A TOPOGRAPHICAL

HISTORY OF STAFFORDSHIRE.

A

TOPOGRAPHICAL

History of Staffordshire;

INCLUDING ITS

AGRICULTURE, MINES, AND MANUFACTURES.

MEMOIRS OF EMINENT NATIVES;

STATISTICAL TABLES;

AND EVERY SPECIES OF INFORMATION CONNECTED WITH THE LOCAL HISTORY OF THE COUNTY.

With a SUCCINCT ACCOUNT of the RISE and PROGRESS of the STAFFORDSHIRE POTTERIES.

COMPILED FROM THE MOST AUTHENTIC SOURCES,

By WILLIAM PITT,

Author of the Agricultural Surveys of the Counties of Stafford, Worcester, &c. under Authority of the Board of Agriculture.

NEWCASTLE-UNDER-LYME:

PRINTED BY AND FOR J. SMITH,

AND SOLD BY HIM, HIGH-STREET, NEWCASTLE; ROGERS, STAFFORD; LOMAX, LICHFIELD; SMART, WOLVERHAMPTON; AND THE DIFFERENT BOOKSELLERS, IN STAFFORDSHIRE; ALSO BY LONGMAN, HURST, REES, ORME, AND BROWN, PATERNOSTER-ROW, LONDON.

1817.

TO

THE RIGHT HONOURABLE CHARLES CHETWYND,

EARL TALBOT,

VISCOUNT INGESTRE;

LORD TALBOT, BARON OF HENSAL;

A PEER OF THE REALM;

LORD LIEUTENANT OF THE COUNTY OF STAFFORD; PRESIDENT OF THE STAFFORDSHIRE AGRICULTURAL SOCIETY, &c.

THIS VOLUME,

IS VERY RESPECTFULLY INSCRIBED,

BY HIS LORDSHIP's

MOST OBEDIENT HUMBLE SERVANT,

WILLIAM PITT.

Wolverhampton, April, 1817.

SUBSCRIBERS.

A.

ANGLESEY, the Most Noble the Marquis of, *large paper*,
ANSON, the Lord Viscount, (10 *copies*,)
Acton, Mr. John, Newcastle, *l. p.*
Adam, Benjamin, Esq. Tunstall,
Adams, William, Esq. Cobridge,
Adams, Mr. William, Hanley,
Adams, Mr. Benjamin, London,
Adams, James, Esq. Brymbo-hall, Wrexham,
Ainsworth, Nicholas, Esq. Newton, near Middlewich,
Allen, Mr. Joseph, Newcastle,
Alsop, George, Esq. Uttoxeter,
Allsopp, Samuel, Esq. Burton-upon-Trent,
Alsop, Mr. Cotes Lodge,
Amphlett, Mr. William, Pepperhill, Salop,
Ashcroft, Mr. Thomas, Barlaston,
Askin, Mr. William, Cheadle,
Astbury, James, Esq. Meaford,
Astbury, William, Esq. Lincoln's Inn, London,
Aston, Mr. Joseph, Wombourne,
Atcheson, N. Esq. F. S. A. London, *l. p.*
Atcheson, Mr. Robert, London,
Atkinson, John, Esq. Maple-Hayes,
Audley, Mr. John, Newcastle,
A. B.

B.

Bacchus, George, Esq. London,
Bacon, Mr. William, Wolverhampton,
Badnall, Richard, Esq. Leek,
Badnall, Mr. Joseph, ditto,

a

Bagnall, Mr. William, Trent-Hay,
Bagnall, Mr. Daniel, Bilston,
Bagshaw, Samuel, Esq. Newcastle,
Baker, William. Esq. Fenton,
Baker, Mr. Hanley,
Baker, Mr. R. Bookseller, Tamworth,
Ball, Mrs. Hougher-Wall,
Ball, Mr. John, Abbots Bromley,
Bamford, Mr. William, Rugeley, (2 *copies*,)
Barber, John V. Esq. Walsall,
Barber, Samuel, Esq. ditto,
Barnes, Edward, Esq. Shareshill,
Barker, Mr. Radford,
Barratt and Son, Messrs. Bath, (2 *copies*,)
Basford, Mr. Stephen, Bilston,
Basnett, Rev. J. B. Cloughs,
Bass, Michael, Esq. Uttoxeter,
Bate, Mr. Thomas, Penn,
Bathwell, Mr. Thomas, Burslem, *l. p.*
Batkin, Miss Eliz. Whitehall,
Bayley, Mr. Joseph, ditto,
Bayley, Mr. Robert, Macclesfield,
Bayliss, George, Esq. Kelsall,
Bedford, William, Esq. Birmingham,
Bedford, Mr. Thomas, Wolverhampton,
Belcombe, H. S. M. D. Newcastle, *l. p.*
Bell, James, Esq. Uttoxeter,
Bennett, John, Esq. Dimsdale,
Bennett, Mr. Penkridge,
Bent, William, Esq. Stonyfields, *l. p.*
Bent, Richard, M. D. Basford,
Berks, Mr. Thomas, Keel,
Best, Rev. Thomas, Uttoxeter,
Bevan, Rev. Mr. Trysull,
Bate, Mr. Newcastle,
Best, Mr. Norris, jun. Bilston,
Birch, Mr. R. Tettenhall Wood,
Bird, Rev. William, Church Eaton,
Bishop, William, Esq. Stafford,
Bishton, George, Esq. Brocton,

Blackwell, Mr. John, Cobridge,
Blagg, John, Esq. Cheadle,
Blair, James, Esq. ditto,
Blount, Edmund, Esq. Bellamont,
Blunt, Rev. John, Blurton,
Booth, Hugh, Esq. Clayton,
Booth, Mr. William, Hanley,
Boughey, Sir. J. F. Bart. M. P. Aqualate-Hall, *l. p.*
Bourne, Ralph, Esq. Fenton, *l. p.*
Bourne, John, Esq. ditto,
Bourne, William, Esq. Cheadle,
Bourne, Mr. James, jun. Dudley,
Bowan, Mr. John, Bilston,
Botham, Mr. Samuel, Uttoxeter,
Bradburne, Rev. T. Pipe-Hill,
Brandon, Thomas, Esq. Cheadle,
Brassie, Mrs. Chester, *l. p.*
Brewster, Mr. Joseph, Brewood,
Brierley, Mr. Thomas, Macclesfield,
Brindley, Mr. Joseph, Longport,
Bristol, Mr. Andrew, Newcastle,
Brittain, Mr. J. ditto,
Broade, Mr. P. B. Fenton-Vivian,
Brocklehurst, Wm. Esq. Macclesfield,
Brocklehurst, John, Esq. ditto,
Brook, Mr. Wolverhampton,
Broomhall, Mr. Newcastle,
Broughton, Rev. H. D. Broughton-Hall,
Brown, Hugh, Esq. Cheadle,
Buckeridge, Charles, D. D. Lichfield,
Buckle, Mr. Wolverhampton,
Bucknall, Mr. George, Albrighton,
Bulkeley, Colonel, Huntley-Hall, near Cheadle,
Burgess, Mr. Joseph, Folly,
Brougham, Mr. Stephen, Bookseller, Burslem,
Beckett, John, Esq. Pipe-gate, *l. p.*

C.

Caldwell, James, Esq. Linley-Wood,

Cale, Mr. George, Wolverhampton,
Calkin, Mr. James, London,
Campbell, Capt. Salop Militia,
Carless, Rev. E. Wolstanton,
Cartledge, Mr. James, Burslem,
Case, Henry, Esq. Shenstone,
Case, Mr. Manchester,
Chadwick, Charles, Esq. Mavesyn Ridware,
Chadwick, Mr. Charles, Leek,
Chambers, Mr. John, Ettingsall,
Chamberlain, Mr. William, Fordhouses,
Chawner, Mr. Thomas, Cheadle,
Child, Smith, Esq. London,
Child, Baddeley, Esq. Lambeth,
Chinn, Henry, Esq. Lichfield,
Chetham, Mr. J. L. Lane-End,
Chetwode, Sir John, Bart. M. P. Oakley-Hall, *l. p.*
Chetwynd, George, Esq. Brocton-Lodge,
Clare, Rev. John, Bushbury,
Clarke, Edward, Esq. Rickerscote,
Clarke, Mr. James, Newcastle,
Clarke, Mr. Aldridge,
Clarke, Mr. George, Wolstanton,
Clayton, Mr. William, Poynton, Cheshire,
Clews, John, Esq. Newcastle,
Clews, Mr. James, Cobridge,
Clive, Mr. J. H. New-Fields,
Clough, Mr. Ralph, Stoke,
Clowes, William, Esq. Port-Hill,
Clutterbuck, Lewis, Esq. Ford-House,
Coare, George, Esq. London,
Collins, John, Esq. Stafford,
Cook, Mr. Newcastle,
Coombe, Mr. T. G. Newcastle, *l. p.*
Cooper, Adjutant, N. S. L. M. Newcastle,
Condliff, Mr. T. Tunstall,
Corbett, Mr. R. Bilston,
Copeland, William. Esq. London,
Cornforth, Mr. Chapel-Ash,
Corser, John, Esq. The Oaks, near Wolverhampton,

Cotterill, Rev. T. Lane-End,
Cotton, Mr. C. Burslem,
Cotton, Mr. Benjamin, ditto,
Coupland, Charles, Esq. Leek,
Coxon, Mr. Hanley,
Coyney, Captain, Weston-Coyney,
Creswell, Mr. Edward, Bilston,
Crewe, Mr. John, Newcastle,
Crockett, Henry, Esq. Little-Onn,
Crockett, J. M. Esq. ditto,
Crook, Mr. Thomas, Manchester,
Crosby, Mr. John Monro, Newcastle,
Crowley, Mr. Wolverhampton,
Cruso, John, Esq. Leek,
Cruso, John, jun. Esq. ditto,
Crutchley, Mr. K. K. Wolverhampton,
Curtis, Mr. Walsall.

D.

Dudley and Ward, the Right Hon. Lord Viscount, *l. p.*
Daintry, J. S. Esq. Macclesfield,
Dale, Mr. John, Stafford.
Daniel, Mr. James, Burslem,
Darwall, C. H. Esq. Walsall,
Davids, Rev. Ince, Aldridge,
Dean, Mr. William, Bilston,
Delves, Court, Esq.
Denman, Mr. Wolverhampton,
Dent, John, Esq. Stone,
Desvoeux, Rev. Henry, Burton,
Devey, Mr. H. Penkridge,
Dicken, Mr. John, Burslem,
Dickenson, Mr. S. Lewis, Brineton,
Dodson, Mr. George, Lichfield,
Downing, Mr. Henry, Dudley,
Drayton, The Book Society,
Drewry, Mr. Joshua, Stafford,
Dudley, Mr. Edward, Wolseley-Bridge,
Duffort, Mr. Stoke,

Dyott, John, P. Esq. Lichfield,
Dimmock, Mr. Joseph, Etruria.

E.

Earp, Mr. Thomas, Uttoxeter,
Earp, Mr. Tettenhall,
Eaton, Mr. George, Newcastle, *l. p.*
Eginton, John, Esq. Bilbrook,
Eld, J. Esq. Haughton,
Ellerton, Rev. Joseph, Stafford,
Emerson, Mr. Robert, Leek,
Evans, Mr. R. Wolverhampton.

F.

Farmer, Mr. Thomas, Wolverhampton,
Farmer, Mr. Edward, ditto,
Farmer, Mr. Richard, ditto,
Faulkner, Mr. Burslem,
Fenton, Thomas, Esq. Stoke-Lodge,
Fereday, Samuel, Esq. Ettingsall Park,
Fletcher, William, Esq. Uttoxeter,
Fletcher, William, Esq. Tean,
Fletcher, Lady, Betley-Court,
Flint, Mr. Charles, Leek,
Floyer, W. H. C. Esq. Hints,
Fithon, Mr. Newcastle,
Ford, Hugh, Esq. Leek,
Forrester, Mr. George, Lane End,
Foster, Mr. Thomas, Wobaston,
Foster, Mr. Marmaduke, Newcastle,
Fowke, Mr. Wolverhampton,
Fowler, Richard, Esq. Birmingham,
Fryer, Mr. John, Wolverhampton,
Finney, John, Esq. Leek,
Fynney, Mr. A. A. ditto.

G.

Gower, Earl, M. P. *l. p.*
Gallimore, Mr. Jesse, Burslem,

Garner, Mr. Hugh, Cobridge,
Gaunt, John, Esq. Leek,
Gater, Mr. Thomas, Norbury, Cheshire,
Gerrard, Mr. Jesse, Uttoxeter,
Gibbons, J. Esq. Wolverhampton,
Giffard, Honourable Mrs. Chillington,
Giffard, Thomas, jun. Esq. ditto,
Gilpin, Messrs. and Co. Wedges Mill, Cannock,
Ginders, Mr. Jeremiah, Ingestre,
Goodwin, Mr. R. Hazlewall, near Cheadle,
Goodwin, Mr. Richard, Uttoxeter,
Gough, John, jun. Esq. Oldfalling,
Granger, Henry, Esq. Tettenhall-Wood,
Green, John, Esq. Essington,
Green, Henry, Esq. Moreton,
Green, Mr. Joseph, jun. Walsall,
Gresley, Sir Roger, Bart. *l. p.*
Griffin, Thomas, Esq. New Inn Mill,
Griffin, Mr. Thomas, Leek,
Grosvenor, T. F. Esq. Pickwood, near ditto,
Grove, Edward, Esq. Shenstone,
Grundy, Mr. George, Lapley.

H.

Harrowby, the Right Hon. the Earl of, *l. p.*
Haddock, Mr. Thomas, Cannock,
Haden, Rev. A. B. Chapel-Ash,
Hales, John, Esq. Cobridge,
Hall, Robert, Esq. Newcastle,
Hall, Thomas, K. Esq. Holly-Bank,
Hall, Richard, Esq. Moss Pit,
Hall, Mr. George, Newcastle,
Hall, Mr. John, Burslem,
Hall, Mr. Joseph, Longport, *l. p.*
Hallam, Mr. John, Newcastle,
Hammond, Mr. Stafford,
Hamilton, Robert, Esq. Fenton,
Hamper, William, Esq. Birmingham, *l. p.*
Hand, Enoch, Esq. Uttoxeter,

Hand, Mr. John, Leek,
Hardern, Mr. David, Macclesfield,
Harding, Rev. Thomas, High-Offley,
Harley, Mr. Thomas, Lane-End,
Hart, Thomas, Esq. Uttoxeter,
Harvey, Mr. John, Lane-End,
Hassells, Charles, Esq. Newcastle,
Hatfield, Mr. George, ditto,
Hatton, Mr. John, ditto,
Hawkins, Mr. William, Stafford,
Hawson, Thomas, Esq. Smethwick-House,
Hayes, Mr. Samuel, Newcastle,
Heape, Mr. Wolverhampton,
Heath, Dr. Fakenhem, Norfolk,
Heath, Rev. William, Inkborough, Worcestershire,
Heath, Mr. Thomas, Hanley,
Heath, Mr. William, ditto,
Heath, Mr. Thomas, Burslem,
Heathcote, R. E. Esq. Bank-House, Warrington,
Heaton, Mr. C. Endon,
Hellier, James Shaw, Esq. Woodhouses,
Henshall, Mr. Newcastle,
Hewitt, Mr. Charles, Lichfield,
Hicken, Rev. William, jun. Audmore,
Hill, Mr. Joseph, Newcastle,
Hill, Mr. Robert, ditto,
Hilliard, Mr. T. H. Leek,
Hinckes, Mr. Samuel, Willenhall,
Hinckes, P. T. Esq. Tettenhall,
Hinckes, Rev. Josiah, Tettenhall Wood,
Hinton, Mr. Hanley,
Hodgkiss, Mr. Joseph, Bloxwich,
Hodgson, Benjamin, Esq. Uttoxeter,
Holdgate, Mr. Thomas, Stoke,
Holland, Mr. R., Streethay, near Lichfield,
Holland, Mr. C. A. Burton,
Hollins, Miss, Shelton,
Holyoake, Francis, Esq. Tettenhall,
Hordern, Mr. Henry, Saredon,
Horton, Mr. Thomas E. Prior's Lee,

Hordern James, Esq. Wolverhampton,
Hordern, Mr. Joseph, Saredon,
Hordern, Miss, Shareshall,
Hordern, Mr. Thomas, ditto,
Howe, Mr. John, Leigh,
Howorth, John, Esq. Cliff-Park, near Leek,
Hubball, John, Esq. Stafford,
Hubbard, George, Esq. Cheadle,
Hulme, J. M. D. Ball-Haye,
Hulse, Mr. James, Newcastle,
Hussey Phineas, Esq. Wyrley-Hall.

I.

Jacob, Mr. Henry, Birmingham,
James, Mr. John, Walsall,
James, Mr. Richard, ditto,
Jefferies, Mr. Coven-Mill Farm,
Jenkins, Captain, Sedgley,
Jesson, Mr. Wolverhampton,
Illidge, Mr. Thomas, ditto,
Johnson, J. Esq. Great-Haywood,
Jones, Mr. John, Burslem,
Jones, Mr. Betley,
Jones, Mr. Edward, Coseley,
Inge, Rev. Charles, Rugeley,
Ingleby, Joseph, Esq. Cheadle,
Johnson, Mr. Ralph, Burslem,
Joynson, Mr. Hanley,
Ironmonger, Mr. Newcastle,
Jones, Rev. Trevor, Lichfield, *l. p.*

K.

Kay, Thomas, Esq. London,
Kealy, Mr. J. F. Newcastle, *l. p.*
Kearns, Mr. Newcastle,
Keates, Mr. James, Eadon, near Leek,
Keeling, Mr. Hanley,
Keeling, Mr. James, Shelton,
Keen, William, Esq. Stafford,
Kemsey, Rev. Matthew, Brewood,
Kempson. Rev. —— Brewood School,
Kilmister, G. R. Esq. Leek,
Kinnersly, W. S. Esq. Newcastle, *l. p.*
Kinnersley, Mr. Clement, Uttoxeter,

L.

Lichfield and Coventry, the Hon. and Right Rev. the Lord Bishop of, *l. p.*
Lander, W. Esq. Rugeley,
Lane, John, Esq. Bromley,
Lane Joseph, Esq. Green-Hill,
Laugharn, Thomas Esq. Leek,
Lathbury, Mr. Burton-upon-Trent,
Lathbury, Mr. Richard, Sudbury-Hall,
Lawton, C. B. Esq. Lawton-Hall,
Lawton, Rev. J. Lawton,
Lawton, Mr. William, Burton,
Lay, Mr. Joseph, Leek,
Lea, Mr. William, Leacroft, Cannock,
Leech, James, Esq. Newcastle,
Leek, Mr. Edmund, Deepmore,
Lees, Mr. John, Longport,
Leigh, J. Esq. Consall-Wood, near Cheadle,
Leigh, Rev. C. Newcastle,
Leigh, Mr. J. Manchester,
Lester, Mr. T. M. Newcastle,
Levett, Rev. Richard, Milford,
Lingard, Mr. John, Wolverhampton,
Lister, J. Esq. Armitage,
Lister, Thomas, Esq. Armitage,
Littleton, E. J. Esq. M. P. Teddesley, *l. p.*
Loch, James, Esq. London,
Locker, Mr. Burslem,
Lomax, Mr. Lichfield,
Lovell, Mr. George, Kings's Sutton, Northamptonshire,
Lowe, Mr. Thomas, Cheadle.

M.

Machin, Mr. Joseph, Burslem,
Mainwaring, Edward, Esq. Whitmore,
Malbon, Dr. Congleton,
Malkin, Mr. Shelton,
Mallabar, Mr. Edward, Uttoxeter,
Mander, Mr. John, Wolverhampton,
Mare, Matthew, Esq. Great-Fenton,
Marsh, Richard, Esq. Lloyd-House,
Mather, Mr. Thomas, Leek,
Maydew, Mr. Thomas, Cobridge,
Mayer, Mr. Thomas, jun. Newcastle,
Mayer, Mr. Joseph, Hanley,
Miller, Mr. William, Dunstall,

Mills, Thomas, Esq. Leek, *l. p.*
Mills, Mr. James, Macclesfield,
Milner, Mr. Edward, Cheadle,
Minshull, Mr. John, Stone,
Minton, Mr. Herbert, Stoke-upon-Trent,
Molineux, Mr. R. Ryton, Shropshire,
Molineux, Rev. William, Wolverhampton,
Molineux, George, Esq. ditto,
Monckton, Hon. Edward, Somerford,
Moore, Mr. Shirleywich,
Mott, John, Esq. Close, Lichfield,
Mottershaw, T. Esq. Silkmore House,
Mousley, Mr. T. jun. Leek,
Morris, Mr. S. E. Burton,
Moreton, Miss, Wolstanton,
Moreton, Mr. John, Wolverhampton,
Mytton, Mr. ditto.

N.

Nairne, Fosham, Esq. Wednesbury,
Newcastle and Pottery Library, The
Newcastle Book Society, The
Newbury, Mr. B. C. W. Hanley,
Nichols, Edward T. Esq. Swithamley-Hall,
Nichols, John, Esq. F. S. A. London,
Nichols, J. B. Esq. F. L. S. ditto,
Nickisson, William, Esq. Newcastle,
Noble, Rev. Mark, Rector of Barming, Kent,
Northall, Mr. W. K. Wolverhampton,
Northen, F. H. M. D. Newcastle,
Norris, Mr. J. P. Westminster.

O.

Oakden, Mr. John, Kingsley,
Okeover, Rev. C. G. *l. p.*
Oldershaw, Mr. John, Tarvin, near Chester.

P.

Paget, Mr. Burslem,
Palmer, Mr. Enoch, Newcastle,
Parkes, Mr. C. Wright, Woodeaton,
Parker, Miss H. Apedale,
Parker, Mr. D. Shrewsbury, *l. p.*
Parker, Mr. W. H. Walsall,
Parker, Mr. John, Rushall-Hall,
Peak, Mr. Keel,

Peak, Mr. Joseph, Newcastle,
Peak, Mr. George, ditto,
Peake, Mr. S. Silverdale,
Peake, Mr. William, Castle-Croft,
Pearson, John, Esq. Tettenhall,
Pearson, G. K. Esq. Macclesfield,
Pearson, Mr. Joseph, Wolverhampton,
Peel, Sir Robert, Bart. M. P. Upper Grosvenor-street, London, *l. p.*
Peel, the Right Hon. Robert, M. P. Stanhope-street, ditto, *l. p.*
Pennant, David, Esq. Downing,
Petit, L. H. Esq.
Petit, Rev. J. H. Hilton-Hall,
Perkins, T. B. Esq. Rickerscote,
Perks, Mr. George, Wolverhampton,
Perry, Mr. Edward, Longport,
Phillips, Mr. Bushbury Hill,
Phillips, Mr. J. E. Newcastle,
Pigot, Dr. Chester,
Plant, George, Esq. Newcastle,
Podmore, Mr. John, jun. Lincroft House,
Poole, James, Esq. Finney Green, *l. p.*
Poole, Mr. Wolstanton,
Porter, Sinckler, Esq. Leek,
Porter, Rev. Robert, Draycott, *l. p.*
Poulson, Mr. John, Wood House,
Pountney, Mr. Richard, Low Hill,
Powys, Rev. Edward, Westwood,
Pratchett, Richard, Esq. Birmingham,
Pratt Mr. John, Saredon Mill,
Pratt, Mr. John, Lane Delph,
Price, W. T. Esq. Wolverhampton,
Proud, J. F. Esq. Wolverhampton,
Proud, Mr. Bilston,
Prowse, Mr. Hanley.

Q.

Quintin, Mr. H. Hillsfield Hall.

R.

Reed, Rev. Joseph, Wolverhampton,
Richards, Mr. Joseph, Coven,
Ridgway, John, Esq. Cauldon Place,
Riley. Mr. John, Burslem,
Riley, Mr. Richard, ditto
Riley, Mr. William, Moat Hall, near Wolverhampton,
Rhodes, Mr. Josiah, Newcastle,
Roberts, Mr. Dudley,

Roberts, Mr. John, Macclesfield,
Robinson, C. B. Esq. Hill Ridware,
Robinson, Mr. T. S. Uttoxeter,
Robinson, Mr. Ralph, Newcastle,
Robinson, Mr. C. Brown Hills Cottage,
Robinson, Mr. Dudley,
Rogers, Spencer, Esq. Longport,
Rogers, Mr. Bookseller, Stafford,
Round, Mr. B. Wednesbury,
Rowley, John, Esq. Bower End,
Russel, Jesse, Esq. Ilam Hall.

S.

Stafford, the Most Noble the Marquis of, *l. p.*
Stamford and Warrington, the Right Hon. the Earl of, *l. p.*
Salt, Mr. George, Betley,
Sant, Mr. Aaron, Burslem,
Savage, Mr. J. Wolverhampton,
Scarlett, Mr. William, Penkridge,
Scott, Sir Joseph, Bart. Barr Hall, *l. p.*
Scott, John, Esq. Great Barr,
Scott, Mr. Newcastle,
Scott, Mr. Bookseller, Burton,
Sergeant, Mr. John, Cheadle,
Shaw, Mr. J. Newcastle, *l. p.*
Shaw, Mr. Charles, ditto, *l. p.*
Shaw, Mr. Job, Uttoxeter,
Shelley, Mr. Newcastle,
Shenstone, Mr. Standeford Mill,
Sherratt Samuel, Esq. Walsall,
Sherratt, Mr. Thomas, Newcastle,
Sherratt, Mr. R. R. Blythford,
Shufflebotham, Mr. Wm. Leek,
Simpson, Mr. Derby,
Simpson, Mr. Thomas, Wolverhampton,
Simpson, Rev. A. Madeley,
Simpson, Mr. Nicholas, Shelton,
Slaney, Rev. Richard, Penkridge,
Sleath, Rev. Dr. Repton, Leicestershire,
Sleigh, Richard, Esq. Leek,
Sleigh, John, Esq. ditto,
Smith William, Esq. Southwell, Nottinghamshire,
Smith, Richard, Esq. Tibbington,
Smith, Robert Esq. Appleby, Leicestershire,
Smith, John, Esq. Fenton,
Smith, Mr. John, Springfields,
Smith, Mr. Sampson, Cheddleton,

Smith, Mr. Thomas, Cheadle,
Smith, Mr. J. P. Burton,
Snape, Rev. William, Newcastle,
Sneyd, Walter, Esq. Keel Hall,
Sneyd, William, Esq. Ashcombe,
Sneyd, Edward, Esq. Byrkley Lodge,
Soars, Mr. Benjamin, Newborough,
Sorton, Mr. William, Newcastle,
Sparrow, John, Esq. Bishton,
Sparrow, Mr. J. F. Burslem,
Spode, Josiah, Esq. Mount, near Stoke, *l. p.*
Smart, Mr. Wolverhampton,
Stanley, Mr. William, Burslem,
Stanley Mr. John, Prestwood,
Steedman, George, Esq. Mill Meece,
Steele, Mr. R. Newcastle,
Stevenson, Ralph, Esq. Cobridge,
Stokes, Thomas, Esq. Bilston,
Storr, Captain, Staffordshire Militia,
Swinnerton, John, Esq. Newcastle,
Swinnerton, Mr. B. ditto,
Sutton, Rev. T. Sheffield, Yorkshire.

T.

Talbot, the Right Hon. the Earl of, 10 *copies.*
Talbot, Mr. John, Wolverhampton,
Tarratt, Mr. Joseph, ditto,
Tayleur, Cresswell, Esq. Burleigh Villa,
Taylor, Herbert, Esq. Uttoxeter,
Taylor, Mr. James, Hanley,
Taylor, Mr. Joseph, Madeley Heath,
Taylor, Mr. William, Macclesfield,
Taylor, Mr. William, Newcastle,
Thacker, William, Esq. Muchall Hall,
Thomas, Mr. Edward, Burton,
Thompson, George, Esq. Rough Hills,
Timmis, Mr. T. Wolverhampton,
Tindall, Rev. William, ditto,
Toft, Mr. Thomas, Hanley,
Tollet, George, Esq. Betley Hall, *l. p.*
Tomlinson, John. Esq. Cliff Ville,
Tomkinson, Mr. S. Burslem,
Tristram, Mr. J. B. Shelton,
Trubshaw. Mr. Little Haywood,
Turner, Mr. H. Oulton Hall,
Turner. Mr. John, jun. Newcastle,
Turner, Mr. Samuel, jun. Stone,

Turner, Mr. James, Leek,
Turner, Mr. Thomas, Newcastle,
Turnock, James, jun. Esq. Stafford,
Turtle, Mr. James, Newcastle, *l. p.*
Twemlow, Thomas, Esq. Peats Wood,
Twemlow, Francis, Esq. Betley,
Twigg, Mr. Joseph, Burslem.

U.

Vincent, the Right Hon. the Earl of, *l. p.*
Vickerstaff, Mr. Ralph, Burslem,
Underhill, Mr. J. Wolverhampton.

W.

Wainwright, Mr. Fradley,
Walford, John, Esq. Hill Hall,
Walhouse, Moreton, Esq. Hatherton,
Walkden, Mr. Joseph, Hall Hill,
Walker, Mr. Randle, jun. Wolverhampton,
Walker, Rev. Thomas, ditto,
Walker, Mr. Thomas, Burslem,
Walker, Mr. Randle, Penn,
Walklate, Mr. Richard, Lane End,
Walthall, Mr. P. jun. Newcastle,
Warburton, Mr. B. Cobridge,
Ward, Thomas, Esq. Newcastle,
Ward, John, Esq. Burslem,
Ward, Mr. J. H. Wolverhampton,
Watson, Holland, Esq. Congleton,
Wear, Mr. William, Hanley,
Webster, Mr. Stafford,
Webster, Mr. George, Macclesfield,
Wedgwood, Mr. Joseph, Basford Bank,
Wenman, William, Esq. Wolverhampton,
Wheeldon, Mr. Wolverhampton,
Wheeler, Trevor, Esq.
Whitaker John, Esq. Macclesfield,
White, Rev. H. Lichfield,
White, Mr. Thomas, Newcastle, *l. p.*
Whitehouse, Mr. Edward, Ettingsall,
Whitgreave, T. F. Esq. Moseley,
Wigan, Edmund, Esq. Lapley, 2 *copies*,
Wilkinson, Mr. Joseph, Macclesfield,
Willatt, Mr. Burslem,
Williamson, H. H. Esq. Longport,
Wilson, Mr. Jonathan, Macclesfield, 2 *copies*,

Winterbottom, Mr. J. K. Stockport, Cheshire,
Wicksted, Charles, Esq. Betley Hall,
Wolverhampton Library,
Wood, Enoch, Esq. Burslem,
Wood, John, Esq. Brown Hills,
Wood, Mr. Joseph, Tean,
Wood, Mr. Thomas, Barlaston,
Wood, Mr. Thomas, Wolverhampton,
Woodford, Mr. John, ditto,
Woolfe, Mr. Benjamin, Leek,
Wolferston, Mr .S. Pipe, Statfold, *l. p.*
Woolston, Rev. W. Adderbury, Oxfordshire,
Wright, Rev. T. Audley,
Wright, Mr. Richard, Lichfield,
Wrottesley, Henry, Esq. M. P.

Y.

Yates, Mr. Blymhill,
Yates, Mr. Attorney, Wilmslow,
Yates, Mr. James, Hanley,
Yates, William, Esq. Shelton,
Yeoman, Rev. Thomas, Stoke,
Yonge, Weston. jun. Esq. Madeley,
Young, Mr. William, Cheadle.

Davenport, John, Esq. Longport.

PREFACE.

It is now two years since the Publisher of the following Work first circulated his proposals; and during that time several individuals have been employed in collecting materials for such a comprehensive History of the County of Stafford, as might embrace every object connected with the ancient as well as the modern state of this division of the Kingdom. The utmost solicitude of the Publisher has been excited by his earnest desire to produce a valuable Record of his native County, which might gratify not only the Subscribers, but the general reader of Topography; and he trusts that his efforts will prove equally satisfactory and successful.

In the regular order of the Work, a brief sketch of the Ancient and General History of the County is first given, comprising Dr. Wilkes's View of Staffordshire from the time when the Romans invaded Britain to the year 1735; the Ancient and Modern Names of Places; and the General Topography of the County. A description of the five Hundreds or Divisions of the County then follows, containing the History of whatever is remarkable either in the antiquities or the modern state of particular places; descriptive sketches of local and picturesque beauty, whether natural or artificial; and occasional brief biographical anecdotes of ingenious or eccentric individuals.

Among remarkable and celebrated places, the City of Lichfield is entitled to particular notice, for its having been the birth-place of many eminent men. The Publisher has carefully collected such documents as might afford a full, satisfactory, and perspicuous record of this metropolis of Staffordshire; with a complete list of the Prelates who adorned it by their virtues and munificence during the long series of nearly twelve centuries. The many important public events which occurred in this ancient city in the course of ages, afford several illustrations of the political and ecclesiastical history of England; and the description of its present state, its magnificent Cathedral, and the fertile and beautiful country in its vicinity, present a combination of instructive and amusing facts to the reader.

From this scene of ecclesiastical magnificence, the History progressively proceeds in the description of those manufacturing towns and districts to which Staffordshire has for ages owed so much of its wealth and celebrity. The iron manufactories of Wolverhampton, Walsall, and Wednesbury, with the Coal Mines at Bilston, and other places, are described; and a brief account given of the curious and valuable manufactures of Soho, near Birmingham.

In the description of Uttoxeter, the reader will find the history of that town curiously illustrated by a chronology of remarkable events during four centuries and a half; and the sketch of Leek contains an account of the present state of the silk manufacture in that town.

The Potteries, productive as they have long been of a manufacture of the greatest national importance, occupy a prominent part of the Work; and their history, from original documents, collected on the spot, will be found worthy the attention not only of the manufacturer and the merchant, but of the man of taste and the philosopher.

The Medallic history of Staffordshire has not been overlooked; and the "Sketch of its Coinage" will be read with interest. It was contributed by a gentleman who ranked high in the numismatic science.

The Botanical History of the County, and that of its inestimable Mineralogy, follow in regular order; its Zoology is also described in detail; and the Corinthian column of its Topographical Architecture terminates in numerous illustrations of MAN, in those "Biographical Sketches of Eminent Natives of Staffordshire," which will doubtless gratify the self-love and the patriotism of every native of the County who reads this authentic record of their ingenuity, wisdom, and heroism.

A very complete History of the Agriculture of Staffordshire follows the Topographical part; and the produce of the soil, under various modes of cultivation, is faithfully detailed from actual observation.

An Account of the different resident Noble Families who derive their honours from names of places in the County, interspersed with historical and characteristic anecdotes, concludes the Work.

Such is the HISTORY OF STAFFORDSHIRE now offered to the attention of the public; and the Publisher cannot conclude, without expressing his warm and grateful acknowledgments to those numerous Subscribers who have so liberally patronized his publication; several of whom have supplied him with documents communicative of much valuable information upon the various subjects which the Work embraces. It would have afforded him great pleasure had he been permitted to give the names of several gentlemen who have thus kindly assisted him in his first effort to obtain public patronage as a Publisher; but that native and retiring modesty, which is the constant attendant of true merit, has left him no appeal from their sentence. It only remains for him to return them his sincere thanks for their favours and their support.

If his Book should be deemed worthy the notice of criticism, the Publisher must plead the general issue. He is aware that it will be found "fruitful in error;" but all who are acquainted with the subject, will allow, that error is inseparable from such a work, than which none is more liable to inaccuracies, that all the diligence of research, or the severest exercise of judgment, cannot possibly guard against. The candid reader will readily enter into the feelings of the Publisher, when he confesses, that, after the lapse of two years, almost exclusively devoted to the production of his County History, and during which time he has encountered many unexpected and formidable difficulties, he is fearful of having led his Subscribers and the Public to form too sanguine expectations of his book, and which all his exertions have not been able to realize. He, however, throws himself upon their indulgence. If even the negative merit of industry be conceded, he must feel himself more than rewarded; but if his book should fall under the lash of a too rigid censor, he will submissively kiss the rod.

Newcastle-under-Lyme, March 10, 1817.

A

TOPOGRAPHICAL

HISTORY OF STAFFORDSHIRE.

GENERAL AND ANCIENT HISTORY.

STAFFORDSHIRE, which lies about the middle of the Island, was formerly much covered with wood, as appears from the remains of the extensive forests of Sutton Coldfield, Cannock-wood, and Needwood Forest, in which, according to tradition, the Druids or priests of the people, performed their religious duties, prior to the arrival of the Romans in Britain.

Barr Hill, situated on the west side of Sutton Coldfield, runs nearly north and south. At the north end stands the town of Aldrich, so called from *ald* (old), and *rice* (a station or jurisdiction); and at the south end is Barr, a village connected with Aldrich. Near this place is a common, which to this day retains the name of Druid-heath, with a pool at the east end of it; and near to this pool is a small area, encompassed with a treble* ditch, which, from not being sufficiently large for an army, is supposed to have been a seat of the Druids. Two or three hundred yards south, is a perfectly-round hill, encompassed with a single ditch, and rising to the height of seven or eight feet; this Dr. Wilkes conjectures to have been a summer seat of the Arch-druid: a similar area, about two miles distant, and adjoining the old Roman road of Street-way, being drier and more lofty, might possibly be his winter residence. The Druids, we are told, gave notice of their days of sacrifice by fires made upon high hills; for this purpose, and for the observations said to be made by the Druids upon the heavenly bodies, Barr-hill was peculiarly adapted: a clump of trees, called

* Neither the Greeks nor Romans ever used treble ditches, so that the above must be referred either to the Britons or their Druids. Although Julius Cæsar asserts that the Druids worshipped a plurality of gods, yet Mr. Cook has pretty plainly shewn that they taught and believed the religion of the Patriarchs; that pillars, a large oak or groves of oak, or large stones, served them instead of temples; and that they believed in one God, and in a resurrection. The Druids and their doctrine undoubtedly originated in the East: they continued in this Island until they embraced Christianity, every attempt of the Romans to destroy them having proved ineffectual.

Bar Beacon, now grows upon its summit. Many remains of Druidical temples are still in existence where they were constructed of stone, as at Abery and Stoneheage; but from the circumstance of those in this county having been formed of wood, they have long ago perished from time or accident. Barrows or Lows there are many, but whether of British origin or not, cannot now with certainty be ascertained.

Few or no places of habitation of the Britons, fenced like the works of Caractacus with large stones, have been found in this county, except indeed those at Wrottesley and Wilbrighton may be deemed such; but there are several fortifications of earth, cast up into high banks, with entrenchments round them, that may be presumed British: and this presumption is strengthened by the testimony of Tacitus, who tells us, that the Iceni chose a place for fight fenced with a bank of earth, having a narrow entrance to keep off the horse: an ancient fortification, situated about a quarter of a mile to the westward of Over Stonall, called Castle Old Fort, nearly answers this description, being encompassed with a double trench 160 paces in diameter, between double entrances, and which Plot thinks a British rampart from its ancient name; and there is no account of any action there of later date. Near Seisdon, on the borders of Shropshire, is Apewood Castle, a very ancient and lofty fortification, situated on a round promontory, and continued for a mile in length: this too, Dr. Plot believes to have been a British work, raised for defence against invaders. To these may be added, the large fortification of earth in Beaudesert Park, which Mr. Pennant asserts to be of British origin.

The great stone standing in a leasow near the Comptons, in the parish of Kinfare, by some called Bolt-stone, by others the Battle-stone, having two chops in the top of it, and measuring above two yards high and near four yards in circumference, may be numbered amongst the British antiquities; as likewise the great stone in a field south of Cannock Church: the Britons erecting such monuments upon civil as well as religious occasions. Cæsar acquaints us that the Britons had iron only in such small quantities, that they made their money of it; and there is reason to believe they often headed their warlike instruments with stone instead of metal, as a dart was found at Leek headed with flint, curiously jagged at the edges, with teeth like a sickle; and flints have likewise been found shaped like arrow-heads: stone axes and hammers, doubtless British, have also been met with on the Wever-hills and on Morredge.

Dr. Plot describes a noble antiquity near Wrottesley, as the foundation of some ancient British city or fortification of great extent: it occupies a circuitous space of three or four miles, and includes parts of Wrottesley, Patshall, Pepperhill, and Bonnegal Parks, and some parcels of the commons of King's-wood and Westbach. Foundations of old walls, large door-hinges, and an antique dagger, have been found here; and from the circumstance of some of the stones being squared, Dr. Plot inclines to think it rather some ruined city than a fortification only.*

The ROMAN *Conquests* in this Island were divided generally into higher or Western, and lower or Eastern Britain, (separated from each other by a line that was carried through the Island), and distinguished by the names of Britannia Prima, Secunda, Flavia, &c. Flavia or Flavia Cæsariensis, comprised all the central counties, including of course Staffordshire, which was possessed before the arrival of the Romans by the Cornavii of Ptolemy, [Camden agrees with Ptolemy,] to whom belonged, amongst other towns, Etocetum or Wall, near Lichfield: Uriconium or Wroxeter, is supposed to have been another principal place of the Cornavii, who occupied Shropshire as well as Staffordshire. This, together with the surrounding country, fell under the dominion of the Brigantes, previous to the Roman invasion. Roman Britain is divided naturally into eastern and western by a chain of hills, running from the Highlands of Scotland and joining the Peak of Derbyshire, the Moorlands of Staffordshire over Cannock-heath and Sutton Coldfield to the range of Edgehill in Warwickshire, and the Chiltern-hills in Buckinghamshire: this extensive range of hills may be termed the backbone of the Island.

There are considerable traces in this county of what are called *Roman Roads*. The principal road of this description is the Street-way, or as Camden, and all authors since his time call it, the Watling-street: it was called Street-way either because it runs in a straight line, or because it was covered with strata of sand and gravel.† Its course is almost due east from Crackley-bank, where it enters this county, and separates it for about two miles from Shropshire, through Weston, and by Stretton (Pennocrucium), over the river

* The same author mentions a stone found in the ruins, that, after ten loads of stone had been hewn from it, required thirty-six yokes of oxen to draw it. The great cistern of the malt-house at Wrottesley was made from it, and which, though left very thick at the bottom and sides, is so capacious, that it will wet thirty-seven strike of barley at one time.

† Bede calls such made roads, strata or covered.

Penk to the Spread Eagle, where it crosses the road from Wolverhampton to Stafford; hence it turns a point to the south, over Calf-heath to the Four Crosses (a very ancient inn,) and continues to Norton and Wyrley Common; it then soon enters Ogley-hay, Cannock-heath, and continues due east two or three miles to Muckley Corner, passing an old entrenchment called Knave's Castle; from Muckley Corner, where it crosses the road from Lichfield to Walsall, it continues due east through Wall (the ancient Etocetum,) about a mile east of which its line seems interrupted, but afterwards continues east by south through Weeford and Hints to Fazeley, where it leaves the county, after having passed nearly thirty miles in a direction east to one point southerly, and in some parts, for many miles together, in a perfect right line.

There were two *Roman Stations* on this road: one of them is placed at Wall, where are now to be found numerous fragments of antiquity, such as Roman bricks, pavements, &c.; and barricades of wood have been lately dug up, evidently military, having the ends cut off by axes (saws not being then in use,) the wood quite black and almost petrified, from the immense lapse of time it had remained in the earth: coins of Nero and Domitian have also been found. The other station has been fixed by some at Penkridge, from its similarity of name to Pennocrucium; but this place being two miles north of the Watling-street, some antiquaries have placed it at Stretton, as agreeing in distance with the old Itinerary of Antonine. This point remains unestablished by the discovery of any Roman vestiges.

Another Roman road, called Ikenild-street, enters this county from the south at Hockley-brook, near Birmingham, and passes through the parish of Handsworth to Sutton-park, which it crosses near two miles, after having gone over part of the Coldfield. It again enters the Coldfield, and passing Little Aston and Shenstone, crosses the Watling-street to the east of Wall, and passes by Street-hay, Whichnor, and Braunston, leaving Burton-upon-Trent about a mile to the east, and continuing by Stretton, it crosses the Dove into Derbyshire, being about twenty-five miles from its entrance into Staffordshire near Birmingham. This is a convex well-formed road, and rectilinear in great lengths: upon Sutton Coldfield and in Sutton-park, it is visible in its original state, not having been much injured by carriages: its course through this county is from north by east to north-east. In Sutton-park is a very ancient bathing-spring called Rowton-well, probably a work of the Romans.

A second Watling-street, also of Roman origin, enters Staffordshire on the north at Rudge-heath, and runs first south and then south-west by the two New Inns across the forest of Kinver to Stourton, and on to Stourbridge-heath, where it enters Worcestershire; but it does not appear whether this road began at Wroxeter or Chester: it was probably made by the Romans under Agricola, when he lay at Chester, or perhaps by Vespasian, when he lay at Wroxeter.*

There are two *Roman Camps* upon this road. One of them, and the largest, occupying an area of twenty-five acres, is situated on a hill at Chesterton; the other is on a flat just under Ashwood Lodge, near Green's-forge: they are both square, very perfect, and lie on the south-east side of the road.

Chesterton Camp is situated on a hill about eleven miles from Wellington, on the old Watling-street, and is perhaps as complete a Roman camp as any on the island: a brook runs close under it on the south and south-west, on which sides the face of a perpendicular rock renders it inaccessible except at the gateway, where a cut has been made for the horse to go to water, and the passage of troops from the road, which is two or three hundred yards distant: the gateways are yet visible, and the bank is generally called the Walls of Chesterton. This place, though now included in Shropshire, is within the ancient limits of this county.†

It is upon record, that Ostorius fortified the rivers Sabrina and Antona before he marched against the Brigantes: these rivers are supposed by some writers to have been the Severn and Avon, and by others the Avon and Tern, which latter river rises at Maer, in

* This road, I presume, crossed the Street-way at the place called the Cock at Watling-street, which therefore gave this name to both roads: part of it goes over a flat called the Wilmers or Wild-moors, and being subject to inundations, is with difficulty traversed in winter. It is now the line of road from Chester or Shrewsbury to Bath and Bristol.

† Frontinus says, the Romans learnt the art of laying out a camp from that of Pyrrhus King of Epirus, which fell into their hands; that they made great improvements upon his plan; and that whenever a new camp was to be formed, it was always constructed upon one uniform plan, and proportioned to the number of soldiers intended to occupy it. The tent of the commanding officer was always in the centre, and a spacious area left about it, where rewards and punishments were publicly distributed: this tent was called Prætorium, it being at first set apart for the Prætor. In Britain the Romans had their summer and winter camps: the former in high and airy situations, the latter in low and warmer parts of the country; they were laid out in a kind of streets, and horse and foot were so disposed as not to interfere or incommode each other.

this county, and running south-west, through a flat, swampy, mossy country, falls into the Severn nearly opposite Wroxeter. The camp at this place, which was twenty times larger than any other in this part of the Island, together with two fortified rivers, would effectually prevent any annoyance from the Britons to the north and west, and the necessity of keeping this post must have been apparent so long as that part of the country remained unconquered. Roman coins of almost every Emperor from Claudius to Valentinian, have been found there.*

Stukeley says: "Two miles beyond Etocetum, on the top of a hill, is Knave's Castle, on the south side of Watling-street, a small tumulus enclosed within three ditches; an entrance on the south side; it has been hollowed on the top. This is a vast moor or common full of heath: the Watling is very fair and straight, and in many places the ridge is perfect for a great length. A little west of the bridge, under which the Penk crosses the Watling, are a few houses belonging to Stretton, on an eminence: this is thought to be Pennocrucium. This village of Stretton is a little north of the road, and a mile south is Breewood, a town on a branch of the Penk, which they say has been an old city, where King John kept his court; a little brook called Horsebrook runs a pasture or two below the road, and parallel to it; the Penk is here a full river; the Watling is here east and west; the old Roman city no doubt was by the road side near here, and perhaps by Horsebrook. Breewood may have been a Roman town, but it is too far out of the road to have any connection with Watling-street, and Penkridge is too miles and a half off, so that it can have no claim. This town must have borrowed its name from the river, as that from the Roman city. Penkridge stands by the side of a large marsh made by the river. The church is built of good stone: a remarkable stone cross in the street."

Mr. Dickenson, who, in November 1796, surveyed this part of the road with the eye of an antiquary, says: "Not a doubt remains with me at present of the site of Uxacona having been at the modern village of Ocon-yate. One great difficulty occurred, viz. the

* A Roman urn, dug up in a gravel-pit at the corner of Stoke-lane, and east entrance into Stone, is now in the possession of Mr. Forster, surgeon, of that town, who purchased it of the labourer that discovered it: it is bell-shaped, being nine inches diameter at the mouth, ten deep, and three and a half diameter at the bottom: it is ornamented with zig-zag work, and contained some ashes and small pieces of human bones. [*For an account of Roman coins found in this county, see the article "Coinage."*]

distance of this place from Wroxeter, which is only eight miles, and the Itinerary distance is eleven; this induced Horsley to fix upon a spot opposite Sheriff Hales for the site of Uxacona, which having carried him three miles too much to the east, he in consequence placed Pennocrucium at Penkridge. To obviate this objection a thought luckily occurred, confirmed since by weighty arguments, that the present road from Watling-street turnpike to Wroxeter was only a *via vicinalis,* and that the grand military road turned off at Watling-street gate to the left, towards Little Wenlock, in order to maintain the high ground on the south side of the Wrekin, and at length fell into the Roman road leading from Wroxeter to Worcester, a few miles short of Wroxeter: this course reconciles the site of Ocon-yate to the Itinerary distance, and removes every difficulty. In the present road between Watling-street and Wroxeter there was formerly a morass, which no carriage could without difficulty pass even at midsummer. My friend, Mr. Pennant, concurred with me in opinion, that a ruin which we both saw at Oconyate many years ago, was the remains of a Roman hypocaust. Ocon-yate being then fixed on as the site of Uxacona, the Itinerary distance of twelve miles brings us precisely to Stretton, where I have discovered a remarkable eminence, called Roley-hill, with a gentle declivity to the meadows on the river Penk, which has the conveniency of a ford in this part. This eminence occupies about five acres, and is crowned with a tumulus, which seems to have served for an exploratory mount: it is scarcely one mile from Street-way."

Dr. Plot was likewise of opinion that Stretton was the site of Pennocrucium; and some Roman vestiges have been found there since his time, viz. a celt, and the brass head of the bolt of a catapulta.

The Romans occupied this country upwards of four hundred years. The Britons, in course of time, became reconciled to their conquerors, were taken into their service, and made their best soldiers: they intermarried, and by degrees became one people.

The SAXONS first arrived in Britain in the year 449, during the joint reign of the Emperors Martian and Valentinian; and about ten years afterwards, (according to the Saxon annals) the Roman soldiers were ordered to leave the Island; but it was not till some years afterwards that they finally quitted it, when they were accompanied by nearly all the British youth that were capable of bearing arms: it was at this period that the Romans collected and buried their money. The Scots and Picts now became troublesome visitors; and Vortigern, the British king, was compelled to apply to his German neighbours for assistance. These invaders, under the general name of Saxons and Angles, after repeated incursions, discovered the weakness of the Britons, at first drove them to the extremities of the Island, and ultimately forced them to retire into Scotland, Wales, and Cornwall, where their posterity has ever since continued. It is very probable, that upon this occasion, the Britons hid or buried their treasure. Some years afterwards, the Danes, encouraged by the success of the Saxons, came hither for the purpose of plunder; but the Saxons, under Hengist and Horsa, after assisting Vortigern to clear his kingdom of foreign enemies, seized it for themselves, and by degrees established the Heptarchy, or Seven Saxon Kingdoms:* Staffordshire was comprehended in Mercia, one of the largest. Egbert, King of the West Saxons, having, about the year 800, conquered the other six Kings, became the first Saxon Monarch; but the kingdoms retained their names much longer, and Mercia continued to have its own governors, though subject to the monarch of the whole.†

The *Saxon Antiquities* in this county are numerous. Dr. Plot commences with Berry-bank (formerly Wolferchester, the work of Wolfer), near Stone, where, on the top of a hill, are remains of the ruins of a large castle, fortified with a double vallum and entrenchments 250 yards in diameter: on the south side was a round conical hill, resembling a tumulus, which, according to tradition, was the seat of Wulferus, king of Mercia, who put his two sons to death for embracing Christianity.‡

Dudley Castle, built upon a lofty hill about the year 700, by one Dudo, an Anglo-Saxon, will be again referred to: it ranks next in antiquity.

In the parish of Mere there is an old fortification, called the Burgh, of an irregular oval form, (like the figure of the hill), and

* Many of the Lows and Camps now visible were undoubtedly the work of those troublesome times; though, for want of historians, we neither know the authors nor cause of their appearance.

† Mercia, (says Dr. Wilkes), whereof Staffordshire always formed a part, and in which some of the Kings had their places of abode, was the best erected, and was so called because it was marked out or limited by several of the other kingdoms: but these people being heathens, were constantly warring against each other.

‡ Wulferus is said to have governed Mercia from the year 657 to 676: the Low adjoining, is probably the place of his sepulture.

fenced in some places with a double trench or rampart; opposite to this, on the heath, is a place called Camp-hills, between which and the Burgh is a large conical hill called Coplow, supposed to be the burial-place of some celebrated commander slain in battle here.*

Upon an eminence, near Alveton, on the north-east border of the county, are the remains of another fortress called Bunbury, similar to, but much larger, than that at Mere: it is of an irregular form, encompassed with a double, sometimes treble trench, from north-west to north-east (the rest being inaccessible by nature), and includes in the whole an area of one hundred acres. This fortification is supposed to have been raised by Coelred king of Mercia (Kenred's successor), when he was invaded, in the seventh year of his reign, by Ina king of the West Saxons, A.D. 716.

The ancient fortifications of Tamworth, and the Lows of Elford and Wiggington, will be noticed hereafter.

Near the village of Billington, in the parish of Bradley, is an old doubly-entrenched fortification, 300 yards in diameter, which, according to ancient writers, marks the place of battle between duke Wada and the other murderers of King Ethelbert on one side, and king Earldulf, his successor, on the other; but this part of our history is very obscure.

The old fortifications on Kinfare-edge, of an oblong figure, 300 yards long and 200 wide, with an artificial bank cast up round them, and fenced with a deep ditch on the N.N.E. and S.S.W. sides (the edge or hill on the other side being inaccessible), are said, by tradition, to be of Danish origin; but the name seems more properly Saxon, and the neighbouring parish of Wolverley is evidently named from Wulfer (Wulfer-ley), whence it seems probable that Kinfare-edge was occupied by Wulfer, the Saxon king of Mercia.

Barrows or *Lows*† are still to be found in many parts of the county: these were heaps or mounds of earth, raised as monuments

* Henry of Huntingdon asserts, that about the year 705, Osrid king of Northumberland, was slain at Meare, by Kenred king of Mercia, whose dominions he had invaded; that the tumulus at Coplow received the body of Osrid, the lows and hillock on Camp-hills those of the soldiers slain, and that the Burgh was the strong-hold which Kenred raised against his enemy.

† Some of these *Lows* were perhaps raised upon civil occasions: those on high grounds may have been merely exploratory or elevated for a prospect or look-out. The following have been observed in different places: one near Swinfen, called Offlow, and supposed to give name to the hundred; a large one

to the illustrious dead, or men slain in battle, and were formerly very numerous; but many of them have been mutilated, or carried away for agricultural and other purposes.

There are many *Saxon Camps* in this county: these are easily distinguished from those of the Romans. The Romans always took care to have a good supply of water, and placed their camps near a road, that the men might always be in readiness to march; but the Saxons generally fixed upon high hills, with a steep precipice in front, preferring security to convenience: the former generally chose a square spot of ground, the latter gave themselves no trouble about the form, but had recourse to ditches. There is one of their camps on Kinver-edge, and another on the top of Berry-bank, near Darlaston-bridge (both occupied by King Wulfer;) and the camp on the top of a hill called Bunbury, near Alveton, in the Moorlands, and referred to King Coelred, about 716, is also Saxon. The Roman camps, and those supposed to have belonged to the ancient Britons, have already been noticed.

The Saxons brought into this island a kind of fortification which they called a *Castle*: this was placed on a high hill, rendered difficult of approach, and was sometimes surrounded by a moat or ditch: it served as a residence for the chief, and a constant garrison being kept, such places were considered, before the use of gunpowder, a good security to their occupiers. Considerable remains of this description of fortification yet exist, as at Dudley, Tamworth, Tutbury, Stafford, &c. At this latter place a castle is said to have been built on the north side the river, by Elfleda, queen of Mercia, about the year 918, besides the one erected long after that time by Randulph, the first Earl of Stafford, about a mile west of the town; though some accounts say, that castle was of older date, and that Ranulph merely re-edified it.

on the Watling-street, near Hints, now a rock of stone, which, with two others at Kingswinford, Dr. Plot thinks Roman petrified barrows; one at Catshill, two on Calf-heath, and one on inclosed grounds east of Great Saredon. The same writer notices a Low near Bushbury, three on Morredge, others near Okeover and Mathfield, one near Colwich Common, and another larger one at the other end of this common called Row-low, probably the place of sepulture of some petty king, Row-low importing *regale sepulchrum*; one in Arbour-close, north-west of Okeover Chapel; three on the Wever-hills, and three others called Queen-low, Gallows-knoll, and Astlow Cross. To these may be added, the Lows on Ribden, Reeden, and Cauldon-hills, also Cock-low, and some others on the hills near Warslow, on Eaton-hill, on a hilly pasture betwixt that and Oncott, those near the town of Leek, and many others in different parts of the county: there were also Lows on Wombourn Common.

It seems pretty certain that the town or Castle of Chesterton-under-Lyme (as Camden calls it), given by King John to Randall last Earl of Chester, must have been a place of note before the Conquest, as, so long ago as the reign of Henry the Third, it was going to decay, when the Earl of Lancaster built another, surrounded by a great pool, which he called the new castle. This castle was built at the north-west of the present town of Newcastle, but of the castle itself there is little or none now remaining.*

The next considerable antiquity of this kind, is Heyley Castle, near Betley, built upon a lofty rock, with the same stone that was dug from the ditches. It was given (says Camden) by Harvey Lord Stafford to Henry de Adithley, in the reign of King John, which Henry seems to have built this castle. Other records say, that the heirs of William de Betley gave the land of Heyley to the said Henry, who was possessed of Aldithlege, now Audley, in this neighbourhood; where also have been traces of a castle, built by some of this family, or of the Verduns, before them.

The origin of Terley* Castle, near Drayton, and of Stourton Castle, in the south of the county, is unknown; but they are both believed to be of great antiquity. Stourton Castle was the birth-place of Cardinal Pole.

The Castle of Chartley is said to have been built in 1218, (2 Hen. III.) by Randall Blunderville, third Earl of Chester; and in the reign of Edward the First. Walter de Longton, Bishop of Lichfield, some authors say built, others repaired, Eccleshall Castle. In the beginning of the reign of Edward the Second, Alveton Castle, east of Cheadle, was built by Theobald de Verdun; and not long after, in the same reign, "the goodly castle of Caverswall, west of Cheadle, was built by Sir William de Caverswall, of masonry:" a house surrounded by a deep moat stands upon the site of this castle.

Entrenchments are to be found in several parts of this county: the small one at Essington-wood, and those called the Moats, near

* Newcastle doubtless had its origin, and was named, after this decayed Castle, there being no account of such town in Domesday-book, or any other records of this county, before the building of that new castle.

† The Lords of Terley enjoy the odd custom of taking an amercement for bastards born within the lordship. The cognizance of the Bishop and ecclesiastical court, and the discovery of the father, is avoided, on oath being made that the bastard was begotten within the manor, and paying ten shillings to the lord: a bastard brought here, on paying 1l. 19s. 11d. to the lord, entitled the parents to the same privileges.

Kinnerston (anciently Rodbaston), are inclosed with double trenches. Indeed, the number of moated houses now remaining (often five or six in one parish,) sufficiently prove the necessity that existed, during the Barons' Wars, and other troublesome times in England, for country families residing in castellated mansions.

Of *Battles* fought in ancient times in this county, the old writers assert that the Danes were totally overthrown at Testenhall or Thestenhall, now Tettenhall, by the Mercians: the site of this battle is unknown, but it is said to have been "so very terrible, that it could not be fully described by the most exquisite pen."

Another bloody battle was fought at Wednesfield, noticed in the account of that place.

A stone is set up on Blore-heath, near Drayton, but in Staffordshire, in memory of a great battle in 1549, when the Earl of Salisbury defeated the King's troops (Henry VI.) commanded by Lord Audley, and killed 2400 men, together with their gallant leader, and all his principal officers. Queen Margaret (as the tradition goes) viewed the battle from the tower of Mucclestone Church, whence she fled (says Leland) upon the defeat, by the advice of John Hales, Bishop of Lichfield, to Eccleshall Church, and he protected her there. The Earl of Salisbury enjoyed his victory but a short time: he was taken prisoner by the Lancastrians at Wakefield, in 1640, and beheaded there, after the battle.

The inhabitants of Checkley have a tradition of a battle that was fought a quarter of a mile east-north-east of their church, in a place called Naked Fields, from the bodies of three bishops slain in the battle lying there naked for some time after the fight: three tall pyramidal stones stand close together in the church-yard, as a monument to these bishops. There are similar monumental stones in the church-yards of Leek, Draycott, and Chebsey, which are supposed to be of Danish origin.

In the year 895 (says Dr. Wilkes) the Danes came up the Severn as far as Bridgnorth, and committed great ravages. We hear no more of them for thirteen years, when they raised a great army, and fought two bloody battles with King Edward, who totally defeated them with the loss of many thousand men: these were the battles of Tettenhall and Wednesfield.*

* It was customary with the Danish kings to address their armies from elevated spots of ground, and it is very probable that the little hill called King's Standing, upon Ikenild street, Sutton Coldfield, was thrown up about this time: although upon a Roman road, it is evidently not Roman.

From the time of the establishment of the Heptarchy, Staffordshire always formed a part of the kingdom of Mercia. Alfred, who may justly be called the first English monarch, about three years before his death, drove the Danes from Bridgnorth, and they never afterwards troubled him. He now proceeded to divide his kingdom into shires or counties, hundreds and tythings: this division was specified in a book called the Roll of Winchester, where Alfred kept his court; but, although often referred to by ancient writers, no copy of this book has been preserved. This part of Mercia then took the name of Staffordshire, from Stafford, a town situated near the middle of it, and so called from a shallow place in the river Sow, on which it stands, that could be passed by the help of a staff only.

The whole Saxon Heptarchy was once divided into hides of land, which division included only the cultivated part. Antiquaries rate the hide of land at one hundred acres of arable (the meadows and woodlands not being included) and a hide of land was as much as one team of oxen could manage in a year: Mercia contained 30,000 hides.* The Hyde-farm, near Breewood, and Hydeley, near Stafford, are so called from this method of dividing land.

From the first division of this county it has always been in five hundreds, Offlow, Totmanslow, Pirehill, Seisdon, and Cuddlestone: the two first had their names from Lows or burying-places; the third is named from a hill near Stone; the fourth from a village, the importance of which is now forgotten; and of Cuddlestone nothing remains but a bridge over the Penk, a mile above Penkridge, called Cuddlestone-bridge.

The NORMAN CONQUEST (1066) forms a most important period in the history of this country. A total change of circumstances took place, and many ancient families had the mortification of seeing their manors and lands distributed amongst Norman favourites, and themselves reduced from affluence to poverty. Besides the royal demesnes in this county, the Conqueror gave Hugh de Montgomery Earl of Arundel, thirty manors; Robert de Stafford, eighty-one lordships; Henry de Ferrers, seven manors, besides the castle and borough of Tutbury; and to William Fitz Ansculph, twenty-five manors: these great barons divided their baronies into lesser

* The hide of land was then valued at about 6l. or 1s. per acre upon 120 acres; when above that price, it was considered as a mark of the extravagance of the times, and the ambition of the laity: but whether this was the annual rent, or fee-simple, does not appear.

that river were strictly guarded, the King was secreted in a barn belonging to a catholic gentleman named Wolf, of Madeley, where he continued till night-fall, and then returned with Pendrell to Boscobel. Next morning, for a better concealment, the King, together with Colonel Carless, who commanded for his Majesty in the late battle, mounted a large oak, in Boscobel Wood, where they sheltered themselves among the leaves and branches for four-and-twenty hours. On Sunday night Pendrell conducted the King, upon a miller's horse, to the house of Mr. Whitgreave,* a catholic gentleman, at Moseley, where he was received with every mark of loyalty and respect: here he remained till Tuesday night, when his Majesty took leave of his faithful host, and was taken by Lord Wilmot to Bentley, the residence of Colonel Lane, about four miles from Moseley towards Walsall. During the King's stay at Moseley he occupied Mr. Huddleston's† (the priest of the family) chamber, from whence was a direct passage into a secret hiding-place that still exists, and in which Mr. Huddleston had been often compelled to take shelter in those times: in this hole his Majesty hid himself upon an alarm of the approach of the rebels, but Mr. Whitgreave had the address to prevent a search for the King. On Monday night, Lord Wilmot left Moseley, and went back to Bentley, to consult with Colonel Lane upon measures to be taken for his Majesty's safety. A scheme was here formed for his journey to Bristol. Colonel Lane having a relation in that city, he obtained a pass for his sister and a servant to visit him. The King, disguised as a servant, rode before Miss Lane, and reached Bristol in safety; but no vessel being ready to sail, his Majesty sought a temporary security at Colonel Wyndham's, in Dorsetshire. At length, a vessel was procured at Brighthelmstone, where, after

* The late T. H. F. Whitgreave, esq. who died in January, 1816, was the great-grandson of the preserver of Charles II.: he has left heirs male. Three generations of this family have lasted from 1651 to 1816, or 165 years, which is 47 years to a generation, and may be reckoned three lives to a century. The late Mr. Whitgreave died in the same mansion in which the King was sheltered.

† It appears, from some anecdotes of the Rev. John Huddleston, lately published, that Richard Pendrell, the honest and loyal protector of Charles II. died February 8, 1671. Mr. Huddleston was the King's constant companion at Mr. Whitgreave's, and when his Majesty departed, he took solemn leave of him, with assurances of his friendship. Mr. Huddleston afterwards retired beyond sea, and became a Benedictine monk; but returning, upon the Restoration, he was appointed one of Queen Catharine's chaplains, and was always excepted by name in proclamations or edicts issued against the Catholics.

various adventures, the King arrived on the 14th of October, took shipping, and next day was safely landed at Feschamp, in Normandy. Upon the King's restoration, 1669, he rewarded all his faithful preservers with pensions.

In the year 1745 an alarming REBELLION made its appearance in this part of the country. About 8000 Scots rebels marched from Edinburgh to Carlisle, which place surrendered to them on the 15th of November. Having greatly increased their numbers, they penetrated south to Lancaster, and afterwards to Manchester: here they formed a regiment of the malcontents, passed the Mersey, and advanced to Congleton, but suddenly turned to the left as far as Derby. The Duke of Cumberland, who was dispatched against the rebels, had his advanced-guard at Newcastle-under-Lyme, and a battle was daily expected in that neighbourhood. The alarm which these movements created was but of short duration: the rebels deemed it prudent to hasten back to Scotland. They were followed by the Duke, and the memorable battle of Culloden, (April 16, 1746), again restored the country to peace and tranquillity.

The domestic history of the present auspicious reign is remarkable for the great improvement of the country by turnpike roads, opening an easy conveyance from place to place; by navigable canals, carried over natural rivers and under mountains; and by the great improvements in manufactures and agriculture, in consequence of these public-spirited exertions. These improvements, and those of the mines, and other resources, compared with the natural and original state of the country, form the object and subject of this work.

MARKET-TOWNS, *(Ancient and Modern.)*

Market Towns.	Market Days.	Market Towns.	Market Days.
1. Stafford ..	*Saturday*	13. Tamworth ..	*Saturday*
2. Lichfield ..	*Friday*	14. Tutbury, *(dec.)*	*Tuesday*
3. Wolverhampton	*Wednes.*	15. Abbots Bromley	*Tuesday*
4. Walsall ..	*Tuesday*	16. Breewood	*Friday*
5. Burton-upon-Trent	*Thursday*	17. Penkridge	*Tuesday*
6. Uttoxeter ..	*Wednes.*	18. Cannock, *(dec.)*	
7. Newcastle ..	*Monday*	19. Betley, *(dec.)*	
8. Leek ..	*Wednes.*	20. Wednesbury	*Saturday*
9. Stone ..	*Tuesday*	21. Burslem ..	*Monday*
10. Cheadle ..	*Saturday*	22. Hanley ..	*Saturday*
11. Eccleshall ..	*Friday*	23. Lane-End ..	*Saturday*
12. Rugeley ..	*Tuesday*	24. Longnor ..	*Wednes.*

A LIST OF THE PARISHES* OF STAFFORDSHIRE,

In the respective Hundreds.

SEISDON HUNDRED CONTAINS TWENTY-ONE PARISHES:

1. Amblecot and Brierley-hill
2. Arley Over
3. Bilston
4. Broome
5. Bobbington
6. Byshbury
7. Codsall
8. Clent
9. Enville
10. Himley
11. Kinfare
12. Kingswinford
13. Pattingham
14. Patteshull
15. Penn
16. Rowley-Regis
17. Sedgeley
18. Tettenhall
19. Tryaull
20. Wolverhampton
21. Wombourne.

* By the term Parish is here meant, a tract of land having a place of worship, and united in some degree by a common or mutual interest, without regarding the ecclesiastical constitution, or dependance upon a mother church.

OFFLOW HUNDRED CONTAINS FIFTY-TWO PARISHES:

1. Alrewas
2. Aldridge
3. Armitage
4. Barr
5. Barton-under-Needwood
6. Bloxwich
7. Burton-upon-Trent
8. Clifton Campville
9. Darlaston
10. Drayton Bassett
11. Edingale
12. Elford
13. Fairwell
14. Hammerwich
15. Hamstall Ridware
16. Haubury
17. Harlaston
18. Haselour
19. Handsworth
20. Harborne
21. Hints
22. King's Bromley
23. Lichfield St. Chadd's
24. Lichfield St. Mary's
25. Lichfield St. Michael's
26. Longdon
27. Marchington
28. Mavesyn Ridware
29. Newborough
30. Norton-under-Cannock
31. Pelsall
32. Pipe Ridware
33. Rolleston
34. Rushall
35. Shenstone
36. Smethwick
37. Statfold
38. Tamworth
39. Tattenhill
40. Thorpe Constantine
41. Tutbury
42. Tipton
43. Walsall
44. Weeford
45. Wednesbury
46. Westbromwich
47. Wednesfield
48. Wittington
49. Whichnor
50. Wiggington
51. Willenhall
52. Yoxall.

CUDDLESTON HUNDRED CONTAINS TWENTY-THREE PARISHES:

1. Acton
2. Baswich
3. Breewood
4. Bednall
5. Blymhill
6. Bradley, *juxta* Stafford
7. Cannock
8. Castlechurch
9. Coppenhall
10. Dunston
11. Forton
12. Gnosall
13. Haughton
14. Lapley
15. Norbury
16. Penkridge
17. Rugeley
18. Shareshall
19. Sheriff Hales
20. Stretton
21. Weston-under-Lizard
22. Wheaton Aston
23. Church-Eaton.

PIREHILL HUNDRED CONTAINS FORTY-SEVEN PARISHES:

1. Abbot's Bromley
2. Adbaston
3. Ashley
4. Audley
5. Barlaston
6. Betley
7. Blithfield
8. Blurton
9. Biddulph
10. Bucknall
11. Burslem
12. Chebsey
13. Colton
14. Colwich
15. Eccleshall
16. Ellenhall
17. Broughton
18. Fradswell
19. Fulford
20. Gayton
21. Hanley
22. High-Offley
23. Keel
24. Lane-End
25. Madeley
26. Maer
27. Marston
28. Milwich
29. Mucclestone
30. Newcastle
31. Norton-in-the-Moors
32. Sandon
33. Seighford
34. Stafford
35. Standon
36. Stone
37. Stowe
38. Stoke-upon-Trent
39. Swinnerton
40. Ranton
41. Talk-on-th'-Hill
42. Thursfield
43. Tixall
44. Trentham
45. Weston-upon-Trent
46. Whitmore
47. Wolstanton.

TOTMANSLOW HUNDRED CONTAINS FORTY PARISHES:

1. Alstonefield
2. Alveton
3. Bagnall
4. Blore
5. Bradley-in-the-Moors
6. Bramshall

Parishes.	Parishes.
7. Butterton	24. Ilam
8. Caldon	25. Ipstones
9. Calton	26. Kingston
10. Caverswall	27. Kinnersley
11. Checkley	28. Leek
12. Cheadle	29. Leigh
13. Cheddleton	30. Longnor
14. Croxden	31. Mayfield
15. Dilhorne	32. Oakover
16. Draycott-in-the-Moors	33. Onecote
17. Endon	34. Meerbrook
18. Elkstone	35. Rocester
19. Ellastone	36. Sheen
20. Flash	37. Warslow
21. Gratwich	38. Waterfall
22. Grindon	39. Wetton
23. Horton	40. Uttoxeter.

TOTMANSLOW	40	*Total in the County:* 183 *Parishes.*
PIREHILL	47	
CUDDLESTON	23	
OFFLOW	52	
SEISDON	21	

ANCIENT AND MODERN NAMES OF PLACES.

[*From Domesday to the present Time.*]

Ancient Name.	Modern Name.	Ancient Name.	Modern Name.
Abetone	*Apeton*	Befecote	*Befcot*
Ache or Acle	*Oaken*	Belintone	*Billington*
Aclei	*Oakley*	Bercheswic	*Baswich*
Acoure	*Oakover*	Bernulvestone	*Barlaston*
Actone	*Acton*	Bertone	*Barton*
Aldidelege	*Audley*	Bertone Abb	*Burton-on-Trent*
Almentone	*Ammington*	Bertone	*Biterton, near Weston*
Alrewas	*Alrewas*		
Alrewic	*Aldridge*	Betelege	*Betley*
Alverdeston	*Alston*	Bidolph	*Biddulph*
Alvidelege	*Alveley*	Bigeford	*Bickford*
Anestanefelt	*Alstonefield*	Bilesroch	*Bilbrook*
Anne	*On (high & little)*	Billestune	*Bilston*
Badehale	*Badnal*	Biscopsberie	*Byshbury*
Baltredelege	*Balterley*	Bispestone	*Bishton*
Barcarderslem	*Burslem*	Blidevelt	*Blithfield*
Barre, Barra	*Barr*	Blocheswic	*Bloxwich*
Bedehald	*Bednal*	Blora	*Blore, near Ilam*

Ancient Name.		Modern Name.	Ancient Name.		Modern Name.
Bradelie	..	*Bradeley*	Comegrave	..	*Congreve*
Bramelie	..	*Gerard's Bromley*	Conton	..	*Comton*
Bramselle	..	*Bramshall*	Cote	..	*Cotes or Coton*
Brantestone	..	*Branston*	Cotewalderstone		*Cotwalton*
Bresmundescote		*Bescote*	Covehale	..	*Copnal*
Bretlie	..	*Bradley*	Cove	..	*Coven*
Breude	..	*Brewood*	Covelau	..	*Cowley*
Brigeford	..	*Bridgeford*	Crachemers	..	*Crakemarsh*
Broctone	..	*Broughton, or Brockton*	Cressvale	..	*Cresswell*
			Crotewich	..	*Gratwich*
Bromelie	..	*King's Bromley*	Cuchesland	..	*Cooksland*
Brotone	..	*Broton*	Cudulveston	..	*Cuddleston*
Brumhelle	..	*Bromhall*	Cuneshala	..	*Consall*
Brunitone	..	*Brineton*	Denstone	..	*Denston*
Brunlege	..	*Abbot's Bromley*	Derlaveston	..	*Darlaston*
Bubintone	..	*Bobbington*	Dochesig	..	*Doxy*
Buckenole	..	*Bucknall*	Dodington	..	*Derington*
Caldone	..	*Caldon or Cauldon*	Dorveslau	..	*Dodsley*
			Draicote	..	*Draycot*
Catespelle	..	*Cospel End*	Draitone	..	*Drayton*
Cavreswell	..	*Caverswall*	Dregetone	..	*Drineton*
Cedda, Cedla	..	*Cheadle*	Dulmesdene	..	*Dimsdale*
Celle	..	*Chell*	Dulverne	..	*Dilhorne*
Cerletone	..	*Charlton*	Dunestone	..	*Dunston*
Certelie	..	*Chartley*	Ecleshelle	..	*Eccleshall*
Cervernest	..	*Charnes*	Ectone	..	*Acton Trussel*
Ceseteford	..	*Seighford*	Edboldestone	..	*Adbaston*
Cetterville	..	*Chatkill*	Edelachestone	..	*Ellaston*
Chebbesie	..	*Chebsey*	Edgersly	..	*Agardsley*
Cheltetone	..	*Chedleton*	Efnefeld	..	*Enville*
Chenestelie	..	*Knightley*	Elachestone	..	*Ellaston*
Chenestetone	..	*Chesterton*	Eleford	..	*Elford*
Chenevare	..	*Kinver*	Elveton	..	*Ellerton Grange*
Chenwardeston		*Kinnerston*	Elmelecote	..	*Amblecote*
Chinesteton	..	*Knightley*	Enedien	..	*Endon*
Chingeslie	..	*Kingsley*	Erlide	..	*Yarlet*
Chrochesdene or Crochestone	..	*Croxden*	Ernelege	..	*Arley*
			Esseningtone	..	*Essington*
Claitone or Clotone	..	*Clayton*	Estendone	..	*Essington*
			Estone	..	*Aston*
Claverlege	..	*Clareley*	Estrenone	..	*Stretton*
Clistone	..	*Clifton*	Etinghale	..	*Ettingshall*
Codeshale	..	*Codsall*	Etone	..	*Water and Church-Eaton*
Coltone	..	*Colton*			

Parishes.

7. Butterton
8. Caldon
9. Calton
10. Caverswall
11. Checkley
12. Cheadle
13. Cheddleton
14. Croxden
15. Dilhorne
16. Draycott-in-the-Moors
17. Endon
18. Elkstone
19. Ellastone
20. Flash
21. Gratwich
22. Grindon
23. Horton

Parishes.

24. Ilam
25. Ipstones
26. Kingston
27. Kinnersley
28. Leek
29. Leigh
30. Lougnor
31. Mayfield
32. Oakover
33. Onecote
34. Meerbrook
35. Rocester
36. Sheen
37. Warslow
38. Waterfall
39. Wetton
40. Uttoxeter.

TOTMANSLOW	40	*Total in the County:* 183 *Parishes.*
PIREHILL	47	
CUDDLESTON..............	23	
OFFLOW	52	
SEISDON	21	

ANCIENT AND MODERN NAMES OF PLACES.

[*From Domesday to the present Time.*]

Ancient Name.		Modern Name.	Ancient Name.		Modern Name.
Abetone	..	*Apeton*	Befecote	..	*Befcot*
Ache or Acle	..	*Oaken*	Belintone	..	*Billington*
Aclei	..	*Oakley*	Bercheswic	..	*Baswich*
Acoure	..	*Oakover*	Bernulvestone	..	*Barlaston*
Actone	..	*Acton*	Bertone	..	*Barton*
Aldidelege	..	*Audley*	Bertone Abb	..	*Burton-on-Trent*
Almentone	..	*Ammington*	Bertone	..	*Biterton, near Weston*
Alrewas	..	*Alrewas*			
Alrewic	..	*Aldridge*	Betelege	..	*Betley*
Alverdeston	..	*Alston*	Bidolph	..	*Biddulph*
Alvidelege	..	*Alvelry*	Bigeford	..	*Bickford*
Anestanefelt	..	*Alstonefield*	Bilesroch	..	*Bilbrook*
Anne	..	*On (high & little)*	Billestune	..	*Bilston*
Badehale	..	*Badnal*	Biscopsberie	..	*Byshbury*
Baltredelege	..	*Balterley*	Bispestone	..	*Bishton*
Barcardeslem		*Burslem*	Blidevelt	.	*Blithfield*
Barre, Barra	..	*Barr*	Blocheswic	..	*Bloxwich*
Bedehald	..	*Bednal*	Blora	..	*Blore, near Ilam*

Ancient Name.	Modern Name.	Ancient Name.	Modern Name.
Bradelie ..	*Bradeley*	Comegrave ..	*Congreve*
Bramelie ..	*Gerard's Bromley*	Conton ..	*Comton*
Branselle ..	*Bramshall*	Cote ..	*Cotes or Coton*
Brantestone ..	*Branston*	Cotewalderstone	*Cotwalton*
Bresmundescote	*Bescote*	Covehale ..	*Copnal*
Bretlie ..	*Bradley*	Cove ..	*Coven*
Breude ..	*Brewood*	Covelau ..	*Cowley*
Brigeford ..	*Bridgeford*	Crachemers ..	*Crakemarsh*
Broctone ..	*Broughton, or Brockton*	Cressvale ..	*Cresswell*
		Crotewich ..	*Gratwich*
Bromelie ..	*King's Bromley*	Cuchesland ..	*Cooksland*
Brotone ..	*Broton*	Cudulveston ..	*Cuddleston*
Brumbelle ..	*Bromhall*	Cuneshala ..	*Consall*
Brunitone ..	*Brineton*	Denstone ..	*Denston*
Brunlege ..	*Abbot's Bromley*	Derlaveston ..	*Darlaston*
Bubintone ..	*Bobbington*	Docheaig ..	*Doxy*
Buckenole ..	*Bucknall*	Dodington ..	*Derington*
Caldone ..	*Caldon or Cauldon*	Dorveslau ..	*Dodsley*
		Draicote ..	*Draycot*
Catespelle ..	*Cospel End*	Draitone ..	*Drayton*
Cavreswell ..	*Caverswall*	Dregetone ..	*Drineton*
Cedda, Cedla ..	*Cheadle*	Dulmesdene ..	*Dimsdale*
Celle ..	*Chell*	Dulverne ..	*Dilhorne*
Cerletone ..	*Charlton*	Dunestone ..	*Dunston*
Certelie ..	*Chartley*	Ecleshelle ..	*Eccleshall*
Cervernest ..	*Charnes*	Ectone ..	*Acton Trussel*
Cesteford ..	*Seighford*	Edboldestone ..	*Adbaston*
Cetterville ..	*Chatkill*	Edelachestone ..	*Ellaston*
Chebbesie ..	*Chebsey*	Edgersly ..	*Agardsley*
Cheltetone ..	*Chedleton*	Efnefeld ..	*Enville*
Chenestelie ..	*Knightley*	Elachestone ..	*Ellaston*
Chenestetone ..	*Chesterton*	Eleford ..	*Elford*
Chenevare ..	*Kinver*	Elveton ..	*Ellerton Grange*
Chenwardeston	*Kinnerston*	Elmelecote ..	*Amblecote*
Chinesteton ..	*Knightley*	Enedien ..	*Endon*
Chingeslie ..	*Kingsley*	Erlide ..	*Yarlet*
Chrochesdene or Crochestone ..	*Croxden*	Ernelege ..	*Arley*
		Esseningtone ..	*Essington*
Claitone or Clotone ..	*Clayton*	Estendone ..	*Essington*
		Estone ..	*Aston*
Claverlege ..	*Clareley*	Estrenone ..	*Stretton*
Clistone ..	*Clifton*	Etinghale ..	*Ettingshall*
Codeshale ..	*Codsall*	Etone ..	*Water and Church-Eaton*
Coltone ..	*Colton*		

SHERIFFS OF STAFFORDSHIRE.

[*From the earliest Accounts to the present Time.*]

HENRY I.

Nicholas de Stafford.

HENRY II.

Milo de Gloucester.

Robert de Stafford, of Stafford Castle: he was allowed 29*l.* 18*s.* for stocking the King's manors: six years.

Alexander Clericus: eighteen years.

Thomas Noel, of Ranton and Ellenhall: eight years.

RICHARD I.

Thomas Noel.

Thomas de Cressewell, of Cressewell.

Hugh, Bishop of Coventry, assisted by others: three years.

Hugh Bardolf: two years.

Hugh de Chaucumb, of Chaucumb, Northamptonshire: two years.

JOHN.

Jeffery, and Hugh de Erdington, of Erdington, near Birmingham: five years.

Thomas de Erdington, and Thomas de Alta Ripa, of Cheshire: two years.

Thomas de Erdington: nine years.

HENRY III.

Ranulph, Earl of Chester, and Hen. de Adithlea, of Audley: four years.

Ranulph, Earl of Chester, and Philip de Kinton: three years.

Ranulph, Earl of Chester, of Chartley Castle.

John Bonet: three years.

Henry de Adithlea, and Robert de Leia, of Leigh.

Henry de Adithlea: four years: William de Bromley, of Gerard's Bromley, being his deputy.

John Extraneus, or Stranger, and Robert de Haga, of Hay: four years.

Robert de Acton, of Acton.

John Extraneus, or Stranger, for ten years successively: Justice also of Chester.

Thomas Corbet, of Barlaston: two years.

John Grendon, of Shenston and Grendon: six years.

Hugh de Acovere, of Okeover: two years.

William Bagod, of Bromley Bagot: three years.

William de Caverswell, of Caverswell, and James de Aldahel.

James de Aldahel, supposed of Audley: seven years.

EDWARD I.

Ralph Mortimer: three years.

Bogo de Knovill: three years.

Roger Springhuse: seven years.

Roger Springhuse, and Lionel Ramescley: three years.

Robert Corbet, King's Bromley.

William Tittley: six years.

Ralph de Schirle, of Shirley, county of Derby: three years.

Thomas Corbet, Barlaston: two years.

Richard de Harlegh: two years.

Walter de Beysin: two years.

John de Acton.

John de Dene.

EDWARD II.

Roger Trumvin.

Johannes Extraneus.

Hugh de Crofts, of Croft's Castle, Herefordshire: two years.

Hugh de Audley, of Audley: 3 years.

William de Mere.

Robert de Cheyne; Sheriff also of Shropshire: but for some offence he was fined, his lands and goods

seized for the king, and his body attached.
Roger Trumwinne: two years.
Robert de Grendon, of Grendon, county of Warwick: three years.
John de Swinnerton, of Swinnerton: two years.
Henry de Bishbury, of Bishbury: three years.

EDWARD III.

John de Hinkele, and Henry de Bishbury: two years.
John de Hinkele: two years.
Henry de Bishbury: two years.
Richard Peshale, of Peshal and Horsley: two years.
John de Hinkele, Stoke, near Stone.
Simon de Ruggeley, of Rugeley.
Richard de Peshale, and Simon de Ruggeley: four years.
Adam de Peshale.
Thomas de Swinnerton: two years.
John de Aston.
Henry, Earl of Derby, of Tutbury Castle: seventeen years.
John de Swinnerton.
Robert de Grendon.
John de Perton.
Philip de Lutteley, Lutteley: 4 years.
Henry Pius, Pyott, or Pye.
John de Perton, of Perton, near Wolverhampton: two years.
John de Gresley, of Colton.
Nicholas de Stafford.
John de Verdon, of Alveton Castle.
John Bassey, or Basset, of Cheadle.
Nicholas de Stafford.
Peter de Careswell, Caverswall Castle.
Walter de Hopton.
William de Careswell.

RICHARD II.

Brean Cornwall, of Hereford.
William Calleson.
John de Verdon.
Roger de Wirley, of Hamstead.
William de Walsall, of Walsall: two years.
Humphrey Stafford, of Grafton, Worcestershire.
William de Walsall.
Roger de Mavesyn, of Mavesyn.
Adamar de Lichfield.
William Chetwynd, Ingestre.
Humphrey de Stafford.
William de Walsall.
John Delves, Apedale or Hilderston.
John Swinnerton.
William de Shareshall, of Shareshall.
Adamar de Lichfield.
Robert Fraunces, Foremark, Derby.
Robert Mavesyn.
William Walsall: three years.

HENRY IV.

William de Shareshall, Shareshall.
Robert Mavesyn, Knight, and William Newport, Knight, of Weston-under-Lizard.
Robert Fraunces, Foremark.
Humphrey Stafford: two years.
William Newport.
William Walsall.
William Newport, Knight.
Robert Fraunces.
T. Aston, Knight, of Haywood.
John Delves.
Thomas Giffard, of Chillington.

HENRY V.

John Bassett, of Blore or Cheadle.
Robert Babthorpe.
John Delves.
Richard Vernon, of Harlaston.
John Meverel, of Throwley.
William Trussel, of Cubbleston.
Humphrey Halton.
John Delves: two years.

HENRY VI.

Thomas Gresley, Knight.
Hugh Erdeswick, Esq. Sandon.
N. Montgomery, Leigh and Cubley.
John Bagot, Esq. of Blithfield.

Roger Aston, Esq.
Richard Vernon, Esq.
Philip Chetwynd, Esq.
Thomas Griffith, Esq. Whichnor.
Nic. Montgomery, Knight.
Roger Aston, Esq.
Ralph Egerton, Esq. Wrinehill.
Thomas Stanley, Esq. Elford.
Robert Strelley, Knight.
Nic. Peshale, Esq.
Philip Chetwynd, Esq.
Ralph Bassett, Esq.
Thomas Stanley, Esq.
Thomas Gresley, Esq.
Hum. Lowe, Esq. Tymmore
Ralph Archer, Esq. Seatfeld.
William Mytton, Esq. Mitton.
Nic. Montgomery, Knight.
Thomas Blount, Esq. Feld.
John Griffith, Knight.
Hum. Blount, Esq.
Thomas Ferrars, Esq. of Tamworth Castle: two years.
Hum. Swinnerton, Esq.
John Stanley, Esq. Elford.
Thomas Astley, Esq. Patshull.
Robert Aston, Esq.
Richard Bagot, Esq. Blithfield.
Th. Cotton, Esq. Hamstall Ridware.
John Delves, Esq.
John Cotes, Esq. Cotes.
William Mytton, Esq.
Hugh Egerton, Esq.
John Stanley, Knight.

EDWARD IV.

Walter Wrottesley, Esq.
John Harcourt, Esq.: two years.
Hum. Peshale, Esq.
John Stanley, Knight.
Thomas Bassett, Esq.
John Harcourt, Esq.
John Acton, Esq. Acton.
John Stanley, Knight, Elford.
Rand. Brereton, Knight.
Henry Beaumont, Knt. Wednesbury.
Walter Griffith, Knight.
William Bassett, Esq.
George Stanley, Esq.
John Stanley, Knight.
John Aston, Esq.
Hugh Egerton, Esq.
Richard Bagot, Esq.
Nic. Montgomery, Esq.
John Aston, Esq.
William Basset, Knight.
Hum. Stanley, Knight.

RICHARD III.

Nic. Montgomery, Esq.
T. Wolseley, Knight, Wolseley.
M. Constable and H. Stafford, Knts.

HENRY VII.

Hum. Stanley, Esq.
Hugh Willoughby, Esq.
William Harpur, Esq.
Hugh Peshale, Esq.
Thomas Gresley, Knight.
Rand. Okeover, Esq.
Roger Draycot, Esq. Paynesley-hall.
Richard Wrottesley, Esq.
Hum. Stanley, Knight.
Richard Harcourt, Knight.
John Mytton, Esq.
John Draycot, Esq.
Thomas Gresley, Esq.
William Harpur, Esq.
John Ferrars, Knight.
John Aston, Esq.
Richard Wrottesley, Esq.
William Harpur, Esq.
John Draycot, Knight.
William Smith, Esq. two years.
Lewis Bagot, Knight.
John Mytton, Esq.
—— Aston, Knight.

HENRY VIII.

J. Giffard, Esq.
Thomas Nevil, Esq. Shenstone.
John Egerton, Esq.
John Mytton, Esq.
Thomas Gresley, Esq.

William Chetwynd, Esq.
Thomas Nevil, Esq.
Richard Wrottesley, Esq.
John Giffard, Esq.
Ralph Egerton, Esq.
Edward Grey, Knight, Enville.
Lewis Bagot, Esq.
John Giffard, Knight.
William Smith, Knight, of Elford.
Edward Littleton, Esq. Pilaton.
Edward Grey, Knight.
John Giffard, Knight.
John Blount, Esq. of Blount's-hall.
John Vernon, Esq.
Edward Aston, Knight.
Thomas Giffard, Esq.
John Giffard, Esq.
William Wrottesley, Esq.
John Vernon, Esq. Harlaston.
Phil. Draycot, Knight.
Edward Aston, Knight.
William Chetwynd, Esq.
John Dudley, Knight.
George Gresley, Knight.
John Vernon, Esq.
Edward Littleton, Esq.
Edward Aston, Knight.
John Giffard, Knight.
William Basset, Knight.
T. Fitzherbert, Esq. Swinnerton.
George Gresley, Knight.
John Harcourt, Knt. Ranton Abbey.
Walter Wrottesley, Esq.

EDWARD VI.

Francis Meverell, Esq.
John Fleetwood, Esq. Colwich Priory.
William Snead, Esq. Bradwell.
Edward Littleton, Knight.
William Basset, Knight.
George Blount, Knight.

MARY.

Thomas Giffard, Knight.
T. Fitzherbert, Knight.
Peter Draycot, Knight.
Edward Aston, Knight.
John Harcourt, Knight.
William Snead, Knight.

ELIZABETH.

Hum. Walls, Esq. Horecross.
Ralph Bagnal, Esq. Newcastle.
John Leveson, Esq. Trentham Priory.
William Gresley, Esq.
Edward Littleton, Knight.
Ralph Okeover, Esq.
John Schrimshire, Esq. Aqualate.
Simon Harcourt, Esq.
John Wrottesley, Esq.
John Fleetwood, Esq.
Richard Bagot, Esq.
Walter Aston, Esq. of Tixall.
T. Trentham, Esq. Rocester Abbey.
George Blount, Knight.
John Giffard, Esq.
Thomas Horswood, Esq.
Ralph Adderley, Esq. Blakehaugh.
Ralph Snead, Esq.
Richard Bagot, Esq.
John Chetwynd, Esq.
Thomas Trentham, Esq.
Walter Aston, Knight.
Edward Littleton, Esq.
John Grey, Esq.
Thomas Gresley, Esq.
Edward Leigh, Esq. of Rushall.
Ralph Okeover, Esq.
Walter Leveson, Esq.
William Basset, Esq.
John Bowes, Knight, of Elford.
Robert Stanford, Esq. of Rowley, near Stafford, and Perry-hall.
Edward Aston, Esq.
Thomas Leveson, Esq.
Fran. Trentham, Knight.
Edward Littleton, Knight.
Henry Griffith, Esq.
Ralph Snead, Esq.
T. Horswood, Esq. Stourton Castle.
William Crompton, Esq. Stone Park.
Walter Wrottesley, Esq.
Walter Bagot, Esq.

William Chetwynd, Esq.
Wm. Skevington, Esq. Fisherwick.
Edward Leigh, Esq.
Walter Bagot, Esq.

JAMES I.

Walter Bagot and Ed. Leigh, Esqs.
William Whorwood, Knight, of Sandwell and Stourton Castle.
Gilbert Wakering, Knight.
Edward Brabason, Knight.
Walter Chetwynd, Esq.
James Schrimshire, Esq.
W. Heveningham, Esq. Aston.
Simon Weston, Knight, Rugeley.
Fran. Trentham, Esq.
Thomas Meverel, Esq. of Bold.
Thomas Littleton, Knight.
Richard Fleetwood, Bart.
John Pershale, Bart.
John Offley, Knt. Madeley Manor.
Hugh Wrottesley, Esq.
Thomas Schrimshire, Esq.
Henry Leigh, Esq.
Edward Winsor, Esq.
Ralph Snead, Esq.
W. Comberford, Esq. of Comberford.
William Skevington, Esq.
Edward Stanford, Esq.

CHARLES I.

Thomas Parker, Esq. of Park-hall.
Hervey Bagot, Bart.
William Bowyer, Knt. Kniperslev.
John Bowes, Esq. Elford.
John Cotes, Esq. Woodcot.
William Wollaston, Esq.
Thomas Broughton, Esq.
Thomas Whorwood, Knight.
Henry Griffith, Bart.
Hum. Wyrley, Esq.
Richard Pyot, Esq. Streethey, and Hum. Wyrley, Esq.
Edward Littleton, Bart. Pilaton.
John Skevington, Esq.
John Skrimshire, Knight, Aqualate
John Bellot, Esq. Moreton, Cheshire.
John Agard, Esq. of Fald.
Edward Moseley, Bart.
William Comberford, Esq.
Simon Rudgley, Esq.: two years.
Edward Mainwaring, Esq. Whitmore.
Thomas Kinnersley, Esq. of Loxley.
Walter Chetwynd, Esq.

CHARLES II.

Edward Seabright, Bart.
John Offley, Esq. Madeley.
—— Jolliffe, Esq. Caverswell.
Gerard Skrimshire, Esq.
Robert Ducie, Esq. of Sheustone: two years.
Brome Whorwood, Esq. Stourton Castle.
Thomas Wilbraham, Esq.
Thomas Chetwynd, Esq.
Rowland Cotton, Esq. or Ralph Okeover, Esq.
Jonathan Cope, Esq. Ranton Abbey.
Humble Lord Ward, Himley, and —— Shilton, Esq. Wednesbury.
Brian Broughton, Bart. Broughton.
John Belloe, Esq.
Joh. Bowyer, Knight and Bart. of Kniperslev.
William Sneyd, Esq. Keel.
Joh. Wyrley, Knight, Hamstead.
Francis Leveson Fowler, Esq.
Walter Wrottesley, Bart.
Joh. Whitehall, Esq. Pipe Ridware.
Edwin Skrimshire, Esq.
Rowland Cotton, Esq. Crakemarsh.
Richard Creswall, Esq.
Henry Gough, Esq. Perry-hall.
William Lawton, Esq.
Thomas Jolliff, Esq.
Joh. Wilson, Esq.
Thomas Allen, Esq.
Simon Unwin, Esq. Clough.
Richard Amphlet, Esq.
Edward Mainwaring, Esq.
Joh. Offey, Esq.
Edward Littleton, Esq.

Charles Skrimshire, Esq.
Matthew Floyr, Esq. Hints.
William Inge, Esq. Thorpe.

JAMES II.

Walter Chetwynd, Esq.
Jonathan Cope, Esq.
Walter Wrottesley, Bart.
Philip Draycote, Esq. of Painsley.

WILLIAM AND MARY.

William Cotton, Esq.
William Ward, Esq.
Joh. Newton, Esq.
Grosvenor Dyson, Esq.
William Jolliffe, Esq.
William Bagnall, Esq.
Joh. Taylor, Esq.
Joh. Chetwynd, Esq.
William Murhall, Esq.
Joh. Chetwode, Esq. Oakley.
Walter Launder, Esq.
Oswald Moseley, Esq. Rolleston.
Benjamin Jolliffe, Esq.
Thomas Nabbs, Esq. Stafford.

ANNE.

Joh. Babington, Esq.
Thomas Okeover, Esq.
Matthew Ducie Moreton, Esq.
Thomas Crompton, Esq. Stone.
William Trafford, Esq.
James Wood, Esq.
Joh. Jarvis, Esq. Darlaston.
Edmund Arblaster, Esq. Longdon.
Walter Moseley, Esq.
Leigh Brook, Esq. of Haselover.
Edward Littleton, Esq.
Harry Grey, Esq. Enville.

GEORGE I.

Oswald Moseley, Esq.
Thomas Birch, Esq.
J. Turton, Esq. Orgreave.
T. Whitby, Esq. Creswell.
Thomas Brown, Esq.
Ralph Sneyd, Esq. Keel.
Hum. Hodgetts, Esq.
John Turton, Esq.
Henry Goring, Esq. of Kingston.
—— Scott, Esq.
—— Townsend, Esq.
—— Hussey, Esq.

GEORGE II.

Edward Wilson, Esq.
Joh. Newton, Esq. King's Bromley.
Joh. Arden, Esq. Longcroft.
William Robins, Esq.
R. Williamson, Esq. Stafford.
W. Faulkner, Esq. Rugeley.
C. Kinnersley, Esq. Loxley.
Joh. Hodgetts, Esq. Prestwood.
R. Adderley, Esq. Coton.
—— Jolliff, Esq.
Richard Fowler, Esq. Penford.
William Inge, Esq.
Richard Davenport, Esq. Ellaston.
Joh. Dolphin, Esq. Shenston.
William Murhall, Esq.
Sir Robert Lawley, Bart. Canwell.
Thomas Webb, Esq. Blakenall.
Samuel Hellier, Esq. Woodhouse.
Ch. Bosville, Esq. Biana.
George Hunt, Esq. Rocester.
Joh. Jervis, Esq.
Joh. Wyrley Birch, Esq. Hamstead.
Edward Busby, Esq. of Barton-under-Needwood.
Henry Vernon, Esq. Hilton.
T. Bradney, Esq. Penn.
Richard Drakeford, Esq. Stafford Castle Church.
Thomas Mills, Esq. Leek.
J. B. Leigh, Esq. Rushall.
J. T. Chetwode, Esq. Oakley.
W. Acton Moseley, Esq.
R. Whitworth, Esq. Batchacre.
Sir Nigel Gresley, Bart. Mere.

GEORGE III.

Joh. Dolphin, Esq. Shenston.
J. Smith, Esq. Fenton.
Sir Edward Littleton, Bart.
J. Sneyd, Esq. Bishton.
William Armett, Esq. Heaton.

J. Hodgetts, Esq.
William Inge, Esq.
Edward Mainwaring, Esq.
F. Eld, Esq. Seighford.
C. Kinnersley, Esq. Loxley.
John Marsh, Esq. Wombourn.
Thomas Hoo, Esq. Barr.
Sir Thomas Broughton, Bart.
Thomas Whitby, Esq.
John Williamson, Esq. of Stafford.
John Turton, Esq. Sugnall.
John Swinfen, Esq. Swinfen.
E. Walhouse Okeover, Esq.
R. Floyer, Esq. Hints.
P. Keay, Esq. Abbot's Bromley.
Ch. Tollet, Esq. Betley.
R. Gildart, Esq. Norton.
Sir John Heathcote, Longton.
T. Stevenson, Esq. Stafford.
T. Parker, Esq Park-hall.
T. Whieldon, Esq. Fenton.
T. Fletcher, Esq. Newcastle.
T. L. Fowler, Esq. Penford.
John Sparrow, Esq. Bishton.
M. Walhouse, Esq. Hatherton.
Simon Debank, Esq. Leek.
G. Molineux, Esq. Wolverhampton.
M. Boulton, Esq. Soho.
T. Swinnerton, Esq. Butterton.
Henry Vernon, Esq. Hilton.
Sir Robert Lawley, Bart.
Richard Dyott, Esq. Freeford.
Jos. Scott, Esq. Barr.
H. Farmer Oakover, Esq.
Thomas Bambridge, Esq.
Robert Parker, Esq. Park-hall.
George Birch, Esq.
Richard Jesson, Esq.
John Heyligier Burt, Esq.
William Phillips Inge, Esq.
John Lane, Esq.
William Cary, Esq. Cannock.
Theophilus Levett, Esq.
Henry Webb, Esq.
James Beech, Esq.
Thomas Mottershaw, Esq.
Walter Sneyd, Esq.
Sir Oswald Mosley, Bart.
Henry Crockett, Esq.
John Smith, Esq. Fenton.

MEMBERS OF PARLIAMENT FOR STAFFORDSHIRE.

1542. John Dudley and Philip Draycot, Knts.
1547. William Paget and John Harcourt, Knts.
1552-3. William Devereux and Walter Aston, Esqs.
1553. Thomas Giffard, Knt. and Edward Littleton, Esq.
1554. John Draycot, Knt. and Thomas Gray, Esq.
1554-5. Thomas Giffard and Edward Littleton, Knts.
1556. Brian Fowler and Francis Meverell, Esqs.
1558. Ralph Bagnall Esq, and Simon Harcourt, Knt.
1563. Simon Harcourt and John Gray, Esqs.
1571. John Gray and Thomas Trentham, Esqs.
1572. John Fleetwood and Thomas Whorwood, Esqs.
1585. Edward Dudley and Edward Leigh, Esqs.
1586. John Gray and William Basset, Esqs.
1588. Walter Harcourt and Thomas Gerard, Esqs.
1592. Walter Harcourt and Christopher Blount, Knts.
1597. Edward Sutton, *alias* Dudley, Esq. and C. Blount Knt.

1601. Thomas Gerard and John Egerton, Knts.
1603. Edward Littleton and Robert Stanford, Knts.
1614. Edward Littleton and William Bowyer, Knts.
1620. William Bowyer, Knt. and Thomas Crompton, Esq.
1623. William Bowyer and Edward Littleton, Knts.
1625. Simon Weston, Knt. and William Bowyer, Esq.
1628. Harvey Bagot and Thomas Crompton, Knts.
1640. Edward Littleton, Bart. and William Bowyer, Knt.
1640. Edward Littleton, Bart. and Harvey Bagot, Knt.
1653. George Bellot and John Chetwood, Esqs.
1654. Charles Wolseley, Bart. and Thomas Crompton, Esq.
1656. C. Wolseley, Bart. Thos. Crompton and Thos. Whitgrave, Esqs.
1658-9. Thomas Whitgrave, Knt. and Thomas Crompton, Esq.
1660. Edward Bagot and William Snead, Esqs.
1661. Sir Thomas Leigh and Sir Randolph Egerton.
1678. Sir Walter Bagot and Sir John Bowyer.
1681. The same, The same.
1685. The same, and Edward Littleton, Esq.
1688. The same, and John Grey, Esq.
1690. The same, The same.
1695. Henry Pagett, Esq. The same.
1698. Edward Bagot and Henry Pagett, Esq.
1701-2. The same, The same.
1705. The same, The same.
1708. Sir John Wrottesley, The same.
1710. William Ward and Charles Bagot, Esqs.
1713. Henry Vernon and Ralph Sneyd, Esqs.
1714. William Ward, Esq, and Lord Pagett.
1722. W. Leveson Gower, Esq. The same.
1727. The same, and Sir Walter Bagot.
1734. The same, The same.
1741-7. The same, The same.
1754. William Bagot and Henry Thynne, Esqs.
1761. Right Hon. Lord Grey and Wm. Bagot, Esq. L.L.D.
1662. The same, The same.
1763-8. The same, The same.
1778. Sir Wm. Bagot, and Sir John Wrottesley, Barts.
1782. Sir J. Wrottesley, Bart. and Viscount Lewisham.
1793. Lord Granville Leveson Gower, and Sir Edward Littleton, Bart.
1796. The same, The same.
1802-7. The same, The same.
1812. The same, Edward J. Walhouse, Esq
1815. Right Hon. Earl Gower, The same (now Littleton).

Dr. WILKES's VIEW OF STAFFORDSHIRE FROM THE ROMANS TO THE YEAR 1735.

RELIGIOUS HOUSES, AND THEIR VALUE WHEN DISSOLVED BY HENRY VIII.

		Founder.	Date.	£	s.	d.
Abbeys.	Burton,	Wulfric,	1004,	356	16	3
	Croxden,	Bert de Verdon,	1176,	103	6	7
	Dieu la Cross,	R. Earl of Chester,	1220,	243	3	6
	Hilton,	Henry de Audley,	1223,	76	14	10
	Radmore,	(Removed to Stoneley),	1154,			
Nunneries.	Blithbury,	Hugh Mavesyn,				
	Black Ladies	Isabel Lander,				
	White Ladies,	Hubert Walter,	1195,	11	1	6
	Fairwell,	Bishop Clinton,	1140.			
Priories and Cells.	Calwich,	Geva Ridel	1125,			
	Dudley,	Ger. Paganel,	1155,			
	Lapley,	Algar, *(a Saxon)*	1146,			
	Ranton,	Robert Fitz Noel,	1180,	90	2	10
	Rocester,	Richard Bacon,	1140,	111	13	7
	St. Thomas,	Gerard de Stafford,	1162,	141	13	2
	Sandwell,	William Offney,	1155,			
	Stone,	Robert Lord Stafford,	1100,	119	14	11
	Trentham,	Robert Earl of Chester,	1218,	106	3	10
	Tutbury,	Henry de Ferrars,	1081,	244	16	8
Colleg. Churches.	Gnosall,					
	Lichfield,	King Oswy,	656,			
	Penkridge,	King Edgar,	964,			
	Stafford,					
	Tettenhall,					
	Wolverhampton,	Lady Wolfruna,	996.			

Furnaces.		
Trescote Grange,	Madeley,	Rushall,
Leek,	Mere-heath,	Teanford.

MANSIONS OF NOBILITY AND GENTRY.

Alveton, ...	.. Earl of Shrewsbury.
Beaudesert Park, ..	.. Earl of Uxbridge.
Caverswall, ...	.. Viscount Vane, *(Irish)*.
Chartley Park, ...	.. Earl Ferrers.
Croxden, ...	.. Earl of Macclesfield.
Drayton Park, ...	.. Viscount Weymouth.
Dudley Park, ...	.. Lord Dudley.
Elford, ...	.. Earl of Berkshire.
Enville, ...	.. Earl of Stamford.
Fisherwick Park, ..	.. Visc. Massareen, *(Irish)*.
Himley Park, ...	.. Lord Dudley.
Hore Cross, ...	.. Earl of Bristol.
Ingestre Park, ...	.. Viscount Chetwynd.
Ingleton, ...	.. Lord Moreton.
Painsley, ...	.. Lord Langdale, *(Irish)*.
Sandon, ...	.. Duke of Hamilton.
Sandwell, ...	.. Earl of Dartmouth.
St. Thomas, . ..	.. Viscount Falconbridge.
Sugnall, ...	.. Lord Glenorchie, *(Scotch)*.
Tamworth, ...	.. Earl of Northampton.
Tixall Park, ...	.. Lord Aston.
Trentham Park, ..	.. Lord Gower.
Weston, ...	.. Earl of Bradford.

Market Towns.	Market Days.	Market Towns.	Market Days.
Alrewas, (d*)	*Tuesday.*	Penkridge, (s)	*Tuesday.*
Breewood, (s†)	*Monday.*	Rugeley,	*Thursday.*
Betley,	*Friday.*	Stafford,	*Saturday.*
Bromley, Abb. (s)	*Tuesday.*	Stone,	*Tuesday.*
Burton,	*Thursday.*	Swinnerton, (d)	*Wednesday.*
Cannock, (d)	*Tuesday.*	Tamworth,	*Saturday.*
Cheadle,	*Friday.*	Tutbury, (s)	*Saturday.*
Eccleshall,	*Friday.*	Uttoxeter,	*Wednesday.*
Kinfare, (d)	*Tuesday.*	Walsall,	*Tuesday.*
Lapley, (d)	*Tuesday.*	Wednesbury, (s)	*Friday.*
Leek,	*Wednesday.*	Wolverhampton,	*Wednesday.*
Lichfield,	*Tuesday & Friday.*	Yoxall, (d)	*Saturday.*
Newcastle,	*Monday.*		

* D. *Disused.* † S. *Small Account.*

Kings of Mercia.	Died.	Kings of Mercia.	Died.	Kings of Mercia.	Died.
Crida,	595	Kenred,	709	Ceolwulf,	822
Wibba,	615	Ceolred,	716	Beornwolf,	826
Ceorl,	625	Ethelbald,	757	Ludican,	828
Penda,	655	Beornred,	757	Withlaf,	840
Peada,	656	Offa,	796	Beortulf,	853
Oswy,	658	Kenwulf,	820	Bertred,	875
Wulfere,	675	Kenelm,	820	Ceolulfe,	886
Ethelred,	704				

Iron Forges.	Iron Forges.	Iron Forges.
Aston Parva,	Cannock-wood,	Wombourn-heath,
Breewood,	Chartley,	Hints,
Bromley Regis,	Congreve,	Oakamoor, 2
Bromwich,	Coven,	Swin,
Burton,	Cunsall,	Whittington.

Arley, ...	Sir Thomas Littleton.
Aston Park, ...	Sir Edward Symons.
Blythfield Park, ...	Sir William Bagot.
Broughton, ...	Sir Bryan Broughton.
Barton, ...	Sir Charles Bromfield.
Cannock, ...	Sir Robert Fisher.
Canwell, ...	Sir Robert Lawley.
Clifton, ...	Sir Robert Pye.
Elmhurst, ...	Sir Theophilus Biddulph.
Knipersley, ...	Sir Thomas Gresley.
Oakley, ...	Sir John Chetwood.
Patteshull Park, ...	Sir John Astley.
Pilaton Park, ...	Sir Edward Littleton.
Rolleston, ...	Sir Oswald Moseley.
Ranton, ...	Sir Jonathan Cope.
Wolseley Park, ...	Sir William Wolseley.
Wrottesley, ...	Sir Richard Wrottesley.
Aqualate, ...	Charles Baldwin, Esq.
Bentley, ...	John Lane, Esq.
Hilton, ...	Henry Vernon, Esq.
Heylins, ...	John Turton, Esq.
Keel, ...	Ralph Sneyd, Esq.
Loxley, ...	Craven Kinnersley, Esq.
Madeley, ...	John Crewe, Esq.
Norbury, ...	T. B. Skrimsher, Esq.
Okeover, ...	L. Okeover, Esq.
Sinai, ...	Earl of Uxbridge.
Stone, ...	W. Leveson Gower, Esq.
Whichnor, ...	Crewe Offley, Esq.
Wotton, ...	John Wheeler, Esq.

GENERAL TOPOGRAPHY.

WITH the exception of Cannock-heath, the middle part of the county is for the most part level: this immense waste swells to a considerable elevation, from which issue many streams, that flow in all directions, and afterwards unite in the Trent.

East of Cannock-heath, and all the way to the Tame, is a tract of light soil, or sandy and gravelly loam; and east of that river, is Elford, Wigginton, Haselor, Harlaston, Sclercott, Thorp-Constantine, and Clifton Campville, on a strong or mixed loam, or useful corn and pasture land. Between the Trent and the Dove is Needwood Forest (lately inclosed), and as far up as Uttoxeter, is a tract of strong or loamy soil; and again, west of Stafford and Penkridge, and on to Eccleshall and the borders of Shropshire, is a strong or mixed marl or loam. The country south of Cannock-heath is a mixed gravelly loam; but the part south and west of Wolverhampton is a lighter soil. To the north of Dudley, and again near Walsall, are considerable tracts of limestone, some in elevated cliffs, and others in strata beneath the surface, either in detached parts or solid rock: the limestone in both places is proved to be nearly upon the same level, by canals communicating between them, and into the heart of the works, at 500 feet above the level of the sea. Some of the cliffs are considerably above that level, and the under-strata beneath it: petrified marine substances are found both imbedded in the solid rock, and in the fissures between the strata.

The *North-west* of the county from Stone, through Newcastle to Talk-on-th'-hill, and thence to Betley and Drayton, is generally loam, with a varied understratum of sand, gravel, marl, or gritstone rock; the soil, being the happy medium between the sterility of sand and the harshness of clay, is adapted either to tillage or pasture. The country is inclosed with quicksets well planted, the timber trees are luxuriant and of flourishing growth, the meadow and grazing lands covered with a good herbage, and the corn crops have generally a promising appearance: upon the whole, this district may be termed a fine country, though not without some spots of inferior land. Near Maer-heath, and Ashley Common, may be found a thin black peat-moor soil, and on the hilly parts, poor sand, gravel, or rock.

The *Southern* part of the county has some peculiarities, and is less level than the middle. The limestone hills, and cliffs of Sedgley and Dudley Castle, rise to two or three hundred feet above the surface, and furnish an inexhaustible supply of that material; while Rowley Regis exhibits a striking singularity, being in itself an insulated mountain, composed of a species of granite or primeval stone, ending in various peaks or summits, but covered with a rich vegetable earth. The loftiest point, called Turner's Hill, is the most elevated ground in the south of Staffordshire: the other highest points of Rowley Mountain, are Oakham and Corney-hills. The base of this mountain is composed of an extremely hard, rusty-blue coloured substance, called Rowley rag-stone, not at all gritty or calcareous: it lies in an infinite number of fragments, some of them of immense size, both beneath, and upon, the surface. Large quantities of it are carried to Birmingham, and other places, for paving and repairing roads. The rock called Rowley hail-stone is of great size, and similar quality. The surface-soil is a good marly loam, retentive of moisture even in its elevated situation, and produces a good herbage of grass. The roads in this parish are rocky precipices, and most of the heavy carriage is performed upon the backs of horses.

The hills of Clent, to the south-west of Rowley, are composed of stone-trash (*traccia arenacea,*) or innumerable small fragments of broken rock-stone intermixed with a red sandy loam: the lower ground partakes more of a marly or loamy quality. The stone-trash apparently goes to a great depth, and by quickly imbibing rain, renders the soil porous and liable to burn in hot summers, except where the moisture is retained by the marly loam abounding in the surface-soil. A considerable part of Clent hills is sheep-walk, though covered generally with a fine turfed herbage: these hill-summits nearly equal those of Rowley in height, and seem capable of the turnip and barley culture. There is a great deal of waste land in this part of the county, producing heath, fern, sheep's fescue, matt grass, white galium, and sheep's sorrel.

Kinver is on a light gravelly soil with a grit-rock bottom, of various quality, more or less fertile: the precipice of Kinver-edge has already been noticed. Some progress towards improvement is making in a considerable tract of land in this neighbourhood, inclosed some years ago. West of Kinver, the soil changes into a strong clayey or marly loam, more or less harsh or friable, including Enville, Bobbington, and Over Areley, (on the Severn.) Here the hills are often rocky, with small fragments upon and near the surface, in some parts quartzose, in others gritty: the surface-soil changes in colour from red to grey, with all the shades between: the cultivation the same as on other strong lands. Enville is famous for black cherries, and the Wake held here is called Cherry Wake. The pleasure-grounds of Enville (the magnificent seat of the Earl of Stamford and Warrington) are extensive, and well-stocked with timber trees and underwood of every description, amongst which the oak preponderates in great profusion, and is of kindly and vigorous growth: these grounds command some beautiful and extensive prospects. Over Areley was formerly famous for fruit, especially apples, but the orchards have been suffered to decline: some recent attempts have been made towards restoring them, and the soil and aspect are well adapted for the purpose.

The *North* part of the county, called the Moorlands, is situated to the north and north-east of a line conceived to be drawn from Uttoxeter to Newcastle-under-Lyme, and includes about one-sixth of the county: it is bounded on the east by the Dove, which separates it from Derbyshire, as the Dane parts it from Cheshire. The face of this portion of the county is various, but in general hilly, with large tracts of uncultivated land. The first market-town from Uttoxeter is Cheadle, situate in a pleasant vale of good mixed soil, but surrounded by large and barren hills, composed of huge heaps of sterile gravel. Cheadle-park, now in cultivation, abounds in mines of coal of great value; but the hills are generally on an understratum of sand or sandy rock, and covered with heath (*erica vulgaris,*) broom, whortle-berries, mountain carex's, cinquefoils, cestus's, matt grass, and knap weeds, all of diminutive growth: these barren wastes are very extensive.

The commons or wastes between Cheadle and Oakamoor, called High Shutt, Ranger, and Alveton-common, consist of an immense number of rude heaps of gravel, with an understratum of soft sandy rock, thrown together without order or form, or rather into every form that can be conceived.*

A little to the north of Oakamoor, so called from being covered with dwarf oaks, the limestone country begins: it extends up Dovedale to near Longnor, and includes, from the Dove to Morredge,

* This barren tract, impracticable to the plough, might be improved into woodland and plantation, and some open spots of the most favourable aspects

a breadth of fifty or sixty square miles. The quantity of limestone here is inexhaustible, and in many places of immense thickness. This is the best part of the Moorlands: the soil seems to have a natural aptitude for producing a fine herbage of grass.*

The Weaver-hills are of considerable extent, and composed of immense heaps of limestone, covered with a good calcareous earth capable of being improved into arable and pasture land: they are inclosed in large tracts by stone walls, but not subdivided, and large breadths have never undergone the least improvement. These hills are very lofty, and nearly as high as the Moorland or Derbyshire Peak-hills, which may be seen from their summits: the fall even from the foot of them to the highest parts of the Dove or the Churnet, is very great; and those rivers are here very rapid.†

Stanton Moor, to the east of these hills, is a considerable waste on a limestone surface: the limestone is intermixed with gypsum or alabaster. Lime being much used here as manure both on ploughed land and turf, large quantities of it are burnt in this neighbourhood: it has been observed, that after liming a coarse turf, white clover has flourished abundantly where that plant had not before been noticed.

The fences, in a large tract of this country, are almost wholly of stone walls, built without cement or mortar, and consequently liable to gaps and breaches: quickset-fences are certainly much superior, both for shelter and appearance, and particularly as these walls do not admit of drains or ditches alongside them.

reserved for gardens to cottage tenements, and cultivated with the spade and hoe. Upon one of these rocky declivities, to the north-east of Oakamoor, is a thriving plantation of Scotch fir, spruce, oak, lime, birch, sallow, and mountain ash: this is perhaps the highest improvement to which the sides of barren precipices can possibly be brought by human industry; and it were to be wished that all public and patriotic societies that offer premiums for planting, would limit such premiums to grounds impracticable to the plough, or of small value, not exceeding per acre a specified sum. The practice of planting extensively upon rich level arable or pasture land, may be considered a national evil, such land being much more useful in culture.

* The breed of cows of the long-horned kind, in this district, is generally superior to those of the south of the county: this may be attributed to the plough and the growth of corn being chiefly attended to in the latter division, whilst here stock forms the principal object: but this general rule, like many others, must be admitted with some exceptions.

† On these summits, in a very red soil, the Upland Burnet (*poterium sanguisorba*) grows in profusion among the limestone: it would be worth while to transplant the young roots, and save the seed for cultivation as winter food for sheep, the plant being remarkably hardy, and good food for sheep or cattle.

The Moorlands, in point of elevation, are far above the south of the county, rising at least from 100 to 200 perpendicular yards above the general level: on these heights are large tracts of inclosed pasturage. This elevation greatly increases the degree of cold, by combining the causes that produce it, and by being more exposed to the elements of air and water in every form and variety of wind, hail, snow, or rain, and the country is almost in a state of nature, without any shelter or covering except stone walls, which have little effect in keeping off the cold winds: the little wheat and barley sown here will not ripen, nor oats in due time, except the season be very favourable.

The plantations at Oakover (partly of oak), are in a flourishing state; and the house, situated in a paddock of excellent turf, commands a fine view of the Dove, with a neat Gothic church: the paddock is well stocked with deer.

At Ilam, farther up the Dove, is a very pleasant seat, which, from its romantic situation, suggests the idea of a glen in the Alps: two considerable rivers, the Hamps and Manyfold, here rush from under the limestone hills, after a subterraneous passage of several miles in separate streams. The precipices which surround the valley in which Ilam stands, are well clothed with oak, and other wood; the pleasure-walks from the seat, on one of these precipices, are diversified and beautiful, and resemble shelves, one almost perpendicularly above another, the sides of which are furnished by nature, with scarcely any assistance from art, with a profusion of flowers of no ordinary appearance.* Several curious specimens of petrified fish are preserved here, said to be of the carp or barbel species, but which, in reality, are fragments of chert *(silex petre)* embodied in the limestone rock when in a state of fluidity; the limestone being softer than the chert, most of these fragments project in the middle, which gives them a resemblance to fish: three or four very good specimens are shewn to those who visit Ilam. The subterraneous rivers here form two very fine cascades.

The bold and romantic hills of Thorp Cloud, and Bunster, on either side the Dove, may be seen from these grounds: the latter (in Staffordshire), is an immense heap of limestone, but covered with a light earth, and well stocked with rabbits: between these

* Amongst others, native geraniums of different sorts, the centaurea scabiosa, and other shewy natives: in a meadow over the water, the ramson *(allium ursinum)* a shewy flower but no desirable pasture-plant, flourishes in great profusion.

hills the Dove falls in abrupt cascades. Large quantities of limestone have fallen from the overhanging precipices into the channel of the river.*

Mill-dale, near Alstonefield, is a long narrow vale or glen of great depth, the sides of which are composed of overhanging precipices of limestone, estimated to be from 100 to 150 yards of perpendicular elevation, and so very steep that they can be ascended only in very few places: the width of this glen, vale or dale, at the top, scarcely exceeds the depth of its sides.

The vale of Manyfold is situated between Wetton and Butterton, where the waters of the Manyfold are absorbed by the fissures under the limestone hills, and discharged again at Ilam, four miles below. The warmest imagination can scarcely conceive a spot more wild and romantic than some parts of this vale. Thor's-house Cavern, is a considerable excavation on the side of a lofty precipice; it has somewhat the appearance of the inside of a Gothic church, and seems to be in part a work of art.

The tract of country north-east of Mole-cop, is the worst part of the Moorlands, and of Staffordshire, the surface of a considerable portion of it being too uneven for cultivation. Large tracts of waste land here, though so elevated in point of situation, are merely high moors and peat-mosses; of this description is a part of Morredge, Ax-edge, the Cloud-heath, High-forest, Leek-frith, and Mole-cop, though amongst the highest land in the county. The calcareous or limestone bottom ends at Morredge, and the understratum, in the tract of country to the west, is generally sandy or gravelly clay or gritstone rock.

The summits of some of the hills in this county terminate in huge tremendous cliffs, particularly those called Leek Rocks or Roches, and Ipstones Sharp Cliffs, which are composed of large heaps of rude and rugged rock piled one upon another, and have a most terrific aspect: in some places single blocks of immense size are heaped together, and overhang the precipices, threatening destruction to the traveller; and some broken masses of prodigious bulk, have evidently rolled from the summits. Leek Rocks or Roches, are

* The sides of the immense precipices of Bunster have a scattering of wood, which might be increased: the ridge of this mountain terminates in some places in a number of conical sugar-loaf hills of bare limestone. The highest point of Bunster is estimated at 300 perpendicular yards above the Dove, and the fall of that river from hence to the Trent is 100 yards more. This hill, though of great elevation, is not equal to the Weaver, nor so high as some of the Peak-hills over the Dove.

composed of a coarse sandy grit-rock; those of Ipstones of coarse plum-pudding stone *(breccia arenacea)*, and seem like sand and small pebbles cemented together.

It is evident, from the huge tremendous cliffs in which these hills, and others in the neighbourhood, terminate, and the immense fragments of broken stones which lie scattered in every direction, that these rocks, at an early period, have been torn in pieces by some violent convulsion of nature. These stone cliffs and fragments cover a considerable extent of country, particularly about Ipstones, Wetley Rocks, Leek Rocks, to the west of Flash, High Forest, the Cloud-heath, and Mole-cop common, with the waste to the north of this common, and in many other places.

Upon Morredge and Axheath commons, large quantities of peat are dug for fuel. The peat-mosses on Morredge are generally three or four feet deep, with an understratum of gravelly clay. The peat here, and on Axedge, is extremely porous and retentive of moisture, and even in a dry season, yields sufficient water from pressure to supply several small lakes and rivulets, and starves the natural herbage: indeed, nothing grows on or near these spots except heath *(erica)*, whortleberries, the cottony rush or grass *(eriophorum vaginatum)*, carex's, and rushes. One great obstruction to the improvement of these lands, is the immense quantities of stone lying on, or of rocks rising out of the surface. The most sanguine friend to agriculture can expect little more here than sheep-walk or plantation, and it would be praise-worthy in the landed proprietors at least to attempt the latter, in which there can be no doubt but their efforts would be crowned with complete success.

OFFLOW HUNDRED.

THIS division of the county of Stafford, is bounded on the east by Warwickshire; on the north-east by the Trent; on the north by the Dove; on the west by Cuddlestone and Seisdon Hundreds; and on the south by Shropshire and Worcestershire. It contains the City of Lichfield, and the towns of Burton-upon-Trent, Walsall, Abbot's Bromley, and Tamworth, with the intermediate parishes.

The general appearance of this part of the county is moderate elevation, with considerable tracts of meadow-land on the banks of the Trent, Tame, and Dove; a light soil from Watling-street to the Trent; and a stronger loam east of the Tame, and between the Trent and the Dove. This latter district includes Needwood Forest, formerly overgrown with wood, but now by improved cultivation a fertile and interesting part of the county.

BURTON-UPON-TRENT.

This town, which is of great antiquity, is situated on the northern bank of the river Trent, in the north-east part of the Hundred of Offlow; and in 52° 53′ north latitude, and 1° 35′ west longitude. It is distant east from Stafford twenty-five miles, north-east from Lichfield twelve miles, and 126 miles north-west from London.

Burton was a considerable place at an early period of the Saxon dominion in this Island. In the Saxon annals it is written Byretun, synonimous to Buryton, a word used by that people to denote places of British or Roman origin. It is therefore probable that a Bury or manor-house, was the residence of some eminent personage here, before the Saxons conquered England. The site of Burton is indeed extremely pleasant. Between an extensive space of level and fertile meadows on the west, and a beautiful range of hills on the east, runs the river Trent; these and other hills about two miles north-west of the town, were formerly covered with wood, thus combining all the attractions of a commanding station, fine prospect, and fruitful soil, the principal requisites for the establishment of a permanent habitation.

The ancient records of Burton commence with the history of St. Medwen, who flourished in the ninth century.—" In this season, one Modwen, a virgin in Ireland, was greatly renowned in the world, unto whome King Ethelwolfe sent his sonne Alfred to be cured of a disease that was thought incurable; but by hir meanes he recovered health, and therefore when hir monasterie was destroied in Ireland, Modwen came over into England, unto whom King Ethelwolfe gave land to build two abbies, and also delivered unto hir his sister Edith to be professed a nun. Modwen hereupon built two monasteries, one at Poulsworth joining to the bounds of Arderne, wherein she placed the foresaid Edith with Osith and Athea; the other, whether it was a monasterie or cell, she founded in Strenshall, or Trentsall, where she hirselfe remained solitarie a certain time in praier and other vertuous exercises. And (as it is reported) she went thrice to Rome, and finallie died, being 130 yeeres of age. Hir bodie was first buried in an iland compassed about with the river of Trent, called Andresey, taking the name of a church or chappelle of Saint Andrew, which she had built in the same iland, and dwelled therein for the space of seven yeeres. Manie monasteries she builded both in England, and also in Scotland, as at Striveling, Edenbrough; and in Ireland at Celestline, and elsewhere."*

In the year 1002, and the twenty-third year of the reign of Ethelred, the Abbey of Burton was founded by Wulfric, Ulfric, or Alfric, Earl of Mercia. The book of Abingdon asserts that Wulfric gave his whole estate, then valued at £700. to endow this Abbey. In the year 1004 this donation was ratified by King Ethelred, and in 1010 Wulfric was slain in an engagement with the Danes.

When Burton Abbey was in its most flourishing state, it possessed very extensive domains, consisting of manors, townships, and lands in this county, and in Derbyshire, Leicestershire, and Warwickshire. In the year 1260, the Abbey was inhabited by thirty monks. There were thirty-five Abbots in succession from the year 1004 to the 4th of November, 1540, when King Henry VIII. after the dissolution of monastic institutions, made the Abbey of Burton a Collegiate Church, dedicated to Christ and St. Mary; to consist of a Dean and four Prebends, and endowed it with the manor of Burton and other possessions. It continued to enjoy this distinction only four years. The seal of the college is beautiful: it is a representation of Our Saviour and his disciples at the Last Supper; the arms of Wulfric, the founder of the Abbey, are engraved at the bottom; and the words in the margin, in Latin, signify "The common seal of the Dean and Chapter of the Collegiate

* Holingshed, b. vi. c. ii. p. 142.

Church of Christ, at Burton-upon-Trent." This Abbey was very extensive; the church was 228 feet in length, and 52 feet and a half in breadth, adorned with a handsome tower at both ends. The cloisters, situated on the south side of the church, were 100 feet square. The fraytor adjoining was 96 feet in length and 30 in breadth, and the principal dormitory on the east 100 feet by 10. The vestiges of the cloisters are still visible in the old wall between the church-yard and the bowling-green. In this wall are several remains of Saxon architecture, particularly some curious arches; and beyond these, on the margin of the Trent, are some other vestiges of the Monastery. The most remarkable is the outline of the east window of a chapel, now filled up with brick and mortar. The old Manor House is the property of the Marquis of Anglesea: what was once the porter's lodge, on the side next the town, is now converted into a blacksmith's shop.

The Church dedicated to St. Modwen was standing in the early part of the eighteenth century; it was taken down, and the present neat edifice, with a fine tower, was erected on its site in the year 1720. In the west gallery is an elegant and well-toned organ, erected in 1771, and the tower contains eight musical bells with a set of excellent chimes. This church contains many monumental inscriptions: in the belfry lies a defaced monument, supposed to be the tomb of the original founder of the Abbey; and within these few years, several stone coffins have been found in the church-yard.

But the most remarkable antiquity in Burton is its bridge, which consists of thirty-six arches, extending 515 feet, and was first built in the time of *Bernard*, Abbot of Burton, in the reign of King Henry II. about the year 1175.*

Burton retains the name of a borough, merely with reference to the translation of *Burgos*, a town, though some of the Abbots were formerly summoned by writ, and sat in Parliament.† The Lord of the Manor has a Court of Record in it, weekly, on Friday, called the Jenter's Court, where pleas can be maintained to any amount, and his Bailiff is also Coroner. A Court of Requests, for the recovery of small debts, is likewise held under the Bailiff. Among other privileges, the inhabitants are exempted from being empannelled on county juries.

The town of Burton consists of one principal street, called High-street, which extends in a parallel line with the river from north to

* Erdeswick's Survey of Staffordshire.

† Shaw's Antiquities of Staffordshire, Vol. I. p. 11.

south. This street is intersected at right angles by another, called Horninglow-street. The Town-hall, which is a neat edifice, is erected on the site of the old market-house. In this hall are held the Courts Leet, and all other public meetings for business or amusement. A remarkable old house is situated near the centre of the town, adorned with a number of wooden pillars, on which several Gothic ornaments are carved.

Among other public institutions, a Free Grammar School, founded and endowed by the Abbot, William Beane, in the year 1520, is worthy of notice. It is situated in the north-west angle of the church-yard. There are two Alms-houses in the town, endowed by Lady Paulett and Mrs. Ellen Parker, for the maintenance of eleven poor women.

In the year 1787, several Sunday Schools for the instruction of poor children, were established in the parish of Burton. An annual subscription of thirty guineas from the Earl and Countess of Uxbridge, the benefactions of the inhabitants of the town, and an annual charity sermon preached for the support of this institution, contributed essentially to its success; and 450 children of both sexes received instruction in eleven schools.

An Act of Parliament has been obtained for lighting and paving the streets, without any expense to the inhabitants of Burton, it being defrayed by the income from certain lands and tenements, under the direction of trustees appointed for that purpose.

There are six Hamlets belonging to the parish of Burton, namely, Burton Extra, Branston, Horninglow, and Stretton, and two in Derbyshire. The commons belonging to these hamlets were inclosed in 1771-2; the arable lands are light and gravelly, and the pastures and meadows level and luxuriant.

The number of freeholders in Burton who voted at the great contested election in 1747* was fifty. According to an accurate survey in October, 1789,

	Houses.	Inhabitants.
Burton contained	636	2926
Burton Extra, ...	102	553
Total, ...	738	3479

In the year 1801, according to the population returns, Burton contained 738 houses, and 3679 inhabitants; and in 1811, the population had increased to the number of 3979.

* At this election, William Leveson Gower, Esq. and Sir Walter Bagot, were returned.

The market of Burton, first granted by King John, and confirmed by King Henry III. is held on Thursday; and in consequence of the trade and manufactures of the town, and the opulence of the neighbourhood, it is well attended and supplied with grain, butcher's meat, butter, poultry, &c.

There are four annual Fairs held in Burton, by a grant of King Henry III. The principal lasts six days; during the first five days, more fine horses, particularly of the black breed, are usually exposed to sale than at any other fair in the kingdom. The sixth day is the festival of St. Modwen, (October 29,) and is appropriated to the sale of cheese, and to a variety of sports and pastimes. A second fair is held on Ascension Day; a third on the 5th of April, for cattle; and the fourth, on the 2d of February, (toll-free,) for horses, cattle, and merchandize.

CHRONOLOGY OF REMARKABLE EVENTS.

In the year 1255, the greatest part of Burton was consumed by an accidental fire.

In the 14th year of the reign of Edward II. Thomas, Earl of Lancaster, was at the head of a formidable insurrection against that Prince; and in the following year the King raised a powerful army, and several of the Earl's confederates deserted from him, and went over to the royal standard. The Earl's faithful adherents marched with him first to Gloucester, and thence to Burton-upon-Trent, and so to his castle at Tutbury, about four miles distant from the town. While on his march, the Earl sent a confidential officer into Lancashire to bring up 500 men to his aid; but he conducted the levies to the King's army, and apprized Edward of the Earl's intentions. The King immediately pursued his enemy, first marching to Coventry, and thence to Lichfield, where the two Spensers joined him with all their forces. On the approach of the royal army to Burton, the Earl placed infantry on each side of the bridge, to prevent the King's passage over the Trent. The King then forded the river at Walton, about three miles above the town, which being made known to the Earl, he sallied forth with his troops from Tutbury Castle, in expectation of being joined by his Lancashire auxiliaries. On hearing of their defection, he fled northwards, and was pursued to Pontefract, where he was taken and beheaded.* The Abbey of Burton suffered so much by the

* Dugdale's Baronetage, Vol. I. p. 781.

depredations of the rebels, that the King, in memory of his victory, gave to it the advowsons of the churches of Tatenhill and Hanbury, in this county, which were forfeited to the Crown by the rebellious Earl of Lancaster.*

The appearance of the *aurora borealis*, or northern lights, is mentioned among the memorable events recorded in the annals of Burton.—"*Mem.* On Monday and Tuesday in the mornynges, being the xvth and xvith dayes of November in the yeare of our Lorde God, one thousand fyve hundreth threescore and fouretene, there were seen in the element, certayne strange lights, whereof some appeared verie fyerie, and terrible to beholde, and stretched downe, as yt seemed, almost to the earthe, and were in the Northe, the Easte, and the Weste. And at those tymes yt was as lighte as if yt had been daye. These lights continued from twane of the clocke in the nighte untyll breake of the daye, and semed most fearfull in the mornynges after foure of the clocke; and hapned the dayes ymediately after the chaunge of the moone, and the ecclipse of the sunne."†

In the year 1611, Edward Wightman, of Burton, was burnt at Lichfield, for holding dangerous and blasphemous opinions.

During the Civil Wars between King Charles I. and the Parliament, Burton suffered greatly, being several times taken and retaken by the conflicting parties in the year 1643.

In December, 1793, a night-watch was first established in Burton.

Burton has been occasionally injured by the inundations of the Trent, particularly one in 1771, which overflowed the greatest part of the town; another in 1792; a third in 1795; and a fourth in 1798.

In 1815, a deputation from Burton presented a loyal Address to the Prince Regent, on his arrival at Beaudesert, the seat of the Marquis of Anglesea, upon which occasion the Bailiff (John Dickenson Fowler, Esq.) received the honour of Knighthood.

MANUFACTURES.

Leland, in his Itinerary, says that Burton was then famous for ornamental works in alabaster; and this fact is confirmed by Camden. This business has, however, been discontinued for more than a century, though alabaster is still to be found in the neighbourhood of Needwood Forest.

* Shaw's Antiquities of Staffordshire, Vol. I. p. 17.

† At the end of Lord Uxbridge's curious rental.

In the 17th century many of the inhabitants of Burton were clothiers, and their kersies were held in great estimation; and some tammies and woollen cloths are now manufactured in this town.

But the chief article of commerce produced in Burton is its Ale, for which it has long been celebrated. The origin of this lucrative business was in the year 1610, when Benjamin Printon began a small brewery, and his success induced others to engage in the same business. Hence the trade has gradually risen to its present importance: there are now nine capital breweries in the town, and great quantities of Burton Ale are annually consumed in this country, besides an export trade to a considerable amount to the northern nations of Europe. Catherine II. Empress of Russia, is said to have been immoderately fond of this beverage. The brewers of Burton, in opposition to the common practice, prefer hard water to soft; and though they have the waters of the Trent running by, they universally use that supplied by their pumps. Yet notwithstanding the singularity of this method, their ale is famous for body and strength.

Three cotton factories erected in Burton afford employment to some hundreds of the inhabitants.

An extensive manufactory of screws has long been established in the parish of Burton; and the manufacture of hats is also considerable in this town.

The mill mentioned in Domesday-book still exists on an islet, about half a mile below Burton-bridge. In the year 1745 it consisted of three parts, one of which was a corn-mill, another for making wash leather, and the third a forge for refining iron. The corn-mill is now rented of the Marquis of Anglesea, by Messrs. Evans and Wilson; the second is converted into a cotton factory by Messrs. Peele, Yates, and Co.; and the forge has been for some years occupied by the Lloyds, Quakers, of Birmingham, who carry on in it an extensive iron manufactory.

An Act of Parliament was obtained in 1698 for improving the navigation of the river Trent. The Burton Boat Company are proprietors of this navigation from Gainsborough. The Grand Trunk Canal from the Mersey to the Trent runs parallel with the river, with which it forms a communication at the south end of the town. Some jealousies arising between the two companies, occasioned the canal to be extended to Wilden Ferry, parallel with the river, to the detriment of both parties, but for the public good, as the price of tonnage was diminished by this opposition.

WHITMORE is a small appurtenance of Burton, and is situated in the meadows about half a mile north of the town. The house, which formerly belonged to the Abbots, was taken down, and a smaller one built for the tenant. Over the parlour-chimney there is still preserved the name of Thomas Feyld, Abbot, curiously embossed in stone.

SHOBNALL, an ancient grange belonging to the Abbey of Burton, is situated about a mile west of the town, upon a bold eminence, above which rises Sinai Park, a summer retreat of the Abbots. The situation is very commanding, and the prospect from it is delightful. There is a chalybeate spring a little below the house.

HORNINGLOW is another hamlet of the parish of Burton. It is an angular hill, which projects into the expanse of meadows near the confluence of the Trent and Dove, and was granted, with the rest of the Abbey lands, at the dissolution of monasteries, to the Paget family. The Marquis of Anglesea is now the principal proprietor. The turnpike-road crosses Horninglow, and the canal passes near it. This hill abounds with gravel, and contains large quantities of very rich marl, variegated with white veins.

STRETTON, a hamlet of Burton, is situated a mile north of the town, on the Ikenild-street, and was a place of some note in the time of the Romans. It is flat, the soil gravelly and light, and the land principally meadow and pasture. It is subject to inundations both from the Trent and the Dove. Near Stretton, an acqueduct of twenty three arches conveys the canal over the Dove, the ground being raised for that purpose to the extent of a mile and a quarter in length.

ROLLESTON is an ancient village, in a low situation near the river Dove, between Burton and Tutbury, with a brook running through it from Needwood. The land is a strong clay loam, with large quantities of excellent marl, which is sometimes used as manure. The church is a rectory in the Deanery of Tamworth; the value of the living, with the tithes, £300 per annum, and the Lord of the Manor is patron. Sir Oswald Mosley, Bart. has a beautiful seat here.

ANSLEY is a small hamlet of Rolleston, about a mile south of it, on the verge of the Forest of Needwood.

TUTBURY is situated five miles from Burton, on the south bank of the river Dove, which separates it from Derbyshire. The elevated and commanding site of Tutbury Castle was occupied by a fortress in the time of the Heptarchy, when it was the principal residence of the Kings of Mercia, who dwelt here in security against any sudden attack of an enemy. But the first certain information on record respecting Tutbury is that immediately after the Norman Conquest in 1066; Hugh de Albrincis was in possession of Tutbury Castle, but was deprived of it by the Conqueror, who gave it to one of his greatest favourites and friends, Henry de Ferrers. In the year 1250, it passed to the Earl of Lancaster, in consequence of Robert de Ferrers having joined Simon Montford in rebellion against Henry III. In 1322, Thomas Earl of Lancaster fortified it against Edward II. but could not hold out, and was obliged to surrender. Tutbury Castle consequently reverted to the Crown. It was afterwards neglected and fell into decay, till it came into the possession of John of Gaunt, who re-built it of hewn freestone, upon the ancient site in 1350.* This castle was now the principal seat of the Dukes of Lancaster, and was for ages distinguished as the scene of festivity and courtly splendour. The number of minstrels which crowded to it was so great, that as an expedient for preserving order among them, the celebrated John of Gaunt, appointed a chief minstrel, with the title of King, with inferior officers under him to assist in the execution of the laws. A charter to that effect was granted to the chief minstrel in 1381, by the Duke of Lancaster.† On the demise of John of Gaunt, the Duchy of Lancaster and its dependencies devolved on Henry Plantagenet, his only surviving son; and when that Prince afterwards ascended the throne by the title of Henry the Fourth, the honour and castle of Tutbury being

* Erdeswick, p. 205.

† Among other sports and pastimes introduced at Tutbury Castle by this eccentric, yet good-natured, Prince, was the annual custom called "Tutbury Bull-running." On the festival of the Assumption of the Virgin Mary, the minstrels assembled to matins at the Priory. In the afternoon they met at the gate, where a bull was given by the Prior. The animal had previously been prepared for this inhuman sport, having his horns, ears, and tail cut off, his body besmeared with soap, and his nostrils blown full of pepper. In this state he was let loose, and if the surrounding minstrels could catch and hold him so long as to deprive him of the smallest portion of his hair, he was declared their property, provided this could be done in Staffordshire, and before sun-set. This custom is supposed by some antiquaries to have originated in an imitation of the bull-fights in Spain; which is not improbable, as John of Gaunt was King of the Spanish provinces of Castile and Leon. In the lapse of ages, the pursuit of the bull, which had been confined to the minstrels, became general, and the multitude promiscuously joined in the barbarous sport, which sometimes terminated in a battle royal and bloodshed. The custom was abolished about forty-two years ago by the Duke of Devonshire, who gave the minstrels four marks in lieu of their former prize.

thus united with the Duchy of Lancaster to the Crown, they have continued so to the present time.

In 1568, Mary Queen of Scots was confined in Tutbury Castle, which was then a solitary spot, neglected and decayed. Here she continued many years a State prisoner, with sufficient time for contemplation, and to repent of her former levity and wantonness. The following account of the state of the castle at the time this unfortunate Queen was confined there, is extracted from the papers of Sir Ralph Sadler, the keeper.

"The whole area, containing about three acres, was encompassed on all sides but one with a strong and lofty embattled wall, and deep foss, as the present ruins plainly shew. The principal entrance was by a bridge under the great gateway to the north, part of which is still remaining. At a small distance to the left of this gateway, or lodge, stood Mr. Dorel's office and bedchamber, and four other rooms. Along the north-east wall, about 100 feet from the entrance, was a lofty tower, embattled, containing four rooms; viz. store-house at the bottom; above that, Curle's apartment; over which was the doctor's; and at the top, the chief cook's. This tower is then said to be very much shaken and cleft, and now very little of it is remaining. At a little distance from this began the principal suite of the Queen's apartments, extending along the east side; viz. the Queen's dining chamber, her closet, and bedchamber, cabinet, place for wood and coal, and then her women's room. These were all above stairs." From this Castle Mary was removed in 1585 to Chartley, and thence in 1586 to Fotheringay Castle, where she was beheaded.

King James the First, in one of his tours through England, visited Tutbury Castle, the former place of his mother's confinement; and such are the strange vicissitudes of life, that these walls which had so long re-echoed the sighs and groans of an unfortunate Queen, were now shaken with the acclamations of a multitude assembled to greet her son, their Sovereign.

During the Civil Wars in the time of Charles I. Tutbury Castle was fortified and garrisoned by the Royalists. The King spent a fortnight at this castle in 1643; and the following items of expence for provisions for the garrison, will shew the prices of those times:

		£	s.	d.
1644.	May 7th, paid for 8 cwt. 2 qr. 17lb. of cheese to Tutbury	17	16	10
	For five pots of butter to ditto	0	12	0
	Dec. For 26 strike of oats, which were sent for by warrant to Tutbury	2	4	10
1646.	Jan. 1. Paid for a standing piece of beef, a quarter of mutton, and a fat pig, and for carrying them to Tutbury	0	8	0

This Castle, after a long siege, being much battered, was surrendered by the garrison to Colonel Brereton in 1648; it was then greatly demolished by the Parliament forces; and when the event of the war was determined, the fortifications were destroyed by an order from the usurpers of power. This demolition, and the dilapidations of time, have finally reduced this once-beautiful and lofty edifice to a picturesque ruin. A considerable part of the gateway remains; and from the few vestiges of the castle, it appears to have been built of hewn freestone, with admixtures of gypsum. A round tower, intended to appear as a ruin, has been erected on a high mound by Lord Vernon, the present possessor, who holds the castle and circumjacent grounds by lease from the Crown. A building has been erected among the ruins, which is the residence of the steward, who entertains the tenants occasionally at wakes, &c. A large room in this house is used for assemblies; and the Minstrel's Court is annually held in it. The green park around the Castle-hill is now a pasture for sheep and cattle; the prospect from the summit of the hill is very extensive, and commands a picturesque view of Needwood Forest.

Tutbury is a pleasant little town, containing about one thousand inhabitants. There was a weekly market held here prior to the Norman invasion, but it is now discontinued. A large cotton factory has been established in this town by Bott and Co.; it is supplied with water from the Dove, and gives employment to a considerable number of men, women, and children. Several woolcombers are also employed in this town.

The Dove fertilizes the rich meadows on its banks, and affords a supply of trout and other fish to the inhabitants of the town. There is an excellent stone bridge over the river at Tutbury; and the turnpike road from Burton-upon-Trent to Uttoxeter passes through the town. The soil of this parish is generally rich, abounding with alabaster and marl; and the meadows are occasionally improved by the inundations of the Dove.

Tutbury Church is a vicarage, dedicated to the Virgin Mary; it constitutes a portion of the ancient priory church, and is a large edifice, with an embattled tower, surmounted by four pinnacles. The principal entrance is an archway of exquisitely rich and beautiful Saxon architecture, and is perhaps the most perfect specimen of the kind in the Island. This entrance is a low semi-circular arch, with a similar arched window above it; it is partly composed of alabaster, richly ensculptured with a variety of grotesque figures;

but no description can give an adequate idea of the beauty of the whole. Part of the arch of the window is beginning to moulder, but the entrance is in a state of perfect preservation.

The Priory was founded by Henry de Ferrers in 1080, and amply endowed by him with lands and revenues. The religious order who resided here were Benedictine monks, and several additional donations were conferred on the establishment by William Rufus and his Queen Maud. Earl Robert Ferrers, grandson to the founder, confirmed to the monks all their possessions, and added the tithes of Newborough to his gift. His descendants also contributed to the wealth and magnificence of this celebrated Priory, scarcely a vestige of which now remains. Little is known of the original extent of this monastery; but among the few facts recorded by antiquaries respecting it, we are informed that it contained a splendid monument to the memory of the founder, with a Latin inscription upon it. A general account of the annual revenue from the possessions belonging to this Priory, is preserved in the First Fruits Office, by which it appears that in 1538 they amounted to the annual sum of £244 16s. 8d. a very considerable income, according to the value of money in the middle of the sixteenth century. On the dissolution of the monastic orders by Henry VIII. Arthur Meverel *alias* Throwley, with eight monks, surrendered this Priory to the King's Commissioners on the 14th day of September, 1538, as appears by the original deed of surrender in the Augmentation Office. This prior afterwards received a pension of fifty pounds a-year.

The site of Tutbury Priory was granted in the sixth year of the reign of Edward VI. to Sir William Cavendish, who pulled down the Priory, and part of the church, to build a large mansion, which was the residence of his eldest son Henry. He dying without legitimate issue, it devolved to his next brother; and it has descended from successive possessors to the present owner, the Duke of Devonshire

A pretended instance of *total abstinence*, in the case of a woman named Ann Moore, an inhabitant of Tutbury, for a long time engrossed a very considerable share of public attention. The imposture was carried on for several years with such extraordinary art and success, that it obtained, in regard to the supposed validity of the woman's assertions, the sanction of a large number of medical, philosophical, and other visitors of every description from all parts of the kingdom. The laudable exertions of a Committee of gentlemen, formed for the avowed purpose of investigating this extraordinary case, at length discovered the cheat; and the wretched woman completed their labours by a formal and unequivocal confession of her guilt before Thomas Lister, Esq. one of the Magistrates for this county.*

Needwood Forest.—This ancient forest, or chace, is the most beautiful part of the honour of Tutbury, and is situated in the northern extremity of the hundred of Offlow, and in the four parishes of Tutbury, Hanbury, Tatenhill, and Yoxall. It was formerly resorted to by the ancient nobility, whose principal diversion was the chace. Camden says, "between the rivers Dove, Trent, and Blithe, Needwood, a spacious forest and full of parks, extends itself, in which the neighbouring nobility eagerly pursue the cheerful sport of hunting.†

* The following account of this imposture, derived from unquestionable authority, is copied from the *Staffordshire Gazette* of May 11, 1813:

"This unparalleled delusion, which has made so much noise, has been pursued but too successfully for nearly six years. The persevering subtilty with which this woman has carried on her imposture, is without example, and baffled all attempts to detect her, and but for the unconquerable scepticism of some of the faculty (particularly Dr. Henderson) her case would have been handed down to posterity, as an indubitable instance of a human being living without receiving the least sustenance. The watch, which had been publicly advertised, commenced on Wednesday the 21st ult. and was strictly carried on until Friday the 30th, a period of nine days, during all which time, these gentlemen verify, that she had not received any nourishment. She, however, gradually grew feebler, her pulse was almost imperceptible, and she at length became so ill, as on the latter named day to induce the gentlemen to suspend the watch. At this time she begged to have her mouth moistened with a wetted cloth, and her desire was complied with by applying to her lips a cloth dipped in vinegar and water; this was done several times, and the gentleman who administered it, declared he perceived her to swallow, although she even then strenuously denied it. The physician attending upon her, at the same time, gave it as his opinion that she could not survive an hour; and yet, at this period, with the immediate prospect of eternity before her, she, *by her own desire*, took an oath, drawn up in the strongest and clearest terms, that for more than four years past, she had not taken sustenance of any description! Her daughter was now admitted to see her, and she in a short time, very much revived.

"From the weighing machine, upon which she was placed during the watch, it appeared that she had lost daily nearly 14 ounces in weight. Nothing now remained to convince every one of her imposition but her own confession of her guilt—and this last proof she voluntarily made before a Magistrate. After this confession, she took milk, in the presence of several of the gentlemen, and now seems fast recovering."*

* See the Rev. Legh Richmond's "Statement of Facts, &c."

† Needwood Forest formed a portion of the Duchy of Lancaster, and after the accession of Henry the Fourth it belonged to the Crown for centuries.

In the survey of Tutbury, taken in the first year of the reign of Queen Elizabeth, the following account is given of the extent of Needwood Forest. "The forest, or chace of Needwood, is in compasse by estimation 23 miles and a half, and the nearest part thereof is distant from the castle of Tutbury but one mile. In it are 7869 yards and a halfe, and very forest-like ground, thinly set with old oakes and timber trees, well replenished with coverts of underwood and thornes, which might be copiced in divers parts thereof, for increase of wood and timber, lately some decayed and spoyled. It is divided into four wards, viz. Tutbury ward, Marchington ward, Yoxall ward, and Barton ward, each containing five miles or more in compasse."

According to a survey made in 1656, this forest was found to contain 9220 acres of land, and by an examination of the timber in 1684, it contained 47160 trees, and 30,000 cord of hollies and underwood, valued at £30,700.

Needwood Forest presents to the eye a great variety of picturesque beauty. In the more fertile part, or the middle and southern divisions, it is diversified by hills and valleys, watered by clear brooks, with here and there a bolder eminence. In the northern part, it is broken into deep glens and lofty precipices, covered with a variety of beautiful trees, particularly the oak, which here flourishes in great perfection. It has been asserted that the forest does not contain less than one thousand acres occupied with oaks,* among which an oak named the *Swilcar*, is considered the monarch. This tree is of great antiquity, but still fruitful in acorns; it measures twenty-one feet round the trunk, at the height of five feet; the lower stem is ten feet high; the whole height 65, and the extent of the arms 45 feet. It contains at least one thousand feet of solid timber, and has been celebrated by poets and botanists.

The uneven part of the forest, comprising upwards of 1200 acres, consisting of abrupt hills and dells, will probably be appropriated to the growth of timber, and the remaining 8000 acres added to the cultivated land of the country. One-eighth of this quantity consists of light sound loam, adapted to the culture of turnips; and the rest,

The owners and inhabitants of some of the circumjacent villages, however, enjoyed certain privileges of common, of which they were eventually deprived by an Act of Parliament, obtained and acted upon for enclosing this extensive and fertile tract. In consequence of this Act, part of the oak timber, and the holly, has been sold and cleared off, and the deer destroyed.

* Marshall's Rural Economy of the Midland Counties, Vol. II. p. 360.

which is generally a red, or whitish-grey marly loam, will produce excellent pasturage, or be productive of good beans.

In the northern extremity of the forest is a singularly romantic valley, called Bertram's Dingle, which, if private property, might be made uncommonly beautiful. Besides the natural beauties of Needwod Forest, it is adorned with four handsome lodges, which have for ages been held under the Crown on a lease of three lives, and transferred from time to time to different owners.

Byrkley Lodge is situated on an extensive plain on the east side of the forest, and is surrounded with a beautiful variety of wood. It is the residence of Edward Sneyd, Esq. who has made many improvements. A stream, called Linbrook, rises here, and runs southward.

Yoxall Lodge is the residence of a poet and divine, the Rev. Thomas Gisborne. It is built in a delightful situation, in a recluse valley on the banks of Linbrook, and is well adapted to the studious pursuits of the owner.

Sherholt Park and Lodge, in Barton ward, is the property of Lord Vernon. The views towards Lichfield, &c. are extensive and picturesque; and near this mansion a rivulet runs into a fine sheet of water, called Lush Pool.

Eoland Lodge is situated on the north-west side of the forest, and is surrounded by woods and lawns. It is the property of Lord Bagot, and was formerly the residence of Francis Noel Clarke Mundy, author of "Needwood Forest," a poem. The poet thus describes his favourite retreat:

"On this green unambitious brow,
Fair mistress of the vale below,
With sloping hills enclos'd around,
Their heads with oaks and hollies crown'd,
With lucky choice, by happy hands,
Plac'd in good hour my dwelling stands;
And draws the distant trav'ler's eye,
Enamour'd of its scenery;
Where all things give, what all express,
Content and rural happiness.
Where far retir'd from life's dull form
Comes no intruder but the storm;
The storm, that with contrasted low'r
Endears the fair, the silent hour.
Thus their wise days our fathers led,
Fleet ran their hounds, their arrows sped,
And jocund health with rosy smile
Look'd on, companion of their toil;

Till tyrant Law usurp'd the land,
Stretch'd o'er the woods his iron hand,
Forbade the echoing horn to blow,
Maim'd the staunch hound,* and snapp'd the bow."

HANBURY.—This village is situated to the north of Needwood Forest, upon a hill that commands an extensive prospect of the meadows on the banks of the Dove, the Moorlands, and Peak-Hills. From its bleak situation it is cold in winter, but delightful in summer.

Hanbury is a place of great antiquity. In the year 680 the Saxon princess St. Werburgh, remarkable for her piety, became Abbess of a nunnery, founded here by her brother Ethelred, King of Mercia. She was buried in this nunnery; and in the year 875 her bones were removed to Chester, where an elegant shrine was erected to her memory. No vestige of the nunnery is now visible; but it doubtless stood to the east of the present church, human bones having been frequently dug up in the ground now occupied by Mr. Hunt's garden, and in an adjacent gravel-pit.

A family, who took their name from the place, were lords of this manor at a very early period. At present, it belongs to the Villiers family, who appoint a game-keeper, and claim common rights on Needwood Forest. The manor-house commands a most extensive prospect. Hanbury Church is an ancient stone edifice, with a square tower; it stands on the edge of a steep declivity, and was formerly a rectory, but is now a vicarage in the deanery of Tamworth. It is dedicated to St. Werburgh, and was founded long before the Conquest. The presentation is vested in the Bishop of Lichfield and Coventry. In the year 1793 the Rev. Hugh Bailye pulled down the old vicarage-house, and built a new one on the opposite side of the church, which commands a charming and extensive prospect.

Extract from the Parish Register.—"On Sunday 14th September, 1777, about eleven o'clock in the forenoon, a smart shock of an earthquake was felt in several places of this neighbourhood (but not in the village.) In some places indeed, particularly Cheshire, and Lancashire, it was so violent, that the people fled out of the churches in great terror."

Felde, or *Faulde*, is a hamlet in this parish, about half way between Hanbury and Tutbury, and situated on a fine natural terrace above the meadows of the Dove. This hamlet was recorded, by mistake, in Doomsday-book for Hanbury.

* This alludes to the order for *lawing*, or cutting-off a claw of all dogs kept within the purlieus of the Royal forests, to prevent their destroying the deer.

In the year 1792, the manor of Faulde was purchased by Mr. Thomas Hunt, of Castle Hay. The old hall is now inhabited by a son-in-law of the present Lord of the Manor.

Coton-under-Needwood is another small hamlet in the parish of Hanbury: it belongs to Charles Bowyer Adderley, Esq. The ancient hall was pulled down, and the present commodious mansion built on its site in 1790, with suitable offices, and extensive and picturesque gardens and pleasure-grounds. English hospitality prevailed here some years ago, when the social friends Messrs. Adderley and Scott made it their residence. In the summer months they pitched a tent upon a circular hill above Coton, where a flag was hoisted, when they were at home, as a signal to their friends:

> "O mark upon yon round ascent
> The social flag and open tent,
> Where life's smooth paths with flow'rs are strewn,
> And mirth makes every hour its own."

The flag-staff was struck in the year 1796.

Draycot-under-Needwood was another of the manors comprised in the gift of William the Conqueror to Henry de Ferrers. It is recorded in Doomsday-book in the following terms: "The said Henry holds Draicote, which contains half a hide. The arable land is one carucate in demesne, and four villans, and four bordars, have two carucates. There are 12 acres of meadow, a wood half a mile in length, and the same in breadth. The whole being valued at fifteen shillings."

This manor has for ages been in the possession of the ancient family of Vernon, who came originally from a town of that name in Normandy. The present possessor is Lord Vernon. In the meadow beyond Draycot Mills are the ruins of an old mansion, encompassed by a moat, with a small drawbridge on the eastern side. This was doubtless the manor-house, or residence of the ancient proprietors. There are several extensive and extremely fertile meadows in the vicinity of Draycot: the turnpike-road from Lichfield to Sudbury passes through this manor.

MARCHINGTON is a small village, situated at the northern extremity of the Hundred of Offlow, with the woody eminence of Needwood gradually rising on its southern side, and the winding stream of the Dove as its boundary to the north. The manor now belongs to Earl Talbot. The earliest record of this manor is in the will of Wulfric Spot, the founder of Burton Abbey, dated A. D. 1004, in which he grants it to Wulfag. It was afterwards part of the demesne lands appertaining to the honour of Tutbury, given by the Conqueror to Henry de Ferrers, and is thus mentioned in the Survey recorded in Doomsday-book: "Henry holds Merchametone, in which are two hides, and in Edgarsley one virgat of land. Uluric formerly held it; and was a freeman. The arable land is seven carucates. In demesne there are two, with one servant, eighteen villans, and nine bordars, who have three carucates. There are forty acres of meadow, a wood affording pasture for cattle or deer, three miles in length, and one mile and a half in breadth." The whole was then valued at one hundred shillings.

Marchington Woodlands is another hamlet in this parish. It consists of a number of scattered houses and farms, on a rising ground, for a mile or two westward of Marchington Chapel.

Hounhill is a manor in the chapelry of Marchington. It consists of about 500 acres of rich pasture-land, mostly of a stiff soil. On the north-west side is a considerable quantity of alabaster. It belongs to Lord Vernon.

Newborough is another township belonging to the parish of Hanbury, to which, like Marchington, it has a chapel of ease, situated two miles west of the mother church, on a small stream called Swerbourn, which runs alongside the borders of the forest, through Yoxall into the Trent. This place was first established by Robert, son of Henry de Ferrers, about the commencement of the eleventh century, who granted certain parcels of land and several immunities to 101 burgesses. In consequence of these privileges, Newborough was inhabited by handicraftsmen, and soon became a flourishing place.

The manor of *Agardsley*, within which this township stands, passed with that of Marchington to the Chetwynd family, and so to the present proprietor, Earl Talbot. Several of the present inhabitants of Newborough are weavers of linen and checks. The chapel is a small modern building, and contains no monuments, for the inhabitants of the township bury their dead at the mother church in Hanbury.

Thorney-hill is an ancient hamlet in the chapelry of Newborough.

Yoxall.—This village is situated in a pleasant valley on the south-west border of Needwood Forest, at the distance of seven miles from Lichfield, and four from Burton-upon-Trent. It was formerly a market-town, and was a member of the honour of Tutbury from the time of the Conquest. Yoxall is supplied with water by the rapid stream of Swerbourn, and good turnpike-roads pass through it to Lichfield, Burton, Uttoxeter, and Ashbourn. The church is dedicated to St. Peter, and is a rectory in the deanery of Tamworth and Tutbury, and Archdeaconry of Stafford. It is an ancient gothic fabric of stone, with a square tower, and contains many monuments.

The principal hamlets belonging to this parish are Woodhouses, about three-quarters of a mile eastward, near which is the mansion of High-wall Hill, where there is a small manufactory of tape established by Mr. Wright; Bond-end, half a mile south; Hadley-end, about the same distance west; opposite to which stands Longcroft; beyond which is Sale; and about three miles north-west from Yoxall church, the hamlet of Horecross.

Longcroft is an ancient mansion and park, originally granted by charter to Roger de Yoxhall, clerk, in the first year of the reign of Henry III. The house has still an appearance of antiquity, and was formerly surrounded by a moat, over which was a stone bridge in front. In 1796 the moat was filled up, the bridge removed, and the mansion greatly improved. It is situated on an eminence which commands an extensive view of Needwood Forest, and the opposite country towards Lichfield. Several groups of full-grown firs and other trees, amongst which is a remarkable old oak, shelter the house from the north and east winds, and add greatly to the picturesque beauty of the scene.—Longcroft was purchased, in the time of Queen Elizabeth, by Simon Arden, who, in the 18th year of her reign, found one light horse, and paid to her Majesty £1. 6s. 8d. for his lands in Yoxall, then valued at £10.*

Horecross is an ancient residence, three miles from Yoxall, on the north-west. In the reign of Henry VIII. it was possessed by the Welles, and the estate then consisted of 200 acres of arable land, 200 of pasture, 40 of meadow, and 50 of wood. The site of the ancient mansion is now occupied by a hunting seat, erected by the present possessor, Lord Scarsdale.

TATENHILL.—This village is situated in a deep narrow valley between two high hills, which gradually descend from the eastern border of Needwood Forest. It is in the deanery of Tamworth, and about three miles from Burton. This parish is extensive, comprising the manors of Callingwood, Dunstall, Barton, Newbold, and Blakenhall. The soil is a strong clay marl. The church is situated on an eminence eastward of the village, and is a large ancient stone edifice, with a tower. It is dedicated to St. Michael,

* Shaw's History and Antiquities of Staffordshire, Vol. I. p. 102.

and valued at £36. 1s. 8d. in the King's books. Tatenhill church is remarkable for an echo, by which four or five syllables spoken in a low voice, on the opposite hill, are distinctly reverberated by the tower. A brook runs through the middle of the village, and formerly turned a corn-mill, which has since been converted into a screw-mill, where 700 gross of screws of various sizes are manufactured weekly.

Callingwood is situated on the east side of Needwood Forest, a mile north-west from Tatenhill church, and in the survey made in the time of Queen Elizabeth, it is called a hamlet within Barton, granted out of the forest to Radulpho de Bosco Cabumpsiato, one of Earl Ferrer's servants. A number of Roman coins were found in this manor, towards the close of the eighteenth century; a presumptive proof that the Romans had been on the spot, which is the more probable, as the Ikenild-street passed to the east of this place. There are some vestiges of an ancient moated house here: the present manor-house is occupied by a farmer. In the year 1650, Sir Edward Mosley, Bart. by virtue of an ancient charter, claimed a messuage and 120 acres, called Barley-fields farm, in Callingwood.

Dunstall is an ancient manor in this parish, situated close to the border of Needwood Forest, and midway between Tatenhill and Barton. This manor is not mentioned in Doomsday-book; but it appears to have been a member of Tutbury from the Conquest, and was given with Newbold, by William de Ferrers, Earl of Derby, in the reign of Henry III. to Walter de Somerville, in exchange for Barton. The present proprietor of this estate is Mr. Meek, by whom it has been much improved.

Newbold.—This manor is situated on the west side of the ancient Roman Ikenild-street, which is here a turnpike-road, leading from Burton to Lichfield. The present proprietor is Mr. Webb, who resides in a commodious modern-built farm-house, not far from the navigation warehouse at Barton.

In this manor, between Burton and Branston, there are salt-springs; and in a place called the Clots, about a mile east of Dunstall, if cattle of a black, red, or brown colour, be put to feed, they will certainly change their colour to a whitish dun, which must be ascribed to the saltness of the soil. As for horses, they rapidly improve upon these grounds, but they soon become dappled, be they of whatsoever colour.*

Barton-under-Needwood.—This village is situated on the

* Dr. Plot's Natural History of Staffordshire.

south-east side of the forest, half a mile west of the Roman road called Ikenild-street, and the same distance from the Grand Trunk Canal, four miles from Barton, and eight from Lichfield.

The original name of this village was *Berton*, a Saxon word, signifying a farm-yard, with the usual barns and other buildings. In Doomsday-book it is expressly mentioned as the property of the Crown. "The King holds Bertune: Earl Algar formerly held it. Here are three hides, with their appurtenances; the arable land is 18 carucates; in demesne are two carucates, and two servants and 17 villans, and eight bordars have nine carucates; there are 20 acres of meadow; a wood two miles in length and one in breadth; and a mill of six shillings rent. At the time of Edward the Confessor the whole was worth six pounds, and now seven pounds."

This manor was granted by the King to Henry de Ferrers. Afterwards it was for some time in the tenure of Walter de Somerville, who gave it to William de Ferrer, Earl of Derby, in exchange for Dunstall and Newbold.

It continued in this family till it was forfeited by Robert, Earl of Derby, who rebelled against Henry III. in 1263; in consequence of which it was given to Edmund, Earl of Lancaster, the King's youngest son, and continued with other estates in the Duchy of Lancaster. It was forfeited to the Crown by Thomas, Earl of Lancaster; and Edward II. in the 19th year of his reign, leased this manor for seven years to Philip de Somerville, at the rent of £55. per annum to be paid into the Exchequer.

This manor continued in the Crown, in the right of the Duchy of Lancaster, till the time of Charles I. who, in 1629, sold it with the park to the city of London, in trustees' names; and they soon afterwards passed to Sir Edward Bromfield, Alderman of London. Dr. Plott, in his map of Staffordshire, in 1682, gives the arms of Bromfield, then living at Barton; and Hardman, who wrote in 1692, thus describes the village: "Barton is a good country town; hath several gentlemen and freeholders in it. The present owner of it is Charles Bromfield, Esq. who hath a copyhold court here. The most remarkable thing is a very fine chapel, built by John Taylor, LL.D. 29 Henry VIII. this being the place of his nativity." From this period it was possessed by several proprietors in succession: the present owner is Eusebius Horton, Esq.* The soil of this parish is generally light and gravelly.

On the north side of the village stands Barton-hall, a good

* Shaw's History and Antiquities of Staffordshire, Vol. I. p. 112.

modern mansion. In the street leading to the church is a curious ancient building, with yew trees in front. This mansion was formerly inhabited by a gentleman of the ancient family of Sanders.

Barton Church is a chapel of ease to Tatenhill, and is dedicated to St. James. The origin of this chapel is curious. In the sixteenth century, a man named Taylor dwelt in a small cottage near the place where the chapel now stands. His wife was delivered of three sons at a birth, and the infants were shewn as a curiosity to Henry the Seventh, who accidentally passed that way. The King ordered that care should be taken to have the boys educated: they all lived to be men, and as the tradition goes, all came to be *doctors*, and to good preferment. The eldest son, John, however, not only rose to eminence as a scholar and divine, but gratefully founded this chapel in the place of his nativity. The chapel is neatly built of durable stone, and contains several monuments.

Barton Free School was founded in 1595, by Thomas Russel, of London, who bequeathed £50. for that purpose, and a certain land-charge for the endowment of the same. The annual produce is £19. and the trustees the Draper's Company. The school-house is an ancient fabric, situated at the eastern extremity of the village. By the benevolence of the Rev. Thomas Gisborne, the school-master's salary has been increased to £50. a-year; and the present teacher, the Rev. Mr. Kirk, obtained a subscription towards repairs, and engaged a classical assistant. His public spirit deserves commendation, and he will probably be enabled to establish a respectable seminary of education.

A little southward of Barton stands Blakenhall, once the principal seat of the ancient family of Mynors.

WHICHNOR.—This small village is situated about half way between Burton and Lichfield. It stands on an eminence on the northern bank of the Trent, about a mile above the confluence of that river with the Tame.

The manor was held by Sir Philip de Somerville, in the 10th year of the reign of Edward the Third, under John of Gaunt, Duke of Lancaster, lord of the honour of Tutbury. During the Duke's residence at Tutbury Castle, to please the people and gain their affection, he established several curious customs, and none more singular than the conditions on which Sir Philip held this manor, Sciercot Ridware, Netherton, and Cowlee. The most remarkable condition of the tenure was to keep a flitch of bacon hanging in his hall at Whichnor, at all times in the year, except in Lent, that it might be delivered to any man or woman who should come and demand it, and at the same time swear that he or she had been married a year and a day without repenting; and that, if they were single, and to be married again, the demandant would take the same party again, before any other in the universe; as appears by the following extract from an ancient roll of parchment, translated from the original copy of the grant which was written in French:

"Here ye, Sir Philippe de Somervile, Lord of Whichnoore, mayntener and gyver of this baconne, that I A. sithe I wedded B. my wyfe, and sythe I hadd hyr in my kepyng, and at my wylle, by a yere and a day after our marryage, I wod not have chaunged for none other, farer ne fowler, rycher ne pourer, ne for none other descended of greater lynage, slepyng ne wakyng, at noo tyme. And yf the said B. were sole, and I sole, I wolde take hyr to be my wyfe, before all the wymen in the worlde, of what condicions soever they be, good or evylle, so help me God and hys seyntis, and thys fleshe, and all fleshes."

Two neighbours were required to testify the truth of this deposition; and if the claimant was a freeman, there was to be given to him half a quarter of wheat and a cheese; and if a villan, half a quarter of rye, without cheese. These things, with the bacon, were to be carried before him, with trumpets, tabernets, and minstrels, &c. past the lordship of Whichnor, and then without music to his abode.

In the year 1661, the manor of Whichnor was sold to the Offley family. William Offley, an ancestor of this family, mercer, and twice bailiff of Stafford, was uncommonly fortunate in his children, having seven sons, all prosperous, and five daughters, all well married. His eldest son, Thomas, was Lord Mayor of London in 1556, and afterwards knighted. His great grandson, John Offley, Esq. was High Sheriff of the county of Stafford in 1680. John Offley, Esq. who resided at Whichnor, sold the manor in 1765 to John Levett, Esq. from whom it has descended to his nephew, the present possessor. Leland, who wrote in the time of Henry VIII. says that Whichnor was the site of a very ancient mansion, which was then in ruins, and that the spot on which it stood was subject to inundations from the Trent. Traces of this mansion are still visible in the meadows at a small distance south-west of the church. The moat is square, encompassing about an acre of ground.

The present house, called the Lodge, was built by Mr. Offley. It is a neat brick building, faced with stone, and commands a beautifully picturesque prospect across the park to the Forest of Needwood. In memory of the singular tenure by which the manor is held, a piece of painted wood, in the form of a flitch of bacon, hangs over the hall-chimney. But it appears that the married people are as averse to the flitch itself as if they were all Jews; not a single individual having carried off the prize from the first day of the institution to the present time.

The church dedicated to St. Leonard, is a small gothic structure of stone, situated on an eminence near the river Trent. Whichnor parish is but small, and contains few inhabitants. At the contested election for the county in 1747, Mr. John Offley was the only freeholder belonging to it who voted on that occasion.

King James the First, in the course of his tour through this part of the kingdom, visited Whichnor on the 21st of August, 1621, and held his Court at the Hall. On the 19th of August, 1624, the King re-visited the Lord of the Manor at Whichnor, and dined with him.

On the 3d of July, 1255, a very remarkable storm of hail fell along the vale of Trent, from the bridge at Wychenofere down to Repriadon, such as had not happened before in the memory of those who saw it. This hail-storm was succeeded by a very great voràge, which swallowed up the earth, together with trees, houses, and corn, and carried them away. There was such an universal destruction of hay in the valley of Trent by the inundation, as had not happened a long time before.*

On Good Friday, March 25, 1596, the mills of Whichnor were burned by a fire, which consumed a large quantity of corn, and destroyed the mill-stones.

A handsome stone bridge, of three large arches, has been erected over the Trent at Whichnor; and a corn-mill, situated between the bridge and the church, is supplied with water by a branch of the river. On the same stream, in the meadow below the bridge, a large forge and slitting-mill has been established by a company at Lichfield. The Justice's meetings for the northern division of the Hundred of Offlow, are held at a large inn on the turnpike-road, near the bridge. Opposite to the inn, and parallel with the road, the Grand Trunk Canal passes, and communicates with the Trent by a lock a little below Alrewas.

Alrewas is a considerable village, in a low situation on the southern bank of the Trent. It is in the deanery of Tamworth, seven miles from Burton, and five from Lichfield. The Grand

* Gales's Annals of Burton, p. 342.

Trunk Canal runs through part of this village, which will doubtless obtain many advantages from the facilities thus afforded for a commercial intercourse with other parts of the kingdom.

The parish church of Alrewas existed at an early period in the ninth century, and was one of the prebends instituted by the Bishop of Lichfield, in 822.*

In the time of the Saxons this manor belonged to Algar, Earl of Mercia. He left this with his other estates to Edwyn, his eldest son; but, after the Conquest, William I. disposed of most of them to his followers and friends, and reserved Alrewas to himself. It seems, as described in Doomsday-book, to have been nearly in the same state at that period as at present; for it had a church, a mill, and twenty-six families of the common people.

"The King," says this ancient and authentic record, "holds Alrewas. Earl Algar formerly held it. The arable land is eight carucates. In demesne there are two; and one bondman, twenty villans, and six bordars, with a priest, have six carucates. There are twenty-four acres of meadow, and a fishery that yields a thousand five hundred eels. Also a wood one mile in length, and half a mile in breadth. In the time of King Edward the Confessor, it was worth ten pounds: it is now worth eleven pounds."*

Thus at the time of the general survey of the kingdom, Alrewas belonged to the King, and continued demesne of the Crown till the 5th of John, when Roger de Somerville obtained a grant of this manor, to hold it in fee-farm for the old rent and 100 shillings increase.†

From this period till the time of Charles II. the manor of Alrewas came into the possession of several individuals, and by an indenture, dated January 3, 1660, John Turton and his brother Philip Turton, purchased of Sir Francis Boynton, Bart. the manor of Alrewas, with the appurtenances, for the sum of £3130. John Turton, gent. who purchased this, and succeeded to the other estates, mentioned as appurtenances, was a Knight, and one of the Barons of the Exchequer, in the reign of William and Mary, and afterwards one of the Justices of the King's Bench. He died in 1707, and was succeeded by his grandson and heir John, who possessed the manor nearly fifty years, and was one of the Justices of the Peace for this county. In this capacity he was distinguished for profound knowledge of the laws of his country, his equal distribution of justice, and his exemplary life. He died August 18, 1771, aged 84.

* Magna Britannia, Vol. V. p. 123. † Dugdale's Baronetage, Vol. II. p. 106.

In 1752, John Turton, Esq. father of John Turton, of Sugnall Esq. sold the manor of Alrewas and Orgreave Hall, with the estates, to George Lord Anson, second and youngest son of William Anson, Esq. of Shugborough, in this county. Lord Anson, so justly celebrated for his nautical skill as a circumnavigator, and his heroism as a warrior, who, to use the energetic language of the poet Thomson, bore his country's name "in thunder round the world," after his various toils and perils, naturally wished to enjoy peace and tranquillity in his native county. On the demise of his Lordship in 1762, his elder brother Thomas Anson, Esq. who was then one of the representatives in Parliament for the city of Lichfield, succeeded to the handsome house and estate at Shugborough, which the great navigator had so much enriched and adorned by the treasures of his voyage. By his Lordship's will, great part of his fortune devolved to his sister's son, George Adams, Esq. whose son, Thomas Anson, Esq. now Lord Viscount Anson, is proprietor of Alrewas and the estate at Orgreave. These possessions, according to a survey made by Mr. Wyatt, in 1793, consist of 5745 acres, besides the common of 2191 acres and a half. But this is a small part of the property his Lordship possesses in this county. He has lately enlarged the magnificent mansion at Shugborough, where his Lordship at present resides.

The old manor-house at Alrewas is situated a little south-west of the church, near the navigation bridge. At a little distance stands a cotton factory.

The church, the antiquity of which has already been recorded, is a fine specimen of Saxon antiquity. It is built of stone, with a low tower, and adorned with a Saxon arch at the west end. It is a vicarage, dedicated to All Saints, and contains several monumental inscriptions.

Extracts from the Register-book of the Parish of Alrewas:—

"1577. This year, the 11th of November, appeared a blazing starre, which was seen forty days after.

"1585. This year, July 6, the second bell and the greatest bell weare caste at Nottingham, by Henry Oldfield, bell-founder.

"1593. This year, in the summer time, there was a great plague in England, in divers cities and townes, as in London their died in one week to the number of 200, and in Lichfield their died to the number of eleven hundred and odde, and as at this time of wryting not clean ceased, being the 28th of November.

"1601. This year, the 1st of July, the bridge upon Tame, called

Salter's-bridge, being greatly in decaye and broken downe, was of new begunne and made broder by two foote, which cost the workmanship tow hundred poundes, whereof this hundred of Offeley payed one hundred poundes, the other four hundreds payed the rest: the 17th of July the foundation began to be layed.

"1613. Robert Nevell the father, and Robert Nevell the son, being in Salter-holme-field the 26th day of June, tending of the towne beastes, in the herdman's walk there, after many furlonge side, about four o'clock in the afternoon of the same day, their was a mightie great tempest of rayne, lyghtning, and thunder, and the father and sonne standing under an oke tree, to save themselves from the rayne, where both of them stricken to death, the barke of the oke tree rent a great length, the leaves of the tree smitten and blowen away the most parte of them. One other younge youth of ten yeares age, Thomas Fraunces, being their to fetch or help the herdman, being within the compasse of tenne yeardes of the same place, was saved, and nothing hurte as the other, the heares of their head singed with the lyghtning and in some parte of the body, and face blackened.

"1614. The mill in Alrewas-hey was reared and sette up the 20th daye of November, but it did not grind until the 21st of December after, being the feast of Saint Thomas the Apostle.

"1621. This year the fallow fields of Alderwas were first sowed with pease.

"1621. August 21st, King James at Whichnor: the Court held at the hall there.

"1643. Upon the 13th of August, being the Lord's-day, there happened in this town of Alrewas a suddaine and fearful fire, in the house of George Thornixarke, an alehouse near the church, occasioned by shooting off a horne-gun; which, in one hower's tyme, consumed the said house and most of his goods, and the barn and hay of John Fitchet, and the house and all the goods of John Francis, and burnt up the trees about them, and had like to endangered the whole towne. God give us grace hereby to amend our lives.

"1675. On January 4, between the hours of seven and eight at night, a terrible earthquake was felt at Alrewas.

"1711. All the bells at Alrewas church were re-cast this year.

"1793. There has been erected, within a few years past, a cotton manufactory at Alrewas, belonging to which there are 200 persons daily employed; the present proprietors of the said manufactory are Messrs. Dickins and Finlow, of Burton-upon-Trent.

"1794. The annals of agriculture for the last century have not exhibited an instance of so heavy a harvest having been housed, in this kingdom, within so short a period as that of the present year; the corn and grain of every kind were housed and completed in the parish of Alrewas by the 16th of August.

"This year six troops of yeoman cavalry were raised for the use of the county of Stafford, and they were trained and disciplined on Fradley Heath.

"1795. On Tuesday, February 10th, a great flood succeeded a sudden thaw: great damage was done by the amazing quantity of ice and waters to several bridges. It is very remarkable, amongst such a variety of damage done to bridges in several parts of England, that Salter's bridge over the river Tame, in the parish of Alrewas, sustained very little damage, which is supposed to be preserved by the extensive breadth of the centre arch, and the great sweep that it commands.

"1795. On Thursday, November 5th, about midnight, a violent storm of rain and wind happened in the parish of Alrewas, and places adjacent; during the course of the night, or rather next morning, before day-light, several trees were blown down in the said parish.

"November 18th. Another violent storm, which was followed by an earthquake."

Among the casualties recorded in the parish register of Alrewas, it appears that no less than fifteen persons were drowned in the Trent, Tame, and Canal, in the course of the eighteenth century.

Croxall is a small parish, situated mostly in Derbyshire, and therefore properly belongs to the history of that county.

Bromley Regis, *or* King's Bromley.—This village is situated higher up the river Trent than Alrewas. According to Doomsday-book, it belonged to Earl Harold, in the reign of Edward the Confessor. At the Conquest it became the property of the Crown, and continued so till the reign of Henry III. when it passed to the Corbets, several of whom were knights and high sheriffs of the county.

In the sixth year of the reign of Edward the Fourth, this manor was sold by Robert Corbet, Esq. to William Praers; on his demise, it devolved to Alice, his sister and heir, who was married to Mr. Patriche. It was sold on the 2d of May, 1569, by Edward Patriche, to Francis Agard, of Ireland, for the sum of £1,240.

In 1573, the manor of Bromley Regis comprised 100 messuages,

20 cottages, two water-mills, one dovecote, 1000 acres of arable land, 200 acres of meadow, 200 of pasture, 200 of wood, and 1000 of heath or furze.

This manor continued in the Agard family till the year 167*, when it was sold by the trustees of Charles Agard, Esq. to John Newton, of the Island of Barbadoes, and was bequeathed in 1794 by Sarah Newton to her cousins, John and Thomas Lane, Esqs.

The soil of this parish is of a gravelly and sandy nature. The Grand Trunk Canal passes to the south of the village. The common, comprising about one thousand acres, has lately been inclosed, and is now in cultivation. At the mill, a manufactory of bar-iron, and another of tin, is carried on to a considerable annual amount.

The church, which is built at some distance from the village, is dedicated to All Saints. It is a fine gothic building, with large and beautiful windows, and contains several monuments of the Agards and Newtons.

Among other charitable donations recorded on a large board fixed to the south wall of this church, the following are most memorable:

"In 1692, Edward Cross, of this parish, gave £5. a-year, charged upon an estate at Bromley-Hurst, in the parish of Abbot's Bromley, to the poor of this parish, to set out an apprentice yearly for ever.

"In 1699, the Rev. Richard Cross, Rector of Baggington, Warwick, erected a Free School House (John Newton, Esq. late lord of this manor, contributing an acre of land whereon to build it), and endowed the said school with an estate in Marchington Woodlands, in the parish of Abbot's Bromley, valued at £30. per annum, (now £60.), the rent to be paid to the master for teaching poor children therein for ever."

Hamstall Ridware.—This manor is situated on the northern bank of the Trent; it is in the deanery of Tamworth, about a mile north-west of Bromley Regis, and two miles west of Yoxall. The river Blythe runs through the middle of the parish and falls into the Trent.

The etymology of Hamstall is derived from the Saxon word *homestall*, which signified a fixed habitation. According to a record in Doomsday-book, Ridware was held in Edward the Confessor's time by Edmund, a Saxon freeman, under the Earl of Mercia; but within twenty years of the Norman Conquest, it had passed into the hands of Walter, a vassal of the Earl of Shrewsbury.

In the reign of Henry I. William de Ridware was lord of this manor, where he resided; it passed during successive ages to several possessors, and about the commencement of the seventeenth century it became the property of the Leigh family, and on the demise of Edward Lord Leigh, his sister, the Hon. Mary Leigh, became lady of this manor and of Yoxall.

The parish of Hamstall Ridware contains 2,521A. 2R. 16P. of tenantable land, and 36A. 2R. 32P. of waste land and waters.

The ancient manor-house is an extensive building, and was formerly fitted-up in a style of neat magnificence. Near it stands a watch tower, about fifty feet high, open at the top, from which there is a richly-varied prospect of the surrounding country.

Among other curiosities, the head of a stone hammer is shewn, to visitors, by the present resident of the mansion. The remains of a coat of mail and other accoutrements, provided for Charles I. in the time of the Rebellion, and a bridle for a scold, are also preserved here. The bridle is made of narrow thin plates of iron, with vacancies for the nose and eyes. When fixed on the offender, a flat piece of iron projects into the mouth and presses on the tongue. There is a ring in the centre, through which a cord was put to lead the offender into the church-yard, where she was obliged to remain till she gave signs of reformation, which, tradition says, this forcible deprivation of speech soon produced. Dr. Plot mentions two other similar bridles in his time, which were considered great curiosities; one at Newcastle, and the other at Walsall. There is another exhibited at Mr. Green's Museum, at Lichfield.

The church is a rectory, dedicated to St. Michael. It is a small structure of stone with a spire; has some painted glass in the windows, and contains several handsome monuments, and among others one to the memory of Thomas Allestree, M.A. who was a minister of the Church of England 54 years. He composed 600 sermons, and preached above 5000 times. He died the 30th of June, 1715, in the 78th year of his age.

Pipe Ridware.—This manor is situated on the northern bank of the Trent, six miles north-west of Lichfield. This is the first of the three Ridwares mentioned in Doomsday-book, where it is thus recorded. "Riduuare: the arable land is five carucates, which Alric holds." It came into the possession of the family of Pipe, in consequence of the marriage of Robert de Pipe, with Maud, the daughter of Thomas de Thamenhorne, in the reign of Edward the First. The manor was then distinguished by the name of *Pipe* Ridware, doubtless from the desire of the possessor to perpetuate his name. It remained in this family till 1443, when the heiress was married to Sir William Vernon, whose successor enjoyed possession of this manor till the ninth year of the reign of Queen Elizabeth, A. D. 1567, and left three daughters co-heiresses, of which Margaret became the wife of Sir Thomas Stanley, Knight, second son of the Earl of Derby, who in her right became seized of Pipe Ridware. In 1606, Edward Stanley, Esq. of Tonge Castle, sold this manor and lands, and also a fishery in the river Trent, to John Chadwicke, of Hill Ridware, gent. for the sum of £[illegible] 6s. 8d. In 1614, John Chadwicke gave this manor and fishery to Henry Agarde, of King's Bromley, in exchange for certain lands in Mavesyn Ridware, and a fishery in the Trent, between Armitage and Handsacre. In 1677, John Whitehall, Esq. of Sharpcliffe, near Leek, purchased this manor and settled here. He was succeeded by his son James Whitehall, Esq. who left two daughters, co-heiresses, one of whom was married to the late Sir Thomas Parker, Knight, and the other to Fisher Littleton, Esq. in consequence of which this manor and estate, with other lands in the Moorlands, were divided.

The ancient manor-house of Pipe Ridware was situated at Lismore. It was called Pipe-hall, and built in the year 1319. There are yet some vestiges of an ancient moat, on the spot, with extensive adjoining orchards, along the extremity of which the principal road to Lichfield formerly passed.

After a succession of ages the old manor-house fell to decay, and the present mansion was built on the bank of the river Trent by Mr. Whitehall, the before-mentioned purchaser of the estate. It was some years ago the habitation of the Rev. John Arden, and has since been repaired, rough-cast, and whitewashed, which has destroyed its antique and picturesque appearance.

The village of Pipe Ridware is but small, with few inhabitants. The extent of the whole parish is only about 700 acres; the soil in some parts of a cold and stiff nature, and in others mixed with gravel and sand.

The church, dedicated to St. James, is a very ancient structure; it contains monuments of the Whitehall family; and a curious circular font, about two feet and a half in diameter, en-sculptured with circles interlaced. The cupola contains two bells, and the church-yard, about an acre in extent, is very dry, and pleasantly situated.

Mavesyn Ridware.—This manor is bounded on the south by the southern bank of the Trent; on the east by Hamstall Ridware; on the north by the river Blythe; and on the west by the manor of Colton, in the hundred of Pire-hill.

After the Conquest, this manor was given by William I. to Roger de Montgomery: another adventurous follower of the Conqueror, named Azeline, held it under him. It probably passed from Azeline to the family of Mavesyn, in right of blood; for tradition asserts, that "the Mavesyns, *sithens the Conqueste*, were, and have been ever, in possession of Ridware."*

This manor continued in the possession of the Mavesyn family during ten generations. In the year 1403, Sir Robert Mavesyn, Knight, was slain at the battle of Shrewsbury. A feud had previously existed between this gentleman and Sir William Handsacre; and their animosity was influenced by the civil war. An authentic account of the termination of the quarrel between these Knights is preserved in Latin, in the British Museum, to the following purport: "The river Trent flows with a clear stream by Mavesyn Ridware, so called, because on its northern bank is situated the ancient inheritance of the Malvesyns. The inhabitants say that a jealousy subsisting between the families of Mauvesin and Handsacre, it so happened when Henry IV. had obtained the crown of England from Richard II. and it was rumoured that Percy, of Northumberland, was in arms against the King, Mauvesin had ridden forth with six or seven of his vassals on the part of King Henry; it chanced also that Handsacre, who espoused the opposite cause, had left home the same day with an equal number of attendants, to join Percy. These rivals met, and inflamed with rage, rushed furiously to battle. Handsacre was slain; and the victorious Mauvesin, proud of his conquest, marching to Shrewsbury, there lost his life, fighting valiantly for the King." Mavesyn and Handsacre fought on an open flat meadow, just above High Bridge, in Mavesyn Ridware, lying on the side of the Trent, between their respective mansions, which are in sight of each other. Sir Robert left behind him two daughters. Margaret, the younger daughter, became the wife of Sir William Handsacre, Knight, and thus terminated a feud which had been so fatal. Sir John Carvarden, Knight, married Elizabeth, the eldest daughter and co-heiress of Sir Robert Mavesyn; and the manor devolved on their descendants for seven generations. In the eighth year of James the First, A. D. 1611, this mansion and its dependencies were possessed by Gerard Stanley, gent. of Harlaston,

* Manesyne Ridware brieft, temp. Edward IV.

in this county, and John Chadwicke, Esq. of Wade-lane House, in Mavesyn Ridware. Mr. Chadwicke, in right of his wife, became possessed of the ancient manor-house, with five parts in eight of the manor, and the whole of the fishery. In 1600 he became sole lord of Pipe Ridware, with lands, and the fishery of the Trent down to King's Bromley, by purchase. In 1616, he gave the manor of Pipe Ridware to Henry Agarde, Esq. of King's Bromley, in exchange for lands in Mavesyn Ridware, and a fishery in the Trent, within Armitage and Handsacre, above High-bridge, so that he extended his right to both sides of the river.

Mavesyn Ridware has continued in the possession of the family of Chadwicke for upwards of two centuries. The parish is of considerable extent, and contains 1660 acres of arable land, 405 of meadow, and 375 of pasture. Charles Chadwicke, Esq. is the present owner.

The road from Lichfield to Uttoxeter extends through the whole length of this parish, High-bridge on the Trent being nearly five miles from the former, and Blythford-bridge, eight miles from the latter. The face of the country is agreeably diversified with an intermixture of corn-fields and pastures, well fenced with flourishing hedges of thorn, and sufficiently thickened with forest trees. The soil towards the north and north-west is an intermixture of marl and clay; about the middle of the parish it becomes loamy and light; and ends towards the south in a vale of sound light land, bounded by open meadows, enriched by the inundations of the Trent.

The crops of wheat and barley are generally excellent; and a variety of oats, the white Poland being sometimes cut in July.

The village of Mavesyn Ridware is small, consisting of the manor-house, church, and parsonage, two farm-houses, and eight cottages.

The manor-house is situated between the church and the Trent. A gatehouse of stone, 87 feet in length, is all that remains of the ancient mansion. The gateway faces the church: over it there is a large room, which, tradition says, has been an oratory. The walls above the gateway on the north side, and from the ground on every other side, are of brick. On the spot where once stood the antique and magnificent abode of the Mavesyns and the Cawardens, now stands a small house, built in 1718 by Charles Chadwicke, Esq. as a convenient box, pleasantly situated for a summer residence.

This beautiful part of the vale of Trent presented numerous attractions to the admirer of rural amusements. Needwood Forest afforded the sportsman a variety of game: the chace was also pursued with avidity and delight in its ample space, while the Trent offered to the skill of the angler great varieties of fish. This river abounds with pike, perch, greyling, eel, gudgeon, and craw-fish; trout and burbot are more rare; barbel, roach, dace, and chub, may be seen in large shoals. A pike was caught in the Trent at Mavesyn Ridware, in 1785, which measured one yard six inches and a half, and weighed twenty-four pounds and a half. An enormous pike was caught in 1772, above four feet long, weighing thirty-one pounds and three-quarters. The otter is the most formidable enemy of the pike in this river, and will attack, kill, and devour those of the largest size. Above fifty brace of pike have been killed here in a season, it being the rule to take none but those that measure twenty inches in length; so that out of 179 caught, 113 have been thrown into the river, as being under size. Though the neighbouring brooks abound with trout, there are few in the Trent, yet these few are the finest fish in the river. Several have been caught of the weight of four pounds a-piece and more.

"This and the neighbouring royalties have had '*game of swans*' immemorially; swans and their aeries are named in Handsacre very early, by deed s. d. and are the great ornament of the river, where they are in a sort of wild state, having no food given to them even in the most severe winters. In pairing-time they are very vicious, and fight furiously, the weaker bird being driven to a distance, and one is frequently killed, or has a wing broken if he escapes; so that in the extent of three miles there are not often seen here more than two nests. The hen lays seldom less than five, generally six, very often seven or eight, and even nine eggs, making her nest in the same spot for years together, unless disturbed; and on a rise of water has been seen to raise her nest, yet the nests are often destroyed by floods. At this time the cob jealously guards his mate: a hardy dog has been stunned with the force of his wing, and trampled under water till nearly exhausted; and a lusty farmer was unhorsed and ducked, as he attempted to cross the river incautiously too near the nest. The cygnets are not marked and pinioned till after Michaelmas, when they begin to take a daily flight."*

The swans belonging to eight royalties on the banks of the Trent, are distinguished by peculiar marks made by the swan-marker, who annually catches them for that purpose with an iron crook.

* Shaw's History and Antiquities of Staffordshire, Vol. I. p. 189.

Quails, which were formerly numerous in the meadows on the banks of the Trent, are now rare; as the quail is a solitary bird, perhaps it has abandoned its former haunts here in consequence of increased population.

The old Church was dedicated to St. Nicholas, and is called the parish church of St. Nicholas in 1691: sometimes it was styled the manor church, which may shew that the parish and manor were co-extensive. This ancient building was of stone, consisting of the nave, north and south aisles, chancel, steeple, and a roomy porch; the whole covered with shingles of oak.

In 1782 the whole of the old church, then very damp and nearly in ruins, was taken down, except Trinity aisle and the steeple, and a new one erected on its site, at the expence of £730.

This modern edifice is elevated above the church-yard, and built partly of the old stone, faced with red brick. The pointed gothic door and windows are ornamented with stone, and a stone cornice, and the roof covered with blue slate. There are only three memorials of the dead in this church, the most remarkable of which is a blue stone in front of the pulpit steps, with the following inscription:

"In full assurance
of a happy resurrection,
are here deposited the remains
of the Rev. Joseph Hollinshead, Clerk,
late Curate of this Church.
In him were eminently conspicuous
the good christian,
the diligent pastor,
the dutiful son,
and the sincere friend.
Suddenly,
and early in life, he obeyed
his great Creator's summons,
and left his parents to deplore
the hope of their old age.
He died April 3, 1753,
Aged 29."

Trinity aisle is a venerable relique of antiquity, and has been the cemetery of the lords of the manor, from the time of its erection in the twelfth century. It was dedicated to the Holy Trinity, and in the days of Popish superstition mass was celebrated on particular days.

The rectory-house is an indifferent old brick building. The glebe is about eleven acres; the tithe is moderate, 2s. 1½d. an acre being paid as a composition; and both together produce about £230. per annum, valued in the King's books at £7. 2s. 11d.

The mill is a small building well supplied with water from the Trent.

According to the Parish Register, the first entry in which is dated November 12, 1538, it appears that from 1538 to 1547 inclusive, the marriages were 20, births 69, burials 55; and from 1787 to 1796, there were 30 marriages, 122 births, and 79 burials.

Curious Extracts from the Churchwarden's accounts:

		£	s.	d.
1651.	For one our glasse	£0.	6s.	9d.
1652.	For one dyall	0	1	4
	Post to set a dyall on	0	1	0
1653.	Paid Gaunt, the clocke-maker, for the clocke	2	10	8
1662.	For the booke of comon praier and the booke of artickles	0	16	0
1663.	For the church bible	3	0	0

The following Extracts from the Constable's accounts, are illustrative of the unsettled state of the country during the Civil war:

		£	s.	d.
1642.	May 16, Paid to Collonell Bagot towards the mentenance of his Majestie's garrison at Lichfeild in money and retornes	7	16	8
	Sept. 12, The same payment, &c.	7	10	0
	Dec. 29, Spent at Mary Edwarde's amongst oure neghbors and Collonell Cromwill's soldiers	0	2	0
1643.	Assessment to pay for five horses that were *lost*, (*i. e.* taken away by Col. Cromwill's forces)....			
	March 17, Provisions went to Lichfeild for Prince Rupert and his foarces	4	5	6
	April 1, Ditto	4	12	10
	Contribucion weekly paid for the garrison at Lichfeild, from Nov. 24, 1643, to Oct. 25, following, inclusive. In money 76 8 6			
	In reternes ...... 33 14 6—	109	18	0
	Weekly payment paid towards the mentenance of the garrison at Stafford, from Feb. 17 to July 20, 1644. In money 21 5 0			
	In reternes .... 19 18 10—	41	3	10
1645.	The whole disbursement this yeare in monny and retornes	289	3	2
1647.	Soldiers under Sir Thomas Fairfax quartered upon them, by ordinance of Parlyment, from Oct. 7, 1646, to Jan. 25, 1647, after the proportion of 372 men in every division:			
	Aug. 28, Quartering 400 Scotes prisoners, and 20 soldiers with horses, theire garde	4	19	8
	Oct. 11. One moneths towards mentaininge the county troops	8	8	0

In the 25th Henry VIII. there were thirty-six householders of Mavesyn Ridware.*

* Harl. MS. p. 594.

At the contested election in 1747, ten freeholders voted from this parish.

Hill Ridware is a small village, situated about half a mile north-east from the church of Mavesyn Ridware. The principal public road in the parish passed through this village: the stocks and whipping-post stand where three roads meet. A may-pole was erected near the cock-pit, at the south end of the village, when George I. was crowned.

In 1796, the village consisted of 30 houses, containing 143 inhabitants. Here are two day-schools for girls, and one for boys.

Rake End is situated between Hill Ridware and Blythbury, and almost contiguous to the former, insomuch that it is often reckoned part of it. In this neighbourhood are two of the most pleasant situations in the parish: Cawarden-spring, near a mile to the west, and Bentley, above half a mile to the east.

Blythbury.—This hamlet contains about one-third of the parish, extending from Bentley Pool-bridge to the river Blythe. Here Hugo Malveysin settled in the reign of Henry I. and founded a church and priory on this demesne.

The priory dedicated to St. Giles was situated on the southern bank of the Blythe, the northern arm of which here bounds the manor, flowing through a pleasant valley. All the wants of the monks were supplied by the fertile tract around them, and the translucent stream. A fruitful garden and orchard, and the river Blythe, which supplied their corn-mill with water and their table with fish, rendered this spot peculiarly agreeable and convenient to a religious confraternity; and as the priory was in the vicinity of a public road, the monks had frequent opportunities for the exercise of their benevolence, by relieving the poor who made daily application at their gate, and the weary and houseless traveller. These monks were of the order of St. Benedict, and were commonly called Black monks from the colour of their outer garments. Part of the priory was appropriated to nuns; but the male and female inhabitants repaired occasionally to the same church to sing requiem for the dead.

The old priory has been destroyed; in 1789 there was a good farm-house on the spot, but scarcely a vestige of antiquity remained. It is now a manor-farm, called Blythbury-farm, of about 140 acres of land, of which 116 are tithe-free, and it has a fishery in the river Blythe.

Armitage.—This village is situated on the southern bank of the

Trent, opposite Mavesyn-Ridware. It is about two miles east of Rugeley, and five north-west of Lichfield. The parish is divided into two hamlets, Handsacre and Armitage.

Armitage was formerly called *Hermitage*, from a tradition that a hermit resided in a sequestered spot between the church and the river Trent. The hamlet and manor of Armitage contains the church, and gives name to the parish of which it forms about three-fifths. It is said to have been formerly "famous for tobacco-pipes made from clay found at Stile Cop."*

When hearth-money was collected in this county, in 1660, the constablewick of Handsacre and Armitage paid for 68 hearths £6. 16s.

At the contested election for the county, in 1747, nine free-holders voted from this parish.

The church, which is dedicated to St. John the Baptist, is situated on a dry, pleasant, rocky eminence, which commands a delightful prospect of the vale of Trent northward, while to the east are visible the lofty hills of Charnwood Forest, in Leicestershire. This church is a most beautiful and picturesque object, built of stone, very ancient, and somewhat ruinous in its external and internal appearance.

The steeple is fourteen feet square, and forty-three feet high. It contains three bells, with the following inscriptions:

> 1st. "Gloria in excelsis Deo, 1627."
> 2d. "Mr. Joseph Lycet, churchwarden, 1627; Mr. William Bryan."
> 3d. "I to the church the living call,
> And to the grave do summon all, 1727."

Hawksyard was an ancient manor within Armitage, held under Handsacre. The mansion was the seat of gentlemen for several ages, but it is not known to a certainty who were the successive possessors. It does not appear at what period the family of Rugeley first became lords of Hawksyard. Hawksyard-hall stood about three furlongs west from Armitage church, in a low swampy situation, not far from the Trent.

Hawksherd Park is now called Armitage Park, which is adorned with a neat hunting-seat, built in the rustic style, and surrounded by a beautiful and picturesque domain, to which the Grand Trunk Canal is no inconsiderable embellishment. Thomas Lister, Esq. is the present possessor.

Handsacre, another hamlet in the parish of Armitage, is thus mentioned in the Conqueror's survey. "Hadsacre: the arable

* Plot's Staffordshire, p. 131.

land is five carucates. Robert holds it." Hubert de Handsacre lived in the reign of Henry I. and was lord of Handsacre. It continued in the possession of this family till 1462, when it devolved by marriage to other families.

Near the turnpike-road, a little southward of the village of Handsacre, stands some remains of the ancient manor-house, surrounded by a moat. The structure was originally large, raised on a foundation of stone, and what is termed by antiquaries "half-timbered," exhibiting many gable ends, with heavy stacks of chimneys.

A farm on the opposite side of the road, is said to have been Handsacre-park; and a small cottage by the side of the old road, is reported to have been the Park-lodge.

About two hundred yards north from the hall is Church Croft, nearly two acres; and in the adjoining Mill-croft there is an oblong square, 187 yards long by ten or eleven wide, evidently the site of some building pointing east and west. On this spot some sacred edifice formerly stood, from which the adjacent church croft took its name.

Handsacre Mill stood on Longdon brook, about 200 yards north-west from the hall, and gave name to the present Mill-close, where the foundations of the building, with one of the millstones and a part of the wheel, still remain.

Verdon's Manor was a part of Handsacre, which formerly belonged to the ancient family of Verdon, but was granted in 1318 by Sir Robert de Verdun, to Robert, son of Sir Henry Mauveysyn, Knight, and Margaret his wife.

LONGDON.—This large village is situated on the turnpike-road between Lichfield and Stafford; four miles from the former, and twelve from the latter, in the deanery of Tamworth. It is remarkable for its great length, and consists of a number of scattered houses. This gave rise to the couplet,

> "The stoutest beggar that goes by the way,
> Cannot beg through Long' on a summer's day."

This village was formerly remarkable for the number of gentlemen's seats in its vicinity, no less than seven families of rank and opulence having formerly made this place their residence. The manor of Longdon is of great extent. It belonged to the church in the time of the Saxons, and in the 20th of William the Conqueror, A. D. 1086, to the Bishop of Chester.

In the time of Henry II. Longdon was part of the forest of Cannock, and given by that King to the church of St. Chadde, and

Walter Bishop, and his successors. It is recorded in the curious hundred-roll in the time of Henry III. "that the Bishop of Chester held this manor with its members, and the town of *Lyth* (Lichfield), in barony, of the King, and had there a free Court, and held plea of all things usual, together with that of forbidden distress; and had waif and view of frank pledge, without the cognizance of the sheriff. And the Bishop and his bailiff were wont to attend at the two great hundred courts, and there to demand his own free court, till the coming of the Bishop Alexander (about 1224), who withdrew himself from thence."*

In 1292, the ancient family of Tromyn, of Cannock, held Beaudesert under the Bishop. On the 29th of September, 1546, the Bishop of Lichfield and Coventry surrendered to Henry VIII. his manors of Longdon and Heywood, with their appurtenances, and accepted in their stead £183. a-year. The King, by his letters patent, bearing date the 26th of October the same year, granted the same to Sir William Paget, Knight.

According to a curious rental of Sir William Paget's estate, made in 1549, this manor of Longdon produced a clear rent of £78. 18s. 8d.; Longdon itself being about £40. which then consisted of the manor of Beaudesert, with appurtenances, valued at £13. 10s. particularly a mine of coal then valued at £4. per annum, and a smith-mill situated in the ancient park, and 48 messuages, with the appurtenances, and a mill called Longdon-mill.

Beaudesert, formerly the palace of the Bishops of Chester, and now the mansion of the Marquis of Anglesea, is situated on the side of a lofty eminence, in an advanced part of the forest of Cannock, about a mile south-west from Longdon church. The mansion is a magnificent edifice of stone, built in the form of a half H; it is sheltered in the rear by rising grounds, adorned with a variety of trees, and enveloped in groves of the most perfect and luxuriant growth. The greatest part of it was re-built by Thomas Lord Paget, in the reign of Queen Elizabeth, and various additions have been since made by his successors. The principal entrance is under a light gothic portico, which leads into a large and handsome hall, eighty feet by twenty-one, with a lofty arched ceiling, a large music gallery at the east end, and a beautiful gothic window at the west end, and adorned with the arms of the first Sir William Paget, and Preston, whose daughter he married.

* Tenure Roll of the Hundred of Offlow.

On a pillar in a corner of the hall, is fixed a piece of copper ore of a conical shape, from the Anglesea mine, upwards of 1200lb. weight.

The dining-room, large drawing-room, and lesser drawing-room, are spacious, and proportionate to the hall. In the lesser drawing-room is a portrait of the first Lord Paget, by Holbein, and a painting of the House. The library contains a valuable collection of books and MSS. and among others, the register of Burton Abbey. Among the modern improvements belonging to this mansion, the stables and coach-houses are most worthy of observation. They are built of beautiful white stone, in the form of a crescent, and situated at a convenient distance from the mansion, in a romantic valley. Beyond these buildings, the extensive gardens display all the beauties of horticulture; both the soil and situation are well adapted to the production of the most valuable herbs, roots, fruits, and flowers, and the walks and pleasure-grounds around the house present that charming variety of scenes to the rambler's eye, which are so grateful to the imagination. The park, where herds of deer ramble at will, contains an extensive succession of hills and valleys, abounding with excellent springs, and reminds the observer of the magnificent and picturesque scenery of Needwood Forest.

Upon the very summit of Castle-hill, in the rear of the mansion and on the verge of Cannock-chace, are traces of an ancient camp. It is encompassed with a rampart and two ditches, which are nearly circular, except on the south-east side, where it is straight. The extent is 970 paces in diameter; the two entrances are opposite to each other; and before that on the east are several advanced works. From its elevated situation, and the advantage of commanding an extensive view, it was well calculated for the purposes of observation and temporary defence. From Castle-hill, in clear weather, may be seen the counties of Stafford, Derby, Leicester, Warwick, Worcester, Salop, Chester, Montgomery, and Flint. The boundary line which separates the hundreds of Offlow and Cuddlestone, runs across part of this hill, and divides Beaudesert-park into two equal parts.

The parish of Longdon abounds with coal, particularly that valuable and elegant species termed *cannel*, which admits of a fine polish, and is worked into a variety of useful and ornamental articles, particularly inkstands, candlesticks, &c. It also is excellent fuel, and when placed edgewise on the fire, it so easily admits combustion that it presently flames as bright as a candle. The veins of coal are 20, 30, and even 40 fathom deep; the upper stratum is generally cannel, and the lower inferior coal raised for common use.

The family of Paget is ancient. Lewis Paget, Esq. was one of the gentlemen of Staffordshire who, in 1486, signed a certificate relating to the office of Master of the game at Cankewood; but the first who attained to the dignity of the Peerage was William Paget, a man of eminent abilities. He was born in London, was educated in St. Paul's school, and received his academical education in Trinity Hall, Cambridge. He afterwards went to study in the University of Paris; and in consequence of his learning and accomplishments, he was afterwards employed by Henry VIII. in important public missions. That Sovereign bequeathed him a legacy of £300. instituted him one of his executors, and appointed him one of the Council to his successor, Edward VI. In 1549, he was called by writ to the House of Peers, by the title of Lord Paget, of Beaudesert, in Staffordshire, and took his place in Parliament on the 3d of December, being then Comptroller of the Household, and Chancellor of the Duchy of Lancaster. But so uncertain is human greatness, that this meritorious Nobleman was committed to the Fleet in 1551, and soon afterwards removed to the Tower by the influence of the Duke of Northumberland. The Duke of Somerset, the friend of Lord Paget, was also committed to the Tower, and soon afterwards tried, condemned, and beheaded, being charged with a design to murder the Duke of Northumberland, and some other Noblemen at Paget-house. It was expected that Lord Paget would be implicated in this charge, but the Duke of Northumberland contented himself with depriving him of his public employments. On the 22d of April, 1552, he was divested of the ensigns of the Garter, on pretence of defect in blood. On the 16th of June following, he was fined £6000. in the Star-chamber, but in February ensuing he had a discharge for the payment of £2000. and in March, 1553, he obtained a grant from the King to him and his posterity for ever of the arms borne by the family.

On the accession of Queen Mary, he was sworn of the Privy Council. She also restored to him the Order of the Garter, and bestowed on him divers grants.

When Queen Elizabeth ascended the throne in 1558, Lord Paget, at his own request, was permitted to quit the public service, for "he was a strict zealot of the Romish Church."* He died on the 9th of June, 1563, aged 57, and was buried at Drayton, in Middle-

* Camden's History of England, Vol. II. p. 394.

sex. A stately monument to his memory was erected above the choir in the cathedral of Lichfield, which, together with that beautiful church, was destroyed during the Civil war.

In 1688, William Lord Paget was Lord Lieutenant and *Custos Rotulorum* of the county of Stafford, and appointed Envoy Extraordinary to the Emperor of Germany. His Lordship died at an advanced age in London, on the 26th of February, 1713, and was succeeded by his son Henry, who, on the 19th October, 1714, was created Earl of Uxbridge by Queen Anne. His only son and heir, Thomas Catesby (Lord Paget), was elected to two Parliaments in the time of George the First, and is classed among the few noble authors who have distinguished themselves by their productions in the English language. In 1734, he published a poem, in quarto, entitled an Essay on Human Life, which was more remarkable for good sense than poetic beauty. He also published a political pamphlet in 1740, and several poems and essays, in one volume octavo, of which only a few copies were printed, to be given away to his friends. Perhaps it would be much to the advantage of modern literature, if the Noble authors of the present day were equally modest and circumspect. His Lordship died at Drayton, near Uxbridge, in January, 1742.

Henry, the second Earl of Uxbridge, was born in 1719, and dying unmarried in 1769, the barony of Burton and earldom of Uxbridge became extinct, but the Barony of Paget descended to Henry Bayley, in right of his mother. Henry Bayley Paget, the late Earl of Uxbridge, was born June 18, 1744; married April 11, 1767, to Jane, daughter of Arthur Champagne, dean of Clonmacnoise, in Ireland, and succeeded to the barony of Paget on the 17th of November, 1769. The issue of his Lordship's marriage has been seven sons and five daughters, of whom the eldest son, Henry William, was born May 17, 1768.

The late Earl was exalted to the earldom of Uxbridge May 19th, 1784; and his conduct as a public character was eminently conspicuous and patriotic. He was Colonel of the Staffordshire Militia, and devoted much of his attention to the discipline of that fine regiment, while his example incited his sons to distinguish themselves by deeds of heroism in the service of their King and country.

At the commencement of the war with the French Republic in 1793, his Lordship's eldest son, Lord Paget, raised a regiment of infantry, and was distinguished for his bravery as its leader in several engagements on the Continent. His second son, the Hon.

William Paget, served in the navy, and rose by his merit to the rank of Captain of his Majesty's ship Romney. In an engagement between that ship and the French frigate La Sibylle, fought on the 18th of June, 1794, and which lasted an hour and ten minutes, this gallant young officer evinced the most undaunted courage. He not only compelled the French to strike their colours, but captured the frigate and three merchantmen. The untimely death of this naval hero was soon afterwards a subject of public regret.

On the decease of the late Earl of Uxbridge, 13th March, 1812, his eldest son, Henry William, succeeded to his titles and estates. But notwithstanding the extensive domains and ample revenue which invited him to ease and security, this heroic Nobleman, the enthusiastic votary of glory, continued in his career of military honour and success, and by a series of important services, achieved amid perils and privations, he realized that renown which shall illumine a page in his country's history to the latest records of time:

"Patriots have toil'd and in their country's cause
Bled nobly, and their deeds as they deserve
Receive proud recompence, we give in charge
Their names to the sweet lyre, th' historic muse
Proud of the treasure marches with it down
To distant time; whilst Sculpture in her turn
Gives bond in stone and ever-during brass
To guard it and t'immortalise her trust."

Among a variety of brilliant achievements by which the military life of this great warrior has been distinguished, it would be difficult to point out the most memorable. When in the disastrous retreat of the unfortunate Sir John Moore, his Lordship charged a superior number of the French pursuing cavalry, dispersed their squadrons, and captured their commander, General Lefebvre, he demonstrated the superior prowess of British cavalry, led by a skilful and intrepid officer; but when at the sanguinary and decisive battle of Waterloo, he engaged in the gallant enterprize of assailing Buonaparte himself, surrounded by his bravest troops, he signalized his heroism, and though he fell mutilated on the field of honour, he established his claim to public esteem on the imperishable basis of superior merit. A grateful country has, by the voice of its Senate, testified its high sense of his worth, while his Prince, by additional honours, enrolled him among the ornaments of the State. He was created Marquis of Anglesea on the 23d of June, 1815. His titles are, Henry William Paget, Marquis of Anglesea, Earl of Uxbridge, Lord Paget, Lord Lieutenant and *Custos Rotulorum* of Anglesea, Keeper of Carnarvan Castle, Ranger of Snowdon Forest, Vice-Admiral of North Wales, Pembrokeshire, and Carmarthen. His principal seats are Beaudesert, and Sinai Park, in this county; Drayton, in Middlesex; Stabridge, in Dorsetshire; and Place Newyth, in the Isle of Anglesea, where he is possessed of an immense property in his valuable copper mines.

On Monday, November the 6th, 1815, His Royal Highness the Prince Regent, accompanied by the Duke of Clarence, arrived at Beaudesert, on a visit to the Marquis of Anglesea, and next day the Royal Brothers were joined by the Austrian Archdukes John and Lewis, who were then making the tour of this country. An intimation of His Royal Highness's intention to visit the Marquis of Anglesea having been received at Beaudesert, the Earl of Uxbridge, Lord Yarmouth, Lord Graves, Sir Arthur Paget, and Gen. Sir E. Paget, with nearly two hundred of the Noble Marquis's tenantry on horseback, proceeded to Lichfield to make the necessary arrangements for the reception of the Royal Visitors. The Staffordshire regiment of Yeomanry Cavalry, under the command of the Hon. Edward Monckton, formed at the entrance of the city, and awaited the arrival of His Royal Highness. It having been 125 years since the neighbourhood of Lichfield was honoured with the presence of a reigning Prince of England, the most lively interest was excited in all classes of people, who seemed anxious to afford another proof of that loyalty and attachment to the reigning family, which has ever distinguished the ancient and respectable city of Lichfield. His Royal Highness reached Lichfield about six o'clock, changed horses at the George Inn, and proceeding rapidly through the city, was met at Longdon by the Marquis's tenantry, headed by his keepers, (a particular form of the ancient system, and probably indicative of his Lordship's right of free-warren over Cannock-chace), and conducted to the Hall, amidst the acclamations of assembled thousands. A detachment of the Yeomanry Cavalry escorted His Royal Highness from Lichfield to Beaudesert. Addresses to His Royal Highness the Prince Regent, from the City of Lichfield, and the Town of Burton-upon-Trent, having been previously prepared, the Deputations were introduced by the Noble Marquis into a spacious dining-room, where they were most graciously received by His Royal Highness, who stood up in the centre, with General Bloomfield on his right, in waiting. On the right, were the Duke of Clarence, with the Marchioness of Anglesea, and the whole of the female part of the Marquis's family: in the back-ground was the Marquis of Anglesea, having on his left the Earl of Uxbridge, Lords Yarmouth and Graves, &c. &c. the whole standing. The Addresses having been read, His Royal Highness was pleased to return most gracious answers, and the Deputations withdrew to partake of an elegant repast prepared by the Marquis. During their stay, the Royal Visitors joined in the sports of the field, or the joys of the banquet, with all the amenity of private life; and on departing, expressed themselves highly gratified with their reception at Beaudesert. Thus terminated a scene of princely hospitality, which reflected equal honour upon the Illustrious Guest and the Noble Donor:—a Prince who, by his wise councils, had raised himself and his country to the highest pinnacle of earthly glory—a Warrior, who had fought and conquered on the deathless plains of Waterloo, and whose gallant services mainly contributed to the success of that ever-memorable struggle for the independence of nations.

Chestall, situated to the east of Beaudesert, was formerly the residence of the Rugeleys. At present the only mansion on the spot is a farm-house, which is the property of the Marquis of Anglesea.

Stonywell is situated about a mile south-east of Longdon Church. This mansion was formerly the residence of an ancient family, one of the descendants of which was memorable for his learning and piety. "John Stanywell or Stonywell," says Wood, "was born in the parish of Longdon, in Staffordshire, within which parish is a small hamlet, called Stonywell, from a well wherein is at the bottom a large stone, which seems to be no more than a little rock, whence springs the water that supplies that well. This person being much addicted to learning and religion when a youth, was taken into a monastery, and was bred a Benedictine monk. Thence he was sent to Gloucester College, in Oxon, where the monks of Pershore had an apartment for their novices to be trained up in academical learning; of which college he was, when in his elder years, Prior for a time, and was then noted among those of his profession for his learning and strict course of life. Afterwards, being Doctor of Divinity, he became Lord Abbot of the said Pershore, and at length a Bishop, (suffragan only as it seems), under the title *Episcopus poletinsis*. He paid his last debt to nature, after he had arrived to a great age, in the beginning of 1553, and was buried according to his will, in a new chapel built by him within the parish church of St. James, in Longdon, for the use of which chapel and parish he bequeathed all his books, his two chalices, his crewetts, holy water stock, vestments, albes, altar clothes, with other things belonging to his private chapel in Longdon."*

The Stonywells continued to reside here for several ages: the ancient mansion is now destroyed, and a modern brick farm-house built for the use of the tenant.

Liswis-Hall was originally the property of an ancient family of the name of Liswis, the first record of whom is an old deed, without date. In the time of Henry III. Roger de Lisewis held a sixth part of a Knight's fee in Longdon, of the barony of the Bishop of Chester. "In 1356, William de Lysweys de Longdon, granted to his son Richard, and Margaret his wife, in frank marriage at the church door, all his lands, tenements, rents, and services, &c. in Longdon, and all his demesne there, with suit of court, amercements, wards, marriages, reliefs, &c. belonging to that manor, and all the liberty in the forest of Cannock, granted to his ancestors, rendering to the said William, during his life, one mark of silver."†

Richard Leswys left a daughter and heiress, Margaret, who carried this estate in marriage to John Legydd, with whose daughter it devolved to James Arblaster. In this family the estate continued till 1769, when Edward Arblaster, of Over Stonall, sold it to Francis Cobb, Esq. of Lichfield. Liswis-hall is a neat mansion, of brick, faced with stone, at a convenient distance from the turnpike-road, with a lawn and meadow in front, through which a serpentine river glides within its banks.

Haneh-Hall is situated on the turnpike-road leading from Lichfield to Abbot's Bromley, and about a mile east of Longdon-green. This ancient mansion was built in the reign of Edward I. by one of the Astons, of Haywood, and was the residence of his successors for several generations. It afterwards came into the possession of a family of the name of Orme, the first of whom, William Orme, of Haneh-hall, married Grace, daughter of Nicholas Hurt, of Casterne, in this county. On his demise, in 1623, his son William succeeded to the estate, and suffered greatly for his loyalty to Charles I. during the Civil war.

"William Orme, of Longdon, gent. to settle the tithes of Upper Mayfield, of the value of £36. per annum for ever, upon such place as the Parliament shall appoint, for which he is allowed £360. so his fine of £1,395. is reduced to £1035. Yet afterwards, upon

* Athenæ, Vol. I. p. 580.

† Dr. Wilkes's Antiquities of Staffordshire.

review, was increased £500. more, besides decimation; the reason whereof was, that he did in effect garrison and furnish with provisions the close of Lichfield, (at the request of the church) upon the first besieging thereof; by which, and this being plundered before the relating, and oft marching with the King at his own charge, a personal estate of £2000. was consumed, he having no profits of his real estate of £500. per annum for four years, though betwixt 1637 and 1654, he had by one wife fourteen children. And farther, in revenge of his steady loyalty, was barbarously taken out of his bed, when sick of a palsie, and imprisoned at Stafford, so long as to bring him to distraction."*

The Parkhursts afterwards possessed Hanch-hall, and Dormer Parkhurst, Esq. voted at the contested election for the county, in 1747. It was afterwards repaired and inhabited for a short time by Thomas Fowler, Esq. of Pendford.

Longdon Church is a large ancient gothic edifice of stone, with a handsome tower. On the south side is Bishop Stoniwell's chapel. It contains some mural monuments, particularly one with a long inscription, recording the loyalty and military services of the above-mentioned William Orme.

Dr. Plot mentions three remarkable instances of longevity in this parish, William May, aged 108, and his wife Joyce 98, who were buried on the same day; and good wife Nip, who lived to the age of 109.

FAIRWELL.—This small village is situated on the side of a woody vale, on the north-east border of Cannock Forest. In 1140, the church dedicated to St. Mary was given to a Priory of Benedictine nuns by Roger Clinton, Bishop of Lichfield and Coventry, who endowed it with the mill, and all the lands lying between the brooks called Chistals and Blackesiche; likewise six husbandmen, with their tenures, horses, and services. King Henry II. confirmed this grant, to which he superadded forty acres of land cleared from wood in the forest of Cannock. Upon the suppression of the lesser religious houses, in 1527, this priory-church was given to Lichfield cathedral to maintain the choristers.†

This manor now belongs to the Marquis of Anglesea: the soil is principally gravel, clay, and sand.

In 1747, the old nunnery chapel was taken down, when three

* Scarce and Valuable Tracts, printed in 1752.

† Monasticon, Vol. I. p. 441.

rows of coarse earthen vessels of different dimensions were found in the south wall, six feet from the ground. These vessels were laid on their sides, and their mouths, which were covered with a thin coat of plaster, were placed towards the chapel.

Fairwell Church is a picturesque object, situated on the side of a hill; the body of the edifice and the tower are of modern brick, but the chancel, with its large gothic window, is a fine piece of antiquity. It is a vicarage in the patronage of the Marquis of Anglesea.

Charley, a hamlet in the western part of this parish, is situated on the verge of Cannock Forest. It consists of a few houses, and an ancient hall.

LICHFIELD.*

This ancient City and Bishop's See is situated in a fine and fertile valley, 119 miles from London, and nearly in the centre of England. The origin of this city has occasioned much learned controversy, and a great difference of opinion among antiquaries, whose vague conjectures afford no conclusive arguments to the mind intent upon truth.

Bede, one of our most ancient writers, calls it *Licidfeld*, or the Field of Carcasses, from a tradition that one thousand British Christians suffered martyrdom here in the reign of the Emperor Dioclesian. This account is treated as a mere legend by sceptical writers, and, among others, by Dr. Stukeley, who traces the etymology of the name from *lich*, a Saxon word, which signifies a morass. But whatever may have been the origin of the name of this city, it derived its first importance from the Saxon Kings. When Peada, son of Penda, King of Mercia, married the daughter of Oswy, King of the Northumbrians, and restored tranquillity to the Mercian kingdom, he at his return from Northumberland brought four priests with him, and made Lichfield his residence. A few years afterwards, Cedda or St. Chadd, came to Lichfield, and devoted to a life of solitude, fixed his cell at or near Stow Church, where Bede says he was buried. St. Chadd was a Bishop, but at this period his habitation was more like an hermitage than a palace. It was surrounded with woods, a brook ran close to the entrance, and the spot was well adapted to prayer and contemplation.

* The *Close* of the Cathedral of Lichfield is in the southern division of the hundred of Offlow, but as the Militia Lists, Taxes, &c. &c. are never included in that division, we have placed Lichfield in Offlow North.

M

Lichfield was erected into a Bishopric by Oswy, in the year 666, and undoubtedly it owed this distinction to the superior sanctity of Cedda.

In the year 1075, Lichfield was by the synod held at London, considered too small for the residence of a Bishop. Prior to this period, the English Bishops were remarkable for their humility and self-denial. They generally travelled on foot about their diocese to instruct the people, and see that the inferior clergy lived according to the precepts of their great and humble Master. But the invasion and conquest of England by a Norman adventurer, and his banditti, changed the face of things. Instead of ancient simplicity, the luxurious manners of the Continent were introduced, and pride extended its influence from the court and the castle, to the hitherto sacred institutions of religion itself.

When the general survey of England was taken by the order of William I. and recorded in Doomsday-book, the lands of this See were not said to belong to the Bishop of Lichfield, but the Bishop of Chester, "who then held Lecefelle, with its appurtenances. The church formerly held it. Here are 25 hides and a half, and one virgat of land. The arable land is 73 carucates. In demesne there are 10 carucates, and 10 servants, and 42 villans, and 12 bordars, having 21 carucates. There are 35 acres of meadow, and two mills, rented at four shillings. It was formerly, and is now, worth 15 pounds. To it belongs a wood, eight miles and a half and seven furlongs in length, and six miles and a half and eight furlongs in breadth. To this manor belong three members, Padinton, (now Packington), the arable land is four carucates; and the two Hamerwiches, the arable land is five carucates; and Tichebroc, the arable land is one carucate; and Nortone and Werelia, four carucates of land; and Rowleia, one carucate of land: all are waste."

Lichfield first became a place of importance in the reign of Henry I. when Roger de Clinton was made Bishop. This rich and powerful prelate was a man of genius and enterprize, zealous to promote the glory of the church, and emulous of distinction. He added to the extent and beauty of the cathedral, environed the city with a ditch, fortified the Close, and enrolled and mustered a garrison. He also increased the number of prebendaries, but made no great addition to the buildings of the city, for his zeal for Christianity incited him to embark in the Crusade, and he died at Antioch.

In the reign of King Stephen, according to William of Malmsbury, "Lichfield was a small village, much inferior to a city. The country about it was covered with trees, and a rivulet ran near it. The Cathedral stood on a narrow neck of land; it was formerly famous for the abstinence of its members, but our Bishops are ashamed of the place, because the episcopal dignity is now to be supported in a different manner."*

In 1296, Walter de Langton was advanced to this See, and to his munificence Lichfield owes much of its present splendour. At this period, the city was intersected by three large pools of water, and Bishop Langton built the bridge over the Minster pool, cleaned the ditch which encompassed the Close, and fortified it with a strong stone wall. He also laid out £2000. in a monument for St. Chadd, the first Bishop of Mercia. The vicars' houses in the Close were a donation by him, and he made an addition to their salaries. Among other public-spirited efforts to improve the city, he obtained a grant from Edward I. to lay an annual tax on the inhabitants for one-and-twenty years, to pave the streets. A short time before his death, he began to build St. Mary's Chapel at the Minster, where he proposed to be buried; but not living till it was finished, he left money for that purpose. This great and good man died on the 16th of November, 1321, and was interred in the Cathedral. Among other benefactions, he gave the old house belonging to the Bishop to the vicars' choral. In the eastern part of the Close he erected a new palace for his successors, and along the north side of it he built proper houses for the Dean and Chapter. He gave a large quantity of plate for the use of the church, and more to the vicars, to whom and the canons he was a great benefactor.

In 1547, by a charter of Edward VI. Lichfield was elevated to the dignity of a City and Corporation. This Prince also restored the burgesses the privilege of sending representatives to Parliament, of which the city had been deprived from the year 1354, the twenty-seventh of Edward III. This charter was confirmed, and many additional immunities conferred on the citizens of Lichfield, by Queen Mary, Queen Elizabeth, James the First, and Charles the Second. When James II. in his arbitrary experiments on the feelings of Englishmen, required the surrender of the ancient charters of Lichfield, he granted them a new one, with some reservations. But in 1688 he published a proclamation, restoring to Corporations their ancient privileges: consequently, the new charter was rescinded, and the corporation reverted to the ample and satisfactory charter of Charles II. This charter is dated No-

* De Gestis Pontificum Lichfeld.

vember 5, 1664. By it the King confirmed all the former charters granted by his predecessor, for the good government of the city of Lichfield, and conferred upon the citizens new and extensive privileges. In this charter, he appointed that two bailiffs should be elected annually on St. James's day; the senior bailiff to be nominated by the Bishop, and his colleague by the Common Council, consisting of one-and-twenty brethren, elected by the citizens. That the bailiffs, or any of the Common Council, should, upon ill behaviour, be liable to be removed by a majority of the body corporate, and upon the removal or death of any of them, others should be elected by the burgesses. That the bailiffs and Common Council should be empowered by law to buy and sell, to sue and be sued as a body corporate, by the name of the bailiffs and citizens of the city of Lichfield, and have a common seal: that they should have power to make bye-laws, and when reduced into writing, to punish offenders against them. This charter further empowered the Corporation to elect a recorder and a high steward, who should be magistrates during life, and have power, together with the bailiffs and justices, to hold courts of gaol delivery, and award judgment of death, or other punishment, according to the nature of the offence, and that no other judges or justices should have power in the city. That the bailiffs and Common Council should elect one of their own body corporate to be coroner, who should have the custody of all records and writs, &c. That there should be four fairs held in the city, yearly; on Ash Wednesday, the Friday in the week after St. Simon and St. Jude, the Friday after Twelfth-day, and on the first day of May, old style; and that the bailiffs should have all tolls. That the bailiffs and Common Council should annually, on the feast of St. James, (July 25th,) elect any one of the citizens and inhabitants of this city, to serve the office of Sheriff of the city and county of Lichfield: and that the sheriff should, on the feast of the nativity of the Blessed Virgin Mary, (September 8,) yearly perambulate the boundaries of the city and county of Lichfield, and the precincts thereof.

The sheriffs have from time immemorial perambulated the city and county, being a circuit of sixteen miles, with numerous attendants, as specified in the charters. Besides the court of gaol delivery, there are other inferior courts held in Lichfield, particularly the court of quarter sessions, pie-poudre court, and a court of record, held by the authority of the bailiffs, for the recovery of debts amounting to forty shillings and upwards. There are also three annual courts held in the city, namely, the court of the view of frank-pledge, the court of array, and the great portmote-court. Of these the court of array is the most remarkable, and worthy of record, as illustrative of the manners of our ancestors.

Early on the morning of Whit-Monday, the high constables of the city, attended by ten men armed with fire-locks, and adorned with ribbons, preceded by eight morrice-dancers, a clown fantastically dressed, and drums and fifes, escort the sheriff, town-clerk, and bailiffs, from the Guildhall to the Bower at Green-hill, a mount situated at the south-east extremity of the city, in the parish of St. Michael, where a temporary booth is erected for their reception. On this mount the title of the court is proclaimed by the common cryer; the names of all the householders in the twenty-one wards of the city, according as they are enrolled, are called over; and all persons owing suit and service to this court, called "the court of array, or view of men and arms," of the manor and lordship of Lichfield, are required to appear, under pain of fine and amercement. After this ceremony is ended, the constables, attended by the armed men, &c. take their leave, and march through the streets to the opposite extremity of the city, and summon the dozener, or petty constable of that ward, to attend. He immediately comes forth, bearing a flag or ensign, joins the procession, and the armed men fire a volley over every house in the ward. The inhabitants, on this salute, invite the constables into their houses and present them refreshments, while drink is given to their attendants. The dozener then, bearing his pageant, attends them to the Bower, and the town-clerk from a roll calls forth the name of every householder in the ward. Those who answer to their names are invited into the booth, and regaled with a cold collation, and those who neglect to appear are fined one penny each. In this manner the constables go through the twenty-one wards, and perform similar ceremonies in each; consequently, it is late in the evening before they have performed their toilsome task. The Court of Array is then concluded, and the constables, preceded as before by the martial music, dancers, and armed men, and attended by the petty-constables, formed in two lines, with their gorgeous banners, conclude the festivity by a procession through the principal streets to the market-place, where on their arrival the town-clerk, in the name of the bailiffs and citizens, delivers a charge to the high-constables. The purport of this address is to thank them for their attendance, and inform them, that in consequence of the firm allegiance of their predecessors, several charters and immunities were granted to this city, which it is hoped will stimulate them to the performance of their duty to the King and their fellow-citizens. He concludes with advising them to retire peacably to their homes, and pursue the paths of industry and virtue, that they may always be worthy of the peculiar privileges they enjoy as free-born Englishmen, and inhabitants of this loyal and respectable city. The dozeners then deposit their colours under the belfry in the adjacent church of St. Mary.

The origin of the court of array has excited the inquiry of antiquaries, some of whom suppose that it was first instituted by Oswy, King of Northumberland, in memory of a victory obtained by him over Penda, King of Mercia, in a battle fought near this city. Others, with greater probability, trace its origin to an act passed the 27th year of the reign of Henry II. in the year 1181, by which it was enacted, that the high-constable of every town should oftentimes view the arms and armour of the men in their jurisdiction.

The Rev. Dr. Falconer throws great light on this institution, and on the manners and customs of our ancestors in the remote ages of ignorance and superstition. "By having recourse to the history of former times," says he, "it will be no difficult matter to account for this uncommon ceremony; it is a mixture of religious and civil institutions; the bailiffs are lords of the manor, and hold a court-leet at their Guildhall, on Whit-Monday, annually; but in former times, that room being too small for the accommodation of the number of inhabitants who were bound to pay suit and service to them, they adjourned the court to the more spacious place on Green-hill. The town was then divided into distinct wards; the names of the inhabitants in each ward were enrolled, by which means it was known with certainty the number of inhabitants each ward contained; and by appearing personally, it was discovered how many of them were capable of bearing arms in defence of their country. To this was joined the religious institution: a representation of the tutelar Saint was formerly exposed to the view of the populace annually, to excite veneration, and likewise it was exhibited on some particular emergent occasions to ward-off some threatening evil. This being a time when all the inhabitants were assembled, it was deemed most proper to exhibit the effigy of their favourite Saint; and after it had answered the purposes of the day, what more proper place could be thought of than the belfry of the church, and where they might have an opportunity of holding a synod? However absurd and ridiculous these ceremonies may appear in the eyes, and to the enlightened minds of the present age, they had undoubtedly their advantages; pageants and parades struck the minds of the populace with reverential awe and respect; they made them submit patiently to subordination; they afforded them an opportunity not only of discovering the number of inhabitants, but the number of effective men, to protect their rights and properties; and they promoted mirth and hilarity. Nor were these the only benefits to be derived from them; the neighbouring inhabitants partook of the advantages; the nobility and gentry resident within several miles round the town, before they had adopted the pernicious custom of deserting their native mansions, and mispending their time and substance in the debaucheries of the metropolis, thought it sufficient entertainment and recreation for themselves and their children to attend this Whitsun-bower gala. Crowds of them were used to flock in, and to increase their amusement there were displayed a variety of other exhibitions adapted to the taste of these times, such as bear and bull-baitings, interludes, flying-chair, legerdemain practitioners, wild beasts, &c. Uncouth as these amusements may be deemed by our modern refined taste, they had their charms and their utility; the novelty and dexterity of them excited admiration, they did not tend to promote vice and immorality, and they afforded an opportunity to all ranks of people to assemble and spend their time in innocent mirth and hilarity."*

An inferior court, called Maudlin's Court, is annually held in Lichfield. It was anciently instituted for punishing drunkenness, and for the purpose of administering the oath of office to the dozeners or petty constables.

The inhabitants of Lichfield first sent representatives to Parliament in the 33d year of the reign of Edward I. 1304; and in the 4th, 5th, 6th, 7th, and 20th of Edward II.; and the 1st, 14th, and 27th of Edward III. From that period they were discontinued till the year 1547, the 1st of Edward the Sixth, when Lichfield was raised to the dignity of a city and corporation.

Since that era, so auspicious to the civil and religious liberties of Englishmen, Lichfield has been favoured with many immunities conferred by successive Sovereigns.

On the 10th of March, 1701, it was determined by the House of Commons, that the bailiffs, magistrates, freeholders, and all that

* The ceremonies of Greenhill Bower, after having been discontinued for many years, were revived on the 3d of June, 1816, (Whit Monday,) with some eclat, but stripped of much of their former pomp.

held by burgage-tenure, have a right to vote in the election of citizens to serve in Parliament for the city of Lichfield. That such freemen only of the said city as are enrolled, and pay scot and lot, have a right to vote, &c. December 18, 1718, it was further determined by the same authority, that the right of election of citizens to serve in Parliament for the city of Lichfield, is in the freeholders of forty shillings per annum; and all that held by burgage-tenure, in such freemen only of the said city as are enrolled, paying scot and lot there, and in the bailiffs and magistrates. The number of electors is about six hundred.

Lichfield has given the title of Earl to several families. The city is divided into three parishes, St. Chadd's, St. Mary's, and St. Michael's. The central part of Lichfield is occupied by St. Mary's parish; the parish of St. Chadd occupies the north-east part of the city and suburbs; and the parish of St. Michael includes the south-east division and its suburbs.

A distinct part of Lichfield is the Close, in which the Cathedral stands. This district is a separate jurisdiction, governed by the Dean and Canons, who are the only magistrates within its bounds, and entirely independent not only of the city but the county of Stafford.

In 1781, a plan of the City and Close was made from an actual survey by John Snape. The names of the wards, with the number of houses and inhabitants, were then as follow:

	Houses.	Inhabitants.
Bacon-street ward	71	372
Bird-street ward	41	230
Sandford-street ward	47	292
Sandford-street below the Water ward	35	185
St. John-street ward	47	260
St. John-street above the Bars' ward	38	262
Sadler-street ward	55	291
Boar-street ward	53	223
Wade-street ward	45	190
Dam-street and Butcher's-row ward	46	256
Tamworth-street ward	66	234
Lombard-street ward	25	164
Stow-street ward	64	278
Green-hill ward	79	367
Total within the City	722	3555
——— Close	43	216
	765	3771

Sir Frederick Morton Eden, Bart. collected many valuable facts illustrative of the state of the poor, &c. in England. His account of Lichfield, in 1797, is equally correct and interesting, being the result of the strictest inquiry and actual observation. "Lichfield," says this intelligent writer, "contains three parishes; viz. St. Mary's, St. Chadd's, and St. Michael's. The first has most houses and inhabitants, but no land; the other two have few houses, but a considerable quantity of land. In 1781, the number of houses in Lichfield was 723; and of inhabitants about 3556. It is supposed that, since that period, the population has considerably increased. In the whole city 408 houses pay the window-tax; the number exempted could not be ascertained. The prices of provisions are, beef and mutton, 5d. the pound; veal, 4½d.; bacon, 9½d. and 10d. the pound; milk, ¾ of a quart for a penny; butter, 11d. the pound; potatoes, 4s. the bushel; bread-flour, 5s. the stone; coals, 6d. the hundred weight.

"The wages of labourers are from 9s. to 12s. a-week, according to the season; women for weeding, are paid 8d. a-day; for hay-making, 10d. a-day and victuals. Those who spin lint, earn from 4d. to 6d. a-day. There are 46 alehouses in this city; and five friendly societies for men, and two for women: from 100 to 200 members belong to each society.

"Land near the city lets from £3. to £4. an acre: the average is about 30s. an acre. Farms are generally small: the principal articles of cultivation are, wheat, barley, oats, turnips, and clover. Tithes are mostly taken in kind. The total land-tax for the city is £420. 19s. 4d. The poor are maintained at their own houses; about 23 pensioners at present receive £2. 17s. 6d. a-week; six of these are bastards. Several house-rents are paid, and casual reliefs are given to many of the necessitous.

"The rates in St. Mary's are about 3s. in the pound on the nominal rental; in St. Chadd's and St. Michael's half the rack-rents are assessed. St. Mary's and St. Chadd's have each a work-house. In St. Mary's work-house there are at present 41 paupers. They manufacture a little blanketing for the use of the house."*

St. Mary's Church is situated on the south side of the market-place, in the central part of the city. According to an inscription on the steeple of the ancient structure, this church was first founded in the year 866, and this date is also recorded in the parish register. The authenticity of this date, however, is questioned by Mr. Shaw,

* Sir F. M. Eden's State of the Poor, &c. in England, Vol. II. p. 661.

who proved a similar claim to antiquity in Wednesbury church to be a most egregious error. Indeed this desire to enhance the antiquity not only of places but of families, is among the common expedients of pride to impose upon ignorance.

Leland, in his Itinerary, says, "St. Marie's is a right beautiful piece of worke in the very market-place." It was certainly a beautiful gothic structure, with a lofty spire. The master and brethren of the gild of the Blessed Mary had a chantry here, which was dissolved in 1545 by Henry VIII. Afterwards the members of the Cathedral officiated at this church, but at length founded a distinct vicarage here, reserving to the Dean and Chapter the right of appointing the vicar.

During the persecution of the Protestant clergy in the short and sanguinary reign of the bigotted Queen Mary, William Langeley, M. A. then minister of this church, was sequestered for preaching on Christmas-day, and administering the sacrament by the common prayer-book. John Butler, M. A. of Edmund-hall, Oxford, was ejected here after the Restoration.

In 1717 the old church being decayed, was taken down, and the present elegant edifice erected on its site. It was opened on the 30th of December, 1721, by the Rev. William Baker. The body of this church is neat, fitted-up with oak pews, and a spacious gallery. The altar-piece is handsome, and on the north side of it is the burying-place of the Dyott family, where stands a monument to the memory of Sir Richard Dyott, Knight, renowned for his fidelity to Charles I. during the Civil war.

St. Michael's Church is situated on the summit of Green-hill, at the south-east extremity of Lichfield. It is an ancient edifice of stone, with a high spire, erected in the reign of Henry VII. who rebuilt several churches, which had been demolished during a long and destructive civil war. The elevated and solitary spot on which this church stands, with the great extent of its cemetery, which is the principal burial-place belonging to the city, and includes a space of nearly seven acres in extent, have a tendency to tranquillize the mind of the observer, and prepare it for serious meditation. A walk, paved with white pebbles, and shaded by a grove, leads to the principal entrance. In 1593, part of the spire was blown down by a tempest. The interior of this edifice is adorned with many monuments and inscriptions. St. Michael's is a perpetual curacy, in the presentation of the vicar of St. Mary's. The church-yard commands an extensive prospect, and the public walks from it through the fields on the south-east side of the city are very pleasant, to the summit of that celebrated mount Borrowcop Hill, of undoubted Saxon origin, and described by most antiquaries as a *tumulus*, and the site of a Saxon fort. A small edifice, with seats, has been erected by subscription on the top of this hill; and from this lofty eminence may be seen, in a clear day, many delightful prospects of the circumjacent country, which presents the diversified scenery of magnificent villas, neat farm-houses, cheerful villages, fertile and highly-cultivated hills and valleys, and all the beauties of the English landscape, where

> "Town and village, dome and farm,
> Each gives to each a double charm."

Borrowcop-hill also commands a full view of the race-ground on Whittington Heath, and the stand erected for the accommodation of the nobility and gentry who visit Lichfield races annually in the month of September. Indeed the prospect is so extensive, varied, and picturesque, as to excite the admiration of every visitor, who will be amply repaid for the toil of ascending this eminence, by a view from its summit of a considerable part of the counties of Warwick, Salop, Stafford, Derby, Nottingham, and Leicester; the churches of Barton, Seckington, Lallington, Clifton-Camville, Whittington, Shenstone, and Lichfield, especially that superlatively grand object in the fore-ground, the magnificent Cathedral, while the south-west view of St. Michael's on the opposite eminence, the tower of St. Chad's, and the turret of St. Mary's, harmonize in presenting a highly-picturesque view of the city and its environs.

The parish of St. Chadd, or as it is commonly called Stow, comprises a considerable part of the north-east division of Lichfield and its suburbs, and several adjacent hamlets. The church is a very ancient fabric of stone: it is dedicated to St. Chadd. It derived its name from the circumstance of St. Chadd having had his cell here in the year 669.

In a small garden near this church is St. Chadd's Well, where many Popish devotees have been known to resort. Leland, in his Itinerary, mentions, "Stowe Church, in the east end of the towne, where is St. Chadd's well, a spring of pure water, where is seen a stone in the bottom of it, on which, some saye, St. Chadd was wont naked to stand in the water, and praye. At this stone St. Chadd had his oratory in the tyme of Wulphar, King of the Merches."

In the times of Popery, when rites and ceremonies were considered essentials of religion, the clergy and populace assembled at this Well on Holy Thursday, and adorned its border with green boughs and flowers. This custom is still observed, for the clergyman, on Ascension-day, attended by the churchwardens of St. Chadd's, and a multitude of children, with green boughs in their hands, visits this well, when the ceremony of strewing the spot with flowers takes place, and the gospel for the day is read. The water of this well is whitish, and supposed to possess medicinal virtues. The interior of St. Chadd's church was completely repaired about twenty years ago; in the chancel and aisles there are several monuments; and in the north aisle, where the shrine of St. Catherine formerly stood, there is now placed a very ancient font.

On the summit of Stow-hill, there is a handsome mansion erected by Mrs. Elizabeth Aston, and lately occupied by her sister Mrs. Gastrel. The lake in front of the house is called Stow-pool: it abounds with fish, and belongs to the corporation of Lichfield.

"There are many pleasant gardens and agreeable walks in this part of the city and its vicinage. Passing near Minster-pool, in the Mill-croft, on the left hand, is seen a stupendous Willow tree. The wide-spreading branches of this noble tree adorn the verdant fields, and form a pleasing entrance to the gardens of the Parchment-house, which are greatly ornamented by Mr. Saville's curious botanical and flower garden, containing many scarce and very valuable plants, the produce of various climates. Adjoining are two cold-baths, erected at the expence of the Rev. Dr. Falconer.

"The large Willow tree has been generally supposed to have been planted by the late Dr. Samuel Johnson, or his father; but the Doctor never would admit the fact: it is probable that the vicinity of a building known by the name of 'the parchment-house,' occasioned such a supposition. The business of parchment-making was for many years carried on by old Mr. Johnson, at that place, until he had greatly enriched his servants, and injured his own fortune. There are now no vestiges of such a manufactory; the pits are filled up, and the yard occupied in part by a gardener, and by Mr. Saville.

"Dr. Johnson never failed to visit this tree whenever he came to Lichfield. During his visit here in 1781, he desired Dr. Jones to give him an account of it, saying it was by much the largest tree of the kind he had ever seen or heard of, and therefore wished to give an account of it in the 'Philosophical Transactions,' that the size might be recorded. When in Lichfield in 1784, he begged to have another copy of the letter, having mislaid the former, and not being able to recover it; but he was so ill during his stay that it was forgotten."*

Dimensions of the Lichfield Willow from the admeasurement of Dr. Jones. The trunk rises to the height of twelve feet eight inches and five-tenths, and is then divided into fifteen large ascending branches, which in very numerous and crowded subdivisions, spread at the top in a circular form, not unlike the appearance of a sturdy oak, inclining a little towards the east. The circumference of the trunk at the bottom is fifteen feet nine inches and five-tenths; in the middle, eleven feet ten inches; and at the top, immediately below the branches, thirteen feet. The entire height of the tree is forty-nine feet; and the circumference of the branches, at their extremities, upwards of two hundred feet, overshadowing a plain not far short of four hundred feet. The surface of the trunk is very uneven, and the back is much furrowed. The most moderate computation of its age is near fourscore years; and some respectable authorities strongly incline to think a century has passed over its head.†

MEMORABILIA.—In 1301, Henry Champanar, son of Michael de Lichfield, bell-founder, granted for the "use and solace of the fryars minor of Lichfield, his fountains in Frontewell, near Alreshaw."

Queen Elizabeth visited Lichfield, July 30, 1575, and remained in this city until the 3d of August.

In 1593, March 21st, the top of St. Michael's steeple, and St. Mary's by the market-place, were blown down by a great tempest, which did the like damage at Stafford. This year, during the summer, there was a great plague in many cities and towns in England. In Lichfield there died upwards of eleven hundred, and the disease had not ceased its ravages on the 28th of November. At least one-half of the population of Lichfield must have been carried off by this memorable visitation; for it appears that in the taxation of the city during the reign of Henry VIII. the number of householders amounted to four hundred, which, at the rate of five inhabitants to a house, would give a total population of 2000.

The Alms-house in Bacon-street, was founded in 1504. The

* Shaw's History and Antiquities of Staffordshire, Vol. I. p. 346.

† On Monday, August 14, 1815, a violent gust of wind carried away the greater part of this venerable tree.

houses in this street were mostly burnt down in the war between the King and Parliament.

Bridge-street, by mistake now generally called Bird-street, is at present the principal thoroughfare of the city, the road from Chester to London passing through it, which was formerly carried round by Stow church, through what was then the principal part of the town, now an obscure lane, called Frog-lane.

Modern Lichfield is an open handsome city; the houses in general are well built, the streets regular and spacious, with an excellent pavement and convenient foot-ways, and kept very clean. The principal inhabitants are gentry, mostly persons of small independent fortunes; the remainder consist of tradesmen and artificers. The general appearance of the city affords an idea of snugness, cleanliness, and elegance, and it is delightfully situated in a fertile spot, abounding with the most valuable productions of the agriculturist. Some of the public buildings have already been described, and others are still more worthy of attention, especially the CATHEDRAL, the history of which will form a separate article.

The principal manufacture carried on in Lichfield is sail-cloth, which gives employment to a considerable number of hands.

A society or company of saddlers and glovers was established here in 1620.

There are two weekly markets held in Lichfield; one on Tuesday, which is inconsiderable; and another on Friday, for the sale of corn, cattle, meat, poultry, butter, and every useful article from the farm. At the market held on Friday, specimens of the improved live stock of several neighbouring gentlemen-farmers are occasionally exhibited.

The citizens are supplied with excellent water from the fountains at Aldershaw, according to the grant before-mentioned. This indispensable article of life is conveyed by pipes, free of expence, to the inhabitants, to four conduits; namely, 1st, the Crucifix; 2d, the Market-cross; 3d, Butcher's-row; and 4th, the Stone-cross.

The land in the vicinity of Lichfield, in a low situation on the banks of the rivulet which intersects the city, is generally meadow and pasture. The arable or garden-land is mostly on the sides of the hills, or in a high situation, and is a light sandy soil, producing plentiful crops of early pease; besides turnips, potatoes, and other roots and herbs belonging to horticulture. This soil is also productive of good wheat, barley, oats, and clover.

The Wyrley and Essington Canal passes near the city, affording a ready medium for the conveyance of coal, lime, and other heavy articles, and opening an extensive communication by the Grand Trunk, with distant inland towns and sea-ports.

A short tour through the principal streets, for the purpose of noticing remarkable objects, shall terminate this description of modern Lichfield.

The market-place is in the centre of the city. Opposite the market-house, at the corner of Market-street, stands a stuccoed house, supported in the front by pillars, and memorable for being the birth-place of Dr. Johnson.

The market-house is a light and convenient building of brick, erected on the spot formerly occupied by the market-cross. In Boar-street stands the Guildhall. It is a neat edifice of stone, with a pediment adorned with the city arms, an escutcheon representing a landscape with three Kings, and many other martyrs slain, and a view in bass-relief of the Cathedral. The hall in front is spacious, and in the rear are apartments in which the members of the Corporation transact public business. Underneath is a gaol, in which debtors and felons apprehended within the limits of the *county* of Lichfield are confined.

The play-house was erected in Boar-street in 1790. It is a small structure, with some ornamental stucco-work in front, and is now the property of a society of gentlemen. At the south-west corner of this street is a free-school, founded and endowed by Thomas Minors, Esq. in 1670, "to teach thirty-four boys of this city to read the psalter and bible in English."

Westward of this school are the gates leading to the Friary, which was formerly a monastery belonging to the Franciscan Friars. It was founded in the year 1229, by Alexander Stavenley, Bishop of Lichfield and Coventry. This monastery was destroyed by a fire, which consumed the greater part of the city, in 1291, but the church belonging to the institution was preserved. In 1545, the monastery and church were totally demolished, and the present mansion erected. It has since been the residence of several of the most respectable gentlemen in this part of the county. In the reign of George II. this mansion was the seat of Michael Rawlins, Esq. and the Duke of Cumberland had his head-quarters here, during the time the army was stationed at Lichfield, in the Rebellion, 1745. On the east side of this mansion a curious monument was discovered, in 1746, of which Dr. Wilkes gives the following account: "October 14, 1746, Mr. Michael Rawlins, living at the Friars in Lichfield, having occasion to build a wall, and set down a pair of gates, as he was sinking for the foundation of the former, and making holes for posts, on which to hang the latter, found a grave-stone, about six feet under the surface, with a cross-fleury standing on four steps, cut in the middle, and round the edges the following Monkish verses. There is no date; but a coffin, with bones in it, lay immediately under the stone, which he removed and fixed in an old niche of the building, which he has converted into a stable. By the letters, this stone seems to have been made not very long before the dissolution of the house, for they are an old English letter, and very legible. I took down the words thus:

"Ricardus Mercator, victus morte noverca,
Qui cessat mercari, pausat in hac ierarca:
Extulit ephebus, paucis vivendo diebus,
Ecclesiam rebus sic & variis speciebus;
Vivat et in Cœlis nunc mercator Micaelis!"

The following translation appeared in the Gentleman's Magazine:

"Richard the merchant here extended lies,
Death like a step-dame gladly clos'd his eyes;
No more he trades beyond the burning zone,
But, happy rests beneath this sacred stone.
His benefactions to the church were great,
Though young he hasten'd from this blest retreat.
May he, though dead, in trade successful prove,
Saint Michael's merchant in the realms above."

Bridge-street, now called Bird-street, is a spacious handsome street, with several good private houses, shops, and the two principal inns. A bridge, at its northern extremity, affords a passage across the pool, and separates it from the precincts of the Close.

Next to Bridge-street is Bacon-street, which was nearly all burnt to the ground during the sieges of the Close in the Civil war.

St. John-street is remarkable for two public institutions, both founded by Bishop Smith, in 1492. The first and most important of these is the Free Grammar School, with appropriate buildings, on which Henry VII. settled the revenues of the Hospital of Denhall at Wyrehall, in Cheshire. This school was further endowed by Edward VI. and is memorable for having been the *Alma Mater* in which Addison, Woolaston, Ashmole, and Johnson, received the rudiments of their education. The hospital of St. John, formerly a monastery, is situated opposite the school. It was re-built by Bishop Smith, who endowed it as an hospital for the support of a master, two presbyter chaplains, and ten poor men.

In Bacon-street, there is an hospital for fifteen poor women, rebuilt and endowed by Dr. Milley in 1504, on the site of an original structure erected by Bishop Heyworth, for the same purpose.

THE CLOSE AND CATHEDRAL.

On the northern side of the city, beyond a beautiful sheet of translucent water which forms the boundary, the Close gradually rises into a gentle eminence, presenting to the admiring spectator, in all the majesty of gothic magnificence, the extensive Cathedral, with its ensculptured western front, and its three "heaven-directed spires." The admirer of human ingenuity cannot view this vast and beautiful pile without emotions of enthusiasm; and however we may ridicule the imposing pomp of Popery, we must confess that such costly and elegant edifices, erected to the glory of the Deity, were well calculated to excite sublime devotional ideas in the minds of the uninstructed multitude.

The Close, in which Lichfield Cathedral or Minster stands, was originally inclosed by a ditch, but by whom, or at what period, is not ascertained. Dr. Wilkes dates the origin of the Cathedral as far back as the year 657; but Fuller is of a different opinion, and asserts that Offa, King of Mercia, in 785 exalted Lichfield to an archiepiscopal See.

"At this time (785) the archbishoprick of Canterbury was in part removed to Lichfield; five essential things concurring to that great alteration.

"1. The puissance and ambition of Offa, King of Mercia, commanding in chief over England. He would have the brightest mitre to attend the biggest crown.

"2. The complying nature of Pope Adrian, except any one will call it his thankfulness, to gratify King Offa for the large gifts received from him.

"3. The easy and unactive disposition of Lambert, Archbishop of Canterbury; unless any will term it his policy, that finding himself unable to resist (a pope and a prince overmatch for a prelate), he would not strive to keep what must be taken away from him.

"4. The commodious situation of Lichfield, almost in the navel of the land, and where should the highest candlestick stand (the metropolitan Cathedral) but in the midst of the table? whereas Kent itself was but a corner (whence it taketh its name), and Canterbury seated in the corner of that corner, a remote corner thereof.

"5. The antiquity of Lichfield in christianity, where the British church suffered a massacre from the Pagans three hundred years before St. Augustine's coming to Canterbury, witness the name of the place, being another Helkath-hazzurim, or Field of Strong Men, where so many worthies died for the testimony of the truth.

"On these and other considerations, Adulph was made the first and last Archbishop of Lichfield (though others make Humbert and Higbert his successors in that dignity), and six suffragans, viz. Worcester, Hereford, Leicester, Sidnacaster, Helmham, and Dunwich, subjected to his jurisdiction. Yet was not the archiepiscopal See removed (as some seem to conceive), but communicated to Lichfield, Canterbury still retaining its former dignity, and part of its province.

"Anno Domini 799. Offa being dead, down fell the best pillar of Lichfield church, to support the archiepiscopality thereof. And now Canterbury had got Athelard, a new Archbishop, who had as much activity to spare as his predecessor Lambert is said by some to want. Wherefore he prevailed with Kenulph, King of Mercia, and both of them with Leo, the new Pope, to restore back the archiepiscopal See to Canterbury, which in the next century was perfectly effected."*

From the ninth to the twelfth century little is recorded respecting Lichfield Cathedral, and the See during three centuries appears to have been of so little importance as to be sometimes united to that of Chester, and at others to that of Coventry, according to the will or caprice of successive Bishops. When Bishop Clinton succeeded to this See in 1127, he not only enlarged the Cathedral, but fortified the Close, and conferred many immunities on the prebendaries. His adventurous spirit, however, hurried him among other enthusiasts into the Holy war, and his untimely death deprived Lichfield of the benefits which it would have obtained from his munificence had he continued a resident.

At that remote period, the Cathedral was built of timber; but in 1269, Henry III. granted permission to the Dean and Chapter of Lichfield to get stone for building their church out of the forest of Hopwas. There is no record of the time employed in the erection of this fabric, but it must have been several years. Indeed, from the obscurity of Monkish historians, who were then the annalists of the age, there is no certain account in existence by which we might learn when and by whom the present beautiful edifice was erected. It

* Fuller's Church History, Book II. p. 104.

was completely finished in Bishop Heyworth's time, who was appointed Bishop of Lichfield and Coventry in the year 1420. The principal contributor to the embellishment of the Cathedral was Bishop Langton, who laid the foundation of St. Mary's Chapel, and left money to complete it. Several other zealous friends of the church, whose names are unrecorded, doubtless contributed to promote the completion of this stately and beautiful edifice, and it was only by the efforts of successive prelates during the lapse of several ages that it was brought to perfection.

All the pomp of Popish superstition was subverted by the Reformation; and at the Dissolution, the costly and venerated shrine of St. Chadd, and several images of exquisite workmanship, were removed from Lichfield Cathedral. A celebrated writer on Ecclesiastical history, says, "in the time of William Heyworth, the Cathedral of Lichfield was in the vertical height thereof, being (though not augmented in the essentials) beautified in the ornamentals thereof. Indeed, the west-front is a stately fabric adorned with exquisite imagerie, which I suspect our age is so far from being able to imitate the workmanship, that it understandeth not the history thereof. But alas! it is now in a pitiful case indeed, almost beaten down to the ground in our Civil dissentions. Now, lest the church should follow the castle, and vanish quite out of view, I have at the cost of my worthy friend (Elias Ashmole) here exemplified the portraiture thereof."*

Whatever zeal the Protestants, in the first era of the Reformation, might evince in the removal or destruction of shrines, crucifixes, images of saints, and other objects of Popish idolatry, they did not attempt to deface the exterior of Lichfield Cathedral, which continued for upwards of two centuries in all its magnificence, an object of admiration to travellers, and the principal architectural ornament of Staffordshire. But when Charles the First, by unconstitutional and arbitrary measures opposed the Parliament, and levied imposts without controul, a most sanguinary Civil war ensued, and the Close and Cathedral of Lichfield were exposed to three destructive sieges. These events are of too much importance in this history to be overlooked, and the following authentic particulars will not only demonstrate the deplorable state of ruin to which this noble pile was brought, but prove illustrative of the manners of that age.

"The situation of the place on an eminence, surrounded by water

* Fuller's Church History, Book IV. p. 175.

and by deep ditches, and fortified with walls and bastions, rendered it unhappily a proper place for a garrison. In the year 1643, Sir Richard Dyott, and some of the principal nobility and gentry of the country, under the Earl of Chesterfield, more remarkable for their loyalty to their sovereign than their experience in the arts of war, garrisoned this Cathedral, when it underwent the attack rendered memorable by the death of Lord Brook, commander of the Parliamentary forces. The loss, however, of that furious fanatic, and professed enemy to the church, gave but little respite to the garrison, the siege of which was continued, and soon after taken by Sir John Gell, as appears in the following curious account:

"The attempt upon Lichfield Close made by Robert Lord Brook, wherein he lost his life, the manner whereof is not a little remarkable, which was thus. This Lord being strangely tainted with fanatic principles by the influence of one of his near relations, and some schismatical preachers, though in his own nature a very civil and well-humoured man, became thereby so great a zealot against the established discipline of the church, that no less than the utter extirpation of episcopacy, and abolishing all decent order in the service of God, would satisfy him; to which end he became the leader of all the power he could raise for the destruction of the Cathedral of the diocese of Coventry and Lichfield, in order whereunto, when he had marched within half a mile of Lichfield, he drew up his army, and there devoutly prayed a blessing upon his intended work, withal earnestly desiring that God would, by some special token, manifest unto them his approbation of their design, which being done, he went on and planted his great guns against the south-east gate of the Close, himself standing in a window of a little house near there to direct the gunners in their purposed battery: but so it happened, that there being two persons placed in the battlements of the chiefest steeple to make shot with long fowling-guns at the cannoneers, upon a sudden accident which occasioned the soldiers to give a shout, this Lord coming to the door (compleatly harnessed with plate armour cap-a-pie) was suddenly shot into one of his eyes; but the strength of the bullet so much abated by the glance thereof on a piece of timber, which supported a pentiss over the door, that it only lodged in his brains, whereupon he suddenly fell down dead.* Nor is it less notable that this accident fell out upon the second day of March, which is

* The spot on which he fell is now distinguished by a pavement of white pebbles, and a marble tablet with an inscription in memory of the event.

the festival of that some-time famous Bishop St. Chadd, to whose memory Offa, King of the Mercians, first erected this stately church, and devoutly dedicated it."*

After the fall of Lord Brook, Sir John Gell assumed the chief command of the besieging army, which continued the siege with such vigour that the garrison surrendered on the 5th of March upon condition, "that their Earl and Governor, (the Earl of Chesterfield,) should surrender the garrison, called the Close of Lichfield, forthwith, upon condition of free quarter to all in general within the said Close; and for any other terms to cast themselves upon the mercy of the Parliament."

Accordingly, the Royalists were made prisoners of war, and the Parliamentary troops took possession of the Close, where they left a strong garrison. These soldiers were guilty of the most indecent outrages in the Cathedral, where they demolished the monuments, pulled down and destroyed the carved work, broke the windows, and destroyed the records. Colonel Russel, the Governor of the garrison, carried away the communion-plate, and whatever moveables he found of value in the Cathedral.

In the month of April following, Prince Rupert, after the reduction of Birmingham, came with a strong body of forces to Lichfield, and besieged the Close, which was bravely defended by the garrison. But Prince Rupert erected batteries in Gayfield, an eminence north of the Cathedral, and which commanded the Close, and he also sprung a mine under the wall and effected a breach, which he entered, but after a sanguinary conflict was repulsed with great loss. In consequence of the loss suffered by the garrison in defending the breach, Colonel Russel agreed to capitulate on honourable terms on the 21st of April, 1643. Prince Rupert then appointed Colonel Henry Bagot the Governor of the Close, who kept possession till the 10th of July, 1646.

The unshaken fidelity of the citizens of Lichfield to the cause they espoused was truly memorable and praise-worthy; and when Charles I. was in his greatest adversity, they still maintained their loyalty, and sympathized with their Sovereign. Of the truth of this assertion there is a sufficient record, for after the battle of Naseby, in 1645, in which the royal army was defeated, the King came to Lichfield on Sunday June 15, when the following Address was immediately presented to him by the Corporation of Lichfield:

* Shaw's History and Antiquities of Staffordshire, Vol. I. p. 238.

'MOST GRACIOUS SOVEREIGN, 'June 15, 1645.''

'Though the sad report of the late ill success hath so oppressed our souls with grief, that we are rendered more apt to expresse our loyall affections in tears than words, yet the safety and presence of your sacred person (as dear to us as our lives) hath so much revived and restored us, that we have taken the boldness, though suddenly and rudely, in a few words, to present to you the most zealous affections and loyall services that a most obliged and gratefull people can possibly bear to a most gracious Sovereign.

'And as we are not so stupid as not to be sensible of God's corrections, where he is pleased so sharply to punish us for our sins, so we are not so unchristian-like as to despair of God's finall blessing upon a most just and righteous cause, nor so unmanly as to lay down our courage and confidence for one cross event, as knowing that man's necessity is God's opportunity, and that God's power is most glorified in man's weakness.

'And albeit the sunne may be for a time eclipsed, even by that plannett which itself enlighteneth, and for a time be obscured even by those clouds which itself drew from the earth into an higher region, yet the light and virtue of the sunne is not thereby made lesse conspicuous: eclipses and clouds last not always. *Mendacia diu non fallunt:* men will not always be worked into this sin of rebellion. Truth at last prevails; right never dieth; but will shortly, by God's blessing (all clouds being dispelled), restore the sunne of this our firmament to his former splendour and glory, and therein his faithfull subjects to their former peace, plenty, and happiness, which is our daily prayer, and shall be our incessant endeavour to the utmost expence of our estates and blood. *Nec plus obire possumus, nec fas est minus.* More we cannot undergoe, less we may not.

'Sir, your most humble and loyall subjects, the bayliffs, sheriffe, and their maces, &c. citizens of this your citty of Lichfield, doe humbly, according to their duty, surrender into your Majesty's hands these ensigns of their authority, which they are resolved to beare from you and under you, or not at all, and, whether they live or die, to live and die your Majestie's most faithful and loyall subjects.'

"His Majesty was pleased to answer that they were eminent for their loyalty, and required Richard Dyott to give them all thanks in his name.

"Then he gave his hand to the head officers to kiss, and they with their chief gentlemen of the towne, waited upon his Majesty to his quarters."*

In March, 1646, the Close was invested by the Parliament's army, under Major-General Lothian, and continued to harass the garrison till the 10th of July, when Colonel Bagot being satisfied that the King had not an army in the field, and that all the fortresses and garrison-towns in possession of the Royalists were besieged, he surrendered on very honourable terms.

As this was the first Cathedral that was seized by the republicans, from the circumstance of its being situated in a fortified place, it was damaged more than any other, by the cannon of the besiegers and the fanaticism of the soldiers. According to the most moderate computation, the damage done to the Cathedral alone was estimated at £14,000. It continued in a ruinous and neglected state till the Restoration, when Bishop Hackett was appointed to this See. This magnanimous and pious prelate came to Lichfield in the beginning of the year 1662, and found his Cathedral in an indescribable state of ruin; 2000 cannon-shot and 1500 hand-grenadoes having been discharged against it. He resolved to restore this venerable edifice to its former magnificence; and the very morning after his arrival he roused his servants by break of day, and set his own coach-horses with teams, and hired labourers, to remove the rubbish. By his own contributions, the benefactions of the Dean and Chapter, and the money which he collected by a personal application to every gentleman in the diocese, this excellent and public-spirited man in eight years completely restored the magnificence of the Cathedral to the admiration of the country.

When the Cathedral was finished, the Bishop consecrated it with great pomp and solemnity, and composed a service for the occasion. The following account of this memorable ceremony is truly interesting:

"His Lordship being arrayed in his episcopal vestments, attended by the Dean, Dignitaries, Prebendaries, and other members of the church, accompanied by many of the nobility and gentry, the bailiffs, citizens, and civil officers of the city and county of Lichfield, with an immense concourse of people, entered at the great west doors of the Cathedral. The Vicars, Choristers, &c. first walked up the south aisle of the church, when the Bishop, with a loud voice, repeated the first verse of the 144th psalm. Afterward, the whole choir alternately sang the psalm to the organ. In the same order they pro-

* Green's MSS.

ceeded to the north aisle. The Bishop sang the first verse of the 100th psalm, which was repeated by the whole company. Then the train passed to the body of the church, where the Bishop began the 102d psalm, which, when the Vicar's choral had concluded, he commanded the doors of the choir to be opened, and, in the same form, first encompassed the south side. The Bishop began the first verse of the 122d psalm; the company finished it, and, with the like ceremony, proceeded to the north side, and sang the 131st psalm.

"At the conclusion of the procession, Bishop Hackett went to the faldisdroy, in the centre of the choir, and, after praying privately, he with a loud voice called upon the people to join him in the Lord's Prayer, and other devotions, suitable to the awful occasion. He pronounced a solemn blessing upon the act in which they were engaged, and upon all that were present. Then followed the usual service of Morning prayer, two especial anthems being added, contributions made for the poor, &c."*

The expences of the repairs of the Cathedral amounted to £9092. 1s. 7½d. of which Bishop Hackett contributed out of his own purse the sum of £1,683. 12s.

DESCRIPTION OF LICHFIELD CATHEDRAL IN ITS MODERN STATE.

The extent of this edifice from east to west is 411 feet in length, and from north to south 67 feet in breadth. It is adorned by three light and elegant spires; the central spire is 258 feet 7 inches high, and the two others in the west front 183 feet. The south steeple contains ten bells.

The west front is adorned with a rich variety of sculpture, particularly several figures the subjects of which were taken from Sacred history. They are now rather indistinct, several of them having been injured during the sieges of the Close. It is to be regretted that the reddish stone of which the western front and its ornaments are composed, is perable and of a very perishable nature. It was, however, easily wrought by the sculptor, and as the stone was to be procured in the neighbourhood, any mutilations might be the more readily replaced. Bishop Hackett repaired such figures as had been defaced; and although time has again considerably injured them, they present a variety of curious and interesting objects, the effect of which is peculiarly grand and impressive. On

* Shaw's History and Antiquities of Staffordshire, Vol. I. p. 243.

the top of the pointed roof, between the two spires, a statue of Charles II. was placed by Bishop Hackett on the spot where a figure of Adam formerly stood, and on both sides of the steeples were the figures of the Patriarchs. The next two rows of figures represent Judges, Prophets, and Prophetesses, and below them the Kings of Judah and Israel in various attitudes, and amongst others that of King David playing on the harp. Over the centre of the beautiful porch is the figure of a Bishop, supposed to be St. Chadd. Within the porch the four Evangelists are represented, holding the Gospels in their hands. On the two sides of the porch stood Moses and Aaron, and between the two great doors stands the Virgin Mary with the infant Jesus in her arms. On the top of the central pillar, there is a figure of Christ between two Cherubs, with his arms open, inviting all those that are heavy laden to come unto him; and he would give them rest, as it is said in the gospel of St. Matthew, ii. 28. On the outside, the vacant spaces in the walls between the doors were filled with the statues of the twelve Apostles, but they were removed in 1749 by Dean Penny. The ornaments in sculpture around the doors were originally very elegant and beautiful, but they are now much defaced by violence and time. Over the west doors, a magnificent circular window was made at the expence of James II. In 1776, the Rev. Dr. Addenbrooke, who was Dean of Lichfield upwards of thirty years, adorned this window with painted glass.

The northern and southern entrances are also adorned with a variety of ornaments, particularly the former, which is beautified with sculptured mouldings, three of foliage, and three of small figures in ovals. In one of the lowest is the representation of a Bishop baptizing a person kneeling before him. Antiquaries suppose that the former was intended to represent St. Chadd, the original founder of the Cathedral, and the latter Wulferus, whom he converted to Christianity. The roof of this Cathedral was formerly covered with lead, but it grew so out of repair that the Dean and Chapter were obliged to substitute slates on account of the smallness of the revenue left to maintain the grandeur of this ancient pile. The western doors open in the main body of the Church, which is lofty and spacious, supported by pillars formed by a variety of slender columns with foliated capitals. Along the walls of the aisles are rows of false arches in the gothic style, with a seat beneath each. The upper rows of windows are triangular. The length of the body of the Cathedral from the west door to the choir, is 213 feet, the breadth 153 feet; the breadth of the side aisles 67 feet, and the height of the nave 60 feet.

In the year 1788, the present Bishop of Lichfield and Coventry, and the Dean and Chapter, circulated an address throughout the diocese, inviting individuals to contribute by subscription to the internal improvement of this Cathedral. Upwards of £5000. was soon collected, but as the sum requisite, was, according to the estimate of Mr. Wyatt, the architect, nearly £6000. the borrowed sum of £1000. was added as a debt upon the edifice. The improvements suggested by Mr. Wyatt were then speedily realized. The pulpit and pews in the nave were removed, the decayed brick floor taken up, and a new floor of Derbyshire Hopton stone laid instead of the former. The choir was also enlarged, by throwing it and St. Mary's Chapel into one, and it is now sufficiently capacious to contain the whole congregation. In the upper part of the choir are nine windows, three on each side, and three at the end. In the central window at the east end is a fine painting of the Resurrection, from a design by Sir Joshua Reynolds. The chaste but brilliant effect of the preternatural light, the graceful form of the ascending Saviour, and the animated expression of the countenance, have excited the admiration of all visitors, and the approbation of connoisseurs.

In 1797, the groins, walls, and roof of the north transept underwent a thorough repair, the whole edifice was completely pointed, the middle spire taken down a considerable way and rebuilt with renovated beauty, and new buttresses erected to support the south transept.

The Chapter-House is an octagon of two long and six short sides, ornamented with gothic arches: the approach to it is from the north choral aisle, through an arched passage: above it is the library, of the same form. It contains some curious manuscripts, the most remarkable of which is called St. Chadd's Gospels, not because it was used by that venerable man, but because it was used in his church. It is said to be 1000 years old, but no antiquary has hitherto been able to ascertain the fact: the writing is in ancient Saxon characters, illuminated with several drawings. There is also a manuscript copy of Pope Nicholas's Valor, a fairly-written folio of Chaucer's poems, illuminated, and a beautiful heraldic list of the Knights of the Garter, &c. This library contains a copy of the Koran, taken from the Turks at the siege of

Beda. There is also a considerable collection of printed books, chiefly on divinity.

The Close contains several houses, built at a convenient distance from the Cathedral, so as to allow a sufficient space for passengers. The Bishop's palace is situated at the north-east corner of the Close: it is a spacious edifice of stone, adorned with the arms of the bishopric in front, with the date 1687. This palace is generally occupied by tenants, the Bishops having for ages chosen Eccleshall Castle for their residence. Westward of the palace stands the Dean's mansion, re-built in the reign of Queen Anne. The houses of the Prebendaries are situated in different parts of the Close. The vicarage consists of two small quadrangles of houses, situated in the north-west corner. Several other handsome buildings occupy the rest of the Close, the most remarkable of which is a building of brick, faced with stone, at the west entrance, erected by Andrew Newton, Esq. and liberally endowed by him, for the reception and support of twenty aged and necessitous widows, or unmarried daughters of clergymen. This structure is supposed to occupy the scite of the ancient Castle.

From the first foundation of Lichfield Cathedral in 655 to the present year, 1816, or more than eleven centuries and a half, eighty Bishops, many of them men eminent for learning and piety, have been seated in the episcopal chair.

The members of Lichfield Cathedral are a Dean, Precentor, Chancellor, and Treasurer; a Sacrist, Sub-sacrist, Organist, eight Choristers, and two Vergers. The Cathedral is governed by the Dean and six Canons, who constitute the Chapter, and hold their court every alternate Friday to hear and determine causes.

The Close is a peculiar jurisdiction, distinct from and independent of the city of Lichfield, and the county of Stafford, the Dean and Canons being sole justices within its precincts.

According to the population returns of 1801, the three parishes of Lichfield contained:

	Houses Inhabited.	Families.	Males.	Females.
Saint Mary,	444	496	1058	1364
Saint Michael,	187	237	450	457
Saint Chadd,	242	264	537	646
Total,	873	989	2045	2467

In the census for 1811:

	Houses Inhabited.	Families.	Males.	Females.
Saint Mary,	450	487	1069	1318
Saint Michael,	214	237	476	519
Saint Chadd,	291	317	614	791
Total,	955	1041	2159	2628

Before the Civil war between Charles I. and the Parliament, there were many handsome altar-tombs of Bishops, and other eminent characters, interred in this Cathedral, and above one hundred coats of arms, and other heraldic ornaments, in painted glass, in the windows, all of which were defaced or destroyed by the licentious soldiery belonging to the republican army. The Cathedral is now adorned with several modern monuments to the memory of individuals, among which the following are most remarkable.

On the north side of the west door, the large mural monument of marble, erected to the memory of Lancelot Addison, reminds us of his eminent son, whose writings improved the morals and the literature of Englishmen. On the south side a monument of mixed marble is inscribed to the memory of Gilbert Walmsley, the friend and early patron of Dr. Johnson.

A marble monument, to the west of the north door, is remarkably elegant. It represents a female figure leaning upon an urn, or Beauty weeping over the ashes of her preserver. The urn is inscribed with the cypher M. W. M. and the following epitaph:

Sacred to the memory of
the Right Honourable
Lady Mary Wortley Montague,
who happily introduced, from Turkey
into this country,
the salutary art
of inoculating the small-pox.
Convinced of its efficacy,
she first tried it with success
on her own children,
and then recommended the practice of it
to her fellow-citizens.
Thus, by her example and advice,
we have softened the virulence,
and escaped the danger, of this malignant disease.
To perpetuate the memory of such benevolence,
and to express her gratitude
for the benefit she herself received
from this alleviating art,
this monument is erected by
Henrietta Inge,
relict of Theodore William Inge, Esq.
and daughter of Sir John Wrottesley, Bart.
in the year of our Lord M.DCC.LXXXIX.

Near the south door, a large and beautiful mural monument, erected by conjugal affection to the memory of a virtuous and patient wife, attracts the attention of the curious visitor. It is adorned with a celestial crown, and other emblems of immortality, with the following inscription :

Lucy Grove,
eldest daughter of Edward Sneyd, Esquire,
and wife of William Grove, Esq. LL. D.
late of Coventry, now of Lichfield Close,
died the sixth of December,
M.DCC.LXXXVII. in the
fortieth year of her age,
and the twentieth of her marriage, leaving
two sons, and two daughters.

Grief, love, and gratitude, devote this stone
To her, whose virtues bless'd an husband's life,
When late in duty's sphere she mildly shone,
As friend, as sister, daughter, mother, wife.

In the bright morn of beauty, joy, and wealth,
Insidious palsy near his victim drew :
Dash'd from her youthful hand the cup of health,
And round her limbs his numbing fetters threw.

Year after year her Christian firmness strove
To check the rising sigh, the tear repress ;
Soothe with soft smiles the fears of anxious love,
And Heaven's correcting hand in silence bless.

Thus tried her faith, and thus prepar'd her heart,
The awful call at length th' Almighty gave ;
She heard—resign'd to linger or depart—
Bow'd her meek head and sunk into the grave.

In 1793, two monuments were erected against the east wall of the south transept, to the memory of Dr. Johnson and David Garrick ; the former at the expence of the gentlemen of the Close, and the latter by the widow of the English Roscius. They are of mixed marble, alike in size and form : the busts, of white marble, are not considered good likenesses of the great originals.

Upon the sarcophagus, under the bust of Garrick, is the following inscription :

Eva Maria, relict of David Garrick, Esq.
caused this monument to be erected to the memory
of her beloved husband,
who died the 20th of January, 1779, aged 63 years.
He had not only the qualities of private life,
but such astonishing dramatic talents,
as too well verified the observation of his friend,
'His death eclipsed the gaiety of nations,
and impoverished the public stock of harmless
pleasure.'—*Johnson.*"

The inscription under the bust of Dr. Johnson is equally comprehensive and panegyrical:

The friends of SAMUEL JOHNSON, LL. D.
a native of Lichfield,
erected this monument,
as a tribute of respect to the memory of
a man of extensive learning,
a distinguished moral writer, and a sincere Christian.
He died the 13th of December, 1784,
aged 75 years.

BISHOPS OF LICHFIELD,

[From their Institution to the present Time.]

Mercia, the largest kingdom of the Heptarchy, followed the Pagan idolatry, till Penda, one of the most powerful of its kings, was defeated and slain by Oswy, King of Northumberland, who having obtained quiet possession of his conquest, continued Peada, the son of Penda, in the government of the parts of Mercia south of the Trent, on condition of his marrying Oswy's daughter, which proposal was acceded to by Peada. Oswy now began to convert the Mercians to Christianity, in which faith Peada, soon after his marriage, was baptized. Peada afterwards built a monastery at Medeshamstead (Peterborough), to shew his personal respect for religion; and in 655, appointed DWINA first Bishop of the Mercians. The Bishops had then no certain dwelling-places, but contented themselves to live in the monasteries. Dwina ruled about two years, and was succeeded by

Date.

657. CELLACH, a Scotchman by birth: but Wulfer having established himself King of Mercia, removed Cellach, the 2d Bishop, and constituted in his stead,—TRUMHERE, 3d Bishop of Mercia: he died, and was succeeded about

662. by—4, JARUMANNUS: whilst he presided, the Cathedral Church at Lichfield began to be built in the year 666: he died, and in

669. Wulfer made CEDDA 5th Bishop of Lichfield. This is the St. Chadd who gave name to the Cathedral and parish: he died March 2, 672, and was buried in the Cathedral. He was succeeded by—6th, WULFRID, who had been his Deacon: he was deposed in 674, and SEAXWULF was made 7th Bishop of Lichfield by Theodore, Archbishop of Canterbury: he governed this See 20 years, and then died, according to the Saxon Chronicle. In

691. HEDDA was consecrated Bishop of Lichfield. The Church of Lichfield was dedicated by this Bishop, December 30, 700, and the

Date.

body of St. Chadd removed into it: he survived this dedication 21 years.

721. Aldwine, 9th Bishop, succeeded him, and remained until 737: in his time lived Bede, the ancient historian.

737. Wiota succeeded Aldwine: he died the 10th Bishop in

752. When Hemele succeeded him, who governed 12 years.

764. Cuthred succeeded him, and died 768, by the Lichfield Register.

768. Berthunus, the 13th Bishop, governed till 785, when, in

785. Sigibert was chosen into his See; soon after which Offa, an uncontrollable monarch, made this an Archiepiscopal See, and, in

786. Adulphus was by Pope Adrian created Archbishop by the influence of King Offa, and remained until 812, when, in

812. Herewin succeeded him in the See of Lichfield only.

817. Athelwald succeeded as Bishop of Lichfield: he instituted prebendaries in his Cathedral Church in 822, when Ceolwolfe was King of the Mercians. He governed this See 35 years.

852. Humbert was chosen 18th Bishop of this See: he died 867. Kenferth governed this See after him, and sat 23 years, dying in 890: his successor was

890. Tunfrith: he governed this See 30 years. Alfred was, part of this time, King of the West Saxons.

920. Ælwinus succeeded and ruled this See in King Ethelstan's time, and remained 24 years.

944. Alfgar was elected into this See, and having sat 16 years,

960. Kinsius succeeded, being the 23d Bishop: he sat 14 years.

974. Winsius succeeded him: he is said to have confirmed King Edgar's charter, called the Golden Charter. He presided 18 years, and was the 24th Bishop.

992. Ælfegus was soon elected Bishop of Lichfield: his name is found affixed to two charters of King Ethelred. He governed 15 years, and was the 25th Bishop,

1007. Godwin succeeding: he governed 13 years, and was the 26th Bishop.

1020. Leofganus was his successor, presiding only 7 years: he died 1027, and was the 27th Bishop.

1027. Brithmarus was elected into this See, and governed it 11 years: he was the 28th Bishop.

1038. Wulfius was chosen: he governed 16 years. In his time, 1044, Leofric, Earl of Hereford, enlarged and almost new built the monastery at Coventry, and endowed it with large possessions. Wulfius died 1045, and was the 29th Bishop.

1054. Leofwin was made Bishop of Lichfield by King Edward the Confessor: he ruled 12 years, and died 1066. At the latter end of his life William the Norman invaded England. He was the 30th Bishop.

Date.

1067. PETER was by him appointed: he translated this See to Chester, and was called Bishop of Chester and Lichfield. He died 1086, and was the 31st Bishop.

1086. ROBERT DE LYMSEY succeeded him in the united See, and was consecrated by Archbishop Lanfranc: at this time Leowin, Abbot of Coventry, died immensely rich; and this Bishop, desirous of gaining the riches, got that Abbey conferred on him by the King, and removed the See itself to Coventry in 1095, and so became the first Bishop of Coventry. He died September 1, 1116, and was buried at Coventry: he was the 32d Bishop.

1121. ROBERT PECCAM was named by Henry I. Bishop of Coventry, after the See had been vacant from four to five years: he was (says an old historian,) that King's butler, but more truly his chaplain, and was consecrated at Abingdon by Ralph, Archbishop of Canterbury: he died August 20, 1127, and was buried in his monastery at Coventry. He was the 33d Bishop.

1127. His successor was ROGER DE CLINTON, Archdeacon of Buckingham, appointed by Henry I. Bishop of Coventry; what he did for Lichfield has been before noticed: he increased or founded the Prebends, viz. the two Gaias, Freeford, Hansacre, Curborough, Darnford, Offelow, Statfold, and the two Ulvetons, and settled the principal dignitaries. To this Roger, King Stephen granted the churches of Pencris and Stafford; also the Church of Wolverhampton, with all its manors and appurtenances; which grants were confirmed by Pope Lucius in 1144. He built a monastery for Cistercian monks at Buildwas, in Shropshire, and King Stephen confirmed all his revenues and gifts to it: he is supposed to have founded the Priory of St. John's, Lichfield, and was the 34th Bishop.

1149. WALTER DURDENT succeeded him: he had been Precentor of Lichfield, was raised by King Stephen, and was consecrated at Lichfield, October 2, by Theobald, Archbishop of Canterbury: he obtained from King Stephen the liberty of coining money. Henry II. by charter, 1154, also granted to the Church of St. Chadd, this Bishop and his successors, all lands of the grubbed-up parts of the forest of Cannock, of Longdon, &c. as expressed in the curious deed: he died 1161, and was buried at Coventry. He was the 35th Bishop.

1161. RICHARD PECHE or PECCAM, son of the former Robert, Bishop of Coventry, was unanimously chosen to succeed him, and by the consent of Henry II. made Bishop. His father had before made him Archdeacon of Coventry, which office he held till he was Bishop. He was sent over into Ireland anno 1181, one of the Justices, but weary of the troubles of life, he returned and put on the habit of a canon regular: he died October 6, 1183, and was the 36th Bishop.

Date.

1183. GERARD DE PUELLA succeeded to this See of Coventry, Lichfield, and Chester: Robert Montensis calls him a man of great learning and honesty. He was one of the Archbishop of Canterbury's domestic chaplains, and famed for his eloquence. He sat but 16 weeks, and died at Lichfield, (not without suspicion of poison,) January 13, 1184: he was the 37th Bishop.

1188. HUGH DE NOVANT, a person eminent for eloquence and piety, was elected by the monks into his place, but not consecrated till 1188, by Baldwin, then Archbishop of Canterbury: between his election and consecration he was appointed the Pope's Legate. He was an incessant adversary to the monks, whom he opposed deservedly, for they vexed the Bishops continually with their contests and quarrels. Richard I. in 1190, gave this Hugh authority to remove the monks of Coventry, and put secular priests in their place; and when the monks refused to obey this order, he made way by the sword, wounding some, and putting the rest to flight. He was said to be wounded in the conflict as he was standing by the altar, died March 27, 1199, and was buried at Caen, in Normandy. He was the 38th Bishop.

1199. The monks chose in his stead JEFFERY DE MUSCHAMP, who was consecrated by Hubert, Archbishop of Canterbury: he died October 5, 1208, and was buried at Lichfield. The nation being then under an interdict from the Pope, in his quarrel with King John, no legal new election took place for some time. He was the 39th Bishop.

1215. WILLIAM DE CORNHULL, Archdeacon of Huntingdon, was chosen Bishop of Coventry and Lichfield, and consecrated January 23, 1215, in the reign of King John. He granted the Chapter of Lichfield a free power of choosing themselves a Dean, which privilege was confirmed to them by Pope Honorius III. Before this the Bishop always elected the Dean, as he did the Canons: this Bishop died September 14, 1223, was buried at Lichfield, and was the 40th Bishop. After his death there arose a dispute between the Churches of Coventry and Lichfield, about the election of a Bishop to succeed him, which was referred to the Pope, who appointed, in

1224. ALEXANDER DE STAVENBY, and consecrated him, with the Bishop of Paris, upon Easter-day. He was a very learned man, and thought to excel most of the philosophers and divines of his age. He is said to have had many visions and strange dreams: he increased the Prebends of the Church, and enriched it with many things. While he sat, it was settled by the Pope, that the election of a Bishop should be by the monks of Coventry and the Chapter of Lichfield in rotation. He was the 41st Bishop.

Date.

1238. NICOLAS DE FARNHAM, a very excellent man, was chosen Bishop, but he declining the honour, it was bestowed on HUGH DE PATESHULL, Treasurer of St. Paul's Church and of England: he had not sat three years, when he was taken away by death in his full strength, December 7, 1243: he was a man of upright life, and was buried at Lichfield. He was the 42d Bishop.

1245. ROGER DE WESEHAM, Dean of Lincoln, was made Bishop of Coventry and Lichfield, by the interest of the Bishop of Lincoln, and consecrated by Pope Innocent IV. at Lyons. Henry III. was not consulted about this election, but he confirmed it, and the Bishop enjoyed it 11 years. About this time the monks of Coventry and canons of Lichfield agreed that in electing a Bishop the number of electors should be equal on both sides. In 1256, the Bishop being old and afflicted with palsy, resigned his Bishopric, and died at Brewood, 1256. He was the 43d Bishop.

1257. ROGER DE MAYLAND was a Sub-deacon, Pope's Chaplain, and a Prebendary of Lichfield, when he was chosen Bishop. He was consecrated by Boniface, Archbishop of Canterbury, in the Cathedral of that city, (with other Bishops assisting,) March, 1258. He died December 16, 1295, and was buried in his Church at Lichfield, having presided 38 years. He was the 44th Bishop.

1295. WALTER DE LANGTON was unanimously chosen into this See; and being much in favour, Edward I. did many things for the advantage of his Church, as before shewn. He died at London, November 16, 1321, and was buried in St. Mary's Chapel, erected by himself, at Lichfield: he was the 45th Bishop. After his death the monks of Coventry and the canons of Lichfield disagreeing, John XXII. then Pope, put an end to their controversy, by conferring this bishopric on ROGER DE NORTHBURGH.

1322. He was consecrated, June 20th. He sat 38 years, and did nothing memorable, except causing the body of his predecessor to be deposited in a more magnificent tomb, for the many benefits he had rendered to the See: he died in 1359, and was buried under a sumptuous tomb. He was the 46th Bishop.

1360. To him succeeded ROBERT DE STRETTON, a very eminent person, canon of this Church: he was chosen at the earnest request of the Prince of Wales, whose Chaplain he was, and consecrated in 1360: it is recorded that this Bishop could not read. He died at his manor-house at Haywood, March 28, 1385, and was the 47th Bishop.

1386. WALTER SKIRLAW was legally elected Bishop of this See, but was translated by the Pope's bull, against his own consent, to that of Bath and Wells: he was the 48th Bishop. To him succeeded, in

Date.

1386. Richard Scrope, by the Pope's appointment. In 1398 he was made Archbishop of York, (being the 49th Bishop,) upon whose translation, in

1398. John Burghill, a Dominican Friar, who was Confessor to Richard II. was translated from Landaff to Lichfield and Coventry: Walsingham says, the Pope did this by his mere arbitrary power. He was the 50th Bishop.

1415. John Ketrick, Notary at the Court of Rome, Archdeacon of Surrey, and Bishop of St. David's, was translated to this See, and was sent to the Council of Constance, in 1416. He was the 51st Bishop.

1420. William Hayworth, Abbot of St. Alban's, a person of good judgment and learning, was consecrated Bishop of Lichfield and Coventry. He went to the Council of Basle, in 1434; and died April 10, 1446. He was the 52d Bishop.

1447. William Booth, a prudent man, Prebend of St. Paul's, London, was made Bishop of this See. After he had sat five years he was translated to the Archbishoprick of York, in 1552. He was the 53d Bishop.

1452. Nicholas Close, Bishop of Carlisle, upon his removal, was translated by Pope Nicholas to the See of Lichfield, and died in November following. He was the 54th Bishop.

1453. Reginald Bolars, soon after Bishop Close's decease, was translated from the See of Hereford to the See of Lichfield, by the Pope's provision, dated February 7, and governed it six years. He was the 55th Bishop.

1459. John Hales obtained this See after the death of Bishop Bolars, and was promoted to it by Henry VI. He was consecrated at Coventry, November 25. This holy father found his Church in a bad condition, and took every opportunity of preferring learned and discreet men. His Chancellor and Registrar, were Dr. Salter and Thomas Mylles; his eminent Dean was Dr. Yotton, and many other learned men were members of his Church. This Bishop died October 3d, 1490, aged 90: he was a person of great learning, and well respected by all Christians. He was the 56th Bishop.

1492. William Smith, Archdeacon of Surrey, was preferred, through the solicitation of Prince Arthur, by Henry VII. This good Bishop erected in the city of Lichfield an hospital for poor people, and a grammar-school. After he had sat four years at Lichfield, he was translated to Lincoln. He was the 57th Bishop.

1496. John Arundel obtained this See after the above translation. In his time the Library was finished. He sat about six years, being the 58th Bishop, and was then translated to Exeter.

1503. Jeffery Blythe, Doctor of Laws. He built a house for the

Date.

Choristers, and gave 50 oaks from his lands towards repairing the Church: he governed 32 years with great reputation. Many eminent men lived in his time. He died in London, in 1534, and was the 59th Bishop.

1534. Rowland Lee obtained this See: he was a person of great eminence. Being highly in favour with Henry VIII. he was made Governor of the Marches of Wales. He cleared the Marches of robbers, who much infested them, and united Wales with England as one body. In this Bishop's time the tenths, &c. of all church livings were translated from the Pope to Henry VIII. who, about this time, erected the bishopric of Chester, lessening that of Lichfield, and seized the images and rich shrines of saints, with their jewels and ornaments; but at the humble request of this Bishop, the King gave up the shrine of St. Chadd. Bishop Lee died Jan. 24, 1542, at Shrewsbury, and was the 60th Bishop.

1543. Richard Sampson was translated from Chichester to this See. He was a learned man, and wrote many books. He died at Eccleshall, in 1554, and was the 61st Bishop.

1554. Ralph Bane succeeded him. He was Doctor in Divinity, a good Hebrew scholar, and lived in difficult times. Dr. Bane was the 62d Bishop.

1559. Thomas Bentham succeeded Dr. Bane. In Queen Mary's reign he was ejected from his Fellowship for his zeal against Popery in the reign of Edward VI.; and on Queen Elizabeth ascending the throne, he was nominated to this See on the deprivation of Dr. Bane. He died February 21, 1578-9, and was the 63d Bishop.

1579. William Overton succeeded him. He was born in London, and brought up by the charity at Glastonbury. He took holy orders in Edward VI.'s reign, but upon the accession of Mary he left College. In 1565 he took his degrees, and in 1579 was made Bishop of Lichfield and Coventry; where he was much commended for his hospitality, and the good repair in which he kept his palace. He died at an advanced age, in 1609, and was buried at Eccleshall, being the 64th Bishop.

1609. George Abbot, Doctor in Divinity, and Dean of Winchester, was promoted to this See. He had been a celebrated preacher in the University, and was consecrated December 3, but removed in February to the See of London. He was the 65th Bishop.

1610. Richard Neale was his successor in this See. He was of low parentage, but of admirable parts from his childhood: he made great proficiency in academical learning, and being in holy orders, was at once Schoolmaster, Curate, and soon Vicar of Cheshunt, Herts, Prebend and Treasurer of Chichester, Master of the Savoy, Dean of Westminster, &c. He became Bishop of

Date.

Rochester in 1608, of Lichfield and Coventry in 1610, and of Lincoln in 1613. He was the 66th Bishop.

1614. John Overhall, Doctor in Divinity, was raised to this See, to the great joy of all good men, and was consecrated on the 3d of April. He remained but four years, and was then translated to Norwich. He was the 67th Bishop.

1618. Thomas Morton, upon Bishop Overhall's translation, was raised to this See by James the First, having before been Bishop of Chester. He sat in this See 14 years, and was then translated to Durham, in 1632. He lived till 1659, being then 95 years of age, and was the 68th Bishop.

1632. Robert Wright, D. D. of Trinity College, Cambridge, became Bishop of Lichfield and Coventry. He had been Chaplain to Queen Elizabeth and James I. and Bishop of Bristol. When the Bishops were excluded, in the Long Parliament, he signed the Protestation against it, and suffered 18 weeks' imprisonment in the Tower. Being released, he retired to Eccleshall Castle, (then a garrison for the King), where he died in 1643, and was buried in the Church there. He was the 69th Bishop.

1643. Accepted Frewen, Dean of Gloucester, was nominated by Charles I. to succeed him, with this title: without power or profit, he retired to London and lived there till the Restoration, when he was translated to the archiepiscopal See of York. He was accounted a good scholar and orator: he was the 70th Bishop.

1661. This See was conferred upon John Hackett, D. D. of Trinity College, Cambridge, one of the most eminent persons of his time for learning and public spirit. He governed this See about nine years; and though he was intent upon the improvement of the Cathedral, he did not neglect his diocese. He was a constant preacher himself, and used his utmost endeavours to oblige his Clergy to follow his example. He died soon after he had completed his buildings, October 28, 1670, and was buried under a sumptuous monument set up by his son, Sir Andrew Hackett, ancestor of the present family of that name at Moxhull, in Warwickshire. He was the 71st Bishop.

1671. Thomas Wood, Dean of Lichfield, succeeded to this See, July 2, but was suspended for non-residence by Archbishop Sancroft, and died April 18, 1692, aged 85. He was the 72d Bishop.

1692. William Lloyd, who had distinguished himself by writing against Popery, was made Bishop of St. Asaph, in 1680, and afterwards translated to this See, where he sat about 7 years: in 1699 he was promoted to Worcester, and died at Hartlebury, August 30, 1717, aged 90. He was the most eminent Chronologer of his time, and the 73d Bishop. Bishop Lloyd was succeeded, in

Date.

1699. By John Hough, whose mother was the daughter of John Byrch, Esq. of Leacroft, in this county. He was educated at Walsall-school, and admitted of Magdalen College, Oxford. He was Bishop of Oxford, in 1690, and translated to Lichfield, in 1699, and from thence to Worcester, in 1717. He was a very amiable and distinguished character, and died in 1743, in the 93d year of his age. He was the 74th Bishop.

1717. He was succeeded by Dr. Edward Chandler, an Irishman, who had been Chaplain to Bishop Lloyd, and was translated from this See to Durham, for which it was said he gave £9000.: he was a Prelate of great erudition. He published "A Defence of Christianity," and three Sermons. Bishop Chandler died in 1750, and was the 74th Bishop.

1750. His successor was Richard Smallbrooke, born, it is believed, at Birmingham, in which town is a street of that name. He was Chaplain to Archbishop Tenison, and a considerable author. His Vindication of Our Saviour's Miracles, against the attack of Woolston, is written with much ability and acuteness: he wrote 23 books, and filled his Church with his relations. He married and had issue three sons and four daughters.—1. Thomas Smallbrooke, A. M. Archdeacon of Coventry, Precentor of Lichfield, and Chancellor of this Diocese; 2. Richard Smallbrooke, L.L.D. Advocate in Doctor's Commons, who, in 1742, became joint Chancellor with his brother. He was the 75th Bishop.

1750. Frederick Cornwallis was the next successor. Having presided over this Diocese 18 years, with wisdom, diligence, and benevolence, he was translated to Canterbury, in 1768. He was the 76th Bishop.

1768. John Egerton, son of Henry, Bishop of Hereford, succeeded him, but was translated to the See of Durham, July 8, 1771. He was the 77th Bishop, and his successor was

1771. The Hon. Brownlow North, Dean of Canterbury. Lord North was at this time Minister. He was translated to Winchester, and was the 78th Bishop.

1771. The next who presided over this See, was Richard Hurd. He was born at Congreve, near Penkridge, but afterwards removed to Penford, where his father occupied a farm, and was educated at Brewood-school, under Dr. Budworth: he afterwards pursued his academical studies with great ability, and becoming Fellow of Emanuel College, Cambridge, was appointed Archdeacon of Glocester, and Assistant Preacher of the Rolls Chapel with Bishop Warburton. After this, he was preacher of Lincoln's Inn, and Lord Mansfield soon cultivated his acquaintance: he was Clerk of the Closet, and had the honour of being tutor to the Prince of Wales. He was an elegant and accomplished scholar, as well

Date.

as an excellent and highly-esteemed Prelate, and was author of several masterly and classical performances. In 1781, this learned Prelate was translated to the See of Worcester, to the heartfelt satisfaction of the whole country. He was the 79th Bishop.

1781. His successor, the present Prelate, is the Hon. Dr. JAMES CORNWALLIS, who was born in 1743, and received the rudiments of his education at Eton school. He afterwards studied divinity at Christ Church, Oxford, and in August, 1791, was appointed Dean of Windsor, to which is annexed the Deanery of Wolverhampton: in 1794 his Lordship resigned them for the Deanery of Durham. In his public functions as a Protestant Bishop, his Lordship has ever been exemplary in the performance of the important duties of that office. He has also contributed much to the improvement of the Cathedral, and the Episcopal Palace at Eccleshall. The Bishop is the patron of the following livings in his Diocese, viz. the Rectory of St. Philip's, Birmingham; the Vicarage of Dunchurch, Warwickshire; the Vicarage of Duffield, Derbyshire; the Vicarage of Prees, Salop; the Vicarages of Eccleshall, Hanbury, and Penn; the perpetual Curacy of Gnosall, Staffordshire; and Patron of Towcester, Buckby, and Pightesley Vicarages, in Northamptonshire; Belgrave, Leicestershire; Burton, Wybunbury, and Coppenhall, Cheshire; and Towen, Merionethshire. The Diocese contains 557 parishes, comprising the whole counties of Stafford, (except Broome and Clent,) Derby, the major part of Warwickshire, and nearly one-half of Shropshire.

A BRIEF HISTORICAL ACCOUNT

Of the Villages, &c. in the Parishes of ST. CHADD *and* ST. MICHAEL, *which are without the boundaries of the City of Lichfield.*

Curborough is a hamlet in the parish of St. Chadd, about a mile north of Lichfield; it was anciently a member of the Bishop's barony of that city, and was afterwards held at the manor of Longdon. The present proprietor is Theophilus Levett, Esq.

Elmhurst is a hamlet situated two miles north of Lichfield, near the turnpike-road to Uttoxeter. In the time of Edward I. Richard Puer held one-fourth part of a Knight's-fee at Elmhurst, of the Bishop of Lichfield. The subsequent owners of this estate were the Biddulphs; it was sold in 1764 by Sir Theophilus Biddulph, to Samuel Swinfen, Esq. whose nephew afterwards sold it to Francis

Perceval Eliot, Esq. late a major in the army. This gentleman was a skilful agriculturist, as appears by his various letters to Arthur Young, Esq. When the Staffordshire Cavalry were embodied, in 1794, he published a pamphlet for the use of the corps, which he dedicated to their colonel, Earl Gower Sutherland.

In March, 1797, the freehold and leasehold estates of Elmhurst and Stichbrook, containing about 856 acres, were upon sale by auction, and valued at 50,000 guineas.

Stichbrook is only remarkable for being in the vicinity of Christian-field, where, tradition says, one thousand British Christians were massacred.

Abenhall is an ancient mansion in the parish of St. Michael. It stands in a low situation, and was built in 1294 by Thomas de Abenhall.

Towards Fairwell, the ancient mansion of *Ashenbrook* attracts the attention of the traveller; it is the property of a Roman Catholic family, and several curious relics of painted glass adorn the windows.

Pipe is a manor about a mile south-west of Ashenbrook. It is a very ancient lordship, and an extensive constablewick, no less than nine villages being subject to the jurisdiction of the constable of Pipe.

Edial-Hall is a good square brick building, with a cupola and ballustrades at the top, the whole being inclosed by a court and garden-wall. This house was once the residence of the celebrated Samuel Johnson, who opened an academy here in 1736, for the instruction of young gentlemen in the Latin and Greek languages.

HAMERWICH is a small village about four miles from Lichfield, on the right of the Walsall road, adjoining the forest of Cannock. Hamerwich Chapel is a small structure pleasantly situated on an eminence.

At a public-house called Muckley-Corner, the turnpike-road from Walsall crosses the Roman Watling-street.

Pipehill is a hamlet on the road from Lichfield to Walsall; it contains two houses, and a third to the east of the road, the residence of Mr. Bradburne, where some curious specimens of a Roman *vallum* are to be seen.

Wall, a hamlet and manor, is divided by the Watling-street, and is the ancient Roman station of Etocetum. The chief remains of this remarkable antiquity are walls which encompass two acres of land called the Castle Croft, but the part most visible, with the ancient cement, is that which forms the garden of Mr. Jackson, where pavements of Roman brick, and great quantities of foundation stones, have been dug up.

Aldershaw was formerly a manor. The mansion is neat, with walled gardens, canals, groves, and other rural ornaments. The situation is pleasant, on a rising ground which commands a fine view of Lichfield Cathedral and part of the city. Aldershaw-hill is the residence of Charles Bournes, Esq.

Freeford is an ancient manor, in a delightful situation, about two miles south of Lichfield: it formerly belonged to a family who took their name from the place. In the reign of Edward I. William de Freeford held half a knight's fee in Freeford of the Bishop of Chester. In the 40th of Edward III. it appears, by a French deed, that Sir John de Freeford was lord of the manor. According to the records of this manor, preserved at the family mansion, it devolved to the family of Dyott by marriage. This respectable family suffered much for their attachment to the Royal cause during the Civil wars.

The manor-house is distinct from the rest of the hamlet, and is extra-parochial. This house is in a retired situation, in a pleasant and highly-cultivated domain: it contains two spacious rooms, appropriated to that genuine hospitality which has long been characteristic of the proprietor, (Mrs. Dyott, widow of the late Richard Dyott, Esq.) and verifies the inscription over the hall-door: *Nil nisi bonum, portus amicis.*

Among the antiquities in this house, may be mentioned a curious silver tankard, which holds two quarts, on which is engraved the figure of an Oliverian, in the costume of the age, pointing at a gnat and swallowing a camel. On the lid is the following inscription:

"The tankard says to Presbiter'an Jack,
Th' art lyer lusty; in deceit no quack;
Yett sipp, nay drinke, none sees or hears,
Come we'll agree like savage beares."

Underneath the figure are the following lines:

"No ignoramus hypocrites drink here:
Ye lips of saints with hands of hell forbeare,
Profain not (Regicides) our loyal bowles,
Sacred to Charles his health, and all true souls."

Fulfen is a small manor about a mile north of Freeford: it was for several ages in possession of the family of Fulfen. In the 21st of Henry VI. Thomas Fulfen granted it to Oliver Chatterton and his

R

heirs, and in the 14th of Charles I. it was purchased by Sir Richard Dyott, of Lichfield.

Streethay is a manor about two miles east of St. Michael's Church, Lichfield. It derives its name from its situation on the Roman road between Lichfield and Burton.

The family of Streethay were in possession of this manor for many generations. In 1591 Philip Streethay, of Lichfield, sold this manor, with its appurtenances, to Richard Pyott, citizen and alderman of London, for £1,530. His eldest son, Richard Pyott, resided at Streethay, and was Sheriff of the county of Stafford in the 11th of Charles I. During the time he served the office of High-sheriff, King Charles made a tour through this part of the kingdom, as appears by the following letter, the subject of which must be left to the discussion of natural philosophers.

"To my very loving friend the High-sheriffe of the county of Stafford:

"Sir,—His majestie, taking notice of an opinion entertayned in Staffordshire, that the burneing of ferne doth draw downe raine, and beeing desirous that the country and himselfe may enjoy faire wether as long as hee remains in these parts, his majestie hath commanded mee to write unto you to cause all burneing of ferne to bee forborne untill his majestie be passed the country. Wherein, not doubting but the consideration of their own interest, as well as of his majestie's, will invite the country to a ready observance of this his majestie's command, I rest

"Your very loving friend,
"Pembroke and Montgomery."

"*Belvoir, August* 1, 1636."

Richard Pyott, Esq. having no issue, devised the manor of Streethay to Edward Wilmot, Barrister-at-law, the father of Edward Wilmot, Esq. the present possessor.

Fisherwick is a small manor in a low situation on the banks of the Tame. It was held by the Bishop of Chester, and in the reign of Edward III. it came into the possession of Sir Roger Hillarye. In 1521, Fisherwick was the property of Sir John Skevington, Knight, alderman of London, and in this family it continued till 1756, when Lord Viscount Massareene sold it to Samuel Swinfen, Esq. by whom it was conveyed, in 1758, to the Right Honourable Arthur Chichester, Earl and afterwards Marquis of Donegal.

The old manor-house at Fisherwick, as represented in an engraving in Dr. Plott's History of the Antiquities of Staffordshire, appears to have been a magnificent specimen of the style of architecture in the reign of Queen Elizabeth. In 1766, this antique fabric was demolished, and in 1774 a modern mansion was erected of white freestone, by the Marquis of Donegal. A view of the elevation of this truly-magnificent building is given in Shaw's History of Staffordshire, accompanied with a florid description; from which it appears, that the noble founder erected it on such an extensive scale, and of such durable materials, that it might perpetuate his name. A light portico of six lofty columns, of the Ionic order, supported a pediment, adorned with his Lordship's arms in alto-relievo, with the initials of the founder's name, and the date of the erection inscribed in large characters on the frieze. The extent of the front was 180 feet; the pilasters and decorations of the principal windows were of the Corinthian order; and the interior of the fabric and its furniture were correspondent to the dignity of its exterior. The surrounding pleasure-grounds were laid out with great taste by Brown, the celebrated projector, so severely satirized for his fanciful decorations of the landscape, by Cowper, in his Task. But after this profusion of embellishment, and extravagance of expenditure, this *earthly paradise*, as it was foolishly termed by vulgar minds, was sold by the Marquis in 1810 to Richard Howard, Esq.; and this splendid mansion has since been sold by auction, and demolished by other architectural projectors for the value of the materials, which have been carried off to decorate the *paradise* of some other fanciful mortal.

WHITTINGTON is a distinct parish, in the deanery of Tamworth. The village of Whittington is situated on a dry gravelly soil, about three miles south-east of Lichfield. This manor passed with the rest of the members of Longdon to Sir William Paget, Knight, from whom it has regularly descended to the present possessor, the Marquis of Anglesea.

The land of Whittington is generally a good deep sandy or gravelly loam, productive of excellent turnips, barley, and grass. Whittington-heath, where Lichfield races are held, contains about 690 acres of land; it is now a sheep-walk, and would, if inclosed, and properly cultivated, be some of the best turnip and barley land in the county.

Tamehorn, an ancient manor in this parish, devolved, by marriage, to the late Richard Dyott, Esq. who greatly improved it by his excellent turnip and barley culture.

ELFORD is a pleasant village in the deanery of Tamworth. It

in succession? But so it is; nor do I know that in any part of the kingdom better are got upon the whole; the land is good, yet there are many parts in this county, its equal, where the crops are much inferior. The usual mode with the farmers in this part, is to lay all or the greatest part of their manure upon the land for the barley crop, amongst which are sowed the grass seeds, which put the land down in as good condition (or better) than when took up from grass, which enables it to carry a great stock of cattle, &c. while in grass seed, &c. and in its course comes up again in such condition that is sure of gaining good crops. You will remark, that in this course of husbandry, he must be a bad farmer who does not give all the land ploughed one good dressing of lime in the course of the three years' cropping, which is mostly put upon the land at Michaelmas, or soon after, on the barley fallow, which dressing varies in quantity per acre; but, I observe, those that use the most lime have their land in highest condition. The writer of this has been in the farming line forty years, consequently has had the opportunity of seeing different culture of land of the same quality; his has been for many years upon land of nearly the above description, after the wheat is got (if dry weather) to plough the stubble under directly; he has sometimes ploughed the spaces betwixt the shock rows before the wheat has been got, finding by experience the earlier it is ploughed the better; and, if the weather suits, puts his lime on, harrows it in, and cross ploughs, and draws the furrows before he begins to sow his wheat; when done that seedness, draws it up and lets it lie in that state until seed-time in the spring, then ploughs it down and oft up again the same day, not chusing to have rain fall upon it betwixt these two ploughings; by this means of early ploughing at Michaelmas, the seeds that have been shed (and some there will be, let him be ever so good a farmer) have time to vegetate and to be totally destroyed again before spring ploughing; and I have by this mode put a farm of strong deep soil in good culture from its open field state. I have this year, 1797, about 20 acres of turnips growing, not upon such as before described, but that loose soil much addicted to couch, lakeweed, and chickweed, &c. and there is scarce a weed sprung up amongst them, which I attribute to a winter fallow. My principal system is never to plough strong land wet, particularly loose fallows, nor cart my manure upon the land (if possible to avoid it) when dirt adheres to the wheels. Many are the times when we farmers had better let our men be at play, than daubing themselves and horses at plough; much worse at harrow, which never ought to be used but when the land is dry.

"The different opinions upon the mode of culture of land has employed many pens, and volumes have been written upon the subject; I believe, notwithstanding the improvements made in the present century, that farming is in its infancy.

"*October 23, 1797.*" "W. Bourne."

There is a handsome stone bridge across the Tame at Elford, and a little above it on the river is situated a paper-mill and corn-mill, formerly the property of the late Mr. Bage, the celebrated author of 'Harmsprong, or Man as He is Not,' and other popular novels. Mr. Bage resided for many years at Elford, and afterwards removed to Tamworth, where he lived respected, and died regretted, in the year 1801. He was a native of Derby, and the late Mr. Hutton, in his history of that town, gives the following quaint account of the author and his publications:

"Wherein is an excellent picture of life, a full display of character and sentiment. These have travelled to the Continent, passed through the Frankfort press, and appeared to the world in a German habit.

"Although fortune never made him conspicuous in the great world, she gave him what is preferable, affluence and content. In directing a paper-mill, may be found that head which is able to direct empires; that judgment, which can decide in difficult cases; a penetration, which can fathom the human heart, and comprehend various systems of knowledge; a genius, which constitutes the companion for Newton in philosophy; for Handel in music; for Euclid in mathematics; a master of the living and dead languages; and all, like the wealth of a merchant who rises from nothing, acquired by himself.

"That rectitude which is rarely found, is here obscured from the public eye; but is a pearl of great price and a credit to our species. Though a diminutive figure, yet one of the most amiable of men; and though barely a Christian, yet one of the best."

This high eulogium on the genius and character of one of our best modern novelists, will probably excite curiosity in the reader to learn some further particulars respecting him. The following biographical sketch will probably prove sufficient on the subject:

Robert Bage was born at Derby, in the year 1728. His father was a paper manufacturer, and the son being intended for the same business, received a common school education. In his youth, however, he was remarkable for the vigour of his intellectual powers, and his love of knowledge. He married and settled at Elford, where he conducted a paper-mill to the end of his life.

His desire of knowledge increased with his years; and at intervals of leisure from his business, he studied and became a proficient in the modern languages. He afterwards turned his attention to the more abstruse branches of the mathematics, and engaged a teacher at Birmingham, with whom he spent an evening every week till he obtained the requisite instruction.

From his retired situation, he was but little accustomed to the manners of elegant society; hence, his productions are rather those of a man of reflection, than a close observer of men and manners. He wrote five novels, namely, Mount Kenneth, Barham Downs, The Fair Syrian, James Wallace, Harmsprong, or Man as He is Not, and Man as He Is.

Mr. Bage was happy in his matrimonial connection, and left two sons; one promising youth died before him. He died in the year 1801, aged 73 years; his character for benevolence and integrity was high, and his friends, who were much attached to him, describe his temper as open, mild, and social. He was kind to his domestics, and his humanity even extended to the domestic animals around him, particularly his horses when past work. Such a man deserves a brief memorial, yet the boast of his friend Hutton, that he was "*barely a Christian*,"[a] might better have been omitted in his panegyric, for much greater and wiser men than either Mr. Bage or Mr. Hutton, have gladly and gratefully acknowledged themselves humble believers in that most merciful dispensation of the Deity, the revelation of his will by the Saviour of the World.

Elford Church, dedicated to St. Peter, is a fine old structure, with windows in the pointed style of gothic architecture. There are some fine paintings on glass in the windows, and several ancient monuments, particularly an alabaster tomb of an Arderne and his wife. The statue of Sir John Stanley lies under an arch, in armour, with both hands supplicatory: his head rests on a helmet, with the eagle and child, the cognizance of the Stanleys. Under another arch is the eldest son, a child with curled hair, in a long gown, recumbent: one hand points to his ear, the other holds a ball, the unfortunate instrument of his death.

A rich altar tomb is ornamented with three recumbent figures, namely, that of Sir William Smith, in full armour, and his two wives, Isabel and Anne; the former of whom wears a coronet.

HASELOR is a small village about a mile north-east of Elford, and contains only a few houses and a chapel, which is a prebend of Lichfield Cathedral. This manor passed from the Staffords to the Stanleys, and by marriage to John Brook, Esq. It afterwards became the property of Joseph Girdler, Serjeant-at-Law.

Haselor-Hall is a curious specimen of those ancient mansions, so common in Staffordshire, Cheshire, and Lancashire, built chiefly of wood and plaster, with several gable ends, transom windows, &c.

EDINGALE is a small village, situated on the eastern verge of the county, and the banks of the river Meese. It is about six miles distant from Tamworth, and seven from Lichfield. The church is a perpetual curacy in the patronage of the prebendary of Alrewas. It is a small building with a tower, and is situated on an eminence, but contains nothing remarkable.

CLIFTON CAMVILLE is a large village and manor, in a pleasant situation. It fills up the eastern angle of the county, and the hundred of Offlow, the point of which angle has been commonly supposed to meet the western point of Leicestershire, in the centre of a cross, dug in the turf of a small common called No-man's Heath. This manor takes part of its name from the situation on a bank with the Meese to the north-east; the latter part of its name is derived from the Camvilles, a family who were in possession of it from the year 1200 to 1316.

At the time of the general Survey, after the Conquest, it was in the King's own hands, and was then a place of almost as much importance as at present, having 46 families, with a church, and a mill.

It afterwards belonged to Hugh Earl of Chester, and his descendants, till carried in marriage to William de Ferrers, Earl of Derby. This manor was held from the time of Henry II. by Marmion, and afterwards by the Camvilles. It was for some time in the possession of General Severne, and from him it came by a female branch of the Pye family to the Rev. Richard Watkins, of Rock, Worcestershire.

Clifton and *Haunton* form a valuable manor of several thousand acres: the soil is similar to that of Elford; the valleys on the banks of the brooks are good meadow-land, and the uplands are a strong and mixed loam. Considerable dairies are kept in this parish, and much cheese made. The sheep are generally of the Leicestershire breed, the cows are of the long-horned kind, and give abundance of excellent milk.

Clifton Camville Church is dedicated to St. Andrew, and is

adorned by one of the finest spires in the kingdom, and which may be seen at the distance of several miles. The interior of the edifice has two chancels, separated by a handsome screen. In the large gothic window of the north chancel is the representation of St. Mark. In the middle of the south chancel is a magnificent monument of alabaster, in honour of Sir John Vernon and his lady, both of whom died in the year 1545.

The north chancel contains a large handsome mural monument, executed by that celebrated sculptor Rysbrach. It is of mixed marble, with two Ionic columns; the inscription between them is to the memory of Sir Charles Pye, Bart. and supports a pediment adorned with the arms of the family.

Harlaston is a considerable hamlet in this parish, about two miles west of the mother church. There is a mill here, on the river Meese, which is here the boundary between this county and Derbyshire. The chapel is a small ancient structure, with a low wooden spire.

THORP-CONSTANTINE is a small village, situated about two miles to the south of Clifton. The church, dedicated to St. Constantine, is small and neat: it is a rectory in the deanery of Tamworth.

Statfold is a compact manor, classed in the south division of Offlow hundred, though separated from the rest of it by intervening parts of the north division.

In the year 1550 this manor came by marriage into the Wolferston family, Humphrey Wolferston being then united with Catherine, daughter and heiress of John Stanley, the former proprietor. The present owner of this estate is Samuel Pipe Wolferston, Esq. The manor-house was built in 1571, by Francis Wolferston, Esq. who also built an octagonal turret as an observatory. The grounds are pleasant, and adorned with flourishing trees; and though the situation is comparatively flat, yet not less than 38 churches and chapels may be seen from the top of the observatory. The soil is in general a strong clay loam, inclined to wet, and difficult to drain; yet when properly managed it is productive of good wheat, barley, oats, and beans: the oak timber is excellent. The church is a small stone structure without a steeple.

TAMWORTH.

Tamworth is situated at the confluence of the Anker and the Tame. The latter river, which is much the largest of the two, gives name to the town, and divides it nearly into two equal parts, one of which is in Warwickshire, and the other in Staffordshire.

The situation of Tamworth is pleasant; the south-side of the town is bordered with meadows, and on the north a fertile well-inclosed country, rising by a gradual ascent towards Wigginton. An ancient foss, called the King's Dyke, is visible on three sides of the town, being some hundred paces in length, and forty-five feet broad. This is supposed to have been the boundary of the castle.

This town, which is of undoubted antiquity, was almost totally destroyed by the Danes about the commencement of the tenth century; but Ethelfleda, the daughter of Alfred the Great, a princess who inherited her father's public virtues, rebuilt Tamworth, and added much to the strength and beauty of the town; for in 914, having been successful against the Danes, and driven them out of the country, she raised the circular mount, and built a castle on the top of it, which was a watch-tower and a place of defence. Her brother, Edward the Elder, convinced of her abilities for public affairs, appointed her governess of Mercia, in which situation she behaved with great prudence and vigilance.

Leland, in his Itinerary, mentions some curious particulars respecting Tamworth. "The towne of Tamworth," says he, "having a celebrate market, is of ancient memory, and after the Danes had razed and defaced it, Ethelthleda, lady of the Merches, and sister of King Edward sen. repayred it. The towne in respect of the bottom, where Tame and Ancre runne, is sett on the declive of a small hill, and the principall street and buildings of it lye by west and east.

"The north part and side of the principal street is in Staffordshire, and on this side is the paroch-church of Tamworth. The south side, and part of this street, lying towards the right ripe of Anker, is in Warwickshire, and the castle also, which standeth at the very point of the confluence of Anker and Tame. I saw but three notable things, the paroch-church, the castle, and the bridge.

"The Castle of Tamworth standeth on a meetly high ground, on the southe part of the towne, hard upon the ripe of Anker at the mouth of it. The Marmions, Freviles, and Ferrers, have been lords of it since the Conquest."*

When Ethelfleda rebuilt Tamworth, she erected a strong tower, called the Dungeon, upon an artificial mount. The present castle was built on the site of this ancient tower, but by whom is uncertain.

Tamworth Castle, at the time of the usurpation of the crown by

* Itinerary, Vol. IV. f. p. 189.

William the Conqueror, was granted by him, with the adjacent territory, to Robert Marmion, lord of Fontenoy, in Normandy. It continued in this family till the year 1291, when it passed by marriage to William Mortein, and from him to the Freviles. In little more than a century it was carried by marriage into the family of Ferrers, and afterwards to the Comptons. In 1751, Lady Charlotte Compton carried Tamworth Castle, and the estates belonging to it, to her husband, the Hon. George Townshend, afterwards Marquis Townshend.

"The architecture of the Castle now remaining seems to have been of various periods. The hall is large, but exceedingly rude and comfortless. This, and some other parts, appear of a very ancient date. The rooms and stair-cases are most of them irregular and uncouth. Yet, by Leland's account, it seems, the greater part was built since his time. 'The base-court,' says he, 'and the great ward of the castle, is clean decayed, and the wall fallen downe, and therein be now but houses of office of noe notable building. The Dungeon-hill yet standeth, and a great round tower of stone, wherein Mr. Ferrers dwelleth, and now repaireth it.'* Such was its state in Henry the Eighth's time.

"There are two noble rooms, comparatively modern, fitted-up with oak wainscot, and round the cornice of the largest, the arms of the family, impaling every match to the earliest periods. From the windows of this room are pleasing and rich views over the river (which runs at the foot of the Castle mount,) to the meadows and woodlands, where formerly the park was. Lord Leicester, it is said, once had thoughts of making this castle his residence, and for that purpose had Wyatt down to survey it; but finding, from the antiquity of the greater part of the building, and the neglect of inhabitation here for nearly a century, the scheme not very practicable, he gave it up: it is still, however, kept in external repair.

"From the Castle leads is a noble circular view of a rich woodland country in Staffordshire, Warwickshire, Leicestershire, and Derbyshire. Canwell, the elegant new mansion of Sir Robert Lawley, Bart. appears to great advantage from hence.†" Tamworth Castle has lately been fitted-up as a cotton factory.

The beauty of the situation of Tamworth is seen from the Castle

* Leland's Itinerary, Vol. IV. p. 181.

† Shaw's History and Antiquities of Staffordshire, Vol. I. p. 419—20.

to great advantage, varied with rich meadows, two bridges over the Tame and the Anker, and the rivers wandering picturesquely along the country. Michael Drayton, born on the banks of the Anker, thus celebrates that river and his mistress:

"Clear Anker, on whose silver-sanded shore
My soul-shrin'd saint; my fair idea lies:
A blessed brook, whose milk-white swans adore
Thy crystal stream, refined by her eyes;
Where sweet myrrh-breathing zephyr in the spring
Gently distils his nectar-dropping showers;
Where nightingales in Arden sit and sing
Amongst the dainty dew-impearled flowers.
* * * * * * *
Fair Arden! thou my Tempe art alone;
And thou, sweet Anker, art my Helicon."

Tamworth is now a handsome well-built town; even in the reign of Henry VIII. Leland describes it as "all well builded of tymber." Lichfield-street is particularly elegant, and the gardens on the bank of the river large and pleasant.

The borough of Tamworth belonged to the Crown till the reign of Henry III. when it was declared a free Corporation; but the inhabitants afterwards forfeited their charter, which was restored by Edward II. The market was held weekly on Saturday, by prescription, from the time that the Kings of Mercia resided here. But the town fell into decay, and lost the name of a borough, till the reign of Queen Elizabeth, who granted it a charter, according to which it is governed by a high-steward, two bailiffs, a recorder, a town clerk, and twenty-four principal burgesses, and sends two members to parliament.

In the year 1796 the Most Noble George Marquis Townshend was chosen high-steward for Tamworth. The parliamentary interest was till lately divided between the Marquis Townshend, proprietor of Tamworth Castle, and the late Marquis of Bath, by the influence of his neighbouring seat and estate at Drayton-Basset; but Sir Robert Peel having purchased the latter seat, was consequently elected one of the representatives of the borough in Parliament.

There are five annual fairs held in Tamworth. The first, the Monday before January 25th; second, St. George's-day (old style,) May 4th; third, St. Swithin, 26th July; fourth, the first Monday in September; and fifth, on the feast of Edward the Confessor, the 24th of October, and the four following days.

Tamworth is a town of considerable population, and a variety of manufactures are successfully carried on in it; particularly a manufactory of superfine narrow woollen cloth, the printing of calicoes, a linen manufactory, and several tanneries and ale-breweries. In 1801 the population of this town, in the Staffordshire division, was 1123 persons, of whom 642 were employed in handicraft trades and manufactures.

According to the return made to parliament in 1811, the population of Tamworth amounted to: males 1258; females 1581; total 2839: but whether this return was made for the Staffordshire division, or the whole town, is not expressed.

The Church, dedicated to St. Editha, is very spacious, and has undergone material alterations and repairs at different periods. It is adorned by a tower, the stair-case of which is mentioned by the wonder-loving Plott as a great curiosity in architecture. The image of St. Editha was destroyed at the commencement of the Reformation. After the Dissolution, this Church, which had been collegiate, came to the possession of Edward VI. and the right of patronage descended to Queen Elizabeth, who, in the 22d year of her reign, by her letters-patent, granted the right of patronage to Edward Downing and Peter Aynhofon, and their heirs for ever. This right being transferable, has gone through different hands into the family of Repington.

There are also several meeting-houses for Dissenters in Tamworth.

In the year 1588, Queen Elizabeth restored to the town the free grammar-school, and granted an annuity for the salary of the schoolmaster. This excellent seminary is now in a flourishing condition.

In 1678, Thomas Guy, citizen and stationer, of London, erected and endowed an alms-house in this town, for six men and six women.

The Right Hon. Lord Viscount Weymouth settled upon two schools in Tamworth a yearly sum of five pounds each; one to teach poor children on the Staffordshire side, and the other poor children on the Warwickshire side of the town, to read English.

Thomas Guy, the celebrated bookseller and stationer, who built the noble hospital in Southwark, which bears his name, was a great benefactor to Tamworth; for, besides the alms-house already mentioned, he, in the year 1701, rebuilt the town-hall, which stands in Warwickshire.

The moat-house is a curious mansion, situated on the north bank of the river Tame, in the county of Stafford, at the extremity of Lichfield-street. It was built in 1572, by the Comberfords, as a town-residence or secondary seat to the principal family mansion, about two miles distant.

In the parish register are the following memoranda: "In 1563, and in 1606, the plague in Tamworth.—1597. Dyvers died of the bleuddie flixe, at which tyme the darthe of corne somewhat abated by reason of deathe.

"1598. mem. that the 20th day of this Aprill, Robert earl of Essex went from Drayton-Basset toward Ireland with an hoste of men to make warre againste the earl of Tyreone, an Irishman."

The bailiffs and burgesses of Tamworth are lords of the manor; the rest of the parish is divided into several lordships, and belongs to different persons, viz. Siersoote, Wigginton, Comberford, Coton, Hopwas, Bonehill, and Bitesscote, in the county of Stafford; there are also several hamlets belonging to the borough of Tamworth in Warwickshire.

Siercote, an ancient manor in the parish of Tamworth, is situated a mile north-east of the Church. It is bounded on the south-east by the turnpike-road and Amington, in Warwickshire; on the east by Statfold; and on the north by Wigginton.

This manor and estate now belongs to Joseph Grundy, gent. It is a capital farm; the soil is a strong loam, productive of good crops of grass and corn.

Wigginton is a hamlet in the parish of Tamworth, situated about two miles north-east of the Church. It is thus recorded in Domesday: "The King holds Wigetone, consisting of two hides. The arable land is six carucates. There are eight villans, and one servant, and one border, and eight burgesses in Tamworde. In all they have six carucates. Here is also a meadow, six furlongs in length and two in breadth. In the time of Edward the Confessor, it was valued at thirty shillings, but at the time of this survey £4."

South-west of Wigginton, near a piece of land called the Low Flat, is a remarkable eminence, which is called Robin Hood's Butt. On the north-west of the hamlet, is a flat space called the Money Lands, where many ancient coins, and several human bones, have been ploughed up. Dr. Plott, in his History of the Antiquities of Staffordshire, mentions a remarkable instance of abstinence: "Mary Vaughton, of Wigginton, in this county, from her cradle to this day has lived with so small a quantity both of meats and drinks, that all people admire how nature is sustained without any sensible exhaustion, she not eating in a day a piece above the size of half-a-crown in bread and butter; or if meat, not above the size of a pigeon's leg at most. She drinks neither wine, ale, or beer, but only water or milk, or both mixed; and of either of these scarce a spoonful in a day. And yet she is a maiden of a fresh complexion, and healthy enough, very piously disposed, and of the Church of England, and therefore the less likely to put a trick upon the world; besides, it is very well known to many worthy persons with whom she has lived, that any great quantities, or different liquors, have always made her sick."

When hearth-money was collected, in the year 1660, Wigginton paid for 114 hearths, £11. 0s. 8d.

In 1811, the population returns, for Wall, Weeford, and Wigginton, were: 438 males, 507 females; total 945.

The Chapel is a small structure of brick, where divine service is performed once a fortnight.

Comberford, the ancient seat and manor of the Comberfords, is in a low situation on the east bank of the river Tame, and on the left hand of the turnpike-road, midway between Tamworth and Elford. The manor and estate were lately the property of the Marquis of Donegal.

Coton, a small hamlet in Tamworth parish, is situated near to Hopwas-bridge.

Hopwas is situated on the west bank of the river Tame. In Domesday-book it is thus recorded: "The King holds Opewas; there are three hides; the arable land is six carucates. In demesne there is a mill of thirteen shillings and four-pence rent; and eleven villans and two bordars employ five ploughs. Here are thirty acres of meadow, a wood six furlongs in length and three in breadth. The whole was valued aforetime and then at 40s." Hopwas is a small hamlet situated at the bottom of a hill, the most remarkable object on which is a house on the summit, environed by a wood called Hopwas Hayes. The turnpike-road from Lichfield to Tamworth passes through Hopwas; and in an opposite direction, the Birmingham and Fazeley canal passes through it, to join the Grand Trunk at Fradley-heath. The old stone-bridge, of sixteen arches, over the Tame at Hopwas, was washed down by the memorable flood in February, 1795, and was again repaired at considerable expence by the county.

Dunstal is a good old farm, situated on the west side of the Tame, between Hopwas and Fazeley. It belongs to the Levett family, of Lichfield.

Bitterscote is another small hamlet, in the parish of Tamworth.

Bonehill is another small hamlet, which has passed, with Fazeley and Drayton manor, from the late Marquis of Bath, to Peel, Wilkes, &c. Bonehill is now remarkable for the cotton-factory of Sir Robert Peel and Co. and a neat modern mansion, the residence of the present minister of Tamworth.

Fazeley is remarkable for nothing so much in antiquity as being the termination of a very spacious and beautiful part of the Roman street, or Watling-street-road, where it leaves this county, in its way through Warwickshire to London.

On the fine stream of Tame, which runs through this village, stands an ancient corn-mill, and the large cotton-mills of Sir Robert Peel and Co. who accidentally, in digging for other purposes, discovered a large spring of the purest water, which, upon analyzation, was found to be of a much superior quality to any before known in Fazeley or its neighbourhood. This spring-water has been conveyed in pipes to the works, where it has been used for the purposes of bleaching, printing, &c. and has contributed much to the prosperity of the establishment. Several large factories, and numerous tenements for the accommodation of the workmen and their families, have been erected at Fazeley, and a very considerable increase of population has kept pace with the success of the manufacturers. The facility of inland navigation from this village to different parts of England, by the canal from Birmingham, which divides into two branches, one uniting with the Trent and Mersey, and the other extending to Coventry and Oxford, has greatly contributed to the prosperity of the place.

According to the population returns in 1811, Fazeley contained 508 males, and 657 females: total, 1165.

WALSALL.

This ancient borough is pleasantly situated on a branch of the river Tame. The town gradually rises from the eastern bank of the brook, and stands on the side of an eminence or cliff of lime-stone, on the summit of which the church is built: Walsall is considered the second market-town in the county. The government is vested in a mayor, recorder, twenty-four aldermen, a town-clerk, two serjeants at mace, and a beadle.

The mayor, with the senior alderman, and late mayor, are in the commission of the peace, and hold a Court of Quarter Sessions, and award judgment in cases of petty larceny and similar offences.

Walsall consists of twelve large and regular streets: the houses are in general well built and convenient, and the population very considerable.

According to the returns in 1811, Walsall contained, males 2686, females 2855: total 5541.—The foreign of Bloxwich then contained, males 3009, females 2639: total 5648. Grand total 11,189.

The staple manufacture of Walsall was formerly shoe-buckles and chapes; but that trade has declined. Considerable manufactories of bridle-bits, spurs, stirrups, and all sorts of hardware connected with saddlery, are now carried on, and the town has enjoyed its full share of prosperity.

The market, which is held on Tuesday, is well supplied with provisions.

The parish of Walsall is large, including the foreign of Bloxwich, Harden, Birch-hills, Walsall-wood, Windmill, Caldmore, and Lime-pit-Bank.

The Church is situated on a bold and commanding eminence, the streets gradually descending from it on every side: it is a fine ancient gothic structure, with a lofty spire. This church, which is dedicated to St. Matthew, formerly belonged to the Abbey of Hales Owen, and was conferred on that religious establishment by William Rufus. It is built in the form of a cross, but does not exhibit any traces of Saxon architecture. The interior is spacious, the roof lofty, and each side of the chancel has seven stalls, the seats of which are ornamented with a variety of grotesque figures carved in *basso-relievo*. Under this part of the church is an archway of massy gothic workmanship. It is a common passage through the eastern part of the church-yard. The window of St. Catherine's Chapel is adorned with the figure of that Saint painted on glass; and St. Clement's Chapel has several empty niches in its walls. The tower of the Church, which is built at the south-west angle, is of limestone; it is strong and plain, and surmounted by a handsome spire. About the year 1775, when the tower was finished, eight good bells were put up; and two years prior to this, a fine-toned organ, built by Green, of London, was placed in the east gallery. The church-yard being too small for the parish, in 1756, a spacious cemetery was enclosed by a wall on the south side of the town, and consecrated for the interment of the dead. At an agreeable distance beyond this burial-ground, the Vicarage, a neat rural mansion, rises amid embowering shades, and forms an interesting object.

Limestone of excellent quality is found in the hills near the town.

A canal has been made to the town, communicating with the lower level of the Birmingham canal, and iron mines opened on its banks. The Wyrley and Essington Canal also approaches within a short distance of Walsall.

There are several places of worship in Walsall appropriated to dissenters. There is also an excellent Free Grammar School.

Bloxwich, in the foreign of Walsall, is situated to the north-west, between two and three miles distant from that town. This populous hamlet has a small chapel of ease, and maintains its own poor. According to the following inscription on a tombstone in the chapel-yard, the inhabitants of Bloxwich have been strenuous in the maintenance of their independence :

"To the memory of Samuel Wilkes, late of this parish, lock-smith, who died 6th November, 1764. Reader! if thou art an inhabitant of Great Bloxwich, KNOW, that the dust beneath thy feet (when overseer of the poor of this parish) was imprisoned in thy cause, because he refused to surrender thy rights, and to submit to an arbitrary mandate, by which it was intended to incorporate the poor-rates of the foreign with those of the borough, and thereby to compel the foreign to the payment of a greater proportion of parochial taxes than is warranted by law; his resistance was attended with success. The benefit is thine."

The foreign of Bloxwich consists of two villages, called Great and Little Bloxwich. Great Bloxwich is situated on a lofty eminence, and is a large and populous village, inhabited by manufacturers of saddlers' ironmongery. There is a large open common, and a common field of good sound arable land, belonging to Great Bloxwich. The chapel is a donative in the gift of the inhabitants. Little Bloxwich is a mile to the north, and farther from Walsall; the Wyrley canal passes through it.

The road from Walsall to Stafford passes through Bloxwich; a road also branches off to Lichfield, and another to Wolverhampton.

RUSHALL is an ancient manor and village, situated on the road to Lichfield, between one and two miles distant from Walsall, on the north-east. At the time of the Conquest, a Saxon family called Neel, of Rushall, had been long settled here, and its representative did fealty to the Conqueror, and his descendants remained in possession of the manor for some generations afterwards.

The Leigh family have been in possession of Rushall for more than two centuries. The ancient mansion, which is now in ruins, is described by Erdeswick as "built about with a wall, and a gatehouse of stone, all embattled castlewise." This castellated mansion was strongly fortified during the contest between the houses of York and Lancaster; and in the war between Charles the First and the Parliament, it was defended by a numerous garrison. A Mr. Pitt, of Wolverhampton, endeavoured to bribe Captain Tuthill, governor of Rushall, to betray the garrison for £2000.; but the captain discovered the treachery, and Pitt suffered death for it in the year 1640. There are two cannon-balls shewn here, as relics of the Civil war. One of these, twelve pounds weight, was found in a piece of timber; the other, only three pounds, was found at Ryecroft, on the opposite side of the road.

This fortification, which is now in ruins, was built of rough limestone, and encompassed about one acre of ground; near the centre is a detached building, containing a large room on the ground-floor, and a bedchamber over it, adorned with several family portraits of the Leighs. The offices on the north side are now inhabited by a tenant. The extent of the manor is about 1500 acres, part in tillage, and part grass. It is in general too cold for barley, but is excellent wheat land. By a singular agreement with the lord of the manor, the tenant is bound to manure with lime only for corn crops, and to lay all his compost on grass-land. The consequence has been a great improvement of the estate, and the circumstances of the occupier.

A Roman *fibula*, with several ancient coins, were found by some workmen in May, 1795, who were employed in making a pit for burning limestone, on an estate called Linley-farm, belonging to Mr. Leigh.

The most memorable individual of the family of the Leighs, was Edward Leigh, who was educated at Oxford, and in 1623 obtained the degree of M.A. He finished his education in the Middle Temple, London, and was the author of *Critica Sacra*, and several other useful books. He was chosen member of the Long Parliament in the time of the Civil wars: but at the mandate of Cromwell, he was confined, with many other members, at the King's Head, in the Strand. He lived till the year 1671, and was buried in the chancel of Rushall Church.

On the ruins of Rushall, a singular plant, with a beautiful flower, grows in great profusion. It is the *Antirrhinum majus*, or Snap Dragon, and is in full bloom about Midsummer.

Rushall Church is a vicarage, dedicated to St. Michael, and appears to have been a chapel of ease to Walsall. On the south side of the chancel there still remains the aperture of a confessionary, and vase for holy water. It is an ancient and mutilated fabric of rough stone in the pointed style of architecture, and contains monuments of the Leighs and other families. The steeple contains five bells. In the south part of the cemetery stands the base of an ancient cross. According to the population returns in 1811, there were 324 males and 289 females, making the total inhabitants of Rushall amount to 613.

The Leigh family, or their successors, are lords of the manor, and patrons of the church.

DARLASTON is situated three miles south-west of Walsall, four east-south-east of Wolverhampton, and twenty from Stafford. The parish is about two miles long, and one and a quarter wide. It is populous, and the manufactures carried on by the inhabitants are principally of gun-locks, stirrups, buckles, and nails. A considerable number of men are employed as miners, and a small proportion in agriculture. The Birmingham canal crosses the parish of Darlaston, and is continued to Walsall. According to the population returns in 1811, the inhabitants in this parish were: 2622 males; 2259 females: total 4881.

The parish contains about 800 acres, of which only about 30 acres are meadow, and the rest arable and pasture. It abounds with coal and iron-stone. The coal at one of the mines is seven yards thick, and the iron-stone, which is found under the coal, is three-quarters of a yard thick. The mines are damp, and the miners are subject to asthmatic complaints, few of them living to 70 years of age. The parish abounds with excellent springs, and abundance of clay, which is made into bricks and tiles.

Darlaston Church is an oblong brick edifice, re-built in 1721: the tower was built in 1606, by Thomas Pye, an author of some repute. The chancel contains a monumental inscription to the memory of Walter Wilkes, and Joyce his wife, who lived together 62 years and saw their great-grand children. He died April 15, 1692, aged 82; she out-lived him 13 years, and died April 22, 1705, aged 97 years. There are two meeting-houses in this parish; one for the Methodists, who are very numerous, and another for the Independents, who are but few.

A Sunday School was erected in Darlaston church-yard in the year 1793, where a great number of children of both sexes are instructed in the principles of useful knowledge and the Christian religion. This institution is supported by voluntary subscriptions, and a collection after a charity sermon annually preached on the Sunday after St. Lawrence.

Bentley-Hall is an ancient manor belonging to Wolverhampton. It is memorable as the residence of Colonel Lane, who, with his sister Jane, concealed Charles the Second after his defeat at the battle of Worcester, and assisted him in effecting his escape out of the kingdom. The manor is now the property of Lord Anson.

WILLENHALL is a considerable and populous place, situated mid-way between Wolverhampton and Walsall. It is in a low situation, on the banks of a brook. The town consists of one long street: the principal house was formerly inhabited by Dr. Wilkes, who was born here. It did not begin to flourish till the iron manufacture was introduced in the reign of Queen Elizabeth; and it has since become populous. It is conjectured that a greater number of locks are manufactured here than at any other town in Europe of the same size; and in consequence of prosperity, many good houses have been built by the opulent manufacturers.

On the 13th of June, 1776, James Lees, of Willenhall, aged 63 years, exhibited a padlock and key manufactured by himself, which was not the weight of a silver twopence; and he said he would engage to make a dozen locks with their keys, that should not exceed the weight of a sixpence.

According to the population return for Willenhall, in 1811, the males were 2014; the females 1509: total 3523. Hence the males exceed the females in number in the proportion of 4 to 3. Similar returns were also made for the parishes of Wednesfield, Darlaston, Wednesbury, and Tipton. This disproportion of number can only be accounted for by the situation not being very congenial to the habits and general feelings of the sex.

The Church of Willenhall is considered as a chapel of ease to Wolverhampton; the patronage is in the principal inhabitants that inherit lands here, but the curate is also to be approved by the lord of the manor. Some time ago, there was a serious dispute between the Marquis of Stafford and the inhabitants, about the nomination of a curate, but the townsmen persevered, and carried their point. The foundation of Willenhall Church was laid, May 6, 1748; and the edifice was opened for divine service October 30, 1749.

It contains a monument, erected to the memory of Dr. Wilkes, in 1800; and we shall conclude the account of this place with a brief biographical account of that beneficent and ingenious man.

Richard, the eldest son of Mr. Richard Wilkes, of Willenhall, and

Lucretia, the daughter of Jonas Astley, of Wood Eaton, was born March 16, 1690, and received the rudiments of his education at Trentham. On the 13th of March, 1709, he was entered of St. John's College, Cambridge, and was admitted scholar in 1710. After passing through several gradations, he took the degree of M. A. in 1717. He entered into orders, and preached at Wolverhampton, and for sometime at Stow, near Chartley; but in consequence of being disappointed in his expectation of preferment, he retired from the Church, and began to practise physic at Wolverhampton in the year 1720. On the 24th of June, 1725, he married Miss Rachael Manlove, of Lees-hill, near Abbots-Bromley, with whom he had a handsome fortune. He then retired from Wolverhampton, and returned to his natal mansion at Willenhall, where he became very eminent in his profession. He published a Treatise on the Dropsy; and during the time a destructive distemper raged among the horned cattle in Staffordshire, he published a pamphlet entitled, "A Letter to the Gentlemen, Farmers, and Graziers of the County of Stafford," on the subject of the prevention and cure of that disorder.

In the year 1747, during his recovery from a severe fit of illness, he amused himself with writing his own epitaph, which he calls a true picture from the life. It is rather to be admired for the originality and independence of the sentiments, than its poetic beauties, and is as follows:

"Here reader stand awhile, and know
Whose carcase 'tis that rots below.
A man's who work'd by reason's rule,
Yet sometimes err'd and play'd the fool;
A man sincere in all his ways,
And full of the Creator's praise;
Who laugh'd at priestcraft, pride, and strife,
And all the little tricks of life;
Who lov'd his King, his Country more,
And dreadful party rage forbore,
He told nobility the truth,
And wink'd at hasty slips of youth.
The honest poor man's steady friend,
The villain's scourge, in hopes to mend;
His father, mother, children, wife,
His riches, honours, length of life,
Concern not thee; observe what's here,
He rests in hope, and not in fear."

His wife died in 1756, and in October the same year, he married Mrs. Frances Bendish, who died in 1798, at a very advanced age.

Dr. Wilkes was a skilful physician, and his prescriptions were often attended with success. His general knowledge was considerable, and his mind active, and always employed in some praise-worthy pursuit. He died at Willenhall, of the gout in his stomach, on the 6th of March, 1760, in the 70th year of his age, and was buried in the Church, where the monument before-mentioned was erected to his memory. His death was universally lamented by his friends and townsmen. He was an indulgent landlord, a good master, a friend to the poor, to whom he always gave gratuitous advice as a physician, and pecuniary aid; and that he was unassuming, the following brief account of himself and his ancestors will demonstrate: "My family," says he, "came out of Hertfordshire, and settled here about 300 years ago, and lived much in the same manner, for if one spent a little of the estate, it was recovered again by another. But as none of my brethren married, and I have no son living, the name of Wilkes will end in this town, and be forgotten with me." Dr. Wilkes was also a great antiquary, and collected materials for a History of Shropshire; but to the antiquities of Staffordshire his attention was principally directed.

Willenhall Spa rises on the north side of a brook which runs to the east. About 200 yards up the brook are several springs, one of which was formerly consecrated to St. Sunday. Below the spa, on the opposite side of the brook, white clay with yellow veins is found, which, when mixed and made into cakes, is sold to glovers by the name of ochre cakes. The whole country abounds with mines of coal and iron-stone.

Wednesfield is a hamlet, about two miles north-east from Wolverhampton, on the western verge of the hundred. It is memorable as the scene of a battle between Edward the Elder and the Danes, in the year 910, of which Dr. Plott gives the following account: "King Edward, with an army of West Saxons and Mercians, overtook the retreating Danes at the village of Wednesfield, and overthrew them in a bloody battle, wherein he killed Eowills and Halfden, two of their kings, and Ohtea and Scurfer, two of their earls, and nine other noblemen, of which great slaughter there are no remains but a low in a ground called South Low-field, which once had a wind-mill set upon it; another field is called North Low-field, doubtless from lows in it, since removed; and such was likely Stowman's-hill, on the road betwixt Wolverhampton and Walsall, half a mile s. w. of Nechels." Stowman-low or Hill, has been since removed to mend the roads, and nothing remarkable discovered.

The parish of Wednesfield was a common-field within the last half century, and a gate on the road at the end of the manor was called Wednesfield-gate. The Wyrley and Essington Canal passes through the whole length of Wednesfield liberty, close by the church and moat-house, and is an accommodation for lime, coal, and manure. On digging into the earth about Wednesfield, the stones are often found encrusted with hard shells, a proof that they increase in bulk by the petrifaction of the surrounding matter.

In 1811, Wednesfield contained 1248 inhabitants; namely, 702 males, and 546 females. The principal manufacture is traps of all sizes and kinds, and among others man-traps.

The Church is a neat structure of brick and stone, erected in 1760, chiefly at the expence of Mrs. Martha Gough. It cost £2000.: the patronage is in Mr. Gough, and the present minister is the Rev. John Clare.

WEDNESBURY.

This town stands at a short distance from the source of the river Tame, five miles from Wolverhampton, and three from Walsall. It is a place of great antiquity; the name is Saxon, from *Woden*, their god of battle, and *Wodensborough* has been modernized to Wednesbury.

In the year 912, according to the Saxon annals, Duke Ethelred, son-in-law of Alfred the Great, and Viceroy of Mercia, died; and his wife, Adelfleda, continued to govern the kingdom with great wisdom. She fortified Weadesburg, now Wednesbury, in the year 916, and built a castle on the hill on which the Church now stands. A few traces of the foundation of this fortress are still perceptible. After the Conquest, it was demesne of the Crown. Henry the Second bestowed it on the Heronviles, in exchange for the town of Countsfield, in Oxfordshire, so that it is now a parcel of the honour of Woodstock.

In 1794, Thomas Hoo, Esq. who was lord of this manor, died without issue, and his estates went in the female line to Mrs. Whitby, and the Hon. Mrs. Foley. The present lord of the manor is Sir Joseph Scott, Bart. of Barr, in the right of Lady Scott, daughter of Mrs. Whitby.

The old manor-house, which is situated about a quarter of a mile N. E. of the Church, is now a farm-house. On the opposite side of the hill is a curious ancient house, called Oakeswell Hall, surrounded by high garden-walls and lofty trees: it formerly belonged to the family of Hopkins, but was purchased by Mr. John Howe, and is now inhabited by some of the family, who are Quakers.

The parish of Wednesbury has long been celebrated for its valuable mines of coal and iron-stone of excellent quality; and the inhabitants have the advantage of a general inland navigation, which has contributed to the prosperity of Wednesbury and its vicinity.

In 1811, the population of Wednesbury was 2806 males, 2566 females; total 5372: and since that period many new houses have been ereeted.

Wednesbury is distinguished for its numerous and valuable manufactures, the principal of which are, coach-harness, saws, trowels, edge-tools, bridle-bits, stirrups, hinges, nails, iron axletrees, wood screws, a variety of cast-iron work, and though last not least, gun-barrels, and locks. The finest enamel paintings are likewise among the productions of its artists. The coal in the neighbourhood, which is considered the best in the kingdom for the smith's forge, on account of its peculiar intensity of heat, is found in separate veins from three to fourteen feet in thickness, and is productive of an ample revenue to its proprietors. A peculiar species of iron ore is found here called Blond metal, which is chiefly used in the manufacture of nails, horse-shoes, hammers, axes, and other heavy tools. Reddish earth, called kip, is also found in the neighbourhood of Wednesbury, which is used in glazing vessels of different kinds.

Dr. Wilkes speaks of the wildfire which is discoverable in some of the old coal-pits near this town. "We have," says he, "long had a wildfire in the old coal-pits in Wednesbury Field. It breaks out spontaneously among the vast heaps of slack left in the coal-works, and which contains a great quantity of sulphur, and frequently smokes out through the surface, and acts upon the several strata, some of which are reduced to cinders. It hardens clay into what is called pock-stone, which is good for repairing the roads, or laying the foundation of buildings. Another kind of fire in these mines goes off with a tremendous explosion, driving every thing before it; but when the proper means are used, this is prevented.

"Mr. Savary, the original inventor of the steam-engine, set one of these engines down in Wednesbury, near a place called the Broad-waters, but the water was too powerful for his machinery, and he was forced to give up the undertaking: thus he had discovered a power sufficient to produce any effect, but was unable to form machinery for using it. This was reserved for after-ages; but the world is highly indebted to him for the principle, since so highly improved, and so generally applied."

The most powerful machinery in the world for raising water has been erected in this neighbourhood at Oakenhill, by the Birmingham Canal Company, upon Boulton and Watt's application of Savary's principle.

The art of manufacturing iron with pit-coal being perfected, furnaces are now very numerous in the vicinity of Wednesbury. They are plentifully supplied with the materials on the spot; and the utmost facility for the conveyance of the manufactured goods afforded by branches from the canal in all directions.

In the year 1742, when Lady Huntingdon sent John and Charles Wesley to Wednesbury, to preach the doctrine of Methodism, the miners and iron-manufacturers, who were then in a rather uncivilized state, rose in a mob, and broke the windows of the house where some people had assembled to hear the preachers. Some of the rioters being brought by a warrant before Justice Pershouse, he reprimanded the methodists. The minister of Wednesbury then joined with the magistrates, (according to Mr. Wesley,) and the mob was encouraged to persecute and insult the methodists, and all who joined them.

These riotous proceedings being noticed by Government, one of the Middlesex justices, in an interview with Mr. John Wesley, informed him, that he had orders from the King to do him justice, his Majesty being determined that no man in his dominions should be persecuted for conscience-sake. Thus by the beneficent influence of the Sovereign, the rioters were intimidated, and the spirit of persecution subsided.

Wednesbury Church is an elegant gothic edifice in the pointed style of architecture, situated on the summit of a bold eminence. The tower is handsome, and supports a lofty and beautiful spire. The interior is divided into a chancel, nave, and north and south aisles. There are several monuments in this church, particularly of the families of Lord Dudley and Lord Harcourt, with several ancient tombs, the inscriptions on which are not legible. It is now a vicarage in the King's books, and the gift of the Chancellor. The present vicar is the Rev. A. B. Haden.

WESTBROMWICH is a village situated to the south-east of Wednesbury, on a gravelly hill. The road to Birmingham passes through the village. The population of the parish of Westbromwich has rapidly increased in consequence of the various manufactures iron, particularly locks, gun-locks, and nails. In 1811, the number of inhabitants was 7485; viz. 3773 males, and 3712 females. A large street in this parish, called Hill-Top, contains several good houses: and near it is a villa belonging to Mr. Elwall; and a good inn on the Dudley road, called the Swan inn.

Dr. Plott mentions an excellent sand for the manufacture of glass, found near Westbromwich. The learned Doctor, whose propensity to the marvellous is remarkable, gives an account of Walter Parsons, who was born at Westbromwich, and was a man of gigantic stature and extraordinary strength. In his youth he was bound apprentice to a smith, and was so tall that they were obliged to dig a hole in the ground for him to stand in up to his knees when he struck at the anvil. He afterwards was employed as porter to James the First, and was courageous in proportion to his strength, but so good-natured that he scorned to take any advantage of it. On being offended by a man of ordinary stature as he walked London streets, he only took him up and hung him by the waistband of his breeches upon a hook for the amusement of the passengers. He would sometimes, by way of merriment, take up two of the tallest yeomen of the guard, and carry them about the guard-chamber in spite of their resistance. Such is the tale of Dr. Plott.

The valley of the Tame, which bounds this parish to the north-east, is rich, fertile, and pleasant, and the surrounding country picturesque. The prospect is agreeably varied with the villas and pleasure-grounds of George Birch, Esq. at Hamstead, and of Sir Joseph Scott, Bart. at Great-Barr.

The Church of Westbromwich is an ancient building, surmounted by a tower.

Sandwell park and mansion, the seat of the Earl of Dartmouth, is in this parish, and situated to the south and south-east of the Church. It takes its name from *Sancta Fons*, or the Holy Well. The present mansion is built on the site of a priory of Benedictine monks, dedicated to St. Mary Magdalen. Some vestiges of the original foundation are still visible to the observant eye, in the rear of the house, and among the offices, where a stone coffin was dug up a few years ago. At the time of the dissolution of monasteries, in the reign of Henry the Eighth, it was given to Cardinal Wolsey. The estate afterwards came into the possession of the present family of Legge.

The present mansion is of brick stuccoed white, and forms a

square; and in the extensive lawn in front is the holy well, which is now enclosed by iron rails.

Sandwell, though situated in a populous neighbourhood, and only four miles from Birmingham, and close to the turnpike-road from that town to Wolverhampton, is enclosed by a high park-wall and a thick plantation of trees so effectually as to be totally secluded from the busy world. The elegant mansion, besides every accommodation for convenience, contains a handsome library and a neat chapel; and the principal rooms are adorned with landscapes and portraits by the best masters, particularly Kneller, Housman, Gaspar, Vivian, Highmore, Sir Joshua Reynolds, Jansen, Richardson, and Vandyke. A variety of picturesque prospects are to be seen from different points of view in the park.

HANDSWORTH is an extensive and agreeable village, situated at the south-east extremity of this hundred, at the distance of about two miles from Birmingham. The parish is extensive, and includes several manors and hamlets. The principal object which it contains is Hamstead House, a noble and ancient mansion, the residence of the Birch family, to whom the manor of Handsworth now belongs. The grounds on the winding banks of the Tame are romantic, and adorned with full-grown and beautiful trees. A lime, which grows on a rocky point, is a magnificent object, and equally remarkable for its elevated situation and luxuriant growth. Three feet from the surface of the ground the trunk is twenty-two feet in circumference, the height of the tree is seventy-five feet, and the shade which it throws extends one hundred and eighty.

The Church of Handsworth, dedicated to Saint Mary, is an ancient gothic structure of brown stone, with a tower. It contains a few neat monuments, and has a variety of coats of arms painted on the windows. It is a rectory in the deanery of Tamworth, and in the patronage of the Birch family.

The population of Handsworth has more than doubled in 72 years. According to the return in 1811, there were 1141 males; 1310 females: total 2451.

But the most remarkable improvement in the parish of Handsworth is the extensive manufactory of Messrs. Boulton and Watt, at the Soho. The waste-lands called Handsworth-heath, which about sixty years ago was a barren waste and rabbit-warren, is now a populous village or town. A beautiful garden, with pleasure-grounds and a piece of water, now covers one side of this heath; five spacious squares of building, erected on the other side, contain workshops for 600 people. The extensive pool gives motion to a large water-wheel, and by its means to a great number of different tools; and the mechanic inventions here employed, are superior in number, variety, and utility, to those of any other manufactory in the world.

When the projector first commenced business here, his productions were only such as are usually made by the artists of Birmingham; namely, buttons, buckles, watch-chains, and trinkets. But with increasing prosperity his plans of utility were extended, and the manufacture of plated wares on a large scale was likewise introduced; and when these substantial and useful branches had been fully established, the proprietors began to bring forward works of elegance in stone bronze and or-moulins. This new manufacture immediately received the sanction of his Majesty, and the principal nobility, while the invention and activity of the manufacturer kept pace with such distinguished patronage.

Messrs. Boulton and Watt afterwards turned their attention to the manufacture of wrought plate, and, after a struggle, obtained the establishment of an office of Assay in Birmingham, for regulating the purity of the metal. Wrought plate is now a prominent article among the many rich productions of this great establishment, to the extension of which, the steam-engine has contributed in a very considerable degree.

The coining mill, erected in 1788, works eight machines, each capable of striking from 70 to 80 pieces of money per minute, the size of a guinea, which is between 30 and 40,000 per hour. And at the same blow which strikes the two faces, the edge of the piece is also struck either plain or with an inscription, and thus every piece becomes perfectly round, and of equal diameter.

In a national point of view, the manufactures produced at Soho are highly important, by collecting together ingenious artists, in consequence of which rival talents have been called forth, competition excited, and the different branches of the business brought to the greatest possible perfection.

The mansion of Soho, erected at a short distance from the manufactory, is elegant, and surrounded by beautiful pleasure-grounds.

Near to the Soho is the residence of Mr. Eggington, who has brought the art of staining glass to higher perfection than was ever known before. This artist's productions have met with the warmest approbation from persons of distinguished taste and judgment.

Hamstead (from Homestead,) has been the ancient seat of the Wyrleys for many generations, but went in the female line, about the commencement of the eighteenth century, to the Birches, the present owners. The present mansion is modern, and erected about a quarter of a mile from the old site. It is delightfully situated on an eminence, and commands a rich prospect of the vale of Tame, and the Aston and Sandwell estates. The interior of the mansion is handsome and commodious; and the dining-room is adorned with a full-length portrait of Judge Birch, an ancestor of the family. In the rear there is a large walled garden, with excellent hot-houses.

Perry, (formerly Pyrrie,) is an ancient manor, on the N. W. adjoining Sutton Coldfield, the river Tame passing along the side of it from Hamstead. Perry Hall and Park, and other estates, and a moiety of the manor, was purchased by Sir Henry Gough, Knight, in the year 1669, and it has continued in that family down to the present owner, John Gough, Esq.

Harborne, situated to the west of Birmingham, is surrounded by the counties of Warwick, Salop, and Worcester. The soil is good corn and grass-land, and very productive of potatoes. The church is in the deanery of Tamworth, and is a vicarage dedicated to St. Peter: it is a neat structure, with a tower. A charity-school for the education of fifty children, has been established in this parish.

Dr. Plott, ever in quest of marvellous facts, informs us, that "James Sands, of Harborne, died December 6, 1588, aged 140, having outlived five leases of a farm of 21 years each, after his marriage; his wife lived also to 120 years of age."

Population of Harborne in 1811: males 626, females 658: total 1284.

Smethwick is situated to the north of Harborne, to the south of Westbromwich, and two miles to the west of Birmingham. This manor was formerly the joint property of Mr. John Reynolds and Mr. John Baddeley. The latter gentleman was a most ingenious self-taught mechanic, and mathematical-instrument-maker. He was a native of Shropshire, and consequently his biography belongs to the history of that county.

A neat chapel was erected in Smethwick, in the year 1719, by a donation from Mrs. Dorothy Parkes, and voluntary contributions in this and the neighbouring parishes.

The population of this parish has rapidly increased since the Birmingham Canal passed through it. The iron foundry belonging to the Soho manufactory is erected at Smethwick, on the bank of the canal; and there is also a large manufactory of gun-barrels here, which are forged and bored by the aid of the steam-engine.

According to the population returns to Parliament in the year 1811, Smethwick contained 631 males, 697 females: total 1328.

Tipton is situated on the main line of the Birmingham Canal, about two miles to the north of Dudley Castle. It is celebrated for the richness of its iron mines, and the number of its iron-works. At four of these it is estimated that 90 tons of pig-iron are made weekly.

There are also three slitting-mills, and one rolling-mill, belonging to Tipton iron-works.

The manufacture of hinges is also carried on to a considerable extent, and gives employment to fifty workmen. It is supposed that nearly one-half of the inhabitants of this parish are nailors, and the nail trade, in all its varieties, gives employment to young and old, women and children, who begin to work at six years of age.

There are also manufactories of shovels, tongs, augers, and edge tools, wood screws, awls, and fenders.

Besides these, various manufactures of iron, a very extensive manufactory of a different kind has been established here by James Keir and Co. on the banks of the canal; namely, the manufacture of alkali, soap, white lead, red lead, and metal sashes for windows.

In the year 1811, the population of Tipton was 8407, namely, 4351 males, and 4056 females.

Tipton Church is a large edifice, with a gallery; the steeple contains six bells. The Church is a curacy, in the gift of the Prebends of Lichfield.

Great-Barr.—This pleasant village is situated on the declivity of a lofty hill called Barr-beacon. It is a place of great antiquity, and the derivation of the name has been the subject of disputation and conjecture among antiquaries. The family of the Scotts have long been proprietors of this manor, and the mansion, which was formerly called the Nether House, is now the seat of Sir Joseph Scott, Bart. It stands in a romantic and fertile vale, surrounded by an extensive lawn, and a great variety and abundance of trees. The hills in the rear of this villa, gradually sloping, and covered with trees, present a truly picturesque scene; they are intersected with gravel walks, and at different points rustic seats are fixed, on which the rambler may rest and gratify the eye with various views of the subjacent landscape. The ascent to Highwood and Barr-beacon, however fatiguing to the lover of rural magnificence, amply repays him by a prospect of fifteen counties, visible from the lofty summit.

Barr-beacon is supposed by antiquaries to have been the elevated point from which the Druids gave notice to the people of their quarterly sacrifices. In the time of the Saxons, a beacon was placed here to alarm the country on any invasion of the Danes.

One of the most interesting objects of this rich and charming rural scenery, is an Urn erected to the memory of Miss Mary Dolman, the cousin of Shenstone the poet. It stands near the flower-garden, and is of statuary marble, and fluted, with a medallion of Miss Dolman in the centre, and an elegant epitaph in Latin, from the pen of the poet, engraved on the pedestal.

The Chapel of Great-Barr is an elegant edifice, adorned by a stately spire, containing six musical bells. The interior is very complete; the pews neat, and the gallery is adorned with a handsome organ. This chapel was founded by Mrs. Bromwich, and the bells were the gift of Mrs. Whitby. An exquisite painting on glass adorns the east window ; it is the production of Mr. Eginton, who has borrowed it from the Rev. William Peter's Spirit of a Child, but has improved on the original design.

According to the population returns in 1811, Great Barr contained 436 males, and 360 females : total 796 inhabitants.

ALDRIDGE is a large village situated on a high ground, about two miles south-east of the turnpike-road from Walsall to Lichfield. It is bounded on one side by the extensive waste of Sutton-Coldfield, and on the other by Cannock-heath. The soil of this parish in general is a sound gravelly loam; the higher ground well-adapted for turnips and barley: the lower, being more cool and moist, is better suited to wheat and oats, or for meadow and pasture.

The manor of Aldridge belongs to Edward Croxall, Esq. of Shustoke, Warwickshire. The Wyrley and Essington canal has been cut through a considerable part of this parish, terminating in this direction near Hay-head lime-works.

Every Christmas-day the rector used to give to every person, great and small, of this parish, that would come to his house, as much bread, beef, mustard, and vinegar, as they could eat. Of late years, instead of this entertainment, the rector pays to the inhabitants of Aldridge and Great-Barr, £26. 13s. 4d. being about 1s. 6d. to each householder, which the master of the family disposes of according to his pleasure.

X

The Church of Aldridge is a rectory. It is an ancient stone structure with a tower, is dedicated to St. Mary, and contains several monuments of the Scotts and other families.

The village contains several good houses, and an extensive distillery. The population is 425 males; 422 females: total 847.

SHENSTONE is a pleasant village situated on a gentle eminence, three miles south of Lichfield. Almost every house in this village has a small garden in front, which contributes at once to the beauty of the place and the accommodation of the inhabitants. The parish is of considerable extent, well watered by several rivulets which abound with excellent trout. It contains much good sound up-land, suitable for turnips, and barley, and other grain. There are also fertile tracts of meadow and pasture-land on the banks of the rivulets, where sheep, cows, and oxen are fattened in abundance. The air of Shenstone is wholesome, as is proved by the longevity of the inhabitants, of which several instances are recorded: Henry Lea died in Shenstone in the year 1660, aged 107; and Susannah Southwell, who married at the age of 112, could relate occurrences to which she had been an eye-witness 100 years before.

There are several gentlemen's seats in the parish.

In 1769, Shenstone parish was surveyed and valued at £4940. per annum. The Church, dedicated to St. John, stands near the centre of the village, and is a very ancient edifice.

Shenstone-Hall, which is in the village, was some time ago the property of Lord Berwick.

Shenstone-Park is a low tract about a mile east of the village. The mansion, which is called the New Hall, stands on a gentle eminence on the margin of Black-brook. It has also been the property of Lord Berwick, but was sold in 1797, to Edward Grove, Esq. of Stretton.

Wood-End and Park-Hall stand upon the turnpike-road leading to Birmingham, half-a-mile from Shenstone Church. The court in front of Park-Hall is surrounded with a high brick wall, and adorned with yews, laurels, and other evergreens.

Shenstone Moss, at the foot of St. John's Hill, is situated a quarter of a mile to the south of the Church, where there is a neat mansion called the Moss.

Footherley is a hamlet, beginning half-a-mile south-west of the church, and extending a mile and a half to Radley Moors. Footherley-Hall was the property of John Dolphin, Esq. of the Moss, but was afterwards sold to Charles Simpson, Esq. of Lichfield.

Little Aston, situated about a mile south of Footherley, and two miles from Shenstone Church, is so called to distinguish it from Aston, near Birmingham. The mansion, which was built by William Tenant, Esq. lord of the manor of Shenstone, is very elegant, and the domain much improved. An extensive lawn surrounds the house, shaded with trees, and adorned with a noble lake, over which a handsome bridge has been built, opposite to which a conservatory of hewn stone adds to the general beauty of the scene.

Upper Stonall is a hamlet, between three and four miles west of Shenstone Church. Near the hamlet, on a small hill, are vestiges of an ancient fortification, called the Castles, or Old-Fort. There were formerly two large inns in this hamlet, called the Swan and the Welch Harp, the great road from London to Chester and Holyhead then passing through it. But the road being now through Birmingham and Wolverhampton, the inns and the population of the place have declined.

Lynne is an estate extending nearly two miles from Upper Stonall to Shenstone. It contains about twelve houses, of which the most ancient is Lynne-Hall, now a farm-house. The estate is now occupied by the Owens, a branch of the Owens of Albrighton.

Chesterfield, on the Watling-street-way near Wall, consists of farms and tenements, and was formerly part of the Roman station.

WEEFORD is a village situated to the east of Shenstone, and the south of Watling-street, in a pleasant valley on the banks of Black-brook. It takes its name from the London-road formerly passing through a ford of Black-brook here, and hence called *Wayford*. There is a beautiful tract of narrow but rich meadows, on the banks of the brook in the vicinity of this village, bounded by low and fertile eminences. The parish is distinguished, as containing the low called Offlow, which gives name to the hundred. Dr. Plott says that the etymology is probably Saxon, though not the sepulchre of Offa, who was buried at Bedford; but it certainly contains the bones of some chieftain who fell in battle near this place. Weeford has been the scene of civil warfare. A Purefoy was slain here by Sir Henry Willoughby, in the cause of Edward IV. and Sir Henry was desperately wounded by Lord Lisle.

The Church of Weeford is a small ancient building: the living is a prebend in Lichfield Cathedral.

Thickbroom is a small hamlet and manor on the banks of Black-brook, which takes its name from the broom, with which it formerly abounded.

Swinfen, another hamlet in this parish, is pleasantly situated on the road to Lichfield, and has been in possession of the Swinfen family for ages. The mansion-house was built by Mr. Wyatt the architect, for Samuel Swinfen, Esq. It is a magnificent structure, and the lawn, lake, and surrounding pleasure-grounds, present a beautiful variety of rural scenery.

The Church, dedicated to St. Mary, is a prebend in Lichfield Cathedral. It is a small ancient fabric, in a low situation, near the turnpike-road.

Hints is situated to the east of Weeford, and the west of Drayton, on the side of an eminence which commands an extensive and beautiful prospect of the subjacent vale, and the surrounding country. The manor and estate have for two centuries been in possession of the Floyer family.

The Church is a modern structure, and is an elegant specimen of Grecian architecture. From its lofty situation, it is a very beautiful and picturesque object. The interior contains several monuments of the Floyers and Lawleys.

In the year 1792, as some labourers were digging for gravel on Hints common, they found a pig of lead twenty-two inches and a half long, and 150lb. weight, with the following inscription upon it in basso-relievo:

IMP. VESP. VII. T. IMP. V. COS.

This curious piece of antiquity was perfect in all its parts; and the inscription is to be read, "*Imperatore Vespasiano septimum, Tito Imperatore quintum, Consulibus.*;" and plainly betokens the year of Christ 76, U. C. 828, when Vespasian and Titus were consuls; the first, the seventh, and the other, the fifth time. Titus the son, is called Emperor as well as his father, having been associated in the government of the empire.

Canwell is a hamlet situated at the south corner of Hints parish. It was the site of a small priory of Benedictine monks, founded in 1142, by Geva Ridell, daughter of Hugh, Earl of Chester, and dedicated to St. Mary, St. Giles, and All Saints. It was one of the monasteries seized by Cardinal Wolsey; and in the time of Queen Mary, it was in possession of Vesey, Bishop of Exon, who died seised of this manor, four messuages, and 1500 acres, held of the Crown.

The present possessor is Sir Robert Lawley, Bart. and the magnificent mansion was erected by his father, under the inspection of Mr. Wyatt, the celebrated architect. The expence of the building, improving the grounds, and making large gardens, amounted to £60,000. The prospect from different parts of the grounds is extensive and beautiful.

Drayton-Bassett is situated two miles south of Tamworth, and was a place of some distinction in ancient times. The latter part of the name is derived from the family of the Bassets, who were for some time lords of the manor. It afterwards became the property of the Earls of Essex.

The Church, dedicated to St. Peter, is a rectory: the King is the patron, and the Lord Chancellor presentist. This edifice is modern, and built in the gothic style of architecture. The interior is adorned with a variety of tombs and heraldic blazonry of the Bassets.

Pelshall is a manor held by the Earl of Darlington, and is situate on the Walsall water, about eight miles distant from Wolverhampton, in the direct line for Lichfield. The Wyrley canal from Wolverhampton passes upon its long level, the whole length of Pelshall, and will facilitate the future working of its mines of coal and iron.

The Chapel is a small ancient brick building. On the north wall is a rudely-painted figure of an old man holding a purse in his left hand and money in the right, and underneath this inscription:

> "Richard Harrison gave 20 groats to the poor of Pelshall, to be paid yearly on New-year's day. The land charged to pay the money is the Old Croft. Aged 101."

According to the returns to Parliament in 1811, Pelshall contained, 252 males; 219 females: total 471.

Norton-under-Cannock, is situated on the south of that extensive waste called Cannock-heath, and about half-a-mile north of the Watling-street. At the time of the Conquest, it belonged to the Bishop of Chester. Two-thirds of this manor were purchased in 1760, by Richard Gildart, Esq. of Liverpool, and the other third has since been purchased by Phineas Hussey, Esq. of Little Wyrley.

The Church is a neat gothic edifice, with a small tower, and contains several monuments of the Fowke, Hussey, and other families. Norton wake is held on the first Sunday in August, when it is customary to adorn the interior of the Church with flowers and laurels.

Little Wyrley is a hamlet of this parish; it was held by the Bishop of Chester, at the time of the Conquest, and has passed from the family of Fowkes, by marriage, to the family of Hussey.

The old manor-house, called Little Wyrley Hall, is, from its situation amid groves of elms and other full-grown trees, a curious and picturesque specimen of the architecture of our ancestors. The Hall is decorated with several pieces of ancient armour, which, with the old-fashioned wooden chairs, give it an appearance of venerable antiquity. The rooms are adorned with heraldic emblems of its owners, and several family portraits. The mansion has been much improved by Mr. Hussey, the present possessor, whose hospitality is well known. The groves are inhabited by numerous rooks, which, after breeding-time, retire from the place, and the ground is then swept and remains clean till their return. The Brown-hills Colliery, an old and profitable establishment, on this estate, is the property of Mr. Hussey.

SEISDON HUNDRED.

THIS Hundred, which takes its name from a small village six miles from Wolverhampton, includes the south-west part of the county, and contains twenty-one parishes.

WOLVERHAMPTON.

The first mention of this place upon record, is, that the pious Wulfrune, relict of Athelme, Duke of Northampton, in 996, built and endowed the church or monastery here, which was called from her name and title Wulfrunes-hampton, whence its present name.

Wolverhampton stands on a rising ground, is a considerable manufacturing town, and the largest in the county: it has a weekly market on Wednesday, and one annual fair on the 10th of July. The population of Wolverhampton has increased more than two-fold since 1750; in that year the inhabitants were 7454, in 1801, 12,565, in 1811, 14,836, and now considerably exceed that number. The town is surrounded by gardens, and the air generally healthy: a navigable canal from Birmingham passes through it on its eastern side, and joins the Staffordshire and Worcestershire canal at Atherley. The parish is nominally very large, but, including only so far as its poor-rate extends, it does not exceed four or five square miles, and about 3000 acres: the country around it is well inclosed. The fine and highly-productive tract called Broad-meadows and Whitmore-ends, was, in the sixteenth century, literally a morass.

Sir Simon Degge says, "this town is pleasantly situated upon a sandy (or rather gravelly) hill, where, upon the Dean's-land, one Richard Best, first a stationer in London, and after a farmer of excise to the Rebels, built a very handsome brick house, and walled about a garden and orchard near the church." This house is now called the Deanery-hall, and is leased, with the other Deanery estates, to the Earl of Darlington: it is at present occupied by James Hordern, Esq. A little north-west from this, in Tup-street, is a large handsome house, built by the Giffards of Chillington, with a chapel attached for the use of the Catholics in this town. Beyond this, is another spacious mansion, with excellent walled gardens, which commands a beautiful prospect to the west, and is the residence of the Molineux family: the Molineuxs came into England with Isabella, wife of Edward the Second, from Flanders, in 1307: George Molineux, Esq. an opulent merchant and banker, who was High-sheriff for this county in 1791, is the present owner.

The *old* Church is a deanery annexed to Windsor, the Dean of Windsor being likewise Dean of Wolverhampton: the revenues of the deanery are held by lease at a reserved rent of £38. per annum, but are supposed to exceed that sum by £300. The service is performed by the Dean, a sacrist, and seven prebends: at the lower end of the chancel are sixteen stalls for the Dean, the prebends, sacrist, three curates, and four lay singing men: six boys have been added. The Rev. Thomas Walker is the present sacrist or perpetual curate, who has held that situation since 1788. The prebends are,—Kynvaston, Rev. Peter Thoroton; Fetherstone, Rev. Thomas Walker; Hilton, Rev. Dr. Muckleston; Willenhall, Rev. Robert Ellison; Monmore, Rev. Robert Fell; Wobaston, Rev. George Fieldhouse and Rev. W. Molineux; Hatherton, Hon. and Rev. Augustus Legge.

This Church was anciently dedicated to the Blessed Virgin Mary, but in the time of Henry III. was altered and re-dedicated to St. Peter: it is pleasantly situated on a gravelly hill, and commands a fine prospect towards Tettenhall, Shropshire, and Wales. It is a handsome fabric of stone, with a fine embattled tower (40 yards high) of rich gothic architecture, and consists of a lofty nave, with two aisles, and a chancel. The north and south porches are of ancient stone-work: the nave is supported by five pointed arches resting on octagonal pillars. One of the most curious features of the church is the richly-sculptured stone pulpit, with a circular flight of steps, and a large well-executed figure of a lion at its base.

On the north side of the tower (in which are ten good bells), is a chapel, anciently called St. Catherine's; now Lane's chancel, and the cemetery of the Lanes, formerly of Bentley, which contains several monuments and inscriptions of that family: they have been recently repaired and beautified at the expense of John Lane, Esq. of Abbot's Bromley, in this county.

The great chancel was suffered to go to decay till Dr. Turner became Dean of Windsor, who expended £500. in repairing it. This chancel contains a full-length statue of brass, in honour of

Admiral Sir Richard Leveson, who commanded against the Spanish Armada. Here is likewise a curious stone font, beautifully embellished with figures and flowers, and evidently of great antiquity.

Memorabilia.—In 1529, the Church was robbed, and the churchwardens went to a wise man.

1533. The high altar cost £95.

1572. Seats first allowed to be built in the church.

Plott says, "in the church of Wolverhampton are seven bells rung together in peal, which must needs be very unmusical."

Nov. 6, 1796, about two o'clock in the morning, a violent gust of wind blew down the whole range of battlements on the south side of the church, together with part of the south transept, and did great damage in the neighbourhood.

In the church-yard, near the south porch, is a round column, about twenty feet in height, exhibiting a variety of rude carvings divided into compartments, containing a bird and beast looking back at each other, dragons with fore feet and long tails in lozenges, birds and roses, a band of Saxon leaves, beasts or griffins, and other grotesque representations: the whole is surmounted by a plain capital. There are similar monuments in Leek church-yard, at Checkley, Chebsey, and Draycot-in-the-Moors, but it is doubtful whether they are of Danish or Saxon construction.

Among the charitable donations recorded in the Church, is the following: "Sir Stephen Jennings, Alderman of London, born in this town, founded one free-school, and purchased the manor of Rushock, in the county of Worcester, now (1707) of the yearly value of £300. at least, for the perpetual maintaining of a schoolmaster and usher, the care and government of which he did leave to the Worshipful Company of Merchant Taylors, of London." Sir Stephen Jennings was Lord Mayor of London in the year 1508.

The new Church, dedicated to St. John, is a handsome stone fabric, with a beautiful and lofty spire, and pewed and painted according to modern taste. The Act for erecting this Church was obtained in 1755, but the damage which the building sustained by a fire in 1758, and a deficiency of funds, prevented its completion till the year 1776. It is in the gift of the Earl of Stamford, who contributed liberally towards its erection, and the present minister is the Rev. Joseph Reid.

Besides the two Churches, there is a Scotch kirk, a Roman Catholic chapel, and four meeting-houses for the Methodists and Independents.

In the south-east part of the town is a large brick mansion, with a moat round it, called Turton's Hall, built for the Leveson family, but now converted into a manufactory of japanned goods by Messrs. Wrighton: the moat is in part filled up.

Plott says, "they have but four weak springs to supply so large a town, which are altogether behind the Cock Inn, called Pudding-well, the Horse-well, Washing-well, and the Meat-well; from which last they fetch their water for use all over the town, in leather budgets laid across a horse, and bringing to the other wells their tripes, horses, and linen." The town was for many years supplied with water from these springs by the water-works erected by Dr. Wilkes, which have been long discontinued; and a plentiful supply is now obtained from numerous wells sunk a considerable depth through the solid rock upon which the town is built. There is likewise an excellent and never-failing spring called the Culwell, close to the town, on the road leading to Wednesfield, which is one of the sources of the little river Smestall. Wulfruna's-well, on Dunstall estate, a medicinal spring at Chapel Ash, another called Spa-well, between this town and Sea-well, a fourth at Monmore-green, a fifth called St. Lawrence's-well, at Grazeley-brook, and an ancient arched well at Waddam's-hill, called Meg-a-doodle's-well, all noticed by Dr. Wilkes, are now neglected.

The first steam-engine that ever raised water in considerable quantity was erected near this town, a short distance from the half-mile stone on the Walsall-road.

On the 22d of April, 1590, a fire broke out in Barn (now Salop) street, which burnt 104 dwelling-houses and 30 barns, with a considerable quantity of corn: the fire was not extinguished till the 27th.

In the year 1743, wheat sold in Wolverhampton market at 3s. barley and pease 1s. 6d. and oats 1s. 2d. the bushel of 38 quarts. In the year 1800, wheat was 21s. per bushel, and hay £12. per ton; and in 1812 and 1813, wheat was as high as 25s. and barley 16s. the bushel.

An elegant structure, comprising a Public Library and News Room, has been recently erected in Queen-street, and many other improvements are in progress under the authority of an Act of Parliament.

A permanent Library was instituted in the year 1794, and has been progressively increasing in number of books and subscribers.

In the year from Easter 1790 to the same period in 1791, the poor-rate expenditure amounted to £2175. 18s. 3d. but it has since considerably exceeded that sum.

Wolverhampton is governed, under the county magistrates, by two constables; one chosen annually by the Dean's tenants, the other by the inhabitants at large.

Few towns possess greater local advantages than Wolverhampton for carrying on an extensive trade and manufacture; situated nearly in the centre of the kingdom, in the midst of the most productive coal and iron-mines, and having a free and easy access to the great rivers, Thames, Severn, Trent, and Mersey, by means of the different canals which surround it, every opportunity is afforded of conveying and receiving materials and merchandise.

The skill and ingenuity of its mechanics, especially the locksmiths, have been known and established for ages, and though the manufactures of this town are very distinct from those of Birmingham, they have, owing to the superior wealth and fame of the latter, been generally (though improperly) classed under the head of Birmingham wares.

The articles peculiarly the manufacture of Wolverhampton, are principally those of heavy ironmongery, having the materials so essentially requisite for their cheapness, coal and iron, upon the spot. In this class are to be reckoned locks of every kind and description, in iron, brass, and wood; hinges, bolts, latches, screws, &c.; all kinds of edge-tools used by carpenters, wheelwrights, cabinet-makers, smiths, &c.; as axes, hatchets, hammers, gimblets, files, &c.; also articles for domestic use, as box-irons, candlesticks, chafing-dishes, coffee and malt mills, fire-irons, rat-traps, &c.; fine steel toys, as cork-screws, snuffers, nut-cracks, sugar-nippers, buckles, watch-chains, &c.

Another branch of manufacture, for which there is a great and permanent demand both for home consumption and foreign supply, is tin and japan wares: several hundred hands are employed in these various articles, and in forming the most elegant and expensive tea-trays, both paper and iron, plate-warmers, tea and coffee pots, caddies, inkstands, &c.

To these may be added another class of manufacture in the brass-foundry line, which employs a great number of workmen in an almost innumerable variety of brass articles.

Amongst individual establishments of importance, is the chemical elaboratory of Mander, Bacon and Co. for the manufacture of oil of vitriol, aquafortis, and other acids, spirits of hartshorn,

Glauber and Epsom salts, and every other chemical preparation connected with medicine, manufactures, and commerce, for British and foreign supply.

A glass-house has been established within these few years, where are made all descriptions of the best white glass.

There are several steam-engines, working various mills, one of which is upon a large scale, for the spinning of woollen yarn for carpets; at which, and also at another, considerable quantities of musket-barrels were ground and polished during the late war. A mill, called the Union Mill, was established about the year 1812, containing about 14,000 shares at £1. each, for the purpose of grinding corn and baking bread, and is found to answer extremely well. An indictment was preferred at the Assizes of 1814, at Stafford, against the proprietors, as an illegal combination; a verdict was, however, given in their favour, after a trial before a special Jury of fourteen hours.

In the adjoining villages and hamlets are made, in addition to the articles before-mentioned, great quantities of nails, chains, gun-locks, and many other similar articles.

One of the most lucrative branches of trade formerly enjoyed by this town is now nearly extinct, viz. that of rich and elegant articles in steel, consisting of watch-chains, buttons, buckles, sword-hilts, &c.: watch-chains have been made of the value of 20 guineas each, and buckles at 10 to 15 guineas a-pair, and not unusually sword-hilts at 50 guineas. These articles were generally sent to France, Spain, and Italy, but the French Revolution, and the wars which arose out of it, seem to have put a period to this business.

Wolverhampton is not remarkable for public buildings worthy of attention. The free grammar-school, founded in 1668 by Sir Stephen Jennings, Lord Mayor of London, is an excellent institution, and very liberally endowed, its revenues being upwards of £1200. per annum. In it are taught the Hebrew, Greek, Latin, French, Dutch, and German languages, besides writing, arithmetic, drawing, &c.; there are several masters, and though education may be obtained here gratis by the children of the inhabitants, but very few comparatively are sent. In 1626, the inhabitants of Wolverhampton obtained a decree of Chancery against the Merchant Taylors Company, which declares this school free for all England: in a subsequent cause, the management was vested in the Bishop of Lichfield, Sir E. Littleton, Bart. the Hon. Edward Monckton, and others, as trustees. The present school (in John-street), is a handsome brick building, erected by the Merchant Taylors Company in 1794, with houses for two masters: the Rev. William Tindall is the present head-master. There are also two charity-schools, and an hospital for a priest and six old women.

The increasing communication between England and Ireland has of late years rendered Wolverhampton a great thoroughfare, it lying in the direct road from London to Holyhead, and about mid-distance. The intercourse with Manchester and Liverpool is also considerable, owing to constant conveyances between those towns and Birmingham.

Grazeley, situated near the road leading from Wolverhampton to Stourbridge, is the property of James Perry, Esq. The field called Ab-low-field took its name from a low or tumulus near Grazeley-brook, on which was planted a bush called Iseley-cross.

Near Grazeley is a very ancient place called the Lea, occupied for many generations by the family of Waring, but now part of the Grazely estate. A short distance from Lea is Merrydale, the seat of William Hetrick, Esq. and nearer Wolverhampton, on the road to Worcester, stand the newly-erected mansion of the Rev. G. W. Kempson, one of the Magistrates for this county, and an elegant villa lately built by Mr. Jno. Mander, of Wolverhampton.

Dunstall Hall, an ancient castellated mansion, with a moat and gateway, one mile north-west of Wolverhampton, was formerly the property of the Wightwicks, but has been many years occupied as a farm-house: James Herdern, Esq. of Wolverhampton, lately purchased it with the demesne of 260 acres.

Stow-heath Manor, which comprehends the east part of the township of Wolverhampton, with Bilston, and part of Willenhall, was an ancient domain of the Crown, but now belongs to the Giffard and Leveson Gower families.*

BILSTON is a large and populous place, three miles south-east of Wolverhampton, and though generally comprehended within that parish, is a distinct township as to all parochial purposes: it is principally inhabited by manufacturers of japanned and other wares, and colliers and workmen employed in the extensive iron-works established here. A branch of the Tame passes through Bilston, and a long line of the Birmingham canal intersects the parish, by which an active and constant communication is kept up with all

* The detached hamlets and prebends generally connected with Wolverhampton, are merely so by some manorial and ecclesiastical ties, having distinct places of worship, parochial and other assessments.

parts of the kingdom: the great London road to Holyhead passes through it. Bilston has been long celebrated for its vast mines of coal, iron-stone, quarry-stone, and clay: here are also numerous furnaces for smelting iron ore, forges, and slitting-mills. It has been asserted that more iron is made in Bilston-fields than in the whole kingdom of Sweden.

The Chapel, dedicated to St. Leonard, is a perpetual curacy within the exempt jurisdiction of Wolverhampton, but the right of nomination is vested in the inhabitants at large: it is a neat modern edifice, fitted-up with much taste. A sharp contest for this curacy took place in 1813, which was decided in favour of Mr. Leigh by a great majority. Here are also two Dissenting chapels, and an excellent charity-school.

Mines.—The strata in the mines now working at Bradley, near Bilston, are as follow:

No.	Strata.	Depth. Ft.	Depth. In.
1.	Surface soil,	1	6
2.	Clay and ratch,	9	0
3.	Clunch,	2	6
4.	Iron-stone,	0	2½
5.	Clunch,	2	0
6.	Iron-stone,	0	2
7.	Soft clay,	0	2
8.	Dark batty clunch,	3	0
9.	Gray jointy rock,	4	0
10.	Iron-stone,	0	1½
11.	Rockbinds with iron-stone,	4	0
12.	Soft parting,	0	1
13.	Strong black rock,	4	0
14.	Dark clunch,	7	0
15.	Iron-stone,	0	5
16.	Dark clunch with iron-stone,	5	0
17.	Ditto fuller of iron-stone,	0	10
18.	Soft clay,	1	8
19.	Batt,	2	3
20.	Brooch coal,	3	6
21.	Fire clay,	0	4
22.	Black iron-stone,	0	1
23.	Black earth,	1	6
24.	Iron-stone,	0	2
25.	Black earth & iron-stone,	1	6
26.	Iron-stone,	1	5
		56	5
	Brought over,	56	5
27.	Rockbinds with iron-stone	10	0
28.	Dark earth, with iron-stone,	6	0
29.	Rockbinds, with iron-stone	9	0
30.	Peldon,	4	0
31.	Gray rock,	23	0
32.	Dark clunch,	2	0
	To the main Coal,	110	5
33.	White coal,	3	0
34.	Tow coal,	2	3
35.	Benches & brazils,	4	6
36.	Foot coal,	2	3
37.	Slip batt,	2	3
38.	Slips,	2	3
39.	Stone coal parting,	0	4
40.	Stone coal and patchels,	4	6
41.	Penny coal,	0	6
42.	Springs and Slippers,	4	6
43.	Humfry batt,	0	4
44.	Humfries,	2	3
	Main Coal Stratum,	28	11
	Upper Stratum,	110	5
	Whole Depth,	139	4

The surface soil is a gravelly loam: No. 2, is generally rubbish, but in some places the clay will make brick; No. 3, clunch, a mouldering argillaceous schistus, of no value, which falls into a slack-coloured powder; No. 4, and the other stratum of iron-stone, as may easily be conceived, will not pay for getting, unless incumbent or appendant to some other useful strata which are got for use; the rocks, No. 9 and 31, are of no particular value, except to mend roads, and for burr walls; No. 20, brooch coal, is useful, and sometimes got; the thicker strata, No. 15, 16, 17, 24, 25, 26, 27, 28, 29, are got so far as they will pay for getting; No. 30, peldon, contains nodules of hard, basaltic, durable stone, too hard to cut, and of no use except for rough walls, or to mend roads, or for pavements.

Of the Coal Stratum.—The white coal is very good, either for shops or other use; the tow coal and brazils are good furnace coal, and the other strata are characterized further on.

In working these mines, the coal is generally raised up the shafts in skiffs or skips, by the force of a small steam-engine called a whimsey: these skiffs, when loaded, weigh from half a ton to fifteen hundred weight each, more or less. The shaft is sunk through the different strata, and to the bottom of the main coal. The opening, called the gateway, is the first work of the miners after sinking the shaft: it is made in the under-stratum of coal, from 8 to 10 feet in height, 9 feet wide, and is carried through the whole extent of the work: excavations, called stalls, are then opened from the gateway, from whence the coal is got for use, leaving sufficient pillars between the stalls to support the roof. The skiffs are loaded in the stalls, and drawn to the shafts by horses, which are let down the shafts and kept in the mines for that purpose; the under-stratum of coal is got first, after which the incumbent strata are let down more easily in large blocks, and afterwards broken smaller by pikes or wedges: long iron prongs are used to force down the upper coal when it cannot safely be come at with the pick, and scaffolds are raised to bring the workmen in contact with the upper coal.

The thickness of coal here, including the brooch coal, exceeds 10 yards, and a cubic yard of coal weighs about a ton, so that this mine contains, in its perfect parts, 10 tons of coal to each square yard of surface on the earth, which would be upwards of 48,000 tons per acre; about one-third is left in pillars, and a good deal lost in rubbish, besides which there are often faults, and defective parts, containing extraneous matter, so that it is reckoned very good work if 20,000 tons are raised for use from an acre, which is supposed to be sometimes done, and more generally from that down to 12,000 tons per acre; but this is to be understood of the old mine only: the new mine, taken in all its practicable parts, is supposed to be quite equal to the old mine in coal, and much superior to it in iron-stone, as strata of good coal are asserted to have been discovered lower down.

The following is an accurate account of the strata in a mine at Catch-em's-Corner, near Bilston, in a coal-pit-field upon the estate of Joseph Lane, Esq. for which we are indebted to Mr. George Jones, an experienced miner and engineer,

No.		Depth. Ft.	In.
1.	Surface soil, gravelly loam,	1	0
2.	Red clay, rubbish unfit for brick,	51	0
3.	Sandy and gravel,	6	0
4.	Clunch, a friable argillaceous schistus of no use, mouldering into a dark blue earth,	3	3
5.	Rockbinds, harder than clunch, but somewhat similar,	15	0
	Depth to the Coal,	76	3
6.	White coal, very good shop coal,	3	0
7.	Tow coal, best furnace coal,	2	0
8.	Brazils, very good furnace coal,	3	6
9.	Batt, hard useless rubbish,	0	2
10.	Foot coal, very good,	2	6
11.	Hob and jack, rubbish with nodules of iron-stone,	10	0
12.	Slip coal, moderately good,	2	0
13.	Stone coal, good, but with stone intermixed,	3	0
14.	Patchels, moderately good coal,	2	0
15.	Sawyer coal, very good,	2	6
16.	Slipper coal, equal to the last,	3	4
17.	Batt, a useless clayey substance,	0	4
18.	Humfries, good coal, with waste matter intermixed,	2	6
	*Depth to the bottom of the main Coal,**	113	1

* This is the old mine, which has been worked many years: the different coal strata are in thickness 26 feet 4 inches at this place. At Bradley they are, including the brooch coal, upwards of 30 feet; and towards Tipton, where the flying reed or upper coal joins the main coal, the strata of coal united considerably exceed 10 yards in thickness, but it is there deeper beneath the surface of the earth.---Very important discoveries have been recently made beneath the main coal, and inexhaustible supplies of iron-stone have been found, equal in quality to any before known, and more abundant in proportion to its surrounding extraneous matter: different strata of useful coal have been also found.

No.			Feet.	In.
	Brought forward,—Bottom of the main Coal,		113	1
19.	Blue and white clay, useless,		5	3
20.	Gubbin-stone measure, iron-stone with rubbish intermixed,		2	9
21.	Table batt, a rubbishy schistus,		5	0
22.	Gray clunch, some iron-stone intermixed,		16	9
23.	Heathen coal, good,		2	6
24.	Fire clay, and clunch,	These strata are applied to no use; the fire clay might be used, but is inferior to that of Amblecot, near Stourbridge.	7	7
25.	Rockbinds, and rubbish,		7	0
26.	Soft clunch, and batt,		2	0
27.	Rubble coal, intermixed with batt,		2	4
28.	Rockbinds, an useless schistus,		3	2
29.	Strong white rock of no particular value,		5	1½
30.	Peldon, a hard basaltic durable stone, too hard to cut, but fit to mend roads,		5	1
31.	Rockbinds, same as No. 28,		17	3½
32.	New mine iron-stone measures in two strata,		3	3
33.	Dark clunch, useless,		9	0
34.	Light clunch and penny-stone, iron-stone intermixed,		6	0
35.	Sulphur coal, not used,		2	6
36.	Fire clay and clunch rubbish, of no value,		16	9
37.	Rockbinds, similar to Nos. 28 and 31,		2	9
38.	Strong white durable rock,	of no particular value	6	1
39.	Gray rock,		3	4
40.	Strong hard rock,		16	6
41.	Gray rock,		17	0
42.	Clunch binds, an useless schistus,		0	9
43.	New mine top coal, good coal intermixed with rubbish.		5	8
44.	Parting batt, clay and rubbish,		0	6
45.	Coal, olders, inferior coal, (holers from *hole*,)		1	10
46.	Baby fire clay, rubbish,		5	9
47.	Strong gray rock, useless,		7	11
48.	Rockbinds, of no value,		10	0
49.	Dark clunch, ditto,		3	7
50.	Fire clay coal, inferior,		4	11
51.	Batt, useless,		0	8
52.	Coal, but inferior,		2	9
53.	Clunch and fire clay, a little used,		10	2
54.	Thin coal, inferior,		1	4
55.	Getting rock, containing iron-stone,		4	9
56.	Strong gray clunch, useless,		1	4
57.	Black basses, an inflammable substance, not used,		7	7
58.	Poor Robin, iron-stone measure, good iron-stone,		3	6
59.	Black basses, same as 57,		5	0
	Carried forward,		356	1

No.		Feet.	In.
	Brought forward, ...	356	1
60.	White iron-stone measure, in nodules, part good, ...	3	6
61.	Bottom coal, moderately good, ...	1	6
62.	Parting batt, or fire clay, ...	0	4
63.	Lower coal, tolerably good, ...	5	3
	Total Depth, ...	366	8

Thus to this depth have been found below the main coal nine different strata of useful coal, being in thickness put together 28 feet 1 inch, and seven strata of iron-stone, generally of good quality; three other strata of iron-stone have also been explored beneath the above, of very good quality, viz. the ball-stone, the black-stone, and the blue flats, all very good iron-stones: there are, no doubt, other mines below, from the promising appearance of the intermediate strata.

The mines of Bilston and its neighbourhood contain, in the old mines and the new, an abundant supply of iron-stone and coal, not to be exhausted in any reasonable given time, and new sources of supply have been ascertained, that will last beyond calculation: the long upper level of the Birmingham Canal passes through them in a zig-zag lengthened course, with collateral branches, and other canals in all directions, so as to accommodate every part at a small distance. The working of these mines has been carried of late years to a great extent; not only the coal-trade, but the number of iron furnaces has been increased, and the business done at each of them much extended.

Towards the latter end of the year 1815, and the beginning of 1816, considerable derangement and stagnation took place in some of these iron-works and mines, occasioned by a combination of unfortunate causes and circumstances. The orders for iron by government were slackened, or suspended, in consequence of the general peace; the war price of provisions had been high, and the miners' and colliers' wages raised in consequence to five shillings per day, and the iron sold too low to indemnify the masters' expences; large issues of copper and silver tokens and paper-money were made to pay workmen's wages; the tokens and paper-money fell into disrepute, and some bankruptcies ensued, and others were distressed for means to go on, till many workmen were thrown out of employment, and were obliged to beg their bread. Some riots ensued, which were obliged to be quelled by the military.

Owing to the continued depression of the coal and iron-trades, and consequent discharge of workmen, a number of colliers determined amongst themselves to make their case known to Government, and adopted the novel plan of *yoking* themselves to several waggons laden with coal, having previously drawn up a Petition signed by several Magistrates, which they intended to present to the Prince Regent, together with the coal. Accordingly, three *teams* of this description set out for the metropolis, each waggon having about 50 men yoked to it; whilst others took different directions. One of the waggons proceeded by the route of Worcester, another by Coventry and Birmingham, and a third by Stourbridge. They proceeded at the rate of about 12 miles a-day, and received voluntary gifts of money, &c. on the road as they passed along, declining to *ask* alms: their motto, as placarded on the carts, being—"Rather work than beg." To prevent their progress to the metropolis, police magistrates were sent from the Home-Department Office, who met two of the parties, one at St. Alban's and another near Maidenhead, and told the men that they had adopted wrong means to procure relief, and persuaded them quietly to return. They were allowed the value of their coals, which were left to be distributed to the poor, and sufficient means were given them to reach their homes. The conduct of these distressed men was most exemplary: they listened with the greatest attention and respect to the advice of the Magistrates, and after obtaining a certificate of their good behaviour, returned with the waggons to their families and friends. A similar proceeding took place a few miles from Chester: the Magistrates of that city met the third *team*, dissuaded the men from further persisting in their ill-advised undertaking, and gave them £20. for the coal, with which they were perfectly satisfied, and immediately returned to their own neighbourhood. The novelty of this affair created some sensation in London.

In 1811, Bilston contained 1,848 houses and 9,646 inhabitants.

A particular species of sand is found at Bilston, of an orange colour, and so extremely fine as to be scarcely palpable: it is much used in the casting of metals. Here is also dug an excellent grit-stone, in great repute for setting the finer sorts of edge-tools.

The quarry-stone lies in beds one above another (according to Plott) twelve beds deep, "every bed being thicker than that above it: the lower bed is about a yard thick, of which they make troughs, cisterns, &c. Some of the tables rise so large and even, that Mr. Hoo got one eight yards long, and not varying an inch in thickness:

some of the stone is also curiously streaked black, whereof there are elegant patterns at Mr. Gough's, Perryhall."

Bradley is a hamlet situated to the south of Bilston. Bradley-hall, an ancient possession of the Hoo family, is now converted into a farm-house: here is abundance of excellent coal and iron-stone. In opening a colliery at Bradley, about four years since, the roof of the mine fell, and incarcerated a number of men and boys, all of whom, except one man, were extricated from their perilous situation, and providentially recovered, though they had been without food for several days. Sir Joseph Scott, Bart. of Barr, is one of the proprietors of this estate, in right of his wife, a daughter and co-heiress of Mrs. Whitby.

Bradley Moor is remarkable for a curious phenomenon called the *wild-fire*, occasioned by a vein of coal having taken fire in the earth, which has continued to burn for a number of years, but it is now nearly extinct.

Near Bradley are the extensive iron-works of Fereday, Smith, and Co. supposed to be the largest establishment of the kind in the world. The powers of the steam-engine, and other mechanical improvements, are here employed to great advantage in the wielding of ponderous hammers of two or three tons weight, and huge rollers are acted upon to separate the dross from the metal while in an almost fluid heat: castings of iron, weighing from ten to twelve tons each, are made in one piece, and bars of one to four inches thick, are *sheared* off with astonishing facility. The iron is here wrought from the ore to the nail-bar, and afterwards manufactured into a great variety of articles of convenience or commerce: many of the boats employed upon the canal are constructed of plates of iron. The hissing of the blast furnace, the clanging of hammers, the dusky appearance of the workmen, and the various operations upon unwieldy masses of red-hot iron, combine to excite an idea of terror in the spectator:

"The ponderous hammer falls,
Loud anvils ring amid the trembling walls.
Strokes follow strokes, the sparkling ingot shines,
Flows the red flag, the lengthening bar refines.
Cold waves immersed, the glowing mass congeal,
And turn to adamant the hissing steel."

BYSHBURY, or *Bushbury*, is a pleasant though retired village, about two miles north-east of Wolverhampton: its ancient name, Biscopesburie, seems to imply that it was once the seat or residence of some of the Mercian Bishops. In the time of William the Conqueror, this manor was held of William Fitz Ansculf, whose posterity assumed the name of Byshbury: it afterwards belonged to the Grosvenors, and the Goughs, ancestors to the well-known antiquary of that name. Peter Tichborne Hinckes, Esq. of Tettenhall, is the present lord, who purchased the estate and mansion in 1790. The parish of Byshbury is divided into two distinct parts, Byshbury and Essington, each maintaining its own poor, and paying its quota to the church-rates: the liberty of Byshbury, including Moseley, is very extensive, and comprises nearly 6000 acres. At Byshbury-hall is still preserved a chair called the King's chair, in which Charles the Second sat during his concealment in this mansion. In 1770, the skeleton of a man in complete armour, supposed to have been one of Charles the First's cuirassiers, was found by some workmen employed in cleansing a pit in an adjoining field.

The Church, dedicated to the assumption of the Blessed Virgin, is an ancient stone edifice, surmounted by a massy embattled tower. The nave is spacious, and rests upon two pointed arches on short hexagon pillars on the north, and four similar arches on the south side: the chancel, more modern than the rest of the building, is large, lofty, and handsome, particularly the roof, which is constructed of oak, supported by curiously-carved flying buttresses.

This church contains monuments of the families of Moseley, Huntbach, Hellier, Leacroft, Whitgreave, Gough, and others, including that to the memory of Thomas Whitgreave, Esq. the celebrated protector of Charles the Second. About fifty years ago, the monument of Hugh Byshbury was opened, and found to contain a stone coffin with a skeleton, and a silver chalice, now used in the communion service. On the north side of the chancel is a mural monument erected by Mr. Huskisson, M. P. to the memory of William Huskisson, Esq. of Oxley, in the parish of Byshbury; Elizabeth, his wife; and Richard, their second son, a naval surgeon, who died at Guadaloupe in the year 1794. It is inscribed as follows:

Brave, sensible, humane, thy mind and heart
Completely fashioned for the healing art,
Led thee, young HUSKISSON, with guardian care
To rescue valour from perdition's snare,
Teaching the mariner in noxious seas
To foil the hovering harpy of disease,
Thy bolder pity labour'd to sustain
The soldier drooping with contagious pain,
Where death, that Nature to her rage might bend
Robb'd her of thee, her salutary friend.

Humanity and honour join'd in grief,
Where they dejected lost thy dear relief.
Paying to virtue their acknowledg'd debt,
Gave thee a foreign grave with fond regret.
A brother who in thee could once rejoice,
(His bosom-friend by Nature and by choice,)
Feels and records on this thy vacant tomb,
Pride in thy worth, and anguish in thy doom.

Byshbury is a vicarage in the patronage of some respectable families in the neighbourhood: in the year 1800, it was presented to the Rev. John Clare, A. M. who is also a Magistrate of the county: the Rev. Gentleman resides in the vicarage-house, erected by himself in a pleasant part of Byshbury-field.

This village is sheltered on the east by a lofty hill, covered with a profusion of yew and other trees, over which, tradition asserts, the great London road to Chester once passed: from the summit of this hill (which is 650 feet above the level of the sea), is a very rich and extensive prospect. Byshbury has been much improved of late years: the soil is gravelly, with a dry and pure air.

Byshbury liberty is chiefly occupied by opulent freeholders, who cultivate their own estates, which are in a high state of improvement. Richard Phillips, Esq. has a handsome house and offices on the upper or south part of the hill; Oxley is the property of James Hordern, Esq. who purchased it of the Huskisson family in 1793; Wobaston, which gives name to one of the prebends of Wolverhampton, has been a possession of the Forsters for several generations; Ford-houses, a newly-erected mansion, the residence of the owner, Lewis Clutterbuck, Esq.; and Low-hill, a neat and modern villa, the property and residence of Richard Pountney, Esq. all deserve notice as so many objects of rural beauty, independent of their pre-eminence in agricultural improvement. There are several other ancient seats within the lordship of Byshbury, and amongst them Old-fallings, occasionally occupied by John Gough, Esq.

A great Low or tumulus (says Hunthach) is still visible here; and Plott has fixed another of these places of sepulture (of Roman erection), in a field called Gun-birch, anciently Birchen Leasow, from the circumstance of part of the bolt of a catapulta having been found in its vicinity, before alluded to.

Moseley is a constablewick in the parish of Byshbury, and contains between five and six hundred acres. Here is a neat mansion, the property of Mrs. Horton, who owns part of the estate. The other part of Moseley belongs to T. H. F. Whitgreave, Esq. in whose ancient seat Charles the Second found an asylum after the battle of Worcester: Moseley is still the family-mansion. Mr. Whitgreave retains a domestic chaplain, and a small building is appropriated as a place of worship for the Catholics in the neighbourhood.

Essington.—This manor comprehends the north-east part of the parish of Byshbury, and chiefly belongs to the Vernons, of Hilton. Some remains of an ancient residence of the De Essingtons, formerly seated here, may still be traced. Essington-wood and the Snead-commons (about 1000 acres) have been inclosed, and are now under cultivation. A branch of the Wyrley and Essington canal passes over these commons: a coal-mine has been worked upon the former for time immemorial. The uplands command some good prospects. A very fine species of sand was some years ago discovered upon an estate in this neighbourhood, which possesses some of the qualities of the Lemnian earth. Some curious pebbles have also been found at different times: Dr. Plott describes a "valuable" one picked up on Coven-heath. An attempt to establish here a manufactory of emery from ground pebbles, failed.

TETTENHALL lies about two miles west of Wolverhampton: Camden says this place before the Conquest was called Theotenhall, *i. e.* the House of the Pagans. A great battle is said to have been fought between Tettenhall and the Wergs, where, in a field called Low-hill-field, a large tumulus is still to be seen. In 911, Edward the Elder defeated the Danes at Tettenhall: Plot notices some extensive remains of a Danish encampment in Wrottesley-park.

The Church, dedicated to St. Michael, is a Royal free chapel, and enjoys all the privileges of such peculiars: it contains several monuments of the Wrottesley and other families. Sir John Wrottesley, Bart. is patron, and the Rev. Charles Wrottesley, his brother, the present officiating minister. In the steeple is a peal of five bells.

The following is an accurate account of the civil state of Tettenhall parish in 1780. Tettenhall is divided into four prebends: 1. Tettenhall prebend; containing the village, with Tettenhall-wood, Compton South, and Autherley East. 2. Pirton prebend; containing Pirton, Trescote, and Wightwick. 3. Wrottesley prebend; containing Wrottesley, the Wergs, Dipton's Farm, and Kingswood. And 4, Pendeford prebend; including Bilbrook, Lane-green, Barnhurst, Cronk-hall, Palmer's-cross, and part of the Birches. The whole parish is now inclosed, and contains about 8000 acres of inclosed land, with upwards of 200 of woodland. The number of houses was then 332, and of inhabitants about 2000, which may be increased from 5 to 13 per cent.

The register commenced in 1602: in the first ten years there were 226 births and 169 burials, and in the ten years ended 1804, 526 births and 425 burials, the increase in that time being as 2 to 5. The population principally consists of persons employed in agriculture, with a few locksmiths: there is a good resident gentry. Tettenhall may be considered extremely healthy, the births to the deaths being as 5 to 4: here are no dissenters or Roman Catholics.

There are two rivulets o[illegible]ll rivers. The Penk rises in Penkridge-well meadows, and [illegible] through Wrottesley and the Wergs to the Dam-mill and Pen[illegible], near which it meets a stream from Chillington: these strea[illegible] form a pool to Pendeford-mill, and pass through the Pendeford estate to Breewood parish. The Smestall comes from the Culwell, Wolverhampton, and the Showell moat at Byshbury, takes a south-west direction through Tettenhall and Compton, where it works a mill, and another at Wightwick, and then passes on for the Severn. The Staffordshire and Worcestershire canal summit is north of Compton, where the lockage begins with a fall of about 350 feet to the Severn.

The face of the country is generally nearly level, with gentle eminences and easy valleys, excepting a cliff or precipice, which runs through the village and along Tettenhall-wood to Wightwick; the land to the west keeping the level of its summit, and that to the east being the valley of the Smestall and of the Canal, whence it rises gradually to Wolverhampton.

The perspective from Tettenhall-wood is extremely picturesque: the ridge of the cliff above Tettenhall church-yard commands a bold and extensive view of Wolverhampton, and the surrounding country. The prospect from the newly-erected mansion of Joseph Pearson, Esq. Barrister-at-Law, is diversified and beautiful; and the seats of Francis Holyoake and P. T. Hinckes, Esqs. are not inferior in point of situation, or the extent of charming scenery which they embrace.

The upland of this parish is generally a sound gravelly loam, more or less strong, with an admixture of pebbles, and an understratum of clay, sand, gravel, or rock: the lowlands generally a four-foot stratum of peat upon gravel, but drained into meadow-land.

There is a fine square grove of 80 elms upon Tettenhall-green, of 90 years' growth, some of which contain upwards of 100 feet of timber: the Wrottesley estate has some good coppices of oak.

COMPTON is a small village to the east of Tettenhall-wood, containing several farm and other houses, with a wharf and some warehouses on the Canal. The mansion and demesne of Wightwick, to the south-west, have been purchased by P. T. Hinckes, Esq. of Tettenhall-wood. Aldersley consists of two farms, also the property of this gentleman. Here the Birmingham Canal communicates with the Staffordshire and Worcestershire, forming what is called the Aldersley junction.

PENDEFORD is situated upon the Penk, and consists of the mansion of Thomas Fowler, Esq. three farm-houses, a few tenements, and about 1000 acres of land: the estate is well timbered, and contains free-stone.

WROTTESLEY, about two miles west of Tettenhall, is the seat of the Wrottesleys, whose ancestors have possessed it for many centuries. The house stands upon a rising-ground, and was erected about 1696 by Sir Walter Wrottesley: it is a large and magnificent structure. This family trace their descent from Sir Hugo de Wrottesley, Knt. (1 Henry III.) to the present Sir John Wrottesley, Bart. M. P. for the city of Lichfield in 1799. He married, in 1795, Caroline, eldest daughter of the Earl of Tankerville, by whom he has issue.

Nurton and Perton are small hamlets belonging to the Wrottesley family: Trescote contains nothing remarkable.

A dreadful thunder-storm, with hail, happened in this neighbourhood, on the 18th August, 1742, which completely threshed-out the growing corn, and occasioned much damage: the hail-stones remained under the roofs of buildings for several days.

On Thursday, December 27, 1799, at Tettenhall-wood, Fahrenheit's thermometer stood 8° above zero, or 24° below the freezing point: the cold was equally intense in January, 1814.

CODSALL is a small village and parish, picturesquely situated on an eminence, to the south of Tettenhall: the manor forms part of Tettenhall Clericorum, and belongs to the Wrottesleys. The church, dedicated to St. Nicholas, is a curacy in the patronage of that family, several of whom are buried here. In the church-yard is a handsome monument to the memory of Nathaniel Barrett, Esq.

On the road leading to Tettenhall, is the Birches, belonging to J. T. Stubbs, Esq.; about half a mile to the west of the village, the Stockings, a good house and estate, the property of Mr. Bedford; and near it, Wood-hall, an ancient family-seat of the Deanes, but now tenanted.

At BILBROOK, in the parishes of Codsall and Tettenhall, Mr. J. Eginton has lately erected a handsome house, and introduced upon his estate all the improvements of modern agriculture.

Oaken, a hamlet in this parish, contains two elegant mansions, occupied respectively by the Dowager Lady Wrottesley and Christopher Wood, Esq.

Codsall-wood is an extensive waste adjoining Chillington-park-wall and Shropshire. Near Codsall-wood is a remarkable sulphur or brimstone well, which springs up through the hollow stump of a tree, and runs down the road, leaving a yellowness on the moss resembling flour of brimstone: in warm dry weather it emits a sulphureous exhalation. It is reckoned salutary in scorbutic cases.

OVER PENN, a pleasant village, two miles south of Wolverhampton, commands some extensive prospects of the surrounding country. The church, dedicated to St. Bartholomew, is a vicarage in the patronage of the Bishop of Lichfield: it was repaired in 1765, when the tower, of brick, was erected. Here is a charity-school for the poor children of Penn, founded by the Rev. Charles Wynn, vicar of this place, in 1714; and alms-houses for five poor persons, endowed by Mrs. Ann Sedgwick, 1761. The Marquis of Stafford is lord of the manor. South of this village, is The Lloyd's (in old deeds Luyd and Lyde), the property and residence of R. B. Marsh, Esq. Nether Penn comprehends the west side of the parish, which extends to Trescote-grange. Finchfield, and Castlecroft on the Canal, are also included in this parish: the former is the property of the Perehouse family, and the latter has been purchased by Mr. Joseph Tarratt, of Wolverhampton. The Bates, of Penn and the Leasowes, possess considerable property in this parish, mostly in their own occupation. Here is Muchall Hall, the property and residence of Wm. Thacker, Esq. with that of Mr. Burne, and several other respectable occupants. The modern system of farming is attended to, and the neighbourhood is in an improving state.

WOMBOURNE is an ancient village, in a low and recluse situation, four miles from Wolverhampton: Dr. Wilkes derives its etymology from *Won* and *Bourne*, and thinks that a victory was obtained over the Danes on the adjacent common, from the lows still remaining there. The manor was purchased by the Wrottesley family in the reign of Charles the First, and has lineally descended to its present representative. The church is a vicarage, dedicated to St. Benedict: the presentation has lain dormant since the death of the last vicar, the Rev. Joseph Honeybourn, in 1760. It is an ancient structure, with a spire, surmounted by the figure of a large dragon: the interior was repaired and modernized, and an organ added, by Sir Samuel Hellier, Knt.

The land of Wombourne is generally a light sandy or gravelly loam, remarkable for the early produce of vegetables and corn; and extensive garden and nursery-grounds have long been cultivated here. The Staffordshire and Worcestershire Canal passes through Wombourne parallel to the Smestall water, on which has been erected an iron-work called the Heath-forge, with a genteel mansion. The Woodhouses were long settled in this parish: they took their name from an estate so called, which afterwards passed to the Helliers, one of whom, Sir Samuel Hellier, greatly improved the estate and mansion; but dying, in 1792, they devolved upon the Rev. Thomas Shaw Hellier, whose son, James Shaw Hellier, Esq. is the present occupier and owner. The house is situated in a very retired and picturesque valley.

Overton or *Orton*, a hamlet in Wombourne parish, contains two or three good farm-houses, and other tenements.

TRYSULL is a small village on the banks of the Smestall. The church, dedicated to All Saints, is an ancient edifice with a small stone tower, on which is sculptured the figure of a bishop. The manor belongs to the Wrottesley family. The waste land in this parish, formerly very considerable, has been inclosed: the upland is a light sandy soil.

SEISDON is a small village or hamlet on the Smestall water, and remarkable for giving name to this hundred, for which no adequate authority can now be adduced. Here is a narrow bridge of several arches over the Smestall.

In this parish is the ancient fortification called Apewood Castle: it stands on a lofty round promontory, the whole extent of the ridge for a mile together having hollows cut in the ground, and apparently forming one continued line of defence, the two hills at each end being the principal bastions. Plot conjectures it to have been a British work.

Near Seisdon-common is a large triangular stone called the War stone; and at a short distance, a small square camp with a single trench.

PATTINGHAM is a pleasant village, situated to the north of Trysul and west of Tettenhall. The church is a vicarage in the patronage of the present incumbent: it is a neat Gothic fabric, with a stone tower. The interior was repaired and newly-pewed a few years since: near it is an ancient stone cross, and a good parsonage-house. The manor, and great part of the land, belong to the Pigot family. The village stands on a rocky bottom, and the lands about it are for the most part light and sandy. Pattingham is supposed to be a very ancient place, from the circumstance of several Roman relics having been discovered here at different times, particularly a valuable torques of gold, found in the year 1700, and a small pig of gold in 1780.

PATTESHULL is a small village about two miles north-west of Pattingham. The church, a curacy in the gift of the Pigots, is an elegant Grecian fabric, erected by Sir John Astley, and contains several good monuments. The parish consists of Patteshull, Burnhill-green, and Westbach, and contains several farm-houses and small cottages: it is in general level flat land, sandy, gravelly, or marly, producing all kinds of grain and pulse, and is chiefly manured with lime. The air of Patteshull is pure and healthy, the deaths not exceeding one in eighty annually.

The mansion and demesne at Patteshull have always been considered as pre-eminently picturesque, and art has been successfully employed in heightening and embellishing the beauties of nature: the aspect is various, the soil fertile, and trees of every species vegetate in great luxuriance. The park contains within its walls 341 acres. Sir John Astley sold this estate, with Pattingham, to Lord Pigot, for £100,000. whose celebrated diamond paid for it. Snowdon pool, a very fine piece of water, is on this demesne.

HIMLEY is a pleasant village, delightfully situated six miles south of Wolverhampton, on the road to Stourbridge. Here is the splendid seat of Lord Viscount Dudley and Ward, who is lord of the manors of Sedgley, Himley, Swindon, King's Swinford, and Rowley Regis.

The mansion is a noble structure, situated in the midst of an extensive and rich park, abounding with deer, having at a proper distance in front a magnificent sheet of water, and in the back ground, close on the left, a steep rocky hill, finely clad with ancient and modern foliage; while, to the right, the woods and hills gradually arise towards the horizon, where the spire of Sedgley church, and windmill towards Dudley, raise their lofty heads amidst the picturesque scene, charmingly secluded by a barrier of hills from the manufacturing country beyond. From the upper part of the park, the woods and water on the right appear to great advantage. Over the house the village church is a pleasing object, and the rising back-ground over Swindon common and Ashwood is richly terminated by the distant hills and woods of Enville.

The interior of the house consists of a spacious hall or dining-room, well furnished with pictures, &c. on the left of which is a billiard-room, and beyond that the library. The opposite wing consists of a large and elegant music-room, decorated with full-length portraits of the late Lord and Lady Dudley, &c. and furnished with one of the best private organs in the kingdom.

Himley has frequently been the scene of rejoicing and festivity upon public occasions. The King's escape from assassination, in 1786, and several of our great naval victories, were celebrated here by illumination, fire-works, &c. But the noble owner is not more conspicuous for his loyalty than for benevolence of heart. In 1792, no less than 1000 poor persons in the neighbourhood were regularly relieved by him; and his Lordship's princely donations to the subscriptions in London, and in this county, (1816), for the relief of the poor, besides innumerable acts of private charity, entitle him to the thanks of his countrymen, and the enviable appellation of the Poor Man's Friend.

The following lines, written by W. T. Fitzgerald, Esq. are fixed against an old yew-tree, in a walk near the house:

This stately yew, which has for ages stood,
The gloomy monarch of its native wood,
Perhaps some Norman Baron planted here,
Who liv'd by rapine and who rul'd by fear;
The tree a symbol of its master's mind,
Emblem of death, and fatal to mankind!
Beneath its boughs no verdant plants are seen,
Its baneful branches poison every green:
And thus the feudal tyrant's hated reign,
Oppress'd the village and laid waste the plain.
To these dire scenes a happier age succeeds,
No despot threatens, and no vassal bleeds.
At Himley now, the poor man finds relief,
Forgets his poverty, and checks his grief;
Raises his languid eyes and drooping head
To bless the liberal hand that gives him bread;
While in the mansion mirth and song attend,
To cheer the stranger, and delight the friend.
But still the yew, though hastening to decay,
Retains the venom of its pristine day;
Its branches still their gloomy nature shew,
And frown upon the cheerful scene below.

Besides this aged yew, there are many other remarkable trees of this and other species, particularly some large oaks, and several fine specimens of forest scenery.

The beautiful and extensive wood called Baggeridge, adjoining, is intersected with commodious carriage roads, and well stocked with pheasants.

The church, dedicated to St. Michael, is a rectory in the patronage of the Dudley family: it is a plain fabric of brick whitened and was built in the year 1764 by the late Lord Dudley and Ward. The parsonage-house is an excellent brick building, with a good garden.

The soil in the neighbourhood of Himley is a light sandy loam, in high cultivation.

To the west is the village and hamlet of Swindon, also a light sandy soil: here are an iron-works, some forges, and a blade-mill, where, by a peculiar temperament of the iron, it is formed into scythes, sickles, axes, &c. The navigable canal passes through Swindon. In 1796 a common of 1100 acres in this hamlet was inclosed, and it is now good turnip and barley land: the hilly parts have been planted, and are very picturesque.

Dudley Castle, the ancient residence of the Dudleys, was built by Dudo, an English Saxon, about the year 700, and stands on a lofty hill of lime-stone rock, the sides of which were formerly so well covered with trees, as to obtain it the appellation of "the Castle in the Woods. The prospect from the summit of the hill on which the Castle stands is extensive and beautiful, including the counties of Worcester, Stafford, Derby, Leicester, Warwick, Salop, Hereford, and part of Wales, the woods of Enville and Himley, with the Castle and obelisk at Hagley: the Malvern-hills, the hills of Rowley, Clent, Abberley, the Cleys and Wrekin, are noble features in the scene:

> "Mountains on whose barren breast
> "The labouring clouds do often rest."

The buildings of the old castle encompassed an area of about an acre, surrounded by a strong wall flanked with towers: under the chapel is a large vault called the prison.

During the Civil Wars this Castle was garrisoned for the King, and was one of the last that held out for Charles. In 1644, it was gallantly defended, during a three weeks' siege, by Colonel Beaumont, till relieved by a detachment of the King's forces from Worcester. In May, 1646, Colonel Levison surrendered it to Sir William Brereton, the Parliamentary General. In the valley are still the traces of an entrenchment thrown up against it in Cromwell's time, and several cannon-balls, some of them 32-pounders, have been found among the ruins. It was for several years after this inhabited by the lords-keepers, the plantations kept in order, and the park within the walls well stocked with deer. In 1750, a set of coiners took refuge in the buildings, which were either wilfully or accidentally destroyed by fire. It is now quite neglected, and forms a very picturesque ruin.

William, the present Lord Viscount Dudley and Ward and Baron of Birmingham, was born January 21, 1750, succeeded to his honours and estates October 8, 1788, and married Miss Julia Bosville, by whom he has issue one son, John William, who is seated in Parliament for Ilchester.

In the lime-stone quarries near this place are found a variety of fossils and marine substances in a state of petrifaction, particularly the Dudley locust, a water insect; the cornu ammonis, a worm coiled up; astroites or star stones; cockles, &c. which are frequently found in large masses. Here is that astonishing work, the Dudley Tunnel, one mile and three quarters in length, cut through a hard rock; and some stupendous caverns, where the lime-stone has been got, and conveyed away by means of this subterraneous canal.

The village of SEDGLEY stands on high ground mid-way between Wolverhampton and Dudley. The parish of Sedgley is large and populous, including Gornall, Coseley, Ettingsal, &c. in all nine villages, containing 2000 houses and 12,000 inhabitants, chiefly consisting of lime-men, colliers, and nailers. Here are immense heaps of lime-stone, and skirtings of coal and iron-stone in abundance. The facility of water-carriage has greatly contributed to the increase of the coal and iron trade in this neighbourhood: the Dudley tunnel passes under the lime-stone hills.

The church is a vicarage, dedicated to All Saints: it is an ancient structure of rough stone, with a lofty spire, which forms a fine object in the surrounding landscape.

Between Wolverhampton and Sedgley, is Sedgley Park, an ancient seat of the Dudleys, which has been long used as a Roman Catholic seminary. There is a fine eminence to the south-east, on which was formerly a beacon, the summit of which still bears that name: the stratum of this hill is principally lime-stone. Here is a considerable estate belonging to the Petit family, called Ettingsal Park, which contains great quantities of lime-stone and iron-stone: it is occupied by Samuel Fereday, Esq.

On the east side of the parish, adjoining the Canal, is a very extensive bed of excellent coal, in some of the mines fourteen yards thick, and intersected with different strata of iron-stone.

At Cotwell End and Gornall, excellent grinding stones are dug: those at the latter place possess a coarser grit, and are generally used for setting thick-edged tools.

The number of freeholders from this parish who voted at the contested election in 1747, was 753. There are three chapels for dissenters.

On Monday, April 23, 1797, during a violent thunder-storm, a fire-ball fell into the chimney of a house occupied by Stephen Cox, a nailer, of Sedgley, whilst he and his wife were sitting by the fire-side, and their child was sleeping on a chair near them. Hearing an extraordinary noise in the chimney, the mother snatched up the child, and thereby saved its life, as the pillow upon which it had lain was scorched, and a great quantity of bricks fell near it. The electric fluid passed through two houses adjoining, and, attracted by a nail, penetrated into a back kitchen, and forced out a piece of the door, which appears as if cut with a knife: it partially melted in its passage the wards of a lock and the key that was in it; part of the lead in the windows of all the three houses was also melted, and a piece of one of the window-frames struck off.

Ellers, on the west side of this parish, is a pleasing specimen of the taste of S. Fereday, Esq. in ornamental gardening: several varieties of the most delicious and highly-flavoured fruits are cultivated in great profusion.

Swinford Regis, or King's Swinford, is so called from having been a domain of the Crown till the reign of John, who gave it to the Baron Dudley, in which family it has ever since continued. It is supposed to have taken its name from Sweyn, the Danish King, who was crowned King of England about the year 1000.

The church is a rectory in the patronage of Lord Dudley. It is an ancient structure, with a massy square tower: over the principal south door is some curious rude sculpture. The interior contains several monumental inscriptions of the families of Corbyn, Hodgetts, and Bendy. The old parsonage-house stands at the east end of the village, and is in part moated.

Near the middle of the village, encompassed by lofty walls, stands Bradley-hall, an ancient half-timbered mansion, with gable ends, transom windows, and other features of the grotesque architecture of 1596, which date appears on its front. To the west is a pleasant eminence called Summer-hill, where are several good modern houses; and at a short distance, is Ashwood, an excellent house, erected by Lord Dudley, and formerly inhabited by Sir Joseph Scott.

At Holbeach, Stephen Lyttelton, and others concerned in the Gunpowder Plot, were taken in 1605.—About half a mile to the south, is Shutt-end, an ancient structure, the seat of the Bendys for many generations. Beyond this is Corbyns Hall, which took its name from the owners.

In Pensnet Chase, near Bromley, is a brine spring, but the brine is too weak to make salt.

Between Flotheridge-pool and Green's-forge, are vestiges of a Roman camp or entrenchment: it is situated on a flat, had only a single ditch, and nearly reaches down to the Smestall. There are also two uniform stone barrows or tumuli, at Barrow-hill, in this parish, which, according to Plot, are petrifactions.

On the north bank of the Stour, is *Prestwood*, formerly the seat of the Hon. Edward Foley, and now of his successors: it is a handsome Gothic mansion, fitted-up with much taste. The grounds were laid out by the ingenious Mr. Repton: the situation is enriched by hill and dale, wood and water, and possesses all the varieties of the landscape.

Near Prestwood is a wire-mill, where the manufacture of iron-wire is carried on to a great extent.

AMBLECOTE and BRIERLEY HILL have generally been considered as part of Swinford, but having distinct places of worship, it may not be improper to describe them as one parish. Erdeswicke says, "being past Rowley, Stour receives a little brook which comes from Hales-Owen, and is from that place the bound between Worcester and Staffordshire until it comes to Stourbridge; but before it cometh there by a mile hath Amblecote standing on the north bank thereof."

This place is described in an inquisition taken April 6th (2d Eliz.) with 10 messuages, 6 cottages, two water-mills, 200 acres of land, 100 of meadow, 1000 of pasture, 100 of wood, and 2000 of heath land. By an old deed, dated 46th Edw. III. it appears that coal and iron-stone were then raised here, and what is now known by the name of pit-coal was at that time called sea-coal.

The coal-mines hereabouts are very extensive, and the strata ten yards thick: supposing 120,000 tons to be raised every year, from 20 pits, they would not be exhausted in 1000 years. One half of this valuable property belongs to Lord Dudley.

Plot says, "the clay that surpasses all others is that at Amblecote, on the banks of the Stour, of a dark bluish colour, whereof they make the best pots for the glass-houses of any in England; and it is so necessary to be had that it is sent to London both by land and water-carriage; the goodness of which clay, and plenty of coal hereabout, no doubt has drawn the glass-houses into these parts, there being divers set up at Amblecote, Old Swinford, Holloway-end, and Cobourn-brook."

This valuable mineral, or glass-house pot-clay, supposed to be found no where else in the known world, is about 150 feet below the surface, and 45 feet below the coal, to the extent of nearly 200 acres, but the best sort is only found upon about 48 acres: the stratum is about two feet and an half thick, of which the middle is the finest. The outside is carefully picked off, and used in copper mines. The principal proprietors are Lord Foley, who has three acres, Edward Hickman, Esq. twenty-three, and Mr. Waldron, twenty-one: Lord Dudley has also some clay in the neighbourhood.

This clay is taken up in lumps of 100lbs weight each, which are afterwards carefully washed, picked, and scraped by women. One yard is calculated to produce a ton, and 4000 tons a-year are got: it will fetch from 34s. to 44s. per ton. It is sent to most of the manufacturing towns in England, Ireland, and Scotland; and considerable quantities are exported for the use of chymists, and furnaces requiring very strong fires. The exportation of it as fuller's-earth is prohibited unless manufactured; on which account it is shaped like bricks, which may be ground down and used as clay. It possesses this peculiar excellency, that a pot made of it, with a proper heat, will melt almost any thing into glass, provided it be fluxed with proper salts. The largest pots made of this clay are for crown glass, plate glass, broad glass and bottles, and hold from 15 to near 30 cwt. each; those for flint glass and phials, from 6 to 10 cwt. each. The largest will last one or two months, the smallest from nine to twelve months.

Broad glass has been manufactured here from the period of its first introduction into England from Lorraine and Normandy. The art of making glass is very ancient. Pliny* says, glass was first discovered by accident in Lydia, by certain merchants making a

* Natural History, lib. 36.

fire on the ground where there was plenty of the herb kali, the ashes of which mixing with sand, produced glass. Different vegetables and minerals make different sorts of glass, and of various colours: the most valuable is a transparent red glass, which, some writers assert, contains gold. Crystal glass has long been made here; but the art of cutting and engraving it, is a recent introduction. The glass trade has considerably increased of late years, and is carried on with great spirit. A number of lofty and spacious glass-houses have been erected between this place and Stourbridge; and the vicinity contains many handsome houses and villas, belonging to manufacturers, most of whom have acquired large fortunes.

Brierly-hill Chapel, a neat small structure, was built by subscription, and finished in 1767, Kingswinford being the mother church: it stands on the summit of a hill, and forms an interesting object in the distance. The Rev. Thomas Moss, A. B. was the first minister: he was afterwards minister of Trentham, and domestic chaplain to the Marquis of Stafford. This gentleman was author of the celebrated little piece, entitled "The Beggar," which, in its original state, and not as it has been given by Dr. Enfield, and copied from him into other works, we here insert. The alterations* certainly take much from the force and beauty of the poem.

Pity the sorrows of a poor old man,
Whose trembling limbs have borne him to your door,
Whose days are dwindled to the shortest span,
Oh! give relief, and Heav'n will bless your store!

These tatter'd clothes my poverty bespeak;
These hoary locks proclaim my lengthen'd years;
And many a furrow in my grief-worn cheek
Has been the channel to a stream of tears.

Yon house, erected on the rising ground,
With tempting aspect drew me from my road;
For plenty there a residence has found,
And grandeur a magnificent abode.

Hard is the fate of the infirm and poor!
Here craving for a morsel of their bread,
A pamper'd menial forc'd me from the door,
To seek a shelter in an humbler shed.

Oh! take me to your hospitable dome;
Keen blows the wind, and piercing is the cold!
Short is my passage to the friendly tomb,
For I am poor and miserably old.

* See Gent. Mag. Vol. LXX. p. 41.

Should I reveal the source of every grief,
If soft humanity e'er touch'd your breast,
Your hands would not withhold the kind relief,
And tears of pity could not be represt.

Heav'n sends misfortune—why should we repine?
'Tis Heav'n has brought me to the state you see;
And your condition may be soon like mine,
The child of sorrow and of misery.

A little farm was my paternal lot,
Then like the lark I sprightly hail'd the morn;
But, ah! oppression forc'd me from my cot,
My cattle dy'd, and blighted was my corn.

My daughter!—once the comfort of my age!
Lur'd by a villain from her native home,
Is cast abandon'd on the world's wide stage,
And doom'd in scanty poverty to roam.

My tender wife!—sweet soother of my care!
Struck with sad anguish at the stern decree,
Fell—lingering fell, a victim to despair,
And left the world to wretchedness and me.

Pity the sorrows of a poor old man,
Whose trembling limbs have borne him to your door;
Whose days are dwindled to the shortest span,
Oh! give relief, and Heav'n will bless your store!

A vast trade in glass, &c. is carried on in the neighbourhood, which has been greatly promoted by the Stourbridge canal.

KINFARE or KINVER is a pleasant village on the west bank of the Stour, and chiefly consists of one spacious and well-paved street, containing many handsome and well-built houses. Kinver stands at the foot of an high hill or mountain called Kinfare-edge, and was formerly a market town of some importance: there is still a market-house or town-hall, though the market is now discontinued. Two fairs are annually held here on the 1st of May and 15th of December. The parish is situated near the south edge of the county, and adjoins Worcestershire: the Stour, and the Staffordshire and Worcestershire canal, pass through it. The soil is generally light, sandy, or gravelly, with a range of meadows on the Stour. The Foleys, of Prestwood, are lords of the manor. Here is a well-endowed free school, but the name of the founder is unknown.

On the south of Kinfare-edge is a small plain covered with sand, where are the remains of a camp of an oblong form, 300 yards long and 200 wide, with a single graff or ditch: tradition attributes it to the Danes. Mr. Shaw thinks it the work of Wulfer King of Mercia (659 to 675), on account of its position with respect to the adjoining country, the ditch securing it from any attack on the south, and the north side being naturally inaccessible: the glacis from the camp to the village, about a mile, forms an easy and regular descent.

Bishop Lyttelton says, "the camp on Kinver-edge, which Dr. Plot imagines a work of the Danes, I make no doubt was British. The name *Kin* and *Vaur* signifies the great edge or ridge in the old Welsh or Irish. Just below the camp, on the east side, is a tumulus or barrow, environed with a little ditch: the tradition here is that this was the burial-place of an eminent commander." Dr. Stukeley describes a tumulus on Salisbury plain very similar to this, which he supposed to be Celtic. Near it, is also a large stone of a square figure, and tapering towards the top, about two yards in height, and four in circumference, having two notches on the summit: this stone is called *Bastone* or *Boltstone*.

On the north side of this hill is a remarkable cavern, called Meg-ô-fox-hole.

The church is an ancient fabric, dedicated to St. Peter, and contains some good monuments to the families of Grey, Hampton, Hodgetts, Foley, Talbot, &c.: it is a perpetual curacy, the presentation being vested in trustees. From the form of a window on the north side, Bishop Lyttelton thinks that part of the edifice was erected prior to the conquest: the chapel adjoining the chancel he refers to the time of Henry the Third, when the Hamptons were lords here, and to whom the building is ascribed.

The woollen manufacture is here carried on pretty briskly in narrow cloths, both coarse and fine, the latter of which are but little inferior in goodness to the western broads.*

There are five other villages or hamlets in this parish, viz. Compton, Stourton, Hastoote, Dunsley, and Whittington.

About a mile above Kinver (says Dr. Wilkes) is a place called the Hyde, where was the first mill for rolling or slitting iron that was erected in England. One Brindley, whose posterity long enjoyed it, went into Germany, and there acted the part of a *fool*, and by this means obtained this excellent machine, which has been so serviceable, and brought so much money into this country.

At Stourton is one of these rod-mills: the village stands on an eminence, having Stourton Castle to the south-east.

Stourton Castle was anciently the property of the Hamptons, lords of Stourton, one of whom died possessed of it in 1472. The

* Wilkes's MSS.

celebrated Cardinal Pole was born in this castle in 1500. His descent was illustrious, being a younger son of Richard Pole, Lord Montague, cousin-german to Henry the Seventh. His mother was Margaret, daughter of George, Duke of Clarence, brother to Edward the Fourth. The early part of this prelate's education was conducted by a private tutor, from whose charge he was removed, at a proper age, to Magdalen College, Oxford. Having finished his studies here, he went into orders, and soon after proceeded abroad to attend the foreign universities; during which time he was allowed a very handsome pension from Henry the Eighth, who likewise conferred upon him several benefices *in commendam*. In 1525 he returned to England, and was received by the king with distinguished marks of favour. His court influence, however, was but of short duration; for having vigorously opposed the divorce of Catharine of Arragon, he became so obnoxious to Henry, that he was compelled to seek shelter in Italy, where he wrote his celebrated piece, intituled "*De Unitate Ecclesiastica*." This work exasperated the English monarch so highly, that he not only deprived him of all his preferments, but even caused an act of attainder to be passed against him. He did not however sustain very material loss by this harsh conduct of his kinsman; for the court of Rome immediately preferred him to several benefices in Italy, and raised him to the dignity of a cardinal. Upon the death of Pope Paul the Third he was twice elected to the vacant throne, but declined the honour, because one election was too hasty, and the other made in the night-time. This truly commendable delicacy so much disobliged his friends, that they no longer afforded him their support, and in consequence the bishop of Paletrina obtained the papal see. Immediately after the bigotted Mary had ascended the throne of England, the attainder against the cardinal was repealed, and he returned with distinguished honour to his native country. His first act, upon his arrival, was to absolve the kingdom from the papal inderdict, under which it laboured on account of the apostacy of Henry the Eighth. He was now advanced to the archbishoprick of Canterbury; but enjoyed this dignified station only a few months, having died on the 17th of November 1558, the same day on which the Queen herself expired.*

Stourton Castle was garrisoned during the Civil Wars, and surrendered to the King in 1644: it is now tenanted.

* Gen. Biog. Dict.

The manor and castle of Stourton, with the hamlets of Compton and Dunsley, belong to the Foley family, of Prestwood.

At *Whittington*, an ancient hamlet in this parish, is an old mansion and a grove of fine elms.

ENVILLE.—This parish, manor, and estate, has for two centuries belonged to the family of Grey, Earls of Stamford, who have here a magnificent residence. Erdeswicke calls it "a goodly manor and a park, standing north-west from Whittington and Kinfare, something more than two miles, where Thomas Grey, who died in the 20th year of Elizabeth, built a very proper brick house." Dr. Wilkes thus describes it in 1759: "Harry, the present Earl of Stamford, having purchased several estates, so as to make the whole neighbourhood his property, and having married the Lady Mary Booth, by whom he had an immense fortune, has laid out many thousand pounds in beautifying these premises. This he has done with so much art and elegance, that few places in England can rival, and none exceed it, but has not yet meddled with the house."

Since that period the mansion has been modernized. It is a handsome white structure, and though the greater portion of it is of modern erection, it still retains an air of antiquity. It consists of a centre, and two wings, the former receding considerably, and having an octangular tower at each end. The windows in this part of the edifice are formed by Gothic arches in the pointed style, and round the top runs an embattlement, which completely prevents the roof from being seen. The wings, which stretch themselves out from the towers, appear as modern erections; and behind are several later additions, which, with the brick offices, are judiciously concealed from the view, so that the whole possesses an agreeable and uniform appearance.

In front of this mansion extends a beautiful sloping lawn, which rises boldly on the left, and is adorned by a charming lake, skirted with foliage, and a few ornamental buildings. From the side of the water a path lies through a neat shrubbery, and leads to a fine cascade, formed by the celebrated Shenstone, who indeed originally designed the whole of this delightful scenery. At a little distance below the cascade, is a rural bridge, composed of only one plank, which crosses the stream, and is a very fine and picturesque object. Near this spot stands a small chapel, dedicated to Shenstone; and having its windows embellished with various paintings on glass. This circumstance, together with the thick and gloomy umbrage in which it is enveloped, impresses the mind with a sentiment of peculiar solemnity. From hence the path extends through the wood, till at last it arrives at an open level, from which there is a view up a gently-ascending lawn, on whose summit is erected, with singular advantage, a handsome rotunda, overshadowed by a bold and lofty wood. The path now entering a part of this wood, leads to a verdant alley, opening into a sheep-walk, from a rising point of which, under a lofty yew, there are some of the richest and most enchanting prospects imaginable. At the extremity of the walk, stands the shepherd's lodge, a neat white Gothic edifice, shaded by a few trees, and partly used as an observatory by the noble owner,* from whence may be seen Bardon-hill, the Malvern and Cotteswold hills, the Wrekin, Clee-hills, and the Welsh mountains.

The church, dedicated to St. Mary, is a rectory in the deanery of Lapley and Trysull: the Rev. Richard Wilkes is the present incumbent. The east window is adorned with several shields and coats of arms; also with portraitures of St. John the Baptist, St. George and the Dragon, and the Blessed Virgin and Child. It contains many monuments and inscriptions in memory of the Greys, Moseleys of the Meer, and other families. The lid of a stone coffin with a cross, and the words ROGERUS DE MORF, inscribed upon it, was dug up in 1762, from beneath the west end of the church, whence it is supposed that the Morfe extended to this place, since a small village in the parish still retains the name of Morfe-town. A similar stone, much defaced, with a fleur-de-lis and a cross, likewise lies at the entrance of the porch.†

Enville is famous for a variety of the black cherry, which possesses a peculiar vinous flavour.

Lutteley, in this parish, is supposed to have formed part of Kinfare forest at the time of the Conqueror's Survey, as no mention is made of it at that period.

The Meer, now the seat of Acton Moseley, Esq. is within the manor of Lutteley. The water from which this place takes its name, has been reduced by art and industry, and the part reclaimed converted into good arable and pasture land.

The village and parish of BOBBINGTON is situated on the western edge of the county, adjoining Shropshire, and about seven miles from Bridgnorth. Here are two large farm-mansions called Blacklands (formerly the residence of Mr. Levingston, the anti-

* Beauties of England and Wales.

† Shaw's Hist. and Ant. of Staffordshire.

quary), and Whittemore-hall; and Bobbington-hall, the property of the Earl of Stamford. The soil is a stronger and moister loam than Kinver or Enville. The late Mr. Carwell bequeathed Bobbington-farm (charged with 10l. annually to the curate of Bobbington), for the support of a certain number of scholars at Christ Church College, Oxford.

The church is a perpetual curacy within the peculiar of Bridgnorth: it is in the patronage of the Whitmores.

OVER ARELEY is bounded on the north, west, and south by Shropshire, and on the east by Worcestershire, except a small neck to the north-east, where it joins Kinfare, being about four miles in length and one and an half in breadth: the Severn passes through this village, but visits no other part of the county.

Areley continued in the possession of Sir Thomas Lyttelton, and his male descendants, from the time of Henry the Sixth to the year 1779, when Thomas, son of George Lord Lyttelton, the tenth in descent, gave it by will to his nephew, George Viscount Valentia, (son of his only sister, Lucy,) who resides here, and is the present owner. The family-mansion was re-built by the Lytteltons, about the year 1650, and is delightfully situated on the richly-wooded banks of the Severn. It has received much external improvement from the taste of Lord Valentia, who is known to the literary world by a well-written volume of Travels. He married a daughter of the late Viscount Courtenay, by whom he has one son, George Arthur, born October 2, 1793: his lordship is divorced from his lady. The church of Areley is an ancient structure, dedicated to St. Peter, first erected by Henry de Port in the time of Henry I. and re-built in the reign of the first Edward. The nave is divided from the chancel by a range of pillars: the tower contains a peal of bells, and is a modern erection. The windows are decorated with some old paintings, and there is likewise a modern one by Mr. Eginton. In 1793, the interior was thoroughly repaired and beautified at the expence of Lord Valentia, who is patron. There is a very ancient monument to one of the Heckstans, and several of the Lytteltons. The ecclesiastical jurisdiction is vested partly in the Dean and Chapter of Lichfield, and partly in Lord Valentia: the former annually visit the church, prove wills, and audit the churchwardens' accounts, whilst the latter, by his right of presentation, institutes and inducts; so that Areley is strictly a donative, and exempt from episcopal or archidiaconical jurisdiction.

On this part of the river (Severn), and higher up to Welsh-pool, the fishermen use a light kind of boat called a *coracle*, in which one man being seated will row himself with great swiftness with one hand, whilst with the other he manages his net, angle, or fishing tackle: it is of an oval form, composed of sallow twigs interwoven round at the bottom, and covered on the outside against the water with a horse's hide. It is probable (says Mr. Shaw) these coracles were in general use among the Celtic nations in the earliest ages; and there is great reason to believe that the first inhabitants of Ireland had no other vessels to transport them either from Britain thither, or from thence to Scotland.

The soil of Areley is in general a good clay loam, and from its situation and aspect well adapted to the cultivation of fruit. The hills are often rocky, with small fragments upon and near the surface: the surface-soil varies in colour from red to grey. Hops were formerly much cultivated here, but have lately been neglected.

The names of the most remarkable buildings, are Areley-hall, Heckstones, Bromley, Bannut-tree, and Hawkbach: here is likewise the eminence called Shatterfoot, and Seckley-wood, containing about 600 acres.

Excellent grindle-stones are dug at Heckstones, and quarries of lime-stone, and mines of coal, are in course of working. On the eastern verge of the village is an ancient Roman vicinal road called the Port-way, which probably led from *Brennogenium* (Worcester) to *Uriconium* (Wroxeter), and now forms part of the post road from Worcester to Shrewsbury. In Areley-wood are the remains of a large Roman camp, which is an exact square, with double, and on one side treble, ditches: it was probably the work of Ostorius, who fortified the banks of the Severn during his conflicts with the Silures and Ordovices.

Castle-field was perhaps so named from the Romans having encamped there; and a town and bridge are said to have once existed at Hawkbach. Roman coins have been found on the spot.

Rowley Regis forms a peninsulated promontory to this hundred, and consists of an extensive range of hills, bounded on the east and west by Worcestershire, on the south by Hales-Owen, and north by Tipton and West Bromwich, in Offlow hundred, except a small neck of land between Bradley and Dudley, which joins it to that of Seisdon. The manor belongs to Lord Dudley.

This parish is principally composed of an insulated mountain, terminated by various peaks, pikes, or summits, the loftiest of which, called Turner's-hill, is the highest ground in the south of Staffordshire: the other points of Rowley are Oakham and Corney-hills. This mountain has for its basis a singular species of quartzose stone of the basaltic or granite kind: it is called Rowley rag-stone, devoid of grit, and not at all calcareous.

The church is a rude fabric of stone, and contains nothing remarkable: it is a chapelry annexed to Clent.

In 1794 was found at Rowley, in pulling down a stone wall, an earthen pot of a globular form, which contained about 1200 Roman silver coins of 40 different sorts, many of them fine impressions of the Roman Emperors, and some of Galba and Otho.

CLENT is composed of a group of lofty hills, bounded north and south by Worcestershire, east by Hales-Owen, Shropshire, and west by Brome, in this county: some particulars of Clent-hills, and the Rowley-stone, are given in the "General View of the County."

The parish contains two manors, Upper or Church Clent and Nether Clent. Tradition asserts that the Romans and Britons had an engagement on Clent-heath, where are tumuli or lows.

Clent is noticed by ancient writers as the place where Kenelm, King of Mercia about the year 820, was murdered by the orders of his elder sister Quendreda, but who failed in the object of her ambition, the Mercians having placed Cenulph, her uncle, on the throne. Shenstone pathetically recounts the fate of the "princely boy," in his 23d Elegy.

On the east side of Clent-hill is St. Kenelm's chapel, an ancient fabric, erected in memory of the royal youth of that name, who received the honour of canonization: the interior is plain, and consists of only one aisle. The tower is a very elegant piece of Gothic architecture, richly adorned with niches and pinnacles: it is a donative of Hagley, and generally given to the rector of that place. At the east end of the chapel-yard is a fine spring, much resorted to in former times for its medicinal virtues. It is probable that Kenelm was murdered in the field called Cowbach, but buried here. Although this place consists of only a few farm-houses, and the roads to it are deep and bad, there is an annual fair for cheese, &c. on the 29th of July. At the wake here, called Kenelm's wake, there is a singular custom among the people of throwing crabs at each other.

The church, dedicated to St. Leonard, is a living in the presentation of the Crown, but the tithes of the rectory are in possession of J. Amphlett, Esq. of Clent-hall, the vicar being entitled to small tithes: it consists of one small aisle to the south, with a handsome stone tower (containing six bells) at the west end. The whole body of the church is covered with wooden shingles, a sort of decoration common to the Saxon and early Norman churches.

Brome.—This village is situated on the west side of Clent, and takes its name from the great quantity of broom formerly growing there. The church, a small neat structure, is a rectory, dedicated to St. Peter, in the patronage of Sir Joseph Scott, who owns an estate here, and the ancient mansion called Harborough, to which the Rev. Thomas Dolman, rector of Brome, succeeded by marriage with Mary, a younger daughter of Wm. Penn, of Harborough, gent., Ann, the eldest, being wife of Mr. Shenstone, of the Leasowes, and mother of the celebrated poet of that name, who spent many of his juvenile hours in this village, and was a warm admirer of Miss Dolman, his cousin, of whom he sings in one of his earliest pieces, beginning "In Brome so neat in Brome so clean," which Mr. Shaw has inserted at length.

TOTMANSLOW HUNDRED.

This Hundred is the northern division of the county, and contains that elevated district called the Moorlands, invaluable for mines of copper, lead, and coal, and remarkable for containing the sources of six rivers.

Totmanslow hundred contains twenty-nine parishes and parish-churches, besides eleven chapels of ease belonging to the Establishment, and several meeting-houses or chapels for the Dissenters. It is bounded on the north-east by Derbyshire; on the south-east by the Hundred of Offlow; on the south and south-west by Pirehill Hundred; and on the north-west by Cheshire.

This hundred contains 37,844 inhabitants, principally employed in agriculture, the silk manufacture, and working in the mines.

This division of Staffordshire contains not only a great diversity of surface but of soil; including the highest hills, some of them chiefly composed of rock, or overgrown with heath, and others rich in mines, or covered with trees; and it also contains part of the fertile fields and meadows of Dove Vale, which is the boundary between this county and Derbyshire.

UTTOXETER

Is a handsome market-town of Totmanslow South, situated on a gentle eminence, near the western bank of the Dove. It is 14 miles distant from Stafford, and 136 from London. In the year 1811, Uttoxeter contained 605 inhabited houses, 628 families; 1376 males, and 1779 females: total of inhabitants 3155.

This town is undoubtedly a place of great antiquity, and from its inviting situation on the bank of a river, was probably inhabited before the invasion of the Romans. The most ancient record in existence, however, is dated in the year 1252, when Earl Ferrers granted a Charter to the Burgesses of Uttoxeter.

The situation of this town, on a dry and moderately-elevated soil, is equally pleasant and healthful; it is considered favourable to longevity, and is seldom visited by epidemical diseases. It contains fourteen streets and lanes. The houses in general are well built of brick, and commodious. The wharf belonging to the Grand Trunk Canal Company, with several large warehouses enclosed by a brick wall, and situated at the northern extremity of High-street, has contributed much to the prosperity of this small but flourishing town, which is thickly inhabited, and exhibits undoubted proofs of the opulence of its merchants, tradesmen, and inhabitants in general. There are several neat modern mansions of brick, built in the vicinity of the wharf, and about a quarter of a mile westward of it, stands the work-house, a neat, clean, and comfortable asylum for the indigent, with a large and well-cultivated garden.

The market is held on Wednesday, according to a Charter obtained by Thomas Earl of Lancaster, in the second year of the reign of Edward the Second, A. D. 1308. It is a very considerable corn-market; the grain is not pitched, but sold by sample, and large quantities of corn are sent to different parts of England from this town by the canal. It is also a great market for butter, poultry, bacon, and all kinds of provisions; and several higlers, and other chapmen, come from the Potteries, Walsall, and Birmingham, and purchase and carry away large quantities of the produce of this fertile neighbourhood. The clerk of the market regulates the weights and measures, and endeavours to prevent forestalling, by ringing a bell at half-past ten o'clock; but the farmers evade this by taking their goods to the inns, where the chapmen from distant places privately purchase them.

There is a fair held in the town every fortnight, for the sale of cattle and merchandize.

Uttoxeter is the residence of the High Constable for the Hundred, and a meeting of the Justices is held every fortnight, alternately, at the White Hart and Red Lion inns. There is only one public brewery in this town, but abundance of excellent ale is also brewed by the different innkeepers. There are two well-established banks in Uttoxeter, besides a branch of the Burton and Staffordshire bank, where business is transacted two days in the week.

The Church is an ancient fabric of stone: it consists of a nave, chancel, and two aisles; the roof is covered with lead, but the exterior part of the walls is much out of repair. The steeple and spire is about 170 feet high, and is a conspicuous object from the eminence between Mayfield and Ellerston above Dove-vale, at the distance of seven miles. During a thunder-storm, about two o'clock on Sunday, February the 6th, 1814, the spire of Uttoxeter church was struck by lightning, which forced a large stone out about half way down, passed into the belfry, and penetrated through the wall of the chancel on the south side of the window. In consequence of this accident, part of the spire which had been damaged by the shock was taken down and re-built: when it was finished, and the gilt globe, cross, and vane set up, Margaret Allporte, then a chamber-maid at the Red Lion inn, ascended, stood on the ornamental stone-work under the globe, and kissed a young man (one of the masons), who stood on the opposite side, in the presence of a multitude of spectators. This female adventurer afterwards descended in safety, though it may truly be said that she ventured her neck for a kiss.

Uttoxeter Church is dedicated to St. Mary. It is a vicarage: the patrons are the Dean and Chapter of Windsor, and the present incumbent is the Rev. —— Fowler.

There are several ancient monuments in this church, particularly an altar-tomb, with a mutilated figure in alabaster.

At the east side of the south aisle, a mural monument of marble is erected, with the following inscription:

"Thomas Oldfield, Esq. Major in the Marines, who fell during the memorable defence of St. Jean d'Acre, in Syria, by Sir Sydney Smith, against General Bonaparte and the army of Egypt, while leading a sortie made by the garrison on the 7th April, 1799, for the purpose of destroying the enemy's approaches. Ætat. 43."

The east side of the north aisle is the cemetery of the Kinnersley family, of Loxley, and contains several monuments and inscriptions: Part of the chapel of the Minors family, of Hollingbury Hall, is now used as a vestry, by permission of the present descendant. A vault in the church-yard was the burying-place of the Degge family. Sir Simon Degge, the antiquary, was a native of Uttoxeter; he died at the age of 92, and was buried in this vault. The last male of the family, who was a pauper, died and was buried here about five years ago. The inscriptions on the tomb-stone have long been effaced by the footsteps of boys, and by the hand of time.

The vault of the Gardner family is in the church-yard, and a tomb-stone over it contains the following epitaph:

"In memory of Lieutenant-Colonel Gardner, late of his Majesty's eleventh regiment of dragoons, in which he served with honour from a cornet, and died lamented, Aug. 1, 1762, aged 71 years. His widow, for the sincere affection she had for him, caused this stone to be erected."

We are informed in the manuscripts of Sir Simon Degge, that Uttoxeter was remarkable for the longevity of its inhabitants. In his records, dated 1726, he states that "in three weeks three men and two women were buried here, aged from 82 to 94." The only instances of longevity worthy of notice in the monumental inscriptions in the church-yard are, Samuel Bell, aged 96; John Hill, who died in November, 1814, aged 91; and Catherine his wife, who died in the same month and year, aged 86. But undoubtedly several old people have from time to time died in the town and neighbourhood, who have been buried in Uttoxeter church-yard, without a "frail memorial" of their longevity.

Besides the parish-church, Uttoxeter contains a meeting-house for the Quakers, who are both numerous and respectable in this town; a large chapel for the Calvinists, and another for Methodists.

There is also a free-school in Uttoxeter, founded and endowed by that eminent mathematician, Thomas Allen.

The inhabitants of Uttoxeter and its vicinity, derive much of their opulence from the fertile pastures and meadows on the banks of the Dove. They include many hundred acres of land, composed of deep rich mellow loam, impregnated with a fertilizing sediment of mud and calcareous earth, deposited from time to time by the inundations of the river. The herbage is very fine, without any intermixture of rushes or other aquatic plants, and principally consists of grasses of the common sorts.

The plain on the Staffordshire bank of the Dove, opposite Uttoxeter, is nearly a mile in breadth, and comprises several thousand acres of luxuriant pasturage for black cattle, sheep, and a few horses. A very small proportion of this extensive space is fenced-in for hay, in consequence of the uncertainty and suddenness of the inundations of the Dove, for a great fall of rain, or the sudden thaw of snow in the Moorlands, causes a rapid and resistless flood, which soon overflows the banks of the river, and covers the level fields to a great extent; insomuch, that it requires much vigilance in the proprietors of flocks and herds to preserve them from drowning. The graziers, on an average, pay two pounds an acre for this excellent pasturage.

A CHRONOLOGY OF REMARKABLE EVENTS AT UTTOXETER, FROM THE YEAR 1252 TO 1689:

Transcribed from the Church Wardens' and Constables' Accounts, and other authentic Papers.

1252. Earl Ferrers's Charter granted to the Burgesses of Uttoxeter is dated August 15, 1252.

1308. In the second year of King Edward the Second, Thomas, Earl of Lancaster, (eldest son of Edmund, second son of Henry the Third), obtained a charter for a market every Wednesday at Uttoxeter, and a yearly fair on the eve, day, and morrow of St. Mary Magdalen, July 21st and 22d.

1642. Paid to them that swept Mr. Ward's hall for king Charles the First, *one shilling*.

——. Trained soldiers' pay here, and to Stafford to wait on the King, £34. 6s.

——. Charges when the country went against Stafford the first time, 3s. 4d.

——. Doveridge men's charges when they came to guard the town, *one shilling*.

——. Paid for match, powder, and coals, for some of the town ends, in February, 1l. 5s. 1d.

——. John Sherratt, for leading clods five days to the bulwarks, 16s.

——. Bestowed on the countrymen when they came to guard the town, when the soldiers went to Lichfield, 2s. 3d.

——. Paid to carpenters and labourers at the bulwarks, 5l. 13s. 2d.

——. For watching Lord Stanhope and his son at the Crown, eleven shillings.

——. To a prisoner who came from Hopton battle, *four pence*. N. B. Fought near Stafford.

——. Paid for match, powder, candles, bullets, and coals, for some of the town ends, in March, 2l. 10s. 9d.

——. For removing one of the bulwarks, and for entrenching, 12s. 4d.

Bestowed on Loxley men when they came to trench, *four pence*.

——. November. Paid for drink bestowed on Loxley and Tean men when they came to help the town against Worthley, 2s.

——. Paid for drink for Marchington, and Crakemarsh, and Creighton men, 2s.

1642. Paid to Johnson for prisoners which were Werthley's men, 6s. 7d.

1643. January. Fire and candle for the watch, being Derby men, 1s. 5d.

——. Paid for butter, cheese, five quarters of pease, and oats, to Tutbury and Lichfield, 19l. 2s. 8d.

——. Paid two men for going to Tutbury with provisions, 1s. 4d.

——. Paid William Ferrall and others who watched the ordnance at the Crown door, 1s. 6d.

——. Paid Richard Cartridge for watching on the church, 8d.

——. February. Charges when the country went against Stafford twice, 5l. 11s. 4d.

——. Paid to the ringers when king Charles the First was here, five shillings.

——. June. Paid to a townsman when he went to guard a field-piece to Tutbury, 8d.

——. Charges to Wooton-Lodge with a horse-load of bread, one shilling.

——. July. Paid to workmen for pulling down the bulwarks and taking them away, 10s.

——. October. Paid down when Colonel Hastings was in town, 5l.

——. December. Ale for the Captain and his men, who brought a warrant from Lord Loughborough for 50l. three shillings.

——. December. For 25 strikes of oats, which were sent for by warrant to Tutbury, 2l. 4s. 10d.

——. December 27th. Provisions sent to Ashburn by warrant, from Colonel Dudley, in the king's army, 3l. 0s. 0d.

——. For ale, tobacco, wood, coals, and hay for guards and centinels, 2l. 10s. 4d.

1644. January. Paid Captain Vernon what he laid upon the township, 11l.

——. February 26. Paid to the Committee's Treasurer at Stafford, for weekly pay in money and returns, 100l.

——. March 11. Paid to Tutbury Castle in money and returns, 50l.

——. April. For a rope to hang the man who killed John Scott, and for a cord to pinion the prisoner, 1s.

——. May 7th. For 8 cwt. 2 qrs. 7lbs. cheese to Tutbury, 7l. 15s. 10d.

1544. September. Levies upon the town of Uttoxeter, 608l. 13s. 2d.

1645. January 1st. The parishioners of Uttoxeter paid 6l. 2s. 4d. weekly, besides extraordinary expences to a considerable amount.

——. Paid to Prince Rupert's cook for his fee, 5s.

——. August 4th. Hay, oats, beer, tobacco, wood, and coals, for the guard three nights, when the Parliament forces went against Tutbury Castle, 1l. 19s. 8d.

——. For guides to go a scouting three nights, 9s.

——. October. For a sheet, making the grave, ringing, beer, and for burying the soldier that was slain in the street, 4s.

1646. February 5th. The Constable's accounts amounted to 975l. 7s. 1d.

——. October 6th. Paid to two men for blocking up the town ends with carts, *sixpence*.

——. Beer for soldiers for barricading the town ends, 1s.

——. February 8th. Paid Commissary Ward three weeks' contribution, 14l. 13s. 8d.

——. Paid General Egerton at Tutbury, 30l.

——. February 22d. For carrying two soldiers to Caverswall, who were maimed in the High Wood beyond Uttoxeter, 2s. 6d.

"On the 15th of February, 1646, a sharp encounter betwixt a party of the King's troops from Tutbury Castle, and a party of the Parliament's from Burton-house, in Derbyshire."—*Wharton's Gesta Britannica.*

——. February 14th. For two horses and a man to carry bread and cheese to Tutbury *in the night, being in great want*, 3s. 4d.

——. February 22d. Paid to parson Langley's soldiers in bread and beer, 3s.

——. March 30th. For provisions to the leaguers (or Parliament forces), at Tutbury, 7l. 4s. 6d.

——. April 8th. Ditto, 11l. 2s. 9d.

Tutbury Castle was taken by Sir William Brereton, in 1646.

——. April 20th. Paid Captain John Cloyd for pulling down the bulwarks at Tutbury, 3l.

——. August 26th. Paid to Ashburn by the Churchwardens of Uttoxeter when the town was infested with the plague, 3l.

——. Paid to the inhabitants of Clifton when the plague was there, 5l.

——. October. For quartering General Fairfax's soldiers, 20l.

1646. October 13th. For quartering Colonel Cromwell's soldiers, 20l.

——. December. Quartering Colonel Oakley's men, 13l. 2s. 6d.

1647. May 12th. To 15 men for pulling down Tutbury Castle, 2l. 10s. 4d.

——. October. To 46 travellers, or Egyptians, with a pass from Parliament, to travel by the space of six months together to get relief, 4s.

1648. May. For two men watching in the steeple when the town was fearful of an insurrection, 1s. 4d.

——. For quartering 16 of Colonel Monk's soldiers, 13s.

1651. For a warrant to fetch in and search for Papists' and delinquents' arms, 4d.

——. August 11th. To Robert Goodall for horses, saddles, and bridles, towards *horsing* the militia, 5l. 15s. Five horses and five men were provided by the town of Uttoxeter. Each of the men were paid 7s. 6d. Their names were, John Clarke, Francis Allen, Phillip Needham, Richard Wilkinson, and Thomas Vernon.

——. August 20th. To ale, bread, and pottage, to relieve the Scotch prisoners, taken by Lieutenant-Colonel Downes, whilst in custody, 2l. 8s.

——. To another body of Scotch prisoners, 1l. 10s.

1658. Note: 67l. 10s. was paid to the army yearly for several years.

——. November 5th. Paid the ringers 5s.

——. And for proclaiming the Lord Protector, 1s.

1660. May. Paid the ringers when king Charles the Second was proclaimed, 5s.

——. For painting the king's arms, 19s. 2d.

——. Ringing on the Coronation-day, 8s.

1662. We paid for 622 fire-hearths.

——. May 16th. A warrant for Quakers, 1s.

1667. Royal aid at Christmas, 44l. 1s.

——. At Midsummer, 44l. 1s.

1672. One month's pay to the Government, 14l. 13s. 6d.

——. April 8th. Two month's pay, 29l. 7s.

——. July 30th. A great fire happened in the rear of Richard Cladd's house, which consumed most of the lower part of the town.

——. Paid Mr. Edge and Mr. Chamberlayne for going to Stafford with a petition to the Judge about the fire, 15s.

1632. October 6th. Repairing stocks and cuckstool, 10s.

1688. For wine and ale at the Crown the thanksgiving-day for the Prince of Wales (son of James the Second), and the day after, when the Bishops were released, 2l. 2s. 6d.

A load of coals that day for bonfires, 8s. 6d.

Ale at High-street bonfire, 4s.

Ditto at the Church-yard bonfire, 4s.

Ditto at the Cross bonfire, 4s.

Ditto for the watchmen and others the night after the Bishops were released, 9s. 6d.

Given to two drummers at the same time, 2s.

——, June 8th. King James the Second sent seven Bishops to the Tower; but they were afterwards released.

——. William Prince of Orange, afterwards king William the Third, came to England to oppose the Popish proceedings of king James the Second, whose daughter, Mary, he had married. Great exertions were made by the Protestants to dethrone James. They employed a number of active, intelligent, and resolute men to run or ride through all parts of the kingdom in the day-time, crying "*Fire and Sword! the French are coming!*" which filled the people's minds with terror. Uttoxeter shared the alarm with the rest, as may be seen by the following record:

1688. December. Given to four lads for gathering coals, and for making fires when the alarm was, 4d.

——. December 26th. Paid Mr. Moreton for ale for strangers and townsmen when the alarm was, 1l. 1s.

When the Prince of Orange landed in 1688, king James fled from London: on the 12th of December he was seized and brought back to Whitehall, but on the 23d of the same month he fled out of England. King William and Queen Mary were crowned April 11th, 1689.

The late distinguished *Admiral Lord Gardner* was born here on the 12th April, 1742. He was the eighth son of Lieutenant-Colonel Gardner, of the 11th regiment of dragoons. Having at an early period shewn a strong bias towards the naval service, he was rated, when 14 years old, as a midshipman, on board the Medway of sixty guns, then under the immediate orders of captain Sir Peter Denis, an officer of distinguished merit. In this vessel Mr. Gardner remained for two years, during which time he was present in an action, in which the Duc d'Aquitaine French ship of the line was taken. From the Medway, our young midshipman afterwards accompanied his captain, first on board the Namur, and afterwards into the Dorsetshire. In the former he served under Admiral Hawke, during the expedition against Rochfort; and, while on board the latter, was present at the capture of the Raisonable, on which occasion Captain Denis put in practice the plan now adopted by the new school, of not firing a single ball till within a few yards of the enemy's ship. He likewise bore a share in the general engagement which took place off Belleisle in 1769, between the British and French fleets, commanded by Sir Edward Hawke, and the Marshal de Conflans. Mr. Gardner, having now been five years in constant service, was appointed lieutenant on board of the Bellona, after the customary examinations. In this station he distinguished himself at the capture of the Le Couragenx, whereupon he was raised to the rank of master and commander, and appointed to the Raven of sixteen guns. After the lapse of four years, he was made post in the Preston of fifty guns, which had been fitted-out as the flag-ship of Rear-Admiral Parry, whom he accompanied to Port Royal in Jamaica. During the whole time of his being stationed here Great Britain was at peace with all the nations of Europe, so that the only circumstance which occurred, requiring notice in this sketch, was his marriage with Susannah Hyde, only daughter of Francis Oale, Esq. a West India planter. This lady having soon brought him a numerous family, and being himself ambitious of rising in the service, he made every effort to obtain an appointment as soon as the American contest began. Accordingly he was nominated to the command of the Maidstone frigate, in which he sailed for the West Indies early in 1778; and, in the course of that year, was so fortunate as to make a rich capture on the coast of America. On the fourth of November he fell in with the Lion, a French man of war, having on board fifteen hundred hogsheads of tobacco, and after a severe action compelled her to surrender. With this prize he sailed for Antigua; and was, soon after his arrival, promoted by Admiral Byron to the command of the Sultan of 74 guns. In the drawn battle which was fought some time subsequent with the French fleet under Count de Estaing, off the island of Grenada, Captain Gardner led the van, and greatly distinguished himself. His ship, however, suffered so much, that he was ordered to Jamaica, from whence he shortly after sailed for England, when the Sultan was discharged. He did not, however, remain long out of commission, having been appointed to the Duke in the course of a few months, with which ship he sailed to join the fleet in the West Indies, then under the orders of Sir George Rodney, and was fortunate enough to arrive in time to participate in the glorious victory of the twelfth of April, 1782. On that memorable day, his ship was the first to break through the enemy's line of battle, according to the new plan of attack, suggested by Mr. Clerk, of Elden, and then for the first time put in practice. At one period of this action the Duke, in conjunction with the Formidable and Namur, had to sustain the fire of eleven of the enemy's ships. Soon after this triumph the American war terminated, and peace continued for several years to shed her benignant influence over the several nations of Europe. During this period, Captain Gardner was employed in different capacities. For some time he acted as Commodore on the Jamaica station, and in 1790 was appointed a lord of the Admiralty, when he likewise obtained a seat in Parliament.

In the year 1793, having been raised to the rank of Rear-Admiral of the blue, he hoisted his flag on board the Queen of 98 guns, in which he sailed as Commander-in-chief to the Leeward Islands. Soon after this event, finding the disputes between the republicans and royalists in the colony of Martinico to run very high, and being earnestly pressed by the latter to effect a descent on the island, accordingly Major-General Bruce landed with 3000 men; but that officer judged it expedient to re-embark again, almost immediately, being satisfied that the republican party was too strong to afford just hopes of success in the royal cause. Admiral Gardner now returned to England, and the following year bore a part in the action of the 1st of June, under the gallant Earl Howe. On this occasion his conduct was conspicuous in the extreme, his ship having suffered more than any other in the fleet, with the exception of the Brunswick. In consequence, he not only was particularly thanked by the Commander-in-chief, but was appointed Major-general of Marines, and created a Baronet of Great Britain. On the 22d June, 1795, Sir Alan was present at the action off Port l'Orient, when the French fleet only saved itself from total destruction by a timely flight. Two years after this event, when a dangerous mutiny had broken out at Portsmouth, he manifested a degree of firmness and resolution, during that trying period, worthy of his high character as a British naval officer. From this time he continued to serve in the Channel fleet till the close of the year 1799, when he was sent with sixteen sail of the line, to reinforce the fleet off Cadiz, and in the Mediterranean. Perceiving, however, that little danger was to be apprehended in these quarters, he returned, with nine sail of the line, accompanied by the convoy from Lisbon.

In 1800, we once more find him serving in the Channel fleet, but he was soon afterwards appointed to succeed Admiral Kingsmill, the naval commander in Ireland, being previously raised to the dignity of an Irish peer. This command he continued to hold till the year 1807, when he hoisted his flag as Admiral of the Channel fleet; which ill health, however, soon compelled him to relinquish. He died in 1810, and was buried in the Abbey church of Bath, with the grandeur and solemnity due to his rank and merit.

Lord Gardner's political career was not distinguished by any circumstance of great moment. He sat in three successive parliaments. His first election took place in 1790, when he was returned one of the representatives for the town of Plymouth. In 1796 he was colleague to Mr. Fox in the representation of Westminster. On this occasion he was opposed by Mr. John Horne Tooke, whose wit, satire, and eloquence, were more alarming to the Admiral than a shower of cannon-balls from an enemy's fleet. Notwithstanding this circumstance, however, he once more offered himself as a candidate for the same city, and was again successful. At this time Mr. Fox, in addressing the electors, said, "A noble Admiral has been proposed to you. I certainly cannot boast of agreeing with him in political opinions; but whom could the electors pitch upon more worthy of their choice than the noble Lord, in his private character universally respected, and a man who has served his country with a zeal, a gallantry, a spirit, and a splendour, that will reflect upon him immortal honour."*

Bramshall is a small parish and village, about three miles w.s.w. of Uttoxeter. The number of houses in the whole parish, in 1811, amounted to 29, inhabited by 33 families. Males 76, females 79: total of inhabitants 155.

The village is situated on the summit of an eminence, and contains some good substantial farm-houses, and smaller tenements. The Church is a Gothic structure of stone, with a small wooden belfry, containing three bells. The edifice itself might easily be mistaken for a barn when viewed from the turnpike-road, and its belfry for a dove-cote. It is dedicated to St. Lawrence, and is a rectory under the patronage of Lord Willoughby de Broke. The

* Imperial and County Annual Register, for 1810.

present rector is the Rev. —— Sneyd, and his curate, the Rev. Thomas Astle, resides at Uttoxeter. This parish is fertile, and productive of abundance of corn and grass.

KINGSTON is a parish in the s.w. extremity of Totmanslow South, and the village is situated about three miles and a half from Uttoxeter. The parish contains 67 houses, 67 families; 152 males, 183 females: total, 336 inhabitants.

The Church is a small ancient structure of stone, with a low square tower of brick. This fabric is in a neglected, if not in a ruinous state, and supported on the southern side by three buttresses of brick. The interior is plain: the pews of oak. There are no remarkable monuments in the church-yard, but such is the mischievous, and it might be added, the devilish disposition of boys, that they have recently defaced several of the inscriptions on the tomb-stones.

This church is a donative in the gift of the Chetwynd family: the Rev. John Hilridge is the present curate.

The land in this parish is in general fertile, particularly in the vale through which the river Blithe flows in its course to Blitheford, and thence to the Trent. The upland is sound gravelly loam.

The country between Kingston and Gratwich is pleasant and well enclosed. The valley through which the Blithe winds is beautiful, and productive of abundance of corn and grass; but the meadows were much injured, and some of the hay spoiled, by the inundations of the river during the wet summer of 1816. There is a corn-mill about mid-way between the villages of Kingston and Gratwich.

GRATWICH is a small parish and village, containing 18 houses, 19 families, and 110 inhabitants. It is situated south-west of Uttoxeter, and about four miles distant from that town. The village is built on a plain; and the soil of the parish, which is a gravelly loam, is fertile. The Church is small, built of brick, with a wooden belfry; and the interior is without a gallery, or monuments. It is dedicated to St. Mary, and is a rectory in the patronage of Earl Talbot, of Ingestre. The Rev. Edward Lewis, the present incumbent, lives at Cotton.

There is one instance of longevity recorded on a tomb-stone in Gratwich church-yard: to the memory of Hannah Woolfe, of Shafferlong, who died in 1793, aged 91 years.

LEIGH is an extensive parish, containing five villages and hamlets. In the year 1811, this parish contained 181 houses, 184 families, 471 males, 466 females; total of inhabitants 937: and since that period there has not been a great increase of the population. The parish of Leigh is situated about four miles westward of Uttoxeter; the soil a good loam, and very fertile. The river Blithe, which passes through this parish, contributes much to its fertility. Lord Bagot is possessed of a considerable estate here, and Mr. Evans, of Derby, has a pleasant rural mansion and a considerable estate, with a plantation of fir trees. But the principal part of the land in this parish belongs to several independent free-holders, who have their farms in the highest state of cultivation.

Church Leigh, as the principal village is called by way of pre-eminence, contains several good houses, particularly the residence of the curate.

The Church is a very beautiful specimen of Gothic architecture; built of stone raised from a quarry in the neighbourhood. It is built in the form of a cross; the arched roof of the nave is supported by ten lofty arches, and massy columns, each indented in four divisions. The square tower, which rises in the centre of the edifice, is supported by four noble Gothic arches, twenty feet high. This has a very fine effect, which is much heightened by the admission of light from the two principal windows of the north and south aisles, through the body of the church. The chancel is large, with three windows on each side, and a large square window at the east end, adorned with paintings on glass. All the windows in the chancel are embellished with figures painted on glass, representing Apostles, Martyrs, and Bishops. There are no ornaments in the chancel: the communion table is of oak and plain, surrounded by a screen of carved wood.

The south aisle contains three monuments of the Ashenhurst family, and a very ancient altar tomb, with two figures in alabaster as large as life; one of a knight in armour, with his head resting on a helmet, and his feet on a lion. His hands are raised in supplication. A female figure lies by his side, in a long robe; her head rests upon a pillow, and her hands are also raised as in prayer. This aisle, which is called Ashenhurst's aisle or chapel, is much disfigured by mortar, and pieces of broken forms and pews, insomuch that it is like a lumber-room.

The north aisle, which is called Wood's aisle, contains the tombs of the Woods, of Fole-hall, in this parish. Here is also a neat mural monument of white marble, in memory of William Harrison, of Over Tean, who died in the 84th year of his age.

A small board fastened to the wall, contains the following record, in black letter:

"In this church lieth the body of THOMAS HARVIE, late of Withington, who died and was buried the 26th of June, Anno Domini 1630. Who gave, by his will, 160 pounds to buy land, and the rent thereof to buy Medley cloth for ye poor of ye parish of Leigh yearely for ever, with and for such uses as are more at large expressed in his will. Which was performed and done by his executors."

Another benefaction to the same purport, dated in 1641, provides 24 yards of Medley cloth yearly, to be given alternately, one year to six men to make coats, and another year to six women to make gowns. This Medley cloth, as it is called, was striped, and seems to have been fashionable in England in the time of Shakspeare for Jaques, in "As you Like it," exclaims, "*Medley's* the only wear." The benefaction of clothing is given yearly, but in plain cloth. Among other charitable legacies, is that of John Blurton, who, in 1743, bequeathed £6. 4s. to be given to the poor of the parish of Leigh, to purchase shoes at the rate of 3s. 9d. a-pair; and that sum is now given in money to each individual entitled to it.

Mrs. Elizabeth Whitehall gave ten pounds per annum to the parish of Leigh for ever, "to set out two poor children apprentice."

The following curious caution to ringers is painted on a board, and attached to one of the columns in the church:

"If that to ring you do come here,
You must ring well with hand and ear;
And if a bell you overthrow,
Fourpence is due before you go;
And if you ring in spur or hat,
Two pots of ale are due for that;
And for a pledge to make it sure,
Your hat shall be the forfeiture;
Our laws are old, they are not new,
Therefore the clerk must have his due."

The Church is dedicated to All Saints, and is a rectory, under the patronage of Lord Bagot, whose brother, the Hon. and Rev. Richard Bagot is the present rector, and the Rev. Henry Thomas, curate. This benefice was formerly estimated at £300., but it is now, from the improved state of the parish, worth £700. a-year.

The parish of Leigh is extensive, and contains Church Leigh, the principal village; another hamlet called Lower Leigh, on the banks of the Blithe; Upper Leigh, which contains several farmhouses, and Park Hall, a mansion belonging to Walter Evans, Esq.

of Derby; Dodsley, a hamlet of this parish, south of the Blythe; and Field, a village containing nine houses.

Church Leigh contains a handsome and commodious school-house, originally founded and endowed by a Mr. Allen, for the instruction of boys who are the sons of parishioners. About 70 boys are instructed in reading, writing, and arithmetic. They are admitted at eight or nine years of age, and continued till they are fit to be put apprentice to some trade, or to return home and assist their parents in the farm. The funds of this institution are now sufficiently ample, amounting to about £100. a-year. The old school-house was taken down, and the present enlarged building erected, in 1806.

Another free-school was founded by the beneficence of Mr. Walter Evans, of Derby, on his own estate here, for the instruction of 60 boys and girls or upwards, who are not the children of parishioners, yet reside in the parish of Leigh: and thus ample means for the education of the children of the poor are provided by the philanthropy of two individuals. The Hon. and Rev. Richard Bagot, the rector, supplies this latter free-school with books, paper, pens, &c.: and a Sunday-school for all the neighbouring children, is also opened weekly, at this institution; where from 80 to 100 children receive the benefit of sabbathical instruction.

As a proof of the fertility of the soil in this parish, it is recorded that in the year 1680, an elm was felled on the estate of Sir Harvey Bagot, near the village of Field, which produced the following quantity of timber: it contained 96 tons of solid timber, and 61 loads of firewood. Two able and experienced woodmen were employed five days in felling it. The height of this tree was 120 feet, and it was 25½ feet in circumference in the middle. The workmen were obliged to put two saws together, and place three men at each end to cross-cut this tree, and the size seemed so incredibly enormous, that it was thought necessary to attest the particulars by the signature of Sir Harvey Bagot, the proprietor, and all the other persons who were eye-witnesses of the fact. Perhaps there would be little difficulty in finding even a larger tree in the vast wilds of America, but we have no record of any elm of equal growth being felled in Europe.

CHECKLEY AND TEAN.—This parish is large and populous: the soil is good in general, particularly the land on the banks of the small river Tean, which flows through it. The village of Checkley is situated four miles and a half north-west of Uttoxeter; it consists of a few farm-houses and smaller tenements, but the most conspicuous object is its ancient and lofty church and high square tower of stone. The tower contains a clock and six bells, and the grey appearance of the exterior of the whole edifice, with its windows strengthened by stone-work, and its high flat roof with battlements, impress the idea of antiquity and durability. The interior, though much out of repair, presents a variety of interesting objects to the observer. The roof of the nave is high, of oak, and supported by eight Gothic arches, and round pillars. A gallery on the north side contains several pews of oak, and the pews and forms below are made of the same material, but are now in a ruinous and neglected state. A very high arch supports the roof at the west end of the chancel. This part of the church is lighted by two windows on each side, and a large east window adorned with curious figures painted on the glass. There is an altar tomb in the chancel, with the recumbent figure of a knight in armour, and a female lying by his side. They are of alabaster, and part of the inscription on the tomb is broken off. A tombstone near the communion table bears the following inscription:

"To the memory of the Reverend James Whitehall, rector of this place twenty and five yeares, who departed this life the second daie of March, 1644.

"White was his name, and whiter than this stone,
In hope of joyfole resverrection;
Here lies that orthodox, that grave divine,
In wisdom trve, virtve did soe clearly shine;
One that covld live and die as he hath done,
Svffer'd not death bvt a translation;
Bvt ovt of charitie I'le speake no more,
Lest his friends pine with sigths, with teares the poore!"

Three stones which stand in the church-yard, have for ages puzzled our antiquaries, none of whom have been able to discover why they were set up.

Checkley Church is dedicated to St. Mary and All Saints. It is a rectory: the Rev. William Langley, the present rector, is also the patron of the benefice.

Upper and Lower Tean are two populous villages in this parish. They contain several new tenements, some of which are handsome mansions. The principal building, however, is the extensive tape-manufactory of John and Nathaniel Philips. Tean-hall, which was the ancient mansion of the Ashley family, now belongs to Messrs. Philips, with the whole extensive manor of Tean.

The tape manufactory was first established at Tean in the year 1747, when an ancestor of the present proprietors brought over two Dutchmen who were skilled in the manufacture. It continued for several years in a very limited state, for many obstructions lay in the way of success; particularly the skill and enterprize of the Dutch, who had then one thousand tape-looms in full employment. British perseverance, and superiority of fabric, at length prevailed; and the manufactories of tape established in this village, and at Cheadle and Kingsley, are now the most extensive in Europe. About 800 looms are employed at Tean, and in the vicinity; 120 at Cheadle, and 50 at Kingsley. The yarn is imported brown from Russia, Germany, and Ireland; and considerable quantities are also brought from Scotland and Yorkshire. It is bleached at the extensive bleach-works on the banks of the river Tean, near the village, and it is computed that from two to three thousand individuals, including women and children, are employed in the different branches of the manufacture.

A chapel for the Calvinists, with a Sunday-school, has been erected in Tean by Messrs. Philips. The chapel contains a good organ, and the Rev. Thomas Pritchard, of Cheadle, is minister.

In the year 1811, the parish of Checkley and Tean contained 341 houses, 368 families; 763 males, 935 females: total of inhabitants 1,698.

Heath-house, the residence of John Philips, Esq. is situated on a gentle eminence in this parish, and commands an extensive view of the subjacent vale of the Tean, and of Leigh Church, and the country to a considerable extent. This mansion is built of stone: it is two stories high, with neat hidden attics, and has a southern and a western front, each adorned with an elegant portico of eight Ionic columns. The house is surrounded with extensive plantations of oak, elm, fir, and ash trees. The agriculture of the estate is carried on under the improved Scotch system.

Beamhurst is a small ancient village to the south-east of this parish; and only remarkable for containing a gentleman's mansion, with a fine grove of lime trees in front. The Moorlands commence to the north-west of the parish, which contains the village of Hollington, situated on an eminence. The houses are built of stone, and thatched.

Rocester is a parish situated near the Dove, about four miles N.N.E. of Uttoxeter. It was the demesne of Algar Earl of Mercia, in Edward the Confessor's time. He left it to his son Edwin, who forfeited it by rebellion to the Conqueror, who gave it to Robert de Stafford, one of his followers. A monastery of Canons regular of St. Augustine, was founded and endowed with large possessions at Rocester, by Richard Bacun, nephew to Ranulph Earl of Chester. These possessions and liberties were confirmed to the monks by Henry the Third, in the 30th year of his reign, A.D. 1246. Some of the Stafford family, of Sandon, who settled here, were great benefactors to this monastery; and in the parish church, which formerly belonged to the monks, there are monuments of the Staffords. At the time of the Dissolution, this house was valued at £100. 2s. 10½d. per annum.* Thomas Trensham, a favourite of Henry the Eighth, obtained it of that prince, and made it his residence. It was held by his descendants in the 17th century, but not a vestige of it now remains.

Rocester Church is an ancient structure, with a square tower of stone; it stands in the middle of a field, in which there is part of a broken cross, which probably belonged to the abbey. The church is dedicated to St. Michael, and is a vicarage. The present vicar is the Rev. —— Jell: the Rev. George Hake is curate.

The village of Rocester contains several good houses, and a cotton manufactory, the property of Mr. Bridone. The confluence of the Dove and Churnet in the meadows below this village, contribute to the fertility of the soil.

The parish contains 155 houses, 168 families; 376 males, 497 females: total of inhabitants, 873. Of this number nearly one-half are employed in the cotton manufactory.

Cowbridge is a hamlet belonging to this parish: it is situated to the south-west, upon the banks of the Churnet; and the branch of the Grand Trunk Canal, which terminates at Uttoxeter, passes through it. Near this hamlet there is a wharf on the canal, with large warehouses, and a few tenements.

Mince-pie Hall is a curious mansion, situated on an eminence, with a turret and observatory.

ELLASTONE is a parish of considerable extent to the west, and contains the high Weaver Hills, with the township of Woolton to the north.

The village of Ellastone is ancient and irregularly built. It is situated about two miles north of Rocester, at a considerable distance from the west bank of the Dove. The soil is fertile, being a loamy surface, with a bottom of lime-stone.

* Dugdale's Monasticon.

The parish of Ellastone, including the townships of Calwich, Stanton, Woolton, Prestwood, and Ramshorn, contains 196 houses, 178 families; 445 males, 471 females: total of inhabitants, 916.

Ellastone Church is a handsome Gothic structure of stone, situated on an eminence near the village. The stone tower is square, and adorned with four pinnacles and battlements, with a vane in the centre. It is dedicated to St. Peter, and is a vicarage, in the patronage of Davies Davenport, Esq. The present vicar is the Rev. —— Davenport, and the present curate, the Rev. George Hake.

Between the two small windows at the east end of Ellastone church, on the outside, a very elegant mural monument bears the following inscription:

"Here lies interred the body of BERNARD GRANVILLE, who trusted in the mercy of Almighty God for the forgiveness of his sins, through the merits and mediation of Jesus Christ the Saviour and Redeemer of mankind. He was the son of Bernard Granville, and great grandson of Sir Bevil Granville, who was killed in the Civil wars, fighting for king Charles the First, on Lansdowne, near Bath, in Somersetshire. He died at Calwich, July 2d, 1775, aged 76."

Near this monument is another tombstone, with this inscription:

"Here lie the remains of the Rev. DAVID MILES, A.M. 50 years Vicar of this church; highly respected by his parishioners and acquaintance. He died in the year of our Lord 1783—of his age 87."

There are several instances of longevity recorded on the tombstones in this church-yard, particularly Samuel Banks, of Stanton, who died Dec. 19th, 1786, aged 89 years; Thomas Horobin, died in 1802, aged 93 years; and Ann Parker, aged 92.

Near the church, on the brow of a hill, is a neat modern structure, with two small wings, and a Gothic window in each. It is a national school, built by the voluntary contributions of the neighbouring gentry, and is at once ornamental and useful.

MAYFIELD is a fertile parish, in a pleasant situation in Dove vale. The village of Mayfield is two miles north-east of Ellastone, and two miles distant from Ashbourn, in Derbyshire. The river Dove, which is the boundary of the two counties, flows near the village; and on the opposite bank there is a very extensive cotton manufactory, belonging to Messrs. Cooper & Co. An ancient stone bridge of five arches near the village, presents an easy communication between Derbyshire and Staffordshire: the whole country affords a beautiful view of well-built houses, and well cultivated fields, enlivened by a considerable population. The number of houses in Mayfield parish, including the township of Butterton, is 195, inhabited by 202 families, of which 405 are males, and 531 females: total of inhabitants, 936.

The village of Mayfield contains several well-built modern houses of brick and stone. The Church is a handsome Gothic structure of stone, with a strong and high square tower, adorned with eight pinnacles. It was built by Thomas Rolleston, A. D. 1616, as appears from an inscription above the arched entrance at the west end. This church is dedicated to St. John the Baptist, and is a vicarage; the Rev. —— Belcher, who resides at Ashbourne, is curate. The roof of the edifice is supported by six large Gothic arches; the east window is beautiful; and a number of elder trees, which grow close to the church walls, give it a rural appearance. There are but few records of longevity in the church yard. One tomb-stone is inscribed with the name of Benjamin Thompson, who died in 1809, aged 88 years.

From the eminence about midway between Ellastone and Mayfield, the prospect of Dove-vale is beautiful beyond description. The country exhibits all the variety of meadows, pastures, corn-fields, and woodlands; numerous herds and flocks grazing securely; the river Dove gently flowing and winding amid the verdant meadows; a village, with its picturesque church and tower, skirted by a wood on an eminence, on the Derbyshire side of the river, and the extensive vale stretching for several miles to the south, with a distant but indistinct view of Uttoxeter; the lofty spire of its church; and the smoke of the town carried away like a large cloud by a brisk north-west breeze.

CROXDEN is a parish to the west of Rocester. The village of Croxden is about five miles north-west of Uttoxeter; but the most remarkable object in this parish is Croxden Abbey, the venerable ruins of which are to be seen about half a mile from the village.

This Abbey was founded and endowed with lands and revenues by Bertram de Verdon, in the 22d year of the reign of Henry the Second, A. D. 1176, for monks of the Cistercian order. It is situated in a fertile valley, near a small rivulet called the Peake, and was for many ages the burying-place of the Verdons.

Croxden Abbey is a very picturesque and extensive ruin. The principal entrance at the west end is yet entire, and consists of a most magnificent Gothic arch of excellent workmanship, as the ornamental stone cornices are yet undecayed, although exposed to the weather nearly six centuries and a half. Above this noble arch, part of the abbey wall rises to the height of forty feet, overgrown with ivy. This is a beautiful picturesque object, worthy of the highest efforts of the painter and engraver. About 90 feet from this ruin, another part of the abbey wall stands bare and grey, distinctly marking the length of the edifice. A small arch, ready to crumble into ruins, stands on the north side, between these two high walls. At the west-end, besides the large and perfect arch before-mentioned, there are four small arches of exquisite workmanship, and perfectly entire. The interior has the remains of a large fire-place, and an arched door-way leads to the cemetery, or court, where there is a stone coffin preserved on a level with the surface of the ground. Several other stone coffins have been discovered in these ruins, and one of lead, all of which were covered up again by Mr. Carrington, whose large and commodious farm-house is situated near the western end of the abbey, and the ruin itself is part of his farm-yard. In a recess in Mr. Carrington's garden, a stone crucifix is preserved. It is about two feet and a half high; the left arm is broken off, but preserved in the recess, and on the back of this curious relic, there is a rude representation of the Virgin and Child ensculptured, but now almost worn out by time.

Croxden Church is a small stone fabric, with a belfry of wood, containing one bell. The window on the north side is built up, and the church is lighted by three small windows on the south side, and one at the east end. This church is dedicated to St. Giles, and is a donative. The Earl of Macclesfield is patron, and the Rev. Walter Eddowes the present curate.

The parish of Croxden, including the township of Great Yate, contains 43 houses, 43 families; 137 males, 126 females: total inhabitants 263. The soil is a fertile loam, though a considerable part of the land is hilly.

Bradley-in-the-Moors is a small parish and village to the north of Croxden. It contains 15 houses, 15 families; 42 males, 41 females: total inhabitants 83.

The Church is a small structure of stone, with a tower. It is dedicated to All Saints, and is a curacy. The Rev. Walter Eddowes is curate.

Alveton is an extensive and populous parish, including the townships of Cotton, Denston, and Farley. The village of Alveton is situated to the north of Bradley, four miles N.E. of Cheadle.

This village is built upon very uneven ground, in a most romantic situation near the river Churnet. It contains several well-built modern houses, and is remarkable for the ruins of an ancient Castle, built and inhabited by Bertram de Verdon, the founder of Croxden Abbey.

After the lapse of several years, this castle devolved by marriage to the Furnivals. It continued two successions in this family, when, falling to Joan, the only heir, who married Thomas Neville, brother to the Earl of Westmoreland, he was in her right created Lord Furnival. He left by her only one daughter, Joan, who having married to John Talbot, afterwards Earl of Shrewsbury, she brought the manor and castle of Alveton into his family, and it is now the property of the present Earl of Shrewsbury.

The site of Alveton Castle is very commanding, and naturally strong, built on a rocky precipice on the south bank of the Churnet, the base being elevated about eighty yards above that river. On this side it was inaccessible, and it was strongly fortified by an entrenchment on the side next the church. The ruins consist of two towers, the most perfect of which is overgrown with ivy, with a small vane in the centre; the other is partly fallen in. A covered archway, and fragments of the thick outer wall, also remain.

Alveton Church, which stands near this Castle, above the village, is an ancient and low Gothic edifice of stone, with a strong square tower: the roof is covered with lead. The interior is heavy, and ill lighted with very small windows: the pews are of oak. It is dedicated to St. Peter, and is a vicarage, in the patronage of R. Williamson, Esq.: the Rev. Thomas Blackey, the present vicar, resides in a neat white mansion near the extremity of the village.

The parish of Alveton contains 357 houses, 358 families; 955 males, 943 females: total of inhabitants 1,898.

There is an extensive wire manufactory established at Oakamoor, on the Churnet, and another at the village, near Alveton, by Messrs. Patten and Co. of Cheadle. About 30 workmen are employed at the former, and 60 at the latter place.

A cotton manufactory was begun on this river about thirty years ago by a company of adventurers, but it was afterwards relinquished, and the building is now occupied as a corn-mill. There is also a large tan-yard in the village.

There are two places for divine worship in Alveton, namely, the Church, and a chapel for Calvinists. The Rev. Mr. Tallis is minister of the latter. Several Methodists reside here, but they have not yet erected a meeting-house, and they consequently meet at private houses.

Near the north bank of the Churnet, opposite the Castle, stands Alveton Abbey, the summer residence of the Earl of Shrewsbury. His Lordship has made great improvements on his manor, and given employment to masons, bricklayers, labourers, and different artists connected with architecture. He is somewhat fanciful, and has built and pulled down several ornamental temples on his grounds; and he is now engaged in the erection of a tower on the summit of a hill, which will command an extensive prospect of the circumjacent country. Lord Shrewsbury, who is a zealous Catholic, married a Miss Hoey, the daughter of a bookseller in Dublin, who was also a votary of the Holy Mother Church, and a considerable number of Catholics are retained in the family mansion.

Alveton is a constablewick over thirteen townships, the inhabitants of which owe suit and service to Lord Shrewsbury, who holds a court leet and a court baron, with the view of frank pledge, in the Moot-hall belonging to the manor. It is situated opposite the Church, and is occupied as a free-school, with an annuity of £12. for the instruction of twelve boys and girls. A branch of the Grand Trunk and Caldon canal passes through the valley near Alveton, parallel with the river Churnet, and is continued to Uttoxeter.

On Sunday, April 20, 1806, a great shock of an earthquake was felt at Alveton, and the places adjacent, about one o'clock in the morning. During the concussion several chimneys were thrown down, particularly in the hamlet of Prestwood, but the inhabitants received no injury.

August 31st, 1810, Alveton and the neighbourhood was visited with the most dreadful storm of thunder, hail, and rain, that ever was remembered by the oldest inhabitant. Hail-stones of more than five inches in circumference, lay three inches thick upon the ground, and broke many of the windows, which were also forced-in by a violent west wind. Much glass was broken in the hot-houses of the Earl of Shrewsbury, and those of the neighbouring gentlemen, and some damage sustained in the farms and orchards. This storm lasted three quarters of an hour.

CHEADLE

Is a market-town in Totmanslow South, situated in a vale near the river Tean, and 145 miles from London. This town was the ancient seat of the Bassets, of Drayton, but it contains no remains of antiquity. The hills near the town are bleak and barren; many of them are covered with heath, and have lately been planted with firs and other trees. Cheadle Park, which is three miles in circumference, contains thirty-three inclosures, and has been much improved. Here are some valuable coal-mines; and a high barren hill belonging to it, which rises near the town, is the principal public walk of the inhabitants, as it affords an extensive prospect of the surrounding country. Lichfield Cathedral, which is twenty-seven miles distant, is perceptible on a clear day.

The parish of Cheadle contains 626 houses, 640 families; 1535 males, 1656 females: total inhabitants 3,191. About one-third of the population is employed in the tape-manufactory, and other handicraft arts; and the remainder in agriculture, and as shopkeepers, innkeepers, and various professions connected with mechanical arts. The weekly market is held on Friday, and is plentifully supplied with provisions. There are two annual fairs for horses and cattle; one held on Holy Thursday, and the other on the 21st of August.

There are extensive brass and copper-works, belonging to Messrs. Patten and Co. situated in a valley on the Tean, about half a mile south of Cheadle. Part of the copper is supplied from a mine at Mixon, but the principal part is brought from the great mines at Ecton. The copper and brass are smelted in the furnaces here. The brass is made into ingots for the brass-founders; and copper and brass are rolled into sheets.

Cheadle Church is an ancient Gothic structure of stone, roofed with oak. It is situated on an eminence, and the square stone tower, which contains a clock and six bells, is adorned with four pinnacles and vanes. The interior of the church is neat. The nave is supported by six Gothic arches, and it contains two galleries. When viewed from the chancel, the effect is simple and sublime: the windows "casting a dim religious light," and the large organ at the west end, in a very elevated situation, rising almost to the high roof, gives an air of magnificence to the whole. This church is dedicated to St. Giles, and is a rectory in the patronage of Trinity College, Cambridge. The Rev. Dellaber Pritchett, A.M. is rector.

The following epitaph in the church-yard is unique:

" In sixteen hundred and ninety-three,
George Wood, of Cheadle, set this tree;
Which was alive, but now he's dead,
Up to this stone here lies his head.
Be sure you have account to give,
When you are dead, how you did live."

He was buried at the root of a yew-tree of his own planting, which is now rooted-up.

There is a chapel belonging to the Calvinists in Cheadle. A free-school has long been established in this town, endowed with twenty pounds a-year.

Draycott-in-the-Moors is a small parish about three miles s.w. of Cheadle, and situated on the northern bank of the river Blithe. The village of Draycott is small and irregular. The only manufacture carried on here is that of tape: about eight looms, belonging to Messrs. Philips, of Tean, are employed in this village.

The Church is situated on a hill, in a rural spot near the village. It is a neat structure of stone: the tower and the body of the church are modern, and the principal entrance is adorned with two lofty Doric columns, and a pediment; but the chancel, which is very ancient, is Gothic. The nave of the church is spacious, the roof is supported by columns of the Doric order, and the pews are of oak. In the chancel, and two small chapels, there are several monuments of the Draycott family, with figures in alabaster, recumbent on altar tombs. At the south side of the church, close to the wall on the outside, is the recumbent figure of a knight in armour, with his right hand on the hilt of his sword, and a dragon at his feet. It is dedicated to St. Peter, and is a rectory. The present rector, the Rev. Robert Porter, purchased the living, the rectory-house, and estate, from the late Mr. Bill, some years ago.

Longevity.—" Hannah, the wife of Thomas Barnes, departed this life the 17th Dec. 1777, aged 100 years."

The parish of Draycott contains 92 houses, 97 families; 262 males, 274 females: total inhabitants 536.

Totmanslow is an ancient but obscure hamlet in this parish, consisting of two or three farm-houses, and a few smaller tenements: this place, equally insignificant with Seisdon, gives name to the hundred.

DILHORNE is a parish of Totmanslow North: it is of considerable extent and population. This parish, including Fosbrook township, contains 242 houses, 249 families; 581 males, 603 females: total of inhabitants, 1,184. Some of the lands in this parish were greatly improved by the judicious management of the late John Holiday, Esq. of Dilhorne Hall. In 1792, the gold medal given by the Society for the Encouragement of Arts and Manufactures was adjudged to him, for having planted at Dilhorne 113,000 mixed timber trees. Previously to this, his extensive and romantic winding vales and fertilized hills, were adorned with hedge-row elms, oaks, and other trees. There are also excellent coal-mines in this estate. But the most valuable coal-mines hitherto discovered in this parish, are those on the estate of Samuel Bamford, Esq. who has a handsome mansion near the village of Dilhorne, with extensive pleasure-grounds.

The village of Dilhorne is but small. There is a free grammar-school here, endowed by the Huntingdon family, who formerly held property in this parish, which came to them by marriage with one of the co-heiresses of Sir John Port, the founder of a noble institution for the instruction of youth, at Repton, in Derbyshire. The present master of the grammar-school at Dilhorne, is the Rev. John Smith, M. A. who resides in the vicarage-house.

Another free-school is established at Blithemarsh, in this parish, for the township of Fosbrook; Mr. James Dunn is the present schoolmaster. A school for girls has also been established, by the beneficence of Lady Buller, Mrs. Holiday, widow of John Holiday, Esq. and Mrs. Willatt, who are all residents at Dilhorne Hall. In this school eighteen girls are maintained, clothed, furnished with books, and instructed gratuitously.

Dilhorne Church is an ancient structure of stone, with a very curious octagonal tower, which contains five musical bells. The interior is neat, the pews are of oak, and the roof is supported by eight Gothic arches. The timber of the roof is oak, adorned with curious carved work, and the outside is covered with lead. There are several monuments in the chancel, but none of the inscriptions are remarkable. The parish registers are entire for about two centuries and a half, and the most curious circumstance recorded in them is, that, during the Protectorate, the banns of marriage between a man and a woman of Dilhorne were published at the market-cross of Cheadle by the town crier, and the parties were afterwards married by a Justice of Peace. The church is a vicarage, dedicated to All Saints, in the patronage of the Dean and Chapter of Lichfield Cathedral, and the present vicar is the Rev. Henry White.

CAVERSWALL is a parish at the western extremity of Totmanslow north. It contains, besides the village of Caverswall, the township of Weston Coyney and Hulme. The village is situated near the source of the river Blithe, and consists of several good houses. But the most remarkable object in this village is Caverswall Castle, built by Sir William de Caverswell in the reign of Edward II. It is a very strong and stately structure of stone, consisting of the high keep, with an inner court and garden, and four lower towers, one at each angle of the main building. A deep moat, supplied by water from two springs, prevents access. The castle has two fronts: the north-west front towards the village, is concealed by an outer wall, which extends along the side of the street to the church-yard. On entering the outer gate, another strong high wall and gate exclude the intruder, for it is now consecrated ground, being a nunnery.

In the year 1811, a number of nuns, who had emigrated from France, and settled at Preston in Lancashire, removed to Caverswall Castle, as a more secluded place. This castle, which was sold by the Hon. Booth Grey to Mr. Brett, a banker, of Stone, in this county, was taken on lease by Walter Hill Coyney, Esq. of Weston Coyney, for the nuns; and the sisterhood, amounting to sixteen in number, with their confessor, came hither. This priest, who is a man of very agreeable manners, has taken much pains to convert several of the peasantry of the parish to the principles of catholicism. His success, however, has not been commensurate to his zeal; when he has any thing to bestow upon them, they are mean and willing enough to receive it; but he has discovered that the majority of his converts come to the chapel in Caverswall Castle more for the hope of gain than the hope of salvation. In the mean time, the nuns are sufficiently active in the good work of instructing young ladies in the principles of their faith, and they have at present about 30 pupils in progress. Their discipline is sufficiently strict; the pupils wear an uniform of buff-coloured cotton; they are not suffered to ramble beyond the bounds of the gravel-walk which surrounds the moat, and two or three small fields; they walk two and two, like other boarding-school girls, and in their half-hour's exercise along the walk in the garden, are required, as a religious duty, to utter their ave-marias and pater-nosters in a low voice. The nuns themselves may be termed the Black Ladies. Their dress is entirely sable, with long thick black veils thrown over the right shoulder. Their demeanour is grave, and they generally walk with some book in their hands. Their countenances are pleasing and pensive; and if a man approaches them, they turn away as if they feared the imputation of vanity, or were in danger of weakening their principles by gratifying the eye of curiosity.

Weston Coyney, the residence of Walter Hill Coyney, Esq. is about three quarters of a mile distant from the castle. Mr. Coyney was brought up a Protestant, and was introduced to a Miss Coyney, the inheritor of a considerable estate, and an orphan. This lady is a zealous Papist, and Mr. Hill, when he married her, was obliged to adopt her family-name, and agree that if they had any daughters, they should go to mass with their mother, and the sons might go to church with their father. To this agreement they now scrupulously adhere.

The Church, which is near the Castle, is a small Gothic structure of stone, with a low tower. It was built by Matthew Cradock, Esq. about two hundred years ago. The chancel contains monuments of the family of Cradock, and the family-vault of the Parkers, of Park-hall. The late Countess of St. Vincent was buried in this vault in the year 1816, and a splendid monument, emblazoned with the most memorable achievements of Earl St. Vincent, is to be executed by a sculptor in London. It is a vicarage, in the patronage of the Parker family. The Rev. William Eddowes is the present vicar.

The parish of Caverswall, including the township of Weston Coyney and Hulme, contains 170 houses, 171 families; 450 males, 450 females: total of inhabitants 900.

KINGSLEY is a parish of considerable extent and population, the principal part of which is in Totmanslow South; it also includes the township of Whiston, in Totmanslow North. This part of the country has been greatly improved by inclosures and plantations of trees. This parish contains 213 houses, 216 families; 566 males, 572 females: total of inhabitants 1,138. The Caldon Canal passes through this parish.

The village of Kingsley is irregularly built, but contains some good houses, and a tape-manufactory, belonging to the Tean Company, in which 30 looms are employed.

The Church is a large and very ancient structure of stone, in a decayed state, with a low tower, and a flat roof. Three small ash trees grow on the roof at the east end, and have a most picturesque effect. They have existed there more than forty years; the seeds are supposed to have been sown by birds who alighted on the roof, and though the trees were frequently cut down, they always sprang up anew: they are now about five feet high. There is a free-school attached to this church; it stands at the northern angle of the east end, and the children of the village are instructed there in reading, writing, and arithmetic. The church is dedicated to St. Peter: it is a vicarage in the gift of the Duke of Devonshire. The Rev. John Wood is the present vicar, and the Rev. William Carlisle curate.

Longevity.—There is an inscription in the church-yard to the memory of Mary Davenport, aged 97 years.

To the north-east of Kingsley stands Cotton-hall, the seat of Mr. Gilbert. It is in a low situation, sheltered by high hills and extensive plantations of fir. The hamlet of Cotton contains several rural tenements, and a small chapel.

IPSTONES is a parish in an elevated situation, about three miles to the north of Kingsley, and five to the south-east of Leek. It comprises the township of Morredge and Foxt, and contains 248 houses, 250 families; 631 males, 604 females: total of inhabitants 1,235. The cliffs in this parish are high and picturesque.

The village of Ipstones contains several well-built houses, and two large inns. The Church, which is about a mile from the village, is situated on a gentle eminence, and is a neat modern structure of stone, in the Gothic style of architecture, with a square tower, adorned with four pinnacles and vanes. The interior is roomy and clean, and the pews are of fir. This edifice was built in the year 1790.

Every Saturday morning the church-bell is rung at eleven o'clock to summon the out-poor of the parish to attend at church, where they receive their weekly pittance from one of the overseers.

Among other instances of longevity recorded in the church-yard, is the following: "Here lie the remains of Robert Clowes, interred Nov. 7, 1771, aged 93 years." The church is dedicated to St. Leonard. Mr. Littleton is patron, and the Rev. William Carlisle curate.

Belmont, the seat of the late John Sneyd, Esq. is in this parish. It is situated on a gentle eminence facing the south-east, and surrounded by woods of oak, elm, ash, and other trees. The underwood is mountain ash, hazle, alder, and salixes of several species, which are cut once in six years to make crates for the Potteries.

Caldon is a small parish, about four miles east of Ipstones, and eight miles from Leek, situated to the south of the turnpike-road leading from Leek to Ashbourne, and in a barren and dreary part of the Moorlands. This parish contains 59 houses, 59 families; 165 males, 152 females: total of inhabitants 317. There is no manufacture carried on in this parish; but Caldon Lowe furnishes an abundant supply of lime-stone to the neighbouring parishes, and to a great extent, through the medium of the canal.

The Church, which is small, is dedicated to St. Mary, and is a chapel of ease to Church Mayfield, which is about five miles distant. The Rev. Thomas Blackey is the present minister.

The principal food of the inhabitants of this parish, and of the Moorlands in general, is oatmeal baked into thin cakes on a bake-stone; but this comparatively coarse fare is certainly conducive to health, for the people who are reared on it are hardy, and long-lived.

Calton is a parish, or rather a township, two miles to the north-east of Caldon, and about nine miles from Leek. This township extends into four parishes, namely, Waterfall, Mayfield, Blore, and Crokesden. It contains 44 houses; 44 families; 112 males, 108 females: total of inhabitants 220.

The Church is a small structure, dedicated to St. Mary, and is a curacy, in the patronage of Mr. Wheeldon. The Rev. Mr. Newton is curate.

Okeover is a small parish, situated on the west bank of the Dove, northward of the road from Leek to Ashbourne. It contains 10 houses, 10 families; 33 males, 27 females: total of inhabitants 60.

This parish is remarkable for the barrows of Hallsteds and Arbour Closes, being deep square entrenchments of stone, and supposed by antiquaries to have been formed by the Romans.

Okeover-Hall, the elegant residence of Haughton Farmer Okeover, Esq. is a magnificent mansion of brick and stone, with an Ionic portico in the centre, consisting of two columns and two pilasters, with two wings. The principal front is towards the deer park, where herds of deer quietly graze, or repose at will. The hall and dining-room belonging to this mansion are adorned with several excellent paintings, particularly the Holy Family, by Urbino, valued at 1500 guineas; the Magi at the Tomb of Christ, with an Angel sitting upon it, by Carlo Dolci; Christ bearing the Cross, by Titian; the Unjust Steward, by Rubens; the Baptism, by Titian; a Venus, by the same artist; St. Paul, by Rubens; and several others.

The situation of Okeover-hall is peculiarly delightful, with a lawn in front, and flourishing trees on the eminences; the Dove and its fertile banks in view; and a neat Gothic church within a few yards of the principal entrance.

The exterior of the Church, which is built of stone, is truly picturesque and agreeable. Its Gothic east window, with ivy climbing on each side; its wall, with neat battlements, and the roof covered with lead; and its handsome square tower, with four pinnacles, and a light vane rising in the centre, present a variety of pleasing objects. The interior harmonizes with its exterior, combining convenience with elegance. It is small; the pews and reading desk are painted; the communion-table is of marble; and an organ, adorned with carved wood in mahogany, harmonizes with the simplicity of the whole.

There are two monuments of the Okeover family in the chancel; the inscription on one of them is partly defaced, but the other bears the date of 1624.

On a mural monument of white marble, an angel is represented as extinguishing the torch of life with the left hand, while the left arm rests upon an urn. The right hand holds a serpent, in a circle, with the tail in the mouth, emblematic of eternity; underneath is the following inscription:

"Sacred to the memory of MARY, the wife of LEAK OKEOVER, Esq. near 40 years the daughter of John Nicholls, Esq. She left this life Jan. 30, 1764, aged 63 years.

So clear a reason, so refin'd a sense,
Such virtue, such religious confidence;
Manners so easy, and a mind so even,
On earth must needs anticipate a heaven!

Her husband soon followed, Jan. 31, 1765, aged 60 years.

Thrice happy pair! in mutual love so tied,
Whom death but for a moment could divide!
Knowing this world is not our proper home,
Their wish was for that happier world to come."

There is another mural monument of white marble, behind the reading desk. It represents an urn, with the crest of the family above it, the arms below, and the following inscription:

"Sacred to the remains of E. W. OKEOVER, Esq. who died 30th of June, 1793, in the 41st year of his age. He knew no happiness equal to that of conferring it upon others; and the many good qualities of his heart, will never be forgotten, and can only be estimated by those who were best acquainted with him."

The church is dedicated to All Saints.

One of the ancestors of the present proprietor of Okeover-hall and manor, left an estate at Atlow, in Derbyshire, the rent of which is appropriated to the support of three clergymen's widows, each of whom receives £30. per annum, and have a house rent-free, near the family-mansion. He also left an annuity to support an organist in Okeover church, and money to provide clothing for twelve children in the parishes of Okeover, Atlow, Mappleton, and the hamlet of Swinescote, in Staffordshire, and to place them, in rotation, as apprentices, with a premium.

Blore is a small parish, including Swinescote township, situated about a mile and a half to the N.W. of Okeover, on the west side of the Dove. The village is situated on high ground, and in a fertile tract, within a mile of Ilam, which may be termed the boundary of the comparatively barren Moorlands. This parish contains 39 houses, 39 families; 83 males, 81 females: total inhabitants, 164.

The village of Blore is chiefly remarkable for having formerly been the residence of the ancient family of Bassett, whose mansion stood here. Not a single vestige of which now remains; and the antiquary, in his researches for some fragment of this ancient scene of baronial splendour, may exclaim, "*Sic transit gloria mundi!*" The few houses now in the village are inhabited by farmers and people employed in agriculture.

The Church is an ancient Gothic edifice, with a large tower; the roof is covered with lead, and there are two small windows in the east end.

The only monument worthy of observation is one to the memory of a young woman, with the following inscription:

"Here lieth the body of Mary Hall, who departed this life, Dec. 12, 1764, aged 23 years:

Though short my days, my wit and beauty bright,
I was my parents' joy and hearts' delight;
In me *alone* they comfort *thought* to find,
God called me hence—it was his blessed mind."

The church is dedicated to St. Bartholomew, and is a rectory. The Rev. William Bailey is the present curate. It contains several monuments of the Bassett family, now mouldering in ruins.

Ilam, *with* Throwley, *and* Casterton, *and* Okeover.—This parish, which is situated on the banks of the Manifold, may with propriety be termed the boundary of the sterile part of the Moorlands, the vale through which the river flows presenting on the side towards Blore rich woodlands and cultivated fields, while near the eastern bank of the river, several barren hills rise in frowning majesty, bidding defiance to the skilful hand of the cultivator, the fertilizing dews, the showers, and sunshine.

This parish, exclusive of Okeover, which has been correctly estimated before, contains 26 houses, 26 families; 83 males, 94 females: total of inhabitants, 177.

The village of Ilam contains about eight or ten houses, including the stately mansion of Jesse Russell, Esq. situated on a gentle eminence, with two verdant terraces, and a pleasant lawn, in front, terminated by the church-yard. Behind the house, on the s.w. bank of the Manifold, a flourishing wood of oak, elm, ash, &c. rises along the side of the hill, and spreads to a considerable extent, in form of an amphitheatre, and above it the ground gradually rises in a cultivated eminence to a great height, terminated by a coppice, which is a conspicuous and pleasing object, and may be seen for several miles. The river falls in two natural cascades near the mansion, and their regular and constant sound, has a tendency to tranquillize the mind. Eastward of the church, which is itself a most interesting object, the village appears peeping amid the embowering trees; the white house of the vicar is most conspicuous; while beyond it rise two barren hills or mounts, with the grey rocks above the surface, which seem to forbid cultivation.

The Church is a small but very beautiful Gothic edifice of stone, situated about two hundred yards in front of the principal mansion. The date of its erection, as marked over the entrance, is 1618. The tower is of stone, with four pinnacles, a vane, and battlements. Two of the pinnacles, and almost the whole of the tower, to the ground, is covered with flourishing ivy in full verdure, which has a most picturesque and pleasing effect. Part of the west end, and of the body of the church, is also overgrown with ivy. The interior is plain, and the pews of oak; but there are several monuments in the chancel, particulary some tombs of the descendants of Oliver Cromwell, who lived at Throwley Hall, in this parish.

The following inscription on a tomb-stone in the church-yard is worth notice:

"Richard Ensor, A. M. formerly of Magdalen College, Cambridge, was instituted vicar of this church, Anno Dom. 1669, departed this life, Nov. 6, 1714, aged 79. And on his right hand lies Hannah, his wife, who died Dec. 29th, 1740, aged 95."

The church is dedicated to the Holy Cross, in the patronage of John Port, Esq. The present vicar is the Rev. Burnet Port.

WATERFALL is a parish in the Moorlands, situated between three and four miles from Ilam. Its name originated in the curious phenomenon of the river Hamps, which, after having flowed about eight miles from its source, suddenly disappears among the lime-stone rocks in this parish, where it holds a subterranean course to the neighbourhood of Ilam, where it rises again and forms a confluence with the Manifold. The village of Waterfall is an obscure and inconsiderable place, of no great antiquity, not being mentioned in any ancient record. The parish contains 92 houses, 92 families; 231 males, 224 females: total of inhabitants, 455. The principal employment of the men is in agriculture; some of them work in the copper mines at Ecton.

The Church, which is dedicated to St. James, is a small fabric: it is a curacy.

GRINDON is an ancient manor, which was possessed by Robert de Stafford, in the 20th year of the reign of William the Conqueror. In the reign of Henry the Third, it was held, with Blore, by William Audley, of the Baron of Stafford. In the 9th of Edward the Third, Joanna le Strange was lady of this manor; she was afterwards married to Sir Henry Brailiford, whose daughter and heir, Joan, married Sir John Bassett, descended from the Bassetts, of Cheadle. In this family it continued till the close of the seventeenth century, when it devolved to Christopher, Duke of Albemarle.

The parish of Grindon contains 69 houses, 74 families; 206 males, 197 females: total of inhabitants, 403. The village consists of a few good farm-houses and smaller tenements.

The Church is a small modern structure of stone, in the Gothic style of architecture, with a square tower. The free-school is at the east end, where the children of the parishioners are taught to read gratuitously; but the schoolmaster is paid for instructing them in writing and arithmetic, as he only receives £8. a-year from the funds of the institution.

Among the monuments in the church-yard, the following are the most remarkable:

"RICHARD BULLOCK, of Ford, died 21st June, 1740, aged 92."

"In memory of CHARLES SMITH, who died 25th Jan. 1814, aged 56 years,
He taught at Grindon school near five and thirty years,
And it was his desire that his bones might rest here till Christ appears."

The church is dedicated to All Saints, and is a rectory, under the patronage of the Marquis of Stafford. The Rev. James Whitaker, of Alstonefield, is the present incumbent.

ONECOTE is a township belonging to the parish of Leek. It is situated to the north-west of Grindon, and contains 74 houses, 80 families; 246 males, 218 females: total number of inhabitants 464. Many of these people are employed in the copper and lead-mines of Ecton.

Narrowdale, in this township, is remarkable for the high rocks with which it is surrounded, and the inhabitants of which seldom behold the sun, and when it is visible, about one o'clock, they call it the Narrowdale noon.

Onecote Chapel is a chapel of ease, belonging to the vicarage of Leek.

BUTTERTON is a township and village in Mayfield parish, situated to the north-east of Onecote. It contains 68 houses, 72 families; 161 males, 194 females: total of inhabitants 355.

The village of Butterton is irregular and ill-built, consisting of about twenty farm-houses and smaller tenements, in a low situation, and surrounded with trees, which relieve the eye in these dreary and naked Moorlands.

The Church is a small modern fabric of stone, in the Gothic style of architecture, with a small square tower: it is a chapel of ease to Mayfield. The Rev. James Whitaker, of Alstonefield, is the present minister. There are two white paper garlands, such as are carried in the funeral procession at the burial of young unmarried women, suspended from the roof of the church.

Longevity.—Elizabeth Stubbs died 28th August, 1799, aged 93.

WETTON is a parish situated to the west of Butterton, and adjoining to Alstonefield. This parish is remarkable for Thor's-house, a cavern in a rock which extends into a hill the distance of 44 yards, and is about 30 feet high in the middle. The roof is supported by a large and rough natural pillar, and the cave is divided into several partitions or rooms with rock-work, which has the appearance of human figures. According to tradition, the Druids performed their sanguinary rites in this cavern, and sacrificed human victims enclosed in wicker-work, on the altar of their idol Thor; but there is no historical document in existence to prove the fact, and the story probably originated in the warm and inventive imagination of some fanciful rustic, or hair-brained poet. Dr. Darwin, whom nobody will suspect of credulity, has contrived to conjure up a host of demons in illustration of the supposed enormities of the Druids, but to sober reason the whole appears an absurd fiction.

Ecton-hill is in this parish, near the village of Warnslow. This hill is remarkable for its productive mines of lead and copper, and is upon the estate of the Duke of Devonshire. It was first wrought by the Earl of Devonshire, in the seventeenth century, and afterwards by Sir Richard Fleetwood, and some Dutchmen. But it proved unprofitable, and copper from Sweden was imported at a lower price than it could be produced here.

In 1720, the hill was again examined by a Cornish miner, a company was established, a lease of the mine obtained from the Duke of Devonshire, and £13,000. expended before any returns could be realized. Some of the company then sold their shares at a great loss, but the remainder persevered in their researches, and on sinking a shaft of two hundred yards deep they discovered a rich vein of copper ore, which repaid their expenditure, and enabled them to acquire considerable profits before the expiration of their lease. It then reverted to the lord of the manor, and in the year 1780 it produced from eight to ten thousand pounds, clear of all expences.

It is asserted that the ore annually obtained in the Ecton mines is productive of *one hundred and ninety tons of pure copper!* How far this statement is accurate has not been ascertained, but such a produce is in itself an ample revenue.

Ecton-hill is about 230 yards high, and the diameter of its base half a mile. The entrance to the mine is at the base, near the bank of the Manifold. From this spot a passage is made 400 yards in length, and about six feet high, into the centre of the hill, strongly walled on each side. In the centre there is a large cavity, by which there is a passage to the summit of the hill, with a lodgment of timber for receiving the raised ore. From this place it is conveyed on small four-wheel carriages, which run in grooves, by boys of twelve or fourteen years of age.

The immense mass of copper ore with which this hill abounds, does not lie in regular strata, but is found in a perpendicular direction, widening towards the lower end. About seventy miners are constantly employed here, night and day, and relieved every six hours by a fresh set, so that the total number of men is nearly 300, besides a considerable number of boys.

When the copper ore is brought from the mine to the bank of the river, it is broken by hammers, and divided by girls into three sorts. It is then washed under the superintendance of an experienced miner, and exposed to sale in the open air.

On the opposite side of the hill there is a rich lead mine, the veins of which extend nearly to the copper ore, and a very considerable population is employed and maintained in these works.

The parish of Wetton contains 102 houses, 102 families; 282 males, 311 females: total of inhabitants 593.

The village of Wetton is considerable, and contains about forty houses. A few of the inhabitants are employed in weaving cotton. The Church is a small ancient fabric, in a ruinous state: the tower is of stone, and contains three bells. Money which had been collected for building a new church at Wetton, was deposited for security in the hands of Mr. Thomas Gould, of Ashbourne, who unfortunately became bankrupt.

Over the entrance to the church there is a very curious old arch, with emblems engraved on stone, representing the Lamb and Cross, with a Dove underneath, opposed to the figure of a Dragon, over which Christ appears triumphant. The church is dedicated to St. Margaret: it is a curacy, and the Rev. William Matthew Ward is the present curate.

Alstonefield is a parish on the banks of the Dove, situated about two miles to the east of Wetton. This parish partakes of the general appearance of the Moorlands, abounding with hills and dales, and interspersed with several fertile spots in a high state of cultivation. It extends to the length of fifteen miles, including the townships of Fairfield-Head, Heathy-Lee, Hollinsclough, Longnor, Quarnford, and Warslow with Elkstones, and containing 855 houses, 872 families; 2,073 males, 2,143 females: total of inhabitants 4,216.

The village of Alstonefield consists of several good substantial farm-houses of stone. The Church is a neat structure, in the Gothic style of architecture, with a large east window partly blocked up, and a strong tower adorned with four pinnacles, battlements, and a vane. It contains three bells and a clock.

Near the east window, on the outside of the church, the date 1590 is engraved on a stone, but the edifice seems of greater antiquity. The architectural ornaments of the east window are beautiful.

The interior is worthy of observation. It consists of a nave, chancel, and two small aisles. The roof is supported by six large

Gothic arches. The pulpit and reading-desk are curious; they are painted blue, with gilded cornices, and remarkable as being the gift of the celebrated Charles Cotton the poet, who resided many years at Beresford-Hall, on the banks of the Dove. The date carved in the wood is 1637. Opposite the pulpit is a pew which belonged to the Cotton family; it is also painted blue.

The church is dedicated to St. Peter, and is a vicarage in the patronage of Sir Harry Harper, Bart.; the Rev. James Whitaker, who resides in the vicarage-house, near the church, is the present vicar.

Among the monuments in the chancel, there is one to the memory of the Rev. Roger Farmer, vicar, who died in the year 1682; and another of the Rev. Peter Parr, who was instituted vicar of this church in 1713, and died in 1764, aged 74.

WARNSLOW is a small township belonging to Alstonefield parish, situate to the north-east of Butterton.

The village of Warnslow consists of several good houses; the Church or Chapel is a small modern structure: the Rev. —— Richardson is curate.

UPPER ELKSTONE is a township and village about two miles to the west of Warnslow. The hamlet of Upper Elkstone consists of about six houses, and that of Lower Elkstone of twelve. The surrounding country is barren and uneven, in many places overgrown with heath, and altogether an uncomfortable district to reside in.

The Church is a small modern structure. It is a curacy, the Rev. —— Richardson, who also officiates at Warnslow, is curate.

The following inscription on a tomb-stone in Elkstone churchyard, is worthy of notice:

"WILLIAM GRINDON, of Stonyfold, died Dec. 26, 1791, aged 62 years.

In hopes of future bliss content I lie,
Tho' pleas'd to live, yet not displeas'd to die:
Life has its comforts, and its sorrows too,
For both to all-wise heaven our thanks are due.
How far my hopes and views are founded well
God only knows---but the last day will tell."

SHEEN is a small parish, situated near the west bank of the Dove, and bordering on Derbyshire. The face of the country here is wild and romantic, but the soil about the hamlet of Sheen is fertile.

This parish contains 63 houses, 73 families; 205 males, 209 females: total of inhabitants 414.

The hamlet of Sheen consists of a few large farm-houses. The Church is a small ancient structure of stone, with a low tower. There are several ancient inscriptions of the seventeenth century on the tomb-stones in the church-yard; the letters and figures are large, and cut in basso-relievo, like the carving on wood at that period. There is one remarkable instance of longevity recorded by the sculptor, namely, to the memory of Isaac Gilman, who died Feb. the 8th, 1805, in the 99th year of his age.

The church is a curacy: the Rev. Matthew Beetham is the present curate.

LONGNOR

Is a township of the parish of Alstonefield, and the population is 467 persons, as has already been mentioned. The small market-town of Longnor is situated eight miles north-east from Leek, and contains 100 houses, including two large inns. The market-house, which stands on a gentle eminence, is very convenient: the market is held on Tuesday. There are eight annual fairs held in this town, namely, on Candlemas-day, Easter Tuesday, May the 4th and 17th, Whit Tuesday, August the 6th, the Tuesday before Old Michaelmas-day, and the 12th of November.

Longnor Church is a neat modern structure of stone, with a high square tower adorned with eight pinnacles. The interior is plain, and the pews of fir painted to resemble oak. The church-yard contains several monuments, the most curious of which is the following biographical record of a veteran:

"In memory of William Billinge, who was born in a corn-field at Fairfield-head, in this parish, in the year 1679. At the age of 23 years he enlisted into his Majesty's service under Sir George Rooke, and was at the taking of the fortress of Gibraltar in 1704. He afterwards served under the late Duke of Marlbro' at the ever-memorable battle of Ramillies, fought on the 23d of May, 1706, where he was wounded by a musket-shot in the thigh; afterwards returned to his native country, and with manly courage defended his Sovereign's rights, at the Rebellion in 1715, and 1745. He died within the space of 150 yards of the place where he was born, and was interred here the 30th of January, 1791, aged 112 years.

Billeted by Death, I quarter'd here remain,
When the trumpet sounds, I'll rise and march again!"

There are also some other instances of longevity recorded on the tomb-stones in this church-yard, namely, "To the memory of Sarah Wain, who died Jan. 28th, 1809, aged 91 years." Another, aged 90; and several between 80 and 90.

FLASH is a village situated towards the northern point of the county, near the road leading to Buxton, and seven miles north-east of Leek. It contains about twenty houses, three of which are ale-houses, and three with shops.

The Church is small: it is a curacy, and the present curate is the Rev. Robert Richard Balderstone.

There is also a small meeting-house for Methodists in this village.

Longevity.—Though the situation of this village is one of the most elevated in Staffordshire, and computed to be nearly half a mile above the level of the sea, the air is salubrious, as may be proved from the records of the dead in the church-yard, particularly the following:

"Here lieth the body of JOSEPH BRUNT, late of Calshaw, who departed this life Aug. 2d, 1794, aged 91 years. Also MARY, the mother of Joseph Brunt, who departed this life Feb. 24th, 1782, aged 104 years.

Remember man that thou must die,
And so must all that do pass by;
Our glass is run, our time is past,
For age will bring you down at last."

MEERBROOK is a village situated about four miles to the north of Leek, in a valley, which, though originally bleak and unproductive, has, by skilful cultivation, been rendered pleasant and fertile. It is sheltered on the east by the Roches, two rocky hills which rise about half a mile distant. The parish or township extends to the river Dane to the N.W. and is diversified by hill and dale.

The Church is a small fabric of stone, with a square tower, which contains one bell: a small pinnacle of stone adorns the east end. It is a chapel of ease to Leek, and the Rev. James Turner is minister.

HORTON is a parish situated to the west of Leek: it comprises the township of Blackwood and Crowborough, and contains 152 houses, 156 families; 415 males, 379 females: total of inhabitants, 794.

The village of Horton stands on an eminence: it contains three good farm-houses, and a few smaller tenements. The fields in general are well-fenced, and consist mostly of pasturage. There is also some corn grown, and a good deal of lime used both on the pastures, and the land under tillage.

Horton-common was lately inclosed: it is a thin, black, and meagre soil, which will require much manure and cultivation.

The Church is a handsome stone edifice, dedicated to St. Michael: it is a curacy.

In Rudyard vale, in the neighbourhood of Horton, there is a capacious reservoir, made by the proprietors of the Grand Trunk Canal, to provide an unfailing supply of water for the upper part of their navigation. This reservoir is about a mile and three quarters in length, and more than one-eighth of a mile in breadth, and its depth is about fifteen feet above their guage. It contains when full 2,420,000 cubic yards of water, or sufficient to supply their canal 100 miles in length, or to fill a lock three yards deep 10,000 times: it is consequently a sufficient supply in a dry season.

ENDON, *including* LONGSDON *and* STANLEY, is a large township in the parish of Leek. The village of Endon is about four miles south-west of Leek, and the lower part of it is situated on the turnpike-road, between that town and Newcastle; the upper part comprising some handsome houses.

The church is situated on a hill near the road. A considerable part of this township is hilly, yet fertile and well-inclosed.

The Church is a modern structure of stone, the top of the walls adorned with battlements; the roof is flat, and the square tower is of stone. It is a chapel of ease to Leek, and the present curate the Rev. John Salt, of Burslem, in the Potteries.

Longevity.—The following instance of longevity is inscribed on a tomb-stone in Endon church-yard: "Thomas Plant died June 16th, 1754, aged 99 years."

Another monument bears this singular inscription:

"Beneath lie the remains of WILLIAM MURHALL, Esq. of Bagnall, who died the 4th of Jan. 1762, aged 66 years.

Part of what I possessed is left to others, and what I gave away remains with me."

BAGNALL is a small village, about two miles to the south of Endon, and six miles to the north-east of Newcastle. The Caldon Canal passes through this parish from the vale of Trent, into the vale of the Churnet, being its highest level, and where it is supplied with water from the reservoir in Rudyard vale.

The manor-house in this village, which belongs to John Sparrow, Esq. of Bishton, is now inhabited by a tenant. It is a good mansion of stone, and appears to be of considerable antiquity.

The Church, which is a most miserable structure of stone, and in a ruinous state, stands on an eminence in a pasture field. It might be mistaken for a small barn but for the wooden belfry, containing one small and solitary bell: the east end is actually propped by a piece of timber. The interior is inconvenient: six small windows admit the light.

It is a chapel of ease to Stoke-upon-Trent, and the Rev. John Smith, who resides at Dilhorne, is minister.

CHEDDLETON is a parish situate on the Churnet. The manor, at the time of the general Survey, was held by Roger de Montgomery: it has since devolved to several families, and now belongs to the Earl of Macclesfield.

The Caldon Canal passes through this parish. It commences at the Grand Trunk Canal at Shelton, and passes through the parishes of Bagnall, Endon, Cheddleton, nearly parallel with the Churnet, and is continued down the vale through which that river flows to Kingsley, and as near to Caldon as the level will admit. Vast quantities of lime-stone are conveyed from Caldon Lowe to the Canal, which is continued along the vale of the Churnet to the neighbourhood of Alveton, and thence to Rocester and Uttoxeter, where it terminates.

The village of Cheddleton is situated on a hill, near the Canal and the river Churnet. It contains several good houses; and the Church, a neat fabric of stone, with its square tower, is a conspicuous object. There is one instance of longevity recorded in the church-yard, on the tomb-stone of Sarah Bentley, who died 19th July, 1811, aged 90 years.

The parish of Cheddleton is of considerable extent and population, including the townships of Basford and Cunsall. It contains 261 houses, 261 families; 233 males, 669 females: total of inhabitants 1,395.

The Church is dedicated to St. Edward: it is a curacy, and the Rev. Edward Powys is the present minister.

Ashcombe, the elegant mansion of William Sneyd, Esq. is in this parish.

LEEK and LOWE parish is of very considerable extent, containing the townships of Bradnop, Endon, Heaton, Leek-frith, Onecote, Rushton-James, Rushton-Spencer, and Tittesworth in North Totmanslow hundred, and the township of Rudyard and Cawdery, in Totmanslow South.

This parish contains 1,446 houses, 1,477 families; 3,660 males, 3,968 females: total of inhabitants 7,628.

In a parish of such extent there is a great variety of soil and surface, but the lands in general are not very fertile. Leek-edge is a new inclosure, north of the town of Leek, and is part of the great waste of Morredge. Part of this land has been planted with a variety of trees by the Earl of Macclesfield, who is lord of the manor; and by Dr. Hulme, of Ball-Haye; and Thomas Mills, Esq. of Barlaston; so that the term *Moorlands*, as applied to this tract, may soon give place to the more significant epithet of *Woodlands*. Oats have been sown on part of this enclosed ground, but with little produce, and a part of it has been more successfully improved into pasturage.

Ladder-edge is another new inclosure of land to the south-west of Leek; and though the situation is high, and the soil stony, it has already, by judicious cultivation, and the free use of lime, produced some good grain and turnips. The fences are generally of stone. The Caldon Canal, which comes within half a mile of Leek, is the medium for an abundant supply of lime-stone, which, when burnt into lime, facilitates the improvement of the cold soil, and the purposes of building.

LEEK

Is the principal market-town of Totmanslow North. It is situated on a pleasant eminence near the river Churnet, and contains thirteen streets and lanes. The streets are wide, well paved, and clean; many of the shops large, and several houses of the silk-manufacturers, and professional men, are elegant mansions. There are three large inns in Leek, and several ale-houses. A weekly market is held on Wednesday, and there are seven annual fairs, chiefly for cattle and pedlars' goods. They are held on the Wednesday before Candlemas, Easter-Wednesday, May the 18th, Whit-Wednesday, July 3d, July the 28th, and November the 13th.

Leek has been indebted for much of its present prosperity to the silk-manufactures, which have been successfully carried on in this town for at least half a century, and the population has rapidly increased.

This town, including Leek-Frith township, contains 954 houses, 972 families; 2,023 males, 2,390 females: total of inhabitants 4,413. About two-thirds of the men, women, and children, in Leek, are employed in various branches of the silk-manufacture, which consist principally of shawls, handkerchiefs, ribbons, ferrets, twist, and sewing silks.

The Church of Leek is an ancient Gothic structure of stone with two aisles, and a square tower, which contains a clock and six bells. In the year 1816, eight pinnacles were added to the tower, several parts of the exterior of the edifice were decorated with ornaments in the Gothic style: the interior also underwent a thorough repair. It is asserted that almost as much money was expended on repairs as would have built a new church. There is a mural monument of white marble in the chancel, to the memory of John Daintry, vicar of this church, who died on the 23d of August, 1758, aged 58, and several other individuals of that family.

Leek church is a vicarage in the patronage of the Earl of Macclesfield: the Rev. Thomas Bentley is the present minister.

William Badnall, of Leek, silk-dyer, left by will, dated the 11th of January, 1806, one thousand pounds in the public funds, the interest of which is to be annually laid out in bedding, clothing, or other necessaries, to be distributed among twenty poor widows resident in Leek, who are not less than 60 years of age.

In the church-yard are the remains of an ancient stone pyramid, without an inscription, the origin of which rests entirely upon conjecture.

Among other instances of longevity recorded on the tomb-stones in this church-yard, is one of an individual aged 95, and another aged 90 years.

There are three other places of worship in Leek besides the Church; namely, a meeting-house for Calvinists, a Quaker's meeting-house, and a large meeting-house for Methodists.

There are two Sunday-schools, one supported by the friends of the Established Church, and the other by the Methodists.

An alms-house, for eight poor widows, was endowed in the year 1696, by Elizabeth Ash, widow, of this town. The building is a curious piece of Gothic architecture, situated in the Compton. Each widow is allowed two shillings per week, fuel, and a new gown once in two years.

Leek gave birth to Thomas Parker, the first Earl of Macclesfield, who was the son of Thomas Parker, attorney, of this town. He was born at Leek in the year 1666, and after a grammatical education, applied himself to the study of the law, under the direction of his father. He became so eminent as a barrister, that he was appointed one of Queen Anne's Counsel, and was exalted to the degree of Serjeant-at-Law, and had the honour of knighthood conferred on him on the 8th of June, 1705. He was a representative in Parliament for the town of Derby from 1705 to 1706. In 1710, he was appointed Lord Chief Justice of the Court of King's Bench, and, on the death of the Queen, he was one of the Lords Justices till the arrival of George the First from Hanover. On the 10th of

March, 1716, he was created a Baron of England, by the title of Lord Parker, Baron of Macclesfield; and on the 5th of November, 1721, he was advanced to the dignity of Earl of Macclesfield. In consequence of some notorious mal-practices, his Lordship was impeached by the House of Peers on charges of corruption, tried at the bar of the House, and pronounced guilty, in consequence of which he was removed from his offices, and fined £30,000.

Of this accomplished, but unfortunate Earl, Mr. Noble,* writes as follows: "This every way distinguished character was the son of Thomas Parker, an attorney at Leek, in Staffordshire, in the chancel of which church I have read the inscription on his grave-stone. He left his son about £100. per annum. He received the Great Seal, May 11th,† 1709, which he held till January 4th, 1724-5. It was an extraordinary event, that Lord Macclesfield, one of the great ornaments of the Peerage, who had so long presided at the administration of justice, should himself be arraigned as a criminal, be convicted of mal-practices, and sentenced to pay a fine of £30,000. as a punishment for his offence; that a second Lord Chancellor of England should be impeached by the Grand Inquest of the nation, for corruption of office; and be, like his great predecessor, Lord St. Albans, found guilty of the charge. The prosecution was carried on with great virulence; and though rigid justice, indeed, demanded a severe sentence, yet party zeal and personal animosity were supposed to have had their weight in that which was passed upon him. The whole fine was exacted, and actually paid by his Lordship and his son, notwithstanding the favourable disposition that was shewn in a certain quarter to relieve him in part by a considerable donation. It is certain, there had been gross mismanagement in the offices of the Masters in Chancery, by which the suitors had been great sufferers; and it appeared that these places had been sometimes conferred upon persons, who had evidently paid for them a valuable consideration. The public cry against corruption in high stations was loud and long; and it was not thought prudent to stay proceedings against the supreme Judge in the kingdom. The statute on which the Chancellor was impeached had, indeed, grown into disuse, but it was still a law; a breach of it was proved, and the consequence was inevitable. Lord Macclesfield was a man of learning, and a patron of it. Bishop Pearce, of Rochester, among others, owed his first introduction to preferment

* Continuation of Granger's Biog. Hist. of England, Vol. III. p. 90.

† It was delivered May 12th.

to his Lordship's encouragement. He was also very eminent for his skill in his profession; but rather great than amiable in his general character. He was austere, and not deemed sufficiently attentive to the gentlemen of his Court, to whom his manners are represented to have been harsh and ungracious, unlike the mild and complacent behaviour of his predecessor, Lord Cowper. His Lordship passed the remainder of his life in a learned retirement, much devoted to the studies of religion, of which he had always been a strict and uniform observer."

His Lordship married Janet, daughter of Charles Carrier, Esq. of Wirksworth, in Derbyshire, by whom he had issue, George, the second Earl of Macclesfield, and Lady Elizabeth. He died at his son's house, in Soho-square, April 28th, 1732, in the 66th year of his age.

On Tuesday, the 3d of December, o. s. 1745, the Scotch rebel army, commanded by Charles Edward Stuart, and the Dukes of Perth, Athol, &c. marched through Leek on their way to Derby, and returned on the Saturday following.

Ball Haye, the elegant residence of Dr. Hulme, may be seen to great advantage from the northern side of Leek church-yard.

RUSHTON-SPENCER is a township and parish about five miles to the north-west of Leek.

The village of Rushton-Spencer is situated on a hill to the west of the turnpike-road between Leek and Macclesfield, and near the river Dane, which is the boundary between the counties of Stafford and Chester. The whole township contains 59 houses, 59 families; 193 males, 169 females: total of inhabitants, 362. There is a small manufactory in this parish, where some cotton twist is spun for the Manchester manufacturers.

The Church, which is a chapel of ease to Leek, stands on a hill, at a distance from the village. The church-yard is surrounded by fir-trees; and the church has the appearance of a barn when viewed from the road, to which there is a communication by a path-way across the fields. It is a small fabric of stone, in a very neglected and even ruinous state: the small wooden belfry contains one bell. There is a sun-dial, two remarkably fine full-grown yew trees, and a young flourishing tree of the same kind in the church-yard. The family-vault and chapel of the Michalls family, of Swithamley-hall, is at the east end of the church, near the communion-table.

The following are the most remarkable instances of longevity, recorded in the church-yard: "James Stoddart, aged 90; Daniel

Dale, died the 15th of May, 1794, in the 89th year of his age; and Ann Rigby, died Nov. 15, 1738, in the [illegible]9th year of her age."

There is a singular tomb in this church-yard, of a young man who is said to have been murdered by his villainous master, for the sake of some money which he received from him as a loan. The report is, that a strong opiate was given to the unfortunate victim of avarice, and during the time of its operation he was placed in a coffin and buried. Some suspicion of foul play arising, the coffin was raised, and on opening it, there were evident proofs that the young man had awoke in the grave, and partly turned in the coffin, where he died for want of air! The body was then replaced in the coffin, and buried in Rushton-Spencer church-yard with the *head* towards the east, and the following inscription on a tomb-stone:

"*Memento Mori.* THOMAS, the son of THOMAS and MARY MEAKIN, interred July 16, 1781.

As a man falleth before wicked men, so fell I."

The Church is dedicated to St. Lawrence: it is in the patronage of the Vicar of Leek, and the Rev. James Turner, of Meerbrook, is the present minister.

CUTTLESTONE HUNDRED.

The Hundred of Cuttlestone is bounded on the north by Pirehill hundred; on the south by Seisdon; on the east by Offlow, and on the west by Shropshire. The principal river in this division of Staffordshire is the Penk, which divides the hundred into two parts, flows through it towards the north-east, and after receiving several smaller streams, and its confluence with the river Sow, falls into the Trent. The meadows and pastures on the banks of the Penk are very productive; the uplands to the east of the river are generally a mixed gravelly soil; but most of the lands to the west, are a stronger mixed loam. Near the verge of Shropshire, however, the land becomes lighter.

BREEWOOD

Is a large parish in the south-west part of the Hundred, and contains about 10,000 acres of land, consisting of meadows on the banks of the Penk, and upland of a mixed gravelly loam, which, when well cultivated, yields abundance of the different kinds of grain, clover, and turnips.

The town of Breewood is delightfully situated on a branch of the Penk, about a mile and a half south of Watling Street, and nine miles distant from Stafford. It is a place of great antiquity. A small priory of Benedictine nuns was founded here by Hubert Walter, in the reign of Richard the First, A. D. 1195. At the Dissolution, its revenue was only £11. 1s. 6d. per annum.

Breewood is one of the most ancient market-towns in Staffordshire. A weekly market was formerly held on Friday, but it is now declined, and the decayed market-house pulled down. There is a large annual fair for cattle held on the 19th of September. In the year 1811, this town contained 210 inhabited houses, and 212 families, 69 of whom were employed in agriculture, 87 in trade, manufactures, or handicraft, and 56 not engaged in business; males 475, females 516: total, 991.

The Church is a vicarage, dedicated to the Virgin Mary, in the patronage of the Dean of Lichfield. The Rev. Baptist Preby is the present vicar. The edifice is ancient and large, built of stone, with a tower and lofty spire of the same material. From its elevated situation it is a picturesque object, distinctly visible at the distance of several miles. A new organ was erected and opened on the Wake Sunday, Sept. 24, 1815.

An excellent Free Grammar School was founded at Brewood, by Dr. Matthew Knightley, and endowed with lands to the value of £60. per annum. The Rev. Mr. Kempson, and the Rev. Matthew Kemsey, are now the masters, and receive a limited number of pupils as boarders. Some eminent men have been educated at this school, particularly the late Bishop Hurd, preceptor to the Prince Regent.

The principal manufacture carried on at Breewood, is Joseph Brewster's fixed and portable thrashing machines, kibbling mills, and straw engines.

Chillington is the largest estate in the parish of Breewood, and includes the south and west part. It is the property of Thomas Giffard, Esq. The mansion and offices are extensive; and the gardens, pleasure-grounds, and plantations, are laid out with great taste, and much improved by the present owner. This estate contains several villages and farm-houses, besides Chillington; particularly Gunston, Long Birch, the Hattons, White and Black Ladies, and the Hyde. The Giffard family has large estates in other parts of this county, but their religious tenets have kept them from public employments. Mr. Giffard, of Chillington, is, however, a gentleman of liberal principles, and is married to a Protestant lady, by whom he has a large family. His tenantry are mostly Roman Catholics, who are accommodated with chapels on his estate at Long Birch and Black Ladies.

Somerford is another large estate in the parish of Breewood, including the manor of Coven, Breewood Hall, Engleton Hall, Four Ashes, and Somerford Hall. The present possessor and resident, the Hon. Edward Monckton, is third son to Viscount Galway in Ireland; he has much improved this estate, and Somerford Hall is now an elegant mansion, in a pleasant situation on the eastern bank of the Penk. The estate was nearly stripped of all its timber by the former owners, but Mr. Monckton has improved its general appearance by extensive plantations. The population of the liberty of Somerford, in 1811, was 578 persons, of whom 283 were males, and 295 females.

The parish contains other villages and hamlets, particularly Kiddimore Green, a township situated to the west of Breewood; the villages of Broomhall and Horsebrook, near the Watling Street; the hamlets of Crateford and Standeford; Aspley Farm, and the Laches, hamlets on the eastern side of the parish; and the small village or hamlet called Paradise.

PENKRIDGE.

This is another large parish, containing nine townships, and between 12 and 13,000 acres of land.

Penkridge is an ancient market-town, situated on the banks of the river Penk. Camden, our greatest antiquarian, asserts that this town is built on the site of the Roman *Pennocrucium*. "The military way continues from Wall, very fair and plain, almost without any breach, till it is crossed and interrupted by the river Penk, and hath a stone bridge built over it at Pennocrucium, so called from the river, and standing at the same distance which Antoninus has fixed; which town has not quite lost the name at this day, being for Pennocrucium called Penkridge. At present Penkridge is only a small village, famous for a horse-fair, which Hugh Blount or Flavus, the lord of it, obtained of King Edward the Second."*

This town stands six miles south of Stafford, and ten miles north of Wolverhampton; the turnpike-road to those towns runs through it, and the Staffordshire and Worcestershire Canal passes close to it on the east side, thus affording a ready communication with distant places. In 1811, Penkridge contained 196 houses, and 204 families, consisting of 438 males, 485 females: total 923 persons.

Penkridge Church was made collegiate by a charter of King John, who settled the advowson of it upon the Archbishop of Dublin.

The following particulars respecting Penkridge, are extracted from the manuscripts of the late Sir Edward Littleton, Bart. and communicated by the Rev. Richard Slaney, vicar of the church:

* Camden's Britannia, Vol. I. p. 466.

"SIR EDWARD LITTLETON, BART. *to the* REV. R. SLANEY, *Penkridge.*

"*Teddesley, Nov. 3, 1697.*

"*Sir*,—By 1st Edward VI. 1547, colleges and chantrys are by Act of Parliament vested in the Crown, which puts an end to the claim of the Archbishop of Dublin's jurisdiction over Penkridge; and it remained in the hands of the Crown till granted out to others.

"The grant of King John to the Archbishop of Dublin, bears date 13th Sept. 1206, and is done away by the above 1st Edward VI. 1547, by an Act of Parliament.

"EDW. LITTLETON."

"King Edward II. in the eleventh year of his reign, declared that the chapel at *Pencriz*, and others, were his free chapels, and as such, exempt from all ordinary jurisdiction, impositions, exactions, and contributions, and accordingly ordained, that none should presume to incroach upon their immunities.*"

The Church is a fine Gothic building of stone, with a square tower, and five bells. It is dedicated to St. Michael, being now only a curacy. It is a royal peculiar, having four chapels within its jurisdiction, namely, Dunston, dedicated to St. Leonard; Coppenhall, to St. Lawrence; Shareshill, to the Assumption of the Virgin Mary; and Stretton, to St. John. The official of the peculiar holds visitations, probate courts, and licenses the incumbents to the chapels. In the chancel of the church of Penkridge, there are several monuments of the Littleton family, who are patrons of the church.

There is a charity-school in Penkridge for twelve boys and eight girls.

The principal manufacture of Penkridge is iron, which is, however, inconsiderable. From the situation of the town on the low and flat northern bank of the river Penk, it is subject to inundations when that river is swelled by floods. The market is held on Tuesday. Two annual fairs are held in this town; one on the 30th of April, and the other on the 10th of October. These fairs are allowed to be among the first in England for saddle and draught horses.

A very considerable part of the extensive parish of Penkridge is the property of the Littleton family, who have resided for centuries at Pilaton-hall. The present possessor, E. J. Littleton, Esq. has been twice chosen a representative in Parliament for the county of Stafford.

On the accession of the late Sir Edward Littleton to the estate,

* Dugdale's Monasticon Anglicanum, p. 370.

about the middle of the last century, he found the ancient mansion in a ruinous state. The situation was also repulsive to a man of taste, being low, and without any advantage of prospect. Sir Edward, therefore, resolved to erect a new mansion in Teddesley Park, about three miles from the ancient seat at Pilaton.

Teddesley-hall is built on the verge of Teddesley Hay, which is extra-parochial, and part of Cannock-heath. The waste lands in this extensive manor are now under enclosure, which will contract the enormous waste of Cannock-heath. Sir Edward Littleton married Miss Horton, May 10, 1752, by whom he had no issue. He was chosen five times to represent the county in Parliament, and was remarkable for his patriotism in the senate, and his hospitality at his country residence. When the Staffordshire and Worcestershire canal was projected, Sir Edward warmly patronized this great public work, subscribed for a number of shares, was always one of the committee for managing the business, and passed the canal through his manors and estates from Galey Wharf to Acton Trussell, a distance of upwards of four miles. The canal was cut through the demesne of Teddesley, and within half a mile of the mansion, thus affording a convenient supply of coal, lime, and building materials. During his long residence at Teddesley, Sir Edward promoted general improvements in agriculture, and the comfort of those around him. His mansion was noted as the scene of old English hospitality, where plenty of roast beef and good ale were provided for all decent visitors, without riot, profusion, or wastefulness. The ancient Christmas festivities of Twelfth-night, were kept up for the gratification of the inhabitants of the surrounding country as long as the health of the beneficent owner permitted. Sir Edward died at Teddesley, and was succeeded in his estates by E. J. Walhouse, Esq. his nephew, who took the name of Littleton.

The parish of Penkridge contains several other townships, villages, and hamlets, particularly Coppenhall, Dunston, Lovedale, and Drayton; Mitton, Otherton, Pilaton, and Water-Eaton; Rodbaston, and Whiston. The total population of the parish in 1811, was 2,243 persons.

Cuttlestone-bridge, over the Penk, about three quarters of a mile south-west of the town on the road to Stafford, gives name to the hundred.

CONGREVE, a village higher up on the Penk, is said to be the birthplace of Congreve the poet. It was the birth-place of that eminent prelate, Bishop Hurd, who held the spot in some veneration. It was related by the late Sir Edward Littleton, that in one of his visits to the Bishop at Hartlebury, as he was coming away his Lordship called out, "Is the old house at Congreve, where I was born, yet standing!" This village has also produced a still more modern writer, Miss Barker, authoress of an amusing novel, entitled "A Welch Story," in which she has introduced the character of Sir Edward Littleton, her patron, under the name of Sir Edwin. Mr. Barker, her father, was many years a respectable iron-master at Congreve, where he had one forge worked by the waters of the Penk, and another higher up on the stream at Coven.

Whiston, Mitton, Preston, and part of Bickford, are townships or hamlets of Penkridge, situated on the western stream which falls into the Penk below Cuttlestone-bridge. Lovedale and Longridge are also arable districts of this parish, on a stronger soil. The heaviest clay soil is to the west of the Penk, that on the east inclining to gravel. Drayton, Woolgarston, Otherton, Rodbaston, Kinvaston, and Water-Eaton, are all hamlets of Penkridge, generally upon a gravelly or light loamy soil. The township of Water-Eaton extends south of Watling-street to the common of Calf-heath, lately enclosed, and the principal part of which is in Penkridge parish.

At the small village of Lapley, about three miles west of Penkridge, there was a Priory of Black Monks belonging to the Abbey of St. Remigius, at Rheims, on whom it was bestowed by Aylmer, Earl of Chester, in the time of Edward the Confessor. It was afterwards granted by Henry the First to the College of Tong, in Shropshire. The Church is an ancient building of stone, with a square tower.

Shareshill, which includes Saredon Great and Little, Latherford, and Saredon mill, is a small parish in the south-east angle of Cuttlestone hundred.

The village of Shareshill is pleasantly situated on an eminence, on the summit of which stands the church. It contains several well-built houses; some of the families reside in tenements of their own, and others are tenants. The population of the parish in 1811, was 493 persons, who were chiefly employed in agriculture. The land about the village is a good sound loam, well adapted to grain and turnips, as well as pasturage. The principal land proprietors are Robert Smith, Esq. of Appleby, Leicestershire, who resides on a part of his own estate, and the Rev. J. H. Petit, who resides at Hilton-hall, but who has a good estate here, and is the present minister of the parish. Shareshill Church contains several antique monuments of the Vernon family.

A small rivulet which rises in Essington Wood, runs between Shareshill and Saredon, into the Cannock-heath branch of the Penk, a little above Deepmore Mill. This stream abounds with excellent trout.

Shareshill is a place of antiquity. In the 1st Henry IV. A.D. 1399, Sir William de Shareshall, Knt. of Shareshall, was Sheriff of Staffordshire. On the north and south sides of this village, are vestiges of two encampments, supposed from their square form to have been of Roman construction.

Great and *Little Saredon* are hamlets of Shareshill. The manor, and tithes, and a good estate, belong to the Littleton family; besides which, the Rev. J. H. Petit, Mr. Hordern, banker, of Wolverhampton, Mr. Pratt, of Saredon Mill, and others, possess good estates here. The Cannock-heath branch of the Penk passes near the boundary between Penkridge and Shareshill, and upon it are Saredon and Deepmore Mills, which are very powerful corn-mills.

Fetherstone is one of the places conferred by the Lady Wulfruna, upon the monastery of Wolverhampton, which she founded. It is situated about a mile north of Shareshill, and was formerly the residence of Huntbach the antiquary, from whom it passed in the female line to Sir Samuel Hellier, and is now the property of J. S. Hellier, Esq.

Hilton was another of the townships given to the monks of the before-mentioned monastery by its founder. It probably took its name from an adjacent eminence, which is now adorned by an obelisk. Hilton Hall and Park, the ancient seat of the Vernon family, is situated in the vicinity of Shareshill; and though nominally in the parish of Wolverhampton, the inhabitants go to Shareshill church as their place of worship, and bury their dead in the church or church-yard. The Hall, which is a large structure of brick and stone, was built in the year 1700. It is surrounded by a moat, and the communication to this mansion is by a handsome bridge, which leads to the principal entrance. The park is very pleasant and secluded, abounding with all the varieties of sylvan and picturesque beauty displayed by groves, clumps, and plantations. Hilton Monument is erected on the highest eminence in the grounds, and commands a very extensive prospect of the surrounding country. The plantations are chiefly of oak, and elm, which flourish in the utmost luxuriance, and afford evidence of the planter's skill and attention. The late Mr. Vernon was an early planter, and lived to see some of his first efforts fast approach to maturity. Many of the larches, and Scotch and other fir trees of his planting, now contain from twenty to thirty feet of timber. One plantation of very lofty oaks, is also an extensive rookery.

STRETTON is mentioned in the population returns for 1811, as a liberty belonging to the parish of Penkridge. At that time it contained 47 houses, 48 families; 118 males, 125 females: total, 243 persons.

The village of Stretton is about three quarters of a mile north of Watling Street-way, opposite Breewood. This place, now so obscure, is considered by several learned antiquaries as the real site of the ancient Roman station of *Pennocrucium*, as agreeing in the distance with the account given in the Itinerary of Antoninus. These antiquaries suppose the name of the place to be derived from Street-way, but there are no vestiges of antiquity to confirm the fact. Mr. Dickenson, indeed, in 1796, found a *tumulus* on Roley Hill, in the vicinity of Stretton, which he considered as a very likely place to have been a military station, having the river Penk contiguous. Some small fragments of Roman implements have also been turned up by the plough, from time to time, in the neighbourhood.

The village contains several farm-houses, and smaller tenements, and a family-mansion of great antiquity.

Stretton-hall is evidently a structure of the last century, and is built in a pleasant situation, with an extensive lawn in front, interspersed with plantations, an ancient wood to the north, and the church and village to the south. A high garden wall, sheltered with plantations, extends to the highway. This village formerly belonged to the family of Conolly, of Ireland, from whom it was purchased about twenty years ago by the Hon. Edward Monckton, of Somerford. The upland is a strong marly loam, adapted to wheat and beans: the pasturage is good, and the hedge-row timber grows vigorously. The mansion is now occupied by a tenant. There is a mill on the Penk called Stretton mill. The Church, which was originally erected as a chapel of ease to Penkridge, is dedicated to St. John. It now possesses its own rights of marriage and sepulture, and has no dependence on the mother church.

Dunston is a small township between Penkridge and Stafford, containing 44 houses and 214 inhabitants, mostly employed in agriculture. The soil is an excellent marly loam, with good meadows on the banks of the Peak. Dr. Plot, always on the stretch for wonders, mentions, that at Dunston, west of the chapel, an echo gave a return of seven syllables. The Chapel is a small building of brick, on the west side of the turnpike-road leading to Stafford. It is dedicated to St. Leonard, and is a chapel of ease to Penkridge, in the patronage of the Littleton family, of Teddesley.

Coppenhall is another township about two miles to the north-west of Dunston, but not so populous. The number of houses in 1811 was 16, and of inhabitants 92, almost all employed in agriculture. The soil is a good mixed loam. The Chapel, dedicated to St. Lawrence, is a chapel of ease to Penkridge. It is also in the patronage of the Littleton family.

An enclosure, according to the Act of Parliament, commenced in 1815, and will include all the waste lands in Penkridge, on Penkridge-heath, Dunston-heath, Huntington common, and Teddesley-hay. The two last adjoin Cannock-heath, or are parts of that waste, and include a very considerable extent of land.

CANNOCK.

This extensive parish, which includes the townships of Huntington and Great Wyrley, is in the eastern part of the hundred, and near the middle of the county. Different antiquaries have given different etymologies of this place, some tracing the derivation up to the Cangi, and others to Canute, the first Danish king who reigned in England. The last syllable of the name is conjectured to be derived from the Saxon *ock*, being the same as oak, and denoting a wood or country full of oaks; and it was undoubtedly a forest during the reign of the Mercian kings, and was their favourite chace.

The town of Cannock is not mentioned in Domesday-book, though in the time of king John there was a town here, with a church belonging to the Dean and Chapter of Lichfield. The manor and tithes now belong to the Marquis of Anglesea. Cannock is situated four miles east south-east of Penkridge, eight miles from Walsall, and nine from Stafford, the turnpike-road to these towns passing through it. The town stands on a gravelly soil, in consequence of which water was frequently scarce in summer, so that it has been facetiously asserted that it contained more ale than water. A constant supply, however, was realized by the liberality of Dr. Hough, Bishop of Worcester, who out of regard to the inhabitants, and particularly to the Birches, of Leacroft, his relations, caused water to be brought in leaden pipes from a spring a mile distant, and built a handsome conduit of freestone at the corner of the bowling-green in the middle of the town. This public bowling-green was encompassed with a brick wall in 1753, when the south side of the church was re-built. Cannock is a pleasant little town, and contains several well-built houses and a good inn. According to the population returns in 1811, it contained 224 inhabited houses, 240 families; 532 males, 611 females: total 1,143 persons. This town formerly had a market on Tuesday, which is now discontinued. It was formerly a place of great resort, on account of the salubrity of Reaumore-hill well, which was a fashionable watering-place in its day. Dugdale asserts, that Henry I. resided here in summer, and there are records of a castle having existed here, though no vestiges of it now remain. It was more probably what the antiquaries term a castellated mansion, built in that form to distinguish it as the royal residence. There is some variation in the accounts respecting the supply of water to the town, for it is stated that a Mr. Biddulph *originally* laid pipes from Leacroft more than a mile distant, which conveyed excellent water from the fountain, and that he bequeathed lands for the purpose of keeping the pipes in repair. Perhaps this had been neglected, and the Bishop of Worcester was only at the expence of repairing the pipes, and re-building the conduit.

The parish of Cannock is of considerable extent, containing 25,000 acres; a considerable part of the enclosed land round the town consists of a good light soil, adapted to corn, turnips, and pasture.

Hedgford, an enclosed hamlet on Cannock-heath, contains a good inn, and stabling for blood horses, which are trained and exercised on the excellent turf of Hedgford-hills. The ground being open and elevated, and the soil being a sound gravel, seldom breaks into mire. Hedgford-pool is a lake in a valley of Cannock-heath, the Water from which issues north to the Trent, near Rugeley, and south to the Walk mill, and afterwards west to the Penk, near Standeford and Coven.

Leacroft and *Reaumore Hills* are enclosed hamlets of Cannock, situated to the east and south of the town. At the Walk Mill,

Mr. Gilpin has established a very considerable iron manufactory of edge-tools, augers, and various similar articles, near the Watling-street-way, besides which he has built a good house, and raises coal for his manufactory and for sale. By this public-spirited industry, he has much increased the population, and the prosperity and comfort of the inhabitants of his neighbourhood.

Great Wyrley, formerly called *Wyrley Bank*, and then extra-parochial, is now a populous hamlet of Cannock. This township, according to the population returns in 1811, contained 82 houses, inhabited by 82 families, of whom 51 were employed in agriculture, and 31 in trade, manufactures, or handicraft arts. The number of persons was 361, of whom 187 were males, and 174 females. The village contains a few well-built farm-houses and tenements, with a number of cottages in every variety of rude architecture, from the clay-built shed to the convenient dwelling. It was formerly the reputed home of beggars, who, when asked where they came from, replied Wyrley-bank. It was then a nursery for illegitimate children, and became a nuisance, to correct which, it was annexed to Cannock as a township, and to Chesland Hay respecting the provision for its poor.

Chesland Hay had formerly a large common, which is now enclosed. The Lanes, of King's Bromley, have an estate here, on which an eminence, planted with firs, is conspicuous for several miles to travellers and the inhabitants of the circumjacent country.

Huntington is another township belonging to Cannock parish. It contains 23 houses, and 135 inhabitants. This township chiefly belongs to the Littleton family; the waste lands of Huntington and Teddesley, are now under enclosure, being in a distinct manor from the general part of Cannock-heath.

Hatherton, including the Four Crosses, and the open part of Calf-heath common, is in the parish of Wolverhampton, but the inhabitants attend divine worship in Cannock Church, and bury their dead in the church-yard. This township contains 55 houses, 56 families, and 299 inhabitants, whose only employment is agriculture. The demesne of Hatherton-hall is an ancient residence of the Walhouse family, and now inhabited by Moreton Walhouse, Esq. father to Edward John Littleton, Esq. M.P. This mansion, which is in the Gothic style, is now re-building. The material is stone, and, when finished, it will be a commodious place of residence. The estate consists of several contiguous farms of good sound gravelly loam, well wooded, and of hedge-rows of thriving trees. The hamlet at the Four Crosses, on Watling-street, consists of several farm-houses and smaller tenements, and a capital inn.

Cannock Church is a stone edifice, dedicated to St. Luke; the living is a perpetual curacy, of which the Dean and Chapter of Lichfield are patrons. The present curate is the Rev. Thomas Bradburne.

Cannock-heath, which was a celebrated forest during the reign of the Mercian kings, is now a bleak and dreary waste. At Radmore, within its boundaries, the Cistercian monks founded an abbey, in the reign of King Stephen; but in the year 1154, the monks removed to Stoneley, in Warwickshire.

RUGELEY.

The population of the town and parish of Rugeley in 1811, was 2,213, of which number 1,089 were males, and 1,124 females. The number of inhabited houses was 453, and of families 478. Fourteen houses were then building, a proof of the thriving state of the place.

Rugeley is built in a low situation, in a pleasant healthy country near the Trent. The Grand Trunk Canal, which unites the navigation of the Trent and the Mersey, runs past the north side of the town, between it and the river Trent. It is six miles N.E. of Cannock, seven from Lichfield, and nine and a half from Stafford; the turnpike-road to these towns passes through it, and it consequently is a considerable thorough-fare. The facilities of an inland navigation have contributed to the prosperity of this town. It has a neat cleanly appearance; the houses are in general well-built, and some of them elegant. Rugeley is governed by two constables, who are chosen annually by the inhabitants. The principal manufacture is hats. The market-day is Thursday. Two annual fairs are held in this town, one June 6, for colts and horses, and the other Oct. 21, for horses, sheep, and cattle.

The manor was the property of Simon de Rugeley, of Rugeley. In the reign of Edward III. he was appointed sheriff of the county by that prince, and at the same period another of the family was knight of the shire. The Marquis of Anglesea is the present lord of the manor.

The Church is an ancient structure of stone, with a tower, which contains six bells, cast in 1706. The roof is covered with shin-gies. The interior consists of two low aisles. It is dedicated to St. Augustine, and is a vicarage in the patronage of the Dean and Chapter of Lichfield. The present vicar, the Rev. Charles Inge, resides in the vicarage-house, a neat mansion which stands near the church, and is surrounded with a shrubbery. The church contains monuments of the families of Weston, Chetwynd, Lander, and Sneyd.

Dr. Wilkes traces the etymology of Rugeley, from *Ridgley*, a ridge of hills above the town. Cannock-heath approaches within a mile of Rugeley on the south; and Stile Cop, one of its eminences, covered with trees, is seen at the distance of many miles. From the summit there is a prospect of a great part of Derbyshire, Staffordshire, and Worcestershire, and a part of Shropshire.

The Free Grammar-school of Rugeley was established by Queen Elizabeth, who endowed the institution with lands in and about the town, which had formerly been appropriated for the instruction of boys in a chantry, dissolved by Henry VIII.

John Bamford Cowper also founded a charity-shool, and left four hundred pounds in trusts, the interest of which was to be paid to a schoolmaster, for instructing sixteen boys of the poorest families in Rugeley, in reading, writing, and arithmetic, to qualify them to go apprentices to trades. They were also instructed in the Church catechism, and the principles of the Christian religion.

Among the memorable events recorded in the parish register of Rugeley, there is an account of a fire on Trinity Sunday, May 20, 1646, which destroyed 29 dwelling-houses. The sufferers afterwards received £336. 7s. 10d. On Saturday, Feb. 19, 1708, a fire broke out in the west end of Rugeley, at nine o'clock in the morning, and consumed ten habitations, nine barns and offices, and goods to the value of £808. 3s. 2d. which was paid to the sufferers severally by Benjamin Adie.

Hagley, the magnificent seat of Viscount Curzon, is in the vicinity of this town, enveloped in lofty and luxuriant groves and plantations, enclosed by a park-wall and paling. Above Rugeley, the vale of Trent displays several gentlemen's seats, with surrounding and picturesque sylvan scenery. Among others the modern mansion of Bellamore, belonging to the Blount family, is a beautiful object. The demesne is well cultivated, and variegated with groves and plantations. At a greater distance to the west, the elegant mansion of John Sparrow, Esq. rises near the Trent; and on the opposite bank of the river is Wolseley Hall, the ancient seat of Sir William Wolseley. The park is a most romantic succession of small hills, covered with plantations, and some very ancient oaks, and is a most agreeable contrast to the fertile meadows in the vale of Trent below. At this place is the ancient inn, of Wolseley Bridge, where the Rugeley road divides; one branching off for Shugborough, Haywood, and Stone, and the other for Radford and Stafford.

At *Brindley's-bank*, above Rugeley, the Grand Trunk Canal crosses the Trent upon an aqueduct of five arches. This place takes its name from the eminent engineer who planned the canal, and here carried it over the river; and so perfect has been the workmanship of this aqueduct, that it has withstood the floods of the Trent for nearly half a century.

At *Brereton*, near Rugeley, considerable quantities of coal are raised from the mines, and sold in the neighbourhood, or sent by railways to the Trent and Mersey canal.

Baswich is an extensive parish of Cuttlestone hundred, containing the townships of Acton-Trussel, Bednal, Brockton, and Walton and Milford. According to the population returns in 1811, the number of houses was 224, inhabited by 246 families, consisting of 553 males, and 558 females: total 1,111 persons. About three-fifths of the inhabitants were employed in agriculture.

The Church of Baswich is dedicated to the Holy Trinity. The body of this edifice has been re-built with brick, but its ancient steeple of stone remains. It is situated in a solitary spot at the extremity of the parish, near the Staffordshire and Worcestershire Canal. The present curate is the Rev. Joseph Ellerton, master of the Free Grammar-school at Stafford.

Walton and Milford is an hamlet of Baswich. At Milford there is a capital mansion belonging to the Rev. Richard Levett; and a good mansion and premises occupied by John Collins, Esq. of Stafford.

Weeping-cross is a small hamlet in this parish, with a good inn, situated on the road to Cannock. The arable land here is a good rich mild loam, with meadows on the banks of the Peak.

Radford is another small hamlet or village, with an excellent stone bridge over the Penk, and a wharf on the canal, where much business is done in the coal and lime trades, and other branches of commerce.

Brockton is a village belonging to the township of that name, and situated in the east of this parish. Brockton Hall is an ancient mansion, the residence of Sir George Chetwynd, Bart. It is situated to the north of the road to Cannock, about four miles from Stafford. The village is in a pleasant rural situation, and contains several farm-houses and tenements. Brockton Lodge is an elegant and modern mansion, the residence of George Chetwynd, Esq. Chairman of the General Quarter Sessions for this county, and son and heir to Sir George Chetwynd. A large stone quarry is now worked on the Brockton estate: it produces good free-stone for building and other uses, and excellent bricks are made of the stratum of clay above the stone.

The parish of Baswich contains a considerable tract of Cannock-heath, which extends to Milford, within three miles of Stafford. As the Church is at the northern end of the parish, there are two chapels of ease belonging to it in the southern part, one in Acton-Trussel township, and the other at Bednal.

Acton-Trussel is a township of Baswich, about half-way between the parish church and Penkridge. The Staffordshire and Worcestershire Canal passes along the side of this township; the soil is a good gravelly loam, with meadows on the banks of the Penk. It joins Teddesley on the south, and Bednal on the east. The population in 1811 was 271; namely, 128 males, and 143 females. The chapel of ease is dedicated to St. James.

Bednal, another township of Baswich, lies between Acton-Trussel and cannock-heath. The population is 89 males, 100 females, total 189 persons; all of whom, except one family, are employed in agriculture. The land is a mixed gravelly loam. The Chapel of ease is dedicated to All Saints.

The Rev. Mr. Peploe, a man remarkable for the humility and simplicity of his manners, officiated as curate at both these places for many years. Having a large family to support, he has been known to do extra-duty at Bushbury on Sundays, and to walk thither and back, a distance of full twenty miles. When the late Sir Edward Littleton went to London, Mr. Peploe frequently came to Teddesley during his absence by desire of the Baronet, who thought the presence of a clergyman might keep the servants in order. However humiliating this office may appear, it was attended with the most beneficial effects, for the servants were so regular and well-behaved, that Sir Edward seldom dismissed any of them.

Castle-Church is a considerable parish in the north part of Cuttlestone hundred. It is situated to the south of the river Sow, and contains a small part of the town of Stafford, called Stafford-green; also the Old Castle, Highfields, Burton, Moss Pitt, Silkmore, and Rickerscote. In 1811, the parish of Castle-Church contained 144 inhabited houses, and 162 families; consisting of 247 males, and 319 females: total, 566 persons.

The Church, which was formerly an appendage to the ancient barony of Stafford Castle; stands on the side of the parish, remote from the greatest part of the population. Stafford Castle, which is in this parish, is situated on the summit of a hill, about a mile and a half south-west of Stafford. The ancient keep is raised on an artificial mount, of an oblong form, and the whole castle surrounded by a deep foss, with the additional fortification of a high rampart on the side which was considered the most accessible. During the Civil war between Charles I. and the Parliament, this castle was garrisoned by the Royalists, and was afterwards demolished, according to the following order:

"*Dec.* 21, 1643.—It is ordered by the Committee, that Stafford Castle shall be forthwith demolished. Present: Colonel Rugeley, Mr. Compton, L. C. Chadwick, Mr. Broughton, Mr. Swynfen, Capt. Stone, Mr. Greg, Mr. Pudsey, Mr. Bendy, and Capt. Foxall."

The last effort in the Royal cause at this castle seems to have been made by "the ould Lady Stafford," as she is called by Sir William Brereton, a General in the army of the Parliament, in his letter to the Earl of Northampton, which concludes thus: "there are divers considerable persons in this blocked-up castle, which we resolve to observe and attend as much as possible, until we can recover the same and disperse them."

The Castle, and the land around it, are now the property of the Jerningham family, who have laid claim to the ancient barony of Stafford. The soil is an excellent marly loam, suitable for beans and wheat, and the estate includes several considerable farms. Some years ago, the walls of Stafford Castle were cleared of rubbish, and made of an equal height, by Mr. Jerningham, and in 1815 two new towers were built. When finished, it will, from its elevated and commanding site, be a most delightful residence, and a picturesque addition to the beauties of the landscape. It is situated on the highest eminence of the hills of Billington, where they terminate abruptly, and it is probable that the ancient military station at that place had some connexion with this castle.

All the western part of this parish to Moss-pit Bank, and the Wolverhampton road, is thinly interspersed with handsome mansions. In this district, William Keen, Esq. of Stafford, has just erected an elegant house. The chief population of the parish is about Stafford and Rickerscote, or from one mile and a half to two miles and a half distant from the church.

Silkmore, in Domesday-book written *Selchemore*, is the manor and residence of Thomas Mottershaw, Esq. It is situated in the north-east part of the parish; the upland is a light sandy loam, and the lowland rich meadows, on the banks of the Penk, and a rivulet which comes from the west.

Rickerscote, on the south-east part of the parish, contains good estates belonging to Mr. Clarke, Mr. Perkins, and others. The upland is a good light loam, adapted to either pasture or tillage, with a sufficient proportion of meadow-land on the Penk and its collateral rivulets. These meadows have been greatly improved of late years by drainage and embankment, which protects them from inundations. Several saline springs have been found by digging in these meadows, and a spa has been discovered on the estate of T. B. Perkins, Esq. which is highly saline, moderately sulphureous, and considerably chalybeate. This spa is possessed of properties similar to the waters of Cheltenham and Leamington; containing, as has been found on careful analysis, a proportion of iron, salts, calcareous earth, and hepatic air. It is moderately aperient when taken in small doses, and has a tonic effect similar to that of other chalybeate waters when applied either internally or by immersion.

Castle Church is an ancient edifice, composed of brick on the one side, and stone whitewashed on the other, with a stone tower. It is a perpetual curacy, the patronage is in the Crown, and the Rev. Edward Dickenson, rector of St. Mary's, Stafford, is minister, who keeps a curate to assist him in his duties.

HAUGHTON is a parish, with a pleasant rural village of the same name, situated about four miles W.S.W. of Stafford, on the road to Newport, in Shropshire. The parish consists of this village and a few farm-houses. According to the population returns in 1811, it contained 93 houses, and 98 families, consisting of 223 males, and 232 females: total 455 persons. The principal employment of the inhabitants is agriculture, though tanning and some handicraft arts are carried on in the village.

The Church is a neat edifice of brick, with an ancient stone tower. It is a rectory in the deanery of Newcastle and Stone, and Archdeaconry of Stafford.

GNOSALL is an extensive parish in Cuttlestone west, including

Apeton Liberty, Cowley township, Moreton hamlet, and Knightley township. According to the population returns in 1811, Gnosall parish contained 455 houses, 492 families; 1,174 males, and 1,210 females: total of persons 2,384. The soil in this large parish is various, and the uplands generally a strong loam. There are several valleys with rivulets, and the land on their banks by draining and proper management, form good meadows.

The village of Gnosall is an ancient place, with a large Church, which was formerly a College for secular canons. This edifice consists of a nave and two aisles, a chancel and two aisles, and a cross aisle. The tower, which rises in the centre of the fabric, is in the Saxon style of architecture to the height of the roof; and the western arch is adorned with flat receding chevron mouldings. The principal part of the church is built in the modern style of English architecture, and the five windows in the western end are decorated with pointed arches.

Gnosall Church is a perpetual curacy, dedicated to St. Lawrence; the patronage belongs to the Bishop of the diocese, and the Rev. Mr. Lowe is the present minister.

Dawley Moor, a considerable common, is situated about a mile north-west of the village.

The township of *Knightley* is still farther to the north-west from Gnosall. The estate is the property of Lord Anson. It is of considerable extent, and contains many excellent oaks.

Moreton is a populous hamlet of Gnosall. The estate formerly belonged to Lord Ducie Moreton, but was lately purchased by Henry Green, Esq. who is making great improvements on the land by draining.

Wilbrighton is also a small hamlet of Gnosall.

From Knightley to Chatwell this parish extends about seven miles in length. Chatwell is said to derive its name from a spring called Chad's-well, formerly in repute.

Cowley township and hamlet is also in Gnosall parish. The most remarkable mansion here is Brough-hall, the property of Henry Crockett, Esq. The soil is in some places a strong loam, and in others lighter, and inclining to gravel.

The manors and estates of Walton Grange, Bescot, and Broadhill, are in this parish. They include an extent of one thousand acres of old enclosed land, and are the property of John Gough, Esq. Braunston-heath, near Walton Grange, was formerly a common, but has lately been enclosed. Broadhill is a small common, upon a declivity, interspersed with several cottages on its side and summit, some of which command extensive prospects, particularly a distinct view of Aqualate Mere and Chetwynd Park. Snowden, in Carnarvonshire, was once seen by an observer in a clear winter's morning, in a direction beyond Chetwynd Park, above which it was evidently elevated. The object was covered with snow, and though the distance is nearly one hundred miles, Snowden may be seen in similar favourable circumstances from the summit of Broadhill.

Walton Grange estate contains a coppice of twenty acres, of full-grown oak, equal to any in the county for size and quality.

Apeton is a liberty in this parish, which extends to the brook, near Church-Eaton.

NORBURY is a small parish, situated in the west division of Cuttleston hundred, including the township of Weston-Jones. The population is inconsiderable, amounting to 319 persons, of whom 175 are males, and 144 females, chiefly employed in agriculture. The village of Norbury is four miles from Newport, and about the same distance from Eccleshall.

The manor of Norbury was formerly the property of the Skrymsher family, but now belongs to Lord Anson. Erdeswicke says, "this *goodly manor* formerly belonged to Hugh de Kilpeck, the King's champion at the coronation, and that Joan, his daughter, *temp.* Hen. III. brought it to Sir Philip Marmion, who died 1291, leaving several daughters his heirs, of which Maud, the third, married to Ralph Butler, who made this his seat, and built the manor-house or castle, at the head of the Wild-moors (so called in an old deed there), moating it round. These Wild-moors were a royal waste, that began at the high grounds about Knightley, Gnosall, in Staffordshire, running down by Newport to the rivers Terne and Severne in Shropshire; of which there is a full account in the Philosophical Transactions by the ingenious and learned Mr. Plaxton, whose other MSS. I have before mentioned in my preface to Vol. I. of Staffordshire."*

The soil is a light mixed loam, fit for turnips and barley. The parish is divided principally between Lord Anson and the Rev. Sambroke Higgins. There are two small lakes in this parish, one called Blackmere, and the other Cannegreave. They are both upon a rivulet, which runs through them, by an old mill, and falls into Aqualate Mere.

The Church is a plain old stone building, consisting of a nave,

* Mr. Shaw's letter in Gent. Mag. Vol. 71. p. 381.

chancel, and a modern brick tower at the west end of the nave, in which are four bells and a small one. In the chancel are several monumental inscriptions to the Skrymshers, one of whom was Adjutant General to Prince Rupert, and standard bearer to the Gentlemen Pensioners of Charles II. The armour belonging to one of the Skrymshers was formerly preserved in the church, but nothing now remains except the helmet and part of a gauntlet.

The present incumbent is the Rev. Sambroke Higgins, and the patronage belongs to Viscount Anson.

Weston-Jones is a hamlet in this parish, upon the Loynton estate, which is the property of the resident rector, Mr. Higgins.

Leewoods is a manor in this parish, belonging to Lord Anson.

FORTON is an adjoining parish, and situate to the south-west of that of Norbury, in the western division of the hundred of Cuttlestone, and on the borders of the county. In 1811, it contained 121 inhabited houses, 130 families, 297 males, and 310 females; the population of the parish amounting to 607 persons. The turnpike-road from Eccleshall to Newport, in Salop, passes through the village of Forton, which is about two miles distant from the latter town. The parish contains the hamlets of Warton, Sutton, Mereton, and Aqualate.

Sutton is on the same road that passes through Forton, to the north-east thereof, and contains several farm-houses and small tenements. The upland soil is a light loam. "In this place, Sutton or Southtown, (Mr. Loxdale says,) there is a freehold estate, which Robert de Broughton granted to William de Scavington in frank marriage with Ellen his daughter. Hugh de Scavington gave it with Ellen his daughter to Ralph de Leyton, 1314. John Symonds and Ann his wife were possessed of it in 1406, and in 1442 passed it to their son John Symonds. In the beginning of Queen Elizabeth it was purchased by John Bettenson, whose daughter and heir, Margaret, carried it in marriage to Roger Benbow, rector of Bolas, in which family it continued to —— Benbow, of Sutton, with the deeds from whence the above information was taken about 1740. The Bettensons were a numerous and ancient family in this parish; for, besides the above freehold, there were four more, each of them possessed of a copyhold estate, 1611. Richard B. a younger son of one, then removed into Essex, and from him descended Sir Edw. B. of Chislehurst, in Kent, Bart. who dying without issue, 1723, was succeeded by a near relation of his name, Edw. Bettenson, Esq. Near to Sutton is a small estate, called the Gild or Guild of

Monks'-farm, which "was formerly given to the Abbot of the Benedictine Monastery at Shrewsbury, who had his occasional residence there, as appears from the following words in the Monasticon, Vol. I. p. 683: *Ricardus Dapifer Cestriæ* dedit abbate Benedictinorum de Salop, habitationem hermeticam in Sylvâ de Suttonâ."*

Between Sutton and Forton is a conical building of stone, originally a windmill, or intended as such. It was built in its present form as an object to improve the view from Aqualate-hall by a former possessor. At the bottom of a hilly field called the Yeld, and opposite to this building, is a fountain called Wins-well, which contains several springs of excellent water.

Mereton, or *Meertown*, is an hamlet, which derives its name from the neighbouring lake. The manor is termed the manor of Meer and Forton, and the latter was included in the former at the Conquest. Near to this hamlet is a pool called Moss-pool, respecting which a superstitious notion existed in the time of Plot: the rising thereof was taken for a certain sign of a dearth of corn! † Two other prognostics of a less alarming description are recorded by the same author, (ch. i. sec. 50. 51.) as being a communication made to Edwin Skrymsher, of Aqualat, Esq. by Samuel Taylor, a person belonging to the Severn, but employed by Mr. Skrymsher in making his boats, "who foretold such *rains* as are usual and frequent by the winds backing to the sun, as he called it, i. e. opposing its course: viz. the sun moving from east by south to west and north, and so to east again; and the wind from west by south to east and north, and so to west again: *ex. gr.* suppose the wind now in the north, if it shift thence to the east agreeable to the sun's course, it most times proves fair; but if it back to the sun, and shift westerly and thence southerly, &c. so as to oppose its motion, it seldom fails of bringing rain; and so in all the other cardinal and intermediat points. Much more accurat and certain was the same Samuel Taylor in predicting the *winds* than the *rains* that attend them, though even in this too he made use of the *clouds* themselves, which whenever he perceived to rise in the form of the letter V. jagg'd on each side, and therefore called by the *watermen* the *Hartshead*, he forthwith concluded infallibly that the next point of the compass to which the *wind* would shift, would be either the opposite one to the most patulous part of the V. or Hartshead, (which though happens but seldom) or the point to which the acute angle of the

* Gent. Mag. Vol. 71. p. 231. † Plot, ch. ii. sec. 29.

same Hartshead seems nearest to direct itself; and this most frequently happens, always one of the two."

Aqualate Hall, according to Mr. Shaw,* was first erected by Sir Thomas Skrymsher, Knt. who died in 1633. He was lord of the manor, and patron of the church. An engraving of the original building is in Plot, (see chap. vii. sec. 39) where it is termed Aqualate *House*. From the Skrymshers it descended by marriage to the family of Baldwyn, from whence it came by purchase to the present proprietor, Sir John Fenton Boughey, Bart. one of the representatives for the borough of Newcastle-under-Lyme. The worthy Baronet new modelled, rebuilt, and considerably enlarged the old mansion, before he made it his seat. It is now a magnificent Gothic mansion, ornamented with turrets and battlements. The extensive park and pleasure-grounds about it present a delightful variety of rural scenery, and are adorned with several plantations, laid out with much taste; and in which are some of the largest oaks in this county. An arm from one, approaching fast to decay, extends in an horizontal direction twenty-seven yards from the trunk!

Aqualate-mere, the principal ornament of this demesne, is in front of the Hall, and is the finest lake in the county. Plot states its dimensions to be 1848 yards long, and 672 yards broad; thus making it cover a space of more than 250 acres. The river Meese derives its source from this water; which is the resort of various aquatic birds, particularly wild geese and ducks. The latter are sometimes taken in a decoy. Fish are for the most part abundant in this lake, though in time past the otter has proved a destructive animal amongst them; and the voracity of another of their own tribe or genus, in proportion as their numbers increase, must create a diminution of the other species: pike have been taken out of this water of the weight of 36 lbs. Dr. Plot says that he was informed by Edwin Skymsher, Esq. that carp grew to a vast size in this mere, and two were taken therein "remarkably great; one, a *melter* 33 inches long and 19¾ inches about, weighing 15 pounds, and the other a *spawner*, which, though not above two feet six inches long, or thereabouts, yet was 20¼ inches round, weighing 14 pounds, and sold for 14 shillings, the scales of each being near as broad as one of the milled half-crowns" then in circulation (ch. vii. sec. 38). In the same section he says, pike have been

* See Gent. Mag. Vol. 71, p. 290.

caught in the Lake of Geneva of 80 pounds weight, reckoning 18 ounces to the pound. He also instances, where pike having swallowed their prey, the same has been taken out of them alive. One instance occurred at the Black lake, near Aqualate, and another at Rugeley. Aqualate-mere is supplied with water by several streams, and a small fountain at the foot of a hill planted with firs, called Anc's-hill, on the N.N.E side of the lake, and nearly opposite to the Hall.

Plot conjectures that the Anc's-hills derived their name from some Roman captain that lay upon them, "whose *name*, or at least prænomen, *perhaps* might be *Ancus*." He says that "it cannot be denied that the Romans had indeed some action hereabout, there being a raised work here at *Morton* not farr off, which seems to be of their fashion, and no question the large meera that lyes just below it, had its name of *Aqualat (quasi aqua lata)* from them; * * * * * not to mention that all these are in or near the parish of Forton, and that there is a village also not farr off, call'd Warton, which are both thought to derive their names from some such actions as are presumed by the story to have happened hereabout." (chap. x. sec. 6.)

That a *Battle* has been fought near this place cannot be doubted, though between whom may still remain a conjecture, for stronger evidence than the etymology of words, the traces of camps, or proximity of ancient military roads, can be adduced in attestation of the fact. About 100 years ago were fallen some very old oaks, which grew on or about the grounds called the Anc's-hills; at which time, according to tradition, in pits and excavations then made, human bones, skeletons, and arms, were found. In 1815 the remains of many skeletons were found in rows, and appeared to have been laid in trenches on each other. They were accidentally discovered by digging to form a plantation. At Oulton, about a mile from thence, nearly twenty years ago, were found some arms, which fell into the hands of the late Richard Whitworth, Esq. of Batchacre. Whether the battle was fought by the Romans or Saxons, is it improbable, that Morton, Warton, and Forton, were the stations of the two contending armies prior thereto?

From the field of battle, and the contemplation of contending armies, we pass to another place surrounded by other dead, but more peacefully slain, and more affectionately interred; and that is the parish-church.

Forton Church, situate in the village of Forton, is a stone building, consisting of a nave, north aisle, and square tower at the west end of the nave, in which are five bells. It may hold from two to three hundred persons. At the east end of the north aisle is a tomb to the memory of Sir Thomas Skrymsher, Knt. who died July 13th, 1633; and within the rails of the altar is an inscription on a stone slab, to the memory of Richard Skrymsher, the son of John, and grandson of Sir Thomas S. who is represented as having been the chief instrument in preserving King Charles at Boscobel, inasmuch as he procured an asylum for James Earl of Derby, and the Earl procured one for the king in the house of —— Penderell.

At the west end of the nave, near the font, is a painting of the Royal Arms, inscribed at the bottom:

"The Gift of John Bettenson, saddler, of London, 1723,"

which person, no doubt, was of the same family as that once settled at Sutton, and before noticed. The windows of this Church formerly were of painted glass, and full of arms, and images of saints. During the Usurpation two Oliverian officers on a Sunday evening after divine service, came and destroyed them. This they termed reforming the Church. One coat of arms, being small, escaped their infatuated zeal, and remained in one of the chancel windows. It contained the arms of two families, Botetourt and Somery. "Thomas Botetourt, one of the ancient lords of this manor (by marriage with Joan, sister and co-heir of John Somery, 16 Edw. II.) was a benefactor to the parish-church of Forton; for the roof being then old and ruinous, or (if tradition may be depended upon) beat down by the fall of a very lofty spire, he covered it anew, and ornamented the interior." The Church had a second reparation about the year 1723, and is dedicated to All Saints.

As a church and church-yard are places for solemn contemplation, not merely in respect of the deeds of the rich and great, but of every rank and station in life, it is presumed that the serious and devout mind will not be averse to notice traits of character, and remarkable occurrences amongst those, whose lot has been to move in the middle and lowly ranks of society. Grandeur must be admonished not to

"hear with a disdainful smile
The short and simple annals of the poor;"

and every one should bear in mind, when he walks among the dead, that in many a

"neglected spot is laid
Some heart once pregnant with celestial fire."

In this church-yard, near the east door of the church, is a plain tomb thus inscribed:

Beneath this Tomb are deposited the Remains of
JANE HEWETT and HONOR DARWALL } who died { October 28, 1800, aged 71. October 29, 1800, aged 64.
United by early and sincere Friendship,
They dwelt together upwards of forty-five years,
Redeeming the Time,
In Piety tow'rds God, and in works of mercy to their Fellow-Creatures.
In their death they were not divided.
Reader,
From their Example follow Christian morals,
Hold fast the Christian Faith.
So like Them shall you receive the Hope,
Which maketh not ashamed.

"The above two were maiden ladies, and distinguished themselves by a noble example of sincere and affectionate friendship. They were no relations by birth or family alliance, but became acquainted very early in their youth. A strong and inviolable attachment to each other ensued. When they became possessed of their respective fortunes, and settled in the world, they made the same dwelling their common home in the town of Newport, Shropshire; where they lived together in the strictest amity and friendship with each other, charitable to the poor, and much respected by the whole circle of their acquaintance upwards of forty-five years. During some part of this long period, they experienced a reverse of fortune in the loss of a considerable sum of money, which they had placed in the hands of some gentleman, who became a bankrupt. Their loss was very unequal. The one had to lament the loss of a great part of her all, while the other experienced but a small reduction. However, this unequal change in their circumstances produced no alteration in their dispositions towards each other; but, if possible, was a stronger bond of union. When their fortunes were thus unhappily reduced, they perceived that, in case of the death of either, the survivor would be too much confined by the mere income of her own property. To prevent which, each made her will, by which she bequeathed the interest of her all to the benefit of the survivor, at whose death it was to be distributed among each one's respective relations.

"Their last bed of sickness exhibited a scene truly affecting and interesting to the feelings of every tender heart. She that was first taken ill, had the misfortune to break her leg. When both were confined, each perceived that the awful exit of one, if not both, was approaching; and as long as strength would permit, she that was most able went every day into the room of her friend to take a final adieu! and when her strength was exhausted she was carried by the attendants. At these affecting interviews they bathed each other's hands in tears, and expressed an heartfelt wish, that it might please God to permit them soon to meet again in a happy eternity. Heaven smiled and heard the pious prayer; and that angel that took away the soul of her that first departed, was, a few hours afterwards, dispatched to release the struggling soul of her absent friend, and to re-unite them in bonds of love for ever."*

The living of Forton is a rectory in the patronage of Sir J. F. Boughey, who is lord of the manor and chief proprietor of the estates in the parish. The Rev. Mr. Bright is the present rector.

CHURCH-EATON is a parish in Cuttlestone west, and contains the townships of High Onn, Little Onn, and Marston. In 1811, this parish contained 126 inhabited houses, 156 families; 403 males, 401 females: total 804 persons. The soil is of a good fertile quality, consisting of good loam in the upland and meadows.

The village of Church-Eaton contains one street, with several houses, and some detached farm-houses in its vicinity. It is situated four miles and a half to the west of Penkridge. The Church is an ancient fabric in the Saxon style of architecture. The tower is low, and the interior of the church very neat and clean. This church is dedicated to St. Edith, and is a rectory in the patronage of Earl Talbot, and the present rector is the Hon. Mr. Talbot, of that family.

Wood Eaton is a small hamlet to the N.W. of Church Eaton, and contains a few farm-houses and smaller tenements. The manor is the property of Lord Talbot, and the land fertile, being a good loamy soil, on a marl bottom, and well adapted to either tillage or pasture.

High Onn, a township and hamlet in this parish, contains several good farms and farm-houses. The soil is excellent for corn, beans, or pasture. The principal part of this township is the property of Thomas Giffard, Esq. of Chillington. Matthew Parkes, Esq. also occupies his own estate here, and part of the Giffard estate.

Little Onn is a township east of High Onn, and contains Little Onn Hall, the elegant mansion of Henry Crockett, Esq. The estate adjoining the manor is the property of Mr. Parkes.

* See Gent. Mag. vol. 71. p. 126.

Marston is another township in this parish. It contains some farms belonging to Henry Crockett, Esq. Orslow is a capital estate belonging to Lord Bradford; and Shushions, another productive corn-farm, is the property of the Littleton family, of Teddesley, which has long been occupied by one of the Crockett family. A brook which comes from Gnosall, separates this parish on the N.E. side from Bradley, passing on and falling into the Penk below Cuttlestone bridge.

BRADLEY is a considerable parish of Cuttlestone west, including Billington and Woollaston liberties. According to the population returns in 1811, Bradley contained 106 houses, 106 families; 284 males, 279 females: total, 563. Of these inhabitants 97 families were employed in agriculture, and only nine in handicrafts. This parish is of considerable extent, being four miles in length, and the average breadth full two miles, and it contains 4000 acres of land.

The village of Bradley contains several farm-houses and small tenements. According to an enumeration made in 1800, the parish contained 102 houses, and 620 inhabitants; so that, notwithstanding the boasted salubrity of this district, there has been a diminution of 57 in the number of the inhabitants in eleven years.

Bradley Church is an ancient fabric of stone, with a tower. It is dedicated to All Saints, and is a curacy.

Billington is a township in this parish, which extends within two miles of Stafford. It contains several tenements and farms, and the vestiges of an ancient fortification, which, from its proximity to Stafford castle, might formerly have had some connection with that fortress.

Shredicote is another division of Bradley parish, and contains several farms.

LAPLEY is a parish of Cuttlestone west, situated to the south of Bradley, and containing the township of Wheaton-Aston. The soil of Lapley is various, in some places light and mixed loam, and in others, especially towards Lapley-wood, thinner and colder clay. A rivulet comes from Wheaton-Aston, and another from Marston, which separately and together water the skirts of this parish, and enable the farmers to form good meadows on the banks. According to the population returns in 1811, this parish, including Wheaton-Aston, contained 141 inhabited houses, 154 families; 353 males, 393 females: total 746 persons.

Lapley was formerly a market-town, according to Dr. Wilkes's

Antiquities. The market is stated by him to have been held on Tuesdays.

Algar, a Saxon nobleman, founded a priory in this parish in the year 1146, which afterwards rose to great importance. In the Nomina Villarum, 9 Edward II. the prior of Lapley is repeatedly nominated to manage the concerns of Lapley and Wheaton-Aston. This priory was suppressed in the reign of Henry V. In the Civil wars, this edifice was fortified and garrisoned, Captain Smith being appointed Governor; and in 1643, £8. 13s. 7d. was levied weekly in the surrounding district for the maintenance of the garrison. It was afterwards taken by the Oliverians, and the fortifications demolished by an order dated March 19, 1645:

"Ordered, that the inhabitants of Lapley, Aston, Marston, Mitton, Stretton, Bradley, Whiston, and Bickford, shall forthwith pull down the works and fortifications about Lapley Church, and every constable within the aforesaid places is required to send six men every day till they be finished."

The Church of Lapley, which is a much larger edifice than is necessary for the present population of the parish, is dedicated to All Saints. It is a vicarage in the patronage of the Swinfen family, who purchased it, with a good landed estate, from Sir Theophilus Biddulph, many years ago. The ancient farm-house on the Swinfen estate, is a strong building in a commanding situation, and was doubtless the Governor's residence when the place was fortified.

Bickford is partly in Lapley, and partly in Penkridge parish. The principal land-owners in Lapley, besides the Swinfen family, are Edmund Wigan, Esq. and Mr. Hall.

Wheaton-Aston is a considerable township in Lapley parish, containing 109 houses, 122 families; 273 males, 306 females: total of persons 579. The soil in this township is generally a good mixed gravelly loam, and the land on the banks of a brook which runs through it is well adapted for meadows.

The village of Wheaton-Aston contains several farm-houses and tenements. Nearly one-half of this village was burnt down some years ago, by the thatch which covered the roofs of the houses taking fire. It was soon re-built, and the new houses tiled.

The Chapel, which is a chapel of ease to Lapley, is a brick building, with a house adjoining it at one end. The vicar of Lapley is the minister of Wheaton-Aston.

On a large open space near this chapel, a may-pole is erected, where the country people amuse themselves with rustic gambols on May-day, and at wakes and festivals.

Blymhill *with* Brineton is a parish in the western division of Cuttlestone hundred. It is situated to the west of Wheaton-Aston, and contains 2,433 acres of land, the soil of which is various, but principally a strong mixed loam. This parish contains 83 houses, 109 families; 252 males, 261 females: total 513 persons.

A mansion called High Hall, situated on a beautiful eminence, is supposed to have been originally the residence of William Bagod, who in the reign of Henry II. was lord of Blymhill, and had his seat there.

There were formerly two wastes or commons in this parish, called the Heath and the Lawn; but they are now inclosed, and their improvement commenced.

Blymhill Church is an ancient edifice of stone, with a tower in the Gothic style of architecture, containing three bells. It is dedicated to St. Mary. In 1719, the body of the church, except the chancel, was rebuilt. It consists of a nave and south aisle, supported by four pillars and three Gothic arches. The chancel is very ancient, with three Gothic windows, the principal of which, in the east end, was formerly ornamented with painted glass. On the south side is a Gothic arched recess, inclosing a tomb. The living is a rectory, the Earl of Bradford patron, and the present rector the Rev. Samuel Dickenson, a learned and ingenious naturalist. The principal land proprietor in Blymhill parish is the Right Hon. the Earl of Bradford; but there are also some considerable freeholders, residents on their own farms, and their land is generally under an improved cultivation.

Brineton is a hamlet of this parish, and formerly gave title to a knight, John of Brineton: it is well watered by two rivulets, from one of which the hamlet of Brockhurst is named, being in the Saxon language *Brookwood.* The soil near Brineton is light, with cooler clay loam on the extremities, well stocked with flourishing oak trees.

Weston-under-Lizard is a small parish in Cuttlestone west. The term *Lizard* is applied to distinguish the village from Weston-upon-Trent, and is derived from a hill in Shropshire, situated to the south-west of the parish.

Weston-hall, situated to the south of Watling Street-way, is a magnificent ancient mansion of the Bridgman family, now Earls of Bradford. It is built in the style of architecture peculiar to the era of Queen Elizabeth's reign, with an extensive park and pleasure-ground.

The Church, which is situated near Weston Hall, is a small ancient fabric, dedicated to St. Andrew. The living is a rectory, the Right Hon. the Earl of. Bradford patron, and the Hon. and Rev. George Bridgman is the present rector.

The village is small, and contains a few farm-houses and smaller tenements, occupied by persons employed in handicraft arts, and labourers. The whole parish contains 44 houses, 46 families; 145 males, 130 females: total 275 persons.

SHERIFF-HALES is a parish of considerable extent, situated to the west of Blymhill, in the south-west part of Cuttlestone west, and principally in that hundred, though partly in Bradford hundred, Shropshire.

This parish contains 154 houses, 157 families; 425 males, 384 females: total population, 809 persons.

The village of Sheriff-Hales contains several farm-houses and smaller tenements, and there are also several houses on the different farms in the parish. It also contains the hamlets of Heath-hill, Hilton, and Burleton. The soil of this parish is a light gravelly loam, very productive of barley and turnips. The Marquis of Stafford is the principal land proprietor.

The Church is dedicated to St. Mary. It is a vicarage under the patronage of the Marquis of Stafford.

From the foregoing history of Cuttlestone Hundred, it is evident that this division of the county may, with the utmost propriety, be termed an agricultural district.

PIREHILL HUNDRED.

THIS division of Staffordshire, equally remarkable for fertility of soil, beauty of landscape, numerous population, and valuable productions of nature and art, is bounded on the north-east by Totmanslow hundred; on the east and south by Offlow and Cuttlestone; and on the west and north-west by Shropshire and Cheshire.

Pirehill Hundred is divided into two parts, the North and the South; and contains six market-towns, many populous villages, and 74,603 inhabitants. The extensive district of the POTTERIES is included in this hundred.

STAFFORD,

The principal town, from which the county takes its name, and where the Assizes are held, is in Pirehill South, and situated on the northern bank of the river Sow, 135 miles distant from London, 16 from Lichfield, 16 from Wolverhampton, and 16 from Newcastle.

Much fanciful conjecture has been advanced by antiquaries in tracing the origin of the name of Stafford; but hitherto they have been obliged to derive all their information on this subject from mere tradition. Respecting this place, Plot says (ch. x. sec. 28.) "About this time [705] the place or island where the town of Stafford now stands, anciently call'd *Bethnei*, began first to be inhabited by St. Bertelline, the son of a king of this country, and scholar to St. Guthlac, with whom he tarryed till his death: after which, tho' now unknown to his father, he begg'd this *island* of him, where he led a hermit's life for divers years, till disturbed by some that envyed his happiness, when he removed into some desert mountanous places, where he ended his life; leaving *Bethnei* to others, who afterwards built it, and called it *Stafford*, there being a shallow place in the river hereabout that could easily be pass't with the help of a *staff* only. Now whereabout this desert place should be, that St. Bertelline went to, tho' histories are silent, yet I have some grounds to think that it might be about Throwley, Ilam, and Dovedale; and that this was the St. Bertram who has a *well*, an *ash*, and a *tomb*, at Ilam: for if, as Cassgrave says, the town of Bertamly, in Cheshire, took its name from a miracle that St. Bertelline did there, I know not why the people about Ilam, Throwley, &c. might not corrupt his name as much as they in Cheshire, and call him St. Bertram, instead of *St. Bertelline*."* About the year 1386 "there was a wonderfull miracle wrought at his altar, in Stafford, if we may believe Cassgrave."

At the time of the Conquest, Stafford was undoubtedly a place of some importance, for in Domesday Book it is termed a city, in which the King had eighteen burgesses belonging to him, and there were twenty mansions of the honour of the Earl of Mercia. It then paid for all customs £9 in deniers,† and was governed by two bailiffs. But the earliest record of its immunities as a Corporation, is the Charter of King John, which is rather an exemplification and confirmation of former privileges, than a new grant.

When William of Normandy conquered England, he built several castles, over which he appointed his confidential followers as governors, with troops to keep his English subjects in awe. Robert de Tonei was appointed by him the governor of Stafford Castle, from hence he took his name, *De Stafford*.

The Barony of Stafford, from the Conquest to the reign of Richard III. was very extensive, including sixty knights'-fees, of which nine were in demesne, and fifty-one in services. Eighty villages held of this barony; but by the attainder of Humphrey de Stafford, Duke of Buckingham, who was beheaded at Salisbury for rebelling against King Richard, the barony was dissolved.‡

Edward, the son and heir of this unfortunate Duke, was restored to his honour and great part of his estates, but he also soon afterwards fell a sacrifice to the false accusation of Knevet, his steward, whom he had discharged for unjust exactions on his tenants. The castle and manor of Stafford was restored to the family in the 23d year of the reign of Henry VIII.; but issue male failing, in the

* Stafford was formerly called *Betheney*, and was built by Ethelfleda, the heroic widow of Ethelred, Earl of Mercia, in the year 918, and was a chief town in this county during the succession of the Mercian kings. Ethelfleda also built a castle in the town, for the protection of the inhabitants against the predatory incursions of the Danes from the northern part of England.—*Camden's Britannia*.

† Magna Britannia, No. 62, page 66. ‡ Ibid, p. 67.

death of Henry Lord Stafford, in 1637, his sister Mary became sole heiress of his estate. She married Sir William Howard, who was created a Baron by Charles I. by the title of Lord Stafford. This lord being found guilty of a plot against Charles II. was publicly beheaded on Tower-hill, in 1682, and the title devolved on Henry, his eldest son, who was created Earl of Stafford by James II. On the abdication of James, this earl, who was a zealous papist, attended the fugitive king into France, and married Claude Charlotte, eldest daughter of the Count de Grammont, by whom he had no issue.*

In the reign of Edward VI. the burgesses obtained a confirmation of the charter of King John, with many additional privileges. Queen Elizabeth also extended her munificence to this town, and established the Assizes and Quarter Sessions here by Act of Parliament, in the 17th year of her reign. During her progress through England, in 1575, her Majesty perceiving the town to be rather on the decline, enquired the reason, and was informed, it was partly owing to the circumstance of the Assizes having been removed, in consequence of which she promised that this beneficial privilege should be restored. A record of this memorable visit is preserved in the parish register of St. Mary's church: "*Mem.* That the sixth day of August, 1575, our Sovereign Lady Queen Elizabeth came from Chartley in progress to Stafford castle, and was received upon the Pool Dam, without the east gate, by the bailiff and burgesses, with an oration made by Mr. Lamb the schoolmaster, in the name of the town. And the bailiff delivered to her Majesty a goodly large standing cup of silver and gilt, of white wine, which her Highness respectfully and gratefully received; and so she passed through the East Gate Street, the Market-place, the Crobury-lane, and the Broad Eye, and thence on the road to Stafford Park, in the 17th year of her Majesty's most gracious reign."

According to the Charter granted by Edward VI. Stafford is governed by a mayor, recorder, ten aldermen, twenty common-council-men, a town-clerk, and two serjeants-at-mace.

A weekly market is held on Saturday in the market-place, which is a spacious square, and in the market-hall behind and contiguous to the town-hall.

In the twenty-third year of the reign of Edward I. A. D. 1295,

* Magna Britannia, No. 62, p. 68.

Stafford sent two representatives to parliament, and has continued to enjoy this privilege since that time.

The following memoranda are curious illustrations respecting this subject:

"1519. This yere John Ferrers and Humphrey Barber, being burgesses in parliament for the towne of Stafford, received certaine waightes out of the King's exchequer, which were appoynted to be kept here within this towne as in the Kinge's treasure. 10 Henry VIII."

"1532. This yere Mr. Erdeswicke and Mr. Bickley, being burgesses of the parliament for this towne, were allowed their charges from the towne."

Among the records of the Corporation, is a power of attorney from Mr. Erdeswicke, to a person in the borough, to receive the sum allowed for his expences in attending parliament.

In former times, it was customary for the burgesses of Stafford to nominate a tradesman as a colleague to some neighbouring gentleman in the representation. The right of election is vested in the inhabitants who pay scot and lot, and have been admitted burgesses. This right the sons of burgesses, when of age, and all men who have served an apprenticeship of seven years in the town, are entitled to; and the number of electors is estimated at about six hundred.

The celebrated Richard Brinsley Sheridan, equally famed for his wit, eloquence, and patriotism, was one of the representatives for the borough of Stafford from 1780 to 1806. It is only intended here to give a very brief sketch of the political life of Mr. Sheridan, whose eminent abilities fully entitle him to be thus preminently connected with the parliamentary history of Stafford. In the year 1779, when he stood high in the public estimation as a poet and dramatic writer, Mr. Sheridan first felt an anxious desire to make the House of Commons the theatre of his eloquence. It is believed that an application was made, without success, to the Duke of Portland, then a leader in Opposition, to give Mr. Sheridan a seat in parliament for one of his boroughs. However that may be, a general election took place in 1780, when Mr. Sheridan determined to canvass for himself, and directed his views to, and carried his election for the town of Stafford. He began, as he afterwards continued, through a long political life, to support the views of Opposition; and though his eloquence did not immediately burst forth, but rather reserved itself for the present to shine with more conspicuous lustre hereafter, he was by no means an idle or ineffective assistant of his party. When, upon the Rockingham party coming into power, Mr. Fox was Secretary of State for the Foreign Department, Mr. Sheridan received his first political appointment, as Under-Secretary to that gentleman. The Marquis of Rockingham's death, and the appointment of Lord Shelburne to be First Lord of the Treasury, threw out Mr. Sheridan's party; which, however, again came into power in 1783, and Mr. Sheridan was appointed Secretary to the Treasury. Mr. Sheridan continued in office but a very short time; but it must be observed, that, while employed, he conducted himself, if not with the accuracy and close attention of a man trained to habits of business, at least with integrity and honour.

The Duke of Portland being afterwards succeeded as First Lord of the Treasury by Mr. Pitt, it was Mr. Sheridan's lot almost uniformly to be the defender of unsuccessful men and measures, though this did not in the least shake his attachment to his friends. Mr. Sheridan took little part in the memorable disputes concerning America. The principal efforts of his parliamentary oratory were displayed upon the following occasions: his Defence of Mr. Fox's East India Bill; his observations on Mr. Pitt's Perfumery Bill, in the year 1785; his Speech on the Irish Propositions, in the same year; the part which he took in the question of the proper mode of appointing a Regency on the first appearance of his Majesty's lamented illness; his noble conduct during the Naval Mutiny; and the wonderful display of eloquence that was exhibited to the world in his speech summing up the Begum Charge against Mr. Hastings. Mr. Sheridan was the constant advocate of Parliamentary Reform, though without the dangerous temerity and impetuous violence of reformers in general. He was a staunch friend of the Liberty of the Press, and the bold and constant advocate of religious toleration. Such is a short sketch of Mr. Sheridan's parliamentary career; a career upon which it would be superfluous here to dilate. The facts are so well known, that it would be unnecessary—and they are so involved in the various opinions and prejudices attendant upon the great party questions in which he was engaged, that to express any decided approbation or censure of the motives of himself and his party, would only be to revive animosities and to awaken unpleasant feelings that cannot be too soon forgotten. His speeches, if not so highly finished, and so well calculated for perusal as those of Mr. Burke, produced without all question a much greater impression at the time, and remain unexcelled if not unequalled for strength of argument and beauty of imagery.

Mr. Sheridan having declined the representation of Stafford in 1806, was elected, during the short Parliament, for the city of Westminster; and in 1807 was returned for the borough of Ilchester. At the general election in 1812 he again became a candidate to represent the borough of Stafford, but was unsuccessful. Ralph Benson, Esq. and Thomas Wilson, Esq. are the present Members.

Mr. Sheridan, after a languishing illness, which he bore with great fortitude, died on Sunday, July 16, 1816, and was interred on the Saturday following in Poet's Corner, Westminster-Abbey; in a grave near those of Addison, Garrick, and Cumberland. It was determined, as a means of shewing the most respect to his memory, and giving it the highest degree of interest, that his relations and friends should follow him to the grave *on foot*. Accordingly, a most splendid assemblage of rank and respectability met at the house of Peter Moore, Esq. his warm and constant friend, to pay the last sad rites to departed genius. The pall-bearers were, the Duke of Bedford, Earl Mulgrave, Lord Holland, the Earl of Lauderdale, Bishop of London, and Lord Robert Spencer. The Dukes of York, Sussex, Argyle, the Marquis of Anglesea, (upon crutches!) Viscount Granville, and between fifty and sixty noblemen and gentlemen, formed the procession.

It seems to have been the singular and enviable property of Mr. Sheridan, to unite in himself a number of talents, any of which would characterize the possessor as a man of taste, and some of which would singly serve to immortalize his name. He was at once a poet, an orator, a dramatist, a political writer, and lastly, a most delightful companion.

Ye Orators! whom yet our Councils yield,
Mourn for the veteran Hero of your field!
The worthy rival of the wondrous *Three!* *
Whose words were sparks of Immortality!
Ye Bards! to whom the Drama's Muse is dear,
He was your Master—emulate him *here!*
Ye men of wit and social eloquence!
He was your Brother—bear his ashes hence!
While Powers of mind almost of boundless range,
Complete in kind—as various in their change,
While Eloquence—Wit—Poetry—and Mirth,
That humbler Harmonist of care on Earth,

* Fox—Pitt—Burke.

Survive within our souls—While lives our Sense
Of pride in Merit's proud pre-eminence.
Long shall we seek his likeness---long, in vain,
And turn to all of him which may remain,
Sighing that Nature form'd but one such Man,
And broke the die---in moulding SHERIDAN.
LORD BYRON'S *Monody*.

Stafford was once a fortified place, but from its low situation it could never have been a fortress of much strength. It was fortified, except towards the river Sow, by a wall, and a ditch supplied with water from that river. But it was taken on the first assault, and without difficulty, in May 1643, by Sir William Brereton, General of the Parliamentary army. The walls are now demolished, and the ditch filled up, though some traces of the fortifications are yet perceptible near the old Gaol, and the Grammar School. The ground under the walls, at the outside, was marshy, and might easily be laid under water around the town.

When fortified, Stafford had four gates. That near the bridge over the Sow was called the Green-gate, being at the south and on the London road: it was taken down in the year 1780. The arch of the East-gate was standing a few years ago. The Gaol-gate was in ruins in 1680. The site of the fourth gate is unknown.

Modern Stafford is built in a low but pleasant situation, on a fertile plain. It is of considerable extent, and contains ten streets and lanes. The streets are wide and well-paved, and the houses in general compact and well-built.

Stafford contains several edifices, the most remarkable of which is St. Mary's Church. It is a large ancient fabric in the form of a cross, and was formerly collegiate in the archdeaconry of Stafford. It consists of a nave, two side aisles, a transept, and a chancel of three aisles. The transept is 100 feet in length and 25 in breadth. In the centre of it rises the tower, which is octagonal, and 33 feet square at the base. The aisles of the chancel exceed the dimensions of the nave. "The style of architecture in general is the early pointed. To the north of the door-way there is a fragment of the Saxon billet mouldings with fret under it. This fragment is evidently a part of the first building. The head of the western window has trefoils in circles, and upright mullions, which seem to have been originally three quarter attached columns. The aisles are lighted by three lancet-shaped windows. In the nave are several windows looking to north and south, of much later date than any in the church. They have in all probability been struck out long after the erection of the nave, as there are distinct marks of a sharp pointed roof both at the west end and against the tower. There are, likewise, some remains of the original perpendicular buttresses of the aisles, some of which on the north side come to a point in front.

"In the south transept is a large window, the arch in which has certainly been altered, being flat, as likewise the door-way under it. This window has, in all probability, been formerly divided into two or more of a lancet shape. The windows in the chancel are in the style of the reign of Henry the Sixth, or earlier. That which faces the east has undergone some alterations in the disposition of its mullions and tracery. There seems likewise to have been another over it, if we may judge from the cell that remains. The arches in this division of the church are elegantly light, supported by four clustered three quarter columns, the capitals being composed of thin laminæ. What is singular, in the construction of these arches, is that the columns diminish in height from the transept to the east end, and the two rows do not answer each other, so that no two are equal as to the length of shafts. The present floor, which is horizontal, covers the bases of part of them, and exhibits the foundations of others. The north transept seems at one time to have possessed a very considerable degree of elegance, from a pinnacle that remains on the buttress at the east end. The door-way exhibits some receding mouldings and embossed ornaments. A window which is placed over it appears to be of much later date than the transept itself, on account of its numerous divisions and fanciful ramifications. It is greatly too wide in proportion to its height. The north porch is probably the original one. Round the columns are trefoil heads and bands, and a series of plain leaves constitute the capitals. In the tower, already mentioned, are eight windows, two looking towards each of the four cardinal points.

"The nave is separated from the aisles by five highly-pointed arches supported on four semi-columns clustered against a square somewhat larger than their diameter. The architrave, the abaci of the capitals, which are composed of leaves, some having a small volute, and the flat under side of the arches, are little improved from the heavy massive style of the Norman era.

"The font presents a singular piece of antiquity. It is very large, and of a clumsy construction. The bottom part of it is a square, of two feet diameter, and is ornamented with figures of men or baboons, on three sides; all lying flat on their bellies. On the fourth side is the figure of a ram. Above this square are figures of four lions, which form this part of the font into an octagonal shape. Each of these lions supports an upright figure, between which are four semi-globes. The whole is surmounted by projecting mouldings and fascia measuring three feet six inches across. The height of the font is three feet three inches, and the interior or cavity is sufficiently large for the immersion of infants. In this font are two small holes, one which runs through the centre of it, and the other which penetrates the sides. The inscription appears to be in the Saxon character; but being much plastered with paint, it is not possible to decypher it correctly. All which we could make out of it, was as follows:

CLEX. ME FACIENS
TALE. * * * * IES
* * DISCRETVS.*"

The altar-piece, which is of the Corinthian order, is painted; and the organ is said to be one of the finest in the kingdom. A seat in the nave, appropriated to the three town magistrates, is in the fanciful style called arabesque. It was fitted up in the year 1708, and was the gift of a Mr. Bremley, of this town.

There are several ancient and modern monuments in this church; the most conspicuous of which is an altar tomb in honour of Lady Ann Aston, and her husband Lord Edward, of Tixall. It is in the north aisle, enclosed with iron railing, which includes the burial-place of the Cliffords, of Tixall. A very elegant mural monument of white marble bears the following inscription:

" ✠ "

"MARY, eldest daughter and co-heiress of James, fifth Lord Aston. Born August 14, 1743. Married Sept. 21, 1766, to Sir Walter Blount, Bart. Perished by fire, Jan. 20, 1805.

R. I. P."

Under this inscription there is a finely-executed figure of a phœnix in alto-relievo.

"The monument next deserving of notice in this church is that of Sir Edward Aston, and his Lady Joan; whose figures are represented in alabaster, under a large canopy. Sir Edward was the person who constructed the curious mansion of Tixall. This gentle-

* Beauties of England and Wales, Vol. XIII. pp. 891-3.

man is said to have been a distinguished knight in the reign of Henry the Eighth. He died in 1567.

"Against one of the pillars in the chancel, stands a very handsome antique monument, which appears from the inscription to have been erected to the memory of Lady Barbara Crompton. Besides these there are a variety of other monuments in this church; but neither they, nor the numerous grave-stones which form the pavement of the chancel, merit particular description.

"The church of St. Mary, in the times of Popery, was collegiate. King Stephen bestowed it on the Bishop and Chapter of Lichfield and Coventry some time previous to the year 1136, but the precise year is not known. In 1445 the patronage of this church, having somehow or other reverted to the Crown, was granted by Henry VI. to Humphrey Duke of Buckingham. At the time of the Dissolution, in the reign of Henry VIII. it consisted of a Dean and thirteen Prebendaries, as is stated in Dr. Tanner's Notitia. The living is now a rectory in the gift of the King.*"

The low octagonal tower of St. Mary's Church contains eight musical bells. St. Mary's is a rectory. The Rev. Edward Dickenson is the present rector, and the Rev. —— Maude, is curate.

Westward from the church, at a very short distance, there formerly stood a very ancient building, which Mr. Pennant supposes to have been the Dean's house; and most likely his opinion is correct.

Besides the parish church of St. Mary's, Stafford contains another smaller Church, dedicated to St. Chadd. It is a very old neglected structure. The ancient mouldering tower of stone contains only one bell, and the body of the church was cased with brick in the year 1740, which quite destroyed its venerable appearance. The Rev. Henry Rathbone is minister of St. Chadd's.

There is a chapel for the Calvinists in Smokey-lane, of which the Rev. —— Chalmers is the present minister. A large new chapel has been erected by the Methodists at the Broad Eye: the Rev. William Henry Loxdale Haden, who was formerly a Lieutenant in the 1st regiment of dragoons, is the present minister.

Among the public buildings of Stafford, the *County Hall* is a very conspicuous ornament. It is situated on the east side of the market-square. The principal front is very neat, and built of stone. Above the basement four Doric semi-columns support a pediment,

* Beauties of England and Wales, Vol. XIII. p. 896.

in the centre of which is the dial-plate of the town clock, with the figure of Justice with her sword and balance on the left, and Peace with the sword sheathed and partly concealed by foliage and flowers, and her emblematic dove on the right. These figures are very beautiful pieces of sculpture in alto-relievo. The front is 120 feet in extent. The interior consists of several elegant apartments. An Assembly-room occupies the front: it is lighted by six windows, and leads to the court-rooms on each side. In the centre is a staircase, at the top of which is the Jury-room, and several other apartments.

The *County Infirmary*, which stands on the western side of Foregate-street, is a large building of brick, erected in the year 1772. It is supported by voluntary subscriptions and benefactions, which amount to about £900. annually.

The *County Gaol*, which is situated almost directly opposite to the Infirmary, but at a considerable distance, and at the extremity of the town, is a large structure. It was erected in 1788, contains 150 cells for prisoners, and its internal economy does great credit to Mr. Harris, the gaoler.

A *County Lunatic Asylum* is erecting at a short distance from the Gaol, which, when finished, will form one of the most complete and commodious buildings of the kind in the kingdom. It will be conducted upon the most liberal plan.

The *Alms-houses* of Stafford, some of which exhibit very curious specimens of ancient architecture, were built and endowed at different periods, for the accommodation of the indigent inhabitants of the town, many of whom find a comfortable asylum in their old age.

There is also a *Free Grammar School*, founded by the patriotic Regency who directed the affairs of the kingdom in the time of Edward VI. and bearing the date of 1550. The revenue of this establishment is said to be £250. per annum. The Rev. Joseph Ellerton is the present master.

The parish of St. Mary's, Stafford, is of considerable extent, and includes the township of Marston, and the village of Whitgreave.

In the times of Popery, Stafford contained several monastic institutions. At the north end of the town, without the walls, a monastery of Franciscan or Grey Friars, was founded by Sir James Stafford, of Sandon. A priory of Black Canons was likewise founded by Richard Peakes, Bishop of Lichfield, in the year 1180, on the banks of the Sow, two miles east of Stafford. It was dedicated to Thomas-a-Becket, ten years after the death of that imperious but unfortunate prelate, and the place is yet known by the name of St. Thomas. At the Dissolution, its revenue amounted to £198. a-year. The ruins of this priory are near a corn-mill. The place where the priory stood is now considered extra-parochial, and the ancient church of Baswich is supposed to have originally belonged to the institution.

In the year 1344, Ralph Lord Stafford, bestowed a piece of ground on the Green, south of the river Sow, on the Augustine Friars, where they founded a religious establishment; but since the Dissolution it is gone to ruin.

The principal streets of Stafford, called Foregate-street and Greengate-street, are the great thoroughfares between Liverpool, Manchester, and London. These streets contain many good shops, the three principal inns, and several inferior houses for the accommodation of travellers. The number of ale-houses in the whole town is 72.

According to the population returns in 1811, Stafford contained 850 houses, 938 families; 2,401 males, and 2,467 females: total of inhabitants, 4,868.

The manufacture of leather constitutes the staple of this town, especially shoes, which are made both for home consumption and exportation. In allusion to this circumstance, the late Mr. Sheridan, at an election dinner, gave as his toast, "May the Manufactures of Stafford be trodden under foot by all the world." There is also a considerable manufactory of hats, and another of cutlery, carried on in this town.

A singular custom is said to be still kept up in this borough, but there seems no good authority for the assertion, though it has been repeated by one antiquary after another. "Here the ancient custom called *Borough English* is still kept up, and where the father dieth intestate, the youngest son inherits the lands lying within the liberties of the town, with other privileges belonging to Borough English."*

Among several handsome private mansions in this town, a half-timbered house in Greengate-street, is a remarkable specimen of ancient architecture. It is the property of Mrs. Fielden, who now resides in one part of it; the other part is occupied as a boarding-school for young ladies. This building, which has doubtless stood for centuries, is large, high, and at present in complete repair.

* Magna Britannia, No. 88, page 58.

JOHN STAFFORD, a Franciscan friar of considerable celebrity, was born here, as is generally supposed, towards the close of the fourteenth century. He was, according to Fuller, "no contemptible philosopher and divine." His principal work was a History of England written in Latin.

EDMUND STAFFORD, Chancellor of England in the reign of Henry IV. was likewise a native of this town. He was brother to Ralph, first Earl of Stafford, and consequently son of Edmund Baron Stafford. Richard the Second preferred him to the bishopric of Exeter; and Fuller informs us that some authors assert, he was likewise bishop of Worcester, and ultimately archbishop of York. Godwin, however, makes no mention of him as ever having held any other see but that of Exeter, and his authority we are inclined to regard as decisive of the question. This prelate was a great benefactor to Stapelton's Inn, now Exeter college, in Oxford, having settled two fellowships in it, and furnished liberal endowments for their support.

Another distinguished character, a native of this town, was THOMAS ASHBOURN, who lived in the fourteenth century. He was educated chiefly at the university of Oxford, where he afterwards obtained a fellowship, and entered into orders. Wickliffe met in this divine a most vigorous and active opponent to his new doctrines. Not contented with exerting all his talents and knowledge, in endeavouring to prove their falsity, he caused a convocation to be called at London, in the year 1382, where the writings of the Reformer were solemnly condemned.

THOMAS FITZHERBERT, a very ingenious and learned man, was also born at or in the neighbourhood of Stafford, in 1552, and sent to either Exeter or Lincoln college, in Oxford, in 1568. But having been bred a Catholic, the college was uneasy to him; and though he would now and then hear a sermon, which was permitted him by an old Roman priest, who lived privately in Oxford, and to whom he recurred for instruction in matters of religion, yet he would seldom go to prayers, for which he was often admonished by the sub-rector of the house. At length, seeming to be wearied with the heresy of the times, as he called it, he receded without a degree to his patrimony: where, also refusing to go to his parish church, he was imprisoned about 1572; but being soon set at liberty, he became still more zealous in his religion, maintaining publicly that Catholics ought not to go to Protestant churches; for which, being likely to suffer, he withdrew, and lived obscurely with his wife and family. In 1580, when the Jesuits Campian and Parsons came into England, he went to London, found them out, was exceedingly attached to them, and supplied them liberally; by which, bringing himself into dangers and difficulties, he went a voluntary exile into France, in 1582, where he solicited the cause of Mary Queen of Scots, but in vain. After the death of that princess, and of his own wife, he left France, and went to Madrid, in order to implore the protection of Philip II.; but upon the defeat of the Armada, in 1588, he left Spain, and accompanied the duke of Feria to Milan. This duke had formerly been in England with king Philip, had married an English lady, and was justly esteemed a great patron of the English in Spain. Fitzherbert continued at Milan some time, and thence went to Rome; where, taking a lodging near the English college, he attended prayers as regular as the residents there, and spent the rest of his time in writing books. He entered into the society of Jesus in 1614, and received priest's orders much about the same time; after which, he speedily removed into Flanders, to preside over the mission there, and continued at Brussels about two years. His great parts, extensive and polite learning, together with the high esteem that he had gained by his prudent behaviour at Brussels, procured him the government, with the title of rector, of the English college at Rome. This office he exercised for twenty-two years, with unblemished credit, during which time he is said to have been often named for a cardinal's hat. He died there, Aug. 27, 1640, in his eighty-eighth year, and was interred in the chapel belonging to the English college.

Wood has given a list of his writings, containing ten different works, chiefly of the controversial kind, in defence of Popery, and directed against Barlow, Donne, Andrews, and other English divines. But the treatises which were received with most general approbation by Protestants and Papists, are, 1. "Treatise concerning Polity and Religion," Douay, 1606, 4to. wherein are confuted several principles of Machiavel. The second part of the said treatise was printed also at Doway, 1610, and both together in 1615, 4to. A third part was printed at London, in 1652, 4to. 2. "An sit utilitas in scelere, vel de infelicitate Principis Machiavellani?" Romæ, 1610, 8vo. The language of these pieces is a little perplexed and obscure, and the method, according to the manner of those times, somewhat embarrassed and pedantic; but they evince strong sense, a generous disposition, with much reading and experience, and abound with matter which has served as a fund to several authors, who have since written against Machiavel.*

TIXALL.—The manor and parish of Tixall is situated in Pirehill South, about four miles distant from Stafford. This manor was held by Roger de Montgomery of the King, in the 20th year of the reign of William the Conqueror. Peganus de Gastenois was lord of the manor of Tixall in the reign of Henry II. and his posterity held it for many generations, and in latter times took the name of Wartney. Roger Wartney left only one daughter and heir, who was married to Sir John Mavestone, but having no issue, the estate passed to the Littletons, from whom it came to the Astons. Sir Edward Aston built Tixall-hall, and Sir Walter Aston, about the middle of the 16th century, built before the Hall a stately gate-house of stone.

This old manor-house at length became decayed, was taken down, and the present mansion built in 1780, in a plain style. The ancient gate-way remains, and is a curious pile of Grecian and Gothic architecture.

The manor of Tixall is at present the property of Sir Thomas Clifford, a Roman Catholic, and was let by him to the Right Hon. Lord Viscount Granville.

The hamlet of Tixall consists of a few farm-houses and smaller tenements. The Church is dedicated to St. John the Baptist. It is a small fabric of stone, with a low tower, and a rectory. The Rev. William Corne is the present incumbent, and the Rev. —— Clarke, curate.

The parish of Tixall contains 29 houses, 31 families; 113 males, 93 females: total of inhabitants, 206.

The Staffordshire and Worcestershire canal passes Tixall manor to the south, and forms a beautiful lake to improve the view of the landscape, from Tixall-hall. The upland is a rich loam, and the meadows on the banks of the Trent and the Sow are luxuriantly fertile. A large quarry of excellent freestone on this estate supplies great quantities of durable stone for building locks and bridges, for which it is peculiarly well adapted, having the property of resisting the action of water.

INGESTRE is a parish and manor about two miles north-west of Tixall, and four miles from Stafford. About the reign of Henry II. this manor was held by the Mittons, but in the reign of Edward III.

* Biog. Dict. Vol. XIV. p. 396.

issue male failing, it came by marriage into the family of the Chetwynds.

In the year 1673, Walter Chetwynd, lord of the manor, and patron of the parish church of Ingestre, obtained a faculty of the Archbishop of Canterbury to build a new church in a more commodious place, and to pull down the old ruinous edifice, and convert the materials to that use. "In 1676 it was fully finished, being built in the form of our parish-churches, not very large, but elegant and uniform. The walls were all squared freestone, the chancel paved with black and white marble, the windows adorned with the arms of the Chetwynds, in painted glass, the ceilings with the same in fret-work, and the side walls beautified with the funeral monuments of the family, curiously carved in white marble. The bones he caused to be brought from the old church, and deposited them in a vault made under the church. The body of the church he caused to be separated from the chancel by a screen of Flanders oak curiously carved, the pulpit and seats being made of the same wood, and all of equal height; and a curious font of white marble standing in the entrance. Over the portal at the west end, on a small white marble table, is this modest inscription:

DEO OPT. MAX.
Templum hoc
A fundamentis extructum
WALTERUS CHETWIND,
(*Walt. fil. Walt. Equit. aurat. Nepos.*)
L. M.
D. D. D.
Anno Æræ Christianæ,
1676.

"The church being thus finished at Mr. Chetwynd's charge only, was consecrated anno 1677, by Thomas Wood, Lord Bishop of Lichfield and Coventry, by performing all the sacred offices of the day; which done, the pious and generous founder and patron, Mr. Chetwynd, offered upon the altar the tithes of Hopton, a village hard by, of the value of £50. per annum, as an addition to the rectory for ever; presenting the bishop and dean, who preached at the same time, each of them with a piece of plate, double gilt, as a grateful acknowledgment of their service, and entertaining the nobility and gentry, who came to see the solemnity, with a splendid dinner at his house, together with many of the common people."*

* Magna Britannia, No. 66, p. 94.

Ingestre Hall is a stately ancient edifice, situated on the declivity of a gentle eminence. Behind the mansion the hill is covered with a profusion of trees, among which are many full-grown oaks of a great size. This wood is a part of the pleasure-ground, through which there are several extensive walks in various directions. The mansion, which is built in the style of Queen Elizabeth's reign, is high, with an arched projection at each side, with four windows in each in the second story. Above the entrance a handsome tower projects from the main body of the building, and is adorned by an elegant balustrade, similar to that which extends along the whole front. A high turret rises in the centre, and has the appearance of an observatory. The body of the whole fabric is of brick, but the bows or projections are of stone, adorned with pilasters, which give the whole a most singular appearance. Large sums of money have been from time to time expended in the alterations and repairs of Ingestre Hall. The old front of brick has been decorated with ornaments of stone, so that it is composed in distinct parts of these two materials.

The pleasure-grounds are laid out with great taste; they are very beautiful, exhibiting the varieties of full-grown trees, flowering shrubs, clean well-rolled walks, lawns, groves, the river Trent and the village of Weston in the valley, and a distant view of the ruins of Chartley Castle.

The interior of Ingestre Hall is magnificent, and the apartments furnished in a high style of elegance. Over the fire-place in the great hall, there is a portrait of Walter Chetwynd, Esq. the venerable ancestor of the family.

The Church, which stands near the Hall, is a small fabric of stone, with a square tower. It contains several mural monuments of the Chetwynds and the Bagots. It is dedicated to St. Mary, and is a rectory in the patronage of Earl Talbot. The Hon. and Rev. John Talbot is the present incumbent.

The manor of Ingestre came into the family of Chetwynd, by the marriage of Sir John Chetwynd with the heiress of the estate, in the reign of Edward the Third. His descendants were created Barons of Ingestre and Talbot; and in the year 1784, John Chetwynd Talbot, who had previously succeeded his uncle William in the barony, was raised to the dignity of an Earl of the United Kingdom by the title of Earl Talbot of Ingestre.

All the lands in the manor and parish of Ingestre, are now in the sole possession of Earl Talbot, who, on the expiration of the leases of his former tenants, took their farms into his own hands. The population of this parish is very small, the number of houses being 18, of families 21, consisting of 67 males and 55 females: total inhabitants 122.

Two farm-houses, which stood in the valley below the Hall, where the annual country wake was formerly celebrated, were pulled down on the expiration of the leases, the fences levelled, and the fields incorporated with the park.

The estate of Ingestre extends to Weston, and into St. Mary's parish, Stafford, including Hopton-heath, and the ancient farm of the Beacon. It also contains the site of the Priory of St. Thomas, which is now occupied by a corn-mill, the hills near it being adorned with groves and plantations, which form a conspicuous ornament to the country.

At *Hopton-heath*, a short way to the south-west of Ingestre, a severe action was fought between the King's forces, under the Earl of Northampton, and the Parliamentary army, commanded by Sir John Gell and Sir William Brereton. The Earl of Northampton had purposed the relief of Lichfield; but that town having been compelled to surrender before his troops could arrive, he determined to march upon Stafford, which had been immediately after invested by a detachment of the victorious republicans. Upon the approach of the Royalists, Sir John Gell, who commanded this force, retired with the view of forming a junction with Sir William Brereton, who was collecting his troops with the same intention. This point being effected, both Generals retraced their steps towards Stafford, and encamped at this place, which lies three miles to the north-east of the town. The Earl of Northampton immediately led his forces against them; and, notwithstanding their great superiority in numbers, attacked them with incredible impetuosity. A long and obstinate contest took place, in which, after performing prodigies of valour, the Earl's horse having been shot under him, he was surrounded and slain. Notwithstanding thus unfortunate, the Royalists continued the battle, and, according to their own account, ultimately gained a decided victory. The Parliamentary army, on the other hand, asserted that though defeated at first they were in the end successful; and, if success is to be estimated by its consequences, they certainly had the best of the day.

Coton Clanford, a small village situate about three miles west of Stafford, gave birth to William Wollaston, a distinguished writer, descended from an ancient family in this county, and born on the 26th March, 1660. He was sent to a private school at ten years of age; and upon the 18th of June, 1674, when he was a little past fifteen, admitted a pensioner of Sydney college, in Cambridge. He acquired a considerable reputation for parts and learning; and having taken both the degrees in arts at the proper seasons, left the University in 1681, not without some disappointment upon having missed a fellowship in his college. He had commenced master of arts the summer before; and it seems to have been about this time that he took deacon's orders. In 1682 he became assistant to the head-master of Birmingham school; and in a short time got a small lecture of a chapel about two miles distant. At the end of four years he was chosen second master of the school, and upon this occasion took priest's orders; for the words of the charter were interpreted to require that the masters, of whom there were three, should be in these orders, and yet should take no ecclesiastical preferment. In this situation and employment he continued till the 19th of August, 1688; when, by the death of a rich relation of his name, he found himself possessed of a very ample estate. In November following he came to London; and about a twelvemonth after, the 26th of November, 1689, married Mrs. Catharine Charlton, a citizen's daughter. She lived with him till the 21st of July, 1720; and he had eleven children by her, four of whom died in his life-time.

After his arrival in London, he may most truly be said to have settled there, for he very seldom went out of it; and we are told, that for above thirty years before his death, he had not been absent from his habitation in Charter-House-square so much as one whole night. In this his settlement in town he chose a private and retired life, although his carriage was ever free and open. He aimed at solid and real content rather than shew and grandeur; and manifested his dislike of power and dignity by refusing, when it was offered to him, one of the highest preferments in the church. He was very well skilled in the learned languages, Latin, Greek, Hebrew, Arabic, &c., and thoroughly versed in all branches of useful learning, as philology, criticism, mathematics, philosophy, history, antiquities, and the like. He accustomed himself to much thinking, as well as to much reading: he was indeed of opinion that a man might easily read too much; for he considered the *helluo librorum* and the true scholar as two very different characters. The love of truth and reason made him love free thinking; and, as far as the world would bear it, free-speaking too. He composed a great number of works, the greatest part of which he is said to have burned during the two or three last years of his life; but some imperfect sketches remain.

Not long before his death, he published his treatise intitled *The Religion of Nature Delineated*, a work for which so great a demand was made, that more than ten thousand were sold in a very few years. He had scarcely completed the publication of it, when he unfortunately broke an arm; and this, adding strength to distempers that had been growing upon him for some time, accelerated his death, which happened upon the 29th of October, 1724. He was a tender, humane, and in all respects worthy man; but is represented to have had something of the irascible in his constitution and temperament. His *Religion of Nature Delineated* exposed him to the censure of our zealous christians, as if he had put a slight upon christianity by laying so much stress, as he does in this work, upon the obligations of truth, reason, and virtue; and by making no mention of revealed religion, nor even so much as dropping the least and most distant hints in its favour. It has indeed made him pass for an unbeliever of revelation with others who have not expressed any zeal at all for it; for the late Lord Bolingbroke supposes Dr. Clarke to have had him in his eye when he described his fourth sort of theists. Mr. Wollaston held and has asserted the being and attributes of God, natural and moral; a Providence, general and particular; the obligations to morality; the immateriality and immortality of the soul; a future state: and Clarke's fourth sort of theists held and asserted the same. But whether Mr. Wollaston, like those theists, rejected all above this in the system of revelation, cannot with any certainty be concluded; and though at the same time the contrary perhaps may not appear, because, whatever might have been thought necessary to prevent offence from being taken, it was not essential to Mr. Wollaston's design to meddle with revealed religion. In the mean time, Lord Bolingbroke has treated Mr. Wollaston's *Religion of Nature Delineated* as a system of theism; which it certainly is, whether Mr. Wollaston was a believer or not. His Lordship calls it "strange theism, as dogmatical and absurd as artificial theology," and has spent several pages to prove it so; yet allows the author of it to have been "a man of parts, of learning, a philosopher, and a geometrician." It is written with a degree of elegance far superior to the style of most English writers, and may justly be regarded as one of the best and most classical works in the English language. He was buried at Great Finborough, in Suffolk, (one of his estates, and afterwards the principal residence of his eldest son), close by the side of his deceased wife.

WESTON-ON-TRENT is a small village situated on the banks of the river Trent, about a mile to the north of Ingestre Hall. The Grand Trunk canal passes through this village, and the turnpike-road between Stafford, Stone, &c. This village is apparently prosperous, and contains two inns, and several smaller tenements and shops.

The Church is an ancient structure, in the Gothic style of architecture, and was built in 1685, as appears by an inscription on a stone over the southern entrance. This edifice is of stone, with a low strong tower, ornamented with large open Gothic windows, and partly covered on the southern side with a flourishing ivy tree. A spire about twenty feet high rises from the centre of this tower, and is terminated by a vane. Two large yew trees, one of them with the "high top dry with bald antiquity," stand among the tomb-stones in the church-yard, and are picturesque objects.

Weston church is a curacy: the Rev. Benjamin Johnson is the present minister.

The principal part of the fertile parish of Weston-on-Trent is the property of Earl Ferrers. The population of this parish in 1811, was as follows: 52 houses, 52 families; 209 males, 185 females. Total of inhabitants, 394.

The situation of the village of Weston is particularly pleasant, and affords a delightful view of a variety of interesting objects. The ancient village church with its airy spire; barges constantly passing along the canal, towed by horses; the turnpike-road, with a handsome stone bridge of one large arch over the Trent; a corn-mill, at work in a field on the right of the road towards Stafford, and about a mile distant on the same side; the woody elevations of Sandon Park, with an obelisk to the memory of that great statesman Mr. PITT, on an airy summit environed by trees; on the left near the bridge, Weston-hall, now a farm-house, in rather a neglected state, and the environs of Ingestre Park, covered with clumps and groves.

At a short distance from Weston, the old salt-works of Shirley-wich are worthy of attention. The works are now occupied by Mr. Moore, and a small branch is cut from the Grand Trunk Canal, for the conveyance of coals to, and salt from, the works.

The brine, from which the salt is made, is raised from a saline spring into a large reservoir by a water-wheel, which works a double pump. This wheel is worked by a stream of water which flows from the high ground at Chartley. The quantity of brine daily used is about 600 hogsheads, from which 520 bushels of salt is made, of half a hundred weight to the bushel, or thirteen tons per day. The proportion of salt in the brine is about one-fourth. The whole process of the manufacture is gone through in 24 hours: the water is evaporated, and the salt is formed in squares of about half a hundred weight each. These squares are neatly formed, dried in kilns, and will bear conveyance without packing. The loose salt is packed up in baskets and bags. The pans in which the brine is boiled are made of wrought iron. Eight of these pans are kept boiling. About 20 families of 100 persons are principally employed in this business, and all live at Shirley-wich, which is also the residence of Mr. Moore. The quantity of salt produced every year is about 160,000 bushels of half a hundred weight each; the duty paid to Government is fifteen shillings per bushel, or £120,000 per annum, and the salt is sold at sixteen shillings per bushel. Thus only one shilling is allowed for the owner of the estate, and the manufacturer, including the expence of machinery, the rent of the premises, and the interest of capital. The quantity of coal used is about one hundred tons per week, and the quantity of salt made is about 80 tons weekly, or 4,160 tons a year.

The salt spring is supposed to be inexhaustible, and only varies a few feet with the vicissitudes of the seasons. Some years ago the quality of the brine was weakened, as is conjectured, by the breaking-in of fresh water. A number of men were employed for some weeks to lower the brine in the shaft, and endeavour to discover the cause, but without effect. The brine could not be materially reduced, and it has remained ever since at the present standard. From the incident, however, it seems that the works might be ruined by a sudden influx of fresh water into the reservoir, but the spring has existed in great strength for some centuries, and may continue so for ages to come. This saline water in its natural strength is destructive to vegetation, and if accidentally let into the fresh water stream, will destroy the fish as far as the Trent.

As the duty is remitted on the salt used in bleaching, a similar indulgence might be granted to the agriculturist, and it is very probable that salt properly applied would powerfully promote vegetation. For this purpose foul or impure salt might be used, and the revenue protected from fraud, as there is a regular attendance of revenue officers at the works. Brine and other ingredients might be mixed with dung-hills and composts, and chymistry might improve agriculture.

An elegant mansion has just been erected by Mr. Moore, near Shirleywich.

SANDON is a small village on the turnpike-road between Stone and Rugeley, consisting of a large inn, a few public-houses, and several scattered tenements. The manor of Sandon is large, and the whole parish contains 92 houses, 92 families; 238 males, 242 females: total of inhabitants, 480. The parish contains the hamlets of Great and Little Sandon, Hardwick, Small Vise, Dracote, Leighs Newton, and Cresswell.

Before the Conquest, Algar, Earl of Mercia, was lord of the manor; it was afterwards given by William the First to Hugh, Earl of Chester, who gave it to William Baron Malbank. His son Hugh founded Combermere-abbey; and his son William gave Sandon Church to the Monks. Adilla, the daughter of this William, married Sir William Stafford, and had the manor of Sandon for her portion. She left it to her son James Stafford. He had issue one daughter, who marrying Thomas Erdeswicke, brought this manor into his family. Sampson Erdeswicke was the last of this family. He was a learned and industrious antiquary; and, among other public works, he repaired and glazed Sandon Church. The manor was sold to Mr. George Digby, in the reign of James I. and came by marriage into the possession of Charles Lord Gerard, of Bromley, whose grand-daughter carried it to William, Duke of Hamilton, by one of whose descendants it was sold to Lord Harrowby, whose son, the Right Hon. the Earl of Harrowby, is now the possessor.

The Mansion-house is an elegant building, in a pleasant situation. It was built on the site of the ancient half-timbered house of the Erdeswickes, and traces of the moat which surrounded it are still visible. Beyond the modern mansion, the sides of the hill are covered with plantations of young and flourishing trees; and on the summit of an eminence, the lofty Obelisk erected to the Memory of Mr. PITT, about eight years ago, is a very striking and beautiful object.

The Church stands on a gentle elevation not far from Lord Harrowby's seat: it is dedicated to All Saints, and is a vicarage. The Rev. George Bonney is the present vicar. Sandon Church is a small structure of stone with a square tower, on the western side of which a very beautiful ivy-tree has climbed almost to the pinnacles, rising like two wings, and leaving the aperture or Gothic window of the belfry uncovered. Among the monuments in the chancel the following to the celebrated Erdeswicke is most conspicuous:

"*Ricardus de Vernon* Baro de *Sibroc* 20 Willmi Conquestoris Pater harum ffamiliar 'de Vernon Holgreve et Erdeswick, 1066.

Hoc sibi spe in Xpo resurgendi posuit *Sampson Erdeswich* Armiger. qui gen. recta serie ducit a Ruo de Vernon, barone de *Sibroc* tempore gui' loqst.	Hujus filli et heres Hugo de Verne duxit fillia et haerede Rainaldi Ballioli dm de *Erdeswich* et Holgreve dederat fillio Mattheo uyu' filli' inde dict' fuit de Holgreve.

Vernon. Malbanc.	Vernon—1 Vernon—Vernon. Baliole—Vernon—1 Holgreve. Holgreve. Ricardus fillius junor *Mathaei* de Holgreve tertij cum pater illi Erdeswik dedisset nomen de Erdeswik sibi assumpsit reliquit et ex
Vernon. Manderill.	altera haerede Guil. *dni* de *Leighton Thomam.*—de Erdeswik genuit cujus pronepos Thomas quartus, accepit in uxore Margareta unica filia et heredem *Jacobi Stafford* de *Sandon* militis cujus proava fuit AVda una filliar et heredu Warini ultimi baronis de Sibrei preav.
Stafford's Vernon.	vero *Guil. Stafford* fillius secund' Harvaei Bagod ex Meliceta Baronissae Staffordiae quae fuit proneptis *Roberti* primi baronis *Staffordiae* qui *Anglia Guil.* Conquestore ingressus.
Stafford Walkelin.	Stafforde | Stafforde | Stafforde | Erdeswik Stafforde | Erdeswik | Minshal Erdeswik — Erdesw. Clinton — Erdeswik Erdeswik Erdeswik Basset Erdeswik Harcourt Erdeswik Grey Erdeswik Lee

Sampson Erdeswik. Elisabetha Dikeswell	Sampson Erdeswik. Maria Neale
Elisabetha uxor prima fuit filia secunda et una trium heredum *Humfridi Dikeswell de church Waver* in com. *Warwici* armigeri ex qua quinque suscepit filias *Margaretam* nondum nuptam, Helenam uxorem Thomae Coyne de Weston Coyne in comitatu *Staffordiae* Armigeris Elizabetham, Mariam et Margeriam. omnes superstites necdum enuptas.	Maria uxor secunda fuit fillia secunda genitaetuma heredū Francisci Neale de Kaythorpe in comitatu Lecestrie armigeri quae filij peperit Richardum et Matheaum fillios et Johāna fillia ut priori marito *Everardo Digby* armigero 14 liberos enixa est, e quibus *Everardus*, Joannes, Georgius, Maria, Elisabetha, Ffrancisca, et Christiana, nunc sunt superstites.

Vernon Semper Viret. Anno Domini 1601."

The monument represents a colossal figure of Mr. Erdeswicke, in a recumbent posture, dressed in a jacket, and with spurs on his heels. Above, in two niches, appear the figures of his two wives kneeling. Mr. Erdeswicke died April 11th, 1603. This beautiful memorial has been much injured by repeated whitewashings.

SAMPSON ERDESWICKE.—This celebrated antiquary, to whose researches other writers on the Antiquities and Topography of this county are so highly indebted, was born here, but the year of his birth is not recorded. He was admitted a gentleman Commoner at Brazen-nose College, Oxford, in the year 1553, and after having remained two years at the University, he returned to Sandon, where he amused himself by antiquarian researches, especially such as were illustrative of his native county. He left in manuscript, "A Short View of Staffordshire, containing the Antiquities of the same County," an incorrect copy of which was published in 1717, in octavo, and a second edition in 1723. There are two copies of the original in the British Museum, and one in Gough's MSS. in the Bodleian Library. Mr. Erdeswicke is recorded by Fuller, in his "Worthies of England," as a gentleman descended of a right worshipful and ancient family; and Camden styles him *venerandæ antiquitatis cultor maximus.* He began his antiquarian work in 1593, and continued it till his death, so that it employed him ten years.

Instances of Longevity.—Robert Reeves, of Smallvise, died March 28, 1758, aged 90.—Margaret Done, died February 2, 1781 aged 91.

The lands in the parish of Sandon are fertile, and the trees flourishing.

GAYTON, a village and small parish of Pirehill South, is situated about two miles to the east of Sandon. The village is small and pleasant, on a fertile loamy soil, and from the higher grounds there is a magnificent view of Sandon Obelisk, the plantations of Lord Harrowby, and the still more picturesque ruins of Chartley Castle, situated on the side of a hill, about a mile distant, in the parish of Stone. The parish contains 48 houses, 49 families; 137 males, 124 females: total of inhabitants, 261.

Two rivulets of clear water pass near the village of Gayton, and unite in the adjacent meadows.

The Church, which is of stone, is a place of great antiquity, built in the Gothic style with small arched windows. The window at the east end is adorned with painted glass. The interior is in a ruinous state, the old oak pews and seats mouldering and falling to pieces. There is a very ancient monument of the Ferrers family in a recess on the north side of the chancel.

Gayton Church is dedicated to St. John the Baptist, and is a curacy: the Rev. Thomas Hilditch is the present minister.

FRADSWELL is a village to the north-east of Gayton, and in the parish of Colwich. It is a scattered village on a gravelly soil. The whole liberty contains 38 houses, 38 families; 120 males, 126 females: total of inhabitants, 246.

The Church is a chapel of ease to Colwich. It was re-built in 1764, the lower part of the nave and tower of stone, and the upper part of brick: the ancient stone chancel remains. It is a curacy, the Dean and Chapter of Lichfield being patrons, and the Rev. Benjamin Johnson, who resides at Weston-on-Trent, is the present minister.

MILWICH is a large populous parish of Pirehill South, about two miles north-east of Sandon, on a loamy soil. The village of Milwich is in a low situation, a brook passes near it, and the Church, which is a small modern structure of brick with a tower of stone, is situated at a short distance on a gentle eminence. It is dedicated to All Saints, and is a vicarage: the present vicar is the Rev. —— Rice, and the Rev. John Sell, is curate.

The parish of Milwich contains 109 houses, 111 families; 290 males, 273 females: total of inhabitants 563.

STOWE is a large parish of Pirehill South, to the north of Colwich. It includes Chartley Castle, and the villages of Hixon and Haywood.

Chartley was, at the time of the General Survey, in the hands of the Conqueror, whose successor, William Rufus, gave it to Hugh, Earl of Chester. In this family it continued several successions, and Ranulph, Earl of Chester, built the Castle in the time of Henry III. but dying without heirs, his estate descended to his four sisters, of whom Agnes had this manor, and being then married to William de Ferrers, Earl of Derby, brought it into his family. It is now the property of Earl Ferrers, and is one of the most romantic relics of antiquity in Staffordshire, being an ancient ruin similar to that of Dudley Castle. The site is on the side of a lofty eminence, and the remains of the castle consist of two round towers, partly covered with ivy, rising amid the foliage of nearly one hundred full-grown yew trees, of the age of many centuries. This ancient fortress is surrounded by a deep moat, and doubtless was considered as a place of great strength before the invention of gunpowder. It has long been uninhabited; and Earl Ferrers, in his annual visits to

Stowe, Gayton, and Weston, generally lodges at the house of one of his opulent tenants.

At present, the ruins of Chartley Castle remind the reflecting observer of the description of the Hall of Ossian. The ancient mansion of Chartley was some years ago consumed by fire. It was surrounded by a moat on low ground, about a furlong west of the castle.

Chartley Park consists of one thousand acres of excellent land, in a state of nature, enclosed within an ancient oak paling, with here and there an old tree, and a few modern clumps of plantations. This tract, which has never been cultivated, abounds with fern and furze. The stock is wild cattle, and deer of two species: the wild cattle are mostly white, with the nose and ears tipped with black.

HIXON, in this parish, is a large ancient village, situated on an eminence. The houses are thatched. The post of a pillory remains, which proves the antiquity of this mode of punishment.

The Ferrers family are said to possess 10,000 acres of land, lying nearly together in the parishes and townships of Chartley, Stowe, Gayton, Weston, Colwich, Hixon, Haywood, and Fradswell.

The parish of Stowe contains 125 houses, 156 families; 440 males, 413 females: total of inhabitants, 853.

ABBOTS BROMLEY

Is an ancient market-town of Pirehill South, situated five miles north-east of Rugeley. It is an irregular decayed place, without manufactures or trade, and took its name from the Abbey of Bromley, in the neighbourhood. There are no public buildings in this town worthy of observation, except the Church, which is an ancient structure of stone, with a high tower, which contains six musical bells. It is dedicated to St. Nicholas, and is a discharged vicarage: the Rev. P. P. Neale is the present minister.

The parish of Abbots Bromley is extensive and populous, including the township of Bromley Hurst, and the liberty of Bromley Bagots. This parish contains 279 houses, 291 families; 737 males, 802 females: total of inhabitants, 1,539.

In the town of Abbots Bromley there is a Free-school, founded in 1603 by Mr. Richard Clark; and an Alms-house for six indigent old women, endowed by Mr. Lambert Bagot. The weekly market of this town is held on Tuesday; and there are three annual fairs, namely, the Thursday before Midlent Sunday, May 22, and August 24, for horses and cattle.

A remarkable custom, called the *Hobbyhorse Dance*, is mentioned by Dr. Plot, as having existed in this town within the memory of many persons alive at the period when he wrote. It was a sort of amusement which the inhabitants celebrated at Christmas, or New-year's Day and Twelfth-day. On these occasions a person danced through the principal street, carrying between his legs the figure of a horse composed of thin boards. In his hands he bore a bow and arrow, which last entered a hole in the bow, and stopping on a shoulder in it, made a sort of snapping noise as he drew it to and fro, keeping time with the music. Five or six other individuals danced along with this person, each carrying on his shoulder six reindeers' heads, three of them painted white, and three red, with the arms of the chief families who had at different times been proprietors of the manor painted on the palms of them. "To this hobbyhorse dance there also belonged a pot, which was kept by turns by four or five of the chief of the town, whom we call *Reeves*, who provided cakes and ale to put into this pot. All the people who had any kindness for the good interest of the institution of the sport giving pence a-piece for themselves and families, and so foreigners too, that came to see it; with which money the charge of the cakes and ale being defrayed, they not only repaired their church, but kept their poor too, which charges are not now perhaps so cheerfully born."

This practice seems to have existed at other places besides Abbots Bromley; for we find hobbyhorse-money frequently mentioned in the old parish books both of Stafford and Seighford. It continued in force till the era of the Civil wars between the Parliament and the House of Stuart, at which time Sir Simon Degge informs us, that he saw it often practised. The same author adds, in another part of his work, "that they had something of the same kind to get money for the repair of the church of Stafford, every common council then collecting money from his friends; and whosoever brought in the greatest sum to the hobbyhorse, was considered as the man of best credit, so that they strove who should most improve his interest: and, as he remembered, it was accounted for at Christmas." *

BLITHEFIELD is a parish about two miles west of Abbots Bromley.

* Paroch. Antiq. Stafford MS. Gough's Camden, Vol. II. p. 514.

The village of Blithefield is chiefly remarkable for an ancient mansion of the Bagot family, which contains a splendid collection of paintings in the first style, and by the ablest artists.

Amongst others, there is a portrait of Lord Treasurer Burleigh, equally eminent as a statesman and a christian. He is painted with a white beard, a bonnet, and collar of the Garter, the George, and a white wand. Henry, Earl of Huntingdon, cotemporary with Lord Burleigh, is painted with the collar of the Garter, and a forked beard. The date is 1588, aged 52. But the most remarkable portrait is that of Walter, Earl of Essex, the unfortunate favourite of Queen Elizabeth. He is represented at half-length, and in armour.

Bagot's Park, which belongs to this mansion, is situated two or three miles to the north-east, on the opposite bank of the Blithe. It contains many oaks, of extraordinary growth, some of the trees containing from 300 to 400 feet each. This timber is of great antiquity, being mentioned as full-grown by Dr. Plot in 1686; it is, consequently, verging to decay. The woods extend over many hundred acres, and are almost wholly of oak, in different stages of growth, some of them carrying timber to the length of 60 or 70 feet.

The Church of Blithefield is dedicated to St. Leonard: it is a rectory, Lord Bagot is the patron, and the Hon. and Rev. Richard Bagot, his Lordship's brother, is the present incumbent.

COLTON is a parish of Pirehill South, and situated southward of Blithefield. This parish extends to the river Trent, adjoining to Rugeley. The soil is a mild fertile loam, with a good proportion of meadows on the verge of the Trent, and on a brook which falls into it. The Swedish turnip is cultivated in rows on the Northumberland system in this parish, and the lands in general are productive of abundant crops of grain, and in an improved and high state of cultivation. Colton parish contains 101 houses, 105 families; 266 males, 218 females: total of inhabitants, 484.

The village of Colton is small. It is a pleasant rural place, in a low situation. A private mansion of brick and stone, with a fish-pond and pleasure-grounds, on the opposite side of the turnpike-road, which passes through this village, are the most conspicuous objects.

Colton Church, situated in an enclosure near the road, is small. The nave is modern, and built of brick, and the roof covered with lead; but the chancel is ancient, of stone, and the roof covered with tiles, with a small stone cross at the end. The tower is square, and of stone: it is low, and from its small narrow Gothic windows, and time-worn pinnacles, it must have stood for some centuries. The interior of the Church is neat and clean, and the pews of oak.

There are but few instances of longevity recorded on the frail memorials of the dead in the church-yard. An epitaph on a mason deserves to be mentioned. His tombstone is decorated with the compass and square in basso-relievo, under which is the following inscription:

"Sacred to the memory of JAMES HAYWOOD, who died May 4th, 1804, in the 55th year of his age:

The corner-stone I oftentimes have dress'd,
In Christ, the corner-stone, I now find rest;
Though by the Builder he rejected were,
He is my God, my Rock, I build on here."

The church of Colton is dedicated to St. Mary, and is a rectory in the patronage of the Dean and Chapter of Lichfield: the Rev. Dr. Buckeridge is the present rector, and the Rev. William Robinson, curate.

It is difficult to describe the fine country beween Colton and Colwich, especially the Vale of Trent, and the beautifully picturesque hills on the southern bank of that river. These hills, and the intermediate valleys, exhibit all the varieties of cultivation and the beauties of woodland, with here and there an airy peak rising above the rest, covered with fern and gorse. The whole scene reminds us of Thomson's descriptive lines:

"O vale of bliss! O softly-swelling hills!
On which the Power of Cultivation lies,
And joys to see the wonders of his hand."

COLWICH is a village in a pleasant situation on the turnpike-road between Stone and Rugeley. It consists of several good houses which extend along the side of the road.

The Church is an ancient structure of stone, situated on a gentle eminence, and adorned with a handsome high tower with four pinnacles, surmounted by gilt globes and vanes. The tower stands on the northern angle of the west end, and is a pleasing object on the approach to the village from Rugeley. The principal entrance of the church at the west end is a neat modern portico of hewn-stone, with a battlement and small turrets. Altogether, the exterior of the church is beautiful: the interior is clean, and the pews neat. The organ was the gift of Viscountess Anson. Among the numerous memorials of the dead interred in the church-yard, there is not one remarkable instance of longevity.

The church contains some monuments of the families of Anson and Wolseley. The burying-place of the Ansons is in the form of an Egyptian catacomb. The monument of Sir William Wolseley, bears an inscription descriptive of his accidental death, being drowned in his chariot, on the 8th of July, 1728, in the neighbourhood of Longdon, by the bursting of a mill-dam, which was occasioned by a heavy shower of rain during a thunder storm. His four horses were also drowned, but the coachman escaped, and was carried by the torrent into an orchard, where he remained in safety till the flood abated.

Colwich church is dedicated to St. Michael, and is a prebend of Lichfield, the Dean and Chapter are consequently patrons: the Rev. —— Pitchford is minister.

Captain Michael Clements, a very gallant and brave officer, son of the Rev. Mr. Clements, of Colwich, was born here. He commanded the *Pallas*, at the defeat of Thurot, in 1776, and subsequently up the Mediterranean; and died a superannuated Rear-Admiral in 1796.

The seat of Lord Anson, called *Oakedge-hall*, is a conspicuous object on the southern bank of the Trent. It is in a solitary situation, and has been uninhabited for some years, in consequence of his Lordship retaining the land around it in his own possession, and his principal residence being at Shugborough. A few dependents live in some of the offices, but the windows of the mansion have been shut up some years.

The village of Great Haywood is situated on the north side of the Trent, opposite Shugborough. It contains a few good substantial houses and the remains of a Priory. The Staffordshire and Worcestershire canal, ter having passed the Trent upon an aqueduct, forms a communication with the Grand Trunk Canal near this village.

The extensive parish of Colwich, including the townships of Great Haywood and Wolseley, and the liberty of Fradswell, contains 310 houses, 345 families; 829 males, 869 females: total of inhabitants, 1,688.

Marston is a village and liberty in the parish of St. Mary, Stafford, about three miles to the north of that town, upon a good loamy soil, excellent both for corn and pasture.

The Church, which is only a chapel of ease to St. Mary's Church, Stafford, is a small structure of stone, with a cupola and one bell. The Mayor and Corporation of Stafford are the patrons. This liberty contains 17 houses, 17 families; 52 males, 48 females: total of inhabitants, 100. The Rev. James Ellerton is curate.

Creswell, the family-mansion and estate of the Rev. Edward Whitby, is extra-parochial, and situated two miles to the north-west of Stafford, near the turnpike-road between that town and Eccleshall. The soil of this farm is an excellent loam, with fertile meadows on the banks of the Sow.

SEIGHFORD, including Aston and Doxey townships, is a parish, situated about three miles north-west of Stafford, on a brook which flows from Latchford-pool and falls into the Sow. The village of Seighford consists of a few scattered houses well-thatched. The general face of the country is level, and it is in a high state of cultivation.

The Church is small: the nave is built of stone, and the chancel and a small belfry, of brick. It is dedicated to St. Chadd, and is a vicarage: the Rev. Edward Whitby is the present incumbent.

The parish of Seighford, including the village of Bridgford, contains 138 houses, 144 families; 420 males, 446 females: total of inhabitants, 866.

RONTON is a small parish and village, situated between four and five miles west of Stafford, and three miles east of Eccleshall. It was the demesne of Goderick, a Saxon nobleman, at the time of the General Survey, and afterwards held by him of William de Noel. Camden says, the Harcourts, an ancient and noble Norman family, received it by inheritance from the Noels.

Ronton Abbey was founded by Robert Fitz-Noel, in the reign of Henry the Second, and was a Priory of canons regular of St. Augustine. It was afterwards made a cell to the Abbey of Haghman, in Shropshire, by the founder, who placed the canons in it, under the rule and obedience of the church of Haghman. This Priory paid to the Abbey of Haghman a yearly pension of 100 shillings. At the time of the Dissolution, it was valued at £90. 2s. 11½d. per annum.

The principal remains of this Abbey, which are yet standing, consist of the high tower of the church, part of its walls, and a small part of the cloisters. The church is but small, and thatched. A tenement is erected on part of the site of the ancient abbey. The church is dedicated to All Saints, and is a vicarage, formerly belonging to the abbey. The King is patron. The Rev. Thomas Harding is vicar, and the Rev. William Hicken, jun. curate.

The land in the parish of Ronton consists of a good strong loamy soil, adapted to the cultivation of wheat, and pulse, with good meadow land on the banks of the rivulets. This parish contains 52 houses, 52 families; 144 males, 148 females: total of inhabitants, 292.

ELLENHALL is situated about two miles to the north-west of Ronton. The village consists of several good modern houses, in an elevated situation, on each side of the turnpike-road between Stafford and Eccleshall. The parish is but small, and is chiefly remarkable for containing the manor-house of the ancient family of the Noels, from whom it devolved by marriage to the Harcourts.

The Church of Ellenhall is a small modern structure, situated in a field on the highest part of the eminence near the village. The nave and tower are built of brick, and the chancel of stone. It is dedicated to St. Mary, and is a curacy: the Rev. William Hicken, jun. is the present minister.

The soil of this parish, and the circumjacent country in general, is a strong marly loam, on a deep marl bottom, in some places inclined to gravel. This parish contains 47 houses, 49 families; 127 males, 124 females: total of persons, 251.

ECCLESHALL

Is an extensive parish in Pirehill North, containing nineteen townships, namely, Aspley, Bromley, Broughton, Charnes, Chatwell, Chorlton-hill, Coldmeece, Cotes, Croxton, Horseley, Millmeece, Pershall, Podmore, Slindon, Sugnall Magna, Sugnall Parva, Three Farms, Walton, and Wooton.

The town of Eccleshall is a place of great antiquity, and is said to have been founded by the Romans in the first century. The following record of Eccleshall town and castle was communicated by a correspondent:

"When Nero was Emperor of the Roman empire, Auviragus was King of Britain. Vespasian, then general of the Roman army in Britain, built a town in a large marsh on the banks of the river Sow, adjacent to the forest of Bloor, and ordered one of the Roman standards to be kept there. The standard represented an eagle, and the town was called Eagle's-hall. It was built about the year 60. On the southern side, at the foot of the hill, a temple was erected to Jupiter, and a flameary or place for sacrifices. About the year of our Lord 209, Lucius, King of Britain, gave the town the name of Eccleshall, and the temple of Jove, Haamere. This edifice was consecrated as a christian church, and continued until the year 509, when Creda, a Saxon, the eleventh from Weber, the first king of Mercia, conquered the Britons, and destroyed all the christian churches within the counties of Stafford, Gloucester, Hereford, Chester, Worcester, Oxford, Warwick, Derby, Leicester, Buckingham, Northampton, Nottingham, Lincoln, Bedford, Huntingdon, and part of Hertford. The Christians were obliged to seek an asylum in Worlsor, supposed to be Wales, till about the year 660, when Lawrence, Archbishop of Canterbury, consecrated Devine, Bishop of Lichfield, and afterwards Eccleshall, and the forest of Bloor was given to him by Penda, King of Mercia.

"Eccleshall Church was re-built in the year 661, but was destroyed nine years afterwards by Wulfere, King of Mercia, who, in 670, while at his castle at Ulferoster, or Uttoxeter, was informed that his two sons Ulfred and Rufin, under pretence of hunting, were gone to Eccleshall to Bishop Chadd, to be baptized and instructed in the Christian religion. The King being instigated by his concubine Werebode, hastened to Eccleshall, and finding his sons in the church in divine contemplation, he slew them both with his own hand, and then destroyed the edifice. Queen Erminilda, the mother of the two royal martyrs, took their bodies and buried them in a certain place not far from Eccleshall, and built a monastery over them; and, from the great quantity of stones collected for this building, the place was called Stones, now Stone, a market-town in this county. King Wulfere afterwards repented, re-built Eccleshall Church, and all others which he had destroyed, and was very favourable to the Christians, but died without an heir. His brother Ethelred succeeded him in his kingdom; then Eccleshall began to flourish, and became so famous and populous that it had five parish-churches and two chapels in it. This prosperity continued nearly three centuries and a half; but a sudden reverse took place in the year 1010, when the Danes laid Eccleshall town and castle, and all its churches, in ashes by fire. It lay in ruins till 1090, when Elias de Jantonice, prebend of Eccleshall, re-built the old church, and dedicated it to the Holy Trinity. In the year 1299, Walter Langton, Bishop of Lichfield, re-built Eccleshall Castle."

Eccleshall is a small market-town, situated on the banks of a rivulet that flows into the river Sow. It is seven miles north-west of Stafford, and contains five streets and lanes. The houses are of brick, well-built, and regular. The market-day is Friday, and it is well supplied with provisions. There are four annual fairs: Midlent Thursday, Holy Thursday, August 5th, and the first Friday in November, principally for horses, cattle, and sheep.

According to the parliamentary returns of the population in the year 1801, the parish of Eccleshall contained 594 houses; 1,737 males, and 1,750 females: total, 3,487 persons. In 1811, the town of Eccleshall contained 217 houses, 225 families; 466 males, 550 females: total, 1016 persons. At the same period, the whole parish of Eccleshall contained 693 houses, 711 families; 1805 males, 1813 females: total, 3,618 inhabitants. There are several good inns and large shops in Eccleshall, but no manufactures, except the common handicraft trades. The neighbourhood of the town is pleasant, and the land fertile.

Eccleshall Church is a large and handsome fabric of stone, in the Gothic style of architecture, and consists of the nave, a north and south aisle, and the chancel. The nave is lighted on the south side by nine large Gothic windows, and the chancel by one large and three narrow windows, besides that at the eastern end, which is partly built up. There are also several windows on the north side. The interior is clean and commodious, the pews are of oak, and the chancel contains several monuments, particularly of the Bosvilles, who formerly possessed the estate at Biana, near Eccleshall Castle.

This church and the church-yard are situated at the extremity of the town, on the road to Chester. The square tower of the church is large and high, built of stone, and adorned with eight pinnacles and a vane. It contains six bells and a clock. The rooks have taken possession of this airy citadel, and may be seen in numbers flying in and out of the apertures in the steeple. The general exterior appearance of the church is magnificent, the walls are ornamented with small pinnacles and battlements, and the southern entrance is through a very beautiful and high Gothic arch. This edifice bears strong marks of antiquity, and is evidently verging to decay, for the wall on the northern side is supported by five strong buttresses. This church is memorable for being the place where Queen Margaret was concealed after the disastrous battle of Blore-heath.

Eccleshall Castle is the residence of the Bishop of the Diocese. It was built in 1299, by Walter de Langton, Bishop of Lichfield, and Lord High Treasurer of England, who established it as the principal residence of the Bishops. The manor belonging to this palace was of very considerable extent and value, for it appears that in 1650 it was sold for the sum of £14,224. In the year 1695, the south front of the castle was re-built by Bishop Lloyd, and it afterwards became the principal residence of successive Prelates, and continues so to the present day.

During the Civil war between Charles the First and the Parliament, Eccleshall Castle was fortified, and defended by a Royal garrison. The garrison stood a severe siege, but was ultimately compelled to surrender, the place being no longer tenable, but reduced to a ruinous state by the fire of the enemy's artillery. It continued for several years uninhabitable, but was re-edified by Bishop Lloyd. The beautiful grove which surrounds this ancient fabric was planted by Bishop Hough, and the late Bishop Cornwallis further improved the surrounding grounds.

The Church of Eccleshall is a vicarage in the patronage of the Bishop of Lichfield and Coventry: the Rev. John Henry Powell is the present vicar, and the Rev. Francis Jenkinson is curate.

The Bishop's Woods are situated between four and five miles to the westward of Eccleshall, and contain about 1,500 acres of land, planted with trees, particularly oak. They abound with underwood, a part of which is cut down annually, and sold to the potters to make crates for their earthenware. The timber-trees are left standing in the proportion of from forty to eighty per acre, according to the quality of the soil and nature of the surface. Hence a regular revenue is obtained from these woods, without decreasing their value. In the lower part of the wood, where the soil is best, the under-wood is worth 15s. per acre annually, if well stooled, and the vacant places planted between the stools. The best part of this underwood is made into rakes, scythe-poles, &c. and these extensive tracts are thus productive of several useful articles.

Chebsey is a large parish of Pirehill South, containing the township of Cold Norton. At the time of the General Survey, it was held by Humphredus of Henry de Ferrers. In the time of Henry II. by Euturpius de Hastings, and in this family it continued several successions.

The village of Chebsey is ancient, and consists of farm-houses and smaller tenements. It is situated in a valley, near the confluence of the Eccleshall water with the Sow.

The Church is an ancient Gothic fabric of stone, with a large square tower. It is dedicated to All Saints, and is a vicarage in the patronage of the Dean and Chapter of Lichfield. The Rev. Henry White, Sacrist of Lichfield Cathedral, is vicar; and the Rev. John Jenkinson, the present curate.

The parish of Chebsey contains 72 houses, 73 families; 211 males, 195 females: total, 406 persons.

OFFLEY or HIGH OFFLEY, so called from its elevated situation (which commands extensive views from S.E. to N.W.) is a parish of Pirehill North, containing the townships of High Offley and Woodseves, and adjoins on the north the parish of Eccleshall. The village of High Offley contains four large farm-houses and a few smaller tenements. The Church is an ancient Gothic fabric, consisting of a nave and south aisle, divided by five plain round pillars. It is neatly pewed with oak, and has a capital organ, the gift of John Salmon, Esq. of London, a native of the parish. In the church are two monuments of the Skrymsher family. The vicarage is in the gift of the Bishop of Lichfield: the Rev. Peter Vaughan, D. D. Warden of Merton College, Oxford, is the present incumbent; and the Rev. Thomas Harding, curate. The great tithes of Offley are attached to the prebend of that name in the Cathedral Church of Lichfield.

The manor and parish are co-extensive, and contain (according to the survey by the commissioners for enclosure in 1809) 2450 acres. The manor was for many generations held by the Skrymsher family. It is now the property of the Rev. Josiah Hinckes.

From the foundation of a Roman road (E.N.E. to W.S.W.) traceable on the lands a little northward of the church, and Roman coins in great numbers, flat bricks, armour, fragments of pottery, &c. having been ploughed and dug up on the side of the hill south of the church-yard, it has been conjectured (with great probability) that the Roman station *Mediolanum* was here situated. Knightley and Drayton-in-Hales, both which have been named by antiquaries as the supposed site of Mediolanum, are within a short distance, but are not known to have any evidences of that kind to corroborate such an opinion.

The parish of High Offley is in general fertile: the soil consists of an excellent sound loam, adapted to turnips, corn, or grass. The Batchacre estate, (late the residence and property of Richard Whitworth, Esq. whose remains are interred in High Offley church-yard) in this parish, consisting of about 600 acres, has lately been purchased by John Walford, Esq. who has built Hill-hall for his residence. Mr. Walford is making rapid improvements upon land which had formerly been much neglected, by a very extensive system of under-draining, clearing off rubbish by the irrigation of his meadows, by improved culture on the upland, and by manure. He cultivates the common and Swedish turnip upon the Northumberland system, in small ridges, and horse-hoes between; he has used lime freely, and is making important experiments with peat ashes. The estate abounds with peat, which is dug in squares, laid in heaps, and set on fire when dry, and thus without other fuel burnt to ashes. The ashes can be produced at less than a penny per bushel, and it is supposed that one hundred bushels per acre is equal to a dressing of lime. It may be drilled-in with the seed for any crop, or scattered over grass-land, corn, or any crop, in the manner in which soot is spread, and is of universal application. Large crops of turnips and grain have been thus raised without any other manure.

This parish contains 100 houses, 100 families; 297 males, 289 females: total, 586 inhabitants.

In this parish was a water called *Shebben* or *Shebdon-pool*, which is noticed by Plot as being in his time the resort of a great number of *Pewits*. In one of his plates (tab. 19. in chap. vii.) is a representation thereof, and the manner of taking them. The Pewit is a bird which he considered "the strangest whole-footed water fowle that frequents this county." He terms it the *Larus Cinereus Ornithologi*, the *Larus Cinereus Tertius Aldrovandi*, and the *Cepphus* of Gesner and Turner. In some counties it is called the *Black Cap*, and in others the *Sea* or *Mire-crow*.

Respecting their appearance, habits, and value, he writes thus: "Being of the migratory kind, their first appearance is not till about the latter end of February, and then in number scarce above six, which come, as it were, as harbingers to the rest, to see whether the hafts or islands in the pooles (upon which they build their nests) be prepared for them; but these never so much as lighten, but fly over the poole, scarce staying an hour: about the sixth of March following, there comes a pretty considerable flight, of a hundred or more, and then they alight on the *hafts*, and stay all day, but are gone again at night. About our Lady-day, or sooner in a forward spring, they come to stay for good, otherwise not till the beginning of April, when they build their *nests*, which they make not of sticks, but heath and rushes, making them but shallow, and laying generally but four eggs, three and five more rarely, which are about the bigness of a small hen-egg. The *hafts* or *islands* are prepared for them between Michaelmass and Christmass, by cutting down the reeds and rushes, and putting them aside in the nooks and corners of the hafts, and in the vallies, to make them level; for should they be permitted to rot on the islands, the Pewits would not endure them.

"After three weeks sitting, the young ones are hatch't, and about a month after are almost ready to flye, which usually happens on the third of June, when the proprietor of the poole orders them to be *driven* and *catch'd*, the gentry comeing in from all parts to see the sport; the manner thus. They pitch a rabbit-net on the bank side, in the most convenient place over against the hafts, the net in the middle being about ten yards from the side, but close at the ends in the manner of a bow; then six or seven men wade into the poole beyond the Pewits, over against the net, with long staves, and drive them from the hafts, whence they all swim to the bank side, and landing run like lapwings into the net, where people standing ready, take them up, and put them into two *penns* made within the bow of the net, which are built round, about three yards diameter, and a yard high or somewhat better, with small stakes driven into the ground in a circle, and interwoven with broom and other raddles.

"In which manner there have been taken of them in one morning 50 dozens at a driving, which at 5s. per dozen (the ancient price of them) comes to twelve pounds ten shillings: but at several *drifts* that have been anciently made in the same morning, there have been as many taken as have been sold for thirty pounds, so that some years the profit of them has amounted to fifty or threescore pounds, beside what the generous proprietor usually presents his relations, and the nobility and gentry of the county withall, which he constantly does in a plentiful manner, sending them to their houses in *crates* alive, so that feeding them with livers and other entrails of beasts, they may kill them at what distance of time they please, according as occasions present themselves, they being accounted a good dish at the most plentiful tables.

"But they commonly appoint three days of driving them, within fourteen days, or thereabout, of the second or third of June; which while they are doing, some have observed a certain old one that seems to be somewhat more concerned than the rest, being clamourous, and striking down upon the very heads of the men, which has given ground of suspicion that they have some government amongst them, and that this is their prince, that is so much concern'd for its subjects. And 'tis farther observed, that when there is great plenty of them, the lent-corn of the country is so much the better, and so the cow-pastures too, by reason they pick up all the worms, and the fern-flyes, which though bred in the fern, yet nip and feed on the young corn and grass, and hinder their growth." (See chap. vii. secs. 9, 10, 11, 12.)

These birds are stated to have changed their settlements from time to time. Their first abode, according to Plot, was at a pool about half a mile s. w. of Norbury church. From whence they removed to Offley-moss, near Woods-eves, to Aqualate, and other neighbouring pools. One while they were settled at the pools about Batchacre, where, according to Mr. Shaw, Mr. Whitworth for many years kept a regular and most entertaining journal of all their *singular* proceedings until their gradual departure. Plot ascribes some of their transmigrations to the death of the heads of the Skrymsher family, then the chief proprietors of estates about their places of resort, which he terms a "strange quality." The less credulous reader will probably ascribe their removals from and returns to former abodes, to better feeding and more quiet and abundant accommodation for laying their eggs and protecting their young.

It may be proper here to observe, that, on the above occasions of driving these birds, their young, just before they could fly, were the object of game. According to the information of an aged person now living, (1816), and who, in his juvenile days, was repeatedly present at this annual amusement of pewit-driving, the young were of a grey colour, and the old ones of a beautiful white, with black heads, and the tips of their wings and tails were likewise black. They were a different bird from the Lapwing, sometimes mistaken for the ancient Pewit in these parts. The net placed on the bank of the pool, as a snare, was about two feet high; and when the young were driven and entangled therein, they were put in baskets covered with nets.

The Shebben and Pewit pools are now drained and enclosed, and of course deserted by their former occupiers. The lands, arable, meadow, and pasture, are mixed in due proportion throughout the parish, and are well cultivated.

Adbaston, is a parish in Pirehill North, situate about a mile and a half north-west of the village of High Offley. It contains four townships, and a population of 532 persons. The Church, dedicated to St. Michael, is an ancient Gothic structure of stone, with a tower, and is a curacy in the patronage of the Dean of Lichfield, the impropriation being a part of the endowment of the Deanery of Lichfield and Coventry. The Rev. Thomas Harding is curate. The soil of the village of Adbaston is a mixed gravelly loam, producing turnips, wheat, and other grain: the low land makes good meadows, if well-drained. In this parish died, November, 28, 1714, William Wakeley, of Outlands, aged 125 years.

Adbaston parish contains 92 houses, 101 families; males 272, females 260: total, 532 persons.

Outlands is a small hamlet north of Adbaston, and contains a few farm-houses and cottages. It is generaly a light rocky soil, with some good meadow-land.

Broughton, is a township and chapelry in Pirehill North, and parish of Eccleshall, containing *Broughton Hall*, the property and residence of the Rev. H. D. Broughton. The mansion is a very good specimen of the Elizabethan era, and is surrounded with beautiful plantations, clumps, and shady spreading trees, particularly sycamores: there is also a promising coppice of oak, and a handsome beech grove, near the house. The ancient Gothic Church is enveloped in luxuriant specimens of the yew, holly, and fir, together with a very fine cedar, which promises to vie with those of Lebanon of old. On the road-side are several other cedars. The estate of Broughton consists of good mixed loam on a marl bottom, containing several large marl-pits. It is now principally held in hand, and farmed in a very spirited and improved manner by the owner.

ASHLEY is a parish and village situate about seven miles N. W. of Eccleshall. In 1811 the parish contained 141 houses and 616 inhabitants. The Church is a small neat stone building. It consists of a nave, north and south aisle, a south porch of entrance, a chancel at the east, and a square tower at the west end of the nave. Attached to the north side of the chancel is a cemetery or chauntry. The tower, side aisles, cemetery, and porch, are embattled at the top, and the whole appears to be covered with lead, except the nave. Some of the battlements are ornamented with pinnacles, and at the bottom of the parapets are heads and grotesque figures, as well as on the heads of some of the windows, most of which are or have been adorned with glass paintings, particularly the chancel window. These paintings occupy not more than one or two panes on each of the windows, except that in the chancel. In that corner of the porch, which is on the east side, and which joins the south aisle, is a turret with a pyramidal top, in the inside of which are steps to ascend to the top of the porch from the south aisle. In the interior of the church, the aisles are separated from the nave by three small pointed arches; the north and south sides of the tower range with the arches of the aisles, the east side of the tower is open to the nave through a pointed arch, and indeed the east, north, and south sides of the tower are supported on arches, which are open to the aisles and nave. In the north pier of the east arch is a niche wherein the font is fixed, which is an old stone bason of a size fit for immersion. The chancel is separated from the nave by a pointed arch, on the south side of which is the pulpit and reading desk, and it is separated from the cemetery on the north side by two small pointed arches, the lower parts of which are filled with iron railing, and gate. The east wall under the chancel window is ornamented with wainscoat, on the pannels of which is the Lord's Prayer, Decalogue, and Creed, and paintings of Moses and Aaron. In the cemetery, which is paved with flags, are several mural monuments, and one costly mausoleum, or monument of superior execution, and such as is seldom met with in country churches. The parts may be thus described: On an altar tomb are placed two recumbent effigies, a male and female as large as life, the male in armour, and the female in the dress of the times; his head rests on an helmet, her's on a cushion; at the feet of the male is a lion couchant, and between them near their heads is fixed a talbot sejant; that part of the top of the altar tomb whereupon the male recumbent effigy is placed, is supported on the south or front side by five small square pillars, within which on the floor and immediately under the male effigy, is the image or effigy of a corpse in a winding sheet. Against the north wall and perpendicular to the altar tomb are two niches with semicircular heads, and about four inches deep, within which are four half-length female figures in relievo; they appear in the same kind of dress as the female recumbent effigy, have a ruff, &c. the dresses of the times. Two of these half-length figures are in each niche, and at the upper part of each niche is an escalop. Over the whole of the above-described parts is erected a canopy supported on four Ionian fluted columns, about two yards high, each standing on a pedestal about three quarters of a yard in height. Above these columns is an entablature, about half a yard in height, which is continued over each arch, that is, on the east and west end, and the south side, and is surmounted by other ornaments; upon the corners of the entablature over each of the four columns, is a square pyramid or pinnacle, each side of which is adorned with four roundles; these pyramidal ornaments are about one yard high, and nine inches square at the base. Above the entablature on the front side, or over the south arch, is a shield which can be but imperfectly blazoned: it is quarterly 1 and 4 a lion rampant, 2 and 3, on a bend three roundles, impaling paly of three parted per fess; first division, a bend engrailed; second, a cross moline between four martlets; third, a talbot rampant; fourth, a cross raguled; fifth, two bars, on the first of which is a crescent, and on the chief are three mullets; sixth, a chevron between three vases (or covered cups). Below the canopy, within the west arch thereof, at the head of the recumbent effigies, is a male effigy kneeling on a pedestal, and as large as life, in armour, excepting an helmet; opposite this at the feet of the said effigy, and within the east arch, is also a male effigy in armour, excepting an helmet: this is much smaller than that at the head, but is in the attitude of kneeling and of prayer. These two last have their faces towards the recumbent effigies; and opposite to each other, on the front or south side of this monument, are two small images kneeling, a male and female, and an helmet between them on the floor. This monument occupies a space on the floor of about four by two and a half yards, and may be four yards in height or more. The north side of it is against the north wall of the cemetery, on which same north wall, on the east side of the above monument, on a plain stone fixed in the wall, is the following inscription, in capitals:

"Here lyes the body of Thomas Ld. Gerard, which dyed Ld. President of Wales, Oct. the 7th, Anno 1617.

Gilbert Lord Gerard, dyed May the 11th, Anno 1623.
Dutton Lord Gerard, dyed April the 24th, Anno 1640.
Charles Lord Gerard, dyed December the 28th, 1667.
Digby Lord Gerard, dyed November the 8th, 1684.
Charles Lord Gerard, dyed April the 12th, 1707."

The above inscription, it is presumed, alludes to those represented on the above monument.

Above the last inscription, on the same wall, a plain marble tablet fixed therein, is inscribed:

"To the Memory of Thomas Fletewood, of Gerards Bromley, Esq. and Frances his wife, who was the only sister, and at length sole heir of the

Right Honourable Charles Lord Gerard: the said Thomas Fletewood departed this life the 15th day of July, in the year 1720, and in the 56th year of his age, and lies interr'd near this place. The said Frances Fletewood died at the city of Liege, the third day of February, in the year 1736, and in the 74th year of her age. The said Thomas Fletewood and Frances his wife left issue only one son, Charles Fletewood, now of Gerards Bromley, Esq."

Above this last inscription, on two separate brackets, are the busts of Thomas and Frances Fleetwood.

On the west wall of the cemetery, a plain marble tablet, is inscribed:

"To the Memory of the Right Hon. Charles Lord Gerard, Baron of Gerards Bromley, who died the 12th day of April, in the year of our Lord 1707; aged 48 yeares, and lies interr'd near this place. He marryed Mary, the daughter of Sir John Webb, of Hatheropp, in the county of Gloucester, Barrt. by whom he left no surviveing issue. The said Mary Lady Gerard survived him many years, and died at Joppa in the year of our Lord 1731, in her return from the Holy Land."

On a bracket above the last inscription, is a male bust.

Against the east window of the cemetery, is a mural monument of white marble, which is a plain tablet with a cornice surmounted by a weeping cherub, sitting on a rock, with his glass run down. On a tablet is:

"Sacred to the Memory
Of Hugo Meynell,
Who died May 1800."

Against the north wall of the chancel, over the pillar which is between the two arches whereby the chancel communicates with the above cemetery, is a marble monument to the memory of William Lord Viscount Chetwynd, the top part of which is a niche with a circular head; and within it is placed a large Egyptian black urn, which was made at Etruria in the time of the late Josiah Wedgwood, Esq. F.R.S. A.S.S. Below the niche, on the corona of a corniche, in one line, is the following, in capitals:

"He was a man, take him for all in all, I shall not look upon his like again."

Below the above, a white marble tablet is inscribed, in capitals:

"Sacred to the Memory of William Lord Viscount Chetwynd, who died April 3, A. D. M.DCC.LXX. Aged LXXXVII. years."

Beneath this inscription is a coat of arms, viz.: Azure, a chevron, Or, between three mullets, Or, surmounted by a large Viscount's coronet of white marble.

The founder of the church was a David Kenrick, whose character is thus delineated on a brass plate on the south side of the nave, over the second pillar from the pulpit:

" In perpetuam rei memoriam,
Manubias Deo,
David Kenricus, pietas ejus memoriæ,
Hoc virtutis præmiolum dicavit.

Mira cano, Pietas sævis reperitur in armis.
Ædificat bellum sternere quod soleat!
Hanc sacram struxit miles memorabilis ædem
Principe sub nigro, quem meruisse ferunt,
David Kenricus: Davide beatior illo
Templum cui Superis ædificare nefas.

Fair Virtue's deeds transmit a glorious name
That lives on records of eternal fame.

This holy edifice, arising from the spoils of a conquered enemy, was dedicated by David Kenrick to the Omnipotent.

Hail wond'rous Piety, when seen in arms,
Whose balmy breath a hero's bosom warms
To build, and not destroy with impious ire,
The solemn Temple, and the sacred Choir!
Kenrick, to early scenes of slaughter led,
By Edward's son in toils of battle bred,
This pile erected, by his God approv'd,
His Country honour'd, and his Prince belov'd.
Happier than David he allow'd to raise
An Altar sacred to Jehovah's praise."

The Church is dedicated to St. John the Baptist. The living is a rectory in the patronage of Thomas Kinnersly, Esq. two turns, and Hugo Meynell, Esq. one: the Rev. W. Anwyl is the present incumbent.

The soil of the old inclosed land in the parish is a lightish loam on a sand or gravel bottom.

The parish of Ashley contains 141 houses, 141 families; males 321, females 295: total of inhabitants, 616.

Ashley Common, to the south of the village, is a thin black peat moor, the hilly parts having an under-stratum of sand, gravel, or sandy rock. The herbage is poor, possessing but few of the better grasses, but the soil improves into an inferior friable marl nearer to Eccleshall. Hungershutt, another waste in this parish, is now under inclosure. The principal land-proprietors are Thomas Kinnersly and Hugo Meynell, Esqs.

Blore-in-Tyrley, is a township in Pirehill North, and contains several farm-houses and smaller tenements, the property of Hugo Meynell, Esq. of Hoar Cross, Derbyshire. The inhabitants generally attend Drayton Church, believing themselves to be in that parish. The soil is a gravelly loam.

Blore township contains 29 houses, 33 families; males 97, females 90: total 187.

Blore Heath, about nine miles from Eccleshall, and bordering on Shropshire, is celebrated as the scene of a bloody battle between the Lancastrians and Yorkists in 1459, in which Lord Audley, who commanded for the King, was defeated by the Earl of Salisbury, with the loss of 2400 men. The Earl of Salisbury was at that time on his march from Middleham Castle, with four or five thousand men, under pretence of settling with the King the disputes of the two Houses. Margaret, the faithful and spirited consort of Henry VI., fearing for her husband's safety, directed Lord Audley to intercept him on his way. He posted himself on Blore Heath with ten thousand troops, collected out of Cheshire and Shropshire, whose chieftains were distinguished by silver swans, the badges of their young prince. Salisbury, notwithstanding the disparity of numbers, determined to stand the fortune of the day, but wisely had recourse to stratagem. He encamped at night on the banks of a rivulet, not broad but deep, and in the morning pretended a retreat. Audley following him with the impetuous valour natural to himself and the times, Salisbury made an instant attack on the divided forces of the Lancastrians. The field was long disputed with the animosity usual in civil feuds. Audley fell, with two thousand four hundred of his troops, chiefly the flower of the Cheshire gentry, whose courage led them to the front of the battle. The Queen fled to Eccleshall Castle, and Salisbury joined the Duke of York at Ludlow.*

Rapin's account of this battle is as follows: "Lord Audley was encamped on Blore-heath near a little river; Salisbury posted himself on the other side, as if he meant to guard *the pass*, and hinder his being attacked; then suddenly feigning a fear, he retired in the night, marching so as, at break of day, his enemies could still see the rear of his army. This retreat, which seemed to be with precipitation, inspiring the Royalists with ardour, they began to pass the river in disorder, imagining that they had nothing to do but pursue the flying enemy. But whilst they were in this confusion, some being over the river, others in the water, and others ready to pass, the Earl of Salisbury returned, and fell upon the troops already over, who had scarce time to draw up. The fight lasted, however, four or five hours, because the King's troops were supported by those that were continually passing.

* Pennant, and "General History," of this Work.

But as this could not be done without confusion, the Royal army was at length put to rout, with the loss of 2400 men. Audley himself was slain, with all the principal officers; among whom was Sir Thomas Dutton, Sir John Dunne, Sir Hugh Venables, Sir Richard Molineaux, Sir John Leigh, &c.

"This battle was fought on September 23, 1459, about one mile from Drayton in Shropshire, on Blore-heath, which lies in Staffordshire; where, at the head of the river Sow, a stone is set up in memory of James Lord Audley, there slain."

We subjoin some observations published in the Gentleman's Magazine, from the pen of the Rev. W. Snape, curate of Keel, and the adjoining parish of Maer, upon the probable position of the two armies previous to the commencement of the battle. Mr. S. conjectures, that Salisbury and Lord Audley were posted respectively on the celebrated Byrth and Camp Hills, near and upon Maer-heath, both which will be noticed hereafter.

"In the hurry and confusion of a battle, several transactions, of subordinate consideration or importance, might be omitted or misrepresented, either from want of recollection, or owing to incorrect information; while the principal manœuvres are in their leading points correctly stated. The above account may, probably, be not exempt from such-like errors. In the first place, the stone erected to commemorate the spot where Lord Audley was slain, is said to be placed at the head of the river Sow; which is not strictly correct: it is placed very near to one source of that river near Ashley and Broughton; but the brook by the side of which it is placed, I was informed by persons resident in the neighbourhood, flows into the Tearne or Tern, which joins the Severn, and thus takes a directly opposite course to the river Sow, which unites with the Trent. It is not improbable that Maer-heath and Ashley-heath, both of which are nearly in a direct line to Blore-heath, and at no great distance from each other, were, three centuries and a half ago, considered, on account of the comparatively few inclosures, as one common, and part of Blore-heath. The Byrth and Camp-hills are about two miles from Whitmore, which is very near another head of the river Sow, which flows through that place 'a little river.' Indeed none of the rivers thereabout, nor that by the side of which the stone commemorating Lord Audley's death is placed, can be termed *great*. It appears from the above account of the battle, that the Earl of Salisbury obtained his victory chiefly by stratagem. That he first attempted to defend some pass—the context leads us to suppose a pass over the little river; but the rivers are mere brooks. It is not impossible, therefore, that the pass might be some particular situation in the road, which the armies must or had travelled. After the Earl had made a pretence of guarding some pass, he is said to have feigned a precipitate retreat, and retired in the night; yet marched so as, at break of day, his enemy could still see the rear of his army. Without making arrogant pretensions to military knowledge, I think I may hazard an opinion, that it is very improbable that a General at the head of 5000 men could feign a retreat, so as to induce his enemy, who was twice as strong, to pursue him in disorder, from an apprehension that his retreat was the effect of fear and a conscious inferiority, unless he had continued to retreat to the distance of four or five miles. If we imagine a shorter distance, it is probable that the enemy would have discovered his intentions. Besides, where was the necessity of the Earl retiring *in the night*, yet marching so as, at *break of day*, his rear could be seen by the Royalists, if he had retired only half a mile or so? I think we may be justified in supposing, that the two armies first met, or were within sight of each other, at the distance of five or six miles from the scene of action. No place can be assigned, where they were so likely to approach each other, as somewhere between the field of battle and Heleigh Castle, the seat of Lord Audley; who would, of course, think of defending his castle by placing his troops between it and the enemy. Heleigh Castle is situate about four miles to the north of the Camp-hill; and very near the latter there is a road which leads to Madeley and to Heleigh. Also, at the foot of the Camp-hill, there are, or, previous to the late alterations made on the common by inclosures and cultivation, there were to be seen the vestiges of an old road, which took a southward direction, and passed by the foot of the Byrth-hill, and from thence towards the village of Maer. From the latter place is an old road to Ashley, which lies in the direct way, or nearly so, to Blore-heath, where the battle was fought. It is not improbable that this track was the ancient road, or one of the principal old roads from Heleigh to Drayton. In which case the Byrth and Camp-hills may be considered as stations in the direct road between Heleigh Castle and the scene of action. However, from the above considerations, I have indulged an opinion, that Lord Audley was encamped on that part of Maer-common called the Camp-hill; and that the Earl of Salisbury was posted on the Byrth-hill, opposite to Lord

Audley's army, where he feigned an attempt to guard the pass; for such the road between the Byrth and Little Byrth may be called. As there is no mention made of any skirmish taking place there, we may suppose that he manifested his intentions of guarding the pass, and of preventing an attack by fortifying his situation. Vestiges of a fortification are to be seen on the summit of the Byrth-hill. I may here notice, that on the Camp-hill are two mounds, once thought to have been tumuli; but, when opened, they contained nothing to support that opinion: and they were probably raised merely for stations of observation. Now, as there exists no tradition of any arms having been found, or sepulchral remains discovered, I conclude, that it is not improbable that the two above-named Generals were posted as already observed; and when the Earl abandoned his position, and 'feigned a fear,' we may suppose that he would retreat a few miles, say five or six, which is about the distance from the Byrth-hill to where the battle was decided."

The ancient wooden cross put up to mark the spot where Lord Audley fell having been thrown down by a cow rubbing against it, a stone pedestal was erected in 1765, by the Lord of the Manor, with the cross upon it. On the north side of the pedestal is the following inscription:

ON THIS SPOT
WAS FOUGHT THE BATTLE OF
BLORE HEATH,
IN 1459.
LORD AUDLEY,
WHO COMMANDED FOR THE SIDE OF LANCASTER,
WAS DEFEATED AND SLAIN.
TO PERPETUATE THE MEMORY
OF THE ACTION AND THE PLACE,
THIS ANCIENT MONUMENT
WAS REPAIRED IN
1765,
AT THE CHARGE OF THE LORD OF THE MANOR,
CHARLES BOOTHBY SCRYMSHER.

The stone is now to be seen in an inclosed field adjoining the public road from Drayton to Newcastle-under-Lyme, about two miles distant from the former town. It is a plain cross about 4½ feet high, 10 inches broad, and 8 inches thick; the transverse part may be about 20 inches in length. The pedestal is 3 feet high, so that the height from the ground is about 7½ feet. The field in which it stands is the property of Sir John Chetwode, Bart. M. P. for Newcastle-under-Lyme, whose mansion at *Oakley*, in this neighbourhood, is situated in the township of the same name, parish of Mucclestone, and county of Stafford, which is bounded on this side by a small rivulet running from Maer.

The above cross is within a few yards of this rivulet, which probably the Earl of Salisbury passed; and when part of Lord Audley's troops were over, and others crossing in confusion, the Earl returned, attacked, and obtained his victory.

Mucclestone is a parish and village in Pirehill North, situated on a rising ground about one mile from Blore-heath. In 1066 the manor was held by Kenning, one of the Tanes, and subsequently by the Morgans, till the reign of Elizabeth, one of whom sold it to Thomas Offley, Knt. Lord Mayor of London, in 1556. It is now the property of Lord Crewe.

The Church is ancient, and dedicated to St. Mary: it has a lofty square tower, from which Queen Margaret is said to have beheld the obstinate struggle at Blore-heath, which was so fatal to her cause. The living is a rectory, in the patronage of Lord Crewe. The Rev. Offley Crewe is rector, and the Rev. A. B. Haden, jun. curate.

Mucclestone parish contains 149 houses, 148 families; 388 males, 384 females: total of inhabitants 772.

Maer, is a parish and village in Pirehill North, situate about mid-way between Newcastle and Drayton, and adjoining the parish of Mucclestone. It probably derives its name from a natural lake or mere (covering about twenty-two acres of land) which is the source of the Tearn, a stream that falls into the Severn, and divides this county from Shropshire in the neighbourhood of Market Drayton. The parish comprises the townships of Maer and Maer-way-lane. The soil of the inclosed land is a gravelly loam. The manor anciently belonged to William de Maer and Robert de Stafford, from whom it passed through different hands, till it became by purchase the property of Josiah Wedgwood, Esq. of Etruria, in the Potteries. It comprises upwards of 3000 acres of land.

At the east-end of the lake stands *Maer Hall*, also the property and till lately the residence of Mr. Wedgwood, who has re-edified and converted it into an elegant modern mansion, and embellished the grounds with plantations, and other spirited improvements.

The boggy land has been drained, and the whole of the common called Maer-heath divided and enclosed, and is now in progress of cultivation.

The Church, which is dedicated to St. Peter, is situated a few yards to the east of Maer Hall, and is a small plain stone building, consisting of a nave, north-aisle, a chancel at the east, and a tower at the west end of the nave: it has also a south porch. The church appears to have been erected or rebuilt in 1610, as that date is inscribed on the south side of the tower. It contains monumental inscriptions of the families of Sir John Bowyer, and Robert Macclesfield, Esq.

The cloth for the communion-table is an old Turkey carpet of great antiquity, the gift of Margaret Tether, daughter of Thomas Picken, of Maer-heath, and brought by her from Constantinople and given to Maer Church, in 1639. From 1558 to 1803 there were 24 curates in 245 years.

There are a few benefactions to the poor of this parish. Mr. William Clayton, Gent. of Radwood, in the parish of Maer, left by his last will and testament, bearing date Nov. 8, 1624, the sum of £100, for certain charitable uses, for a limited time. And his son, John Clayton, of the same place, on September 18, 1634, gave the sum of £60, with which sum and his father's legacy, amounting to £160, he purchased from Sir William Bowyer, Knt. of Knipersley, in the county of Stafford, an annual rent of eight pounds, payable every Lady-day and Michaelmas-day, by equal portions, charged upon certain lands, &c. in the parish of Maer. He also appointed trustees to receive the said annual rent, and to pay the yearly sum of *four pounds* thereof, by two equal portions, *to the Minister* of Maer, "*to encourage him* to take pains in preaching, and also in catechizing and instructing the ignorant in the principles of religion." *Also*, to pay the sum of *four pounds*, being the residue of the said annual rent of eight pounds, *for and towards placing poor children*, of the parish of Maer, *apprentices* in some honest and lawful calling. The interest of the sum of £30 was given by different benefactors to the poor of this parish; among whom were Mr. Dale, who gave £20; and Mr. John Cox, mercer, of Drayton, in the county of Salop, who bequeathed in his will, dated April 29, 1691, the sum of £10. The latter (Mr. Cox) bequeathed his charity *to the town of Maer*, and to be distributed in portions of twelve pence each, a little before Christmas.

The lady of the present Lord of the Manor, Josiah Wedgwood,

Esq., very laudably supports a Sunday-school for the poor children of the parish.

The Parish Register commences in January 1558; and the average number of baptisms, &c. each year is nearly as follows for the respective periods stated:

From | to (both inclusive).		Of Baptisms.	Weddings.	Funerals.
1581 to	1610	4	1	3
1611	1640	7	1	6
1641	1667	10	2	7
1701	1730	8	1	7
1731	1770	12	2	6
1771	1807	11	1	6
In 1808	there were	16	0	7

In 1729 were 28 funerals; in 1778, 21 baptisms; and in 1756, 6 weddings; which are the greatest number of each that have yet taken place in any one year. The following is a statement of the amount of the number of funerals that have occurred in the several months of the years from 1701 to 1808, both inclusive, which shews *the order of the months*, as they have been most fatal to the inhabitants of the parish of Maer: April, 87; May, 80; March, 72; February, 71; July, 63; November, 58; January, 53; June, 49; December, 44; August, 42; October 42; September, 39.

Extracts from the Parish Register.—Richard Wilson, died December 24, 1693, aged (*ut ipse dixit*) 138!

Catherine Morgan, died February 6, 1639-40, aged 100.

Mem. 1775.—This year was very remarkable for the wetness of the season, and the lateness of the harvest: corn was not all got in until the middle of November.

The living is a curacy, endowed with a small parsonage-house, about two acres of glebe, and the great and small tithes of the parish, excepting one or two estates, which are tithe-free. It is in the patronage of Josiah Wedgwood, Esq.: the present incumbent is the Rev. Robert Smith; and the Rev. William Snape, is curate.

An annual rent of *six shillings and eightpence*, called Candlestick money, is charged upon certain lands, &c. near the village of Maer, in the said parish, and is paid to the overseers.

The parish of Maer contained in 1811, 76 houses, 80 families; 221 males, 233 females: total of inhabitants, 454.

" About half-a-mile to the north of the village of Maer, is a hill called the *Byrth*;* round a great part of the summit of which has been made a foss and rampart, of an irregular form, corresponding with the figure of the hill, which may be nearly a mile in circumference. At the distance of a mile or more from and to the north-west of the Byrth, is another hill, which is part of Maer-heath, and is called *the Camp-hill*. Between these are two more hills: the one very near to the Byrth, which is named *the Little Byrth*; and the other, which is larger, is called *Coplow*. Dr. Plot is of opinion, that a battle was fought here, about the year of our Lord 705, between *Osrid* king of Northumberland and *Kenred* king of Mercia. He also conjectures Coplow-hill to be the tumulus of Osrid, and the Byrth to be a fortification or 'strong hold that Kenred had raised against him.' His conjectures seem to be grounded chiefly on a quotation from Henry Huntingdon, viz. 'Osrid vero rex belli infortunio juxta Mere pugnans interfectus est.' In taking a view of the ground on Maer-heath, however, there appears to be nothing artificial, except the two small hills on the Camp-hill, and the foss and rampart round the Byrth.

"A few years ago, George Tollet, Esq. a gentleman of learning and an antiquary, the Rev. T. Barlow, and Mr. Poole, of Finney Green, and other gentlemen, went to view Maer-heath, for the purpose of ascertaining whether there were any barrows. They dug into those places that had most resemblance to barrows, but found nothing like sepulchral remains. Coplow-hill is evidently the work of Nature; and as a public road has been made over part of the heath since the time of Dr. Plot, and sections made through some parts of the hills, in which regular strata appear, it is probable that Osrid was slain near some other place called Mere."† This battle, it has been conjectured, might have been fought near Mere or Mere-town, in the parish of Forton, near that fine lake, which ornaments the seat of Sir John Fenton Boughey, Bart. which is called Aqualate Mere, as there is very strong evidence of a battle having formerly been fought there; and which is noticed in a former page (272), under the article 'FORTON.'

Maer-heath was formerly a waste or common of several hundred acres, but it has lately been inclosed, allotted to the different proprietors of the estates in its vicinity, and a considerable part of it planted with young trees. Part of the land was drained before it was planted, at the expense of from £5. 14*s.* to £7. 12*s.* per acre.

* In some maps it is termed *Burgh*, *Bruff*, or *Brough*.

† Gent. Mag. Vol. 89, Sup. p. 603.

Above 600,000 trees have been planted by one individual, and this rugged waste promises to become a beautiful woodland. Grass land with good herbage has already been formed upon the more level and sheltered parts of Maer-heath, some of which has been let at two pounds per acre.

MADELEY is a parish of Pirehill North, situated about six miles west of Newcastle, and bordering on the west upon Shropshire and Cheshire. It consists of the hamlets of Great Madeley and Little Madeley.

The village of Great Madeley contains several good houses, the most remarkable of which is a half-timbered house, with the following quaint inscription on the front: "16— *Walk Knave what look'st at* —47." By the side of the village is a large pool, which supplies a corn-mill with water.

The Church is an ancient structure of stone, and consists of a nave, side-aisles, transepts, and a chancel, which is at the east end of the nave, to the north wall of which latter is attached a cemetery of nearly the same size and plan as the transepts. It has also a south porch, and a square embattled tower at the west end of the nave, in which is a ring of six bells; the sixth having been lately bequeathed by Mr. Samuel Stretch. In the interior the different parts are separated from each other by pointed arches, supported on plain pillars. On the north side of the chancel arch are the reading-desk and pulpit, the latter of which is ancient oak, moderately carved with small Saxon arches. The crimson velvet cloth and cushion were the gift of the Rev. Thomas Barlow, as is mentioned on one of the tables of benefactions. Above the same arch, in the nave, is painted the Pater-noster, the Creed, and the Commandments, on two tables. Between these tables are the Royal arms, painted in the year 1804. A small brass chandelier of 12 lights is suspended from the middle of the ceiling of the nave, and is inscribed :—"Bequeathed by Samuel Stretch, of Madeley." At the west end of the nave there is a small organ. It contains several monuments of the Egerton family, particularly one of Elizabeth, wife of Sir John Egerton, who died in the year 1701. She was daughter and sole heir of William Holland, of Denton Heaton, in the county of Lancaster, from whom descended the Earls of Wilton, who retain the family name of Egerton. There are also several other memorable family monuments, and among others, one in the cemetery attached to the chancel, of Sir John Offley, who died September 6, 1688,

This church stands by the side of a public road in the village; and to the west, on the opposite side of the road, is a school, which is a plain old building, with two apartments, one for the boys, and the other for the girls: it has also a small area enclosed before it. This school was founded and endowed by Sir John Offley.

The Church is dedicated to All Saints, and is a vicarage in the patronage of Lord Crewe. The Rev. Peter Roberts is the present vicar; and the Rev. Alexander Simpson is curate, and Master of the free-school.

The benefactions to the poor of Madeley are rather numerous; amongst them those of Sir John Offley, and Ralph Horton, gent. are most conspicuous.

Little Madeley, is an irregular hamlet, situated about a mile to the east of Madeley church. It contains some good farm-houses and smaller tenements.

Madeley Park, to the south, is the property of Lord Crewe; part of it is in a state of nature, and overgrown with heath, and part of it is now under cultivation. The soil is light, may be quickly brought into culture, and promises to make useful land.

The parish of Madeley contains 177 houses, 186 families; 529 males, 489 females: total, 1,018 persons.

KEEL, is a parish and village in Pirehill North, pleasantly situated about three miles west of Newcastle-under-Lyme, on the turnpike-road to Nantwich, in Cheshire. The village consists of many good tenements, farm-houses, and cottages. The parish is small, containing not quite 3000 acres. In the north and east parts of it, are some mines of iron-stone and coal, both of which are worked at Silverdale, where is also a smelting-furnace. The manor is co-extensive with the parish.

Though Keel is now an independent manor, it seems to have been formerly an appendage to Newcastle-under-Lyme, as appears from the old record called Testa de Nevill, which also shews that it then belonged to the Knights Templars.

The Church stands on a rising ground at the entrance of the village. It is a neat stone building, of an oblong square figure, embattled, with a tower at the west end, containing four bells. At the west end are two doors of entrance, one on each side, opposite to each other: that on the north side is seldom used. Keel Church was rebuilt in 1790. The interior forms an area of about 55 feet by 34. The font stands within a niche in the middle of the west wall, and is an elliptical bason of white marble, about half a yard long, and a foot in breadth; it rests on a square pillar of the same marble, about a yard high. Against the same wall is fixed a wooden screen about seven yards long, and extending to the ceiling in height.

Attached to the south wall are monumental inscriptions in memory of Edward Brett, Esq. and his wife, and Lawrence Cranage, gent. There are also several sepulchral memorials of the Sneyds.

The following is a list of the charitable gifts to the poor of Keel parish:

	£	s.	d.
A Stranger (supposed to be Mr. T. Audley) left	20	0	0
Mr. Abnett, of Audley, left twelve fourpenny loaves, to be given yearly, to twelve widows, every Good Friday.			
Mr. John Heath, jun. left, to be given in bread,	5	0	0
Mr. John Heath, sen. ditto, - - - - -	5	0	0
Madame Brett left twenty shillings a-year, to be given in bread to the poor, on the nearest Sunday to the 19th of February			
John Sneyd, Esq. - - - - - - -	10	0	0
Madam Frances Sneyd left for a Free School, to instruct twelve children - - - - - -	100	0	0
Mrs. Mary Giles, - - - - - - - - -	5	0	0
Mrs. Ann Giles, - - - - - - - - -	100	0	0
Advanced by the Parish - - - - - -	5	0	0
Mr. Samuel Whitehurst gave twenty shillings a year to the poor, to be given in bread, as long as the estate called the Nabbs, in this parish, is free from a parish apprentice.			
Mr. Thomas Breck, gent. left to the poor twenty shillings a-year, charged upon lands in Knutton.			

Most of these benefactors lie buried at Keel.

The Parish Register commences May 1, 1540, from January to December, both inclusive. In 1541, there were six baptisms; and from September 1540 to the end of 1542, three burials. For the first thirty years after the commencement of the Register, the average number of baptisms, each year, was seven, and of funerals three. For the same period, from the commencement of the last century, the average was 17 baptisms and 11 funerals. For the last twenty-eight years the average amount has been increased to 36 baptisms and 18 funerals each year. The whole number of funerals in this parish from the year 1700 to 1808, has also been calculated: these amount to 1546. The following enumeration will shew the months which have proved most fatal to humanity: In January, 164; February, 143; March, 175; April, 171; May, 132; June, 109; July, 160; August, 103; September, 105; October, 91; November, 117; December, 137. Hence it appears that most people have died in March, and fewest in October.

In the tower of this church is a bell, on which is the following inscription:

"Ora pro nobis sancte Johannes Baptista."

By some this may be considered a relique of an ancient custom in the church of consecrating and baptizing bells. "The first fact of this kind on record occurred about the year 968, when Pope John XIII. sprinkled a large bell, which was cast for the Lateran at Rome, with holy water, blessed it, and pronounced it sacred. It is said that a similar custom still exists in the Romish church, and that, on the baptizing of bells, a prayer of consecration is used, which imports, that by its sound the people may be delivered from the assaults of their enemies, and the attacks of evil spirits.

It may also afford some degree of evidence to what Saint the church is dedicated; for in those days of superstition, when the bell was cast, as it was believed that the sound thereof was a shield of defence against the attacks of evil spirits, it may be presumed that the sound, which warded off the bad, would call forth the aid of the good spirits; and as the belief of the validity of the intercession of saints was then more generally prevalent, we may suppose that the people of every parish would be anxious to call more particularly on the protecting Saint of the parish church; and some might inscribe one of their bells after the manner of the above, that such saint may not only hear but see their wishes: and thus it may be presumed that St. John the Baptist is the saint to whom Keel church is dedicated. This hypothesis may be strengthened by the consideration, that, in the time of Henry VIII. "among the injunctions of that king, in the year 1536, it was ordered that the dedication of churches should in all places be celebrated on the first Sunday of the month of October for ever. Yet this order was not enforced, or not obeyed; but, however, most of those jubilees, called wakes, are now celebrated near the time of Michaelmas, when a vacation from the labours of harvest and the plough doth afford the best opportunity for visits and sports.

This transposing of the day, (continues Bourne,*) hath left it more

* See his Eccl. Law, Vol. I. p. 339, of 7th edit.

difficult to know the saint to whose protection the church was committed. Such wakes as are observed on the first or second Sunday after Michaelmas-day, in these we may doubt a translation of time by virtue of the said injunction of King Henry VIII. or by a prevailing custom of postponing such solemnity to the end of harvest: and in such cases the saint may be lost, unless some other way preserved."

Keel wake is kept the first or second Sunday in October, and therefore the bell may be considered evidence of the saint to whom the Church is dedicated.

Extracts from old books belonging to the parish:

" *ANCIENT FEES AT KEEL COURT.*

	s.	d.		s.	d.
Summons	0	1	For swearing each witness	0	4
Ent'ring Action	1	0	For takeing verdict	0	8
If Defendant appears not for ent'ring Judg'ment	0	6	For ent'ring Judg'ment	0	6
Cryer for calling	0	4	For Levari fac[t].	1	0
If he appears for ent'ring Plea	0	4	For serving	1	0
			Juror's fees on tryall	1	0

" The Minister of Keel has six shillings and eightpence by a custom here or by a modus for every corpse that is carried through any part of the parish of Keel to be buried in any other parish; if he the said minister of Keel shall in his own person, or by his deputy in holy orders, offer burial to the said corpse." It may be needless to observe that such custom has long been obsolete.

Origin of Surnames.—" Sarah Legacy, who was left as such to the town, by some sorry person or other on the 5th of November last, baptized February 20th, 1737."

The living of Keel is a small curacy, which has been augmented by Queen Anne's bounty. The patron and impropriator is Walter Sneyd, Esq. of Keel Hall, whose brother, the Rev. John Sneyd, is the present incumbent; and the Rev. W. Snape, is curate. A singularity attaches to Keel Church, which is noticed by Plot as also applying to the Cathedral Church of Lichfield, and the Church of Alveton, in this county, the former of which, he says, is " not placed due east and west, as other churches are, but declines no less than 27° from the true points; whilst the Church at Alveton declines in like manner from the true east 32° northward.

Longevity.—Hannah Wright, died February 5, 1776, aged 96.

The soil of this parish is generally a good loam, adapted either to corn or grass.

The parish of Keel contains 176 houses, 178 families; 461 males, 483 females: total, 944 persons.

Keel Hall, the seat of Walter Sneyd, Esq. stands about half-a-mile south-east of Keel, and has been the residence of the principal part of the Sneyds for upwards of two centuries. It is built principally of red stone, and, according to Plot, was erected in 1581. The south front underwent some alterations by the late Ralph Sneyd, Esq. but the upper part of the wings of that front still retain much of the same figure and plan as that engraved in Plot's History of Staffordshire. Over the entrance-door, on the same side of the house, are two shields, the one containing the arms of Sneyd, the other Sneyd impaling Bagot; and there is affixed a lion passant guardant between them, the crest of the Sneyds. The present occupier has erected a handsome new west front, of the same kind of stone, which is embattled, and adorned with four octagon embattled turrets. The grounds of this demesne are embellished with flourishing groves and plantations, and a building or summer-house, noticed in most maps, called the Pavilion, erected on an eminence, which commands extensive views of the surrounding country. There is a dingle of excellent oak to the south-west on the road to Madeley.

Walter Sneyd, Esq. the present possessor of Keel Hall, was many years Lieutenant-colonel of the King's Own or Staffordshire Militia, a distinction which the regiment acquired from the high state of discipline it attained under his command. He received many tokens of Royal regard from the King during the time the regiment was stationed at Windsor; where it was employed as a kind of body guard for upwards of fourteen years, at that splendid Palace of our Kings. His Majesty stood sponsor to one of his children. Colonel Sneyd afterwards commanded the Local Militia for the division of Pirehill North.

The poor in the neighbourhood of Keel Hall, are supplied with bread, meat, and soup, every Christmas, and in seasons of distress, by the liberality of the benevolent owner.

Betley is a parish of Pirehill North, situated upon the borders of Cheshire, about seven miles west by north of Newcastle, on the road to Chester. It contains the ancient town of Betley, in the vicinity of which the handsome mansion of Lady Fletcher, called Betley Court, and the seat of George Tollet, Esq., called Betley Hall, with their pleasure-grounds and plantations, contribute to the embellishment of the place. Betley had an annual fair for cattle held on the 20th of July.

Near the village is a fine pool, called *Betley Mere*, which belongs to the Earl of Wilton. It is now chiefly remarkable for its excellent gardens, which contribute largely in the supply of vegetables to the neighbouring towns and villages.

The Church is situated on an eminence at the eastern extremity of the town, and has been partly rebuilt. It has a nave, side aisles, a chancel at the east end, and a square tower at the west end of the nave. The most ancient part of the edifice is the nave and aisles; which, as well as the chancel, have plain common tile roofs. The windows of the former are also very plain. The side-walls of the nave are part wood, and part plaster: the timber is framed after the ancient manner, and the spaces between filled with plaster, &c. The aisles were, no doubt, coeval with the nave; but from prior decay, the walls thereof have been rebuilt with common brick. In the interior, the nave is separated from the aisles by four plain pointed arches on each side. The pillars which support them are merely *single trunks of trees*; and the architraves of the arches, (if we may use the expression,) are plain curved pieces of wood. From the three middle pillars or trunks, are turned three similar plain wooden curves across the nave; thus making the nave to exhibit a succession of three pointed arches. The dimensions are as follows:—The length of the nave and aisles may be about 15 or 16 yards. The aisles are narrow, being not quite three yards in breadth. The breadth of the nave is about six yards. There is a small west gallery; and at the east end of the north aisle and nave is a large seat, enclosed by a wooden screen, about eight or nine feet in height from the floor. The nave is separated from the chancel by a wooden partition, on the south side of which is the desk and pulpit, which are adorned with crimson velvet, as is also the altar. The upper part of the partition is neatly ornamented on the side fronting the nave, with three painted tables of the Lord's Prayer and the Belief. Above the Lord's Prayer is painted a dove in glory; and above the Ten Commandments are the King's Arms; on each side of which is painted a group of three angels. The doves and angels are on a blue ground. The chancel is built of stone, in a good style, and seems to have been erected by one of the Egertons.

Upon entering the chancel from the nave, there hangs an hatchment, in memory of Mr. Tollet against the north wall. Arms: checky, argent and azure, on a chevron engrailed Or, three anchors azure; on a chief gules, a lion passant argent (Tollet). It has an escutcheon of pretence argent, bearing on a chevron azure, three garbs Or (Cradocke). Crest: a tower surmounted by a pyramid azure, round which appears, coiled and descending, a serpent proper, langued, gules. Motto: "*Prudentia in Adversis.*"

At the east end of the chancel are mural monuments. That on the north wall is the most ancient: it contains two small Ionian columns, the bases and capitals of which have been gilt. These columns support an entablature, the middle part of which is heightened by a circular arch or round pediment. At the top, above each column, is a shield, each bearing the same arms, viz. Gules, a fess ermine between three arrow heads argent, and between the two in chief is a crescent of the same for difference. The entablature, small columns, and the member or part upon which the columns are supported, form a kind of niche, by projecting about nine inches from the wall, within which are a small male and female image in a devotional attitude, with a desk between them, on each side of which is a book open. Behind the female is a third figure, a female, and smaller than the other two. Above these figures, on the back ground, are three coats of arms.

On the south wall, is a modern marble monument, having the arms of Tollet on a shield at the top; below which, on a white tablet, is inscribed;

"In the Year 1768,
George Tollet, Esq.
Erected this to the Memory of
His Mother Mrs. Elizabeth Tollet, and
His Grandfather George Tollet, Esq.
Commissioner of the Navy in
The Reigns of King William and Queen Anne.
They both died in a very advanced age,
And are buried in a vault beneath
The opposite Monument.
Also in the same vault lieth interr'd
The body of the above-named George
Tollet
The younger, Esq. who died upon the 22nd
day of October 1779, in the 46th year of
his age.

He was learned and charitable."

which took place in the time of Cromwell; when the clergy were not allowed to solemnize matrimony. That Usurper, probably to degrade the clergy, caused all marriages to take place before a magistrate. The following extracts shew the manner in which they were recorded:

"Mr. John Milward, sonne of the Worfull John Millward, of Snitterton, in the countie of Darbie, esqr. and Mrs. Jane Sneyd, daughter of the Worfull Mrs. Jane Sneyd, of Broadwall, within the county of Stafford, widdowe, weare published three severall Lords dayes, in the parish church of Wolstanton, in the countie of Stafford; and likewaise in the parish church of Darley, in the county of Darbie, aforesaid; viz. the 21st and 28th dayes of December, and the 4th day of Januarie, 1656; according to an act of Parliament in that case made and provided; (and noe exception made by any p'son) as by the certificates, under the hands of the Regesters of Wolstanton and Darley, was made playnly to appear unto me; and the said Mr. John Milward and Mrs. Jane Sneyd weare married the 27th day of Januarie, 1656, before me *Edward Brett*, esqr. one of the Justices of the Peace for the countie of Stafford: wittness my hand the day and year last above written.

"*Edward Brett.*"

Again:

"Richard Marsh, son of Thomas Marsh, late of Wolstanton, deceased, and Ann Rowley, daughter of William Rowley, of Broadwall, weare published three severall markett dayes at the Markett Crosse in Newcastle-under-Lyme; viz. the first, the 8th, and the 15th dayes of December, 1656, (and noe exception made by any p'son.) And the said Richard Marsh and Ann Rowley weare married the 23rd day of December, 1656, before Edward Eardley, esqr., one of the Justices of the Peace for the countie of Stafford."

The following is said to have been the Form of Marrying before a Magistrate, as enjoined by the Parliament during the Commonwealth of England, to take place from September 29, 1653.

"MAN.

I, A. B. do here, in the presence of God, the searcher of all hearts, take thee C. D. for my wedded wife; and do also, in the presence of God and before these witnesses, promise to be unto thee a loving and faithful husband.

"WOMAN.

I, C. D. do here, in the presence of God, the searcher of all hearts, take thee A. B. for my wedded husband; and do also, in the presence of God and before these witnesses, promise to be unto thee a loving, faithful, and *obedient* wife."

The Coal in this district differs in quality, thickness, and inclination; some strata burn dull, and leave a considerable residue of ashes, others clear, with a quick consumption: the thickness is from two to ten feet; and the inclination varies from a perpendicular descent to an almost horizontal flatness, but the most usual *dip* is about one foot in a yard. It is gotten at different depths. In some places pits are sunk upwards of 130 yards, and in others coal is gotten within 20 yards from the surface. At Kidcrew, in the north side of the parish, is to be seen some of the most improved kind of machinery for raising coal. The price is according to quality and situation. About one-sixth part of what coal produces at the pit-head, is in some places judged a fair mine-rent to the proprietor of the land.

The water, which drains from the coal-mines, is, in this country, of considerable value to those through whose lands it chances to pass, as it is impregnated with a yellow substance termed Car; which is used by the potters, and is a necessary ingredient in making that kind of ware called Egyptian black. The mode of procuring the car is as follows:—Being of a specific gravity greater than that of water, it forms a sediment at the bottom of the channel of the stream that conveys it from the mine: when a considerable quantity is thus lodged in a certain space, the stream, to that extent, is diverted from its usual course; and the car is thrown out of the channel, from whence the water has been turned off, upon the adjoining banks; where it remains till dry. Sometimes small pits or ponds are made on the adjoining banks, and the car is scooped from the bottom of the channel, and thrown into them, without diverting the course of the water. When it is sufficiently dry, it is sold at the rate of one guinea per cart-load.

A great number of bricks, tiles, and quarries are made for sale in this parish. The price of common bricks is about 28*s.* per thousand; dressed bricks, £1. 15*s.* and floor bricks, £2. 15*s.* per thousand. Three cubic-yards of clay make 1000 bricks; mine-rent, 2*s.* 6*d.* per thousand.

New Chapel, is a village and chapelry in the northern part of this parish, at least five miles north of the mother church. It is situated on a good sound upland of mixed loam. There are several coal-works in this part of the parish.

The Chapel is a modern structure of brick, with two double rows of pews, and a small west gallery. It has no tower, but a cupola with a small bell; and is endowed with land at Norton-in-the-Moors, and Burslem, with some small annuities, and surplice-fees. The Rev. John Lawton is the present minister.

In the chapel-yard is a plain altar-tomb inscribed:

"In Memory of James Brindley, of Turnhurst, Engineer, who was interred here, September 30, 1772, aged 56."

This ingenious man resided at Turnhurst, a mansion in this parish, and is well known as the inventor of the modern system of artificial canal navigation.*

Biddulph, is the most northerly parish of Pirehill North, and borders on Cheshire. The source of the river Trent is in the neighbourhood of Mole Cop, a rocky ridge, near the boundary between Cheshire and Staffordshire. A considerable extent of mountainous land projects from Mole Cop to the north-west.

This is an extensive parish, and is divided into four hamlets, namely, Over, Middle, and Lower Biddulph, and Knipersley. The manor of Biddulph was in the Conqueror's hands, but he soon afterwards bestowed it upon Ormus, who had several sons, the eldest of whom inherited it, but having only daughters, it passed by them into other families. Middle Biddulph was given by Aliva to Roger, her cousin, her uncle Edward's son, who took upon him the surname of Biddulph, and in his posterity it continued to the year 1598; and Over Biddulph to Thomas, the son of Ormus, another of her uncles: but Nether Biddulph went by Aramintia, who married Henry Verdon, to the Bagleys and Mainwarings.†

The Church of Biddulph is an ancient structure of stone, with a low square tower, and is situated on a gentle eminence. In the interior, near the altar, on the north side of an altar-tomb on the south side of the chancel, is the following epitaph in capitals:

"Goe gentle stranger, press not neere
To qvestion who's intoombed heere.
Here lies the man, ah mee noe more
Lest thov this sacred dvst adore.
His honor'd name's conceal'd from thee
To keepe thee from Idolatrie.
Svffice it onely here to find
He was a man of mortall kind.
The composition of his spirit
The exposition of his merit.
Are theames all hvmane thoughts transcend
Matters t' admire not comprehend.

* A memoir of this gentleman will be given in our account of the Grand Trunk Canal.

† Magna Britannia, No. 42, p. 76.

The world knows his noble worth so well
All tongues, all pens, on 's praise shall dwell.
Mvltis ille qvidem flebilis occidit
nvlli flebilior - - - - - - - - E. B.
Ehev: fvgaces - - - -
labvntur anni - - - - - - -."

Sir William Bowyer, Knt. in whose praise the above lines were written, died March 7th, 1640, aged 53.

This church is dedicated to St. Lawrence, and is a vicarage in the patronage of the Gresley and Adderley families alternately. The Rev. James Sewell is the present vicar.

Longevity.—Sarah, wife of William Poole, of Rushton, died Jan. 29, 1806, aged 92 years. Also, William Poole, died January 31, 1811, aged 97 years.

This parish presents some curious remains of antiquity, particularly the Bridestones, consisting of eight upright freestones, two of which stand within a semicircle formed by the other six. Some antiquaries suppose the circle was formerly complete, but the origin and use of these stones is unknown. Westward of the Bridestones there is a kind of artificial cave of freestone.

The parish of Biddulph contains 269 houses, 269 families; 721 males, 739 females: total, 1,460 persons.

Knipersley Hall, in this parish, is the property of Mr. Bateman, of Manchester, and is a considerable farm, of a loamy soil, kept in hand, and cultivated on the modern improved system.

NEWCASTLE-UNDER-LYME.*

This Borough and Market-town is situated within the Hundred of Pirehill, and although entirely surrounded by parishes which are within the Northern division of that Hundred, is considered, for the public purposes of the county, to be in the Southern district. Newcastle is distant from Lichfield 31 miles, from Stafford 16 miles, and from Leek 12 miles.

Monday is the regular market-day, and another market on Saturdays has been recently established to meet the demands of the increased population.

There are well-attended fairs, for horses, cattle, woollen cloths,

* Sometimes spelled *Lime* or *Line*, but all the Charters of the Borough give the addition of *under-Lyme*.

and many other articles, on Shrove-Monday, Easter-Monday, Whit-Monday, and the first Monday in November, besides two other moveable fairs in the months of July and September.

This Borough, and also a Manor of considerable extent adjacent to it, known by the name of the Manor of Newcastle-under-Lyme, derive the former part of their appellation from a Castle, which stood in the midst of a large pool, now nearly surrounded by the town. The Manor appears, from ancient documents, to have extended over the liberties of Penkhull, Wolstanton, Shelton, Hanley, Clayton, Seabridge, Knutton, Dimsdale, Holditch, Hanchurch, Hanford, Whitmore, Keel, Fenton, Longton, Meer-lane, Normacot, Tunstall, Chatterley, Bradwell, and Thursfield.

Antiquaries have generally supposed, upon the authority of Camden, that the castle had its name on account of an older castle, which stood not far from it at Chesterton-under-Lyme; but, supposing this to be correct, the addition of *under-Lyme* still wants explanation, and if it ever formed part of the name of Chesterton, it has long ceased to do so. Camden's account of Newcastle is as follows :*

"The Trent first runs southward, with many windings, not far from New-Castle under Lime, so called upon the account of an older castle which formerly stood not far from it at Chesterton-under-Lime, where I saw the ruinous and shattered walls of an old castle, which first belonged to Ranulph Earl of Chester by the gift of king John, and after, by the bounty of Henry 3d, to the House of Lancaster."

Later authors have been led into error by the ungrammatical construction of the above quotation, for it is certain that Camden must have meant, that the ruinous walls which he had seen of an old castle formerly belonging to the Earl of Chester and afterwards to the House of Lancaster, were the walls of *Newcastle*, and not, as the sentence at first reading seems to imply, the walls of the castle of Chesterton.

Dr. Plot fell into the above-named error, and he proceeds to state,† that the castle of Chesterton went to decay "as long ago as the reign of king Henry 3d, when the Earl of Lancaster built another near by, in the midst of a great pool, which he called the New Castle, that gave original (no doubt) to the town of that name close by it."

* Camden's Britannia, edit. 1695, p. 530.

† Plot's Natural Hist. of Staffordsh. p. 434.

Bishop Gibson, the editor of Camden, follows Dr. Plot in his mistake, and he also asserts* that Newcastle was "built in Henry 3d's time by the Earl of Lancaster."

The authorities which we shall cite fully justify our interpretation of Camden, and they prove, in contradiction to Plot and Gibson, that the town, as well as the castle, of Newcastle-under-Lyme, existed not only before the creation of the Earldom of Lancaster, but even before the reign of Henry the 3d.

In the 5th of king John (1203) the town of Newcastle was amerced, for having changed its market-day from Sunday to Saturday.†

In the 17th of king John (1215) a grant was made from the Crown to Ranulph de Blondeville Earl of Chester of New-Castle-under-Line and its liberties,‡ including, in all probability, the Manor as well as the Castle of that name. The Crown, however, appears shortly to have regained possession of Newcastle and all its appendages. Amongst the inquisitions remaining in the Exchequer,§ there is one of early, though indeterminate, date in the reign of Henry the 3d, stating that our Lord the King held, as forfeited, the New Castle with the Town and Manor; including by name most of the townships which are now considered parcel of the Manor. The Manor appears to have been granted in fee-farm by Henry the 3d, to Gilbert Lord Segrave in 1232,‖ but the castle and town remained in the possession of the Crown. During this period the castle was held for the King under the government of a constable,¶ which appears to have been an office of considerable importance. In the 35th Henry 3d, the constableship was held by one of the Audley family,** and afterwards by William de Fenton, who appears to have been constable in the 37th and 42d years of the same reign.††

Many of the freehold tenants of this manor were bound by their tenures to perform the Castle guard at different periods of the year. The *Testa de Nevill*, which is an ancient record preserved in the Exchequer, and compiled by John Nevill, a justice in Eyre, between the years 1234 and 1238, contains the following particulars of tenures within this manor by knight's service and socage:

* Camden's Britannia, edition, 1695, p. 538.

† From the Record of *Assisæ et Placita Coronæ*. taken at Lichfield on St. Matthew's day, 5th King John.—See the Abbreviatio Placitorum.

‡ Calendarium Rotulorum in the Exchequer, published 1803, p. 39.

§ Calendarium Inquisit. post mort. vol. 1, p. 202.

‖ Magna Britannia et Hibernia, vol. 5, p. 64. ¶ Ibid. p. 64. ** Ibid. p. 64.

†† From old Grants.

KNIGHT'S FEES.

"John de Cnocton holds the vill of Cnocton (*now Knutton*) to fee-farm, paying yearly to the New Castle £4. 11*s.* 6*d.*, also performing the guard of the same castle for forty days, at the cost of our Lord the King.

"Also, William de Erdinton holds the vill of Fenton to fee-farm, paying yearly to the said castle 7*s.* 4*d.*, performing also the aforesaid guard for the aforesaid term.

"Also, William de Hanleg' holds the vill of Hanleg' (*now Hanley*) to fee-farm, paying yearly to the said Castle 6*s.*, and performing, &c.

"Also, Ralph de Bevill holds the vill of Langeton (*now Longton*) to fee-farm, paying yearly 6*s.* to the said Castle, &c.

"Also, William Murell holds one yard-land within the said manor of New Castle, in the vill of Selton, (*now Shelton*) by the serjeanty of keeping *Haiam* (the inclosure or park) of our Lord the King there, which is called the *Haia* of Clive (*now Cliff.*)

"Also, Henry de Audicheley holds within the said manor the vills of Tunstal, Chadderleg,' (*now Chatterley*) Bradewell, Turnedesfeld, (*now Thursfield*) and Normanecot, (*now Normacot*) by the serjeanty of serving on foot, with a bow and arrows, within the said Castle for eight days in time of war at his own cost."

SOCAGE TENANTS.

"The Knights Templars hold Kel, (*now Keel*) a member of New Castle, of the gift of our Lord King Henry, and pay nothing."

In the year 1235, Henry the 3d granted a charter to the burgesses of Newcastle-under-Lyme, erecting the town into a free borough, and granting to the burgesses a merchants' guild (which was the phrase then used for the establishment of a body corporate) and other privileges. Of this charter a copy is subjoined:

"*Henricus Dei gratia Rex Angliæ Dominus Hiberniæ et Dux Aquitaniæ Archiepiscopis Episcopis Abbatibus Prioribus Comitibus Baronibus Justiciariis Vicecomitibus Prepositis Ministris et omnibus Balliviis et fidelibus suis Salutem Sciatis nos concessisse et hac carta nostra confirmasse pro nobis et heredibus nostris Burgensibus nostris de Novo Castro subtus Lymam quod Villa nostra de Novo Castro subtus Lymam sit liber Burgus et quod Burgenses ejusdem Villæ habeant Gildam Mercatoriam in eodem Burgo cum omnibus libertatibus et liberis consuetudinibus ad hujusmodi Gildam pertinentibus et quod eant per totam terram nostram cum omnibus mercandisis suis emendo et vendendo et negociando bene et in pace libere et quiete et honorifice et quod quieti sint de thelonio passagio pontagio stallagio et lestagio et omnibus aliis consuetudinibus salvis in omnibus libertatibus Civitatis nostræ London Quare volumus et firmiter precipimus pro nobis et heredibus nostris quod predicta Villa nostra de Novo Castro sit liber Burgus et quod Burgenses ejusdem Villæ habeant Gildam Mercatoriam in eodem Burgo cum omnibus libertatibus et liberis consuetudinibus ad hujusmodi Gildam pertinentibus et quod eant per totam terram nostram cum omnibus mercandisis suis emendo et vendendo et negociando bene et in pace libere et quiete et honorifice et quod quieti sint de thelonio passagio pontagio stallagio et lestagio et omnibus aliis consuetudinibus salvis in omnibus libertatibus Civitatis nostræ London sicut predictum est. Hiis Testibus, Gulielmo de Ferrariis, Gulielmo de Cantilupo, Radulpho filio Nicholai, Almarico de Sancto Amando, Galfrido Despensatore, Radulpho Tyrel et aliis, Datum per manum venerabilis Patris Radulphi Cicestrensis Episcopi Cancellarii nostri apud Feckeham decimo octavo die Septembris anno regni nostri decimo nono.*"

The Borough appears to have been at this time a place of very inconsiderable size and population; for in the said record called *Testa de Nevill*, which was compiled at the same period, appears the following entry of the number of its burgages:

"In the Town of Newcastle are twenty-eight burgages, each of which pays twelve pence per annum, at the four quarter-days."

The Castle and Borough appear to have continued in the immediate possession of the Crown until the 49th Henry 3d, (1263) when the King was compelled to transfer them by grant to his brother-in-law Simon de Montfort* Earl of Leicester and Chester. The period during which they were in the hands of that rapacious and aspiring nobleman was very short, as he lost his life at the head of the rebellious barons who fought the battle of Evesham,† against Prince Edward, in 1265. All the possessions of Simon de Montfort were forfeited to the Crown; and a great part of them, including this Castle and Borough, was granted by Henry the 3d,‡ in the 51st year of his reign (1265) to his younger son Edmund Plantagenet (surnamed Crouchback) who was created Earl of Leicester in 1265, Earl of Derby in 1266, and in the following year Earl of Lancaster, which was the earliest creation of the latter dignity.§

Among the Barons whose estates were confiscated in consequence of the battle of Evesham was Nicholas Lord Segrave,|| who had pro-

* Calendarium Rotulorum in the Tower, published 1802.

† Life of Simon de Montfort, written by George, second Marquis Townshend, in Longmate's Supplement to Collins's Peerage, p. 312.

‡ Calendarium Rotulorum in the Exchequer, published in 1803, p. 94.

§ Heylyn's Help to English History, ed. 1773, p. 271.

|| Magna Britan. et Hibern. vol. 5, p. 64.

bably inherited the Manor of Newcastle from Gilbert Lord Segrave, for in 1267 we find this Manor again in possession of the Crown, and granted out to Hammon le Strange,* who was one of the Lords Marchers of Wales. The Manor was shortly afterwards re-united with the Castle and Borough, and became part of the possessions of the House of Lancaster.

The said Edmund Earl of Lancaster, &c. died seised of this Castle and Manor, in 1295,† and they descended, with his other vast estates, to his eldest son Thomas, who upon his marriage with Alice, the heiress of Henry Lacy Earl of Lincoln, endowed her, *ad ostium ecclesiæ*,‡ with the Castle and Borough of Newcastle, and all the Hamlets belonging thereto. In the Record called the "Nomina Villarum," returned into the Exchequer in 1315, by the Sheriff of Staffordshire, in obedience to the King's writ, the Earl of Lancaster is named as Lord of the Town of New Castle, and it is stated that the said Earl had the return of all writs there. This Thomas Earl of Lancaster, &c. having been taken in arms against King Edward II. in 1322, was condemned to death by a court-martial, and beheaded near Pontefract. His estates were confiscated at that time, but afterwards his widow had livery of part of them, comprising this Castle and Borough, and the Hamlets belonging thereto. Henry Earl of Lancaster, &c. the brother and heir of the said Thomas, (whose attainder was reversed in 1326) died in the life-time of the Countess Alice, leaving a son, Henry, created Duke of Lancaster, who in time became possessed of all the estates of that great house.§ Duke Henry died of the plague at Leicester, in 1361, whereupon this Castle and Manor, with their members and other appurtenances, were, upon a partition of his estates, between his two daughters and co-heiresses, assigned to his eldest daughter Maud,‖ at that time the wife of William Duke of Zealand and Bavaria. She died seised of these estates ¶ in the following year, without issue, whereupon they became the property of John of Ghent, (commonly called Gaunt) Earl of Richmond, the third son of King Edward the 3d, in right of his wife Blanch, the surviving daughter of the said Duke Henry. John of Ghent, on 13th November, 1362,

* Calendarium Rotulorum in the Exchaq. published 1803, p. 99.

† Shaw's History of Staffords. vol. 1, p. 40. ‡ Ibid. p. 41.

§ Calendar. Inquis. post mort. published 1808, vol. 2, p. 236.

‖ Abbreviatio Rotulorum in the Excheq. published 1805, p. 263.

¶ Calendar. Inquis. post mort. published 1808, vol. 2. p. 247.

was advanced in Parliament, to the dignity of Duke of Lancaster, with great splendour and solemnity. He died in 1398, being succeeded by his eldest son, Henry of Bolingbroke, who, as Henry the 4th, assumed the Crown of England. Ever since his accession to the Throne, the possessions of the Duchy of Lancaster (including this Castle and Manor) have been virtually united to the Crown; though separated therefrom in point of form, by the policy of different Monarchs.*

The Manor and site of the Castle, of which nothing is left but the foundations, are now held by the Marquis of Stafford, as Lessee of the Crown in right of the Duchy of Lancaster. The pool which surrounds the site of the Castle, and the adjoining mills, were granted by the Crown, in the 8th year of James I. to Ralph Sneyd, Esquire, (an ancestor of their present owner Walter Sneyd, Esquire,) subject to a fee-farm rent of £14. 6*s*. 8*d*. The Coal Mines under all the Copyhold Lands within this Manor are the property of the Duchy, and now yield a considerable revenue. They are worked extensively, for the consumption of the Potteries, by the Lessee of the Crown, Lord Viscount Granville. Within the limits of the Borough, all Manorial Franchises have long been exercised by the body corporate.

The town was first incorporated, as above-mentioned, by Henry the 3d, on 18th September, 1235. His Charter was twice confirmed by Edward the 3d, viz. once in 1344,† and afterwards in 1372,‡ and subsequently, in 1378, it received confirmation by Richard the 2d.§ In 1281, Edward the 1st, at the instance of his brother Edmund, added the Grant‖ of a Fair for three successive days, viz. on the Feast of the Holy Trinity, and on the eve and morrow of that day; and other Fairs were afterwards added by Edward the 3d and Henry the 6th.¶

The early Charters do not contain the style of Incorporation; but it appears by the Records of the Borough, which are extant from the year 1368, that a Mayor and two Bailiffs, besides other inferior Officers, were annually elected. Queen Elizabeth, by Charter dated in 1590,** confirmed all the former Grants, and added many

* Blackstone's Commentaries, 11th Edit. vol. 1. p. 119.

† Calendarium Rotulorum in the Tower, published 1803, p. 149.

‡ Original Charter amongst the Borough Records.

§ Ibid., but the date obliterated, and here supplied from the Calendar. Rot. in the Tower, p. 200. ‖ Original Charter amongst the Borough Records.

¶ Calendar. Rotulorum in the Exchequer, published 1803, pp. 171 & 199.

** Original Charter amongst the Borough Records.

new privileges. The Corporate body at that time assumed, and still retains, the style of the Mayor, Bailiffs, and Burgesses, of the Borough of Newcastle-under-Lyme. By this Charter a Common Council was established, for the government of the Borough, and the management of its Revenues and Possessions, consisting of a Mayor, (who is also a Justice of the Peace) two Bailiffs, and 24 capital Burgesses; the three former of which Officers are annually chosen by the Common Council on Tuesday after Michaelmas-day. The office of Clerk of the Market was granted to the Mayor for the time being: The offices of Recorder, Town Clerk, and Serjeants at Mace, were established by the same Charter: under the authority of which several Courts are also held, viz. a Court Leet and View of Frank Pledge twice a-year, a Court of Record held every three weeks for the trial of personal actions arising within the Borough, not exceeding £40. in amount, and a Court of Pie-powder for the trial of Disputes and Offences occurring during the Fairs and Markets: The holding of several other Fairs, and of a weekly Market on Mondays, with many other privileges, were also conferred on the town by this Charter.

Charles the 2d, in 1664, granted a Charter* confirmatory of Queen Elizabeth's. The principal additions made by his Grant are the appointment of two other Justices of the Peace (who are annually chosen out of the Common Council, and, together with the Mayor, hold a General Sessions of the Peace within the Borough four times in every year,) and the extension of the sum for which Pleas may be held in personal Actions from £40. to £50.

This Borough underwent a like fate with many other Corporations, towards the close of the reign of Charles the 2d, at which time a surrender was forced from it of all its Franchises.† His successor James the 2d, in the first year of his reign, granted a new Charter,‡ comprehending nearly the same privileges as the Borough had before enjoyed, but containing an arbitrary and unconstitutional clause which empowered the Crown to remove the principal Officers of the Borough from their situations at pleasure. During the short reign of James the 2d, the new Charter was acted upon, and the obnoxious clause of removal was twice put in force; but after the accession of William and Mary, the former Charters were restored to the Borough, and the Charter of James has never since been considered as of any validity. The office of Recorder is now filled by James

* Original Charter amongst the Borough Records.

† Corporation Records. ‡ Original Charter amongst the Borough Records.

Caldwell, Esq. of Linley Wood, who was elected in 1801, and that of Town Clerk, by Thomas Sparrow, Esq. who has held the office since 1789.

The Guildhall of the Borough is a large and handsome brick edifice, ornamented with stone pilasters, and standing in the centre of the Market-place. Nearly adjacent to it is an elegant stone Cross, surrounded by a flight of steps for the accommodation of the market.

Two Members are returned to Parliament by the resident Burgesses of this Borough. This privilege has been exercised ever since 1352, and probably for a longer period, as the time when the right was first obtained is not known. It appears from a Report of the House of Commons, made by the Committee of Privileges in 1624, that the right of Election was then considered to belong exclusively to the Common Council; but the resident Burgesses at large have long been admitted to share in the Elections, and their right is now for ever confirmed by the last Resolution of the House of Commons, which passed in the year 1792.

The Freedom of the Borough is acquired in four different ways. 1st. by Birth, the claimant having been born within the Borough, the son of a sworn Burgess residing there at the time of the birth; 2d. by Service for Seven Years within the Borough, under an Indenture of Apprenticeship inrolled with the Town Clerk; 3d. by Purchase; and 4th. by Gift, from the Common Council of the Borough. Absence from the Borough for a year and a day successively after admission to Freedom, operates as a permanent loss of the Franchise. The Mayor and two Bailiffs are the Returning Officers.

The following are the total numbers of the Burgesses who voted at the nine last contested Elections for Newcastle:

In 1734, ... 444,
1774, ... 465,
1790, ... 564,
1792, ... 569,
1793, ... 562,
1802, ... 555,
1807, ... 622,
1812, ... 645,
1815, ... 621.

This Borough is at present represented in Parliament by Sir JOHN CHETWODE, Bart. of Oakley, and Sir JOHN FENTON BOUGHEY, Bart. of Aqualate, both in this County. The latter gentleman is a native of the Borough, and was elected for the first time in 1812. The other Representative succeeded in 1815 to Earl Gower, who had vacated his seat for Newcastle in order to be elected for the County of Stafford.

The Church of Newcastle is dedicated to Saint Giles the Abbot. It was re-built in 1720 of brick, with a parapet wall, and is a large and commodious structure, with three galleries, supported by well-proportioned pillars. The pews are of oak and uniform, and there is a good Organ. On each side of the middle aisle are pews for the Corporate Officers of the town, that for the Magistrates being distinguished by a gilt canopy. The Chancel is in the form of a semicircular alcove, and the walls and ceiling of it are elegantly and appropriately painted. Over the middle window is a carved Pelican feeding her young. The font (of white marble) was the gift of Samuel Bagnall Esquire, of Barlaston. A neat clock was lately added to the west gallery by the late Alderman Hill. The Church contains several handsome mural monuments, belonging to the families of Ford, Fenton, Kinnersly, Beard, Bourne, &c. The tower is very ancient, and built of red sand-stone, and the entrance to it is on the western side by a Saxon arch of very elegant architecture. It contains eight bells and a clock with chimes.

The Church was, until lately, only a chapel of ease to the parish of Stoke-upon-Trent, but in 1807, an Act of Parliament passed by which five new Rectories, of which Newcastle is one, were created out of that parish. The Rev. Clement Leigh is now the Rector, and the Patronage is at present vested in the Rev. Charles Simeon, of King's College, Cambridge. The Living is now small, but will be soon considerably augmented by a commutation of the Tithes of the Town-Fields.

There are four Meeting-Houses for Dissenters from the Established Church, one of which, for Presbyterians, has been shut up many years.

On an elevated situation near the north-west end of the town, stands a large brick building, containing Alms-houses for the reception of twenty poor widows. The following inscription on a stone in the north front, explains the origin of this charity:

"Erected
at the expence of the Right Honourable
Grace Countess Granville;
John Lord Gower,
and Bernard Granville Esq.
in completion of the Will of the Most Noble
Christopher Duke of Albemarle.
Anno 1743."

The endowment furnishes a weekly allowance of 2s. 11d. in money to each of the twenty widows, besides twenty shillings, to buy each of them a blue gown and petticoat of cloth or stuff, at Michaelmas. The nomination of the widows is vested in the Marquis of Stafford, Lord Carteret, and the Rev. John Granville, of Colwich, in this county, who now represent the several families that inherit the estates of the original founder, and are chargeable with the payment of the charity.

A Free Grammar School, for teaching Latin and Greek to the sons of the burgesses, and of the poor inhabitants of the Borough, has been long established. The present School, erected in 1722, stands in a damp and inconvenient situation adjoining the church-yard, and it is intended to remove it to another part of the town. The Corporation are the trustees of the revenues of the school; but the right of presentation, to two turns out of three, belongs to the family of William Cotton, Esq. of Bellaport, in Shropshire, and of Etwall, in Derbyshire. The Corporation present upon every third vacancy, and have lately exercised that right on occasion of the resignation of the Rev. John Blunt, the late master.

Another Free School, for teaching children of the poor inhabitants of Newcastle to read, write, and cast accounts, was founded in 1704, by Edward Orme, Clerk. This school, and a house for the master, which is attached to it, adjoin the west side of the church-yard. Sixty boys are educated here, and £5. a-year is appropriated for putting two of them out as apprentices.

The late Thomas Hatrell, Esquire, by his will endowed several other schools for the education of poor children, in this town. Besides the charities above-mentioned, many donations, which have been from time to time given to the Borough, and catalogues of which are fixed in the church, are yearly disposed of in bread and clothing to the poor inhabitants. The interest of £600. given to the Borough in 1726, by Sir Walter Wagstaffe Bagot, Bart. then a Representative of the Borough in Parliament, is appropriated in aid of the church and constable rates; and the interest of a like sum of £600. given in 1730, by another Member for the Borough, John Ward, Esq. afterwards created Lord Viscount Dudley and Ward, is applied in aid of the poor's rates.

The staple trade of the town is the manufacturing of hats, for which it has long been famous. There is a public brewery, on an extensive scale, and considerable manufactories in the silk and cotton trades have lately been established. The streets are mostly well-paved, and remarkably wide, and the houses are in general well-built.

This town is on the direct road from London and Birmingham to Liverpool and Manchester, and it possesses other great advantages in point of local situation, from the abundance and cheapness of fuel, and its vicinity to the important and populous district of the Potteries; and it has an immediate communication by water with the Trent and Mersey or Grand Trunk Canal.

In 1775 an Act passed for enabling the late Sir Nigel Gresley to make a Canal to Newcastle from his Coal Mines at Apedale, distant about two miles, and the town has been since supplied with those coals at a price limited by Parliament. The supply having been defective another Act was obtained in 1812, in consequence of which other mines in the same estates were opened and a small advance was made in the price. As the term, during which the proprietors of these mines are restricted as to price, and are bound to furnish a sufficient quantity of coals to the town, will expire in two years, it seems unnecessary to add the particulars of the prices, &c.

A navigable Canal from the south end of this town to the Grand Trunk Canal at Stoke-upon-Trent, a distance of four miles, has been made in pursuance of an Act which passed in 1795: and a communication by Canal and Railway is also made between the first-named Canal and the late Sir Nigel Gresley's Canal under the authority of another Act passed in 1798. These Canals were made by subscription, and afford great accommodation to the town, although they have not hitherto paid any dividends to the proprietors.

A piece of waste land called the Marsh, lying in this Borough, and containing about twenty-three acres, was inclosed and leased out for building upon by virtue of two Acts which passed in 1782 and 1783. The rents are applied in aid of the Poor's Rates, and produce at present about £200 per annum. An elegant Theatre, and Ladies' Boarding School, both built by subscription, besides a handsome Hotel and Assembly-room, with many excellent houses, are already built on this land.

But the improvement from which the greatest benefit is likely to be derived to Newcastle, is the Inclosure of some open Fields surrounding the town, and containing about six hundred acres, which are the property of many individuals, subject to Rights of Common exercised periodically by the Burgesses of Newcastle. An Act for effecting this Inclosure, as well as the Inclosure of a Common called

Knutton-heath, on which the annual Horse Races are held, received the Royal Assent in June 1816; and as the measure is yet in its infancy, we deem it right to give the leading features of the Act a place in this History.

All the rights of common are to cease at such time after the 25th December, 1816, as the Commissioner shall direct.

205 acres of land in the Town-fields are to be set out in compensation for rights of common, and to be allotted to Trustees for the burgesses.

These 205 acres are to be parcelled out in not more than five allotments, to consist of not less than 30 acres in any one field, lying in one entire parcel, except in the Clayton field, where there may be two allotments not exceeding together 60 acres.

Five acres of the Brampton and Stubbs are to be laid out in Public Walks for the accommodation of the inhabitants.

The residue of the Town Fields is to be allotted amongst the Land-Owners, due regard being had to the present local situations and conveniences of their respective lands. This residue consists of upwards of 400 acres of land, a considerable part of which will be immediately convertible to building and other valuable purposes.

All burgesses possessing a qualification of £40. per annum in freehold or copyhold estates within this county, or being heirs apparent of persons having £100. per annum of like property, are capable of acting as Trustees for the burgesses' allotments.

The Trustees are to let off such portions of the burgesses' allotments as shall be necessary to defray the expenses of keeping up the fences, paying taxes, tithe-rents, &c. of their allotments, and supporting the public walks; and they are to permit the rest of these allotments to be enjoyed by the burgesses as stinted pastures; subject to regulations for limiting the number, and determining the sorts of stock, and the periods of using such pastures. The Trustees have the power of enforcing the observance of their regulations, by penalties not exceeding £5. for any one offence.

All Tithes are to remain payable as heretofore, except from the town fields lying in the parish of Newcastle, where a commutation is to take place, on the principle of giving to the Rector annual rent charges payable out of the several allotments.

Care has been taken in the Act to preserve a continuance of the annual amusement of the Races, by allotting the Race Course on Knutton-Heath, with its appendages, to the Corporation of Newcastle for the purposes of the races, with provisions for the accommodation of the spectators, &c.

According to the returns to Parliament in 1811, this town contained 1354 inhabited houses, and a population of 6175 inhabitants, of whom 2940 were males and 3235 females. The families were classed as follows, viz. 1207 engaged in trade, 47 in agriculture, and 100 not falling in either of these descriptions. At the present period, 1817, the population would not be over-rated at 8000.

An erroneous notion has prevailed, that the town was heretofore greater in size, and that it once had four churches, but that three of them, as well as a considerable part of the town, were reduced in the wars of the Barons. We have already shewn that about the period of its incorporation in the year 1235, which was probably within a century after the first erection of the Castle, the town contained only 28 Burgages. Amongst the "Inquisitiones Nonarum," taken about the year 1341, and preserved in the Exchequer, is a record relating to the Borough, too curious to be here unnoticed. We subjoin a translation of it in the note,* and it shews that at the end of it's second century the town had not advanced much in prosperity. With regard to the tradition of four churches having formerly existed, there is nothing to satisfy us of its credibility. It is said there were formerly churches dedicated to St. Mary and St. Katharine, but the Patent Rolls in the Tower, (vide Calendarium, pages 85 & 226) will justify our assertion, that those were nothing more than *Chauntries* dedicated, it is true, to the abovenamed Saints, but annexed to, and forming parts of, the only Church, or (more properly speaking) Chapel, belonging to the town, which, as we before stated, is dedicated to St. Giles. In fact every authentic record and document proves that the town was not more

* The Inquisitio Nonarum for Staffordshire, taken about 1341, is as follows, so far as regards the Borough of Newcastle:—"Jordan de Lavenden, Ralph "Swanill, Richd. Legowe, Wm. le Grey, Henry Bryan, Henry Kene, John "Pollard, Wm. le Somenour, Richard Roberts' son, Richd. le Glover, Ralph "Lycoris, Philip de Erlyde, being sworn present that the Men of the aforesaid Town have moveable goods, whereof the true value of the ninth of all "the moveable Goods, (except growing corn, wool, and lambs for which they pay "the ninth out of the Town aforesaid,) is worth five marks and two shillings "and no more; and the reason is, because the better men of the same Town and "the greater part of the men thereof live by agriculture and by wool and lambs, "for which they pay to the ninth out of the Town aforesaid;) and the further "reason is, because the community of the whole Town aforesaid is annihilated "and reduced to poverty."

extensive or populous in former times, but on the contrary, that from having at first consisted of a few houses only, it has gradually and regularly advanced, to the present period, in size, population, and importance.

Dr. Plot having already celebrated the curious device of an ancient Iron Bridle, which is kept by the Mayor of this Borough for the punishment of Scolds, we shall close our observations on the town by quoting his quaint remarks upon this subject. "They have a peculiar artifice at Newcastle and Walsall for correcting of scolds, which it does too so effectually, and so very safely, that I look upon it as much to be preferred to the Cucking-Stool, which not only endangers the health of the party, but also gives the tongue liberty 'twixt every dip; to neither of which this is at all liable; it being such a Bridle for the tongue, as not only quite deprives them of speech but brings shame for the transgression, and humility thereupon before it is taken off. This instrument being put upon the offender by order of the Magistrate, and fastened by a padlock behind, she is led round the town by an officer to her shame, nor is it taken off till after the party begins to shew all external signs imaginable of humiliation and amendment."* To the credit of the females of the present age, this whimsical method of correction has not been frequently called for of late years.

Dr. Plot mentions an instance of a stone having been found in a place called *Gallows Field*, near the town, being the place where malefactors were formerly hung, in which stone was an entire skull of a man, with the teeth, &c. in it. Of this fact, an alderman of Newcastle assured the doctor, that he had such an one long in his possession. This curious circumstance Plot endeavours to account for by saying, that it is probable, that the place, when it was used for executions, was nothing else but a sandy land, in which they used to bury the bodies of the persons executed, which, in process of time, turned into stone, about the head of a man, inclosing it in it. This is not at all unlikely; it being well known that sands have been observed to petrify.

This same writer also, in mentioning several instances of men of extraordinary strength living in this county, adduces one in Godfrey Witrings, a butcher of this town, whom he saw take up a form six feet and ten inches long, and fifty-six pounds in weight, by one end in his teeth, and, holding both his hands behind him, lifted up the other end the whole height of the room, striking it thrice against

* Plot's Staffordshire, p. 389.

the floor of the chamber over it; which, the doctor says, by computation according to the centre of gravity, will prove that he lifted up, with his teeth, about one hundred and sixty-eight pounds weight.

Newcastle gave birth to Major-General Thomas Harrison, and to the celebrated John Goodwin, two of the infamous but extraordinary Cromwell's admirers; the one aiding the Usurper's regicide purposes with his sword, and the other with his pen.

These Major-generals, as the Usurper called them, were, according to some, only eleven* in number; according to others† twelve; but Bates says,‡ the districts over which Cromwell appointed the Major-Generals were fourteen. These officers were to keep a strict and vigilant eye over the jarring parties of the Presbyterians, the Independents, and cavaliers, as the loyalists were called; but particularly to watch the proceedings, and curb the factious spirit of the rigid republicans, whom Cromwell had the greatest reason to dread and suspect. The Major-generals had almost absolute power; and they exercised it, as might naturally be expected from the nature of their characters, and the upstart innovations of their master, of whose turbulent spirit they largely partook. So tyrannical did they at length become, that, to prevent worse consequences to himself, Cromwell was compelled to reduce their authority within much narrower bounds. Before this reduction of their power, they could commit to prison all suspected persons; and they chose to suspect, whoever they disliked, (viz.) all moderate, loyal, good, men: they moreover levied money, sequestered those who refused to pay; had power to enlist horse and foot upon any occasion they might think proper to make, or any emergency they might themselves create. From their decisions no appeal lay, but to their regicide master himself.

Of this honourable fraternity was HARRISON, the son of an attorney, of this town. Not having any relish for his profession, he enlisted into the Parliament's army; and, being a person of great volubility of tongue, he soon insinuated himself into the favourable opinion of the army, and became Cromwell's confidant. The Protector knew how to make use of such a person as Harrison; and he did not fail to avail himself of his canting dexterity, and perseverance, in routing the Presbyterians, and bringing his legitimate, though weak, Monarch to the block. These services, for a season,

* Whitelock's Memorials of English Affairs, &c. p. 634.

† Clarendon's History of the Rebellion, Vol. III. p. 458.

‡ Vitæ Selectorum, &c.

procured Harrison a wicked elevation and popularity, among those who admired anarchy, rebellion, and military tyranny, highly seasoned by religious professions; but he did not long enjoy his honours. He was at length executed as a traitor; his head was set up at Westminster hall; and his quarters upon the gates of the city of London. Thus disgracefully perished one of the many persons who, under the most showy pretences, would gladly have entailed to their posterity a form of government, which the people of this country never was, and we hope never will be, long disposed to support: a government which opens the door to every species of oppression, by raising those to govern others who have neither prudence nor judgment to govern themselves—who are clamorous for principles which, wherever they have power, either in their own families, or over their other immediate dependents, they rarely reduce to practice. Monarchy, and that only, suits the genius, and sober and rational character of Englishmen.*

The other person mentioned, as a native of this place, was doubtless honest in his mistaken principles; and, as his personal history is not very generally known, we shall dwell upon it at a somewhat greater length. This person is the celebrated JOHN GOODWIN, a learned divine and most acute and zealous defender of Arminianism, who, as Granger† remarks, "made more noise in the world, than any other person of his age, rank, and profession."

Mr. Goodwin was born in the year 1593, as we are informed by an intelligent gentleman, who is descended from him in this neighbourhood.‡ Every account, hitherto printed, states the place of his birth to be unknown; and it is but just to add, that our own authority is tradition. He received his academical education at Queen's College, Cambridge, where he soon became known by his learning and talents, and for being a smart disputant. Upon his leaving college he was admitted into orders, and became much admired for the erudition and elegance which distinguished his pulpit compositions.

He preached sometime in the country, and removed to London in the year 1632. The year following he was presented to the vicarage of St. Stephen's, Coleman-street. At this time the Archbishop Laud took upon himself the government of the English

* Beauties of England and Wales, Vol. X. p. 1077.

† Biographical Hist. of Eng.

‡ Yet Granger quotes a MS. in Lambeth Palace, which says: "Johannes Goodwin, Norfolc, became Fellow of Queen's College in Cambridge, in 1617." Biog. Hist. of Eng. Vol. III. p. 42. (note.)

church; and John Goodwin, among others, was denounced for a breach of canons, in 1637.*

In the year 1640, the king having allowed the convocation to continue its sittings, after the dissolution of Parliament, the clergy were busily occupied upon two subjects of considerable magnitude. One of these was to grant a subsidy for six years, to meet the exigency of the public affairs. This was proposed to be done, by a tax of four shillings in the pound upon the estates of the clergy. Another object of this prolonged convocation was the enactment of certain canons, or articles, amounting in number to seventeen. These were published on the 30th of June.† The first of these canons, "concerning the royal power," asserts the absolute authority of kings, and the unlawfulness of taking arms, even in self-defence. Many of the other canons bore peculiarly hard on the non-conformists. The fourth, in particular, has been remarked to have a singularly intolerant character. It is there decreed, that no person shall import, print, or disperse, any books written by Socinians, on pain of excommunication, and of being further punished in the Star-chamber. That "no minister shall preach any such doctrines in his sermons, nor students have any such books in his study, except he be a graduate in divinity, or have episcopal, or archidiaconal, ordination; and if any layman embrace their opinion, he shall be excommunicated, and not absolved without repentance or abjuration." Though Mr. Goodwin does not appear to have had any peculiar bias to Socinianism, he, along with others of the London clergy, drew up a petition to the Privy council; and so great was the outcry against the proceedings of the bishops, that the King thought it prudent to issue an order to Laud to soften his severity.

Mr. Goodwin, refusing to baptize the children of the parish promiscuously, and also to administer the eucharist to his whole parish, was ejected from his living in 1645. He then set up a private meeting in Coleman-street parish, on the plan of the Independents. Being thus in a manner freed from the restraints of episcopacy, he attacked his adversaries with considerable warmth; and being a zealous defender of Arminianism, against the rigorous and disso-

* Neale's History of the Puritans, Vol. II. p. 263.

† "Constitutions and canons ecclesiastical, treated upon by the archbishops of Canterbury and York, president of the convocation for their respective provinces, and agreed upon with the king's majesty's license, in their respective synods, begun at London and York, 1640."

tializing dogmas of Calvinism, was attacked with a characteristic and native fury; and he did not fail to defend himself with spirit. But when the factious turbulence of the times had succeeded in destroying for a season, the episcopal government, he hoped to have met with more favour from the Presbyterians. By indulging this expectation, honest John Goodwin by no means shewed the strength of his understanding.

Bishop Burnet accuses Goodwin as a zealous *Fifth Monarchy Man;* but with what justice we know not; he was certainly a friend of Cromwell's, and that is almost sufficient to stamp him a friend to any enthusiastic vagary. The late Mr. Toplady, in his detestation of Goodwin's Arminianism, was glad to have this flaw in his character; and Mr. Wilson observes, of this furious Calvinist, that "he has heaped together whatever he could find to vilify the character of Mr. Goodwin, and stated them with all the acrimony of a party bigot." The author of the Gangræna, however, has not enumerated this among the vices of Goodwin.

On the restoration of monarchy under Charles II. it was naturally expected that Goodwin would meet a severe chastisement; but, as Burnet observes, both "John Goodwin and Milton did escape all censure, to the surprise of all people." All that was done was to call in his book intituled the "Obstructors of Justice," and Milton's celebrated "*Defensio pro populo Anglicano contra Salmasium*," and his *Answer* to "The Portraiture of his Sacred Majesty in his Solitude and Sufferings," and have them burnt by the hands of the common hangman, which was accordingly done, on the 27th of August. The recommendation of the Commons that the king should direct his attorney-general to proceed against the authors of these books was not regarded. It is thought by some that his Arminianism was at length of some use to him.

The Restoration, however, did not restore Goodwin to his former living; and another presentation took place in 1661. The terms of the Act of Uniformity not according with the free spirit of Mr. Goodwin, he continued a nonconformist till his death in 1665, aged 72. He wrote a great number of books, and seems to have had a ridiculous fondness for Greek titles. If our memory does not fail us, some of the late Mr. Wesley's Calvinian opponents used to call him Goodwin Redivivus. Wesley, however, was a much more sober man: had every thing that was benevolent in Goodwin's religious creed, without any of his dangerous political bias.*

* Beauties of England and Wales, Vol. X. p. 1084.

NORTON-ON-THE-MOORS, is a parish of Pirehill North, situated about five miles N. N. E. of Newcastle, on the road between that town and Leek. It contains the townships of Bemersley and Norton; and among other hamlets, and places, are Milton, Whitfield, Ford-Green, Ball-Green, Norton-Green, Badiley-Edge, Brown Edge, &c. The two last are commons, on which are built several cottages, some scattered and some nearly adjoining each other.

The village of Norton is situated on an eminence between two of the upper branches of the river Trent; the land in its vicinity is cold, and the country hilly. This district abounds with coal at various depths, some 80 yards deep and more. The strata are from four to seven feet thick. The price varies: at some pits it is sold at 8s. 4d. per ton.

The Church is a plain small edifice of brick, which was probably re-built in the year 1738. According to a memorandum in the Parish register, "the old Church began to be taken down May the third, and the foundation of the new church was marked out May 6th, 1737."

A square tower at the west end has its top corners ornamented with four balls or small globes, and is remarkable for containing the only doors of entrance, the chief of which is on the west side, and leads into the body of the church. The way to the gallery at the west end is by a flight of steps on the out-side and on the north side of the tower, which also leads to the belfry and station for the ringers. The bottom of the tower may be considered a kind of porch, which contains the following appropriate notice to the parishioners on a table hung against the wall:

"As there is in many parishes a rule made, that, except the corpses be brought to church by fair day-light, precisely at a certain hour, they are not to be buried till the next morning.

"And whereas there is in this parish a growing ill custom of burying at late hours, to the harm and danger of the living, without the least benefit to the dead.

"And whereas many persons attending funerals complain of the inconvenience of being subjected to the cold and damp of the night air, and obliged to return home in the dark.

"Therefore, it is hereby ordered for the advantage of all, and the disadvantage of none, that every corpse shall be at the church at the times hereafter-mentioned, viz.

"In November, December, and January, at half-past three.

"In October and February, at four.

"In all the other months, at five.

"THOMAS MIDDLETON, Curate.

"RICHARD BALL, } Church Wardens,
"FRANCIS HARGREAVES, } 1775."

Benefactions.—The following sums were left by the individuals whose names are here subscribed, as appears from the tables in the church:

1.—*The yearly interest of sums bequeathed for the use of the poor.*

Hugh Ford (*a*)* of the nearer side of the water, 4*l.*

William Ford (*b*), father of H. F. (*a*) of the nearer house to Norton on Ford-Green, 6*l.*†

William Forde (*c*), late of the further side of the water, 20*l.*

Ellen Forde, (*d*) wife of W. F. (*c*) for poor householders, and to be dealt at the discretion of W. F.'s heirs, 30*l.*

Hugh Meare, late of Norton-Hall Houses, 5*l.*

Mr. John Pott, schoolmaster of Norton, (for poor householders), 10*l.*

William Sherratt, late of Bearstone, in the county of Salop, 10*l.*

Mr. Thomas Sherrett, 10*l.*

☞ The principal of this last charity, and for which the curate of Norton is trustee, lies on Mr. W. Sherrett's property in Milton.

2.—*Annuity or yearly sum bequeathed for the use of the poor.*

William Meare, late of Pott-shrigley, in the county of Chester, left charged on a field called Annats-field, at Norton Green, the yearly sum of 2*l.* 10*s.*

3.—*The yearly interest of sums bequeathed and to be distributed to the poor in bread.*

Sir John Bowyer, Knight and Baronet, left 3*l.*

Mr. Moor, 2*l.*

Mr. Josiah Keeling, late of Milton, 12*l.*

This last charity is directed to be distributed, the first Sunday in each month, publickly in the church.

4.—*Poor's pew and seats in the gallery.*

Hugh Forde, of Forde Green, did in his life give the furthermost pew situated in the north end of the gallery, and all the sitting in the arch in the wall in the middle part of the gallery, for the use of the poor inhabitants of this parish for ever.

* The letter after the surnames is used for reference to some of the same name that precede or follow in the list; and where the same letter follows repeatedly the same name, it denotes the same person.

† Forty shillings of which was lost.

5.—The yearly interest of sums bequeathed for the use and towards the maintenance of a schoolmaster.

Ellen Forde (*d*) left 30*l.*

Ellen Forde, daughter of W. F. (*e*), 20*l.*

A School and House for the abode of the parish schoolmaster, is in the church-yard on the eastern side near the Lich-gate.

The Living is at present (1817) a chapel of ease to Stoke-upon-Trent, but by an Act of Parliament passed eight or ten years ago, upon the death, cession, or other voidance of the present incumbent, it becomes a Rectory, and in addition to its present emoluments, (except £50. a-year now allowed by the rector of Stoke) will be endowed with all the great and small tithes of the township of Norton, which at present belong to the rector of Stoke. The Rev. Daniel Turner is the present incumbent.

There are three *M eting-Houses* belonging to Dissenters in this parish, the largest of which stands in the village of Norton.

Several excommunications are recorded in the parish register, in the seventeenth century. The laxity of ancient church discipline in the present day makes such instances and events scarcely known, or very rare.

Dr. Plot* records a singular custom, which formerly was observed in this county, in felling oak trees and stripping the bark; and which he first noticed in this parish. His words are: "In the felling whereof they have this very good custom, that they flaw it standing about the beginning or middle of May, which I first observed in some fences near Norton-on-the-Moors, Milton, Badiley, where there were several oakes stood naked, divested of their bark, which they told me would not be felled till Michaelmass following at soonest, or perhaps not till mid-winter, or the ensueing spring; which I take to be a way of so valuable a consideration, that perhaps it may deserve the debate of a Parliament, whether it might not be worth while to inforce this custome to be strictly observed all over the nation? for tho' by a reserve in the act for due felling oaken timber, it may be done at any time for building or repairing houses, ships, and mills; yet for any other uses none may fell it (in consideration of the tan) where bark is worth but two shillings per load, over and above the charges of barking and pilling, but between the first of April and last of June, when the sap is up, and the bark will run, which causes the out side of the timber to rott

* Chap. ix. sec. 87.

away quickly, and to grow worm-eaten: whereas these being fell'd in or near the winter, and having stood naked all the summer drying in the sun, become in a manner as hard and sound without as within, being as it were all heart, and not so subject to worms; by which means there would be a great deal of good timber saved, and no other could be used; nor would the use of the bark be lost to the tanner, as I suppose is presumed in the present act it would, should it have admitted felling oaks in the winter season, when the bark will not run."

This parish contains, 329 houses, 336 families; 900 males, 861 females: total, 1,761 inhabitants.

BAGNALL is a township in the parish of Stoke-upon-Trent, and in 1811 its population was 288 persons, whereof 156 were males, and 132 females. The number of inhabited houses were 48, and 4 were uninhabited. The number of families were 49, 28 of which were employed in agriculture, 13 in trade, and 8 were of neither description.

The Chapel, which is small, and in a ruinous state, is situate in an open field adjoining to the village, about three miles to the south-east of Norton-on-the-Moors. In the same village is an ancient mansion, now a farm-house, but formerly the residence and property of the family of Murrell, a name which appears in the Testa de Nevill for Staffordshire. Therein a William Murel is recorded as then holding one virgate of land in the village of Selton (now Shelton), and within the manor of Newcastle. [For further particulars see the concluding part of Bucknall.]

BUCKNALL is another township in the parish of Stoke-upon-Trent, whose population in 1811 amounted to 344 persons, whereof 173 were males, and 171 females. The number of inhabited houses were 76, and 4 uninhabited. The number of families were 76, 40 of which were employed in agriculture, 29 in trade, and 7 of neither description.

The Church of Bucknall is situate in the village, about two miles eastward of Hanley, and three south-west of Bagnall. It is a small stone building. In the interior are a few pews, but the seats are chiefly benches or plain forms. On the walls were painted, in fresco, the Lord's Prayer, Decalogue, and Belief, and the two following inscriptions:

Ligna lapis fiunt
Sam. Lea, Curate,
W. Adams, Warden,
1718.

Wm. Adey, Curate,
Matth. Broadhurst, Warden,
1738.

From the first inscription we infer, that the chapel was rebuilt with stone in 1718. The ancient chapel was probably half timber, like the present turret of Whitmore church, and some old churches now to be seen in Cheshire.

Though Bagnall and Bucknall are still townships of the parish of Stoke for civil purposes, such as the maintenance of the poor, &c. and were chapels of ease to the mother church; yet in regard to ecclesiastical purposes, they are now by Act of Parliament become a distinct parish and rectory, which includes the adjoining township of Eaves, belonging likewise to Stoke, and the three townships, Bagnall, Bucknall, and Eaves, are called the Parish of Bucknall and Bagnall. The Chapel of Bucknall is a parish church, and the Chapel of Bagnall is annexed thereto, and a chapel of ease to the said parish church of Bucknall and Bagnall. The inhabitants of the chapelry of Bagnall are not liable to the repairs of the church of Bucknall, nor are the inhabitants of the chapelry of Bucknall liable to the repairs of the chapel of Bagnall.

The *Rectory* is endowed with about 85 acres of glebe, nearly 60 acres of which are situate in the parish of Cheddleton, and commonly called *Lee-house Farm*. Upwards of 21 acres lie in the parish of Horton, and are called *Lask Edge* or *Moor-top Farm*, and a field of about four acres lies near Bagnall, and is called the *Parson's Field*.

It is further endowed with all the great and small tithes of the townships of Bucknall, Bagnall, and the Eaves, and each of them, together with all surplice fees and other emoluments heretofore belonging to the chapelries.

The population of the township of Eaves in 1811 was 232 persons, whereof 115 were males, and 117 females. The number of houses was 40, which were occupied by 40 families, and 3 were uninhabited, so that the population of the parish of Bucknall and Bagnall amounts to 864.

The above noticed act of separation does not exonerate the inhabitants of this parish from the repairs of Stoke Church, nor deprive them of their accustomed right of pews, seats, or sittings therein, or of the right of sepulture in any burial place within the parish of Stoke, to which they were entitled before the passing of the said Act.

WHITMORE is a parish in the hundred of Pirehill North. It was anciently called *Witemore;* and in the Conqueror's Survey the following description occurs:

"Isd' R. (Ricardus Forestarius) ten' Witemore. (de rege) et Nigel de eo. Vlfac tenuit et lib' hō fuit. Ibi ē dimid' hida. T'ra ē III. car'. In d'nio. ē una. et III. uilli et II. bord' cu' I. car. Ibi. I. ac' p'ti. Silua. I. leuu' l'g. et dim' lat. Valet x. solid'."

"The same R. (Richard Forester) holds Witemore (of the king) and Nigel of him. Ulfac held it and was a free man. It contains half a hide. The arable land is three carucates, one is in demesne, and three villains, and two bordars hold one carucate. There is one acre of meadow. A wood a mile in length and half-a-mile in breadth. The whole is worth 10 shillings."

The village of Whitmore is situate on the public road from Newcastle to Market-Drayton, about four miles from the former place. The population in 1811 was 43 families, containing in all 291 persons, whereof 156 were males, and 135 females; 35 families were employed in agriculture, six in trade, and two were of neither description. The number of houses was 44.

The Church* is situate in the village, and was, we apprehend, rebuilt in 1632, as that date appears on a stone over the west door. It is a small stone building of an oblong form, with a half-timber turret on the west end, containing three bells, and is capable of holding about 150 persons. The turret probably exhibits a specimen of the kind of walls erected in the ancient churches prior to the use of stone.

A few monuments are in the Church to the memory of some of the respectable family of Mainwaring, which has been seated at Whitmore for several generations. On the north side of the Church appears a neat walk flanked with rows of trees, forming a beautiful avenue, which is terminated by Whitmore-hall, the seat of Edward Mainwaring, Esq. and which, judging from a date over the front door, was built or rebuilt in 1676.

The Living till of late was a chapel of ease to Stoke-upon-Trent. It is now a rectory endowed with all the great and small tithes in the parish, surplice-fees, and between 30 and 40 acres of glebe; part of which lies near Burslem, and part near Newcastle; but the bulk is situated in and near Whitmore. The reputed value is about

* It is not known to what Saint the Church is dedicated, as no Wake has been kept for a number of years at Whitmore; nor do the bells determine the point—one of them is inscribed in Latin, "St. John," and another "All Saints."

£400. a-year. The Rev. J. S. Brasier, is Rector, and Edward Mainwaring, Esq. Patron.

Benefactions to the Poor, inscribed on a tablet on the north wall in the Church:

"Mr. Joseph Ball, late of this town, left to the poor of the parish of Whitmore, two pieces of land (fifty pounds purchase) the one called Ball's-meadow adjoining to the Town Meadow, and a Meadow belonging to Master John Eardley. The other call'd Holliegrieve [grove-croft] lying between lands belonging to Mr. John Knight and Mr. John Walton.

"Another charitable person left the interest of twenty pounds for ever to the poor of the said parish, in the trust of the Churchwarden and Overseer.

Mr. Middleton, Minister,
Saml. Latham, Churchwarden,
Jos. Hildich, Overseer, Ann. Dom. 1736."

It is conjectured that this last donation was left by one or more of the name of Cleyton, as will be hereafter noticed, and that the first benefactor might be a descendant of the Rev. John Ball, formerly minister of this Church, respecting whom a few memoirs, abstracted from an old printed book, are here subjoined, not out of respect to some acts of non-conformity there recorded, for such a sincere Churchman must regret, but for reasons stated in the conclusion, and because he appears to have been an honest man, generally respected, and an active, pious minister, much revered in his parish and neighbourhood.

John Ball, called in scorn "the Presbyterians' Champion," was minister at Whitmore, near Newcastle, Staffordshire, from about the year 1610 to the time of his death, Oct. 20, 1639. He was born about the year 1585 at Cassington, an obscure village, a mile from Hanborough, in Oxfordshire, and educated at Yarnton school; from whence he was removed to Brazen-nose College, Oxford, where he took the degree of Bachelor of Arts. His parents being of low estate, he was kept at school through the partiality of his master, who admired his pupil's abilities; and was supported at College chiefly by the aid of friends. After he left the University, he was engaged as tutor to the children of Lady Cholmley, at her house in Cheshire, where he also taught other children. His acquaintance being amongst 'the most godly people in those parts,' who much lamented the evils of the times, he often associated with them in keeping private days of fasting and prayer. "Himself, with some other honest Christians, were often convented to Chester, and much troubled there, for keeping *a Fast on Ascension-day*, their fault being aggravated by the Bishop and his officers, because they fasted upon that holy day."

"He was incomparably abstemious and moderate in the use of meat, drink, sleep, and apparel, and therefore had an admirable command over his passions."

When he was judged fit for the Ministry, his friends encouraged him to enter thereupon, and such was his disposition; but he was averse to the subscription to the articles, &c. required by law, and had repeated "conflicts with the then Bishop of Chester about non-conformity." By the intercession of his friends, an Irish Bishop, who occasionally visited London, ordained him and another without requiring their subscription; and "about the year 1610, he was called to be minister at Whitmore, (a small village near unto Newcastle in Staffordshire) where he was entertained into the house of Edward Mainwaring, Esquire, a pious and much-esteemed gentleman."

As a minister he was zealous and active, entertained high and honourable thoughts of God at all times, and ever spoke of his name with reverence. The Sabbath days he endeavoured to keep strictly holy; and the means which he adopted were as follows: Every Saturday afternoon, by way of preparation for the Sabbath, he spent about two hours in his house in praying and expounding some Scripture. On the Sabbath morning he expounded a portion of Scripture in his family, then he preached twice in public, read divers chapters, and catechised also in the afternoon, spending much time therein, especially in the summer season. After the evening sermon, he went unto Mr. Mainwaring's house, being near at hand, where he repeated both his sermons and prayed; likewise after supper he had another exercise in his own family, equivalent to a sermon.

"He was exemplary careful to preserve God's ordinances from pollution and contempt, and therefore did always take great pains with his people to prepare them for the sacrament of the Lord's supper. In his sermons he was wont to tell his auditors, that their persecuting, impoverishing, imprisoning and thrusting daggers into their ministers' bodies, would not be more grievous than their unworthy communicating at the Lord's table, and their unanswerable walking unto the ordinances of grace dispensed amongst them; and when parents presented their children in the congregation to be baptized, he would very affectionately lay open their duty, pressing sacramental ingagements home to their hearts, with strong arguments and authority ministerial."

He frequently preached at Lectures, at Madeley especially, where he was the principal upholder of a weekly sermon, and a monthly fast; and was ever ready to assist others in private fasts and other exercises of religion.

"That he might keep God's servants, of the lowest parts, always high in his own estimation, he would wisely weigh their peculiar serviceablenesse in some kind or other, though they were inferiour in sundry excellencies. Hereupon when one, admiring worthy Master Dod's holy conversation, said, Oh, how well would it be if all ministers were like unto him, Master Ball answered, The Church would be quite undone if all others were like him. For though no man more honoured Master Dod ['s abilities,] (which not many years before his death he witnessed, by listening unto his fatherly counsel bare-headed, with his hat in his hand) yet he knew, that other men, much inferiour in graces, had some special way of advantage to the cause, and the people of God."

Although he conceived that corruptions existed in the church government and in the administrations of holy worship, yet he had a great dislike to the way of separation from our church assemblies on such pretences. He much lamented the first breakings out of independency in England, and indulged a belief, that, if God should give opportunity and hopes of Church reformation, the brethren of the new separation would be found the greatest obstructors thereof. He called the ways of separation a labyrinth, wherein men tire themselves, and grow giddy, as in a maze, but when all is done, there is no way out, but that whereby they entered.

"Being ask'd by one, who much pretended unto immediate inspirations besides scripture, whether he, at any time, had experience thereof in his own heart? His answer was this, 'No, I blesse God, and if I should ever have such phantasies, I hope God would give me grace to resist them."

Other traits of character may be extracted from his life, but we shall confine ourselves to a few remarks in respect of his power of becoming a man of considerable influence with an annuity of twenty pounds, for such was said to be the stipend of his cure, or allowance for his maintenance from the church. This sum might appear to the reader a mean and inadequate allowance, and such it was considered in those days, by his biographer, who terms his outward condition 'mean;' nevertheless it may be affirmed, that his condition was superior to that of many ministers in the present day whose income amounts to no more than £200. a-year; for, in his time, it is not improbable, that land set at four or five shillings an acre. At the time J. B. was minister of Whitmore, land within ten miles of the place set at nearly one-tenth of its present rent. About fifty-three years after J. B.'s death, rich glebe land set in an adjoining parish at about six shillings an acre, which said land now rents for upwards of two pounds; so that when he first entered upon the ministry, it is not improbable that £20. a-year was equal to the rent of a good farm of 100 acres extent and more. And in regard to the comparative price of food, labour, &c. the following statements have appeared in print. Some of them are prior and others subsequent to J. B.'s time, but by taking the average price, a fair estimate may be made.

"In the reign of Henry VIII. it was enacted that butchers should sell their meat by weight: *beef* for one halfpenny per pound, and *mutton* at three farthings. At that time *fat oxen* (for cows were not then slaughtered) were sold for thirty shillings and eight pence each; *fat wethers* for three shillings and four pence, a *fat lamb* for twelve pence."—*Stowe.*

"T. 35 Henry VIII. the fearme of one cow,	8*s*.
The rent of a church yard house for six years,	6*d*.
The rent of one half a close for six years,	6*d*."

"1581.—At that time 200*l*. a-year would have kept a gentleman's or knight's house, when such were wont to keep each 'halfe a score of clean men in his house, and xx or xxiv other persons besides, every day in the weeke; sixteen years before 200 markes would have been sufficient.' In regard to *apparel*, half a crown was thought a great price for a cap, since it could have been had, a few years before, for 1*s*. 2*d*. A pair of shoes cost 12*d*. which used to be 6*d*."

"1594 and 1595.—Butter 3*d*. per lb.; a fore quarter of lamb, with the head, in March, 2*s*. 2*d*.; a capon 1*s*. 2*d*.; a dozen of pidgeons 2*s*. 4*d*.; 28 eggs 8*d*.; beef per stone 1*s*. 6*d*.; a lamb in April 5*s*.; a side of veal in April 8*s*.; a calf's head 10*d*."

In the reign of James I. wheat is stated as having sold as low as 11*s*. 6*d*. the quarter, or 1*s*. 5½*d*. per bushel; and in the succeeding reign of Charles I. at 14*s*. the quarter, or 1*s*. 9*d*. the bushel. In the reign of James I. one full quart of the best beer or ale was one penny.

These prices are in an household book apparently kept in the neighbourhood of Whitmore from 1738 to 1741:

Wheat from 4*s*. to 6*s*. per bushel.
Barley, 3*s*.
Oats from 1*s*. 4*d*. to 2*s*. 6*d*.
[The latter price was sowing grain.]

A Calf from 6*s*. to 9*s*.
A fat Ewe, 5*s*.
Beef 1*d*. to 2*d*. per lb.
Pork 2*d*.
A pair of calves plucks, 6d.

The following is the average price, according to the statement of several old men, some of whom were upwards of, and others approaching to, the age of 80 years in 1813. Two of them had lived from their childhood in the neighbourhood of Whitmore. In their juvenile days,

Bacon was 4*d*. per lb.
Butter, 3*d*.
Cheese, 2*d*.
Beef, 2*d*.
Potatoes, 9*d*. the bushel.
Wheat, 2*s*. 6*d*. to 5*s*. per bushel.
Barley; 1*s*. 4*d*. to 2*s*. 6*d*. ditto.
Malt, 3*s*. to 3*s*. 6*d*. ditto.
Oats, 7*d*. to 15*d*. ditto.

Cow, (barren) 2*l*. 15*s*. to 3*l*.
Ditto, good milch and calf, 3*l*. to 5*l*.
Calf, 4*s*.
Milk, two quarts for 1*d*.
Sheep, (barren) 6*s*.
A Pig (good store) 12*s*.
Labourers' wages 4*d*. a day and meat.
Land, 7*s*. to 14*s*. an acre; and 16*s*. a great price for an acre of meadow.

[A House for 12*s*. which in 1813 would rent for 3*l*. a-year.

Though the prices of the above articles are ever in a fluctuating state, yet they may serve to enable the reader to form an idea of the comparative value of twenty pounds a-year in the days of Mr. Ball, while minister of Whitmore; but this divine possessed other advantages of no mean description besides his annual stipend: the chief of which must be accounted the constant friendship of the first family, and chief proprietor of estates in the parish. Such a connection could not fail of adding weight to his ministry in several respects. It no doubt enabled him to improve his abilities* by a constant intercourse with enlightened minds, and by a free access to a good library: for it appears from his life, that he boarded with the family of Mainwaring, not only while a bachelor, but several years after he was married. Six children were born in their house,

* He is represented as having been a man of great talents in assisting and healing "dejected tempted Christians." His ability to counsel and comfort persons of that description, as he was wont to acknowledge, was occasioned "by his conversing with Mrs. Sarah Mainwayring, (wife to the gentleman in whose house he continued many years) who was much exercised in that kind, and was an unparallel'd gentlewoman for holy tendernesse and exactnesse in religion." She was a Miss Stone, and married Edward Mainwayringe, Esq. at Bowe Church, in Chepsyde, London, Sept. 6th, 1600. She died at Whitmore, where she was buried in July 1648, aged 71 years. The parish register contains an high encomium on her merits.

and when he removed from thence, a house was built on purpose for him in the village of Whitmore, wherein his seventh and last child was born. Moreover, when he boarded with Mr. M. his visitors and acquaintance were kindly received, and hospitably entertained at Mr. M.'s expense. And when he became a housekeeper, a cow was kept for him summer and winter, gratis, by Mr. M.

Other friends likewise contributed to Mr. Ball's comforts, among whom was Mr. William Cleyton, Gent. of Radwood, in the parish of Maer,* who left him twenty shillings a-year during his life; which was one of the temporary purposes to which his charity was, at first, directed to apply, and alluded to in the account of Maer.

When these helps are considered, and the profits of a respectable school are added; for he kept a considerable school at Whitmore, as "some of his pupils were tabled in his own house, and the rest placed in the neighbourhood;" it is no great matter of surprise that with his income "he lived very comfortably, was given to hospitality, and also unto liberality to the poor, (considering his estate) and left a good competency unto his wife and children."

Wordly affairs beyond the bounds of his calling engaged but little of his attention; and it was fortunate for him that there was no necessity; for at Whitmore "he married a godly wife, whom God had provided as an helpmeet for him, he himself being always a meer scholler."

One trait in his character appears highly culpable, which is his mode of entering upon the ministry by the ordination of an Irish Bishop, and contrary to the consent of his own diocesan. However

* In a will, bearing date Nov. 8th, 1625, he bequeathed twenty shillings yearly to Mr. Hodskin, of Meare; and the like sums to Mr. Ball, of Whitmore, and to Mr. Hulme, of Betley, to be paid to each respectively and yearly during their lives, or till they had sufficient maintenance elsewhere; and in case these three ministers mentioned should die, or have competent maintenance elsewhere, his will was that the said sum of three pounds should be distributed yearly to three preaching ministers, at the discretion of his executors. He also left to the poor of Madeley and Whitmore parish ten shillings a-piece yearly. This annuity of ten shillings, we presume, is a part of the last-named charity noticed in the benefactions, p. 376. The other part might have been added by his son John Cleyton, who increased his father's bequests to the poor in Maer. These Cleytons were probably descendants of Mr. Wm. Cleyton, formerly curate of Whitmore, whose death is thus recorded in the Parish register:—"Gulielmus Clayton concionator verbi Dei sepultus fuit XVI Die Junii Anno Domini 1609—fuit iste Gulielmus pastor Ecclesiæ de Whitmore."

for this transgression he may be said to have made some atonement by frequently lamenting, in the presence of others, his own inconsiderate entrance upon the ministry both during his life and at his death.* He considered the minister's office "a weighty calling," as most undoubtedly it is; and the memoirs of this divine are here introduced in confirmation of this important truth. Wherever the officiating minister is rendered weak by poverty of stipend, the influence and assistance of the high and more opulent in his parish, who profess to belong to the Church of England, are indispensably necessary to promote her prosperity: and wherever this assistance is withheld, or influence, in matters respecting religion, perverted by being conveyed through other channels than that of, or unconnected with, the *officiating minister*; whether curate or incumbent, the Church will most materially suffer.

TRENTHAM, is a parish in Pirehill North, adjoining to that of Newcastle. It contains four townships, namely, Blurton and Lightwood Forest, Hanchurch and Butterton, Handford, and Trentham. This parish contains 399 houses, 417 families; 1069 males; 1051 females: total, 2,120 inhabitants.

The village of Trentham is situate on the turnpike-road between Newcastle and Stone, about four miles from the former place. It consists of several tenements and a large Inn. The new Cemetery of the House of Trentham is situated in the village of Trentham, on the eastern side of the road that passes through it. It is a large pyramidal pile of stone, of two stories, the upper of which contains one bell, and is surmounted by a cross. The entrance is at the western side, and a large Gothic window admits the light from the east. The interior of the cemetery contains twenty catacombs on each side, each sufficiently capacious to hold one coffin. The divisions between these catacombs are faced with Derbyshire marble. The funeral service is performed in this Mausoleum on the interment of any of the parishioners, and an extensive burying-ground is enclosed in the rear of this structure, beyond which the minister's house stands in a field: it is now the residence of the Rev. T. Butt, the present minister of Trentham.

The Church, which originally was part of the monastery, noticed hereafter, is an ancient structure of stone situated close to the Hall, and is dedicated to St. Mary.

* His death is thus entered in the register—"Johanes Ball obijt vicessimo die Octobris, 1639, quondam Curatus de Whitmore observatissimus."

The ancient family-vault of the House of Trentham is in the Church, and none of the family have yet been interred in this new cemetery.

Longevity.—Thomas Elkon, of Spratslade, died Feb. 7, 1783, aged 91 years; Ann Lakin, aged 89 years.

There is a remarkable inscription in the church-yard, which points out the antiquity of a family resident in this parish: "To the Memory of the Boulds from the year of our Lord 1558."

"Trentham, heretofore called *Trickingham*, is of no note for any thing ancient, but a little monastery dedicated to the holy and royal virgin, St. Werburga, and erected for canons regular of St. Augustine, in the reign of William Rufus. Ranulph de Gernons, Earl of Chester, is said to be the founder of this monastery."* But as that Earl lived in the time of king Stephen, it is probable he was only the restorer of it, and a benefactor to it. He granted the canons "a yearly rent of 100s. which gift was confirmed to them with divers immunities, by King Henry the Second. It was valued at the Dissolution at £106. 3s. 10d. per annum."†

According to Pope Nicholas's taxation in 1291, the Church of Trentham, then appropriated to the prior and convent of the same, was valued at twenty marks.

Lady Catherine Leveson, daughter of Alice Duchess of Dudley, gave £400. to the township of Trentham, to purchase lands of the yearly value of £20. for the support of a schoolmaster to teach the poor children of the parish, until they be fit to be put forth apprentices to some honest trade and occupation. She also gave to the poor of this parish fifty pounds, to be distributed to them on the day of her funeral. She also gave the yearly rent of £120. for the maintenance of twelve poor widows, whereof three were to be chosen out of this township, by the minister, churchwardens, and overseers of the poor, and to have a gown of grey cloth with the letters K. L. in blue cloth fixed on it. As also one hundred pounds per annum to be paid out of her rents and revenues at Foxley in Northamptonshire, for the placing out ten boys apprentice, two of them to be of this parish, where she died in 1673, but was buried at Lilleshall, by her husband Sir Richard Leveson. His sister and co-heir marrying Sir Thomas Gower, brought this lordship into his family, who made it his seat, and whose descendants were elevated to a peerage. Sir John Leveson Gower, the 5th Baronet,

* Dugdale's Monasticon, Vol. 9, p. 200.

† Magna Britannia, No. 82, p. 154.

was created Viscount Trentham and Earl Gower, July 8, 1746, and his eldest son, Marquis of Stafford, in the year 1786. [See "Peerage," in this Work.]

Trentham Hall, the family residence of the present Marquis of Stafford, is situated on the banks of the river Trent, and is one of the most magnificent seats in the county. It is a modern mansion, and built after the model of Buckingham House, in St. James's Park. It is built of brick, but the front is covered with Egyptian cement similar to stone. The principal front is to the south, and the mansion is sheltered by full-grown trees on the north and east sides.

The pleasure-grounds are extensive and diversified, and the river Trent, which forms a beautiful lake at some distance from the front of the Hall, with the woody hills beyond it, present a beautiful variety of rural scenery. Near the eastern side of the house, an orangery is seen like a mystic temple, and the stranger on approaching it is gratified by the fragrant scent of the snow-white blossoms. A few paces beyond the orangery, and nearer the house, a handsome iron bridge of a single arch, 90 feet wide, affords a ready passage to the rambler over a fine piece of water.

Part of the windows in the front of the lower story exhibit the various green-house plants and flowers of a conservatory. A new eastern wing has been lately completed. The interior of this mansion harmonizes with its elegant exterior. It contains a choice and valuable collection of paintings by the first masters, particularly an exquisite Guido, and some other *chef d'œuvres*. There are also several sketches of Northern scenery by the present Marchioness of Stafford.

Blurton, is an hamlet about two miles to the east of Trentham. The Church is a small structure of brick, with a belfry containing one bell. It is a chapel of ease to Trentham church: the Marquis of Stafford is the patron, and the Rev. John Blunt, minister.

SWINNERTON parish is in Pirehill North, and contains part of Seabridge township. The village of Swinnerton is about three miles to the south of Trentham, and four miles W. N. W. from Stone. This large and fertile manor was possessed for ages by a race of gentlemen who took their names from it. The first of this family on record was Roger de Swinnerton, who in the 34th of Edward the First obtained a Charter of free-warren in all his demesne lands in this manor; and for keeping a market on Wednesday every week, and a fair yearly upon the feast of our Lady's assumption. From this family the manor devolved by marriage to the Fitzherberts in the reign of Henry the Eighth.

The Mansion-house of Swinnerton, the residence of the Fitzherbert family, is situated on an eminence which commands very extensive prospects of the circumjacent fertile country, and of part of Shropshire. In one of the apartments there is a very fine full-length picture of Sir John Fitzherbert, Knt. This house and the demesne land were a few years ago occupied by George Tollet, Esq. of Betley, during the minority of the owner. While under his management, the lands were well cultivated and greatly improved.

Swinnerton Church is an ancient fabric, situated close to the Mansion-house. It is dedicated to St. Mary, and is a rectory which formerly was in the patronage of the University of Oxford, but afterwards came into private patronage. The late Rev. William Robinson, Clerk, was both patron and incumbent. The Rev. Mr. Corns is now rector; and the Rev. Thomas Orritt, is curate.

Swinnerton Park is situated between two and three miles N. N. W. of the Mansion, and to the west of Trentham demesne. It consists of wood land and waste land, of a good quality, but in a state of nature. It is in an elevated situation, capable of being rendered very valuable land, and is somewhat similar in circumstances to Chartley Park, and like it capable of very high improvement. It is of considerable extent, and is much resorted to during the season by many of the poor for the purpose of gathering bilberries, which they sell to the nearest towns and villages. Hanchurch High-heath adjoining is also a considerable waste of similar land. Millstone-Green is another neglected waste contiguous to Hanchurch-heath. Perhaps no part of the county is more worthy of the attention of the agriculturist than these tracts.

Millstone-Green is likewise the name given to an hamlet situate between Newcastle and Whitmore, about two miles and a half from the former place.

Dr. Plot says (ch. ii. sec. 12) "that those places that are situate highest, and enjoy the fewest waters, mines, and woods, must undoubtedly be the healthiest, in proportion as they are free from one, more, or all of them: upon which account (he continues) I take Swynerton, the village of Beech, and all the hill country betwixt that and Trentham (being void of most if not all these) to be the healthiest spot of ground in all the county, which seems amply to be made out by an observation they have there, viz. that they have three christenings for one buryal."

Acton, is an hamlet in this parish, situate about a mile to the east of Whitmore, and is noticed in the Conqueror's Survey. It is therein termed ACTONE, and is stated to be the property of the Bishop of Chester, and held of him by Rob't. The arable land was four carucates, one was in demesne. There were ten villains and eight bondars, holding four carucates. There was likewise a mill of two shillings rent, eight acres of meadow, and a wood three furlongs long and two broad; each furlong being forty perches. It is remarkable that a pinnacle at the top of the gable end of an old barn built of wood is now to be seen, or was within these two years past, having carved thereupon the arms of the See of Lichfield and Coventry; from which is it improbable that the Bishop of Chester's property became vested in the Bishopric of Lichfield and Coventry? To such of our readers, who incline to that opinion, this old barn will be considered as affording a striking proof of the long duration of those kind of buildings, which have been erected with strong timber frames, and covered with boards. The chief, if not sole proprietor of estates in Acton, is Edward Mainwaring, Esq. of Whitmore Hall.

Close to Acton is a large gravel pit, which continues still to be much used by the public.

About a mile from Acton is a water mill, which bears marks of great antiquity; and we presume, in regard to the walls and plan thereof, is the identical mill noticed in Doomsday Book, as rented at two shillings a-year.

The parish of Swinnerton contains 167 houses, 167 families; 461 males, 432 females: total 893 inhabitants.

STANDON is a small parish, situated about three miles south-west of Swinnerton, and four miles north of Eccleshall, near a branch of the river Sow.

The village of Standon is built on an eminence in a pleasant situation, and surrounded by fertile fields.

Standen Church is a small structure of stone, with a tower, containing three bells. It is dedicated to All Saints, and is a rectory: the Rev. William Walker is the present incumbent. The population of Standon is 420 persons.

Cotes, is an hamlet in this parish, situate about a mile to the west of Swinnerton, which, with *Cotes-lodge* and estate, is chiefly, if not wholly, the property of John Cotes, Esq. of Woodcote-hall, near Newport, Salop. This worthy gentleman is now advanced in years, and his virtues will long be remembered when he is no more. No man possesses or is more entitled to the fond esteem of a grateful tenantry. The following testimony of the respect heretofore enjoyed by the family of Cotes will, we trust, prove acceptable to our readers, not only on their account, but from the piety and dignity of the style, and from the affecting description which it gives of the latter days of (Bishop Hough) a venerable Prelate's life.

Bishop of WORCESTER'S *Letter to Lord* DIGBY, *April* 13*th*, 1743, *in the* 93*d year of his age, three weeks before his death.*

"MY LORD,—I think myself very much obliged to your Lordship's nephew for his kind visit, whereby I have a more authentic account of your Lordship's health, than is usually brought me by report; and an opportunity of informing myself in many particulars relating to your noble house, and the good family at Woodcote, which I hear with the uncommon pleasure of one who has been no stranger to them. Mr. Cotes is blessed in his children, all whose sons are not only deserving, but prosperous; and I am glad to see one of them devoted to the service of God. He may not, perhaps, have chosen the most likely employment to thrive by, but he depends on a master, who never fails to recompense them that trust in him above their hopes. The young gentleman will account to your Lordship for Hartlebury; but I fancy you will expect me to say something of myself, and therefore I presume to tell you, my hearing hath long since failed. I am weak and forgetful, having as little inclination to business, as ability to perform it. In other respects I have ease, if it may not more properly be called indolence, to a degree beyond what I durst have thought on, when years began to multiply upon me. I wait continually for a deliverance out of this life into a better, in humble confidence, that, by the mercy of God, through the merits of his Son, I shall stand at the resurrection on his right hand. And, when you, my Lord, have ended those days that are to come, which I pray may be many and prosperous, and as innocent and exemplary as those that are past, I doubt not of our meeting in that state, where the joys are renewable, and will always endure. I am your Lordship's most obedient, and ever affectionate servant,

"JOHN WORCESTER."

Charlton, is a village and chapelry, about five miles north by west from Eccleshall, and near the eastern branch of the river Sow.

The village, in which the chapel stands, is called Chapel Chorlton. The Chapel is a small structure dedicated to St. Lawrence, and is a chapel of ease to Eccleshall. The Rev. F. Jenkinson is the present minister.

At or near to Chorlton is a quarry of good building stone of a white colour, and it was noticed by Dr. Plot (ch. iv. sec. 31) as being in his time "the whitest and freest from ...... stains" in the county, "and of a fine grain enough."

Hill Chorlton, or *Chorlton-on-the-Hill*, is an hamlet a small distance from Chapel Chorlton, and in the same chapelry.

Cold Meece and *Mill Meece* are two hamlets to the south of Swinnerton, about a mile from thence; and nearly the same distance from each other. The latter place, as its name imports, contains a good water mill, and there is an ancient mansion, or a part thereof, which belonged to the family of a Mr. Bill.

BARLASTON, is a village situated about three miles north of Stone, to the east of the road leading to Newcastle.

The Church of Barlaston is situated close to Barlaston Hall, and is a small structure of brick, built in the Gothic style, with four windows on each side, and a handsome east window. The ancient square tower, which is low, and rather in a ruinous state, is of stone. This church is dedicated to St. John the Baptist, and is a curacy, in the patronage of the Marquis of Stafford. The Rev. Benjamin Adams is the present minister.

Barlaston Hall is situated on an eminence, which commands an extensive view of a picturesque part of the vale of Trent. The Grand Junction Canal passes through the grounds below the mansion, which is a large and handsome building. Tittensor-heath and Trentham-woods bound the prospect. Thomas Mills, Esq. is the present proprietor.

The parish of Barlaston contains 77 houses, 81 families; 191 males, 205 females: total, 396 persons.

Darlaston Bridge, over the Trent, is in this parish; and *Darlaston Hall*, the seat of John Smith, Esq. is situated in the meadows, where a very ancient plantation of firs skirts the road.

Tittensor Heath, in this neighbourhood, is a considerable length of high, sound, sandy, waste land. At the extremity of the heath, towards Trentham, there is an Asylum for lunatics, kept by Mr. Thomas Bakewell, facing the vale of Trent, which is on the opposite side of the road.

Clayton, is a liberty in the parish of Stoke-upon-Trent, and is in the northern division of Pirehill. It is situated about two miles

s. w. from Newcastle, and commands a beautiful and extensive prospect. According to Pope Nicholas's taxation, the prior of Trentham held at Clayton (therein termed Claxton) in the deanery of Newcastle, two carucates of land, which was worth ten shillings a carucate; and the profits of gardens, which he had from thence, amounted to one shilling a-year. Clayton gave name to a family who were seated there at a very early period: the last of the eldest branch of the male line, Thomas Clayton, Esq. died about the eleventh of Charles 2d, leaving two daughters. Elizabeth, the eldest, and heiress, married Thomas Lea, Esq. This gentleman was descended in the *direct male line* from the antient Earls and Dukes of Suffolk, and the Earls of Lichfield.

Mary Clayton married William Windsor, Esq. of Rees-hall, Cheshire, and the Grange, Shropshire. Part of the foundation of the ancient Clayton-hall still remains: it was taken down by Mr. Lea, who built the old mansion, which is now in a state of dilapidation, having been inhabited by farmers for more than fifty years, the family residing in Shropshire. In September 1812, the Clayton and other family estates descended by the death of the Rev. John Lea, of Acton Burnell, Shropshire, to the three daughters of his youngest brother, the late Thomas Lea, Esq. of Chester. The Rev. John Lea had long held in abeyance all the honours and titles belonging to his family.

The Claytons, of Clayton, possessed anciently large estates in Cheshire, which passed into other branches of this family settled in other counties.

The seat of Hugh Booth, Esq. is pleasantly situated in the village.

According to the population returns of 1811, Clayton contained 24 houses, 31 families; 87 males, 84 females: total of inhabitants, 171.

STONE,

Is a very extensive parish of Pirehill South, consisting of uneven upland, but in general of a good loamy quality.

The town of Stone is a place of great antiquity, and owes its origin to the inhumanity of Wulfere, King of Mercia, who slew his two sons for embracing Christianity, but afterwards repented, and converted his Heathen temples into Christian churches. The Saxons, according to custom, gathered stones and heaped them upon the place where Wulfere had slain his sons, to preserve the memory of the place, and Queen Ermenilda, their mother, erected a church over their tomb. A town was afterwards gradually built at the place, which in memory of the event was called Stone.

The manor is not mentioned in Doomsday Book, but the Conqueror gave it to Erasmus, who held it till the reign of King Henry the First, and then it being forfeited to the King by his son Enysan de Walton, who had killed two nuns and a priest, who had settled in a small oratory in this town, in honour of St. Wulfad, before murdered there, was given to Robert de Stafford, who out of great devotion to that saint, founded a priory for canons regular of St. Augustine; and after him the family made it their burial-place.*

The Church is a handsome modern structure of stone, built in the Gothic style, with ten windows on each side, and a square tower containing eight bells. It is dedicated to St. Michael, and is a curacy, the King being patron. We believe the Rev. Mr. Buckeridge, of Lichfield, brother or near relative to the Archdeacon of Warwick, is the incumbent; and the Rev. Joseph Smith, is the present minister.

Stone is a well-built market-town, situated on the northern bank of the Trent, seven miles north from Stafford. The Trent and Mersey Canal passes near the town, and has much facilitated its commerce. It is a place of considerable extent, and contains eight streets and lanes. There is a good weekly market for corn and other provisions held on Tuesday.

The principal office for conducting the business of the Trent and Mersey Canal, is at Stone. Some account of this noble undertaking is given in a succeeding part of this Work.

The parish of Stone contains the townships of Beech, Kibblestone, Hilderstone, and Normacott. According to the last population returns, it contained 1,174 houses, 1,227 families; 3,159 males, 3,111 females: total, 6,270 inhabitants.

Stone has four annual fairs: the Tuesday after Midlent; Shrove Tuesday; Whit-Tuesday; and August 6th, for sheep and cattle.

At *Stonefield*, a vast and open tract to the left of the town, the Duke of Cumberland drew up his army in 1745, in order to give battle to the Scotch rebels, who had advanced to Derby. The sequel is well known.

* Magna Britannia, No. 63. p. 78.

About half a mile from the town, a good stone bridge leads to the pleasant hamlet of *Walton*, whence Enysan de Walton, before mentioned, took his name. This Enysan (says Pennant) was the true re-founder of the Priory at Stone; but the Staffords, who were his superiors, assumed the honour of this new foundation. The church of the priory contained many magnificent tombs of this great family at the period of the Dissolution, at which time they were removed for safety to the Augustines of Stafford, but were afterwards destroyed with that place. At a short distance to the left, is a range of hills called *Stone Park*, an extensive farm belonging to the Gower family. This was formerly a place of some consequence, and gives title to the present Viscount Granville.

A little further on, stands the hamlet of *Stoke*; and on the opposite side of the Trent, the remains of the once spacious mansion called Aston House, which originally belonged to a branch of the Heviningham of Suffolk, and was brought by marriage to Sir James Simeon, who rebuilt the Hall. He also erected a Mausoleum in the garden for the interment of himself and family. The vault, with two wings, in Stone church-yard, belonging to the Jervis family, was executed from the model of this cemetery. Edward Weld, Esq. of Lulworth Castle, Dorset, afterwards became possessed of this estate, but it is now the property of the Right Hon. the Earl of St. Vincent.

Burston, a small hamlet about three miles from Stone, was anciently much frequented by the devout, on account of a chapel said to have been erected on the spot where Rufin, the second son of Wulfere, was supposed to have suffered martyrdom. The road from Burston passes, for several miles, along a beautiful vale, watered by the Trent, and enlivened by the Canal, which winds very picturesquely through the meadows. The prospect is bounded by some well-wooded hills called the Orangies, &c. behind which lie the lordships of Hilderston and Milwich. The former was the demense of Robert Bagot in the reign of King Henry I. which he held of the Baron of Stafford, but it afterwards changed its lord, for in 12th Edward I. one Vitalis held this manor of Robert de Stafford, and soon after it came into the possession of Robert Hugglesford, whose family held it till the end of King Edward IIId.'s reign. It afterwards became the estate of Sir John Delves, whose only daughter and heir carried it by marriage to Sir Robert Sheffield, Knt. Recorder of the City of London. It afterwards was sold to Sir Gilbert

Gerrard, of Gerard's Bromley, who was Master of the Rolls in the reign of Queen Elizabeth.*

FULFORD, is a village and parish five miles to the north-east of Stone. The river Blithe passes to the north of this parish; the soil is loamy on a marl bottom; and there is a mansion near the Church called Fulford Hall.

The village is ancient, and irregularly built, being situated on uneven ground.

The Church is dedicated to St. Nicholas, and is a chapel of ease to Stone. It is in a low situation, with a small cupola and one bell. There is also a chapel for Calvinists in this village.

Fulford is noticed by Plot as containing the same kind of white building stone as that at Chapel Chorlton.

* Topographer, Vol. I. p. 121.

THE

STAFFORDSHIRE POTTERIES.

THIS opulent and interesting district of the county, lies to the north-east of Newcastle, and extends about nine or ten miles in length. The POTTERIES, so called, are situated in the Hundred of Pirehill North, and include the towns and villages of *Golden Hill*, *New-field*, *Smith-field*, *Tunstall*, *Longport*, BURSLEM, *Cobridge*, *Etruria*, HANLEY, *Shelton*, STOKE, *Lower Lane*, *Lane Delf*, and LANE END, each of which places will be noticed in this their geographical order.

Golden Hill is within the liberty of Oldcott, parish of Wolstanton, and is chiefly remarkable for its valuable mines of coal, and the fine and extensive prospects which it commands: these include nearly the whole of the county of Chester, with the Welch mountains in the back-ground; and the best general view of the POTTERIES, with the surrounding country.

New-field, was partly formed out of the very extensive town-fields within the liberty of Tunstall, about the year 1613, though William Badyley became seised of a messuage and certain quantity of land there, in the 10th of Edward VI. It is well situated for manufacturing purposes, having abundance of coal.

ADMIRAL SMITH CHILD, who of late years lived on this estate, settled on him by his maternal uncle, Thomas Baddeley, was of Salopian descent, though his immediate ancestry were of Audley, in this county,—his great-grandfather, Smithe Child, having succeeded to Boyle's Hall and other property in the latter township, (of which the Admiral was himself a native) on the demise of Margaret Smithe, widow, about 1657. In the peace of 1763, he married his kinswoman, Margaret Roylance, and resumed, near the commencement of the American war, his naval duties, which were continued to its termination. Whilst on shore, he ardently participated in agricultural and other useful pursuits, and received the honorary freedom of Newcastle-under-Lyme, and of Liverpool.

Upon his death, the following succinct notice appeared in the obituary of *The Naval Chronicle* for 1813:

"Died, at New-field, near Newcastle-under-Lyme, on the 21st January, of gout in the stomach, aged 83, Smith Child, Esq. Admiral of the Blue. He entered the service under Earl Gower's auspices in 1747, as the nautical disciple of Lord Anson, and served at the sieges of Pondicherry and Louisbourg. He commanded the *Europe* in the two actions off the Chesapeake, in 1781, with such credit as enabled him to obtain preferment for most of her officers; but, the following year, his eldest son, (a youth who had evinced great intrepidity on board the *Fame* upon the memorable *Twelfth of April*, and was about to have joined the *Foudroyant*, commanded by Sir John Jervis, then at home,) perished in the unfortunate *Ville de Paris*. In 1795 he took the command of the *Commerce de Marseilles* of 120 guns, and attained his flag on Valentine's day, 1799. The Admiral was, during great part of his life, in the Commission of the Peace for Staffordshire,—a Deputy-Lieutenant, &c. of the county,—and was most eminently and extensively beloved and revered."

His remains were deposited at Wolstanton.

Smith-field possesses many strata of good coal and coarse clay, and embraces some charming prospects. Here is *Smith-field*, a handsome villa, formerly the property and residence of the late Theophilus Smith, Esq.

Tunstall, a liberty in the parish of Wolstanton, is pleasantly situated on an eminence about four miles from Newcastle, and on the turnpike-road from Lawton to that town. It has a neat chapel belonging to the Methodists. There "formerly was a church here, and various human bones have been dug up; but such is the effect of time, that not the least trace of it now remains."* There are several considerable manufactories at Tunstall, particularly of a superior kind of blue tile, the clay found here being favourable for the purpose: it is little inferior, in appearance, to common slate. A turnpike-road runs hence to Bosley, in Cheshire.

The following are copies of two ancient and curious deeds connected with this place:

"TUNSTALL.—At the Court there holden on Tuesday in the week of Pentecost in the 10th Year of the Reign of King Edd. the 4th came William Badyley Son and Heir of Margery Handeson and took Seisin of the Lord of Audley of one Messuage and 20 Acres of customary Land in Tunstall

* Aikin's Manchester.

of which the said Margery died seized To hold to the said William and his Heirs according to the custom of the Manor: and he gives to the said Lord at his entry 20d. In testimony whereof John Harryson Deputy Steward to this Copy hath affixed his Seal. Given as above."

"Be yt knowne to all true m̃e In chryste In the wey of truthe yt I Phythyon of Tunstall dyd purchys a garden place yt lyythe in Tunstal at Hary of Tunstall my broders ye wch garden place I do set at my dysseasse to Margerye my wyffe and to Margyt my doughttd ye wyffe of John Banchcrofte & aftd ye dysseace of my wyffe hytt to remeyne to Margytt my doughttd & to hyr eyres the recorder of this Rychard of rydgwaye & to thys I set to my seale Gyvyn ye last daye of Julye in ye yere & reyne of Kynge Harrye ye Syxt after ye conquest of England xxx. & vij."

Longport lies in a valley between Newcastle and Burslem, and contains some very good houses and first-rate manufactories. This place was formerly called Long Bridge, from a kind of bridge or stepping-stones laid across the whole breadth of the swampy meadows, which bridge was afterwards removed; and from the Trent and Mersey canal passing through it, and its great improvement in buildings and population, the place received its present name. Here is a chapel belonging to the Methodists.

BURSLEM.

THIS extensive and populous town, which claims the honour of being the MOTHER OF THE STAFFORDSHIRE POTTERIES, stands on a rising ground about three miles and a half N.N.E. of Newcastle. It is a parish in Pirehill North, and includes the townships of Hulton Abbey and Sneyd.

The manor of Burslem was in the possession of Robert de Stafford at the time of the General Survey. Henry de Audley was lord of it in the time of Henry the Third, and it continued in the possession of his family to the end of the sixteenth century.*

Burslem is undoubtedly the ancient seat of the Pottery, where earthenwares have been made many centuries, for Dr. Plott, who wrote in 1686, mentions the potteries of this place as the greatest of the kind, and gives a very minute description of the process of making earthenware at that time.† We have devoted a separate article to the description of the various improvements made in this useful and elegant manufacture since Dr. Plott wrote. The vast increase of population, opulence, and knowledge, in this district

* Magna Britannia, No. 60, p. 78. † Hist. of Staff. ch. iii. p. 122.

of the county, affords a sufficient demonstration of its general utility, and the numerous manufactories, the extensive Warehouses, Kilns, and beautiful mansions of the master-potters, with the comfortable habitations of the thousands of industrious individuals employed in this lucrative branch of trade, present a scene of animation truly interesting to the patriotic observer.

Burslem contains several modern streets, and the houses are well built of excellent brick burnt upon the spot. The Market-house is a neat modern structure of brick, situated near the centre of the town. It was erected in the year 1760, upon a piece of waste ground called May-pole bank, by the subscriptions of its inhabitants, to which the then Lords of the Manor, Sir Nigel Gresley, Bart. and Ralph Sneyd, Esq. gave £10. each. It is now a large and excellent market for butcher's meat, vegetables &c.: the market-days are Monday and Saturday, but the former is the principal.

The Church, dedicated to St. John, is a large modern brick edifice, with an ancient stone tower: it stands in a low situation near the canal. The interior is neat and clean, and furnished with a good organ.

This Church is one of the rectories formed out of that of Stoke by an Act of Parliament, passed in 1805. The Rev. Edward Whieldon is rector, and the Rev. John Salt, curate.

The late John Rogers, Esq. of Longport, left £100. towards raising the tower of Burslem Church, provided it be carried into effect within the given time.

According to the Parish Register, there were 451 baptisms and 287 funerals at Burslem church, in 1816.

There are six Meeting-houses for Dissenters of the different denominations, in this parish, including one for Roman Catholics, at Cobridge. The chapel for the Methodists of the Old Connexion, is the largest in the county, and is capable of holding 3000 persons. There are two very extensive Sunday Schools, at one of which 2000 children receive education and religous instruction: a library is attached to it for the use of the scholars. The other school is not so large.

"Few places," says Dr. Aikin, "have so great a diversity of opinion on the score of religion as this; but the effusions of loyalty here upon most occasions may be fairly stated to be general, warm, and sincere."

A large and commodious Subscription Warm Bath was opened at the Bicras Colliery, in August 1816. It is supplied from three powerful steam-engines, the property of Messrs. Wood and Caldwell: the temperature of the water is generally from 85° to 90° Fahrenheit, but it can quickly be either diminished or increased at the option of the bather. It is circular, and sufficiently large for the exercise of swimming. The interior of the circular part is beautifully painted in landscapes, and sea views, by an artist of some eminence.

Burslem is remarkable as being the place where the first clod of that great national undertaking, the Trent and Mersey Canal, was cut by the late Josiah Wedgwood, Esq.; and on July 26, 1816, the 50th anniversary of this memorable event was celebrated by a public dinner, at which all the principal manufacturers of Burslem were present. Enoch Wood, Esq. presided upon the occasion, and after a well-merited eulogium upon the late venerable Father of the Potteries (a native of Burslem), and the inventive genius of Brindley, he exhibited various ancient specimens of Earthenware, descriptive of the progressive state of the manufacture during the last 150 years, which he divided into epochs of fifty years, from the Butter Pot mentioned by Plott down to the time at which the excellent specimens of Queen's or Cream-coloured ware, Jasper, &c. left by the late Mr. Wedgwood, were produced.

It has been often asked, Why the potters fixed themselves here, or the Potteries of this county continued to flourish more than those of any other part of the kingdom, or perhaps of the whole world? The answer to this question appears tolerably obvious—The abundant and almost inexhaustless supply of clay and coal upon the spot, the inland situation of this district, which contributed to render labour cheap, and some other circumstances that will be noticed hereafter, combined to fix and establish this important branch of commerce where it had been so successfully commenced. "The measures or strata, by which the beds of coal are divided, consist most commonly of clays of different kinds, some of which make excellent fire-bricks, for building the potter's kilns, and *saggars*, (a corruption of the German *Schragers*, which signifies cases or supporters), in which the ware is burnt. Finer clays, of various colours and textures, are likewise plentiful in many places, most of them near the surface of the earth; and of these the bodies of the wares themselves were formerly manufactured. The coals being then also got near the surface, were plentiful and cheap. In the time of Plott they were as low as two-pence the horse-load, which, at eight horse-loads to a ton (the usual estimation), amounts to only sixteen pence the ton. In 1795 the price of coals was from four to five shillings per ton at the works. Since that time a regular advance has taken place."* When they first began to get the coals here, it was done by removing the soil and clay which covered them: they were afterwards got in open pits, which, however, soon filled with water. Recourse was then had to draining, by laying soughs or gutters from the lowest part of the land near them, by which means the coals were procured for many years in this neighbourhood, until the upper parts or heads of the mine were generally exhausted, and coals became scarce.

In the year 1719, Lord Macclesfield, who owned an estate of about 150 acres, full of coals, adjoining the town of Burslem, entered into an agreement with the owners of the low meadow lands near the church, for permission to cut a sough or gutter from thence up to his lands, for the purpose of draining his mines. The completion of this gutter furnished an abundant supply for upwards of sixty years, but the coal that lay above this gutter or drain, at length became exhausted, and rendered the article scarce and dear. The proprietors of the coal lands then introduced horse-gins; and steam-engines followed shortly afterwards. The deepest engine-pit in the year 1815 was 111 yards, which, if the owners find sufficient demand for the coal, will give a plentiful supply for many years to come. Coals are now sold here at 8*s.* 4*d.* per ton at the pit.

The coals here range from north to south the whole length of the Potteries (say about nine or ten miles on the east of Burslem), and generally dip from east to west about one foot perpendicular in every four feet in length down the dip. Towards Mole Cop, which is four or five miles north of Burslem, the coals suddenly return along Harecastle Hills, having a greater dip, and range nearly from north to south for the length of four or five miles to the neighbourhood of Red-street, and dip south-east; from whence they again range north and south in the direction of Silverdale, a distance of four or five miles, dipping towards the east, and heading-out to the surface of the earth towards the west.

It has been clearly ascertained that there are 32 different mines of coals, between Burslem and the ridge of hills a little to the east of Norton Church, of various thicknesses, generally from about three to ten feet each, laying in the following order, stratum, super-stratum, &c., and in that neighbourhood known to miners by the

* Aikin's Manchester.

names annexed to each mine. At other parts of this range of mines, they are known by different names, and vary a little both in thickness and quality.

1 Red Shag Mine,	17 Whitfield Mine,
2 Brief Furlong ditto,	18 Church ditto,
3 Bass ditto,	19 Eight Foot ditto,
4 Little Rowe ditto,	20 Ten Foot ditto,
5 Peacock ditto,	21 Bowling Alley ditto,
6 Spend Croft ditto,	22 Sparrow Buts ditto,
7 Great Row ditto,	23 Holly Lane ditto,
8 Cannel Row ditto,	24 Iron Stone Coal ditto,
9 Chalky Row ditto,	25 Flats ditto,
10 Row Hurst ditto,	26 Frog Row ditto,
11 Burn Wood ditto,	27 Cockshead ditto,
12 Little ditto,	28 Lime Kiln ditto,
13 Four Foot ditto,	29 Ridgway Cannel ditto,
14 Easling ditto,	30 Bullhurst ditto,
15 Topmost of Two Little Mines,	31 Badiley Edge ditto,
16 Undermost ditto, ditto,	32 Deep Badiley Edge ditto.

There are also several other thin veins of coal lying between the above mines, which are without names, and have never been got.

In 1811, Burslem town and parish contained 1,658 houses, 1,720 families; 4,119 males, 4,506 females: total, 8,625 persons. Of this population, it was computed that about nine-tenths were employed in, or connected with, the pottery business.

In 1653 the number of houses and population was so small, that, in the register of an adjoining parish it was then termed *parochiella*—" the little parish."

COBRIDGE, is a large village, part in the parish of Burslem, and part in that of Stoke-upon-Trent, and contains several extensive potteries and collieries. The population is considerable, and it is a prosperous and increasing place. Here is a chapel for Roman Catholics.

ETRURIA, is a considerable village, or regular-built street, chiefly inhabited by potters, and situated on the sides of the Canal, about a mile north-east of Newcastle. It is part of the vill or township of Shelton, in the chapelry of Hanley and parish of Stoke-upon-Trent, and contains one large manufactory of earthenware, china, &c. Josiah Wedgwood, Esq. who is the proprietor of this extensive pottery, has a handsome seat and plantations on an eminence near the works. The late Mr. Wedgwood, his father, built and named this place after the Italian Etruria, celebrated for the exquisite beauty of its earthenware, the remaining specimens of which served him as models for the improvement of his productions.

The Methodists have a Chapel at the foot of Etruria Bank.

We copy the following article from the *Gentleman's Magazine*, for February, 1795:

"Died, at Etruria, in Staffordshire, aged 64, JOSIAH WEDGWOOD, Esq. F.R. and A.SS.; to whose indefatigable labours is owing the establishment of a manufacture that has opened a new scene of extensive commerce, unknown before to this or any other country. It is unnecessary to say that this alludes to the Pottery of Staffordshire, which by the united efforts of Mr. Wedgwood, and his late partner, Mr. Bentley, has been carried to a degree of perfection, both in the line of utility and ornament, that leaves all work, ancient or modern, far behind.

"Mr. Wedgwood was the younger son of a potter, but derived little or no property from his father, whose possessions consisted chiefly of a small entailed estate, which descended to the eldest son. He was the master of his own fortune, and his country has been benefitted in a proportion not to be calculated. His many discoveries of new species of earthenwares and porcelains, his studied forms and chaste style of decoration, and the correctness and judgment with which all his works were executed under his own eye, and by artists, for the most part, of his own forming, having turned the current in this branch of commerce; for, before his time, England imported the finer earthenwares, but for more than twenty years past, she has exported them to a very great annual amount, the whole of which is drawn from the earth, and from the industry of the inhabitants; while the national taste has been improved, and its reputation raised in foreign countries. His inventions have prodigiously increased the number of persons employed in the Potteries, and in the traffic and transport of their materials from distant parts of the kingdom; and this class of manufacturers is also indebted to him for much mechanical contrivance and arrangement in their operations, his private manufactory having had, for thirty years and upwards, all the efficacy of a public work of experiment. Neither was he unknown in the walks of philosophy. His communications to the Royal Society shew a mind enlightened by science, and contributed to procure him the esteem of scientific men at home, and throughout Europe. His invention of a thermometer for measuring the higher degrees of heat employed in the various arts, is of the highest importance to their promotion, and will add celebrity to his name. At an early period of his life, seeing the impossibility of extending considerably the manufactory he was engaged in on the spot which gave him birth, without the advantages of inland navigation, he was the proposer of the Grand Trunk Canal, and the chief agent in obtaining the Act of Parliament for making it, against the prejudices of the landed interest, which at that time stood very high, and but just before had been with great difficulty overcome in another quarter by all the powerful influence of a Noble Duke, whose canal was at that time but lately finished. Having acquired a large fortune, his purse was always open to the calls of charity, and to the support of every institution for the public good. To his relations, friends, and neighbours, he was endeared by his many private virtues; and his loss will be deeply and long deplored by all who had the pleasure of knowing him intimately, and by the numerous objects to whom his benevolence was extended; and he will be regretted by his country as the able and zealous supporter of her commerce; and by the steady patron of every valuable interest of society."

The handsome mansions of William Bent, Esq. and Dr. Bent, stand on a rising ground, about half a mile east of Newcastle, on the road to the village of Etruria. They command some charming and extensive prospects.

To the right, on a gentle eminence, between Etruria and Cobridge, is the *North Staffordshire Infirmary*, a newly-erected and handsome piece of building, which, when completed, will be every way adapted to the purposes of such a benevolent Establishment. It was projected, about two years since, by the leading gentlemen and manufacturers of the neighbourhood, whose exertions were seconded by the munificent donations of noble and opulent individuals, and the subscriptions of the manufacturers, tradesmen, and others, in the vicinity. His Royal Highness the Prince Regent gave £600., and the late John Rogers, Esq. of the Watlands, bequeathed the princely legacy of £1000. to the Institution. The workmen belonging to the different manufactories in the Potteries contribute to the funds of the Institution by a small deduction from their weekly earnings, proportioned to the number of children, and the amount of wages, of each individual. The Marquis of Stafford is the Patron.

HANLEY,

Is a large modern town, about three miles east by north of Newcastle. No part of the Potteries can boast of more respectable manufacturers than Hanley, which is next to Burslem in size and extent. However irregular the streets may appear to an observer in the town, Hanley, from its elevated situation, appears to great advantage when viewed from Cobridge, or from the turnpike-road between Newcastle and Leek, which passes near it. The Grand Trunk Canal is close to the town, and affords the utmost facility of inland navigation, for the conveyance of the earthenware to Liverpool, Hull, and London.

There are several collieries and potteries in the vicinity of this town and of Shelton.

A weekly market is held in Hanley, on Saturday, for the sale of provisions of all kinds; it is abundantly supplied, and well attended by purchasers.

In 1812, owing to the increasing population of the town, it was deemed necessary to apply to the Legislature to empower certain trustees to enlarge the market and market-place, and an Act for "establishing and regulating the market, and for enlarging and improving the market-place" at Hanley, was obtained. The Act mentions two market-days, viz. Wednesday and Saturday; but the latter is the principal.

The population of the townships of Hanley and Shelton amounts to nearly 10,000 persons. According to the return in 1811, Hanley contained 911 houses, 930 families; 2,165 males, 2,316 females: total, 4,481 inhabitants. Shelton at the same period contained 1,077 houses, 1,148 families; 2,612 males, 2,875 females: total, 5,487 inhabitants.

Hanley Church was rebuilt in the year 1788. It is a handsome well-built structure of brick, with a square tower 100 feet high. It cost upwards of £5000. and is surrounded by a spacious cemetery, enclosed with a wall and large iron gates. It is a Chapel of Ease to Stoke-upon-Trent. The patronage is vested in certain trustees. The Rev. R. E. Aitkens, is the present minister. He was elected in 1802.

This town also contains several meeting-houses for Dissenters of different denominations.

Shelton, though now generally included with Hanley, in being contiguous thereto, and apparently a continuation thereof, is a distinct vill and township. It contains some of the oldest and most respectable manufacturers. Very excellent porcelain, little inferior to that of the East, has long been manufactured here, and it possesses the advantages of a public wharf upon the Trent and Mersey navigation, which passes this place.

ELIJAH FENTON, an ingenious poet of the last century, and whose life is given among the "Eminent Natives" of this county, was born at Shelton, on the 20th of May, 1683.

STOKE-UPON-TRENT,

Is the parish-town of the Potteries, inasmuch as the most important, and by far the greater part of the whole of the Potteries, are, or were, in this parish. Some portions or districts have since been made distinct parishes by Acts of Parliament. Stoke is situtuated about a mile and a half to the east of Newcastle, on the river Trent, with the Grand Trunk Canal running parallel, and passing through the town, thus affording every facility for the conveyance of heavy articles. There are several wharfs and warehouses in this town and its neighbourhood, and it has many earthenware manufactories. "From this place to Newcastle, on the right, the prospects are extremely beautiful; and nearly at the midway, a view so populous, and at the same time so picturesque, is seldom met with."*

The Church is an ancient edifice of stone, with a stone tower. It was the mother church of most of the churches around it to the extent of several miles,† and is dedicated to St. Peter. It is a rectory, and was in the patronage of the late Rev W. Robinson, who was also incumbent. The advowson, we presume, is now vested in his executors. The Very Rev. the Dean of Lichfield and Coventry, Dr. Woodhouse, is the rector, and the Rev. Thomas Yeoman, is the present curate.

In the Church are some monuments to the memory of the Fentons, of Newcastle, maternal ancestors of Sir J. F. Boughey, Bart. M. P. and the following epitaph commemorative of the virtues of the late Josiah Wedgwood, Esq.:

"Sacred to the Memory of
JOSIAH WEDGWOOD, F. R. S. & S. A.
Of Etruria, in this county,
Born in August 1730,
Died January the 3d 1795:

Who converted a rude and inconsiderable manufacture into an elegant art and an important part of national commerce. By these services to his country

* Aikin's Manchester.

† The following were Chapels of Ease to Stoke, viz. Newcastle, Whitmore, Norton-on-the-Moors, Burslem, Bucknall and Bagnall. Hanley and Lane End still continue such.

he acquired an ample fortune which he blamelessly and reasonably enjoyed and generously dispensed for the reward of merit and the relief of misfortune. His mind was inventive and original, yet perfectly sober and well regulated. His character was decisive and commanding, without rashness or arrogance. His probity was inflexible, his kindness unwearied, his manners simple and dignified, and the cheerfulness of his temper was the natural reward of the activity of his pure and useful life. He was most loved by them who knew him best; and he has left indelible impressions of affection and veneration on the minds of his family, who have erected this monument to his memory."

We insert the following account of former sittings in Stoke Church, in order to preserve an evidence of the primitive distribution of seats in regard to rank, age, and worth, as well as the separation of the men from the women in ancient places of public worship.

April 3d, Anno Domini 1634.

"By virtue of an order made by Robert, (by the Divine Providence of God,) Lord Bishop of Coventry and Lichfield, *John Mainwaring*, rector of the Parish Church of Stoke-upon-Trent, in the county of Stafford, William Allen, and William Hill, Churchwardens; their and others their assistants, with the consent of all or most of the parishioners, have placed the ancient householders of the said parish in the seats in the said church as ensueth.

Lichfield, July 12th, 1664.

"If this be the right copy of the order of my worthy predecessor the Lord Bishop *Wright*, I confirm it, appointing that the elder parishioners, according to their deserving, be seated and preferred before the younger.

"John Lichfield and Coventry.

"The names of the parishioners of *Stoke-upon-Trent*, appointed to place the parishioners in their seats, that have [so done] by common consent, and [whose arrangement was] allowed by my Lord Bishop of *Coventry* and *Lichfield*, Anno Domini 1669.

"*Imprimis.*—Robert Clayton, Thomas Tittensor the younger, Thomas Barratt, Thomas Murhall, John Bucknall of the Gate, John Bucknall of Bentiley, Robert Hunt, John Brown, John Hill, Thomas Ames, Henry Brookes, and Richard Meire, who have placed the parishioners as followeth:

"*The South Side—Men's Seats.*

"1. *Churchwardens* for the time being.

2. Thomas Hunt, Gent. John Brown, Gent. Thomas Fenton, John Lovatt, and Thomas Lea.

3. William Bagnall, John Machin, Richard Lovatt, Sir Richard Leveson's tenement, and Robert Bagnall.

4. William Hill, Thomas Hill, and Mr. Bentiley.
5. Mr. Terrick, Richard Broad, and Mr. Keeling.
6. John Malpass, John Boulton's tenement, and William Simpson.
7. Richard Nichols, Thomas Turner, and Roger Dale.
8. John Brown, John Brown, John Dale, and John Prickett.
9. John Doody, John Bowyer, and Nicholas Lovatt.
10. Randle Woodcock, Thomas Fenton, and Roger Machin.
11. William Knight, John Proctor, and Benson's Heir.
12. Henry Stevenson, Nicholas Lovatt, and John Wright.
13. Thomas Tittensor, Richard Beech, and Randle Bagnall.
14. Roger Tittensor, Randle Woodcock, and Thomas Machin.
15. Roger Wood, Mary Shaw, and George Hales.
16. John Pattison, Wright's house, John Machin de Lane.
17. John Stevenson, Francis Lycett, Hanley's de Hanley.
18. Richard Boulton, John Barratt, and Dawson's house.
19. John Biddulph, William Hall, and John Crockett.

" *North Side—Men's Seats.*

" 1. Thomas Smith, Gent. and Thomas Bucknall, Gent.
2. William Allen, Ridghouse, John Hill, and Roger Machin, senior.
3. Thomas Serjeant, Gent. Mr. Bradshaw, John Murhall, and John Wood.
4. At the *Wall*, Roger Machin, jun. Thomas Rawlins, John Adams, Ottiwell Jolley, and John Machin.
5. George Hanson, John Machin, and Thomas Bucknall.
6. Richard Cartwright, William Beech, and Thomas Pare.
7. Thomas Tittensor, Richard Thorleys, and Brassington's.
8. Richard Meire, Robert Cross, and John Leigh.
9. Thomas Machin, John Boulton, and Robert Whilton.
10. William Allen for tenement, William Allen de Hulme, and John Beech.
11. Sir William Bowyer, for Craddock's, Whatson's, and Anthony Keeling's.
12. Bentiley's de Laund, Richard Walklott, and John Poulson.
13. John Wedgwood, Richard Trevin, and John Smith.
14. John Wood de Ash and tenement, John Bowyer, and John Plant.
15. John Austen, Laurance Naylor, and William Cowap.
16. John Stevenson, Gilbert's house, and John Boulton.
17. Thomas Walklott, John Leese, Jeffry Steel.
18. Lewis o'Land, Hugh Mare, and Francis Pool.
19. Jeffry Meire, Richard Cartwright, and William Beech.
20. John Adams, William Murhall, and Thomas Ames.

" *North Side—Women's Seats.*

" 1. Roger Bradshaw, and John Allen; Roger Bradshaw hath the Form end.
2. Thomas Murhall, Wood de Ash, Austens de Ash, and Allen's de Hulme.
3. John Machin de Bucknall, Boulton's house de Bucknall, William Adams de Bagnall, Hanson's house, John Beech de Bentiley.
4. Richard Poulson, Laurence Sherratt, Randle Booths, Iohn Hill of Shelton, Randle Bagnall of Eaves.
5. Robert Whilton, John Hitchcock, John Bucknall, of the Gate, Thomas Lovat, of the Hole-house, Richard Walklott, of Kerry Hill.
6. Thomas Ames, Thomas Hanley, of Hanley, the whole form.
7. Hitchin of Lane, George Fenton, Robert Hill, and Francis Craddock.
8. Richard Meire, German's house, Edmund Vise, Richard Hewet, and Thomas Pare.

9. Richard Mear, Thomas Turnore, the house of Wright and Hanley, Richard Walklott, for Hanley-heys; Thomas Tittensor, for the Over-house, Thomas Wood claimeth the same.

10. Parker's house, Boulton's house, John Bradshaw and Thomas Bagnall.

11. Thomas Fenton of Boothen, Thomas Ames, John Hanley's of Penkhull.

12. Thomas Barrott, John Rowley, of Shelton, Hugh Wood, and Stephen Fenton.

13. Spooner's house, Richard Tunstall, and John Hankinson.

14. Roger Harrison, William Harrison, Jeffry Meire, and Roger Cowap.

" South Side---Women's Seats.

" 1. Thomas Bucknall, and Richard Serjeant, George Fenton claimeth the same.

2. Thomas Ames, John Turnore, Brassington of Moor-hall.

3. Hen. Brooks, Bartholomew Bowyer, Richd. Lovatt, and Widow Benteley.

4. Robert Clayton, Thomas Dawson, John Lovatt, the house that is holden of Garnitt and Benson a seat, Horne's land a seat, John Trinley claimeth the same.

5. Robert Hunt, Roger Bagnall for Longton house, Whiston's house, Roger Bagnall of Clayton, for Longton house, the 4th seat; Richard Aston claimeth for German's Land.

6. John Kendall, William Machin, and Thomas Lovatt.

7. Thomas Machin, John Proctor, Henry Lovatt, of Eaves, John Kendall, Thomas Tittensor claims the same for Fenton's house.

8. John Brown, T. Broad to hill, and John Brown for his tenement.

9. Widow Hordern, Thomas Tittensor for the lower house, Hugh Machin.

10. Widow Bagnall, Mr. Egerton, Peter Knight, and John Woodcock the fourth seat.

11. Iohn Hammersley, Richard Kendrick, William Barratt, Thomas Machin claims a seat for Bate's house.

12. Iohn Dale, Iohn Hitchin, Iames Hudson, Ralph Bucknall, and Iohn Pulsbury.

13. Hugh Thorley, Roger Fox, Iohn Simpson of Clayton, Widow Febkin, and Iohn Simpkins.

14. Widow Stevenson, Iohn Bourn, and Roger Fixon. Richard Aston is unplaced and Robert Pyler. Also the young maids are to kneel in the short forms.

" Eccleshall Castle, 20th April, A. D. 1634.

" Being fully informed by such as I have caused and required to survey convenient and commodious sitting and placing of the parishioners, of the parish of *Stoke-upon-Trent*, in the county of *Stafford*, and finding no just opposition against the same, by any of the parishioners aforesaid, I, Robert, *(by Divine Providence)* Lord Bishop of *Lichfield* and *Coventry*, do well approve the Order, by parcelling and seating of the said parishioners aforesaid, and by these presents allow, ratify, and confirm the same whatsoever the Rector, Churchwardens, and parishioners have done therein, until just cause shall be shewed to the contrary. In witness whereof I have hereunto set my hand, and caused my seal episcopal to be hereunto placed at the time and place abovesaid.

"ROBERT LICHFIELD AND COVENTRY."

"THOMAS GOODFELLOW, *Clayton*, } Churchwardens for the Year
"JOSEPH BOURNE, *Ford Bays*, } of our Lord 1772."

Some obscurity in the above document appears at the beginning, by the position of the subsequent confirmations in 1664 and 1668, whether the list applies to the period of 1634, and afterwards 1668; or wholly to the latter: and some sentences are not clearly expressed, which we have taken the liberty of filling up between brackets according to our judgment. Whatever obscurity might exist, originally, or by copying, the article is sufficiently explicit for the purposes already stated, and which induced its insertion. By some it may be considered not uninteresting in furnishing a list of the names of the ancient householders in the parish of Stoke; and in affording a partial means of judging of the comparative state of the parish in regard to the population in the seventeenth and nineteenth centuries.

Mr. Mainwaring, whose name appears, as Rector, in the beginning of the above instrument, is remarkable not only as an instance of longevity, but for the duration of his incumbency. The following has been handed to us as being an extract from the Parish Register, though it is evidently imperfect:

"May 1692. Johannes Mainwaring, S. T. P. Rector Ecclesie de Stoke sup. Trent: sepult."

"Johannes Mainwaring cum extitisset Rector Ecclesie Stoke p. spatium Quinquaginta et non'm annor' expiravit die et anno s'pra dict."

If Mr. Mainwaring was 59 years rector of Stoke, his age, at the time of his death, could not be less than 83 years. He was succeeded by the Rev. John Repton, who was likewise perpetual curate of Norton-on-the-Moors.

In the year 1815, a very handsome and commodious *National School*, for the education of 500 children of the poor in the Principles of the Established Church, was erected at this place, at an expense of nearly £1000. which sum was raised principally by voluntary contributions. It is situated at the east end of and adjoining the church-yard.

There are two or more meeting-houses at Stoke for the Dissenters.

The first steam-engine for grinding burned flint for the use of the potters, was established at Stoke.

About midway between Stoke and Newcastle, is *Cliffe Ville*, the seat of John Tomlinson, Esq.; and at Penkhull, is *The Mount*, the seat of Josiah Spode, Esq.

LANE END,

Is a modern and populous town, which has risen in a few years by the almost magic influence of a prosperous manufacture, to a respectable degree of opulence. This town is situated four miles to the south-east of Newcastle, and at the southern extremity of the Potteries. Large quantities of earthenware are manufactured in the extensive potteries of this town and its vicinity.

The township of Lane-End and Longton is populous, and contains 1,032 houses, 1,079 families; 2,277 males, 2,653 females: total 4,930 inhabitants.

Lane-End is a market town: the market is on Saturday, and well supplied with provisions.

The Church is similar to that of Hanley, but does not appear to so much advantage, being in a comparatively low situation. We understand it was rebuilt about the year 1795. The original chapel was principally built and endowed at the charge of the late John Bourne, Esq.* and was consecrated in 1764. The following memorandum may throw some light on the expense of the present building:

* Mr. Bourne was an inhabitant of Newcastle-under-Line, and as our article for that place was too copious to copy the monumental inscriptions of public characters in the Church there, we hope to stand excused by our readers in taking this opportunity of recording a just tribute of respect to a pious, benevolent man, and a benefactor to the people of Lane End. Mr. Bourne was buried at Newcastle, and in the church there, on a mural monument, is the following inscription:

"In Memory of John Bourne, Esq. late a worthy inhabitant of this town, (Newcastle-under-Line) and one of his Majesty's Justices of the Peace for the county of Stafford. A man very remarkable for his public spirit, and extensive liberality. His zeal for the advancement of religion appeared from several new chapels erected and endowed in this neighbourhood, chiefly at his expense. His benevolence towards men was shewn by a constant readiness to assist the needy with whatever their wants required; and by strenuously promoting every measure which promised happiness to mankind. As a magistrate he was active and impartial, distributing justice with temper and judgment. He was kind to his relations, sincere to his friends, and forgiving to his enemies. He died a bachelor, Sept. 13, 1764, aged 73 years."

He was maternal uncle to the Rev. John Fernyhough, B. D. who was forty-three years the respected and beloved minister of Newcastle; and not only

"*Aug 28th, 1792.*

"At a meeting of the Trustees for re-building the chapel of Lane End, the following estimates were received, and Messrs. Leigh and Turner's accepted.

"The different estimates were given in for completing the whole, and in three different valuations of the parts, (for instance):

"1st. For building the chapel with a tower, the height of the roof, for pewing the body and making a chancel.

"2dly, For building the tower 62 feet from the floor, and completing it.

"3dly. For making the galleries complete.

	Messrs. Hulme & Watson's valuation.	*Leigh & Turner's.*	*Mr. Gardner's.*
For making the body complete, with a tower 39 feet high,...............	£2,485.	£2,048.	£2,100.
The Tower complete 62 feet high,............	445.	340.	400.
The Galleries complete......................	396.	420.	360.
For completing the whole, ...	£3,326.	£2,808.	£2,860.

Lane End Church is a Chapel of Ease to Stoke. By the resignation of the Rev. T. Cotterill, the church is now (1817) vacant. The presentation is vested in trustees.

The living is, at present, small: though during Mr. Cotterill's* ministry, from 1805 to 1817, it has been much improved, and, it is said, will be augmented considerably more. In the time of the Rev. Mr. Wright (Mr. C.'s predecessor), it was valued at about £75. a-year. The emoluments arise from upwards of 18 acres of glebe, a good parsonage-house and garden, surplice-fees, and the profits of one or more augmentations from Queen Anne's Bounty. An increased allowance was also made by the trustees, on condition, that

succeeded to a fair portion of his uncle's fortune, but possessed his virtues also. He died March 8th, 1803, at the same age (73), was buried March 15th, at the same place, and was likewise a benefactor to Lane End chapel, inasmuch as he gave £100. on the condition, that another hundred pounds was raised, which was accordingly done by subscription; and thus was procured one of the augmentations from Queen Anne's bounty to Lane End chapel.

* It is but justice to add, that during Mr. C.'s time, and we believe chiefly owing to his exertions and esteemed ministry, an organ has been erected in the church; and a good ring of bells and a clock set up in the tower.

the duty was performed twice every Lord's-day. Its present reputed annual value is about £140.

These instances of Longevity were extracted from the Parish Register:—Lydia Barber, died Oct. 27, 1769, aged 107 years; Rosamond Cook, died Sept. 23, 1774, aged 124 years; Elizabeth Mills, died Feb. 14, 1776, aged 100 years; Sampson Smith, died Aug. 11, 1780, aged 99 years; and Sarah Hollins, died Nov. 10, 1780, aged 100 years.

An English *Charity School* adjoins the Church-yard, which, according to an inscription on a stone in the wall, was built and endowed in 1760. Mr. Bourne is said to have been the founder of this charity.

At Lane End are meeting-houses for different denominations of Dissenters.

Lower Lane and *Lane Delph* present nothing remarkable, and indeed they may be properly incorporated with Lane End, of which place they form a kind of suburbs.

On the left of Lane End, is *Longton Hall*, the seat of Sir John Edensor Heathcote; and near it, *Park Hall*, the residence of Thomas Parker, Esq. a descendant of Lord Chief Justice Parker.

A BRIEF ACCOUNT

OF THE

RISE and PROGRESS of the STAFFORDSHIRE POTTERIES.

IT is at all times pleasureable to trace the progress of a nation in civilization and the arts,—in the attainment of those distinguished excellencies which contribute to the comforts and convenience of society, and the exaltation of its character in the scale of being; but if we admire the united exertions of a people in the cultivation of manners, and the dignified acquirements of moral and intellectual science, we are called to contemplate with feelings of peculiar pleasure, the labours of a class or community of that people, who, under less favourable auspices, associate and form an indivisible interest to promote these important objects, and to gain pre-eminence in some particular branch of commerce or manufactures, in which their industry and ingenuity might enhance the value of the natural productions of the soil, or best improve the advantages of local situation. Such an union, whilst it binds more closely the bond of brotherhood, tends at once to cherish individual merit, and advance the national honour. It was owing to a similar combination of worth and intellect, at no very remote period, that the Staffordshire Potteries make at this day so prominent a feature in the commercial transactions of Britain. Indeed, the advancement of this important division of Staffordshire in its manufactures, and consequent opulence, has been most rapid; and there is scarcely a market in Europe or America, that does not exhibit the envied productions of our Potteries, in all their variety, both useful and ornamental.

The STAFFORDSHIRE POTTERIES present a striking illustration of what may be achieved from a cordial union of the mental and physical powers of man. Little more than a century since, this flourishing seat of commerce scarcely had existence as such. Nature had indeed amply compensated the want of a genial surface, in the rich and inexhaustible mineral treasures which pervade this extensive district; but if the exertions of the Agriculturist were paralyzed, so the sources of future wealth remained unexplored. The Potteries so called, wore a bleak and barren aspect, and consisted merely of several detached and straggling villages, thinly inhabited, and possessing a trade comparatively insignificant. But what a change has public spirit wrought in this interesting tract of country! The ingenuity of the artizan has been seconded by the researches of the mineralogist; and the earth has poured forth her bounties to enrich its owners, and raise the Staffordshire Potteries to the enviable rank which they hold in the commercial world of two Continents. They can now boast their thickly-inhabited towns and villages, public edifices, extensive works, beautiful villas, and handsome habitations, enlivened by a busy and increasing population of 60,000 souls. To point out individuals, who merit the thanks of their countrymen, and the praises of posterity, might in us seem invidious: suffice it to say, all have contributed to this great work. All have deserved the meed of applause for their ceaseless endeavours to promote the well being of all with whom they were more immediately connected. The generation that has passed away, will ever be remembered with grateful veneration for having began the goodly fabric upon a firm and durable basis, whilst their descendants are entitled to equal honour in having raised a superstructure worthy the talents and patriotism of their predecessors.

The exact period of time at which the Manufacture of Earthenware was introduced into this county, cannot now be ascertained: it may, however, be traced with certainty for two centuries back. At that time, its principal seat was the town of Burslem, then called the *Butter Pottery*, from a manufactory established there for making butter-pots. These pots were of a cylindrical form, and, under the provisions of an Act of Parliament obtained for the purpose about 1670, were of a certain size and weight (6lbs.) and made to contain not less than fourteen pounds of butter. This Act, it seems, was obtained in consequence of a fraud practised upon the public by the inhabitants of the Moorlands, by making these pots thick at the bottom, in order to increase their weight.

Burslem is denominated a *Butter Pottery* in some old maps;

but neither Speed nor Camden mention it as such. Dr. Plot (1686) is the earliest writer who notices it, and with whose clear and comprehensive account of the state of the Manufacture of Earthenware at that time, we shall commence our Sketch of the Rise and Progress of the Staffordshire Potteries.

"The greatest *Pottery* (says he) they have in this *County*, is carryed on at *Burslem* near *Newcastle-under-Lyme*, where for making their severall sorts of *Pots*, they have as many different sorts of *Clay*, which they dig round about the *Towne*, all within half a miles distance, the best being found nearest the *coale*, and are distinguish't by their *colours* and *uses* as followeth :

"1. *Bottle Clay*, of a bright whitish streaked yellow colour.

"2. *Hard Fire Clay*, of a duller whitish colour, and fuller intersperst with a dark yellow, which they use for their *black wares*, being mixt with the

"3. *Red Blending Clay*, which is of a dirty red colour.

"4. *White Clay*, so called it seems though of a blewish colour, and used for making yellow-coloured *ware*, because yellow is the *lightest* colour they make any *Ware* of; all which they call *throwing* clays, because they are of a closer texture, and will work on the *wheel*;

"Which none of the three other *clays*, they call *Slips*, will any of them doe, being of looser and more friable natures; these mixed with water they make into a consistence thinner than a *Syrup*, so that being put into a *bucket* it will run out through a *Quill*, this they call *Slip*, and is the substance wherewith they *paint* their *wares*; wherof the

"1. Sort is called the *Orange Slip*, which before it is work't, is of a greyish colour mixt with orange balls, and gives the ware (when annealed) an *orange* colour.

"2. The *White Slip*; this before it is work't, is of a dark blewish colour, yet makes the ware yellow, which being the *lightest* colour they make any of, they call it (as they did the *clay* above) the *white Slip*.

"3. The *Red Slip*, made of a dirty reddish clay, which gives *wares* a black colour; neither of which *clays* or *Slips* must have any *gravel* or *Sand* in them; upon this account, before it be brought to the *wheel* they prepare the *clay* by steeping it in water in a square pit, till it be of a due consistence; then they bring it to their *beating board*, where with a long *Spatula* they beat it till it be well mix't; then being first made into great *squarish* rolls, it is brought to the *wageing board*, where it is slit into flat thin pieces with a *Wire*, and the least stones or gravel pick't out of it. This being done, they *wage* it, i. e. knead or mould it like *bread*, and make it into round *balls* proportionable to their *work*, and then 'tis brought to the *wheel*, and formed as the *Workman* sees good.

"When the *Potter* has wrought the clay either into *hollow* or *flat ware*, they are set abroad to dry in fair weather, but by the fire in foule, turning them as they see occasion, which they call *whaving*: when they are dry they *stouk* them, i. e. put *Ears* and *Handles* to such *Vessels* as require them: These also being dry, they then *Slip* or *paint* them with their severall sorts of *Slip*, according as they designe their *work*, when the first *Slip* is dry, laying on the *others* at their leasure, the *Orange Slip* makeing the ground, and the *white* and *red*, the *paint*; which two colours they break with a wire *brush*, much after the manner they doe when they *marble* paper, and then *cloud* them with a *pensil* when they are pretty dry. After the *vessels* are painted, they *lead* them, with that sort of *Lead Ore* they call *Smithum* which is the smallest *Ore* of all, beaten into dust, finely sifted and strewed upon them; which gives them the *gloss*, but not the colour; all the *colours* being cheifly given by the variety of *Slips*, except the *Motley-colour*, which is procured by blending the *Lead* with *Manganese*, by the *Workmen* call'd *Magnus*. But when they have a mind to shew the utmost of their *skill* in giving their *wares* a fairer *gloss* than ordinary, they *lead* them then with lead *calcined* into powder, which they also sift fine and strew upon them as before, which not only gives them a higher *gloss*, but goes much further too in their work, than *Lead Ore* would have done.

"After this is done, they are carried to the *Oven*, which is ordinarily above 8 foot high, and about 6 foot wide, of a round copped forme, where they are placed one upon another from the bottom to the top: if they be *ordinary wares* such as *cylindricall Butter-pots*, &c. that are not *leaded*, they are exposed to the *naked* fire, and so is all their *flat ware* though it be *leaded*, haveing only *parting-shards*, i. e. thin bits of old pots put between them, to keep them from *sticking* together: But if they be *leaded hollow-wares*, they doe not expose them to the *naked* fire, but put them in *shragers*, that is, in coarse metall'd pots, made of *marle* (not *clay*) of divers formes according as their *wares* require, in which they put commonly 3 pieces of *clay* called *Bobbs* for the ware to stand on, to keep it from sticking to the *Shragers*: as they put them in the *shrangers* to keep them from sticking to one another (which they would certainly otherwise doe by reason of the *leading*) and to preserve them from the vehemence of the *fire*, which else would *melt* them downe, or at least *warp* them. In 24 hours an *Oven* of *Pots* will be burnt, then they let the *fire* goe out by degrees, which in 10 hours more will be perfectly done, and then they draw them for *Sale*, which is cheifly to the poor *Crate-men*, who carry them at their *backs* all over the *Countrey*, to whome they recken them by the *piece*, i. e. *Quart*, in *hollow ware*, so that 6 pottle, or 3 gallon *bottles* make a *dozen*, and so more or less to a *dozen*, as they are of greater or lesser *content*. The *flat wares* are also reckon'd by *pieces* and *dozens*, but not (as the *hollow*) according to their *content*, but their different *bredths*."*

This is admitted to be a faithful description of the state of the Manufacture down to the time when Plott wrote; indeed, there are persons still living, who remember manufactories carried on in Burslem upon a similar principle. Its authenticity is further corroborated by the numerous specimens of the wares here alluded to, which have at various times been dug up in different parts of that town, and now carefully preserved. It is readily discovered, upon minutely examining these ancient pieces, that the Butter-pot claims priority of date, from the circumstance of its being composed of the coarsest clay (then as now used in making the inferior sorts of brick) as well as exhibiting the rudest workmanship; and although many of these pots (weighing 6lbs.) have been lately found without any glaze† whatever, and evidently made at a remote period, others have the appearance of lead ore having been sprinkled on the inside, when in a moist state, and before they were exposed to the heat of the potter's oven.

This partial glazing appears to have been the first attempt at improving the rude article, manufactured, there is every reason to believe, long before the year 1670. Specimens have also been found which exhibit a further improvement in medals of King William and Queen Mary being attached to the sides of the ware, and glazed

* Plott's Hist. of Staff. ch. iii. secs. 23-29.

† The common glaze was produced by lead ore, finely powdered, and sprinkled on the pieces of ware before firing; sometimes with the addition of a little manganese, for the sake of the brown colour it communicates, and where the potters wished "to shew the utmost of their skill," (*Plott*) in giving the ware a higher glaze than ordinary, they employed, instead of lead ore, calcined lead itself; but still sprinkled it on the pieces in the same rude manner.

with the dust of lead ore inside and out; together with a variety, frequently found, *glazed with salt*,* and bearing the initials WM. WR. AR. surmounted by a crown, and others ornamented with medals of Queen Anne, tolerably executed in basso-relievo. These pieces appear to be composed of the clay found in the coal-pits in and near Burslem, and then called Can-marl; whilst others have been found, formed of this clay and a mixture of white sand or pounded gritstone, procured at Mole Cop, and well covered with a salt glaze. This last is known by the name of CROUCH WARE, and proves that the salt glaze had at that time been introduced.

Another description of earthenware, dug up with the above, is made of the same kind of clay, and painted all over with clays of various colours; orange, yellow, red, brown, dark grey, black, and all the shades which a mixture of the native clays could produce. From the dates and paintings upon these pieces, they may likewise be referred to the reigns of William, and Anne.

These clays, it appears, were the only materials then used in the composition of the body, as well as in painting this ware; if indeed we except the black clay, which seems to be a mixture of the ironstone that abounds in the neighbourhood, with a trifling addition of manganese. The glaze, to all appearance, is simply lead ore, or a combination of lead ore with some of the whitest of the before-mentioned materials.

At this early period, a great variety of useful and ornamental articles, in large dishes, jugs, candlesticks, &c. were manufactured at Burslem, which at this day are much admired for superiority of workmanship. Specimens of these are preserved, with dates thereon, down to the year 1707; and from the skill and great variety of colours used in ornamenting them, it is evident that white clays had not then been introduced, and were wholly unknown in the Staffordshire Potteries.

It was not long, however, before a purer clay from Dorsetshire and Devonshire, was brought hither; but this was used only for washing or lining the insides, or ornamenting the outsides, of ware, with flowers and figures of various kinds; and these white ornaments are sometimes found upon a red body, well *glazed with lead ore*.

* This species of glaze, (according to Dr. Aikin) was first brought hither about the year 1690, by the two Elers, brothers and foreigners, who established a small pot-work at Bradley, and met with some success in the manufacture of tea-pots, and a ware in imitation of the unglazed red china from the East.

From these facts it would appear, that the salt glaze was used long previous to the introduction either of flint, or white clay, particularly in making Crouch-ware; the body of it being at first formed of a reddish clay, and afterwards of a dark greyish clay, dug from the coal-pits, which, when exposed to an intense heat, became of a light greyish colour. This clay, mixed with pounded sand from Mole Cop, produced a whitish body, then called STONE WARE, which was for some time generally used, and further improved by several manufacturers, who dipped it in a slip produced from the whitest clay from Devonshire, and which, after drying, was put into the hands of the *Flowerers*.*

Another body was afterwards formed of the whitest clay and a pounded gritstone from Baddeley Edge and Mole Cop; but from the scarcity and expense of this clay, the ware manufactured from it was light and thin, which is apparent from specimens of it now in existence. This variety of ware also passed through the Flowerer's hands, when sufficiently hard and dry, who embellished it with engravings of flowers, birds, beasts, &c. and afterwards took some wool with dry pounded smalts or zaffer, and rubbed the smalts over the engraved figures or ornaments. This, when fired and glazed with salt, produced a fine blue colour upon the parts engraved, and was the only kind of painting or ornament then used, except indeed the raised sprigs of flowers and different animals, &c. that were stuck upon the wares formed in clay or brass moulds, and which raised-work was formed in some instances of a still whiter clay than the body.

Another body, and that indeed the best, was composed of the white clays of Devonshire or Dorsetshire, mixed with a due proportion of calcined flint, previously reduced to a fine powder by pounding it in a dry state, and by passing it through a fine hair sieve, and glazing it by pouring as many bushels of salt as there were mouths or fire-places to the oven (7 or 8) through holes at the top of the oven, at the time the ware had reached the highest degree of heat these materials were capable of sustaining, without fusing the body, which frequently happened, especially if the heat was increased after the salt had been thrown into the oven, as thereby too great a vitrification of the body took place, and an inferior sort of porcelain was unintentionally produced. The thinnest pieces of

* *Flowering* was generally performed by women, who decorated the ware with drawings of flowers, &c. traced with the point of an iron nail, which cut through the thin coat of slip in which the ware had previously been dipped.

this porcelain became transparent, and were therefore of less value than the ware intended to be produced, on account of their being less able when used to resist fire or hot water. This was called WHITE WARE, or WHITE STONE WARE.

At this period commenced what may be termed an era in the history of the Potteries, though at the time it was unperceived, on account of the gradual progress of the manufacture to its then improved state. The articles manufactured from the above materials were very durable, and of neat clear white appearance; and they soon found their way to distant parts of England, and were subsequently exported in considerable quantities to Holland, Germany, Prussia, and other parts of the Continent of Europe. These articles consisted at first of circular pieces, or of such only as were produced from the thrower's wheel, and finished on the lathe; for it does not appear that plaster moulds had yet been used. The only kind of moulds then known seem to have been made of brass, or a kind of clay known among potters by the name of *Tough Tom*, many of which latter have lately been met with in digging up the old shord-rucks,* found in foundations for new buildings in Burslem.

About this time, Ralph Daniel, a potter, of Cobridge, during his travels through France, visited a manufactory where plaster moulds were used, one of which he brought away with him (a table plate), the pattern of which, together with several others, were immediately imitated, by Aaron Wood, whose business it was to make models for the potters. The gypsum mines of Derbyshire furnishing the material in great abundance, complete table services of different fanciful patterns (barley-corn, mosaic, basket-work, &c.) were manufactured. This important acquisition was soon followed by the introduction of a variety of curiously-carved tea and coffee services, &c. Plaster moulds now began to be brought into general use, both on account of the extraordinary absorbent quality of the plaster, when burnt, and its adaptation to the method then practised for making the hollow articles, such as carved tea-pots, coffee-pots, &c. The manner in which casts were taken from the moulds, was by pouring a very thin slip into the plaster mould, and letting it stand a short time before it was poured out, by which a perfect impression was received by the coat of thin slip or clay, which closely adhered to the mould. Then another and thicker coat was added, which, after remaining a few minutes in the mould, was returned, and this second casting more than doubled the thickness of the first

* Rubbish from pot-works.

casting or pouring-in of slip; and these castings were repeated until the required strength or thickness was obtained. The moulds were then placed before a fire to dry, which done, the pieces of ware were easily separated from the moulds, and the seams or marks where the moulds parted, were smoothed or taken off, and the spouts and handles were put on by the person whose branch of the art was then termed "*Handling and Trimming.*" The pieces of earthenware being thus formed, were carried to the oven, in which they were, for many hours, exposed to a considerable degree of heat; and for this purpose they were put or placed in *saggars*, having large holes cut in their sides to admit the fumes of the salt, which thus united more freely with the flint and clay upon the surface of the ware. These saggars, when filled with earthenware, were placed or piled upon each other about ten or twelve feet high, in circular ovens twelve or fourteen feet wide, having seven or eight mouths or fire-places at equal distances. This oven was generally filled once a-week. The *firing* generally commenced on the Thursday evening, and finished on the Saturday following about mid-day.

There were at this time upwards of twenty ovens in the parish of Burslem, all of which cast in their salt or glazing* at the same time, generally on the Saturday morning from nine to twelve o'clock. This occasioned such immense and constant volumes of smoke, as literally to envelope the whole neighbourhood; and it was not unfrequent for passengers to mistake their way, and run against each other, during the continuance of this process. The scene which presented itself upon those occasions, has been not inaptly compared to the emissions of Etna or Vesuvius.

This white glaze soon attracted the attention of enamellers from the china and Dutch-tile manufactories then established in different parts of the kingdom, who began to cover their carved work with fine enamel colours, and soon after made great progress in painting groups of figures, flowers, birds, &c. and in copying the

* The glazing with salt seems to be effected by the following process. The salt poured by degrees into an oven raised to a very high temperature, is immediately decomposed; the alkaline part of it, in a state of vapour, enters into the saggars, and surrounds the pieces of earthenware, dissolving the surface of them, by which means they are covered with a very thin coating of glass, produced by the dissolution of the earths of which the body of the ware is composed, in this alkaline vapour. The muriatic acid of the salt, combined with the caloric in the oven, to which it has a strong affinity, rushes with violence through the chimney of the oven, and through every other aperture of it, and forming the white suffocating smoke above alluded to. Common salt, it is well known, is the crystallised solution of soda in the muriatic acid.

paintings of the richest China from the East upon their tea, coffee, and dessert sets, as well as jars, cornucopiæ, figures, and other ornamental and useful articles.*

Various improvements in the body, glaze, and workmanship of the ware, gradually succeeded; and the manufacture of the *White Stone Ware* had at this time extended itself along the whole range of the line on which the various strata of coal presented themselves, or were found at a depth of from three to six feet from the surface of the earth, at or near Tunstall, Shelton, Cobridge, Hanley, Stoke, Lane End, and Mere Heath, being a distance from one extreme to the other of eight or nine miles.

Many have been the conjectures respecting the introduction of this important acquisition to the Staffordshire Potteries. Some attributed it to two Dutchmen (the Elers, before-mentioned) who were supposed to have brought the art from Holland, and who built a small manufactory at Bradwell, near Burslem, about the year 1690. Their claim to this honour, however, is rendered extremely questionable, from the very improved state of the manufacture of Crouch Ware at Burslem in the time of William and Mary, and Queen Anne, as well as on account of the unfitness of their oven at Bradwell, from its small dimensions, for this manufacture. The foundation of this oven remained undisturbed until within the last twenty years; and on minutely examining the ground about it, fragments of broken red china only could be found. It is therefore very probable that the excellent *red* clay found near the ancient oven at Bradwell, and the contiguity of coal, and clays, induced these foreigners to fix upon that spot, for the purpose of imitating the fine unglazed red china ware of the East, and specimens of these *unglazed* imitations are still in existence. These pieces, it is well known, were manufactured at Bradwell about the year 1690, and were then sold at a high price.

During the time from the commencement of the manufacture of White Stone Ware to the period when it had attained its highest perfection, which was about fifty years, the lead ore glaze continued to be used, in the original way, in numerous small manufactories in every part of the Staffordshire Potteries. The manufacturers, finding themselves in possession of a white body, fired and glazed it in the usual manner; and with the addition of manga-

* *Enamelling* was at first practised by persons who were not manufacturers of the body of the ware, though a few ultimately added this branch of the art to their regular business.

nese rubbed under or upon the glaze with a sponge, produced an article known by the name of TORTOISE SHELL WARE.

Another variety of ware, termed AGATE WARE, was produced by the addition of ground zaffer, in various proportions, and applied either by a sponge or the point of a hair pencil. Hafts for knives and forks, snuff-boxes, &c. &c. were manufactured from these materials, and sent to Birmingham and Sheffield, to be fitted-up with hinges, hoops, and springs.

A similar application of calcined copper, iron, and other metals, gave rise to other descriptions of ware, called CAULIFLOWER WARE, MELLON WARE, &c. &c.

Another and an excellent kind of ware was produced by employing the lead-ore alone, or refined (omitting to use any of the before-named metals), with a small proportion of ground flint. This variety, denominated CREAM-COLOURED WARE, was first made in the year 1750, by Enoch Booth, of Tunstall, near Burslem, and one of his earliest pieces is now in the possession of an inhabitant of Burslem. The colour of this ware was subsequently improved, and considerable quantities of it were manufactured by the late Mrs. Warburton, of the Hot-lane, near Burslem, in 1751.

In the early stages of the manufacture of White Stone Ware, numerous difficulties presented themselves, which to persons without precedent or much chemical knowledge, might have proved insurmountable, had it not happened that different persons were making experiments at the same time with more or less success. Although the native marls or clays (of which the saggars were made) were numerous, it required great judgment in the choice and admixture, to enable them, when made, to sustain a much higher degree of heat than was required in the manufacture of lead ore glaze; and this inconvenience was very generally felt for a considerable time. It was, however, at length ascertained, that saggars composed of Can-marl two parts, and Black-marl one part, were capable of sustaining the degree of heat required. This discovery was followed by that of the due admixture of flint and clay for forming the body of the ware, and the manufacture of White Stone Ware rapidly increased.

About the year 1740, Thomas and John Wedgwood left their father's employ (as lead-ore glaze potters) and commenced the manufacture of White Stone Ware upon their own account; but although very industrious and ingenious workmen (one of them being well skilled in burning or firing the ware, and the other an excellent thrower), they were unsuccessful for a long time, and had actually determined to abandon any further attempt to make the White Stone Ware, when an accidental circumstance encouraged them to proceed. The water with which they prepared their clay, it seems, became highly saturated with salt, owing to the shord-ruck or rubbish from their ovens being placed immediately above their water-pool, and which rubbish contained much salt. The rain, passing through the shord-ruck, dissolved the salt, and carried it into the pool, whence it got into the body of the ware, and, in conjunction with the flint and clay, together with the lime which generally adheres to flint stones, formed a fusible body that arrived at a state of vitrification with a lower degree of heat than was requisite to prepare this body for the salt glaze. This discovery induced them to make another trial with purer water; and in this they succeeded beyond expectation.

The Wedgwoods followed up their success with unremitting diligence; and shortly afterwards built a new and commodious manufactory, where they had a supply of good water. This was near the Windmill, invented and executed by the celebrated Brindley, for reducing flint stones to a fine powder by grinding them in water, and thereby preventing the pernicious effects upon the health of the men employed in preparing the flint according to the old method, by pounding it by hand in a dry state in a mortar. The fine dust of the flint getting into the lungs, produced coughs and consumptions, which frequently proved fatal. This building, censured at the time as having been built upon too extensive a scale, was the first Earthenware Manufactory in the Potteries *not covered with thatch*. In 1750 they erected an excellent and substantial dwelling-house adjoining their manufactory, which so far exceeded the other houses in the Potteries in point of size and elegance, that it then was, and now is, distinguished by the appellation of the "*Big House*;" and about the year 1763, these gentlemen retired from business in the possession of an ample fortune, the just and honourable reward of their industry and integrity.

About this time an improvement was made in the salt glaze by the united efforts of William Littlor and Aaron Wedgwood. Littlor had observed how nearly the White Stone Ware approached to Porcelain; and about the year 1750, he left Burslem, and commenced a Porcelain manufactory at Longton, near Stoke. He so far succeeded as to excite the astonishment of the potters; but it proved an unprofitable article, and the manufacture of it was discontinued. Littler sustained some heavy losses, and sold his estate at Brownhills, in the parish of Burslem. Specimens of this porcelain are preserved, which, in body, glaze, and enamel, may be considered as excellent for that day.

Aaron Wedgwood was a manufacturer of White Stone Ware, and having married Littler's sister, they united their experience, and made repeated attempts to improve the salt glaze. The result of these experiments was the addition of ingredients of easy fusibility with certain proportions of ground zaffer and the flint and clay of which the body of the ware was composed. In this compound, when in a liquid state, the ware was dipped, by which it imbibed a thin coat or covering of the liquid; and this, when exposed to the fire in the usual method of glazing with salt, produced a fine smooth glassy surface, quite free from the small inequalities which are observable on all the pieces of ware glazed with salt alone. Some excellent specimens of this improved species of glaze are preserved, which possess all the beauty of the finest lapis lazuli; and of others, from the admixture of a small proportion of manganese, which have the appearance of the finest Oriental lapis lazuli. These articles were further ornamented by enamelling, gilding, &c.

The following curious instruments or indentures, will be found interesting, as pointing out the importance which was formerly attached to an initiation into the art of manufacturing earthenware, and the price of labour when these agreements were executed.

"This Indenture, made the three and twentieth day of August, in the fifth year of the reign of our Soveraign Lord King George the Second over Great Brittaine, &c. Anno Dni. 1731, Between Ralph Wood, of Burslem, in the county of Stafford, miller, and Aaron Wood his son, of the one part, and Dr. Thomas Wedgwood, of Burslem afores^{d.} potter, of the other part, Wittnesseth that the s^{d.} Aaron Wood, of his own free will and consent, and to and with the direction and appointm^{t.} of his s^{d.} father, Hath put himselfe, and doth hereby put and bind himselfe apprentice unto the s^{d.} Dr. Thomas Wedgwood, the art, trade, mystery, and occupation of a potter to learn, that is to say, turning in the lath, handling, and trimming (throwing on the wheele being out of this indenture excepted), and with him the s^{d.} Dr. Thomas Wedgwood to worke from the eleventh day of November next, being Martinmas day, for during and until the full end and terme of seven years from thence next ensuing and following, and fully to be compleat and ended, during all which time and terme of seven years the s^{d.} Aaron Wood, as an apprentice to his sd. master, will and faithfully shall serve, his secrets shall keepe, his commands lawfull and honest every where shall do, the goods of his sd. mat. hee shall not inordinatly wast, nor them to any one lend, without his sd. mat. lycence, from the business of his sd. mat. hee shall not absent himselfe, but as a true and faithful servant shall, during the sd. terme of seven years, behave and demean himselfe towards his sd. master and all his. And the said Ralph Wood shall, during the sd. terme of seven years, find and provide for his sd. son all sorts of apparrell, whether linnen, woollen, or other, as also meat, drink, washing, and lodging, fitting and necessary for an apprentice to such trade as aforesd. And the sd. Dr. Thomas Wedgwood in consideration thereof, and of the sd. seven years service, doth hereby covenant, promisse, and agree, that hee, the sd. Dr. Thomas Wedgwood, shall and will, during the sd. terme of seven years, teach and instruct, or cause and procure to be taught and instructed, him, the sd. Aaron Wood, his sd. apprentice, in the businesse of the potting trade aforesd. so farr as turning in the lath, handling, and trimming, as much as thereunto belongeth, or the best way and method he can. And the sd. Dr. Thomas Wedgwood doth also promisse and engage to pay unto his sd. apprentice, the sd. Aaron Wood, for every weeke's worke done by the sd. apprentice in the first, second, and third year of his sd. apprentiship, the sum of one shilling weekly, of good and lawfull money of Great Brittaine, and for every weeke's worke done by the sd. apprentice in the fourth, fifth, and sixth year of his sd. apprentiship, the full sum of one shilling and sixpence, and for every weeke's worke done by the sd. apprentice, in the seventh and last year of his sd. apprentiship, the full and just sum of four shillings of lawfull money of Great Brittaine. And the sd. Dr. Thomas Wedgwood doth hereby further covenant, promisse, and agree that he, the sd. Dr. Wedgwood, shall, and will, over and above the weekely wages aforesd. give yearly to the sd. Aaron Wood, his sd. apprentice, one new paire of shoes during the sd. terme of seven years. In wittnesse whereof, the sd. parties aforesd. to these pr.sent Indentures have interchangeably put their hands and seales the day and year first above written :

"Ralph Wood.
"Aaron Wood.
"Dr. Thos. Wedgwood."

"*Sealed and delivered in the presence of*
"Sara × Wood.
her mark.
"Jos. Allen.

Articles of Agreement indented, made, concluded and agreed upon, the twenty-eighth day of September, in the Year of our Lord One Thousand Seven Hundred and Forty Three, and in the Seventeenth Year of the Reign of our Sovereign Lord King George the Second, over Great Britain and so forth, between Aaron Wood, of Burslem, in the County of Stafford, Earth-potter, of the one part, and John Mitchell, of Burslem aforesaid, Earth-potter, of the other part, as follows:

"*First*,—The said Aaron Wood, for the consideration hereunder mentioned, doth covenant, promise, and agree, to and with the said John Mitchell, his executors, administrators, and assigns, by these presents in manner following (that is to say) that he the said Aaron Wood shall and will, for and during the term and time of seven years, to begin and be accounted from the eleventh day of November next ensuing the date of these presents, abide and continue with the said John Mitchell, his executors, administrators, and assigns, as his and their hired and covenant servant, and diligently and faithfully, according to the best and utmost of his power, skill, and knowledge, exercise and employ himself, and do and perform all such service and business whatsoever relating to the trade of a earth-potter which he the said John Mitchell useth, as he the said John Mitchell shall from time to time during the the term aforesaid order direct and appoint, to and for the most profit and advantage of the said John Mitchell that he can, and shall and will keep the secrets of the said John Mitchell relating to the said trade or business, and likewise be just, true, and faithful to the said John Mitchell in all matters and things, and no ways wrongfully detain, embezzle, or purloin any monies, goods, or things whatsoever belonging to the said John Mitchell, but shall and will from time to time pay all monies which he shall receive or belonging to or by order of the said John Mitchell into his hands, and make and give up fair accounts of all his actings and doings in the said employment without fraud or delay, when and as often he shall be thereto required. And in consideration of the premises of the several matters and things by the said Aaron Wood to be performed as aforesaid, the said John Mitchell doth for himself, his executors and administrators, covenant, promise, and agree to and with the said Aaron Wood by these presents, that he the said John Mitchell shall and will well and truly pay or cause to be paid unto the said Aaron Wood, the sum of seven shillings of good and lawful money of Great Britain, by weekly payments, for every six days that the said Aaron Wood shall work with the said John Mitchell as aforesaid during the said term; and also shall and will well and truly pay or cause to be paid unto the said Aaron Wood the further sum of ten shillings and sixpence of like lawful money, upon every eleventh day of November yearly, during the said term: the first payment of the said sum of ten shillings and sixpence, shall be made on the eleventh day of November next ensuing the date hereof. And it is further agreed by and between the said parties to these presents, that the said Aaron Wood shall not be from the service of the said John Mitchell above two weeks in any one year during the said term. And that the said Aaron Wood shall not, and will not at any time or times during the said term, work for any other person or persons at the trade of a earth-potter, but the said John Mitchell, his executors, administrators, or assigns, upon penalty of paying to the said John Mitchell the sum of ten pounds of good and lawful money of Great Britain. And that the said Aaron Wood shall not have person or persons to work with him in the business that the said John Mitchell is to employ him in but himself only. In witness whereof, the said parties to these presents their hands and seals have hereunto put this day and year first above written.

"AARON WOOD.
"JOHN U MITCHELL.
his mark."

"*Sealed and delivered in the presence of*
"J. HENSHALL.
"ANN HENSHALL.

At this time, 5*s.* 6*d.* or 6*s.* was considered the full weekly wages of a journeyman potter, and it appears 7*s.* per week was considered sufficient for a man who was a modeller, and had the full management of the largest manufactory in the Pottery.

"In the year 1763, Mr. Josiah Wedgwood, who had already introduced several improvements into this art, as well with respect to the forms and colours of the wares, as the composition of which they were made, invented a species of earthenware for the table, of a firm and durable body, and covered with a rich and brilliant glaze, and bearing sudden vicissitudes of cold and heat, without injury. It was accompanied also with the advantages of being manufactured with ease and expedition, was sold cheap; and as it possessed, with the novelty of its appearance, every requisite quality for the purpose intended, it came quickly into general estimation and use. To this manufacture the Queen was pleased to give her name and patronage, commanding it to be called QUEEN'S WARE, and honouring the inventor by appointing him her Majesty's potter.

"It is composed of the whitest clays from Devonshire, Dorsetshire, and other places, mixed with a due proportion of ground flint. The pieces are fired twice, and the glaze applied after the first firing, in the same manner as porcelain. The glaze is a vitreous composition of flint and other white earthy bodies, with additions of white lead for the flux, analogous to common flint glass; so that, when prepared in perfection, the ware may be considered as coated over with real flint glass. This compound being mixed with water to a proper consistence, the pieces, after the first firing, are separately dipt in it; being somewhat bibulous, they drink in a quantity of the mere water, and the glaze, which was united with that portion of the water, remains adherent uniformly all over their surface, so as to become, by the second firing, a coat of perfect glass.

"To Mr. Wedgwood's continued experiments, we are indebted for the invention of several other species of earthenware and porcelain, adapted to various purposes of ornament and use. The principal are the following: 1. A TERRA COTTA; resembling porphyry, granite, Egyptian pebble, and other beautiful stones of the siliceous or crystalline order. 2. BASALTES, or Black Ware; a black porcelain buiscuit of nearly the same properties with the natural stone; striking fire with steel, receiving a high polish, serving as a touchstone for metals; resisting all the acids, and bearing, without injury, a strong fire, stronger indeed, than the basaltes itself. 3. WHITE PORCELAIN BISCUIT, of a smooth wax-like surface, of the same properties with the preceding, except in what depends upon colour. 4. JASPER; a white porcelain biscuit of exquisite beauty and delicacy, possessing the general properties of the *basaltes*, together with the singular one of receiving through its whole substance, from the admixture of metallic calces with the other materials, the same colours which those calces communicate to glass or enamels in fusion, a property which no other porcelain or earthen-ware body, of ancient or modern composition, has been found to possess. This renders it peculiarly fit for making cameos, portraits, and all subjects in *basso relievo*, as the ground may be of any particular colour, while the raised figures are of pure white. 5. BAMBOO, or cane-coloured biscuit porcelain. This possesses the same properties as the *White Porcelain Biscuit*, mentioned above. 6. A PORCELAIN BISCUIT, remarkable for great *hardness*, little inferior to that of agate. This property, together with its resistance to the strongest acids and corrosives, and its impenetrability by every known liquid, adapts it for mortars, and many different kinds of chemical vessels."*

The following description of the process used in manufacturing the earthenware, was communicated to Dr. Aikin, by a person on the spot. The practice has varied in but a trifling manner since that time. A piece of prepared mixture of clay and ground flint, dried and prepared to a proper consistence, is taken to be formed into any required shape and fashion, by a man who sits over a machine called a wheel, on the going round of which he continues forming the ware. This branch is called *throwing*; and, as water is required to prevent the clay sticking to the hand, it is necessary to place it for a short time in a warm situation. It then undergoes the operation of being turned, and is made much smoother than it was before, by a person called a turner; when it is ready for the handle and spout to be joined to it, by the branch called *handling*. Dishes, plates, tureens, and many other articles, are made from moulds of ground plaster; and, when finished, the whole are placed carefully in saggars, and taken to the oven. The ware, when removed from the oven, is called *biscuit*, and the body of it has much the appearance of a new tobacco-pipe, not having the least gloss upon it. It is then immersed or dipped into a fluid generally consisting of sixty pounds of white lead, ten pounds of ground flint, and twenty pounds of stone from Cornwall, burned and ground, all mixed together, and as much water put to it as reduces it to the thickness of cream, which it resembles. Each piece of ware being separately immersed or dipped into this fluid, so much of it adheres all over the piece, that when put into other saggars, and exposed to another operation of fire, performed in the glossing kiln or oven, the ware becomes finished by acquiring its glossy covering, which is given it by the vitrification of the above ingredients. Enamelled ware undergoes a third fire after its being painted, in order to bind the colour on.

A single piece of ware, such as a common enamelled tea-pot, a mug, jug, &c. passes through at least fourteen different hands, before it is finished: viz. the slipmaker, who makes the clay;—the temperer, or beater of the clay; the thrower, who forms the ware;—the ball-maker and carrier;—the attender upon the drying of it;—the turner, who does away its roughness;—the spoutmaker;—the handler, who puts to the handle and spout;—the first, or bis-

* Aikin's Manchester.

quit fireman; the person who immerses or dips it into the lead fluid; the second, or glass fireman;—the dresser, or sorter in the warehouse; the enameller, or painter; the muffle, or enamel fireman. Several more are required to the completion of such pieces of ware, but are in inferior capacities, such as turners of the wheel; turners of the lathe, &c. &c.

We cannot better conclude our account of this important Manufacture, than by giving a short biographical sketch of one, who, by his genius and exertions, mainly contributed to the accomplishment of that stupendous undertaking, *The Grand Trunk, or Staffordshire Canal*,—and with which, and the name of BRINDLEY, the best interests of the Staffordshire Potteries must ever be associated. Some account of this, and his other great works, is incorporated in the memoir of their immortal master.

JAMES BRINDLEY, a person of extraordinary genius for mechanical inventions, and distinguished by his skill in planning and conducting inland navigation, was born at Tunsted, in the parish of Wormhill, and county of Derby, in the year 1716. His parents possessed a small freehold, the income of which his father dissipated by field diversions, and by associating with persons above his rank; and thus the education of the son was wholly neglected. Urged by the necessities of his family, he contributed to its support till he was nearly 17 years of age, by that kind of labour which, in country places, is assigned to the poor; but at this age he bound himself apprentice to a millwright, whose name was Bennet, near Macclesfield, in Cheshire, and soon became expert in his business, discovering, at the same time, a singular genius for the extension and improvement of the mechanic arts in general. Accordingly, in the early period of his apprenticeship, he performed several parts of the business without any instructions from his master, and gave so much satisfaction to the millers who employed him, that he was always consulted in preference to his master or any other workman; and before the expiration of his servitude, when Mr. Bennet, by his age and infirmities, became unable to work, he carried on the business with reputation, and provided a comfortable subsistence for the old man and his family. About this time his master was employed in constructing an engine paper-mill, the first of the kind that had been attempted in these parts; but as he was likely to fail in the execution of it, Mr. Brindley, without communicating his design, set out on Saturday evening after the business of the day was finished, and having inspected the work, returned home the next morning, after a journey of 50 miles, informed his master of its defects, and completed the engine to the entire satisfaction of the proprietors. Mr. Brindley afterwards engaged in the mill-wright business on his own account, and soon acquired the reputation of a most ingenious mechanic. The fame of his inventions and contrivances was in a little while spread far beyond his own neighbourhood; and in 1752, he was employed to erect a curious water-engine at Clifton in Lancashire, for the purpose of draining coal-mines, which had before been performed at an enormous expence. The water for the use of this engine was conveyed from the river Irwell by a subterraneous channel, nearly 600 yards long, which passed through a rock; and the wheel was fixed 30 feet below the surface of the ground. In 1755, he constructed a new silk-mill at Congleton, in Cheshire, according to the plan proposed by the proprietors, after the execution of it by the original undertaker had failed; and in the completion of it he added many new and useful improvements. He introduced one contrivance for winding the silk upon the bobbins equally, and not in wreaths; and another for stopping, in an instant, not only the whole of this extensive system, in all its various movements, but any individual part of it at pleasure. He likewise invented machines for cutting the tooth and pinion wheels of the different engines, in a manner that produced a great saving of time, labour, and expence. He also introduced into the mills used at the Potteries in Staffordshire for grinding flint stones several valuable additions, which greatly facilitated the operation. In 1756, he constructed a steam-engine near Newcastle-under-Lyme, upon a new plan, which evinced his genius for invention and contrivance, for which he obtained a patent. The boiler was made with brick and stone, instead of iron plates, and the water was heated by fire-places, so constructed as to save the consumption of fuel. He also introduced cylinders of wood instead of iron, and he substituted wood for iron in the chains which worked at the end of the beam. But in this and similar contrivances for the improvement of this useful engine, he was obstructed by interested engineers; and his attention was diverted from the prosecution of them by the great national object of "Inland Navigation." In planning and executing canals his mechanical genius found ample scope for exercise, and formed a sort of distinguishing æra in the history of our country. However, envy and prejudice, and an attachment to established customs, raised a variety of obstacles to the accomplishment of his designs and undertakings; and if he had not been liberally and powerfully protected by the Duke of Bridgewater, at the commencement of the business, his triumph over the opposition with which he encountered must have been in a very considerable degree obstructed and retarded. The Noble Duke possessed an estate at Worsley, about seven miles from Manchester, rich in mines of coal, from which he derived little or no advantage, on account of the expense which attended the conveyance of this article by land-carriage to a suitable market for consumption. Fully apprized of the utility of a canal from Worsley to Manchester, he consulted Mr. Brindley on the subject; who, having surveyed the country, declared the scheme to be practicable. Accordingly his Grace obtained, in the years 1758 and 1759, an act of parliament for this purpose; and Mr. Brindley was employed in the conduct and execution of the undertaking, the first of the kind ever attempted in England, with navigable subterraneous tunnels and elevated aqueducts. At the commencement of the business it was determined, that the level of the water should be preserved without the usual obstructions of locks. But in accomplishing this object, many difficulties occurred; and it was soon found that it would be necessary to carry the canal over rivers and many deep vallies, and that it would not be easy to obtain a sufficient supply of water for completing the navigation. However, Mr. Brindley, patronized by the duke, and furnished with ample resources, persevered, and at length conquered all the embarrassments, occasioned by the nature of the undertaking, and by the passions and prejudices of individuals. Having completed the canal as far as Barton, where the river Irwell is navigable for large vessels, he proposed to carry it over that river, by an aqueduct 39 feet above the surface of the water. This was considered as a chimerical and extravagant project; and an eminent engineer, who was consulted on the occasion, ridiculed the attempt. "I have often heard," said he, "of castles in the air, but never before was shewn where any of them were to be erected." The Duke of Bridgewater was not discouraged; but confiding in the judgment of Mr. Brindley, empowered him to prosecute the work; and in about ten months the aqueduct was completed. This astonishing work commenced in Sept. 1760, and the first boat sailed over it the 17th of July, 1761. The canal was then extended to Manchester, where Mr. Brindley's ingenuity in diminishing labour by mechanical contrivances, was exhibited in a machine for landing coals upon the top of a hill. It is no wonder, that an object, so curious in itself, and of such national importance, should have attracted general attention.

The Duke of Bridgewater, having found by experience the utility of these inland navigations, extended his views to Liverpool; and obtained, in 1762, an Act of Parliament for branching his canal to the tide-way in the Mersey. This part is carried over the river Mersey and Bollan, and over many wide and deep vallies. Over the vallies it is conducted without a single lock; and across the valley at Stretford, through which the Mersey runs, a mound of earth, raised for preserving the water, extends nearly a mile. In the construction of this mound Mr. Brindley displayed his mechanical genius, by rendering the canal itself subservient to his design, and by bringing the soil, necessary for his purpose, along the canal in boats of a peculiar form, which were conducted into caissoons or cisterns; so that on opening the bottoms of the boats, the earth was deposited where it was wanted, and the valley was thus elevated to a proper level for continuing the canal. Across the Bollan the ground was raised by temporary locks, formed of the timber used in the construction of the caissoons just mentioned. In the execution of every part of the navigation, Mr. Brindley displayed singular skill and ingenuity; and in order to facilitate his purpose, he produced many valuable machines. His economy and forecast, in every part of the work, deserve to be particularly noticed, and they are peculiarly discernible in the stops or floodgates, that are fixed in the canal, where it is above the level of the land. These stops are so constructed, that if any of the banks should give way and occasion a current, the adjoining gates will rise merely by that motion, and prevent any other part of the water from escaping, besides that which is near the breach between the two gates.

Encouraged by the success of the Duke of Bridgewater's undertakings, a subscription was entered into by a number of gentlemen and manufacturers in Staffordshire, for constructing a canal through that county, in pursuance of a scheme which had been suggested some years before; and Mr. Brindley was engaged to make a survey from the Trent to the Mersey. Upon his report that a canal for connecting the two rivers was practicable, application was made to Parliament, in 1765, for an Act to this purpose, which was obtained in the same year.

This canal, which, by its planner, was ingeniously termed the *Grand Trunk*, (in allusion to the main artery of the body, from whence branches are sent off for the nourishment of the distant parts) and which is commonly known by the name of the *Staffordshire Canal*, takes its course from north-west to south-east, across the county of Chester, and thence across Staffordshire beyond its middle, when, turning short in a north-eastern direction parallel to the Trent, it accompanies that river into Derbyshire, and enters it near the place where the high road from Derby to Leicester crosses the Trent over a bridge, substituted to the former Wilden-ferry. In length it is ninety-three miles. Its fall of water from its greatest elevation at Harecastle-hill, is 326 feet on the northern side, and 316 on the southern; the former effected by thirty-five locks, the latter by forty. Six of the most southern locks are fourteen feet wide, adapted to the navigation of large barges, and one of the northern is of the same width. The common dimensions of the canal are twenty-nine feet breadth at the top, sixteen at the bottom, and the depth four feet and a half; but in the part from Wilden to Burton, and from Middlewich to Preston-on-the-Hill, it is thirty-one feet broad at the top, eighteen at the bottom, and five and a half deep. The canal is carried over the Dove in an aqueduct of twenty-three arches, the ground being raised to a considerable height for the space of a mile and two furlongs. Over the Trent it is carried by an aqueduct of six arches of twenty-one feet span each; and over the Dane, on three arches of twenty feet span. There are besides near 160 lesser aqueducts and culverts for the conveyance of brooks and small streams. The cart bridges erected over it are 109; the foot bridges eleven.

For the sake of preserving a level as much as possible, the hills and elevated grounds in the course of the canal have been pierced by five tunnels. Of these, that through the mountain at Harecastle is the principal, and has proved a work of vast labour and expense, in consequence of unforeseen difficulties. Its length is 2880 yards, with a width of nine feet, and a height of twelve, lined and arched with brick; and it runs more than seventy yards below the surface of the earth. The other tunnels are at Armitage 130 yards; at Barnton, in Great Budworth parish, 560 yards; at Saltenford, in the same parish, 350 yards; and at Preston-on-the-Hill, 1241 yards. Each of these is seventeen feet four inches high, and thirteen feet six inches wide. The boats employed upon the canal carry about twenty-five tons, and are drawn by one horse. The tonnage paid to the proprietors for the liberty of navigating is three-halfpence per mile. This great work was begun on July 17th, 1766. It was carried on with great spirit by Mr. Brindley while he lived, and was finished by his brother-in-law, Mr. Henshall, who put the last hand to it in May 1777.

The next object which engaged the attention of Mr. Brindley was the construction of a canal from the Grand Trunk, near Haywood in Staffordshire, to the river Severn near Bewdley, by means of which the port of Bristol was connected with the ports of Liverpool and Hull. This canal, about 46 miles in length, was completed in 1772. His next undertaking was the survey and execution of a canal from Birmingham, which should unite with the Staffordshire and Worcestershire Canal near Wolverhampton. This navigation, which is 26 miles in length, was finished in about three years. Our engineer advised the proprietors, in order to avoid the inconvenience of locks, and for the more effectual supply of the canal with water, to have a tunnel at Smethwick; but his advice was disregarded; and the managers have since been under a necessity of erecting two steam-engines. The canal from Droitwich to the river Severn, for the conveyance of salt and coals, was executed by Mr. Brindley; and he also planned the Coventry navigation, which was for some time under his directions; but a dispute arising about the mode of executing it, he resigned his office. Some short time before his death, he began the Oxfordshire canal, which, uniting with the Coventry canal, serves as a continuation of the Grand Trunk navigation to Oxford, and thence by the Thames to London. The last undertaking, in which Mr. Brindley engaged, was the canal from Chesterfield to the river Trent at Stockwith. He surveyed and planned the whole, and executed some miles of the navigation, which was finished by his brother-in-law, Mr. Henshall, in 1777. Such was the established reputation of Mr. Brindley, that he was consulted on a variety of occasions; and, indeed, few works of this kind were undertaken without his advice.

When difficulties occurred in the execution of any of his works, Mr. Brindley had no recourse to books, or to the labours of other persons; but as all his resources were those of his own inventive mind, he generally retired to bed, and lay there one, two, or three days, till he had devised the expedients which he needed for the accomplishment of his objects. He then got up and executed his design without any drawing or model, which he never used, except for the satisfaction of his employers. His memory was so tenacious, that he could remember and execute all the parts of the most complex machine, provided he had time, in his previous survey, to settle, in his mind, the several departments, and their relations to each other. In his calculations of the powers of any machine, he performed the requisite operation by a mental process, in a manner which none knew but himself, and which, perhaps, he was not able to communicate to others; and after certain intervals of consideration, he noted down the result in figures; and then proceeded to operate upon that result, until at length the complete solution was obtained, which was generally right.

Some have said that Mr. Brindley had been so much neglected in early life, as never to have learned to read and write; and it has also been affirmed, that his aspect was that of an ideot, and his language mean, obscure, and almost unintelligible. But neither of these facts is well founded. Although he read little, and wrote less, he had frequent occasion for correspondence with his friends and employers. His countenance was sensible and animated; but as he was unostentatious in his outward appearance, his dress was always plain. His conversation, on occasions of importance, and among his intimate friends, was instructive and interesting; and enlivened by the singular genius, and the benevolent and patriotic spirit for which he was distinguished, and which gave energy to his pursuits. His want of literature, indeed, compelled him to cultivate, in an extraordinary degree, the art of memory; and in order to facilitate the revival, in his mind, of those visible objects and their properties, to which his attention was chiefly directed, he secluded himself from the external impressions of other objects, in the solitude of his bed. Dr. Wallis, who was eminently distinguished by the tenaciousness of his memory, so that he could extract the cube root of any number to 100 places of figures in his mind, availed himself, for this purpose, of a similar seclusion. Mr. Brindley died of a diabetes, in consequence of sleeping in a damp bed at Ipstones, on the 27th of September, 1772, in the 56th year of his age, at Turnhurst, in the parish of Wolstanton, Staffordshire. He was buried at New Chapel, in this county; and which is a chapel of ease to Wolstanton church.

The talents of Mr. Brindley were of a singular kind; and under the patronage of his Grace the Duke of Bridgewater, they had an opportunity of being unfolded and exercised to their full extent, in the execution of works new to this country, and which will perpetuate his fame to future generations. The brief recital which has been given of his stupendous undertakings, will enable the reader to form some judgment of this extraordinary person. Such was the enthusiasm with which he engaged in all schemes of inland navigation, that he seemed (if we may credit report) to regard all rivers with contempt, when compared with canals. To this purpose, it is said, that in an examination before the House of Commons, when he was asked by a Member, for what purpose he apprehended rivers were created? he replied, after some deliberation, "to feed navigable canals." As to his private character, those who knew him well, and had an opportunity of conversing familiarly with him, respected him in a high degree "for the uniform and unshaken integrity of his conduct; for his steady attachment to the interest of the community; for the vast compass of his understanding, which seemed to have a natural affinity with all grand objects; and, likewise, for many noble and beneficial designs, constantly generating in his mind; and which the multiplicity of his engagements, and the shortness of his life, prevented him from bringing to maturity."

ADDENDA.

Having noticed several Wells, from p. [170] to [174,] which were in former times remarkable in this county, for their medicinal properties, and being holy; we insert the following extract from Mr. W. S. Mason's Statistical Account or Parochial Survey of Ireland, (vol. 1. p. 328.) with a view of illustrating in some degree the superstitious devotion called "well-worship," p. 174. It affords some details of the rites observed on such occasions which might not be generally known, and may afford some amusement to the reader. It occurs, in the statistical account of the parish of Dungiven, in the diocese of Derry, and county of Londonderry, which was communicated by the Rev. Alexander Ross, rector.

"The only Saint particularly respected in this parish is St. Patrick; every remain of antiquity, as the church, the upright stone, &c. is ascribed to him as its author.

"There is a fine spring well, between the town and the old church, called *Tubber-Patrick*, which is the object of peculiar veneration. The absurd and superstitious ceremonies which are here practised to this day, would scarcely be credited without ocular testimony. Early almost every Sunday morning, from Patrick's-day until about Michaelmas, a number of devotees surround this well, and after bowing towards it with great reverence, walk round it a certain number of times, repeating, during their progress, a stated measure of prayer, they then wash their hands and feet with the water, and tear off a small rag from their cloaths, which they tie on a bush overhanging the well—from thence they all proceed to a large stone in the river Roe, immediately below the old church, and having performed an ablution, they walk round the *stone, bowing* to it, and repeating prayers as at the well. Their next movement is to the old church, within which a similar ceremony goes on, and they finish this rite, by a procession and prayer round the upright stone already described.

"How strange that this idolatrous monument of Odin, should still be substituted for the pure sanctuary and spiritual worship of the living Jehovah."

The writer adds, by way of note, that something of the kind, in regard to *bowing* to the stone, is alluded to by the prophet Isaiah (ch. 57. v. 6.) and that it is some consolation to observe, that the dislike to these ceremonies is not confined entirely to Protestants.

The Rev. C. O'Connor, a zealous Roman Catholic, alludes with sufficient contempt to them, when speaking of the massacre in 1641, he boldly attributes it to "the Druidical and well-worshipping excommunications and miracles, by which the people were hoodwinked."—See his Historical Address, published in 1812, p. 244.

In the account of Aghaboe, in Queen's county, and diocese of Ossory, p. 42. by the Rev. Edward Ledwick, LL.D. we have the following account of another well:

"*St. Canice's Well* was in my orchard in Aghaboe; there the pedlars laid down their packs and said their prayers.

"About a quarter of a mile distant from the town was a groupe of *thorn bushes*, called after St. Canice; there, likewise, the poor people performed their devotions.

"These were heathen practices, derived from the earliest ages. At the citadel of Athens stood an olive-tree planted by Neptune and Minerva, and held very sacred. The fathers and councils reprobated such gross superstition; but it was difficult to eradicate it, as among the uninformed it is to be found at present."

We trust that we need not apologize for the introduction of any matter that relates to Public Charities. We wish our information on that part of local history had been more extensive, as it cannot fail of being interesting to many. There is no charity, which, in our opinion, more deserves the attention and support of the affluent and pious than that established at Stafford in 1771, "for the relief of widows and orphans of poor clergymen, &c." and we here subjoin a copy of part of a printed statement sent to the subscribers in 1816, in respect of the chief rules and orders thereof, and the forms of the certificates required by the governors. Such information we hope will be peculiarly acceptable to such of the clergy as are not subscribers; and by adding greater publicity to the charity, will induce some, to whom God has given abundance, to promote and increase the benefits thereof by adding their own contributions, and exerting their influence on its behalf. No more solemn injunction is given in the Bible than the following: "Take heed to thyself, that thou forsake not the Levite as long as thou livest upon the earth." (Deut. xii. 19.)

RULES AND ORDERS.

1. The widows of clergymen so continuing, and their children unprovided for, or not in a way of providing for themselves, are the persons to be relieved by this Society,—or such clergymen, as

shall be disabled by age, sickness, or infirmity, and have not income sufficient for the necessary support of themselves and their families.

2. That the children of necessitous clergymen do, in the life-time of the said clergymen, so far partake of this charity, as to be apprenticed at proper ages.

3. That no widows or orphans be relieved, but those of such clergymen, as at the time of their death or incapacity, were possessed of some ecclesiastical preferment, or *licensed* to a curacy within the archdeaconry, the several peculiars, or county of Stafford.

4. That no person be relieved, until there be a certificate given in, and repeated yearly under the hands of three governors, (or if the person be removed out of the county, of three clergymen living in the neighbourhood of the place, where such person resides) setting forth the age, circumstances, and residence of the person, and that she, or he, is of a sober life and conversation, a member of the Church of England as by law established, (and if a widow) how many children she hath, and of what age they are.

5. That every clergyman subscribing any sum to this charity, and every lay person subscribing one guinea, be a governor and director of this charity for that year; and that every benefactor by a donation of twenty guineas, be a governor and director for life.

9. That the business of the Stewards be more particularly to *promote the charity in their respective districts, both among the clergy and laity, to receive subscriptions, to collect arrears,* and to pay them to the treasurer on or before the annual meeting of the governors.

10. That a general meeting of the governors of this charity be annually held at STAFFORD, on the Thursday next after the twenty-fourth day of June, of which notice is to be given to each district in the Stafford and Birmingham papers.

11. That at this annual meeting applications are to be delivered in and considered, and orders made for the disposal of the money contributed; reserving however a full and sufficient sum in balance to answer exigencies; at the same time the treasurer's accounts for the preceding year are to be received and audited.

FORMS OF CERTIFICATES.

CLERGYMAN'S CERTIFICATE.

"We, whose names are underwritten, do testify that in the Archdeaconry, or county of Stafford, is by disabled from officiating, and has not an income sufficient for the necessary support of himself and his family, the whole not exceeding* and now resides at"

WIDOW'S CERTIFICATE.

"We, whose names are underwritten, do testify that widow of late of in the Archdeaconry, or county of Stafford, is unmarried, aged years, and hath children, of the respective ages of is of a sober life and conversation, a member and regular communicant of the Church of England; has no sufficient maintenance, stands in need of relief, and has not an income in any way of more than* and that her husband was at the time of his death, possessed of some ecclesiastical preferment, or licensed to a curacy, within the Archdeaconry or county of Stafford; and now resides at"

ORPHAN'S CERTIFICATE.

"We, whose names are underwritten, do testify that aged is unmarried, unprovided for, and in no way of providing for self, behaves decently and soberly, and frequents the service of the Church of England, and has not an income in any way of more than* and that his (or her) father was at the time of his death possessed of some ecclesiastical preferment, or licensed to a curacy, within the Archdeacoury or county of Stafford, and now resides at"

"N. B. It is expected that these forms will be adhered to, otherwise the certificate will not be admitted.

"☞ The Governors of this Charity expect, at their next and all future annual meetings, an acknowledgment from every person relieved, specifying the sum.—The acknowledgment may be made, if the application is continued, at the foot of the fresh certificate.

"All persons who are disposed to contribute by will to this charity, are desired to do it in the following manner:

"I give and bequeath to A. B. and C. D. the sum of****, upon trust, that they pay the same to the Treasurer for the time being of the Charity for the Relief of Widows and Orphans of Poor Clergymen, within the Archdeaconry and county of Stafford, which sum I charge upon my personal estate, and desire it may be applied to the use of the said charity."

The total of receipts in 1816, was £426. 4*s*. 4*d*. and of disbursements £426. 2*s*. 8*d*. making a balance in the hands of the treasurer of only *one shilling and eightpence.*

* Here insert the particulars of the income (if any) and how it arises.

In looking over the list of those relieved by the charity, we observe the allowance to each varies from £10. to £36. and the subscriptions vary from 10*s*. 6*d*. to £5. 5*s*. each.

In proportion as the benevolence of the charitable is extended and increased, this and such like charities may be improved.

In the diocese of Derry, in Ireland, is an excellent charity for Clergymen's widows. It is thus noticed in Mason's Statistical Account of Ireland, (vol. i. p. 601.)

"The rector of this parish [Maghera] pays eight pounds a-year, and the curate fifteen shillings, to this excellent institution, by which the widows of those who pay one per cent. per annum of their benefices or curacies, as rated some years ago, are entitled to an annuity of thirty-five pounds a-year, and six of them are accommodated with houses in the city of Derry, rent-free, and kept in repair." In looking over the list of subscribers there recorded, we perceive the subscriptions vary in the first class from £1. 10*s*. to £100. which latter sum appears opposite the name of the Lord Bishop of Derry. The curates' subscriptions in the second class are from 15*s*. to 18*s*. each. In 1813, the fund amounted to £3746. 7*s*. 5¼*d*.

The account concludes thus: "This fund is progressively increasing, and holds forth a bright example to all other dioceses in the British Empire."

Our sincere wish is that the Staffordshire Nobility, Clergy, and Gentry, will enable the Secretary to the Stafford Charity to make a similar report of its "progressive increase."

PECULIARS TO WHICH NO DIOCESE LICENCE IS TO BE DIRECTED.

DERBYSHIRE.

DEAN.

Bakewell,
Ashford,
Beighley,
Longson,
Fairfield,
Chapel-en-le-Frith,
Baslow,
Tiddeswell,
Hope,
Wormhill,
Chelmorton,
Buxton,
Taddington,
Moniash,
Kniveton,
Sheldon.

SAWLEY PREBEND.

Sawley,
Risley,
Wilne,
Longeaton,
Breason,
Calk,
Dale Abbey,
Hartington,
Sterndale,
Peak Forest.

STAFFORDSHIRE.

Cannock,
Rugeley,
Areley,
Brewood,

St. Mary,
St. Chad, al's Stow, } *Lichfield*
St. Michael,
Hammerwich,
Farewell,
Baswick,
Acton Trussell,
Bednall,
Whittington,
Alrewas,
Bromley Regis,
Edingale,
Maveson Ridware,
Pipe Ridware,
Hansacre,
Armitage,

LORD OF THE MANOR.

Hints,
Norton,
Eccleshall,
Charnes,
Millmeece,
Gerards Bromley,
Chadskill,
Chorlton,
Offley,
Colwich,
Fradswell,
Adbaston,
Longdon in com. Stafford,
Weeford,
St. Chad, Stafford,
Tipton,
Prees in com. Salop,
Penkridge,
Copnall,
Sharehill,
Dunston,
Gnosall,
Wolverhampton,
Pelsall,
Willenhall,
Bilston,
Tettenhall,
Codsall,
Oakover.

WARWICKSHIRE.

Harbourne,
Edgbaston,
Itchington, }
Chadshunt, }
Gayton, }
Tachbrooke,
Ufton, al's Oloughton,
Packwood,
Knowl,
Balsall,
Barston,
Merrivale,
Badsley Clinton.

SHROPSHIRE.

St. Mary, Salop,
Cliff,
Ashley,
Albrighton,
St. Mary Magdalen, } *Bridgnorth,*
St. Leonard,
Claverley,
Bubbington,
Alveley,
Quatford,
Buildwas,
Wombridge,
Longdon upon Ternhill com. Salop,
Uppington.

The manor of Hilderstone, on the east side of Stone, belonged to the Hugglesfords, in the reign of King Edward the 3d, after which it became the estate of Sir John Delves, by whose daughter it went in marriage to the family of Sir Robert Sheffield, Recorder of London. It became afterwards the property of Sir Gilbert Gerrard (afterwards Lord Gerrard) Master of the Rolls, and is now the property and residence of Ralph Bourne, Esq.

We have already noticed the village of GREAT HAYWOOD. It is a township situated in the parish of Colwich, and hundred of Pirehill (South) with a population of 700 persons. The manor of Haywood was given by Roger de Meland, in the time of Henry the Third, to Roger de Aston, a descendant of whom carried it by marriage to the Tixals. It afterwards came into possession of the Whitbies, but was subsequently re-united to the house of Tixall, by purchase. The barn belonging to the manor-house was of a most magnificent size, but has been greatly reduced. The horse-bridge over the Trent, adjoining to Haywood, was formerly not less remarkable for extraordinary dimensions. Mr. Pennant says he remembered it to have consisted of two-and-forty arches. Magna Britannia (vol. 5. p. 87) says, "it is longer than any bridge in England, having near forty arches; yet much shorter than the bridge over the Drave, at Essec, which Dr. Brown tells us is at least five miles long, and made all of wood." The tradition is that it was built by the county, in compliment to the last Devereux, Earl of Essex, who resided much at Chartley; and being a keen sportsman, was often deprived of his diversion for want of a bridge. Mr. Pennant, who states this, says he was not clear about the truth of this report, and adds "then there certainly had been a bridge here long before; so that if there was any foundation for such a mark of respect, it could only have been rebuilt, after falling to decay.*

The chief reason for the notice which we have just given of Haywood, and of this bridge, is that we might more conveniently introduce the beautiful Vale of Shugborough, which from the middle of the bridge is seen to great advantage. This vale, varied with almost every thing that nature and art could give to render it delicious, is watered by the Trent and Sow. The first, to use the words of the author last quoted, animated with milk-white cattle, emulating those of Tinian, the last with numerous swans. The boundary on one side is a cultivated slope; on the other, the lofty front of Cannock-wood, clothed with heath, or shaded with old oaks, scattered over its glowing bloom by the free hand of nature. It is more difficult, continues Mr. Pennant, to enumerate the works of art dispersed over this Elysium: they epitomize those of so many places. The old church of Colwich; the mansion of the ancient English baron at Wolseley Hall; the great windowed mode of building in the reign of Queen Elizabeth, in the house of Ingestre;

* Journey from Chester, p. 90.

the modern seat in Oak-edge; and the lively improved front of Shugborough; are embellishments proper to our own country. Amidst these rise the genuine architecture of China, in all its extravagance; the dawning of the Grecian, in the mixed Gothic gate-way of Tixal; and the chaste buildings of Athens, exemplified by Mr. Stuart, in the counterparts of the Chocagic monument of Lysicrates,* and the octagon tower of Andronicus Cyrrhestes.† From the same hand arose, by command of a grateful brother, the arch of Adrian of Athens, embellished with naval trophies, in honour of Lord Anson, a glory to the British fleet. Pennant's friend, as he informs us, the late Thomas Anson, Esq. preferred the still paths of private life, and was every way qualified for its enjoyment: for with the most humane, and the most sedate disposition, he possessed a mind most uncommonly cultivated. He was the example of true taste in this county; and, at the same time that he made his own place a paradise, made every neighbour partaker of its elegancies. He was happy in his life, and happy in his end. The improvements which he began were carried on, with great judgment, by his nephew and successor George Anson, Esq. He was father to the present proprietor, who was created a peer of Great Britain, February 17th, 1806.

The improvements at Shugborough have been farther carried on by his Lordship, the house having been enlarged, and a handsome portico added to it. The highly cultivated state of the demesne marks the laudable agricultural taste of the noble owner.‡

Of the great number of statues which embellish the place, an Adonis and Thalia are the most capital. There is also a very fine figure of Trajan, in the attitude of haranguing his army. The number of which Etruscan figures in the garden shew the great antiquity of the art of sculpture in Italy, long before the Romans became a people. The beautiful monument in the lower end of the garden does honour to the present age. It was the work of Mr. Schemecher, under the direction of Thomas Anson, Esq. just mentioned. The scene is laid in Arcadia. Two lovers, expressed in elegant pastoral figures, appear attentive to an ancient shepherd, who reads to them an inscription on a tomb: "ET IN ARCADIA EGO!"

The moral of this seems to be, that there are no situations of life

* Antiquities of Athens, Chap. IV. tab. 1, 3.

† Ib. Chap. III. tab. 1, 3.

‡ See the account of Shugborough Farm and Demesne, p. 90. *dep.* 2.

so delicious from which death will not at length snatch us. It was placed here by the amiable owner, as a memento of the certainty of that event. Perhaps, also, as a secret memorial of some loss of a tender nature in his early days: for he was wont often to hang over it in affectionate and firm meditation.* The Chinese house, a little farther on, is a true pattern of the architecture of that nation, taken in that country, by Sir Percy Brett.

Opposite to the back-front of the house, on the banks of the Sow, stand the small remains of the ancient mansion, which, according to Leland, originally belonged to "Suckborrow with a long beard," who, as some say, gave it to the mitre of Lichfield and Coventry.† It must have been in very early times; for the manor of Haywood, just mentioned, (in which this is included,) belonged to that See in 1086, the twentieth of William the Conqueror, and so continued till the reign of Edward VI. who gave it to Lord Paget. The house, before that time, was an episcopal palace. The remains still standing, serve to give the appearance of reality and ruin to some beautiful Grecian columns, and other fragments of ancient architecture; which were added to the front by Thomas Anson, Esq.‡

The most powerful machinery in the world for raising water, is employed in Bilston coalpit-field, and in that neighbourhood, the property of the Birmingham Canal Company. They have there erected a double steam engine of twice sixty-horse power, for pumping the subterraneous mine water into their canal. This machinery put in motion the beginning of summer, when the natural supply of water begins to slacken, will, in about three or four months, entirely exhaust and completely empty this reservoir beneath the earth; and some considerable effect may be expected, the strength of 120 horses being employed; they are thus enabled to keep their canal brim-full during the droughts of the driest summer, this subterraneous reservoir losing nothing by evaporation in hot weather.

At Ocker-hill, not far distant from this, is another tier of powerful steam-engines, but of inferior force to the preceding, which raises the water from a lower level of the canal, about sixty feet rise into the main summit; so that their water, after supplying a tier of nine or ten locks, is returned to its original source: and far-

* Pennant, 93. † Magna Brit. V. 89.

‡ Pennant, 94.—For an account of this noble family, see "Peerage," in this Work.

ther to supply this lower level, in dry seasons, they have an engine at Broadwaters, near Wednesbury, to raise the subterraneous water of Wednesbury field into their lower level.

The whole of this hydraulic machinery has, we suppose, cost £50,000. and probably pays 20 per cent. per annum, by enabling the canal proprietors to do as much extra business as will return £10,000. per annum in tonnage.

At the same place they have also some other curious powerful machinery; as 1st, "the Neptune and Vulcan," a floating engine, which will pass through the locks to any part of the canal, and by making a stark or bar of pile plank, connected with the engine across the canal, will pump a mile, or any moderate length, of the canal water in a few hours over the said bar, by which means any foul part of the canal may be laid dry and cleansed out, and the water restored without loss after the operation. The water is here raised by paddles fixed in an horizontal shaft, working in a segment of an horizontal hollow cylinder; the paddles in successive circular motion, scooping or lifting the water over the bar, and raising it no more than is just necessary for its flowing off; by which means no waste or loss of power is sustained, the surface water requiring to be raised only a few inches, and that from the bottom only the depth of the canal. The motion is effected by a steam engine, and the machine requires no more attendance than a canal boat, and is capable of raising the greatest quantity of water to a small lift, of any machine ever constructed.

2d, The Mud Chain Pump, a floating machine fixed to a canal boat or barge, for cleansing the bottoms of canals or rivers, without drawing off the water at all. A number of semicircular iron buckets or scrapers, are fixed to a flattened iron chain, the chain being circular or endless, and drawn round iron rollers at either end, one end being higher and the other lower; a circular motion is impelled by a steam engine, and the scrapers in rotation, drawn or forced along the canal bottom, fill their bucket ends at one scoop; circular holes in these buckets discharge the water, and at the upper roller they are inverted, and discharge their contents in rotation into the boat. The whole machine may be moored or moved at pleasure, and is said to be capable of cleansing out 20 tons of mud in less than an hour, with the assistance of two or three men: the machine is then floated to any convenient place to discharge its contents.

Ancient Coins.—On Friday the 28th of March, 1817, two Urns, containing a great quantity of Roman copper coins, were turned up by the plough on a farm called Little Madeley Parks, in this county, the property of the Right Honourable Lord Crewe, and in the occupation of Mr. Joseph Taylor. The urns were unfortunately destroyed, but many of the coins are in excellent preservation, particularly one of Dioclesian, in the possession of James Poole, Esq. of Finney Green, who has favoured us with the following description of several of the most perfect specimens amongst them. The pieces of Constantine are also very fine. An ancient horse-shoe, and a curious key, were likewise discovered at the same time, and are preserved by Mr. Thomas Mayer, jun. of Newcastle:

Obverse.	*Reverse.*	*Exergue.*
D N Dioclesian O P F S Aug	Quies Augg	P L C
Imp Licinius P F Aug	Genio Pop Rom	P T R
Constantinus Aug	Virtus Exercit Vot. XX.	P T R
Imp Constantinus P F Aug	Principi Juventutis	P T R
Constantinus P F Aug	Marti Conservatori.—[The head of a warrior, with a helmet on.]	
Constantinus	Beata Tranquillitas.—[A Globe on an Altar, and three Stars above the Globe, meant to signify the world preserved by the Gods for the three sons of Constantine 1st.]	S T R
Constantinus Max Au	Victoria	P L N
Imp Constantinus P F Aug	S P Q R Optimo Principi	
Imp Maximinus P F Aug	Genio Pop Rom	P T R
Crispus Nob Caes	Soli invicto Comiti	P L N
Crispus Nobil C	Beata Tranquillitas.—[An Altar inscribed "Votis XX."]	P L O N
Constantinus Jun Nob Caes	Claritas Reipublicae	P T
Constantinus Jun Nob C	Beata Tranquillitas	S T R
Constantinus Jun Nob C	Beata Tranquillitas.—[An Altar inscribed "Votis XX."]	
F L C L Constantinus Jun N C	Principi Juventutis	
Same Obverse	Soli invicto Comiti	P L N
Imp Constantinus P F Aug	Soli invicto Comiti	P L N

Natural History of Blymhill.—The following additions to Blymhill were communicated by the Rev. S. Dickenson.

Papaver Dubium (long smooth-headed poppy).—This is our

common corn poppy, the *papaver rhœas* not being found in Staffordshire.

Salix Russellina (Bedford willow).—This species having the property of fragility in its small branches, which being struck with a finger break off at the shoot of the present; it was generally mistaken for the *Salix Fragilis* of Linnæus, which has not been found in this county. The error was first discovered by his Grace the Duke of Bedford, in honour of whom the species is termed Russellina. Its cultivation is much increased from the rapidity of its growth, and the value of its timber being of easy cleft: it is greatly esteemed for making hurdle bars.

Salix caprea (great round-leaved willow).—Flowers at the beginning of April, the earliest of the willow kind. The catkins are vulgarly and not inaptly called *goslings*, from their striking similarity of colour, and their appearance precisely at the time when goslings are hatched. It is very common, of both sexes, of spontaneous growth.

Salix aquatica (water willow).—The provincial name is Black Withy: this is by far the most common of all our willows, though not noticed by Linnæus. It appears to have been mistaken by him for Salix cinerea, to which, Dr. Smith observes, it approaches very nearly. The wood is greatly esteemed by wheelwrights to make waggon staves: that of Salix caprea is likewise used for the same purpose.

Salix aurita (round-eared sallow).—Inferior in size to Salix aquatica. In hedge-rows in wet soils at Blymhill, very common of both sexes.

At a recent meeting of the Bath Philosophical Society, Dr. Wilkinson gave the analysis of a very strong Saline Spring, situated in the neighbourhood of Stafford. Sixteen ounces of the water contained 214 grains of saline matter, dried at the temperature of 180 degrees. One hundred grains, by analysis, yielded

14 grains	muriate of magnesia,
2	sulphate of lime,
6	sulphate of magnesia,
78	muriate of soda.

The Wedges Mill, and Church Bridge Manufactory, of late years established by Mr. Gilpin, is a considerable addition to the commercial and manufacturing interests of the county: some good houses, and many tenements and workshops, have been erected at the places named in Wyrley liberty. The Wyrley Bank Colliery has been opened, worked by royalty, and raising 10,000 ton of coals annually, employs about thirty families. Attached, is a manufactory employing about fifty workmen. The coals are conveyed on a rail-way, a considerable distance, to the turnpike road for sale and use. The manufactory consists of augers, and edge and plantation tools, which are now in full work, notwithstanding the depression of the times. The screw auger, a modern tool, is made in great perfection; it enters and perforates the hardest wood with facility without any preparation. A steam-mill is in full work at Church Bridge, in the manufactory; and a water-mill on the Hedgford river, called Wedges Mill, is also attached to it, and the whole appears a thriving concern, established within about the last twenty years.

Addition to the Natural History of Insects.—The following, though apparently whimsical, is from high authority. Dr. Hirsch, dentist at the court of Vienna says, the lady-bird (*Coccinella septem punctata*) will cure the tooth-ache, thus: take the living insect, and bruise it between the fingers, then rub these fingers till they become warm at the points, and touch with them the unsound part of the gum, also the diseased tooth; this he did with the happiest effect. A few days after he repeated the experiment with equal success, without bruising a new insect. The following insects have, it is said, the same virtue, used in the same way: the poplar beetle (*Chrysomela populi*) length half an inch, on poplar trees; the wood beetle (*Chrysomela sanguinolenta*) wings red, length five lines, under side blue, shells black with a red border; found in woods; also the *curculio* and *carabus* species, found on various plants. Charcoal is also recommended, from high authority, as a preventative and cure of the same complaint.

Conformably to an ancient custom of the Manor of Newcastle-under-Lyme, within the Duchy of Lancaster, it is essential that the *original* wills of copyhold tenants, dying seised of copyholds, should be produced and proved in the Manor court, within a limited time after the decease of the testator, or testatrix, in default of which the estate is liable to forfeiture.

We subjoin the following observations as illustrative of the General Statistical Analysis: [*see* "*Statistical Tables.*"]

Without much innovation on our present system, the following improvements are obvious: Let every Cottage have a garden sufficiently large to produce vegetables for a family, and particularly potatoes for themselves, and for a hog the whole year. Let as many industrious families as possible have additions of land made to keep a dairy cow. Nothing would add more to the comfort and supply of mankind than an increase of such occupations. Gentlemen of landed estates should promote and encourage small and moderate-sized dairy farms upon all land adapted to grazing; ten or twelve cows would find employment for a family doing their own business, would promote industry, employ moderate capital, and enable the occupier to pay as much rent as a large monopolized farm, and would require no expensive out-buildings; as no horse team should be kept, but potatoes, cabbages, and green crops, raised in plenty by hand-work in the garden style, by the spade and the hoe. An agricultural population might thus be considerably increased to the great improvement of the county.

The principal improvements to be made by the large farmers should be in a more correct culture of beans and other crops, particularly wheat, and in hand-culturing the fallows, and filling up crops. Beans should be universally set in rows and clean-hoed in a garden style. Nothing could possibly be a better preparation for wheat. 30,000 acres of land are now annually summer fallowed in Staffordshire, at a great expense in labour of men, horses, and manure. It is self-evident that this may be occupied by beans, potatoes, and other summer crops, by applying human labour and manure, and fewer horses would be required, if land was once completely clean fallowed. It may be permanently so continued without laying barren through the summer. If two-thirds only of the fallow land were cropped it would add 20,000 acres annually of crops, worth probably £20. per acre, or £400,000. which would pay for much human labour and profit. All land in culture for wheat ought to be examined in the spring and filled up with plants, as it will not bear transplanting and pay for that attention, and if there be a plant in every nine inches, and the land be kept free from weeds it will produce 20 thrave per acre at least. It has been ascertained by experience that this is practicable and will pay for it, and can be demonstrably proved that the general culture of grain and pulse would pay for a more minute attention than is commonly bestowed.

STATISTICAL

STATISTICAL TABLES OF STAFFORDSHIRE.

(From the last Parliamentary Returns.)

It is a remarkable feature in the population of Staffordshire that in places where employment is principally confined to heavy articles, and too laborious for women, the males exceed the females, who appear to have emigrated from time to time to situations more congenial to the habits and temperament of the sex. The facilities afforded to heavy carriage by the opening of Canals has greatly contributed to increase the population of the manufacturing towns, some of which for their extent are almost incredibly populous, and composed chiefly of miners, nailers, locksmiths, &c. a very small proportion being employed in agriculture. The following enumeration is from the Census of 1811:

	MALES.	FEMALES.	TOTAL.
Foreign of Walsall, not including the Town, (Stirrup-makers, and Sadlers' Ironmongery,) ..	3009	2639	5648
Tipton, (in the Iron and Coal Trade,) ..	4351	4056	8407
Wednesbury, (Ditto and Gun-Locks,) ..	2806	2566	5372
Darlaston, (Miners and Forgers of Iron,) ..	2622	2259	4881
Willenhall, (mostly Locksmiths,) ..	2014	1509	3523
Wednesfield, (large Steel Traps,) ..	702	546	1248
TOTAL, ..	15504	13575	29079
Town of Walsall, ..	2686	2855	5541
Fazeley, (a Cotton Manufactory there,) ..	508	657	1165
Tamworth, (ditto,) ..	1258	1581	2839
Lichfield Close, (no Trade,) ..	78	163	241
Lichfield City, ..	2237	2785	5022
Newcastle-under-Lyme, ..	2940	3235	6175
West-Bromwich, (a large Parish, part Agricultural, part Iron-Manufacture,) ..	3773	3712	7485
TOTAL, ..	13,480	14,988	28,468

HUNDRED.	Parish, Township, or Extra Parochial Place.	HOUSES.				OCCUPATIONS.			Persons.		
		Inhabited.	By how many Families occupied.	Building.	Uninhabited.	Families chiefly employed in Agriculture.	Families chiefly employed in Trade, Manufactures or Handicraft.	All other Families not comprised in the two preceding Classes.	Males.	Females.	Total of Persons.
OFFLOW, (NORTH.)											
Alrewas,	Parish,	145	202	1	3	94	97	11	441	680	1121
Alrewas-Hays,	Extra P.	7	7		1	1	6		24	25	49
*Anslow, otherwise **Annesley,*** (a)	Township,	45	45			45			102	123	225
Barton-under-Needwood, (b)	Ditto,	185	209	2	6	106	82	21	544	522	1066
Branson, (c)	Ditto,	53	64		2	58	6		194	179	373
Bromley Regis,	Parish,	82	115		5	73	26	16	256	271	527
Burton Extra, (c)	Township,	156	159	2	1	20	107	32	389	483	872
Burton-upon-Trent,	Parish,	785	842	1	30	60	558	224	1844	2135	3979
Clifton-Campville,	Ditto,	68	68			19	13	36	188	174	362
*Curborough, with **Elmhurst,*** (d)	Township,	44	46			45	1		110	119	229
Draycott,	Parish,	75	77	1	4	63	9	5	194	190	384
Dunstall, (b)	Township,	25	35			30	5		77	80	157
Edinghall, otherwise Edingale, (e)	Parish,	38	38		1	34	4		83	79	162
Fisherwick, (f)	Township,	14	18		2	15	1	2	34	39	73
Fradley, (g)	Hamlet,	56	77			72	5		190	205	395
Freeford, (f)	Ditto,	2	2			2			11	8	19
Hamstall Ridware,	Parish,	60	79			54	12	13	200	228	428
HANBURY PARISH:											
Coton,	Township,	14	14			13	1		37	32	69
Fauld,	Ditto,	12	12			12			34	38	72

Hanbury, ...	Township,	25	26		1	19	6		49	64	113
Hanbury Wood-end, ...	Ditto,	47	47			43	4		121	118	239
Marchington, ...	Ditto,	58	65			47	18		152	172	324
Marchington Woodlands, ...	Ditto,	47	51			51			145	161	306
Newborough, ...	Ditto,	106	124		2	64	22	38	306	317	623
Harlaston, (h)	Ditto,	28	31		2	27	4		79	71	150
Haselor, (f)	Ditto,	5	5			5			20	22	42
Haunton, (h)	Ditto,	38	38			12	4	22	120	109	229
Horninglow, (c)	Ditto,	48	51	1	3	39	12		143	154	297
Mavesyn Ridware, ...	Parish,	75	108			85	18	5	257	291	548
Needwood Forest, (i)	Extra P.										
Oakley, (k)	Township,										
Orgreave, (g)	Hamlet,	25	26			20	3	3	69	80	149
Pipe Ridware, ...	Parish,	16	20			20		:	43	58	101
Rolleston, ...	Ditto,	86	90	1	3	67	12	11	235	240	475
Street-Hay, (f)	Hamlet,	25	25		2	24	1		63	65	128
Stretton, (c)	Township,	73	77		6	50	27		196	174	370
Syerscote, otherwise Sterscote, (m)	Liberty,										
Tamhorn, (l)	Hamlet,	1	1			1			6	3	9
Tatenhill, ...	Parish,	65	73	1	2	52	19	2	195	177	372
Thorpe Constantine, ...	Ditto,	4	4			3		1	30	24	54
Tutbury	Ditto,	231	254	2	9	142	97	15	607	628	1235
Whittington, ...	Ditto,	117	139		3	112	21	6	297	305	602
Wichnor, (b)	Chapelry,	23	23		1	7	15	1	82	77	159
Yoxall and Hoarcross, ...	Parish,	217	278		5	164	98	16	609	736	1345
TOTAL, ..	..	3226	3664	12	49	1870	1314	480	8776	9656	18432

a Anslow is in Rolleston Parish. *b* Barton-under-Needwood and Dunstall are in Tatenhill Parish; Wichnor is a Chapelry to the Rectory of Tatenhill. *c* Branson, Burton Extra, Horninglow and Stretton, are in the Parish of Burton-upon-Trent; Winshill Township, part of the Parish of Burton-upon-Trent, is in Repton and Gresley Hundred, (Derby), and there entered accordingly. *d* Curborough, with Elmhurst, is in the Parish of St. Chadd, Lichfield. *e* Edinghall is partly in Repton and Gresley Hundred (Derby,) and there entered. *f* Freeford, Fisherwick, Haselor, and Street-hay, are in the Parish of St. Michael, Lichfield. *g* Pradley and Orgreave are in Alrewas Parish. *h* Harlaston and Haunton are in the Parish of Clifton-Campville. *i* Needwood Forest is included in the Returns of Barton-under-Needwood, (Tatenhill Parish,) of Draycot Parish, of Hanbury and Marchington (Hanbury Parish,) and of Tutbury Parish. *k* Oakley is in Croxall Parish, (Repton and Gresley Hundred, Derby,) and included therein. *l* Tamhorn is in Whittington Parish. *m* Syerscote is in Tamworth Parish (Offlow North) and there entered.

STATISTICAL TABLES—*continued.*

HUNDRED.	Parish, Township, or Extra Parochial Place.	HOUSES. Inhabited.	HOUSES. By how many Families occupied.	HOUSES. Building.	HOUSES. Uninhabited.	OCCUPATIONS. Families chiefly employed in Agriculture.	OCCUPATIONS. Families chiefly employed in Trade, Manufactures or Handicraft.	OCCUPATIONS. All other Families not comprised in the two preceding Classes.	Persons. Males.	Persons. Females.	Persons. Total of Persons.
OFFLOW, (SOUTH.) *(a)*											
Aldridge, ...	Parish,	137	152	5	3	36	25	91	425	422	847
Armitage, with Handsacre, ..	Ditto,	92	98	1	5	61	18	19	239	244	483
Barr Great, (b)	Township,	120	127		2	78	30	19	436	360	796
Bentley, (b)	Liberty,	17	17			14	3		61	42	103
Biddescote, (c)	Township,	8	8				8		20	19	39
Bonehill, (c)	Ditto,	40	41			17	23	1	120	116	236
Burntwood, Edgehill, Woodhouses (d)	Ditto,	125	125		3	87	38		318	341	659
Canwell, ...	Extra Pa.	3	3			2		1	14	14	28
Darlaston, ...	Parish,	916	1001	6	42	20	914	67	2622	2259	4881
Drayton Bassett, ...	Ditto,	79	88	1	1	70	16	2	224	231	455
Elford, ...	Ditto,	76	76		4	50	18	8	207	190	397
Fairwell, with Charley, ...	Ditto,	29	29		3	24	5		82	83	165
Fazeley, (c)	Township,	125	143		3	35	101	7	368	522	890
Hammerwich, ...	Parish,	41	41			30	10	1	114	101	215
Handsworth, with Soho, ...	Ditto,	484	513	5	12	108	344	61	1141	1310	2451
Harborne, ...	Ditto,	228	281	6	5	26	243	12	626	658	1284
Hints, ...	Ditto,	47	47		1	41	6		148	123	271
Longdon, ...	Ditto,	213	213		6	175	30	8	499	518	1017
Norton-Canes, ... } with	Ditto,	33	33			33			89	74	163
Little Wyrley, ... }	Township,	65	65		3	32	19	14	195	161	356

Ogley-Hay, ...	Extra Pa.	2	2				1	1	5	3	8
Pelsall, (g)	Liberty,	96	105	1	2	13	91	1	252	219	471
Perry-Barr, (e)	Hamlet,	104	106		12	79	23	4	318	258	576
Pipe-Hill, (d)	Ditto,	16	18			12	5	1	45	65	110
Rushall, ...	Parish,	114	117		5	95	12	10	324	289	613
Shenstone, ...	Ditto,	255	261	2	6	163	62	36	666	712	1378
Smethwick, (f)	Hamlet,	242	264	2	6	80	169	15	631	697	1328
Statfold, ...	Parish,	3	3			2		1	14	11	25
Swinfen and Packington, (h) ..	Liberties,	27	28		1	23	3	2	98	89	187
TAMWORTH, (*part of*)	Borough & Parish,	263	277		16	198	48	31	629	698	1327
Tipton											
East Division, ...	Parish,	654	693	8	16	407	129	157	1924	1723	3647
West Division, ...	Parish,	887	961	11	24	589	292	80	2427	2333	4760
Wall, (d)	Hamlet,	13	13		1		13		48	43	91
Walsall, ...	Parish,	1131	1154		19	3	792	359	2686	2855	5541
Walsall Foreign (k)	Township,	1060	1131	4	35	182	919	30	3009	2639	5648
Wednesbury, ...	Parish,	967	1067	5	32	55	924	88	2806	2566	5372
Wednesfield, (g)	Township,	222	207		5	70	137		702	546	1248
Weeford, (i)	Parish,	36	36		2	30	4	2	94	96	190
West Bromwich, ...	Ditto,	1450	1545	16	35	216	1086	243	3773	3712	7485
Wiggington, (c)	Township,	140	145		3	105	25	15	296	368	664
Willenhall, (g)	Ditto,	611	711	9	32	68	571	72	2014	509	3523
TOTAL,	..	11161	11945	82	345	3329	7157	1459	30,709	28,219	59928

a Part of Dudley with Netherton, is in this Hundred, but the whole is entered in Halfshire Hundred, (Worcester.) *b* Bentley and Great Barr are in Aldridge Parish. *c* Bidescote, Bonehill, Pazeley, and Wiggington, are in the Parish of Tamworth, which includes Sciencote, (North Offlow Hundred;) part of Tamworth is in Hemlingford Hundred (Warwick) and there entered accordingly; Wiggington includes Comberford, Coton, and Hopwas. *d* Burntwood, Pipehill, and Wall, are in the Parish of St. Michael, Lichfield. *e* Perry-Barr is in Handsworth Parish. *f* Smethwick is in Harborne Parish. *g* Pelsall, Wednesfield, and Willenhall, are in the Parish of Wolverhampton, (North Seisdon Hundred.) *h* Swinfen and Packington, are in Weeford Parish. *i* Walsall Foreign is in Walsall Parish.

STATISTICAL TABLES—*continued.*

HUNDRED.	Parish, Township, or Extra Parochial Place.	HOUSES. Inhabited.	By how many Families occupied.	Building.	Uninhabited.	OCCUPATIONS. Families chiefly employed in Agriculture.	Families chiefly employed in Trade, Manufactures or Handicraft.	All other Families not comprised in the two preceding Classes.	Persons. Males.	Females.	Total of Persons.
PIREHILL, (NORTH.)											
ADBASTON:	Parish.										
Adbaston	Township,	21	25		1	20	5		65	56	121
Bishop's Offlow, ...	Ditto,	37	40	1	1	35	5		99	97	196
Flashbrook, ...	Ditto,	17	19			18		1	61	56	117
Tunstall, ...	Ditto,	17	17	1	1	15	2		47	51	98
Ashley, ...	Parish,	141	141		2	102	17	22	321	295	616
AUDLEY:	Ditto,						55				
Audley, ...	Township,	112	112		1	43	55	14	321	289	610
Bignall End, ...	Ditto,	40	42			21	18	3	130	106	236
Eardley End, ...	Ditto,	21	23			16	4	3	86	72	158
Halmer End, ...	Liberty,	90	93		1	30	56	7	248	243	491
Knowl End, ...	Township,	35	35		1	30	5		114	117	231
Park End, ...	Ditto,	12	12			10	12		38	37	75
Talk o'th' Hill, ...	Ditto,	159	200		3	39	39	22	418	399	817
Balterley, (a)	Ditto,	46	46		2	42	4		125	124	249
Batchacre	Extra P.	1	1					1	2	2	4
Betley, ...	Parish,	143	151		3	88	31	32	359	402	761
Biddulph, ...	Ditto,	269	269	1	3	250	13	6	721	739	1460
Burslem, (b)	Ditto,	1658	1720	6	49	78	1552	90	4119	4506	8625
Chorlton, ...	Chapelry,	36	36			33	3		97	86	183

DRAYTON-IN-HALES: (c)	Parish,										
Almington	Township,	41	43	1		31	12		101	109	210
Blore-in-Tyrley, ...	Ditto	29	33			30	3		97	90	187
Hales, ...	Ditto,	34	35			31	3	1	103	107	210
ECCLESHALL:	Parish.										
Aspley, ...	Township,	3	3		1	3			15	15	30
Bromley, ...	Ditto,	4	4			2	2		17	15	32
Broughton, ...	Ditto,	2	2			2			9	9	18
Charnes, ...	Ditto,	14	14			10	4		50	41	91
Chatcull, ...	Ditto,	9	9		1	7	1	1	38	32	70
Charlton-Hill, ...	Ditto,	15	15	1		14	1		40	45	85
Coldmeece, ...	Ditto,	7	7		1	5	2		28	14	42
Cotes, ...	Ditto,	33	33			21	11		92	104	196
Croxton, ...	Ditto,	146	146	1		83	63		369	343	702
Eccleshall, ...	Town,	217	225		6	65	97	63	466	550	1016
Horseley, ...	Township,	88	90		1	64	24	2	213	214	427
Millmeece, ...	Ditto,	18	20			14	6		89	54	113
Pershall, ...	Ditto,	16	17			10	6	1	49	49	98
Podmore, ...	Ditto,	9	9		1	9			37	28	65
Slindon, ...	Ditto,	22	23			19	4		61	66	127
Segnall Magna, ...	Ditto,	25	24		1	15	9		58	57	115
Sugnall Parva, ...	Ditto,	9	11		1	9	2		39	32	71
Three-Farms, ...	Ditto,	4	4			4			19	17	36
Walton, ...	Ditto,	21	22			19	3		72	59	1[illegible]
Wooton, ...	Ditto,	31	33		2	23	9	1	84	69	153
Keel, ...	Parish,	176	178	1	10	10	167	1	461	483	94[illegible]
Madeley, ...	Ditto,	177	186	1	10	38	43	105	529	489	1018
Carried forward,		4005	4168	14	103	1408	2348	375	10,467	10,768	21,235

a Balterley is in Barthomley Parish, (Nantwich Hundred, Cheshire.) b Including Hulton Abbey and Sneyd. c Drayton-in-Hales is partly in Bradford Hundred, (Salop), and there entered accordingly.

HUNDRED.	Parish, Township, or Extra Parochial Place.	HOUSES. Inhabited.	By how many Families occupied.	Building.	Uninhabited.	OCCUPATIONS. Families chiefly employed in Agriculture.	Families chiefly employed in Trade, Manufactures or Handicraft.	All other Families not comprised in the two preceding Classes.	Persons. Males.	Females.	Total of Persons.
PIREHILL (NORTH), *Continued.* *Brought forward,*		4005	4168	14	103	1408	2348	375	10,467	10,768	21,235
MAER	Parish,										
Maer, ...	Township,	44	44	1		40	4		122	139	261
Maerway-Lane, ...	Hamlet,	32	36		1	21	10	5	99	94	193
MUCCLESTONE, (a)..	Parish.										
Aston, ...	Township,	36	37		3	29	8		103	106	209
Knieghton, ...	Ditto,	22	26		1	6	17	3	57	56	113
Mucclestone, ...	Ditto,	25	33			28	5		83	84	167
Oakley, ...	Ditto,	10	10			8	1	1	40	39	79
Winnington, ...	Ditto,	36	42			8	33	1	106	99	204
NORTON-IN-THE-MOORS, ...	Parish.										
Bemersley, ...	Township,	33	38		2	22	4	12	89	86	157
Norton-in-the-Moors, ...	Ditto,	296	298	3	1	200	96	2	811	775	1586
OFFLEY HIGH:	Parish.										
Loynton, ...	Township,	7	7			6		1	18	20	38
Offley High, ...	Ditto,	93	93		3	61	30	2	279	269	548
Standon, ...	Parish,	52	68		1	58	8	2	216	204	420
STOKE-UPON-TRENT:	Ditto.										
Bagnall, ...	Township,	48	49		4	28	13	8	156	132	288
Bolleston, ...	Liberty,	9	10		1	8	2		22	24	46

Bucknall,	Township,	76	76		4	40	29	7	173	171	344
Clayton,	Liberty,	24	31		2	24	6	1	87	84	171
Eaves,	Township,	40	40		3	29	6	5	115	117	232
Fenton-Culvert,	Ditto,	337	340	2	11	5	281	54	800	850	1650
Fenton-Vivian,	Ditto,	176	188		3	34	149	5	406	450	856
Hanley,	Ditto,	911	930	14	30	91	817	22	2165	2316	4481
Longton and Lane-End,	Ditto,	1032	1079	17	43	13	778	288	2277	2653	4930
Penkhull and Boothen,	Ditto,	768	780	6	34	44	557	179	1866	1985	3851
Seabridge (b)	Ditto,	28	31			24	4	3	79	80	159
Shelton,	Ditto,	1077	1148	16	36	86	912	150	2612	2875	5487
Swinnerton, (c)	Parish,	167	167	1		150	16	1	461	432	893
TRENTHAM:	Ditto,										
Blurton and Lightwood Forest,	Township,	160	164		4	47	116	1	461	462	923
Hanchurch and Butterton,	Ditto,	42	43		1	36	5	2	132	117	249
Handford,	Ditto,	82	85			59	19	7	202	191	393
Trentham,	Ditto,	115	125		3	75	28	22	274	281	555
Tyrley, (d)	Parish,										
Whitmore,	Ditto,	43	43		1	35	6	2	156	135	291
WOLSTANTON:	Ditto,										
Brieryhurst,	Hamlet,	104	108	2	1	19	5	84	327	319	646
Chatterley,	Township,	34	34			16	18		99	110	209
Chell,	Ditto,	63	69	1	1	12	57		168	188	356
Chesterton,	Ditto,	146	150		5	28	122		406	475	881
Carried forward	..	10,172	10,590	77	307	2798	6546	1246	25,933	27,186	53,119

a Mucclestone Parish contains Bearstone and Derrington, Woore and Cravenhunger, (North Bradford Hundred, Salop.) b Seabridge is partly in Swinnerton Parish. c Including part of Cobridge and Etruria. d Now called the Staffordshire part of Drayton-in-Hales Parish, and contains three Townships, viz. Almington, Blore-with-Tyrley, and Hales.

HUNDRED.	Parish, Township, or Extra Parochial Place.	HOUSES.				OCCUPATIONS.			Persons.		
		Inhabited.	By how many Families occupied.	Building.	Uninhabited.	Families chiefly employed in Agriculture.	Families chiefly employed in Trade, Manufactures or Handicraft.	All other Families not comprised in the two preceding Classes.	Males.	Females.	Total of Persons.
PIREHILL (NORTH), *Continued.*											
Brought forward,		10172	10590	77	307	2798	6546	1246	25933	27186	53119
Knutton,	Township,	133	133		8	51	82		357	356	713
Oldcot,	Ditto,	103	110		4	15	88	7	219	254	544
Rainscliff,	Ditto,	116	128			6	22	100	355	320	675
Stadmoreslow,	Ditto,	45	48			33	13	2	86	65	151
Thursfield,	Ditto,	33	36			8	2	26	91	94	185
Tunstall-Court,	Liberty,	335	350		8	52	250	48	823	854	1677
Wedgwood,	Township,	15	16			6	1	9	57	48	105
Wolstanton,	Ditto,	152	156	10	5	27	107	22	411	437	848
TOTAL,		11,104	11,567	87	332	2,997	7,110	1,460	28,403	29,614	58,017
PIREHILL, (SOUTH.)											
Barlaston,	Parish,	77	81		3	46	28	7	191	205	396
Beech, (a)	Township,	132	137		3	81	47	9	355	326	681
Bromley Abbots,	Parish,	191	201	3	8	97	77	27	470	549	1019
Bromley Bagots, (b)	Liberty,	28	28		1	15	1	12	97	86	183
Bromley Hurst, (b)	Township,	60	62			56	6		170	167	337
Blithfield with Newton,	Parish,	69	79		1	72	7		222	212	434

Chartley-Lodge, ...	Extra Pa.	1	1			1			4	5	9
Chebsey, (c)	Parish,	67	68		2	63	5		184	174	358
Cold-Norton, (c)	Township,	5	5			5			27	21	48
Colton, ...	Parish,	101	105		3	82	19	4	266	218	484
Colwich, (d)	Ditto,										
Crestbell, ...	Extra Pa.	1	1			•		1	7	12	19
Ellenhall, ...	Parish,	47	49			41	8		127	124	251
Fradwell, (d)	Liberty,	38	38			38			120	126	246
Gayton, ...	Parish,	48	49		1	40	9		137	124	261
Haywood Great, (d)	Township,	132	152		3	93	42	17	350	352	702
Hilderstone, (a)	Ditto,	432	432	2	7	165	176	91	1119	1100	2219
Hopton and Coton, (e)	Liberty,	61	62		2	51	7	4	165	167	332
Ingestre, ...	Parish,	18	21			12	8	1	67	55	1122
Kibblestone, (a)	Township,	172	175		12	136	38	1	568	488	1056
Marston, (e)	Liberty,	17	17			15	2		52	48	100
Milwich, ...	Parish,	109	111		1	83	27	1	290	273	563
Ronton, ...	Monastery,	1	1			1			8	6	14
Ronton, ...	Parish,	51	51			46	5		136	142	278
Salt with Enson, (e)	Township,	59	78	2		66	12		195	196	391
Sandon, ...	Parish,	92	92	1	3	70	17	5	238	242	480
Seighford, (f)	Ditto,	138	144		2	128	13	3	420	446	866
Stone, (e)	Ditto,	438	483	3	22	40	399	44	1117	1197	2314
Stowe, ...	Ditto,	125	156		1	106	39	11	440	413	853
Tillington, (e)	Liberty,	3	3			2		1	10	10	20
Tixall, ...	Parish,	29	31			25	1	5	113	93	206
Weston-upon-Trent, ...	Ditto,	52	52		2	28	21	3	209	185	394
Whitgreave, (e)	Liberty,	36	36			33	3		122	118	240
Wolseley, (d)	Township,	140	155		2	96	32	27	359	381	740
TOTAL, ..		2970	3156	11	79	1833	1049	274	8355	8261	16, 616

a Beech, Kibblestone, and Hilderstone, are in the Parish of Stone, which Parish includes Normacot Township. *b* Bromley Bagots and Bromley Hursts are in the Parish of Bromley Abbots. *c* Cold Norton is in Chebsey Parish. *d* Colwich is included partly in the Township of Great Haywood, and partly in that of Wolseley, which Township are in Colwich Parish. Fradwell is in Colwich Parish. *e* Hopton and Coton, Marston, Salt with Enson, Tillington and Whitgreave, are in the Parish of St. Mary, Lichfield. *f* Including Aston and Doxey Townships.

HUNDRED.	Parish, Township, or Extra Parochial Place.	HOUSES. Inhabited.	HOUSES. By how many Families occupied.	HOUSES. Building.	HOUSES. Uninhabited.	OCCUPATIONS. Families chiefly employed in Agriculture.	OCCUPATIONS. Families chiefly employed in Trade, Manufactures or Handicraft.	OCCUPATIONS. All other Families not comprised in the two preceding Classes.	PERSONS. Males.	PERSONS. Females.	PERSONS. Total of Persons.
SEISDON, (NORTH.)											
Bilston, (a)	Township,	1818	1945	14	30	18	1904	23	4887	4759	9646
Bushbury, ...	Parish,	102	104	1	8	64	25	15	272	269	541
Himley, ...	Ditto,	57	72			30	24	18	170	171	341
Kingswinford, ...	Ditto,	1657	1713	7	66	144	1322	247	4139	4128	8267
Moseley, (b)	Hamlet,	13	13			9	1	3	30	32	62
Oaken, ...	Ditto,	37	40		3	31	7	2	104	101	205
Penn, Lower, }	Parish,	40	43			25	7	11	135	118	253
Penn, Upper, }		109	109		2	17	72	20	267	260	527
Perton, with Trescott, (c)	Hamlet,	45	45			29	16		129	112	241
Rowley Regis, .. }											
Lower, ... }	Parish,	410	554	6	23	20	532	2	1170	1157	2327
Upper, ... }		488	521	5	12	35	484	2	1322	1325	2647
Sedgeley, ...	Ditto,	2444	2657	19	36	135	2485	37	6975	6962	13937
Tettenhall-Regis, with Tettenhall Clericorum, (c) .. .. }	Ditto,	245	258		5	103	120	35	573	595	1168
Wolverhampton, ...	Ditto,	2826	6963	47	63	419	2989	3555	7355	7481	14836
TOTAL, ..		10,291	15,037	99	248	1079	9988	3970	27,528	27,470	54,998

SEISDON, (SOUTH.)											
Amblecot, (d)	Hamlet,	188	193	2	4	100	70	23	201	228	429
Areley Upper, ...	Parish,	159	159		16	131	26	2	338	353	691
Bobbington, (e)	Ditto,	72	76		1	63	8	5	180	186	366
Broom, (f)	Ditto,	17	18		1	15	1	2	57	53	110
Clent, ...	Ditto,	159	194	1	4	64	81	49	365	372	737
Codsall, ...	Ditto,	80	112			88	24		266	268	534
Enville, ...	Ditto,	144	148		4	73	35	40	373	373	746
Kinfare, ...	Ditto,	356	371		21	211	132	28	836	832	1668
Orton, (g)	Liberty,	25	29			14	4	11	76	71	147
Pattingham, (h)	Parish,	154	162		3	112	35	15	418	380	798
Patteshull, ...	Ditto,	26	26			25	1		60	82	142
Pendeford, (i)	Township,	37	40		2	28	8	4	115	92	207
Trysull with Seisdon, (k)	Parish,	99	106	1	2	66	26	14	256	235	491
Wombourne, ...	Ditto,	207	252			65	113	74	483	506	989
Wrottesley, (i)	Hamlet,	32	32	2	2	26	5	1	101	97	198
TOTAL,		1755	1918	6	60	1081	569	268	4125	4128	8253
TOTMANSLOW, (NORTH.)											
Alstonefield, (a)	Parish,	123	148		4	87	36	25	320	334	654
Basford, (b)	Township,	42	42		1	21	3	18	141	102	243
Blackwood and Crowborough, (c)	Ditto,	65	69			53	8	8	177	177	354
Carried forward ..	..	230	259		5	161	47	51	638	613	1251

a Bilston is in Wolverhampton Parish, which Parish contains Featherstone, Hatherton, Hilton, and Kinvaston. *b* Moseley is in Bushbury Parish, which Parish contains Essington. *c* Oaken, and Perton with Trescott, are in Tettenhall Parish, which Parish contains Pendeford and Wrottesley, (South Seisdon Hundred.) *d* In Old Swinford Parish, (Halfshire Hundred, Worcester.) *e* Part of Bobbington Parish is in Brimstree Hundred, (Salop.) *f* Broom is locally situated in Worcestershire, east of Kidderminster. *g* Orton is in Wombourne Parish. *h* Part of Pattingham Parish is in Stottesdon Hundred, (Salop.) *i* Pendeford and Wrottesley are in Tettenhall Parish, (North Seisdon Hundred.) *k* Including Woodford Grange, Extra Parochial.

HUNDRED.	Parish, Township, or Extra Parochial Place.	HOUSES. Inhabited.	By how many Families occupied.	Building.	Uninhabited.	OCCUPATIONS. Families chiefly employed in Agriculture.	Families chiefly employed in Trade, Manufactures or Handicraft.	All other Families not comprised in the two preceding Classes.	PERSONS. Males.	Females.	Total of Persons.
TOTMANSLOW (NORTH), *Continued.* *Brought forward,*	..	230	259		5	161	47	51	638	613	1251
Blore, *with* *Swinescote,*	Parish, Township,	39	39		2	33	6		83	81	164
Bradnop, (d)	Ditto,	81	81		2	81			206	214	420
Caverswall,	Parish,	87	87		1	66	20	1	201	246	447
Cauldon,	Ditto,	59	59		2	33	26		165	152	317
Cheddleton, (b)	Ditto,	183	183	1	1	144	37	2	474	478	952
Consall, (b)	Township,	36	36	1		30	5	1	118	79	197
Dilhorne *with* *Fosbrook,*	Parish, Township,	242	249		5	104	132	13	581	603	1184
Endon with Longdon & Stanley, (d)	Ditto,	189	140	1	3	92	39	9	385	381	766
Fairfield-Head, (c)	Ditto,	189	189		2	109	22	58	510	493	1003
Grindon,	Parish,	69	74	1	3	60	7	7	206	197	403
Heaton, (d)	Township,	62	64	1		49	14	1	174	172	346
Heathy-Lee, (a)	Township,	124	141	1	2	80	61		353	353	706
Hollins Clough, (a)	Ditto,	125	128	1	13	29	99		209	304	513
Horton with Horton-Hay, (a)	Parish,	87	87		4	39	10	38	238	202	440

Ilam with Throwley & Casterton,	Parish,	26	26		3	22		4	83	94	177
and Okeover, ...		10	10			9		1	33	27	60
Leek and Lowe, (d)	Parish,	832	836	6	3	125	599	111	1664	2039	3703
Leek-Frith, (d)	Township,	122	137	2	2	107	27	3	359	351	710
Longnor, (e)	Ditto,	100	103		2	38	56	9	230	237	467
Onecote, (d)	Ditto,	74	80		4	65	15		246	218	464
Quarnford, (e)	Ditto,	151	152	1	7	72	74	6	353	346	699
Rushton-James, (d)	Ditto,	61	65	1		35	6	24	171	153	324
Rushton-Spencer, (d)	Ditto,	59	59			48	11		193	169	362
Tittisworth, (d)	Ditto,	45	55			40	12	3	130	143	273
Warnslow with Elkstones, (c)	Ditto,	166	166	1	10	62	24	80	418	410	828
Weston-Coyney and Hulme, (e)	Ditto,	83	84			70	14		239	214	453
Whiston, (f)	Ditto,	64	64	2	3	14	8	42	165	186	351
Total,	..	3545	3652	20	79	1817	1371	464	8825	9155	17,980
TOTMANSLOW, (south.											
Alveton, ...	Parish,	167	167		3	25	139	3	469	465	934
Bramshall, ...	Ditto,	28	33	1	1	22	6	5	76	79	155
Bradley-in-the-Moors, ...	Ditto,	15	15			14	1		42	41	83
Butterton, (g)	Township,	68	72	1	2	54	4	14	161	194	355
Calton, (h)	Ditto,	44	44		2	33	8	3	112	108	220
Calwich, (i)	Ditto,	19	19			19			55	50	105
Cheadle, ...	Parish,	626	640	2	14	65	252	323	1535	1656	3191
Checkley and Tean, ...	Ditto.	341	358	3	1	100	248	10	763	935	1698
Carried forward, ..	..	1308	1348	7	23	332	658	358	3213	3528	6741

a Fairfield Head, Heathy-Lee, Hollins Clough, Longnor, Quarnford, and Warnslow with Elkstones, are in the Parish of Alstonefield. *b* Basford and Consall are in Cheddleton Parish. *c* Blackwood and Crowborough are in Horton Parish. *d* Bradnop, Endon with Longdon and Stanley, Heaton and Leek-Frith, Onecote, Rushton-James, Rushton-Spencer and Tittisworth, are in the Parish of Leek, which Parish contains Rudyard and Candery, (South Totmanslow Hundred.) *e* Weston-Coyney and Hulme are in Caverswall Parish. *f* Whiston is in Kingsley Parish. *g* Butterton is in Mayfield Parish. *h* Calton is in the Parishes of Waterfall, Mayfield, Blore, and Croxden, and extends into the northern division of Totmanslow Hundred. *i* Calwich is in the Parish of Ellastone.

STATISTICAL TABLES—*continued.*

HUNDRED.	Parish, Township, or Extra Parochial Place.	HOUSES. Inhabited.	By how many Families occupied.	Building.	Uninhabited.	OCCUPATIONS. Families chiefly employed in Agriculture.	Families chiefly employed in Trade, Manufactures or Handicraft.	All other Families not comprised in the two preceding Classes.	PERSONS. Males.	Females.	Total of Persons.
TOTMANSLOW, (SOUTH,) Continued, *Carried forward,*	..	1308	1348	7	23	332	658	358	3213	3528	6741
Catton, (k)	Township,	75	76		2	61	7	8	209	199	408
Crakemarsh, (i)	Ditto,	24	27			18	2	7	77	77	154
Creighton, (i)	Ditto,	23	28		1	20	2	1	49	58	107
Croxden with Great Yate,	Parish, Township,	43	43	1	3	25	9	9	137	126	263
Denston, (k)	Ditto,	46	46		3	31	11	4	107	110	217
Draycott-in-the-Moors,	Parish,	92	97		1	81	16		262	274	536
Ellastone,	Ditto,	64	64			53	11		141	144	285
Farley, (k)	Township,	69	69			65	4		170	169	339
Field, (l)	Ditto,	9	9			9			35	34	69
Gratwich,	Parish,	18	19		1	15	4		64	46	110
Ipstones,	Ditto,	180	180	2	16	75	83	22	461	433	894
Kingsley, (n)	Ditto,	149	152	2	3	41	27	84	401	386	787
Kingston,	Ditto,	67	67	3		32	21	14	152	183	335
Loxley, (i)	Liberty,	20	20			18	1	1	83	78	161
Mayfield,	Parish,	127	130		1	80	19	31	244	337	581
Morredge and Foxt, (m)	Township,	68	70			44	8	18	170	171	341

Rocester, ...	Parish,	155	168		10	47	103	18	376	497	873
Rudyard and Caudery, (n) ..	Township,	16	16			15	1		66	49	115
Sheen, ...	Parish,	63	73	1	2	44	13	16	205	209	414
Stanton, (o)	Township,	61	64		3	48	12	4	140	158	298
Stramshall, (p)	Ditto,	30	31		4	18	9	4	64	87	151
Uttoxeter, ...	Parish,	605	628	5	27	136	376	116	1376	1779	3155
Waterfall, ...	Ditto,	92	92		2	51	32	9	231	224	455
Watton, ...	Ditto,	102	102		5	31	67	4	282	311	593
Wootton, (o)	Township,	51	51		6	44	7		109	119	228
Woodlands, (p)	Ditto,	76	77		1	63	10	4	178	208	386
TOTAL,	..	3805	3917	28	114	1646	1539	732	9438	10,48	19,864
CITY OF LICHFIELD:											
Chad St. (q)	Parish,	291	317		4	91	150	76	614	791	1405
Mary St. (r)	Ditto,	450	487	1	9	57	277	153	1069	1313	2382
Michael St. (s)	Ditto,	214	231		2	93	77	61	476	518	994
The Close, ...	Extra Pa.	55	55	4	3	1	5	49	78	163	241
TOTAL, ..	..	1010	1090	5	18	242	509	339	2237	2785	5022
BOROUGH OF NEWCASTLE-UNDER-LYME, ...	Parish,	1245	1354	25	74	47	1207	100	2940	3235	6175
BOROUGH OF STAFFORD: *Mary St. and St. Chad*, ...	Parish,	859	938	17	9	43	851	44	2401	2467	4868

n Rudyard and Caudery are in Leak Parish, (North Totmänslow Hundred.) *o* Stanton and Wooton are in Ellastone parish. *p* Stramshall and Woodlands are in Uttoxeter parish. *q* Containing Crowborough with Elmhurst, (North Offlow Hundred.) *r* Containing Hopton and Coton, Marston, Salt with Enson, Whitgreave and Tillington, (South Pirehill Hundred.) *s* Containing Burntwood, Pipehill and Wall, (South Offlow Hundred.)

HUNDREDS, &c.	Division.	HOUSES.				OCCUPATIONS.			Persons.		
		Inhabited.	By how many Families occupied.	Building.	Uninhabited.	Families chiefly employed in Agriculture.	Families chiefly employed in Trade, Manufactures or Handicraft.	All other Families not comprised in the two preceding Classes.	Males.	Females.	Total of Persons.
Hundred of Cuttleston, ..	East & West,	4097	4299	31	88	2377	1347	575	10391	10664	21055
——— Offlow, ...	North & South,	14387	15609	94	436	5199	8471	1939	39485	38875	78360
——— Pirehill, ...	Ditto,	14096	14723	98	411	4830	8159	1734	36758	37875	74633
——— Seisdon, ...	Ditto,	12036	16955	105	308	2160	10557	4238	31653	31598	63251
——— Totmanslow, ..	Ditto,	7350	7569	48	193	3463	2910	1196	18263	19581	37844
City of Lichfield, ...	..	1010	1090	5	18	242	509	339	2237	2785	5022
Borough of Newcastle-under-Lyme, ...	..	1245	1354	25	74	47	1207	100	2940	3235	6175
Borough of Stafford, ...	..	859	938	17	9	43	851	44	2401	2467	4868
Local Militia,*	..	..	..	..	..	..	..	..	3945	..	3945
Totals,	..	55,080	62,537	423	1537	18,361	34,011	10,165	148,073	147080	295153

* The Eastern Regiment of Staffordshire Local Militia was assembled on the 15th March, 1811, for fourteen days Exercise, to the number of 1344, officers included. The Western Regiment was assembled on the 18th May, to the number of 1296. The Southern Regiment was assembled on the 25th May, to the number of 1309.

PARISH REGISTERS.

Divisions.	Years.	Baptisms. Males.	Baptisms. Females.	Baptisms. Total.	Burials. Males.	Burials. Females.	Burials. Total.	Marriages
Cuttleston Hundred,	1801 1810	282 321	252 290	534 611	187 188	197 164	384 352	105 127
Pirehill Hundred,	1801 1810	825 1229	857 1114	1682 2343	727 726	755 747	1482 1473	413 724
Offlow Hundred,	1801 1810	915 1376	819 1294	1734 2670	957 961	942 1003	1899 1964	452 771
Seisdon Hundred,	1801 1810	653 1122	594 1050	1247 2172	941 861	792 883	1733 1744	468 662
Totmanslow Hundred,	1801 1810	488 617	510 609	998 1226	351 310	380 340	731 650	202 263
City of Lichfield,	1801 1810	88 81	70 86	158 167	59 68	58 63	117 131	80 73
Borough of Newcastle-under-Lyme,	1801 1810	70 124	40 113	110 237	51 91	61 82	112 173	104 17
Borough of Stafford,	1801 1810	75 95	73 83	148 178	77 50	57 54	134 104	53 42

GENERAL STATISTICAL ANALYSIS OF STAFFORDSHIRE.

[*Collected and continued to* 1817, *from the best Authorities.*]

EXTENT OF THE COUNTY, 1250 SQUARE MILES.

SOIL.			Acres.
Cultivated and Waste, in the gross, ...			800,000
Viz. Towns, Villages, Yards, Buildings, Roads, Rivers, Waters, Gardens, or in promiscuous and irregular Occupation, ...			100,000
Waste Land, Woodland, and Miscellaneous, ..			100,000
Remains in regular Farming Cultivation, ...			600,000
At Grass, Meadow and Pasture, ...			400,000
In Tillage, ...			200,000
Viz.	Produce per Acre.	Produce.	
54,000 *Wheat,*	24 *Bushels,*	1,296,000 *Bushels,* (The Seed to be deducted.)	
36,000 *Barley,*	30 ——	1,080,000 ——	
40,000 *Oats,*	30 ——	1,200,000 ——	
20,000 *Pulse,*	25 ——	500,000 ——	
Fallow for Wheat, ...			30,000
Turnips and Green Crops, ...			20,000

POPULATION.	Souls.
Probable Amount of the whole Population, ...	300,000
Viz. Residing in Towns or large Villages, ...	100,000
In smaller Villages and the Country, ...	200,000
Houses, ...	57,000
Families, ...	60,000
Families without Business, Profession, &c.	3,000
Agricultural, Farmers and Labourers, ...	19,000
Employed in Trade, Manufactures, Commerce, & Mines	38,000
Viz. in Handicraft and Domestic Manufactures, ...	12,000
In Manufactures for Exportation, ...	24,000
In the Mines, ...	2,000

A

BRIEF HISTORICAL ACCOUNT

OF THE

Coinage of Staffordshire.

IN our view of the Medallic* History of Staffordshire, we shall, for obvious reasons, commence with the County town of

STAFFORD;

although, within the district to which it gives name, there is one other place to which, in respect to the priority of the enjoyment of the valuable and distinguishing privilege of mintage, it must necessarily concede, viz. Tamworth.

From the discoveries hitherto made, and the historical documents yet existing, it does not appear that the town of Stafford enjoyed the honour of coinage till the time of William the Conqueror. That a mint was established there during his reign, is indisputably evident from three different specimens of his money

* It is somewhat remarkable, that among the very numerous and valuable County and Local Histories which have hitherto been presented to the British public, there is scarcely one, which, however complete and satisfactory in all other points, is not defective in information respecting that ancient and important privilege of Royalty, (the grant of Mints. Bearing this deficiency in recollection, and, anxious as we are to render our labours as worthy as we can of the public eye, we feel much pleasure in laying before our readers the following account of the several Coinages, both ancient and modern, which have originated in this County; which, though perhaps it may in this respect yield to some others, is nevertheless entitled to distinguished notice among them. And this account we submit with the greater confidence, (and consequently with the greater pleasure) as it proceeds from the pen of a Gentleman, who for many years has made the study of ancient English Coins and Medals his favourite pursuit; and whose judgement in this department of our National Antiquities has been repeatedly acknowledged by the most skilful numismatic Connoisseurs.---Ed.

preserved in the Cabinet of the writer of this article, all in the highest state of preservation, and of extreme rarity. Of these, the first exhibits on the obverse, or head-side, a profile portrait of the Conqueror looking to the right, and in the same hand a sceptre, and is inscribed WILLEMVS.* REX. A. (for *Anglorum.*) The reverse has a St. George's cross of double lines extending to the inner circle, with two small crescents attached to each of the lines at the extremity, and a pellet between them. In the centre also is a pellet within an amulet. A quadrilateral compartment formed of semi-circles, and surmounted at each junction by a pellet, terminates in each quarter formed by the double Cross, and at equal distance from each of its limbs. Round the coin, on this side, is read GODWINNE ON STEF, denoting that the piece was struck by Godwinne, the King's moneyer in Stafford; the word ON in the Saxon language signifying "*In.*" This coin is extremely rare, and has never yet been engraven in any numismatic work.

The second of the above-mentioned pennies differs considerably from that which has been just described. On this, the Conqueror appears full-faced, crowned, and with labels appendant to the ears. The inscription runs WILLEMV REX. The type of the reverse is the same in all respects as that of the foregoing coin; but the inscription is somewhat more full, it being GODWINE ON STEFFOR. The third of the pennies alluded to is, on the obverse, precisely similar to the last, excepting that it has, after the word REX an I; but whether this be intended to signify that he was the first Sovereign of England of that name, or is a part of the letter A, (the abbreviation of *Angliæ,*) is a point yet undetermined by medallic connoisseurs; although it is believed the greater part of them incline to the latter opinion. The reverse is similar to that of the two former coins; but the piece, as appears from the inscription, was the production of a different mint-master, being inscribed WVLFNOTH ON STÆF.

A fourth coin of the Conqueror, struck at Stafford, with a reverse considerably different from those of the pieces already described, was some years ago in the splendid collection of the late S. Tyssen, esq. of Narborough Hall, Norfolk; but as we have never seen either the coin itself, or any engraving of it, we are unable to give an account of it.

* As the information conveyed in this article is designed for a popular work, and not for the *Medallist* only, I have written the Conqueror's name WILLEMVS instead of ƿILLEMVS, (employing the Roman W for the Saxon ƿ,) from the idea that it would be far better understood.—W. W.

As the Stafford Mint seems to have originated in the reign of the Conqueror, so also does it appear to have terminated therein; as no coins struck at this town occur till the time of the Commonwealth, when cities, borough-towns, villages, and even private individuals, through the want of small change, were permitted by Government to coin and circulate their own specie. During this period, a Farthing Token was struck by some tradesman of this town; but as Snelling (who quotes the coin) has neither given his name, nor any representation of his currency, nor has the piece itself come to our hands, we are unable to afford our readers any farther information respecting it.

It may neither be improper nor uninteresting here to remark, that while many places, of much less repute and consideration than Stafford at the time above-mentioned, issued common Town Farthings and Half-pence, for the accommodation of their respective inhabitants, not one of that place, of either description, it is believed, has yet made its appearance.

In the large private Copper Coinage which was issued about twenty years ago, Stafford has a place; Messrs. Horton and Co. having struck a Half-penny, bearing on the obverse, the Arms of the Town, over which is its name, and beneath, the date 1797. This coin is circumscribed on the edge with the words: "Payable by Horton and Company." On the reverse is the Cypher of Mr. Horton, with a Cross Crosslet on either side. Above the Cypher is the word "Half-penny;" and below, a small ornamental flourish. The die of this piece was sunk by Mr. Willets; the coin was manufactured by Mr. Kempson, and about a ton weight thereof struck.*

LICHFIELD.

Although this ancient and celebrated City (amongst other marks of Royal munificence) was honoured with that of a Mint, granted by

* It is with great pleasure we are enabled to gratify our readers with an interesting supplement to the foregoing account of the Mintage of Stafford, and which carries it to still higher antiquity than the former discoveries had done. To the Cabinet of the Gentleman to whom we are indebted for these numismatic Memoirs, has recently been added a fine Penny of Edward the Confessor, bearing on its obverse his profile portrait to the right, helmetted, and with a sceptre surmounted with four pellets, diamond-wise, joined. Round the bust is the King's name and style, EADWARD. RE.. The reverse has a cross of double lines, extending to the inner circle only; and in each quarter two lines extending from the same circle, and terminated at their point of concurrence with a pellet. This side of the coin is inscribed, GODWINE ON STAE. being part of the word *Staefford*, which was the orthography of the town's name then in use.—ED.

King Stephen to Walter, Bishop of Coventry, (to which place the see of Lichfield was then transferred,) and his successors, yet not a single coin, issuing from it, has yet been discovered. From this circumstance only it might fairly be inferred, that the quantity of specie (if any) which it issued was extremely small. But when we recollect, that out of more than 400 coins of Stephen (who granted the mint) found near Ashby-de-la-Zouch, not more than fourteen miles from Lichfield, not one was found bearing the name of that city, and moreover, that no coins of any subsequent Monarch struck there have yet been discovered, we may safely conclude, that, from some circumstance or other, now unknown, no Mint was ever established. We must therefore descend for many centuries before we meet with any coin of this place; and the first which occurs, is in the time of Charles the Second, when it appears, that the Corporation struck a Farthing Token, in order to relieve the inhabitants from the scarcity of change under which they then laboured. This curious Coin, which is octagonal, has on the obverse, in Italic letters, within an octagonal compartment, the words "The Citty of," circumscribed "To supply the poore's need." On the reverse, in the field of the coin, is "Litchfield," being a continuation of the first-mentioned legend in the same situation, and round it, "Is Charity indeed;" so that the inscriptions on both sides of the coin, if connected, will run thus: "The Citty of Litchfield"—"To supply the poore's need"—"Is charity indeed." This coin, which is remarkable for having a poetical legend, is dated 1670.

That Farthings were issued by the tradesmen of Lichfield, as well as by those of other places, at this period, there can be no doubt. Snelling, in his list of Towns whose Traders struck Farthing Tokens, mentions, among others, Lichfield; but as he gives no representation or even verbal description of the coin which afforded him authority for placing this city in that list, and as we have, moreover, not been so fortunate as to meet with it, it is not in our power to give a description of it. The Repository, of all others apparently the most likely to exhibit the coins struck at Lichfield, should seem to be the celebrated Museum begun by the late ingenious Mr. R. Greene, of that place, and, as we believe, still preserved entire.

In the year 1794, Mr. Barker, of this city, struck two tokens, of silver, as it should seem, (though not mentioned by Pye) both of which have on the obverse, in a shield, (each side of which is ornamented with a wreath,) the Arms of the City of Lichfield. On the reverse of the former, are the words "Payable on Whit-Monday at the Green-hill Bank," on that of the latter "S. Baxter, Dozener, Sadler's-street Ward," under which is the date 1794.

In the year 1800 was struck at Lichfield, by Mr. Richard Wright, a very fine token of the large or penny size, bearing, on the obverse, the portrait of the late ingenious Mr. Greene, founder of the valuable Museum now in the possession of Mr. Wright. Over the bust is "Richard Greene," and round it a continuation of the inscription in smaller letters, "Collector of the Museum, died June 4th, 1793, aged 77." The edge is circumscribed "Penny Token, payable by Richard Wright, Litchfield." On the reverse is a representation of the beautiful and much-admired West Porch of the Cathedral, circumscribed "West Front of Litchfield Cathedral." In the exergue is the date "1800." The dye of this elegant coin (which is very rare, six dozen only being struck) was sunk by that truly ingenious artist, Mr. J. Gregory Hancock, of Birmingham, and the manufacture of it performed by Mr. Kempson, of the same place.

There yet remains to be mentioned a Half-penny, made at Lichfield for sale, which bears on its obverse a portrait of the celebrated Dr. Johnson; but it is in all respects so unworthy of notice, as to be justly excluded from the Cabinet by every judicious collector.

TAMWORTH.

Although the town of Tamworth be partially situate in the county of Warwick, yet, as a portion of it is in that of Stafford also, and it is not certain whether the Mint, formerly established here, might not have been erected in the latter; we have deemed it right to give a view of the Coinage issued at different periods from that town; desirous, in order to render our work as complete as possible, rather to introduce matter relative to the County (concerning which it may in a slight degree be dubious whether it relate to it or not) than from a similar consideration to abstain from all communication of it. Under these impressions, we confidently anticipate the excuse of our readers for calling their attention to the Numismatic History of Tamworth, which, even though it be entirely excluded from the list of Staffordshire towns, must yet, as a piece of border history, be not wholly unworthy of their notice.

The earliest instance of Tamworth occurring, as a place of mintage, is in the reign of Canute the Great. Kedar, a celebrated Swedish antiquary, has published a coin of this Monarch struck here, having on the reverse, EDRIC ON TAM; *i.e.* "Edric, Moneyer in Tamworth." The name of the same mint-master is found also on a coin in the Bodleian Cabinet at Oxford; but we have not had the good fortune to see either representations of the coins, or the coins themselves,

In the reign of William the Second, we find the Tamworth Mint again employed; a coin of that Sovereign (formerly in the collection of the celebrated Dr. Mead) bearing on its reverse, "IELFWINE ON TAM," but, as we know not where the coin now is, nor have seen either a similar piece, or an engraving of it, we are unable to gratify our readers by a more accurate account of it. We must therefore immediately proceed to the more recent coinage of Tamworth; the first specimen of which is a Town Token, struck during the Commonwealth, of an octagonal shape, and inscribed on the obverse "Tamworth Chamberlains;" and in the field of the coin, "their Halfpenny," being a continuation of the inscription. The reverse is inscribed, "For Change and Charitie," with a Fleur-de-lys in the centre.

In the year 1799, the Rev. F. Blick, of this town, struck a private Token, which does credit to the abilities of every artist employed upon it. On the obverse, it bears a view of the Church and Castle; over which are the words "Church and Castle," and in the exergne, "Tamworth." On the reverse, the initials of Mr. Blick's name, with the motto "*Deus nobis hæc otia fecit.*" In the exergue "Halfpenny Token, 1799."

A second piece (of the same description with the foregoing) belonging to this ancient and reputable place, was issued by a Mr. Harding, and is of the penny size. On the obverse it has a view of the Castle, and above it, "Tamworth Castle." In the exergue "East View, 1799." The edge is inscribed in three compartments, "Penny Token payable at"—"the house of John Harding"—"Calico Printer, Tamworth." On the reverse, a View of the Town-Hall, circumscribed "Town-Hall: Rebuilt by Thos. Guy;" and in the exergue, "Tamworth, 1701."

NEWCASTLE-UNDER-LYME.

We have long been in doubt, whether the Coins of Henry the Second, inscribed NVLAS on the reverse, and of those of Edwards the First and Second, having "Villa Novicastri," were struck at this town, or Newcastle-upon-Tyne, in Northumberland. But even should those of the two last-mentioned Kings be ascribed to the latter town, we are of opinion that Newcastle-under-Lyme may lay a very fair claim to those of Henry. It was, in the time of this Monarch, so far as appears from history, a place of very considerable eminence, having a strong Castle, and no less than four Churches. This circumstance, together with the marks of royal favour bestowed upon it by Henry the First, in its incorporation, incline us to think it probable, that it had a Mint also at the period above-mentioned, viz. in the reign of Henry the Second. We are farther induced to this opinion by an apprehension that Newcastle in Northumberland, however conspicuous afterwards, was not of equal importance and consequence at the time alluded to, with the town of that name in Staffordshire. Whether any documents, relating to either of these places, which might determine this point, are yet in existence, we know not. Placed, therefore, under this uncertainty, we will immediately proceed to notice the Coinage which is indisputably ascertained to have issued from this town; and of this, we regret to say, we can give no further account, than that, so far as appears from Snelling, this town, together with Burton-upon-Trent, Penkridge, Yoxall, Ashley, and Betley, had tradesmen resident in them, who issued their own private tokens; but as he gives neither engravings nor verbal descriptions of these pieces, and none of them have fallen under our inspection, we are deprived of the pleasure of giving a more particular account of them. The Betley Token, indeed, he partially describes; and from hence it appears, that it was coined by "Thomas Rich, in Betley and Lasterley." He remarks also on this piece, that it was one of the very few which were intended for the use and accommodation of *two places*.

LEEK.

Of this town we find no Coins till the year 1793, in which were struck two Half-pennies, in all points alike, except that the letters on the one are somewhat smaller than those on the other. On the obverse is a Chest with a package of goods lying across it, and an erect Caduceus near them, whose point appears fixed in the ground. Around these is the inscription: "Leek Commercial Halfpenny, 1793," and on the edge of the coin, "Payable at Leek, Staffordshire." The reverse bears two Hands joined together, with an Olive Branch behind them, and is inscribed "*Arte favente nil desperandum.*" This coin is singular for having its legend in Latin, the inscriptions on almost all the tokens, struck about this period, being in English.

UTTOXETER.

This town concludes our brief view of the Staffordshire Coinage, and though it be the last specimen of it in order, is yet far from being the least curious and interesting. It was struck by a person of the name of William Wakelin, and is a memorable proof of his loyalty to his Sovereign. It bears on the obverse "Will. Wakelin, 1663," and in the centre, "his Halfpenny." Mintmark, a Rose. The reverse is inscribed "*Vive le Roi in Uttoxetor,*" with a Crown in the centre, and a Rose for the mintmark.

We cannot conclude this department of our work, without observing, that no doubt can be entertained but many other towns, villages, and tradesmen, within the limits of Staffordshire, issued their respective monies, when such monies were allowed by the authority of Government to be in circulation. But if our catalogue of such as we have been able to discover, should appear scanty and defective, let it be remembered that all such Coinages, from being *local,* were extremely limited in their circulation, and consequently are not easily met with, but upon, or near, the places of their respective mintage. Many of these coins, unknown to collectors, are doubtless in the possession of individuals resident in, or near the place in which they were struck; but from the circumstances of their being in the hands of those who are unskilled in the science of Coins and Medals, and having never found their way to the inspection of the connoisseur, remain yet unpublished.

We think it not improper, in conclusion, to give a short account of the several discoveries of ancient coins, which have at various times been made in Staffordshire, which shall be extracted from a communicative letter of Mr. Shaw, the late ingenious historian of this county, published in the Gent. Mag. for Dec. 1796, p. 983.

"At Callingwood, (says Mr. S.) on the north side of Needwood Forest, were found in 1793 upwards of thirty Roman Gold Coins of five Emperors, most of them very perfect and beautiful, eight of which are now in my possession. In 1794, was also found at Rowley Regis, an earthen globe containing about 1200 similar Coins in silver, which, when altogether, formed a complete series of the Roman Emperors; but they were unfortunately soon dispersed at a low price about the country, except about three hundred, which are in the possession of the Rev. Mr. Cartwright, at Dudley. In 1795, was likewise found at Oulton, near Stone, a parcel of near four thousand silver Coins, all Saxon, except some of William the Conqueror; forty of which, of the different sorts, in the highest preservation, are in my possession."

STATE

OF THE COUNTY OF STAFFORD

ABOUT THE YEAR 1680.

[*From Dr. Plot's History, written before that Year.*]

THAT part of the county north of Trent is divided into two parts, called Moorlands and Woodlands; whereof the Moorlands are the more northerly mountainous part, lying between the upper parts of Trent and Dove, to Draycot-in-the-Moors, yielding coal, lead, copper, marble, and millstones; and the Woodlands, the more southerly level part of that country, from Draycot to Whichnor, and Burton, between the lower parts of the said rivers, including Needwood Forest, with all its parks; also the parks of Whichnor, Hore Cross, Bagots, Chartley, Loxley, Birchwood, and Paynsley, which anciently (Plot supposes) were all one wood, producing salt, black marble, and alabaster, besides great quantities of the best timber; and both Moorlands and Woodlands, as goodly cattle, large and fair spread, as Lancashire itself, and such as the graziers say will feed better; the lime-stone hills of the very Moorlands producing a short, but a fine sweet grass, and large oxen, though in an open cold country. Drayton says, in his Polyolbion, of the Moorlands;

> "She from her chilly site as from her barren feed,
> For body, horn, and hair, as fair a beast doth breed,
> As scarcely this great Isle can equal."

And if among the mountains of the Moorlands, much more can they breed and feed cattle too in the rich meadows that adorn the banks of the Trent, Blithe, Tene, Churnet, Hamps, and Manyfold, all in this quarter of the country; and more especially still upon the famous Dove-bank, esteemed by many the best feeding land of England. In the very midst of winter these meadows are adorned with a pleasant verdure, and when the river overflows them in the spring, it enriches them as the river Nile does Egypt, and makes them so fruitful, that the inhabitants thereabout upon such occasions usually chaunte this joyful ditte:

"In April Dove's flood is worth a king's good."

Whereof Michael Drayton renders this very good reason,

"Because the dainty grass
"That grows upon its banks, all other doth surpass."

As he says of Needwood:

"Needwood doth surmount
In excellency of soil by being richly plac't
'Twixt Trent and battening Dove, and equally embrac't
By their abounding banks, participates their store;
Of Britain's forests all (from th' less unto the more,)
For fineness of her turf surpassing."

From which lime-stone hills, and rich pastures and meadows, the great dairys are maintained in this part of Staffordshire that supply Uttoxeter mercat with such vast quantities of good butter and cheese, that the cheesemongers of London have thought it worth while to set up a factorage here for these commodities, which are brought in in so great plenty, that the facters many mercat-days (in the season) lay out no less than five hundred pounds a-day in these two commodities only.

Nor comes this northern part of Staffordshire much behind the south in breeding of sheep, which indeed are but small, have generally black noses, and their wool but coarse; nor in the production of corn; for though the land imployed for tillage be naturally but mean, yet where the industry of the husbandman has any thing showed itself, in marling, liming, or mixing lime with ess, and so laying them together on their heathy grounds, it produces corn of all sorts plentifully enough; the black moorish and gouty grounds of the Moorlands with the best helps are fit indeed only for oates: but the arable lands about Marchington, Draycot-in-the-Clay, Rolleston, Horninglow, and some other townes about Needwood, are of so rich a clay, that they produce as good hard-corne (i. e. wheat and rye, pease, beans, &c.) as any in the south, though not so much; the sheep too of the south bear somewhat a finer fleece, and it produces more and better coal and ironstone.

Besides wool, for the supply of the cloathing trade, and felting, which are chiefly exercised about Tamworth, Burton, and Newcastle-under-Lyme, they sow both hemp and flax all over the county, in small proportions, whereby they are furnish't too in some measure with linens; so that all things considered, this seems, to be a land (*terra suis contenta bonis*) that can subsist of itself. And yet a third part at least, if not half this county, must be confest, when all's done, to be barren heath and gorsy grounds, and woodland: but these yield some of the chief profits as well as pleasures of the country, for though the surface be barren, yet the subterranean riches are usually found in such uncultivated places; and of this sort of land is the Chase of Canck-wood, and most of the warrens and parks of the nobility and gentry, whereof before the late unhappy civil war there were near 50 in this county stock't with deer, and about 33 or 34 yet remaining, so great plenty is there of this kind of land, stored not only with mines, but with all sorts of game, both for hound and hawk, so happily are the profits of the gentry mixed with their pleasures, *utile dulci*.

This heathy, broomy, gorsy, barren sort of soil, for the most part too is a gravelly fast land, whence it is that in Canck-wood, and most of their parks, they have so pleasant and secure pursuit of their game. Hence 'tis too that their highways are so universally good, except in the most northerly part of the Moorlands, where, between the three shireheads and Longnor, the hills and bogs are such, that a horse can scarce pass between those two places; and indeed many of the mountains of that part of the country which they call Roches, Clouds, Torrs, Edges, Cops, Heads, &c. are hardly passable, some of them being of so vast a height, that in rainy weather I have frequently seen the tops of them above the clouds; particularly these of Narrowdale are so very lofty, that the inhabitants there, for that quarter of the year wherein the sun is nearest the tropic of Capricorn, never see it at all; and at length, when it does begin to appear again, they never see it till about one o'clock, which they call thereabout the Narrowdale noon, using it proverbially to express a thing done late at noon.

In his account of tillage, Plot says of Clay ground, if it lye in common fields, as generally it does in this county, they have it always in tillage, sowing it two years, and letting it lye fallow the third; they lay it in ridges, according to the temper of the land, and make their fallows in March or April; after this, a little before the second plowing, which is commonly about the middle of June, they give it its manure, which is cow or horse-dung, unless when folded with sheep, and then immediately spread it, and cast it under furrow with the plow, lest the rain and sun should weaken it;

about the latter end of August they plow it again, to kill the weeds, and turn up the manure, that so they may return it again to their seed, at the last plowing, when they sow, which is usually the week before or after Michaelmas.

The land thus prepared, they sow it with wheat, and if a strong stiff clay or cold land, with red lammas or bearded wheat; otherwise, with white lammas; and sometimes with both mixt; allowing of either two strike to an acre, whereof if they have twenty again, they think it a good increase. The next grain they sow in their common fields after wheat, are usually horse beans, or pease, white or gray, or pease or beans mixt; for these they plow at Candlemas, and sow in the decrease of the moon, having found (as they say) by long experience that they codd better, and are not so apt to run into straw; for seed they allow four strike of pease and five of beans to a statute acre, whereof if they have again 20 strike of the former and 30 of the latter per acre, they reckon it a competent crop. But if either clay or light mould lay out of the common field, so that they may be tilled at pleasure, they are also capable of improvement by marle, especially by the dice or flat marle, which with rain runs like lime, and never bindes the stiffest clay, but rather loosens it, so that after it has afforded eight or nine crops it will yield very good grass; whereas clay-marle laid on the same so bindes the surface, that though they will not fail giving seven or eight crops of corn, yet they are rendered thereby ill-disposed at lest for grass, unless the ground after all be well muck't or manured with muck and lime mixt together, when it will yield two or three crops more. And as for pure sandy gravelly ground, which will naturally bear nothing but rye, French wheat, or oates, nor these neither unless well muck't, and then not above three years together but it must rest again, by the help of these marles used as above, lime, and good muck, they are made as good for all sorts of corn, bearing as many crops as any land whatever.

For the heathy land of this county it is seldom inclosed, but when they intend it for tillage, which is never for above five years neither, and then it is thrown open to the commons again; when they do inclose such, they generally proceed in the manner following: First they stock up the heath with mattocks, and then fallow it in winter, and in the summer ensuing give it its proper manure, which is lime, allowing four loads to each statute acre, each load containing four quarters of lime, which, when slacked, is spread on the ground with shovels, and plowed in under furrow about the middle of September, after which it is sown with rye, allowing for seed two strikes to an acre, which if it yield 25 strike again, they reckon it a competent increase. After rye they sow barley, making their fallows about Candlemas, and giving another plowing in April, then sowing it, affording for seed three strikes to an acre, which generally yields them thirty again; and next barley, or white peas, for which they plow but once in March, and then sow them, allowing three strikes to an acre; fourthly, after peas, they sow oates, either red or white, if it be in good heart, but if poor and weak, black oates, for either of which one plowing is sufficient in March, and four strike of seed; lastly, they sow oates again, (for their common heathy grounds will seldom afford but one crop of barley) and then their inclosures are thrown open to the commons again.

Their broomy, gorsy, or furzy hot sandy land, they first clear of those incumbrances by stocking with the mattock, or drawing them up by the roots by a certain instrument made like a strong lever, armed with a iron hook, which roots up bushes by force; they then manure it with marle, which doth not only so fertilize it as that it shall yield the husbandman 7 or 8 crops, but as some conceive frees it from the annoyance of these plants for ever after. The manure being laid on, they plow and sow this as the heathy lands are, only in some places they first sow with French or buck-wheat, for which they make their fallows in winter, and stir them in May ensuing, allowing one strike to the acre, and so brings their land in order for rye; which they sow after once plowing. Their gouty, moorish, peaty, cold black land, they husband also much after the same manner, only they burne the earth to a good depth; yet these will bear little else than oates, white oates upon the gouty, and black oates upon the cold black land. In marling their lands, upon sound grounds, whether clay or gravelly, they lay eight or nine score or two hundred loads on an acre: on loose wormey ground they commonly lay more.

The white flaxen, and bright red wheat, are the ordinary grains of the country, besides which they now and then sow the Triticum multiplex, or double-eared wheat; Triticum Polonicum, or Poland wheat; and for barley, besides the common long-eared, and spratt barley, they sow sometimes the Triticum spelta, or naked barley. And amongst the oates, besides the white, black, and red oates, at Burton-upon-Trent I found they also sowed the Avena-nuda, or naked oat;* and besides the summer and winter vetches, the wild vetch, Vicia Cracca, is sown in some places; but these only in mea-

down: about Wexford they also sow drills, or lentils, on their poorest land, to feed sheep and cattle in winter; they also sow muncorn or wheat and rye mix't; they matter not how poor or hungry the land be their seed comes off, though to be sown on the rankest soil, in general they chuse corn for seed that grew on land of a different temper from that it is to be sown upon, as from clay lands to sow on sandy, and from sandy, to sow on clay. To avoid blasting and smutting, they steep their grain in brine before they sow it, which they esteem a certain remedy for this disease of corn. To prevent meldewing, the most pernicious of all annoyances that inclosures and rich lands are lyable to, Thomas Cartwright, parish clerk of Wombourn, either mixes his corn with soot before he sows it, or sows soot upon it after the wheat's in the ground, by which means he has preserved the corn from being meldewed in lands lyable to it, and this not for one or two, but for ten years together, which is certain matter of fact. They draw their plows here both with oxen and horses, but rather with the former than the latter, because of their turning to a more certain profit, and of having less of hazard in them, oxen always increasing in price with their fatness: they generally plow with oxen in pairs, but horses in a string, to prevent poaching the land; and so they do in some places with their oxen too, in very wet seasons, being furnished with half-yokes for that very purpose.

When their corn is come up (especially oates and barley,) if sown on a binding land, and it prove a dry time, at and about Church Eyton, they sometimes harrow them again to break the clods and loosen the earth, which will make them flourish much the better; for though it may pluck some up, yet making more spring by half than it destroys, they account it advantagious; after the corn is in the blade, if it grow too ranck, they eat it off with sheep. At Alrewas they mow off the tops of it before it spindles, and in May and June they weed their corn with an iron digger, and another instrument like a pair of smith's tongs, jagged like a rasp on the inner sides to take the firmer hold, with which they pluck up the weeds by the roots.

When the time of harvest is come, they reap their wheat, and bind it, and so they doe their rye; when bound they gather nine sheaves together and sett them upon their butt ends, and cover them with throe, and so let them stand ten or twelve days before they carry them, the corn threshing the better the longer it stands; their barley they mow with the sithe and cador in the south parts of the county, but in the Moorlands they reap it with hooks; their oates they harvest after the same manner with barley, their beans and pease they reap with hooks, letting them lie in reaps ten or twelve days, and turning them once the day before they carry them.

Dr. Plot asserts, that in a light lean soile, wheat being sown, in three years time it degenerated into rye; and that in a strong close soil, rye sown in the same space of time improved itself into wheat, and that many having sown wheat and barley, have yet reaped nothing but darnel; that barley did not only degenerate into rye or darnel, but some times into oates too: hence he concludes that a change of species sometimes takes place.*

Ropes are asserted to have been made of the rind of rushes stronger and more durable than of hemp; also that about Shenstone, they frequently used instead of hops to preserve their beer, the common heath or ling (Erica vulgaris) which gave it no ill taste.

As a proof of the superstition of these times, take the method of curing the disease in cattle called the Foule, which falling into their leggs and feet, causes such impostumes or cores of putrified matter, that they cannot go farr. They strictly observe the turf where the oxe, cow, or heifer thus distempered, sets his sick foot when he first rises in a morning, upon which they usually find some of the sanies, or matter of the impostume, press't out by his weight: this very turf, with the impression upon it, they cut up, and hang upon a tree or hedge towards the north wind, which blowing upon it, the beast becomes cured in three or four days; and this Dr. Plot was told was practised about Tamworth with good success.

I should suppose the learned Dr. Plot was imposed upon in the information he receivedon the following particulars. "It must be ascribed to the saltness of the soile and grass, that if any horned

* This I believe to be erroneous: a mongrel species may accidentally occur by impregnation of the sexual parts of one species upon another, but very rarely, as both male and female organs are in the same plant in all the varieties of grain and pulse. Wheat mixed with rye and sown on a lean light soil, might perish, and the rye succeed, without any change of species, and if darnel abounds either in the soil, or the seed sown, it has a wonderful aptitude to increase and to choke all other plants. I believe, if wheat was sown with one-tenth darnel, little must be expected to be reaped but darnel itself; a bad sort of grain sown may be expected to grow worse, and lean oats sown repeatedly upon hard tilled land, would soon become all husk nearly. To produce the most perfect grain, a good sort must be sown upon land in a proper state to receive it.

cattle of never so deep a black or other colour, be put to feed in a place called the Clots, in Newbold ground, in the parish of Tatenhill, about a mile east of Dunstall, they will certainly change the colour of their coat to a whitish dun, (like a daw's head) in a summer's running; and so they will if put upon Tatenhill, about a mile upon Tatenhill Common; or into Buck stew, another parcel of Newbold grounds; nor does only the grass but the hay of these grounds also turn cattle to this whitish dun, which 'tis said they recover not in two or three years time, tho' put into other grounds; as for horses they are improved upon these grounds at a great rate, only they make them dappled, be they of what colour soever before."

In the Moorlands they cut the turf in the spring time with an instrument called a push plow, being a sort of spade shod somewhat in the form of an arrow, with a wing at one side, and having a cross piece of wood at the upper end of the helve after the manner of a crutch, to which they fasten a pillow, which setting to their thigh, and so thrusting it forward, they will commonly dispatch a large turf at two cuts, and then turn it up to dry, which in good weather is done on one side in 8, on the other in 4 or 5 days; when dryed, if intended for fuel in winter, they pile them up round in manner of a hay-rick 10 or 12 foot high, and let them stand all summer; but if for manuring their land, they heap it up round a good quantity together on the ground, and set it on fire: these heaps they keep burning sometimes three weeks together, still putting on new turf as the old burns away, only giving them vent by air holes, which they make with a stick; the ashes of this turf they call ess, which laid on their meadows, or tillage land, goes further than dung or lime: they also use turf to ridge and head their meaner houses, and sometimes thatch with it. In their moorish boggy grounds they also dig peat, either for fuel or to burn for manure as above.

Of herbaceous plants, the Vicia Sylvestris Cracca, or wild tufted vetch, or tare-grass, has been observed to do so well in meadows, that it advances all starven weak cattle above any thing yet known.*

* This is a perennial plant well deserving a place in the best meadows, being excellent, either eaten green or in hay.

PRESENT STATE

OF THE

AGRICULTURE OF STAFFORDSHIRE.

STAFFORDSHIRE is an inland county, situate near the centre of England, bounded on the north by Cheshire and Derbyshire; on the south-east by Warwickshire; on the south by Worcestershire; and on the west by Shropshire. It lies between 52° 23′ and 53° 13′ north latitude, and 1° 29′ and 2° 27′ west longitude from London. The greatest length from the north part of Ax-edge Common (between Leek and Buxton), to the south part of Woods-Eaves (south of the Severn), from N. by E. to S. by W. is 60 English miles, and the greatest breadth from the junction of the Trent and Dove (north of Burton), to the west point of Terley-heath (near Drayton), nearly from east to west, is 38 miles: the county contains in the gross 1220 square miles, and 780,800 statute acres.

Divisions.—This county is divided into two parts by the Roman road of Watling-street, which passes through it from east or south-east to west, nearly 30 miles, having about three-fourths of the county to the north, and one-fourth to the south. It formed part of the ancient *Cornavii*, under the Saxon Heptarchy, and belonged to the kingdom of Mercia. It is in the Oxford circuit, and diocese of Lichfield and Coventry. Staffordshire is divided into five hundreds: 1. Totmanslow, to the north: 2. Pyrehill, to the north-west; 3. Cuddleston, central, and to the south-west; 4. Offlow, to the east; and 5. Seisdon, to the south: it contains 24 market-towns, and 81 parishes.

Climate.—The air of this county is sharp, and the climate may be termed inclining to wet, the average annual rains being about 36 inches, those of London 21 inches, of Essex 19 to 20 inches, of Lancashire 42 inches, and in Ireland still more. It may be considered an established fact, that the annual rains on the west side of the kingdom are at least double in quantity to those on the east side, as most of our rain comes from the west or south-west, and the clouds are attracted in their course by the high grounds in the midland counties; on which account much more rain and snow falls on high grounds than on low, the source of springs and rivers being generally from high grounds, or valleys supplied from such high grounds by the excess of moisture there falling.

The aspect of this county is various: the middle and south parts are generally level, and of moderate elevation, but with some exceptions, as the lime-stone hills and cliffs of Dudley, Sedgeley, Rushall, Walsall, &c. which contain an inexhaustible supply of that material, and of excellent quality. The quartzoze or ragstone-hills of Rowley, still more elevated, furnish a good material for roads and pavements. The hills of Clent, Barr-beacon, the high grounds on Cannock-heath, and near Rugeley and Beaudesert, the hills of Byshbury and Essington, are formed chiefly of gravel. Tettenhall-wood, Kinfare-edge, and some situations near Enville, command extensive prospects; but much the highest ground in the county is in the northern part, which rises gradually to what is called the Moorlands. It has been ascertained, by tracing the canal levels, that the valleys near Leek are more elevated than the hills near Wolverhampton. The altitude or elevation of a country has great influence on its climate and vegetable productions: it is calculated that 60 yards of elevation is in this respect equal to one degree of latitude. The lowest ground in the county is on the banks of the Severn at Over Areley, where the elevation is about 60 feet above the level of the sea: fruit has been long cultivated here, and succeeds as well as in Worcestershire. The country gradually rises hence to the north to the summit of the Staffordshire Canal, an elevation of 385 feet. The surface of the parishes of Tettenhall and Byshbury, contiguous to this summit, are elevated from this level to 500 feet. Here also fruit succeeds well, particularly the Tettenhall pear, which is indigenous to the neighbourhood. The summits of the Birmingham and of the Wyrley and Essington Canals (elevation 500 feet) extend 40 miles to the north-east and south-east of Wolverhampton: fruit-trees do not succeed well much above this level, without natural or artificial shelter. Byshbury-hill, and the hills of Penn, rise to 650 feet; Barr-beacon, 750; the hills of Sedgeley and Dudley Castle, 800 feet and upwards; and the Rowley and Clent-hills, 900 feet: at these latter elevations the ripening of grain is sensibly retarded, generally two or three weeks later than in the lower land. The summit of the Grand Trunk Canal is elevated about 420 feet, and the vale of Trent varies from that level to about 200 feet at Burton, and is consequently in a mild and temperate climate, though in 53° of latitude. The Vale of Trent may be considered as the Garden of Staffordshire, and the first families of the county have accordingly taken up their residence in this salubrious tract, which includes the noble mansions of Trentham, Sandon, Ingestrie, Shugborough, Blithfield, Hagley near Rugeley, Beaudesert, and many others. This part of the county is beautifully varied in its surface, is well wooded, and exceedingly fertile.

North of the Trent the country rises gradually into hills and mountains, which continue through the northern counties into Scotland, and have been termed the back-bone of the kingdom. The part extending into Staffordshire, called the Moorlands, is a rough, dreary, cold tract, the snow lying long on it. The Moorlands consist in part of gravelly hills, impracticable to the plough from unevenness of surface: other portions consist of high moors, and peat mosses, where peat is dug for fuel, the soil beneath being a gravelly clay. A considerable part of the county is upon a lime-stone bottom, in many parts rising out of the main surface in huge cliffs: the lime-stone is covered with good calcareous loamy earth, and is the best part of the Moorlands, but the elevation is too great to ripen corn in due season: oats have been unripe in November, and no other grain is sown. The elevation of a hill called Bunster, near Ilam, is propably more than 1200 feet, that of the Weaver-hills 1500, and that of the highest hills north of Leek 1800 to 2000 or more: this is equal to 10 or 12 degrees of latitude, and brings the climate to a level with that of the lowlands on the Arctic Circle, where the sun does not rise at all on our shortest day. This land is better adapted to grass than corn; the grass, clinging to the earth, is less exposed, and the humid climate is favourable to its growth.

Soil.—The soil of this county is various, and may be divided into 1. The Argillaceous, constituting the stiff and strong lands; 2. Siliceous, or the loose and light sandy lands; 3. Calcareous, as in the lime-stone districts; and, 4. In the valleys, and again on the highest hills, peat-earth, composed of the roots of aquatic vegetables, saturated with water. The light soils have sometimes a considerable mixture of pebbles, and are then termed gravelly loams; the clay soils are of two kinds: 1. the strong, stubborn, harsh, tenacious clay, or clay loam; 2. the more mild and tractable, consisting of friable marl or loam. There are also various mixed soils, which may be termed loamy, clayey, gravelly, sandy, peaty, &c. according as the different materials preponderate in the composition of them.

The basis of our soils consists of the primitive earths, *(silica)*, sand *(alumina)*, clay, and lime or calcareous earth. If sand most abounds, the soil is dry; if clay preponderates, the soil is wet: lime and magnesia are correctors, increasing the moisture of sand, and diminishing the wetness of clay, and different proportions of these earths will form soils of every degree of dryness or moisture. According to modern chemistry, the fertility of all soils is derived from carbon, which is charcoal in a pure state, freed from its earthy and saline particles. No soil can be fertile without carbon: it should contain one-sixteenth its weight of carbon, but it must exist in the soil in a particular state of combination, so as to become soluble in water, to promote fertility. Dunghill-water contains much carbon: in lime-stone the lime is united with carbonic acid in the proportion of about 43 per cent. Putrescent manures part with their carbon during the process of putrefaction: it is absorbed by the soil, and thence by the plants growing therein.

		Acres.
Division of Soils.—Gross acres of the county, as stated before,		780,800
Deduct for roads, waters, buildings, and yards, one acre in twenty,		39,040
Waste-lands, woods, and impracticable land. These were reckoned, in 1794, at	Acres. 141,760	111,760
Inclosed since, to 1816, (Sutton-Coldfield, Needwood Forest, and different Commons,) probably	30,000	
Deduct		150,800
Remains, cultivated land,		630,000
Of the Waste-lands, the reclaimable part is now probably reduced to		70,000

Of the 630,000 acres of cultivated land, including the pasture part of Parks, 100,000 acres may be meadow and pasture, and 530,000 acres arable: of this may be reckoned,

Two-fifths clay loam, or friable mixed loam,	212,000
Two-fifths gravelly mixed loams, including lime-stone soils,	212,000
One-fifth lighter soils, capable of turnip-culture,	106,000
	530,000

Water.—The county is well watered, principally by the Trent, and its contributary streams. The Trent rises in the Moorlands, near Biddulph: it takes a winding course, first southerly, then south-easterly and easterly, and lastly north-easterly; and, after washing the county in a course of upwards of fifty miles, leaves it below Burton, where it becomes navigable. In its course it receives the Dove, the Manyfold, the Hamps or Hanse, the two last being subterraneous for a considerable length above Ilam. It also receives the Churnet, the Blythe, and the Teyn: these rivers rise in the Moorlands. Afterwards it receives the Sow from Stafford, after the Sow has received the Penk from Penkridge: the Penk has its sources from Cannock-heath, Chillington Pool, Byshbury, and Tettenhall. Below Wichnor it receives the Tame from Tamworth, which has its sources near Walsall, Wednesbury, and Birmingham: the Smestall, which rises near Wolverhampton, runs into the Stour, which passes through the south of the county into the Severn.

The most considerable lake in the county is Aqualate Meer, on the borders of Shropshire, near Newport, said to be 1848 yards long, and 672 broad. There are several other pieces of water of smaller dimensions, and many as ornaments to gentlemen's seats, in different parts of the county.

Navigable Canals have been executed with great spirit, and to great extent: they are 1. the Trent and Mersey; 2. the Staffordshire and Worcestershire; 3. the Birmingham Canal, passing through the Collieries, and Wolverhampton, into the above; 4. Lord Dudley's, or the Stourbridge Canal; 5. the Wyrley and Essington Canal, across Cannock-heath, and uniting the coal and lime-works of the county with the other canals; and, 6. the Dudley Tunnel, and Netherton Canal, are partly in this county. These Canals, in the aggregate, are about 200 miles in length in this county, and cost upwards of half-a-million of money: some of them are so full of business as to pay 30 or 40 per cent. upon the money expended, and have greatly promoted the mining interest, commerce, and population of the county.

Estates are in great variety as to extent and value, from that of the Nobleman, and opulent Commoner, of 15 or £20,000. per annum, to the humble freeholder of forty shillings a-year. The county can boast of many gentlemen of fortune, who farm upon a considerable scale, and a number of proprietors of 200 or 300 acres of land who farm it themselves, and whose lands are cultivated in a first-rate style. It is from this class, and the more opulent farmers, that improvements in stock, and the introduction of new plants, and new modes of culture, are to be expected, and in which, when established, they will be imitated by others.

Tenures.—A large proportion of the county is freehold: considerable portions of landed property are also held on leasehold and copyhold tenures, under the Bishop, Colleges, Deaneries, Prebends, and other Church dignitaries, but seldom by the occupier, being commonly let out again to farmers.

Leases are often granted: those for 21 years are not uncommon, and some for a shorter term. In most covenants fallowing is considered as necessary. The growth of hemp, flax, and rape for seed, are prohibited, or restricted to an acre. Meadow-land is secured from the plough; tenants are restrained from cropping and lopping timber, and from taking more than a stated number of crops, are bound to use all their dung on the premises, and restricted from selling hay, straw, or dung. Sometimes systems of cropping are inserted, but little attended to, if the tenant is thought to be going on well. The tenant is generally bound to keep the building in repair, the landlord finding materials. Many gentlemen who have large estates round their seats, or in the neighbourhood, do not grant leases; and to the honour of many such gentlemen it may be observed, that the possession of their farms upon such tenure is equally secure with a lease, and that whilst the occupier behaves with propriety, he is as little liable to be disturbed in possession, and often rents upon easier terms.

Buildings.—The county of Stafford contains many magnificent and elegant seats of Nobility and Gentry, around which the beauties of landscape have been united with the improvement of the demesne, by taste and attention in planting, and by draining and improving the boggy and unsound spots, and in some instances by ornamental and useful fish-ponds, and sheets of water.—The following are the principal seats:

1. *Trentham,*—Marquis of Stafford.
2. *Sandon,*—Earl of Harrowby.
3. *Ingestrie,*—Earl Talbot.
4. *Shugborough,*—Lord Anson.
5. *Blithfield,*—Lord Bagot.
6. *Chartley,*—Earl Ferrers.
7. *Hagley,* (near Rugeley,)—Lord Curzon.
8. *Beaudesert,*—Marquis of Anglesea.
9. *Weston,*—Earl of Bradford.
10. *Himley,*—Lord Dudley.
11. *Sandwell,*—Earl of Dartmouth.
12. *Enville,*—Earl of Stamford.
13. *Teddesley,*—E. J. Littleton, Esq.
14. *Wrottesley,*—Sir Jno. Wrottesley.
15. *Patshull,*—Sir George Pigott.
16. *Chillington,*—J. Giffard, Esq.
17. *Somerford,*—Hon. E. Monckton.
18. *Hilton,*— —— Vernon, Esq.
19. *Aqualate,*—Sir J. F. Boughey.
20. *Oakley,*—Sir J. Chetwode.
21. *Swinnerton,*—G. Fitzherbert, Esq.
22. *Etruria,*—J. Wedgwood, Esq.
23. *Tixall,*—Sir T. Clifford.

The noble mansion of *Fisherwick,* (Marquis of Donegal,) has been suffered to dilapidate, even to the sale of the materials.

Farm-houses, Offices, &c.—The farm-houses of ancient date are built of wood and plaster, with little design or contrivance for either comfort or convenience. Those of modern construction are better contrived, built with brick, and covered with tile or slate: the barns, cow-houses, stables, and offices, are laid out so as to shelter a compact yard, with some instances of convenient cow-sheds, and feeding-stalls, which are necessary to every respectable farm: most of the old ones are very defective in these particulars.

In the construction of farm-offices, economy ought doubtless to be kept in view, and the money so laid out as to pay an interest to the proprietor, which will always be the case when real conveniences are considered, for which the occupier had better pay interest than go without them.

By the introduction of threshing-mills, less barn-room becomes necessary, as one threshing-floor is enough for any farm. A rick-yard should be in an airy situation, and well fenced, and secured from trespass. A Dutch or hay-barn, on an economical construction, would save much trouble in thatching: the security it gives from showers in hay-harvest, and in preventing the loss which happens from the deteriorated quality of the hay in the tops and bottoms of stacks, would enable an occupier to pay a consideration for it. Stables for horses, cow-stalls, and calf-house, hog-styes, and troughs of durable materials, are all necessary: these conveniences in some places have been well attended to, but in many cases are very defective.

Respecting foddering cribs of brick, the late Sir Edward Littleton observed of their utility, that they shelter cattle from the winds, and their fodder from the rain, and prevent fowls from injuring the fodder: the large and small beasts can feed at the same time. They should not be covered with hipped roofs, or the wet will run down four sides instead of two, and two troughs should discharge the rain from off the cattle: the apertures should be made so high as that the small beasts can but just reach to eat out of them, otherwise as the dung will constantly be rising round them, the cattle will walk into them before winter is over. An opening may be made at either end for the pigs to lie under it, and when the dung gets deep, a trench may be cut through it to drain off the moisture. The apertures for oxen should be wider: Sir Edward erected these foddering cribs in most of the farm-yards upon his estate.

Cottages.—The same worthy gentleman, with great humanity, turned his thoughts to the comforts of the labourer, by erecting upon his estate warm and comfortable tenements for their use. His idea was, that for the sake of economy three dwellings should be put together, with a room to serve them all for washing, baking, &c. and an oven large enough to bake for all three at once, the occupiers heating the oven in rotation, and giving notice of it to their neighbours: if by the side of a common the better, as the furze will supply the oven with fuel, and a few sheep, or geese, may be kept. Every labourer (he said) should have keep for one cow, with meadow-ground for winter-keep, and a garden of one-third of an acre for vegetables, fruit-trees, and hemp; and where it can be done, it will be desirable to have them near a pool of water.

Size of Farms.—The farms of this county are of all sizes, from twenty acres to five hundred and upwards; yet it must be acknowledged that, within the last fifty years, the consolidation of small farms has been prevalent, it having been found expensive to keep the buildings in repair: indeed, small arable farms are unprofitable, as a team of horses will eat up their produce. They should therefore be composed of grass-land, and occupied by a dairy of cows, well managed, and the cultivation should be with the spade, in the garden style. Farms of larger size are best adapted for raising grain for the public supply, as there is room for strength of team, and the greatest surplus of produce will thence be brought to market. But the true system is to have farms of all sizes; to employ the industry of the working farmer, and the capital of the more opulent.

Rents are in this county always paid in money, but something like personal services are in a small degree kept up, as a day's team-work annually for the landlord, or to keep him a dog. Rents have been considerably advanced since the year 1800, land near populous towns being let at five pounds per acre, in some instances at six or seven guineas, and garden-land near towns at 3s. per rod of 64 square yards, which is about £11. 7s. per acre. Many farms in the country have been raised from £1. per acre, to £2. or £2. 10s. The general rent of the county has, perhaps, been raised one-third, and is now from 30s. to £2. and £3. per acre for whole farms.

Tithes.—A considerable proportion of the land of the county remains titheable, though some has been exonerated. Tithes are considered by most people as operating unfavourably upon agriculture, but they must be admitted to be a property equally sacred with any other: no friend to justice can expect an abolition of tithes without an equivalent. Such an equivalent may be found in land, which might be set apart on all titheable estates by proper commissioners; and an equivalent in land must certainly be a more solid property than tithes, as it may be improved by industry, and managed with less trouble than tithes, even by clergymen themselves, or their agents.

Poor's Rates have been gradually and greatly increasing.

	£.	s.	d.
From Parliamentary inquiries, from 1783 to 1785, the annual expence of the whole county, upon a medium of those three years, was	45,404	10	10
Of this, net money paid for the poor,	40,963	13	7
County Rates, including Gaols, Houses of Correction, County Bridges, Militia, Vagrants, &c.	3,261	5	4
Churchwardens' Accounts for repairs of Churches,	1,179	11	11
Total,	£45,404	10	10
The total money raised within the year ending Easter, 1803, by the Poor's-rates in Staffordshire, was	£109,456	0	0

which was nominally stated to be 4s. 2½d. in the pound, and was then probably, upon the real value, a little under 3s. in the pound. In this proportion, the increase of 20 years is nearly as 2 to 5, and this may be considered as permanently fixed, if our present state of things be permanent also.

Weights and Measures.—Those customary in this county differ considerably from the regular standard, and from each other: the custom of Wolverhampton-market being 18 ounces to the pound of butter, 120 to the hundred of cheese, 9½ gallons to the bushel of grain and pulse, and 72 to the bushel of wheat; whilst that of other markets in the county varies, some being more and some less than the above. Malt is generally throughout the county sold by the Winchester bushel of 8 gallons, and wheat-flour by the stone or peck of 14 pounds weight. Much complaint has been made of this variation, and much has been said about a regulation of weights and measures; and it is now understood that the business is in the hands of the Legislature, who intend to establish one universal standard of weights and measures upon the following principle: A cubic foot of distilled water, at the temperature of 56½ of Fahrenheit, weight 1000 oz. avoirdupoise; a gallon is to contain 10lbs. or 160 oz. of such water, which is in its lowest terms 4-25ths of such cubical foot or 276·48 cubic inches: the quart will be 40, and the pint 20 inches, holding of pure water 1¼lb. the quart 2½lbs. and the gallon 10lbs.; the bushel of eight gallons being 2211·84 cubic inches, holding 80lbs. weight of water. An universal standard of weights and measures upon this principle was approved and recommended, in 1814, by a Committee of the House of Commons.

Implements of Husbandry.—These in use in this county for draught, are waggons and carts; for tillage, double and single ploughs,* with and without wheels, scufflers, harrows,† and rollers; for sowing, drill-ploughs and other drill-machines; and for other purposes, thrashing-machines, winnowing ditto, straw-cutting engines, &c. The waggons are either with six-inch wheels, or narrow wheels: the former kept by the larger farmers, in which six horses are drawn double, and convey a load of three tons and a half, or four tons; the narrow-wheeled waggons are drawn by four, and carry two tons and a half, or three tons. Six-inch wheel carts are mostly used about farm-houses, for conveying dung, repairing roads, materials for hollow draining, &c.; some few narrow-wheeled carts are kept by the small farmers for the road, also one-horse carts, and by gardeners, butchers, and people who supply the markets. The wheel-carriages of this county are well constructed, and not capable perhaps of much improvement: they should be made as light as possible, consistent with sufficient strength, as any superfluous weight in the carriage is so much taken from the load. Double or two-furrow ploughs, are much used, and answer well on light soils, where four horses will plough two acres or more per day. The single-wheel plough is a very good tool, requiring no person to hold or touch it, except when turning at the end of the furrow: they require but one attendant, for which a boy of twelve or fourteen years of age is sufficient. The superiority of a plough that requires no holder but a person only to drive the horses, and turn it in and out at the end of the furrow, to one which requires to be held, and the horses of which are guided by reins, must be evident, unless it can be proved that these latter are of easier draught, and that wheels encumber the movement.

The wheels have been much improved by the addition of an iron flay, firmly screwed to the coulter, which, in ploughing leys, takes off the turf and turns it into the furrow, where the plough immediately covers it with earth: by this management, a turf at one

* Swing-ploughs, drawn by two horses a-breast, and guided by the ploughman, chiefly with reins, are in use amongst many of our first-rate agriculturists, who think the strength of two horses so applied equal to that of three when drawing in length. The wheels are considered as adding greatly to the draught.

† The twin-harrow, for eradicating squitch, an excellent implement, is chiefly confined to the adjoining county of Salop.

ploughing has the appearance of a fallow, and harrows nearly as well: this ploughing requires no additional trouble but the strength of an extra horse, and the flay may be used or not at pleasure. The common swing-plough without wheels, is used to plough hedge-sides, ill-formed corners, or any difficult work, when both a holder and driver are required.

Several very ingenious manufacturers of agricultural implements reside near Wolverhampton. Winnowing-machines, of the very best construction, and containing all the newest improvements, are made by Mr. Joseph Cornforth, of Chapel Ash, of the best deal or oak, at from £8. to £10. each, who has made several thousands of them, and sent them to all parts of the kingdom, and even abroad, with the greatest approbation. Mr. Joseph Brewster, of Breewood, makes thrashing-machines either fixed or portable, upon the best construction, and upon the most reasonable terms: they do their work perfectly, and he keeps a team and portable machine in readiness to thrash ricks for any farmer at the shortest notice, and is thus employed all round the country. Mr. Rudge, of Pattingham, makes Northumberland ploughs either with or without wheels, requiring only one person to attend them, with two horses, which will plough any sort of land an acre per day, or more, if required: he has made great numbers for all parts of the country. Mr. James Chambers, of Ettingsal, also makes a great variety of drill ploughs for sowing all kinds of grain, pulse, and seeds, in different numbers of rows and distances; sowing tubes by hand, or mounted on a hand-barrow, with which a strong boy will sow ten or twelve feet wide of turnips, clover, or other seeds, with the greatest accuracy, as fast as he can walk; straw-cutters and turnip-slicers, and various kinds of horse and hand hoes, on the best construction; cultivators and iron-harrows, of various forms, hay-making machines, fly-catchers, to clear turnips of insects, and a great variety of other machinery.

At Ettingsal Iron-works, iron gates for avenues, or for common inclosures, are made on a firm, durable, and cheap construction; also iron hurdles for dividing grass lands or green crops, equally portable with those of wood, and much more durable, with ploughs entirely of iron, at five guineas each, and all kinds of ornamental and useful iron fencing.

Enclosing, Fences, Gates, &c.—The cultivated land of the county is now very generally enclosed, and nothing remains open but the commons or waste lands: the benefit of inclosing common fields has been so apparent, that it has been every where adopted.

In the enclosure of wastes, most new fences are made with post and rail, with a mound or bank of earth raised up near to the lower rail, and a ditch sufficient to drain the land where the drainage is required; in the inside of the bank, a little above the level of the natural surface of the earth, is planted the quicksets, for which purpose the white-thorn or hawthorn *(cratægus monogynia)*, is very generally preferred. In thin soils it is best to dig a trench two feet deep, taking out the under-stratum, and filling it up with soil, to facilitate the growth of the quicksets.

In planting fences on good soils, something better and more profitable than the hawthorn might certainly be introduced. An ingenious friend has suggested that if all hedges had been originally of the most profitable plants, they would have produced time past and to come beverage and materials for spirit to supply mankind, for fattening hogs, with materials for paying the expence of renewing; and all this with every advantage of the hedges now in use.

To effect this, he proposes to plant at regular distances in every new fence a Spanish chesnut; midway between the chesnuts plant a crab, and between the chesnuts and crab, a common plum; in the intermediate spaces, white-thorn; when they have been planted three or four years, and are in a thriving state, cut them off in February with a sharp knife just above the ground, to force out many shoots, and clean them from weeds, leaving the handsome straight plants for fruit: the chesnuts may be grafted with scions from Spain, if such can be procured; the crab stocks with apples; the plum shoots with plums of different kinds; and the white-thorn with medlars. Thus you will have a very productive orchard, without any waste of land, and a permanent and durable fence. The chesnuts would be excellent food for hogs or other cattle; apples every one knows the use of; and plums and medlars, if superabundant, would fatten hogs, or the juice might be converted into spirit superior to malt spirit. He thinks the chesnut, the walnut, and the beech, should be planted for their fruit; but the oak should be preferred to the lime, the fir, the larch, and the poplar, whose fruit is of no value.

In a late account of Sicily, it is asserted that a large tree is common there, called the Carubba tree, bearing a produce like a large strong bean in the husk: it is grown to export, and for feeding cattle. Its fruit is good feed for pigs, horses, and other animals. A tree is said to bear two quintals (350lbs.) of this fruit: if it would stand our climate it might be an acquisition.

It has been proposed, as an economical scheme, to plant willows of the proper kind for making gates, rails, and hurdles, they being of quick growth, and, when cut, soon shooting out again. They will succeed well in moist corners, or by the side of a stream.

Iron gates for common use have been lately introduced, and are purchased at reasonable prices.

Arable Land, Cropping, and Cultivation.—The arable land of the county, as stated before, consists of, 1. clay or marly loam; 2. gravelly mixed loam, including lime-stone bottoms, and thin black soils on gravel; and, 3. lighter sandy soils, adapted to the growth of turnips. The whole of the cultivated land of the county is now inclosed, or at least so little remains in common fields as not to deserve any particular notice. All the common sorts of grain, pulse, roots, and other plants and vegetables, are cultivated on the different soils.

On the deeper and stronger soils a good many beans are grown, generally in the rotation of, 1. fallow; 2. wheat; 3. beans; 4. barley or oats, with clover and other grass-seeds: the clover is often mown for horse-hay, and the land afterwards kept in pasture one or more years. Tracts of excellent land of this description are to be found near Stafford castle, in the neighbourhood of Eccleshall, and in the parishes of Gnosall, Bradley, Church Eaton, and elsewhere. Many farmers here fallow for wheat, and get good crops without manure, then dung the wheat-stubble for beans, which insures the crop, as well as one of barley afterwards, and of clover and grass: on this sort of soil wheat is sown in October, beans the beginning, and oats the latter end, of March, and barley in April. When laid to grass for more than one year, white clover and trefoil are generally sown with red clover and rye grass: if such land has been clean and well fallowed in one tillage, it may certainly pass through a second tillage without fallow. Thus: plough up the ley in a good thick furrow, and set beans in rows by hand, as is the common practice in Worcestershire; keep the beans clean as a garden by the hoe, and a double crop may be expected, and the land will be in as good order for wheat as it can be made by fallow: this is the best system of management for such land. This strong soil will form good permanent pasture, as is proved by here and there a piece of excellent old turf; but the occupiers say it requires many years to turf over, in which it will be much forwarded by top-dressings of good manure.

2. The mixed soils, composed of gravel, sand, and clay. These are often springy; but when well under-drained, some may become

sound and dry enough for turnips; if too cold or wet, they require occasional summer fallows; for when such land is getting foul and stubborn, it is the only mode to recover it, and is certain to make a grateful return for the indulgence. Such fallows should be ploughed up between Michaelmas and Candlemas, and cross-ploughed and well worked in summer, manuring with lime or muck: farm-yard muck is best laid on fallows not too much reduced. After one turning, or when fresh from the fold-yard, spread it immediately, and plough it in. It dresses more land, and meliorates the soil better, than when too much reduced. The couch-grasses are best destroyed by a complete summer fallow, as well as thistles, and the seedling weeds and coltsfoot by early spring culture. The usual culture of the mixed soils is, *first*, fallow as above, with at least four ploughings; *second*, wheat; *third*, the wheat-stubble pin-fallowed in autumn, and laid dry, cross-ploughed in March, and harrowed down, then ploughed up, and sown with barley in April, together with clover and other seeds, and then lying two or more years at grass. When ploughed up again, oats may be sown on the turf, and if the land be clean, you may manure the oat-stubble at Michaelmas, plough and sow winter vetches, to be eaten or cleared off the ground time enough for a short fallow for wheat. Pease drilled in rows, and well hoed, may succeed fallow-wheat, and barley after the pease, and clover and grass-seeds for two years' pasture, as before.

3. The light sandy or gravelly soils, adapted to turnips. On this sort of land the Norfolk system of, 1. turnips; 2. barley; 3. clover; 4. wheat; has been long well known and practised. Some farmers having observed that the same crop every four years is hard tillage, and tires the land, have given it two years' rest under the seeds, and sown the wheat upon a two-year old ley with success; or pease may be drilled in rows upon ploughing up the ley, and well cleaned by the hoe, to be followed by, 2. wheat; 3. turnips; 4. barley, with seeds, and then clover and pasture during pleasure. The land in preparation for turnips, must have at least four ploughings, and sufficient harrowings between. The first ploughing should be soon after harvest, and if possible the land should be laid up in one-bout ridges before Christmas; the second should be in March, and the land well harrowed down; the third in May, and the fourth at sowing; or if the stubble could be ploughed up immediately after harvest, turnip seed and a little rye might be sown and harrowed in upon such ploughing; but, if the growth be worth it, it might re-

main till the following spring for sheep-feed, to be eaten in April, and the ground afterwards worked for turnips. If the growth is not worth preserving, the second ploughing should lay it up in one-bout ridges before Christmas, and the land being exposed to the winter frosts, would harrow down mellow in spring. This last is the true mode of preparation for the Swedish turnip, which requires the land to be got forward enough for sowing by the end of May, and which cannot easily be done if one ploughing only of a stubble be given before winter. The manure in use for turnips is dung pretty well reduced, or lime, or both, or compost of dung and soil, or town manure. Near Birmingham, the parings and shavings of bone and hoof dust have been applied with great success. Turnips must be hoed twice, and hand-weeded, if necessary, as the object of the culture is to clear the land of weeds, which may be thus done effectually. It is now a well-established maxim that land, sound and dry enough for turnips, requires no other fallow or means of cleaning.

The writer once weighed a perch of a very good crop of turnips, and found it 36¼, which is exactly 30 tons per acre, but more acres are grown less than 20 tons per acre than above it; yet, by calculation, this seems very little, for turnips of 6 or 7lbs. weight are a good common size; but if we reckon one upon every square foot of 7lbs. weight, it would be upwards of 136 tons per acre, a weight we must never expect to see grown in reality. It was asserted in the public prints, that near 250 tons of Swedish turnips were grown in 1814, upon five acres of the Earl of Uxbridge's (now Marquis of Anglesea) estate. That such proportion might be grown upon small patches is very probable, it being 66¼ upon a perch, or 23¼lbs. upon a square yard nearly; but such weight is seldom or ever met with upon an average of five acres. If 30 tons be grown upon an acre, it may be considered equal to two acres of the best meadow land; and it is a great acquisition to a farmer, if he can make one acre of sandy or gravelly land equal to two or three of meadow in the support of stock, which may be done by industry and good management. The Northumberland method of getting turnips in drills on one-bout ridges, is extending itself in the county. The manure is deposited under the plants, and an opportunity is given of scuffling between the rows during the growth of the crop.

Wheat Culture, Seed, and Produce.—If turnips are eaten off the land in autumn, it is good management to sow wheat, as there can be no better possible preparation for wheat on light land; or spring wheat may be sown instead of barley, and three or four quarters per acre at least expected, and clover succeeds with wheat better than with barley. The common average crops of wheat are generally stated at from 3 to 4 quarters (Winchester measure) per acre on good land, from two bushels to two and a half sown: this is a very low average produce for good management. But the fact and the misfortune is, that wheat crops are seldom free from a pretty large admixture of grass and weeds, which share with the crop the nutriment from the soil. The following was the result of an examination in September, 1814, upon a twenty-acre field of wheat at Showell farm, near Wolverhampton: the occupier (Mr. Miller), had well drained and fallowed the land, manuring with both muck and lime: the soil mixed loam, of not above average quality. An acre in the middle of the field produced 99 mows of 8 sheaves each, or 33 thraves of wheat; some of the land had a sheaf on six square yards: a sheaf weighed 10¾lbs. and produced 4½lbs. of corn: this is 3 tons 15 cwt. 2 qr. 14 lbs. per acre weight of the crop, and 60¼ bushels at 60lbs. per bushel per acre produce of grain. The grains, when dry, weighed 96 to a half-guinea weight, or three grains of wheat were two grains troy weight, which is 10,500 grains of wheat to 1lb. avoirdupoise: the ears of corn were 276 on a square yard, or 30 and 2-3ds on a square foot, containing on an average 28½ corns each. It will appear by this, that if a wheat crop fill and occupy the whole ground, and be free from weeds, the average produce may be greatly increased.

Wheat is commonly sown at the rate of about 140lbs. weight per acre; this, weighing as above, would be 304 grains on a square yard, or little more than two inches asunder every way: yet it has so great a tendency to tillering, or making lateral shoots, that every grain sown in August might, upon an average, be divided into four plants, and re-planted in October. If taken up the March following, and again divided four-fold, and planted at nine inches, it will make a full crop, if the ground be free from weeds. A single grain would thus furnish the plants of a square yard, or less than half a pound weight of seed wheat would supply an acre; and this method may very probably come into practice when cultivation becomes systematic, and more hands are employed therein.

At present wheat is often drilled in rows at seven inches: this, at one bushel per acre, would be one inch and a quarter distant in the rows, and it is thus more ready to clean by the hoe. The great improvement wanted upon the present practice is, to keep the crop clear of grass and weeds.

Of the common lammas, or winter wheat, there are four varieties: 1. Spike and grain red, or, as the farmers call it, red straw red; 2. spike white, grain red; 3. spike red, grain white; 4. spike and grain both white. The first variety is most generally cultivated, but instances of the greatest produce from the last have occurred.

Spring Wheat Cultivation.—This is a modern introduction to British husbandry not generally known till about the year 1800. It is supposed to be a native of Tartary, and was first introduced under the name of Siberian wheat. The grain is rather longer and thinner than the common wheat, is liable to be hurt by frost, and must not be sown till spring; it has ripened when sown May 15th before barley sown at the same time; but is best sown in April: average value about ten per cent. less than the best autumn wheat. Being a smaller grain, it will ripen with less sunshine, and receives less injury in a wet harvest, drying sooner for grinding.

There are three varieties of the spring wheat, (*triticum æstivum*): 1. Spike and grain red; 2. spike white, grain red; 3. spike and grain white. It has been found by experiment that 9 grains of spring wheat are the same weight with 7 of autumn wheat, and that a 38-quart bushel of spring wheat weighing 72lbs. contained 984,275 grains of corn, which, if equally distributed over an acre of land, would be 203¼ grains upon a square yard, or only 2¼ inches distant asunder every way. As many bushels per acre are generally grown of spring as of autumn wheat; but it is looked upon by the millers as inferior to autumn wheat, particularly when wheat is plentiful. In 1814, Mr. Pratt, of Saredon, an eminent miller, who well knows the value of all the varieties of wheat, grew near 30 acres of it after turnips, which proved well, and yielded per acre 27 or 28 bushels 38-quart measure; which is full four quarters per acre statute measure: some were sown as late as May-day.

Cone-wheat (triticum turgidum), is not much grown in Staffordshire: there are four varieties; 1. White cone; 2. Red cone; 3. Bearded cone; 4. Many-eared cone.

Steeping of Seed.—This is commonly done to wheat before sowing, either by immersion in brine, stirring up and skimming-off whatever swims on the surface; or by sprinkling over it in a heap a sufficient quantity of brine or urine, and afterwards drying it with quick lime: the former method is, doubtless, superior. This, when properly done, has seldom been known to fail keeping the crop free from injury by smut, which is a contagious disease in wheat, and should be rejected as seed when tainted with it: the before-mentioned method is used as a preventative and cure.

Mildew is also sometimes very injurious to wheat: the cause is supposed to arise from a too humid atmosphere, and want of sunshine, which are the proximate causes; and improper tillage, and an abundance of weeds, the predisposing ones. A want of circulation in the air, rendered stagnant by high hedges, may be considered as having a tendency to mildew: hedges should always be plashed against a wheat field. It has been asserted, that minute *fungi* are the cause of this disease; but this is beginning at the wrong end: such, if they exist, being the effect of the distemper, and not the cause, their natural growth being upon putrid and decaying substances.

Rye, for a crop, is not much grown in Staffordshire, but sometimes a little may be sown on light land, or on head-lands. No particular preparation is necessary except a fallow manured upon very poor sands; its produce may be reckoned more than that of wheat on light poor land: in a scarce time of wheat it is a welcome addition to that grain for bread. A valuable use of rye is to form a very early sheep-pasture for ewes and lambs in April; if a pea, oat, or other stubble, be cleared of its crop in August, and immediately ploughed up and sown with rye, and half a bushel of winter vetches per acre to fill up the bottom, it may be half a yard high by the beginning of April following, and form a valuable pasture for ewes and lambs at that pinching season: it may be eaten off in good time to work the land thoroughly for turnips.

Barley is generally sown after turnips on all land where turnips are grown; on loams it is sown after fallow wheat, the wheat stubble being pin-fallowed in autumn, and well worked in spring; or after a clean wheat fallow barley has been grown at one ploughing given in autumn upon the wheat stubble, laid in a proper form for sowing, and drained by proper furrows or gutters, and the barley sown early in the spring following without farther ploughing; as the amelioration of the ground by winter frosts disposes the ground to work kindly under the harrows without more tillage; this upon strong loams: clover, and grass-seeds, are always sown with barley as above.

Drilling-in of barley in rows is practised by some. After drilling, the Rev. Mr. Dickenson defers sowing the grass-seeds for a month, and then hoes them in between the rows of barley; or has harrowed-in grass-seeds by a single-horse harrow without injuring the barley: the grass-seeds late sown succeed equally well, and do less injury to the barley crop in harvest than when more fully grown.

Oats are grown upon one ploughing-up of grass land, and if such land be mellow and in good condition, the greatest crops are thus obtained; 5 or 6 bushels per acre of Poland oats are sown, which upon such land may produce 40 or 50 bushels in return, and on turnip soils should be followed by turnips.

Or the oat stubble may be manured, and immediately ploughed and sown with winter vetches. These may be used in spring, time enough to make a good fallow for wheat or for turnips. Oats are also grown in succession after fallow wheat on cool land (improper for barley), generally upon one ploughing of the wheat stubble, when grass-seeds are sown with the oats. This is too much practised, and an imperfect tillage: it would be much better to sow grass-seeds in spring upon the fallow wheat, and let oats and vetches precede the wheat as above.

The oats adapted to the better kind of land are the white oats, Dutch or Poland, also a variety called potatoe oat, which is an improved variety of the Dutch oat. Red oats are preferred by some for horses, and the black oat is sown upon inferior soils; but the better kind of white oats is alone used for groats and oatmeal, and in that form for human food.

Oats are a considerable article of human food in this county, besides their general use as groats, and in soups and gruels. Oat bread is eaten in the Moorlands over a considerable extent of country; the origin of which custom seems to have arisen from that high mountain-land being too cold to ripen wheat. It is believed that oatmeal is equally nutritious with wheat flour, weight for weight; and two bushels of oats are nearly equal to one bushel of wheat as human nutriment.

For, one bushel of good wheat, Winchester measure, weighing 60lbs. will yield 49lbs. of flour, 10lbs. of pollard or bran: 1lb. waste.

Two bushels of oats, same measure, weighing 80lbs. may yield 50lbs. of oatmeal, 28lbs. of husks: 2lbs. loss.

Advantage on the side of the oats 1lb. meal, (28lbs. of husks being supposed equal to the 10lbs. of pollard or bran); oats therefore to be equal to wheat in cultivation as human food, must be produced in double quantity per acre.

Beans are grown considerably upon the strong loams, and are principally sown broad-cast, and left to chance. The Worcestershire practice cannot be too strongly recommended as much superior, where none but the greatest slovens think of sowing them broad-cast. They are there very generally either set by hand, or drilled by a machine; in the former case women and children are principally employed, who set from 3 to 4 bushels per acre of the large tick bean, or less measure of smaller sorts.

The expence of hand-setting there in 1807 was about 8s. per acre, and a quart of cyder daily to each setter. A gentleman by letter thus informs the writer: "We excel in nothing so much as setting beans; it is superior to the most perfect drilling. They are all set by line, and we prefer setting them north and south, to have the benefit of sunshine between the rows. They are hoed three times with the gardener's common hoe, and they produce sufficient to satisfy the cultivator." The time of setting is February and March.

Various machines are now in use for drilling beans (see "*Implements*"); and the expence may be reckoned at less than one-half of that of setting: it is practicable upon all broken land, particularly after fallow wheat; but turf land should be set by hand, and if clean hoed, must equal any fallow for wheat.

Pease are extremly well adapted for drilling by a machine, their globular form insuring a delivery with great accuracy. They should be grown only upon land in clean culture, and kept perfectly clean by the hoe, which they may be when in rows; the land is then fit for any crop. Turnips after the early sorts, and wheat after the latter, or any other crop, at pleasure. In the neighbourhood of London £50. per acre was made long ago by early pease for podding, immediately succeeded by turnips, also for the market; and both cleaned in a garden style: the same practice is applicable to this county in all dry warm situations near a populous town or neighbourhood. The garden Rouncevel has long been sown in the common fields of Milwich with great success.

Vetches, of the winter kind, are very properly sown in autumn, upon land intended for wheat fallow the next summer, as the crop may be used green or made into hay, in time for working such fallow; if the vetches are well manured for, less manure would be required for the fallow. It would be a good plan to muck for the vetches, and lime the fallow for wheat. Ten tons per acre of green vetches, or from two tons and a half to three tons in hay, may be thus obtained instead of barren fallow. It may be here observed, that vetches mown before they are in full blossom, will shoot again;

but when the full blossom is attained, the ground should be ploughed as the crop is cleared.

Buck-wheat, called here French wheat (*polygonum fagopyrum*), has been grown both for ploughing under as manure, and for a crop. This plant is not considered good enough to cultivate, except on new inclosed or unimproved land, where ploughing it in may increase the vegetable matter; it is, however, good enough for game and poultry, as well as swine. Dr. Withering prefers to it for culture the black bindweed, (bearbine of the farmers, *polygonum convolvulus*), the seeds of which, he says, are produced in greater quantity, and the plant bears cold better.

Hemp and *Flax* are grown in this county, but upon a small scale: hemp generally in small yards, or in the gardens of cottagers, or sometimes an acre or two may be grown by a farmer after turnips; and some few instances have been known of large crops upon a deep black peat, well drained. It should be sown the beginning of May.

Flax is generally grown upon a turf on one ploughing, and should be sown in March or the beginning of April, two bushels and a half of seed per acre. The land should be well harrowed before sowing the seed as well as after, and rolled, and harrowed again very fine: care should be taken that the seed be ripe when pulled, it being a valuable part of the crop. The chaff of flax is excellent food for horses, mixed with oats or beans. A middling crop will produce 30 stone of flax per acre, and a full crop considerably more.

Turnips.—Their usual culture has been before stated, to which may be added, that they have been drilled-in upon the Northumberland system, both common and Swedish, by several public-spirited gentlemen, particularly by Sir John Wrottesley, and Sir George Pigot, who have cultivated considerable portions of their own estates. The system is thus managed, as detailed by a gentleman who practises it. After well working the land, and laying on lime four tons per acre, the lime being spread and well harrowed-in, the land is stricken into two furrow ridges, about two feet from middle to middle; the dung cart is then applied, and a row of well-reduced dung laid along every fifth hollow, at the rate of about ten tons per acre; this is immediately divided and distributed along the hollows, and another plough follows to cover the dung by dividing the ridges upon it in single furrows. The drill machine follows, drawn by one horse, going along one of the furrows: a roller presses down two ridges, one on each side the horse's path.

The seeds are deposited over the manure on the crown of the ridge, and covered in by short rollers following the delivery. Two rows are thus drilled at a time by one horse. They can get on thus at the rate of about an acre and a half per day, with four horses in two teams to plough, one at the drill, and two or three at the dung carts. The whole is immediately done on the fresh soil, by which the vegetation of the seed is promoted, the dung covered up under the seed, and a good crop generally insured. When dung is not used in addition to the lime, the drill immediately follows without turning back the furrows, and five horses do the whole business. The cleaning is performed by a hoe plough between the rows, which are further thinned and cleaned by women and children, no skill being necessary. About 28lbs. of seed is drilled to an acre.

Sir John Wrottesley states, that to form the ridges he uses a double furrow plough with two horses, which turn a furrow either way, and the muck is laid in every eighth row, which one person spreads, and four women or boys follow, each distributing the dung in two furrows: two common ploughs follow with two horses in each, to cover-in the dung, after which comes the roller and drill, which being very light work, is best done by a hackney or blood horse; and that turnip sowing being at a busy season, with the above strength and sufficient horses at the dung carts, according to the distance the muck is drawn from, they generally finish four acres per day.

The Fly.—The insect which most commonly attacks and destroys turnips is of the beetle kind *(chrysomela nemorum)*, wings two, covered by two shells, skippers, hop or take-wing, length half a line, or one-twenty-fourth of an inch: when much disturbed, it takes wing, and is soon out of sight. To keep off this insect, it is recommended to mix an ounce of flour of brimstone with every pound weight of turnip-seed twenty-four hours before sowing: if this be neglected, and the fly be discovered, immediately harrow with light harrows, which will disturb or drive the fly: or if the fly continues, sow eight bushels per acre of dry lime, or dry sifted fine ashes, early in the morning, or in the evening, when the dew is on, that it may adhere to the leaves. Hoeing in time, and repeatedly, is said to disturb and drive the fly: it is asserted that where these precautions have been used, no turnip crop has ever been destroyed by the fly. The following method is said to have succeeded last year, in preserving twelve acres of Swedish turnips. A board was drawn over the field, having its upper surface fresh painted: on the approach of the board the flies hop, and are entangled in the fresh paint, which must be renewed as often as necessary, or tar may be used instead of paint. The following system is proposed by the inventor: Take two boards, half an inch thick, from a deal plank, connect them together by three or four pieces of cord, leaving an interval between them of about a foot; both the boards to be painted, or tarred on the top; some strings or fringes of netting hanging loosely from the hinder edge of the first board will, by playing on the ground, make many of the insects hop and alight on the second: a strip of wood to be nailed under the fore-edge of each board to keep it from warping or catching the ground, will make them mount any clod or stone in their way. A piece of cord, four or five feet long, is to be put through each fore corner of the first board, with a cross stick at its fore end, to be drawn by a person at each end. If the boards be twenty feet long, and laid crosswise, they will thus sweep twenty feet in breadth; and two persons in going less than half a mile will sweep an acre, and in going five miles will sweep more than ten acres: by following the hinder board, you will have the pleasure of seeing the enemy entangled in the paint or tar. The operation to be repeated daily, or twice a-day if necessary, until the turnips get into the rough leaf, when they will be secure from further danger.

Mildew on Turnips,—caused, as is generally supposed, by the *aphis brassicæ*, an insect so extremely minute in its animalculæ state, that fifty or more may be found beneath one pair of seedling leaves of a young turnip plant. In its fly state it has four wings, two long and two short, body black, size of a grain of mustard seed, is extremely prolific, and produces ten generations in one season, each successive generation beginning to breed at ten or twelve days old. First generation *(oviparous)*, hatched by the sun; the succeeding ones, except the last *(viviparous)*, and fifty on an average produced at a time; so that the production of a single fly in a season is 50×50×50, to the tenth power, which amounts to countless millions. Dr. Darwin believed this species to be so wonderful in its increase, that, from their immense numbers, they may in process of time destroy the vegetable world. The top-dressings before mentioned are very probable means of checking the progress of their increase.

Turnip Tenthredo (tenthredo rustica), less frequent on turnips than the former, common on willows, and when they increase beyond the usual numbers, resort to turnip-fields. Great ravages were committed on turnip-fields by the caterpillar of this fly in the summer of 1786; when, after hot weather, they attacked the turnip plants when full half grown, perforated the leaves into net-work, and did not finally disappear till the frost set in; they have not been very formidable since. Many farmers were alarmed, thinking a new enemy to the crop had appeared: it is supposed that some circumstances in the weather had been favourable to their increase, and that their progeny was prematurely destroyed by frost. Ducks were employed in large droves with some success in destroying this caterpillar. They should be pent up all night, and driven early in the morning, fasting, into the midst of the turnip-field, when they will fill their craws as full as they will hold; they will also devour slugs and worms: the former are said to be injurious, and sometimes destructive, to a turnip-crop. Mr. Vagg's grand secret of night-rolling is well known, and still recommended by some persons, as tending to squeeze and destroy injurious insects better than rolling in the day-time.

Some persons are said to have soaked the seed, and made it spurt before sowing, to promote a quick vegetation, which method Dr. Darwin advises to be more attended to. He also advises to prevent or destroy insects, that the ground be strewn with soot, and thinks it probable that if infusions were made in hot water, or for a longer time in cold water, of those leaves which no insects devour, as of walnut, laurel, foxglove, henbane, hounds-tongue, ragwort, or tobacco, and were sprinkled on the ground just after the young plants spring up, it might prevent or destroy insects, without injuring the crop. This, however, can be practised only on a small scale.

Potatoes have been, and are a good deal, cultivated, both with the spade and plough, in various ways, and the culture well understood. They have been applied to the fattening of cattle, hogs, and poultry, as well as for food for mankind. The refuse potatoes, and inferior kinds, are generally applied to feeding hogs, which they will bring on fast, when boiled and mixed with barley-meal, and the inferior or damaged barley will do for this purpose. The markets are principally supplied with potatoes by cottagers who have large gardens, or who rent land for the purpose of growing them: some also are grown for the market by farmers.

The curl is now pretty well got rid of, the sorts liable to it having been rejected, and new varieties raised in their stead. The animal nutriment from potatoes is decidedly superior to that from any other plant or vegetable in common culture and use, from the same breadth of land; this, perhaps, is owing to the starch and mucilage which they contain. An acre of potatoes produces, on an average, as much human or animal nutriment as from three to four acres of wheat; and it has been found by experiment that an hundred weight, properly used in feeding hogs, will produce 4lbs. of pork, or 10 tons from an acre would produce 800lbs. of pork, and so in proportion. Potatoes are thus worth a farthing a pound as hog's meat, when pork is worth 7d. The hog-manure remains to pay for the trouble of feeding, and the labour and expence of taking the potatoes to market will be saved.

Cabbages.—Their cultivation in the field has been common for many years, but has been less attended to since Swedish turnips have become a favourite. They are generally grown on three-feet ridges, manured under the rows, with one plough-hoeing, afterwards hoeing by hand. It may be observed, that they succeed best on deep strong loams, too moist for Swedish turnips; and that they can be come at in frost and snow, when turnips are buried or locked in the ground. To stand the winter, they should be raised from spring-sown plants; but the greatest crop will be raised from plants sown in July or August, planted out the following spring, and having the whole summer for their growth. The excellence of cabbage for milch cows, and ewes and lambs, is well known.

Rape.—This plant has never been known to be cultivated in this county for its seed, and many leases contain restrictions to prevent it. It is properly sown on early stubbles ploughed up as spring-feed for sheep, and followed by turnips. The smooth chadlock is the same plant in its natural state.

Mangel Wurzel, a variety of the beet species, very productive of leaves, with large conical fleshy roots, which appear beautifully variegated when cut across in slices. The author was one of the first cultivators of this plant, having received his seeds from Dr. Letsom, when he first introduced it under the name of the Root of Scarcity. The roots are very sweet, and must contain much sugar, but they have a nauseous flavour. He thought the plant inferior to potatoes, turnips, carrots, or parsnips, and therefore dropped its cultivation.

It has since been considerably cultivated; but has lately sunk into disrepute, upon a charge of paralyzing cattle that eat it. This arose, perhaps, from its having been eaten too freely, it probably requiring to be given only in a limited quantity. Unpleasant effects have also accrued from cattle eating raw potatoes too largely, by hoving and swelling them. Potatoes, when raw, possess considerable acrimony, and should not be given to a beast in larger quantities than a peck at a time, nor more than three times a-day: the acrimonious quality is dissipated by heat. It is probable, the mangel wurzel would be made perfectly safe by baking or boiling.

Its cultivation should be upon good deep soil, well pulverized and manured; the seed should be sown in April, and buried about three-quarters of an inch deep. It may be dibbled to advantage by having a stop on the lower part of the dibble, to prevent the holes being made deeper than here directed. It is proper to form a germination before the seed is planted, by steeping it in soft water twenty-four hours, to forward its growth, and (if necessary) to ascertain its vitality, especially if the ground is dry at sowing. A proper distance would be 12 inches asunder every way, which gives 43,560 plants upon an acre, and would require 2lbs. of seed.

When the plants are fairly above ground, give them a hoeing with a carrot-hoe to kill the weeds, and fill up vacancies by transplanting from a seed bed. A second hoeing should be given with a turnip-hoe when the roots are the size of a radish, and again fill up vacancies. The whole number of plants above-named may thus be preserved, which, at 1lb. weight each plant, would amount to 19 tons 9 cwt. per acre, and so in proportion. When full grown, the leaves may be stripped off and given to cows, deer, sheep, or swine, taking care to preserve the middle leaves, which will soon form another head. The stripping may be repeated.

The crop must be taken up in October or November, (in dry weather if possible), and the leaves stripped for use, leaving the crown of the plant perfect; then put them under cover in straw, in alternate layers, to preserve them from the wet and frost; or they may be laid in ridges the same as potatoes, and covered with straw. They may be given all the winter in limited quantities to horses, cattle, sheep, deer, and swine; and in spring, if the whole are not consumed, the roots may be planted out for seed, or for use; for, after ripening the seed, the roots are still fit for use, and may be sliced and given to cattle as before.

Carrots and *Parsnips* may be cultivated in a similar way, as well as preserved and used; they would be extremely valuable for all kinds of stock in the winter and spring months, and would well repay the trouble of the cultivator.

Weeds.—These are a great plague to the plough farmer, and occasion the necessity of summer fallows. On neglected farms they are known to take away more corn than the tithe-man. The following are those principally infesting the arable land of this county. 1. Couch grass, provincially squitch, which signifies the roots of perennial grasses, as the bent grasses, (*agrostis*); dog's-grass, (*triticum repens*); creeping soft grass, (*holcus mollis*); the tall oat-grass, (*avena elatior*); and some others: the last-mentioned grass has bulbous roots, which breed or shelter grubs, worms, &c. These roots are sometimes so interwoven with the soil in hard tilled land, as to form a perfect matting, and choke up the plough. They are only to be destroyed by repeatedly ploughing the land and working it in hot dry weather, or by forking-out and burning the squitch. 2. Coltsfoot, (*tussilago farfara*); and, 3. horse-tail, (*equisetum arvense*). These are to be destroyed by early spring ploughing, as they blossom and seed in March; and will then, if cut, bleed to death. To extirpate them entirely requires repeated ploughing and hand-weeding, and by perseverance, and laying clean to grass, they will at length disappear. 4. Creeping thistle or saw-wort, (*serratula arvensis*), growing every where, may be weakened by good tillage and weeding; but its seeds fly all over the country, if suffered to ripen, as well as those of coltsfoot and the ranker thistles, and also groundsel (*senecio vulgaris*). They should, by universal consent, be cut before seeding, in hedges, waste places, and elsewhere, and might thus be extirpated. 5. Chadlock, pronounced kedlock: three distinct weeds bear this name, viz. *wild mustard*, *wild rape*, and *wild radish*. If they are suffered to shed their seed in one tillage they appear in the next, and are thus perpetuated, which is the case with all the seedling weeds. They can only be destroyed by pulverizing the land early in spring, and ploughing them up or harrowing them out as they appear, or by hoeing or weeding them out of all crops before they ripen and shed their seed. Other weeds of this class are, 5. Ivy-leaved chickweed, (*veronica hederifolia*); 6. common chickweed, (*alsine media*); 7. bindweed, (*convolvulus arvensis*); 8. pale arsmart or lake-weed, (*polygonum pensylvanicum*); 9. bladder campion, (*cucubulus behen*); 10. corn poppy, (*papaver rhœas*); 11. corn crowfoot, (*ranunculus arvensis*); 11. nettle hemp, (*galeopsis tetrahit*); 12. shepherd's purse, (*thlaspi-bursa-pastoris*); 13. groundsel, (*senecio vulgaris*); 14. tare, (*ervum tetraspermum*); 15. corn marigold, (*chrysanthemum segetum*); 16. corn chamomile, (*anthemis arvensis*); 17. knapweed, (*centaurea*); and, 18. fern, (*pteris aquilina*); and many others.

The weeds common in grass-land which should be extirpated, are, 1. docks, (*rumex*); to be destroyed only by perseverance in rooting up; 2. thistles, (the *carduus lanceolatus*, *palustris*, and *pratensis*, are called boar thistles, from their roughness); these, with the creeping thistle, (*serratula arvensis*), should be cut off close, or rooted up; 3. knapweed, (*centaurea nigra* and *scabiosa*), should be rooted up, being useless and unsightly; 4. goose tansy, silver weed, or feathered cinquefoil, (*potentilla anserina*), refused by cattle, and a mark of surface-water and want of shallow draining; 5. rushes, (*juncus*), where permanent, a general indication of under-water and want of deep drainage; 6. red rattle or louse wort, (*pedicularis sylvatica*), another indication of want of drainage, and a sure mark that the land is unfit for sheep, which soon become scabby and unsound where this plant abounds; 7. ramsons, (*allium ursinum*), common near streams, and give a garlic flavour to the milk of cows that eat it; 8. dog's mercury, (*mercurialis perennis*), common on the hedge banks of moist land, and said by Dr. Withering to be noxious to sheep and deleterious to man; 9. common nettle, (*urtica dioica*), common on rich land in patches, but should be rooted out; and, 10. sedges, (*carex*), on moist land, and to be extirpated by draining and top-dressing.

Care should be taken always to sow clean seed. The following are often sown with the seed, and will, in that case, abundantly increase: 1. white darnel, (*lolium temulentum*); 2. eriff or goose grass, (*galium Aparine*); 3. beggar's needle, (*scandix pecten*); 4. bearbind, (*polygonum convolvulus*); 5. cockle, (*agrostemma githago*); and some of those above-named. Summer fallowing, and well-managed hoe crops, with attention and industry, are the means of destroying all weeds.

Weeding.—For extirpating weeds, the drill husbandry is to be highly recommended. It gives a good opportunity of working between the rows of grain, and particularly of pulse, beans, and pease, and thereby keeping the crops clear. In broad-cast sowing, thistles which cannot be handled should be drawn up with tongs, and grasses and other weeds by hand. Turnips are cleaned first by the hoe, and if any weeds remain they should be drawn by hand, and not suffered to shed their seed.

Every good farmer will root up the docks, and the ranker thistles, which infest his pastures, which may be done by a docking-iron when the ground is moist after rain; if they are only mown, it checks but not destroys them. Other luxuriant weeds, as knapweeds and scabious, should then also be drawn by hand, and this should always be done before their seeds are perfected, as thereby a great increase of these noxious plants may be prevented. To shew the importance of attending to this subject of destroying weeds before they perfect their seeds, it may be observed that the business, if well done, will lessen every year, but if neglected will increase beyond calculation; for

> One year's seeding, makes seven years' weeding;
> But one year's good weeding, may prevent seeding.

As a preventative is much better than a cure, it would certainly be an improvement in our rural polity, if it were made presentable and fineable at quarter sessions, and the offender punished by fine accordingly who should suffer weeds to run to seed on his premises, particularly those of the class *syngenesia*, which having their seeds furnished with feathers and wings, will disperse them over a whole country. Mr. Marshall thinks an indictment at the court-leet might be preferred for such a nuisance, and matters of less importance have engaged the attention of the Legislature.

The following is a list of the most troublesome and pernicious weeds of this class, growing in hedges, road-sides, on heaps of compost and manure, and in waste places, as well as in grass and cultivated land: 1. coltsfoot, (*tussilago farfara*), the seed flies in April and May; 2. creeping thistle, (*serratula arvensis*), the seed flies in July or August; 3. great or spear thistle, provincially boar thistle, (*carduus lanceolatus*), flies in August and September; 4. sow thistles (*sonchus arvensis*), flies in August and September; 5. smooth succory, (*crepis tectorum*), often in foul oat crops, the seed flies from midsummer to autumn; 6. groundsel, (*senecio vulgaris*), the seed flies all the summer. Many other sorts have the same quality, as ragwort, (*senecio jacobæa*); hawkweed, (*hieracium umbellatum*), and many others.

The seeds of these weeds, when fully ripe, will hover in the air, and in windy weather, like winged insects, fly to any distance, till saturated with moisture or becalmed, when they deposit themselves on cultivated ground, or newly-raised banks of earth, where they soon vegetate and establish themselves in vigour. Many persons have supposed, from observing their growth upon newly-raised banks of maiden earth, such as canal spoil, &c. that they had been dug with such soil out of the earth; whereas, nothing can be more simple or natural than their flight from the parent plant, borne up by feathers through a current of air.

Weeds are often suffered to grow on heaps of compost, or manure, so as to perfect their seeds: this is great neglect, as such manure may afterwards do as much harm as good. They should be mown, and not suffered to seed, if the heap cannot be turned over; or if the growing weeds were smothered with fresh muck, or soil, or lime spread over the heap, they would add to the value of the manure instead of fouling it and the country with their seeds.

The true use of summer-fallow is to destroy weeds, and to apply the whole force of the land to the intended crop; but the end is frustrated unless the fallow be well managed. Ploughing in dry weather has a tendency to destroy root weeds, as squitch, thistles, coltsfoot, horsetail, &c. and for the two last the ploughing should be early in the season (March or April); but the ground should afterwards be harrowed down fine, and left for showers to force the vegetation of seedling weeds, and when vegetated, they should be ploughed-in, and the ground again harrowed, to vegetate those seeds which were before too deep: some of the seedling weeds require early pulverization to force their growth, or they will appear in the crop, and sow their seeds there if not rooted out.

Grass: *Natural Meadows and Pastures.*—The natural situation of these is in the valleys, and on their sloping sides, where the sediment of water from the upland has for numberless ages been accumulating and increasing the soil; and the stagnant water, for an equal length of time, has, in the lower parts, accumulated the increasing roots of aquatic plants into regular beds or strata of peat. By the drainage of such peat, a fermentation is brought on, and its surface is converted into vegetable mould, which, by top-dressing, consolidating by rolling, &c. is rendered productive of grass, and becomes meadow-land. By these operations, the vegetable matter of the peat (which before was in an inert state), is brought into action, and an improved peat-moss in length of time becomes valuable meadow-land.

That part of a vale which is composed of sediment of water brought from the upland, is, however, generally richer, and produces a better staple of herbage than that whose origin is from peat. Some of the richest meadow-land in the county has originated from the sediment brought thither by water, accumulating and increasing from the beginning of time. The low-lands adjoining all our rivers and brooks come naturally under this head of meadows and pastures, as well as considerable tracts of flat land, which, by the backing on of water in former times, have acquired a stratum of peat on their surface. Considerable tracts of such land, lying near the rivers and brooks, are improved and kept in heart by the natural overflow of the streams; but this advantage is enjoyed at the risk of having the crop occasionally swept away, or soiled over by untimely floods, which frequently occur.

Very rich meadows are to be found on the Trent, near Burton; and the fertile meadows on the banks of the Dove are an interesting object. This river, which separates the counties of Stafford and Derby, arises from springs under the lime-stone hills of the Moorlands and the Peak, and at times receives an amazing addition from torrents rushing down those hills after heavy rains or the melting of snow. Its upper channel has a great declivity; in many places the river comes tumbling over the rocks in cascades, and in its greatest swell pushes on with astonishing rapidity. The water has a greyish cast from the calcareous earth it contains, to which may undoubtedly be attributed the extraordinary fertility of its banks, which is and always has been proverbial; "As rich as Dove," being an epithet applied to any spot highly fertile. The farmers are accustomed to say, that it is scarcely possible to overstock a few acres of Dove land. This land has an almost perpetual verdure, and the spring-floods of the river are very gratifying to the land-occupiers, who have this proverb: "In April Dove's flood is worth a king's good." It is also said of Dove banks in spring, that a stick laid down there every night shall not be found the next morning for grass. This river fertilizes its banks like another Nile, but sometimes rises so high in twelve hours as to carry off sheep and cattle, to the great alarm of the owners; and in as few hours abates, and returns within its own channel. Below Rocester, where this river receives the Churnet, the plain spreads wide, and continues so with variations to below Uttoxeter, where it is near a mile in breadth, and amounts to several thousand acres, almost entirely pastured with cows, sheep, and some horses, very little of it being mown for hay, the uncertainty and suddenness of the floods making the risk of hay too great. The rent of Dove land was many years ago forty shillings to three pounds per acre; and it would be worth ten shillings per acre more at least, were it not for the inconvenience of floods, which sometimes occur when there has been little or no rain on the spot, from sudden rain or melting of snow on the Moorland or Peak hills. It must be remembered, however, that these floods are, and have been, the cause of the extraordinary fertility of the land.

The following is a list of our principal spontaneous pasture and meadow herbage, ranked according to the author's conception of their merit: 1. meadow-grasses, (*poa pratensis trivialis*), the staple grass of our best meadows, most luxuriant by watering; 2. fox-tail grass, (*alopecurus pratensis*); 3. rough cock's-foot grass, (*dactylis glomerata*): these two last are the most productive of all our grasses, and particularly on grass-land that cannot often be watered; the last is eagerly eaten by all sorts of cattle when young, and would bear mowing three times in the season, if well manured. The fox-tail is also early and very productive. 4. Annual meadow-grass, (*poa annua*), sweet and fine grass, ripening its seed and sowing itself three or four times a-year, after which the old root dies: said to have been sown in Suffolk, and thence called Suffolk grass. 5. Meadow fescue, (*festuca pratensis*); 6. dog's-tail grass, (*cynosurus cristatus*); 7. meadow oat grass, (*avena pratensis*); 8. vernal grass, (*anthoxanthum odoratum*), fine, sweet, and early, but not very productive.

The above grasses are all certainly well worthy of trial in cultivation. The following are also common in grass-lands. 9. Ray-grass, (*lolium perenne*); the seed common at market, and the only grass that has been much cultivated, although seemingly inferior to most of the above-named. 10. Oat grass, (*bromus mollis*); this has also been cultivated, is hardy and productive, but somewhat coarse; it will flourish on poor thin soils. 11. Yellow oat grass, (*avena flavescens*); common in good grass land, and is good herbage. 12. Timothy grass, (*phleum pratense*); this is also sometimes cultivated, and is productive and durable in the ground, but somewhat coarse. The other most common grasses, are bent-grasses of different sorts, (*agrostis alba*, &c.); tall oat grass, (*avena elatior*); a bad couch grass in arable, but a productive meadow-grass; dog's couch or squitch-grass, (*triticum repens*); hair grass, (*aira cæspitosa*); quake grass, (*briza media*); brome grasses, (*bromus arvensis*, &c.); meadow barley grass, (*hordeum pratense*); soft grass, (*holcus mollis*); and many others.

The water meadow grass, (*poa aquatica*), and reed canary grass, (*phalaris arundinacea*), grow very tall, six feet and upwards, are extremely productive, and very common in streams and hedges: cattle eat these grasses, and they are said to be cultivated in the Isle of Ely. They mow it twice a-year for their cattle in Scandia: (*Linnæus*). A variety of the latter, with striped leaves, is cultivated in our gardens, and called ladies' traces. The flote grass, (*festuca fluitans*), generally growing in water, is a sweet good herbage, and very productive. Geese and ducks are very fond of the seeds of this grass. Water hair-grass (*aira aquatica*), generally grows in water, is not uncommon, has a sweet taste like liquorice, and cattle are very fond of it.

Other principal meadow-herbage, valuable for hay and pasturage, is composed of plants of the *diadelphia* class, as follow: 1. The meadow or cow clover (*trifolium flexuosum*), the seed of which is professed to be sold under the name of cow-grass: it abounds naturally in many meadows and old pastures, but the seed bought under this name seldom proves perennial and durable. 2. White clover, (*trifolium repens*), abounds naturally in sound meadows and pastures, is highly valuable, and much cultivated. 3. Hop trefoil, (*trifolium procumbens*), good herbage, abounding in meadows, and sometimes cultivated. The bird's-foot trefoil (*lotus corniculatus*); tufted vetch, (*vicia cracca*); meadow vetchling, (*lathyrus pratensis*); meadow burnet, (*sanguisorba officinalis*); meadow sweet, (*spiræa ulmaria*); and meadow sorrel, (*rumex acetosa*); are common in meadows, and grateful and wholesome to cattle, either green or in hay: the cow-weed, (*chærophyllum sylvestre*), has an uncouth and weed-like appearance, resembling hemlock, but paler in colour; common in meadows; cows eat it greedily when not too old, and it is wholesome food. The crow-foot, (*ranunculus bulbosus, repens*, and *acris*), the butter-cups of the meadows, so very abundant in themselves, acrid and pungent, are, notwithstanding, a grateful and desirable admixture, acting as an astringent, and probably correcting the flatulent qualities of the more luxuriant grasses, and are useful in the hay. The dandelion, (*leontodon taraxacum*,) and yellow goat's-beard, (*tragopogon pratense*), have durable qualities, salutary to cattle; the yellow rattle or penny-grass, (*rhinanthus crista galli*), and yarrow (*achillea millefolium*), are nutritive and wholesome to cattle.

The following may be considered neutral plants, in meadows, neither worthy the farmer's attention to encourage their growth, nor his efforts to destroy them. They ornament the foliage, and (as Dr. Withering has observed,) a display of beauty may have been in some measure the design of the Great Creator. Daisy, (*bellis perennis*); daffodil, (*narcissus pseudo-narcissus*); harebell or English hyacinth, (*hyacinthus non scriptus*); cowslip, (*primula veris*); primrose, (*primula vulgaris*); lady smocks, (*cardamine pratensis*, and *hirsuta*); wood or meadow anemone, (*anemone nemorosa*); goose grasses, (*galium palustre* and *uliginosum*); bistort, (*polygonum bistorta*); cinquefoil, (*potentilla reptans*); meadow rue, (*thalictrum flavum*); valerian, (*valeriana officinalis*); orchis, several species, of great beauty; meadow boot, (*caltha palustris*); ladies' mantle, (*alchemilla vulgaris*); eyebright, (*euphrasia officinalis*, and *odontites*); *veronica*, several species; white saxifrage, (*saxifraga granulata*). The fritillary, (*fritillaria meleagris*), adorns in great profusion some meadows in the parish of Wheaton Aston, and a meadow near Wolseley Bridge, on the road to Stafford.

The pilewort (*ranunculus ficaria*), is one of the earliest flowers in grass-land, pushing out its yellow flowers in April; the water avens (*geum rivale*), grows in the meadows near Shenstone, and shows an elegant flower in June; the purging flax, (*linum catharticum*), and lesser centory, (*chironia centaurium*), are common in dry pastures; the restharrow, (*ononis arvensis*), is a productive plant, but rather to be extirpated than encouraged, though cattle eat it well when young. The roots of it are troublesome in arable land, being strong, and almost stopping the plough. The hogweed or cow parsnip, (*heracleum sphondylium*), though a coarse and weed-like plant, is common in pastures and hedge-sides, and well eaten by cattle when young. If this plant had the advantage of culture, it would bear cutting two or three times in the season, for hogs and cattle.

Artificial Grasses.—The grass-seeds generally sown for laying land to pasture are, red and white clover, trefoil, and ray grass; of the three former mixed, from 10 to 20lbs. per acre, and of the latter from a peck to half a bushel. The meadow or cow-clover is professed to be sold by seedsmen as having the perennial quality of continuing in the land. The natural grasses before-mentioned being in appearance superior to ray grass, it were much to be wished that a more general trial should be made of the best of them, by collecting their seeds for cultivation. The cock's-foot (*dactylis glomerata*) is much cultivated by Mr. Coke, at Holkham, and strongly recommended by that eminent agriculturist, for the laying down of land instead of ray grass.

A great variety of grasses, and other agricultural seeds, were, a few years ago (and perhaps still), professed to be sold by Gibbs and Co. of Half-moon-street, Piccadilly, London. The writer has had some experience of the following:

Burnet, (*poterium sanguisorba*); hardy, and strictly perennial; cows prefer it to clover, but it is less productive: it is a very different plant to the meadow-burnet, being a native of calcareous soils, limestone, or chalk. It grows naturally on the Wever-hills, in the Moorlands, and in Dove-dale, will stand the winter frosts, and be ready for pasture early in spring. It will also bear the greatest drought of the dryest soils, by striking a long tap-root deep into the earth; but it is best adapted for dry calcareous soils.

Lucern (*medicago sativa*), is very productive, equally so with the best vetches or clover, and is perennial and durable, but requires cleaning and nursing for the two or three first years; it will bear transplanting * from a seed bed, which is perhaps the best way of cultivating it: or it may be drilled between rows of barley on very clean land, and might be hoed in autumn and spring, when the barley-crop is cleared, after which it would maintain itself and last many years, and might be cut two or three times each year, being at times manured.

Saintfoin (*hedysarum onobrychis*), is an excellent perennial plant for hay or pasture, equally productive with clover, and may be mown annually for many years. It succeeds best on calcareous soils, with a substratum of chalk, having fissures which the roots can strike through. It will not thrive on thin soils having a hard gravel or clay bottom near the surface, but is quite worthy of culture on loose bottoms, and succeeds well on poor land, having that quality, if in clean culture. It should be sown three or four bushels per acre, with oats or barley, and care should be taken to sow only new seed, as the old seed will not vegetate. It is grown plentifully upon the Coteswold-hills, Gloucestershire, and the Chalk hills of the southern counties.

Many other of our native plants seem well worthy of cultivation, particularly the perennial vetches so common in our meadows; with cichory, which has been highly recommended by Arthur Young, Esq. as food for sheep and cattle. The yellow goat's-beard, before named, seems equally worthy of attention; the common sow-thistle is also very productive, and well eaten by cattle; also comfrey, (*symphytum officinale*), and lettuce, (*lactuca*), might be cultivated for hog-food: the garden-beet is also extremely productive and durable, as well as early in spring.

Hay Harvest, in this county, is mostly in July. Upon regular farms the artificial grasses, clover, and ray-grass, are generally cut first, and the meadow-grasses follow: on the highly-forced lands near large towns, the meadow-grass is somewhat earlier than in

* See Harte's Essay on Transplanted Lucern.

the country. The seasoning or making of clover and ray-grass into hay, is as follows: After mowing, the swathes are suffered to remain till they are well dried on the upper side; they are then turned over, and the other side dried in like manner, and afterwards turned a second time, if necessary; they are then put in larger rows, the ground raked, the rows put in cocks, and carried to the stack: the stalks of clover take a good deal of time to dry throughout. Fair weather and sunshine are very desirable in this business. In making hay from meadow-grass, the first operation after mowing, is spreading it all over the meadow, which is called *tedding*; in the evening it should be raked into win-rows, and afterwards divided into grass cocks; next morning (if fair weather), the grass cocks are cast about into square beds, and the vacant spaces between the beds (if any) raked clean; the beds, after turning, are put in the evening into middle-sized cocks. Next morning (if fair), these cocks are again spread open and turned over, and being sufficiently made, are put into large cocks and carried to the stack. The time for these operations, and their effect, is much facilitated by sunshine and fair weather.

In showery weather, the following is a better process. After mowing, and the swathes being seasoned, rake them clean over; when the other side is seasoned, make them into small cocks; when the cocks are dry on the upper side, turn them over, and when dried on the other side, half a dozen of them may be put together into quarter cocks, and so by degrees into large cocks: this process will require more time than the former, but will keep the hay better protected from the weather. Hay, when spread over all the ground, catches all the rain, but when in cock only a part of it; it will lose its juices and its colour in proportion to this greater or less saturation.

Sir John Sinclair has detailed a Lancashire and a Scottish method of making hay in tipples, adapted to a humid season or climate, thus: The swathes are rolled up in bunches, and set on end like flax, two bunches meeting of ten or twelve pounds weight each; a rope is twisted of the grass, and tied round near the top; the straggling stems are twisted round the top to give it a conical shape; after standing a few hours they become smooth on the outside, so that no rain can penetrate, and they are soon dry after rain: they are never opened or tedded, but carried from the tipple. Not a leaf is lost, and the hay is as green as a leaf dried in a book.

Hay for saintfoin or vetches must be made by the same process as that from clover; but the last-mentioned process seems applicable to them all in bad weather.

Stacking of Hay.—Some people carry tunnels up their hay-rick either by drawing up a bundle of wheat straw, or by a tunnel of wood drawn up and out at the top. These are meant to discharge a part of the vapour generated in fermentation, and thus prevent the danger from the hay firing: those who practice it say they can thus carry their hay a day or two the sooner. A hay stack should not be begun above four yards wide at the bottom, when, if well put together, and the sides sloped outwards, it will hold two waggon-loads in every yard in length or more; and if carried so as to be four yards high, besides the roof, when settled, it may contain from a ton and a half to two tons of hay for every yard in length.

Produce of Meadow Land.—Mr. Curwen has stated that a good meadow may produce annually twenty tons of grass per acre; but this will be seldom realized. To ascertain this point, the fresh-mown grass of two perches of land was weighed in July, 1814, in a meadow of Mr. Miller's, at Dunstall, near Wolverhampton: it was well-grown, and by the side of a floating gutter; the weight was 3 cwt. nearly. This is 12 tons per acre; and reckoning the aftermath at half as much, the whole product would be 18 tons per acre. When made into hay it was a few pounds short of a hundred weight, but the weather had not been remarkable for sunshine. It may be assumed from this experiment, that a well-grown meadow may cut from 10 to 12 tons of grass per acre, and make 3 tons of hay, but that in general not more than half this quantity is obtained; that if a meadow will not give a ton of hay per acre, it is not rich enough to be mown; that a good crop of clover, or vetches, may be 10 tons per acre green, or more at two cuttings: hence will appear the great importance of keeping meadows in good heart. A meadow that will produce three tons of hay per acre is as valuable as an average crop of wheat. It is calculated that an hundred weight of grass per day would maintain a dairy cow; that the hay made from it would do the same in winter; and therefore, that an acre of meadow producing 18¼ tons of grass, would maintain a dairy cow the whole year. But in general it takes two or three acres to do this; one acre of rich meadow is therefore worth two or three acres of average land.

Gardens and Orchards.—Every house in the country should have a spacious garden, as the land will thus produce more human sustenance than in any common course of agriculture. The usual garden-plants cultivated are: potatoes, beans, pease, cabbages, French or kidney beans, broccoli, savoys, turnips, carrots, onions, beets, spinach, lettuce, and many other sorts of pot-herbs and salads. A garden should be well fenced, and sheltered from the north and east winds, and the hedge-rows and corners should be planted with useful fruit trees. A labourer's garden should be from a quarter to half an acre: those sorts of apples and pears which keep sound for a length of time should be planted, as well as those for present use, together with gooseberries, currants, damsons, and other plums: such articles would in pies be a cheap and wholesome food for children, and might at times turn to good account for sale. Suppose a labourer's garden of half an acre thrown into two equal lots, the one for garden vegetables, the other for wheat alternately; let the garden lot be again divided in two, the one-half for potatoes, the other half for garden vegetables: the whole might be cultivated with the spade and hoe without loss of time, by doing a little every day, morning and evening, and in the hoeing and weeding the wife and children might assist. By this plan there would be one-eighth of an acre potatoes, which might produce forty bushels; one quarter of an acre of wheat might (drilled and hand-hoed) produce eight bushels; and a hog might be kept from May to Christmas on the refuse of the garden and wash, and fattened after harvest with boiled potatoes and bran, and ground barley from gleanings: the straw of the wheat would furnish the hog with litter, and dung for the garden would be produced. From these resources many family comforts might be derived; and it were to be wished that gentlemen of landed property would put them in the power of industrious labourers on their estates, by letting them land upon average terms; and if some premium or reward (as a store-pig in May), were given to such as managed in the best style with the least loss of time, it might be a stimulus to industry.

At some intervals it might be advantageous to sow the one-half with hemp, in which case half the wheat stubble, and half the potatoe ground, ridged-up with the spade to lay through the winter, would, on digging down in the spring, be in excellent order for hemp. Half the wheat and potatoes would then be omitted for one season; but the hemp would find employment for the wife and children in manufacturing, to the comfort and convenience of the family.

The Orchards of Staffordshire are inconsiderable; but little fruit-liquor is made, and perhaps the produce is insufficient for its own consumption even at table. This is certainly a great neglect, as many soils and situations are well adapted for fruit, and it appears to have been more attended to formerly, many orchards having been suffered to go to decay. In former times there was a crab-mill in most villages; these have gone to decay with the orchards, and the cottagers, who might have drank cyder and perry (had such been attended to), are now obliged to drink water. It would be a laudable effort to endeavour to restore the orchards, and would add to the comforts of the labouring people in the country.

The parish of Tettenhall has one singular circumstance in the fruit way; it produces a peculiar kind of pear called Tettenhall pear, and known by no other name. Many hundreds of the trees grow in the parish, though scarcely to be found at a distance. The tree is large, and a plentiful bearer, the fruit well-flavoured, bakes and boils well, but will not keep long enough for carriage to a distance unless gathered before they are ripe. The average annual produce of this parish is many thousand bushels more than its own consumption. The pears ripen about Michaelmas, and their whole duration is about one month, during which time the neighbouring markets are plentifully supplied, and they are carried by the canals into Lancashire. They make but a weak perry, and are but little used in that way, there being a demand for them all at market. We must repeat that the culture of fruit-trees in this county has been too much neglected, and that the raising of orchards is a desideratum in its agricultural improvements, as producing an useful article little interfering with its present productions. In such a business due attention should be paid to the proper kinds of fruit, and to those that produce fruit-liquor of the best quality. Fruit-trees might certainly be raised in pasture land, and in hedge-rows, with very little injury to the other produce of such land; and perhaps the increase of human beverage from fruits, instead of grain, is much to be desired, as tending to increase the quantity of grain for food, and being more conducive to health than malt liquor. It has been stated that fruit-liquor is an antidote to the stone and gravel, and that malt-liquor promotes those diseases. Large tracts of this county are well adapted for the growth of apples and pears; deep rich friable loams are their natural soil, or lighter soils of a good depth, and not too much elevated. In thin soils upon gravel, the hard under-stratum should be taken out, and loam put in its place; and bottoms too moist should be under-drained, which would adapt them to grass and corn as well as fruit-trees. It is an unpardonable neglect for a house in the country with land to it to be without an orchard, or for a village on a fertile soil to be without a crab-mill to make verjuice, cyder, and perry, and the means of employing it. The French are far before us in this respect; and though we are superior to them in strong ale and good beef, their peasantry drink wine whilst ours drink water, when at the same time they might drink cyder and perry, if due attention were paid to fruit-trees. But the business can only be done by the land-owner, it not being consistent with the uncertain tenure by which land is commonly held in occupation.

Timber, Plantations, and Woodlands—This county still continues well-stocked with all kinds of timber, notwithstanding the immense quantities that have been cut of late years. Amongst the many well-timbered estates, that of Lord Bagot is most prominent. The woods (in the neighbourhood of Abbot's Bromley), extend over many hundred acres, and are almost wholly of oak of the first quality: many of these oaks are of great length, and contain from 60 to 70 feet of timber. In the Park are many hundreds of extraordinary bulk, containing from 200 to 400 feet of timber each: some of these are of great antiquity, being mentioned by Dr. Plot as full-grown timber in 1686. The succession woods and young plantations are very extensive, and still continuing: they are sometimes made by sowing acorns with fallow wheat, sometimes by planting-out young plants of oak and other wood, and at the end of two years, when such plants have taken well to the ground, they are cut off near the surface, and the second shoot trusted to for the tree, this thriving with more vigour and luxuriance than the first, checked by transplanting. After cutting down a wood, the re-planting is also sometimes effected by striking-in with a pick-axe a sufficient number of acorns, and other seeds of forest-trees and under-wood; and all these methods have been attended with success. The soil is a moist gravelly loam, upon a clay or marl bottom; the young plantations are always well fenced, and carefully guarded from trespass by a woodman appointed for that purpose.

The Chillington estate, belonging to Thomas Giffard, Esq. is extremely well-timbered. From this estate has been sold, within thirty years, timber to a great amount; but the growth of succession wood more than supplies the place, and keeps up the value. Some very fine ripe oaks grow in the woods, and around the pool, a large and beautiful piece of water. The woodlands remain very extensive, and in the modern plantations oak has been particularly attended to; but there are all or most of the other varieties, of extremely promising growth; the young plantations are always well-fenced and secured from trespass, without which planting would be in vain. This branch of agriculture is peculiarly the province of the gentleman landholder, to whom, it may be conceived, nothing can be more pleasing than thus to improve his estate, and at the same time shelter and ornament the country.

The demesne and neighbourhood of Beaudesert, the magnificent seat of the Marquis of Anglesea, is well stocked with fine ripe oaks. The woods upon this estate are of considerable extent. Its situation is upon very elevated ground on the north point of Cannock Forest or Chase, the most extensive waste of the county, said to have been formerly covered with timber, and still called Cannock Wood. The prospect from Beaudesert commands a very extensive view over Needwood Forest, and the Moorland and Peak Hills into Derbyshire.

Upon the pleasure-grounds and estates of Lord Dudley, at and near Himley-hall, are large quantities of well-grown timber, and very extensive woods and coppices of oak, particularly a fine wood called Baggeridge, beautifully intersected with pleasure-roads, cut through it in various directions, and well stocked with pheasants.

Teddesley Park is the seat and property of E. J. Littleton, Esq. one of the representatives of the county in Parliament: the timber and plantations on this estate are very considerable. Manstey-wood is an extensive coppice of fine ripe oak; and there are many other coppices, dingles, and clumps, of full-grown oak and other timber, with spring coppices, and young plantations of different growths, to a great extent. These plantations are generally upon poor cold land, but having a clay or marl bottom, the oak succeeds well, and comes to a large size and the best quality.

Wrottesley estate, belonging to Sir John Wrottesley, Bart. has some considerable woods of ripe well-grown oak, with extensive plantations of modern growth of different varieties. At Patshull, (Sir George Pigot's), the park and estate are very well timbered. At Enville (Lord Stamford's), the woods and plantations within the compass of the pleasure-grounds are of great extent, and extremely well stocked with timber and underwood of every sort, amongst which the oak predominates. The woods of Upper Areley (upon the Severn), are very extensive, and well stocked with oak in every stage of growth. The Bishop's woods, in the neighbourhood of Eccleshall, are said to contain 1300 acres, which are cut in regular rotation, keeping pace with their growth, and producing an annual revenue without injuring them. The woods at Broughton (Sir John D.

Broughton's), abound with plantations, clumps, and particularly fine sycamores, with a fine promising spring coppice of oak. Hilton Park (Mr. Vernon's), contains some very fine ripe oaks, and other timber trees; and a great extent of modern plantation by the late Mr. Vernon, who was a very early and persevering planter: a great many trees of his planting are now fast approaching to maturity. To these may be added, the fine and extensive plantations of oak and other forest-trees at Trentham (the Marquis of Stafford's), and at Whitmore (Mr. Mainwaring's).

Many other estates are extremely well timbered, and the county is not likely to experience a scarcity of that article for ages to come. The estate of Richard Fryer, Esq. at the Wergs, Tettenhall, amongst plenty of various sorts of well-grown trees, contains two of the finest oaks in the county, perfectly sound, and at full maturity; they contain, it is conjectured, near 300 feet of timber, and from one to two tons of bark each.

We shall here make two observations on planting trees; the one on the proper site of ground for plantations, the other on the sorts of timber, respecting pleasure, profit, and utility.

With respect to the proper site of ground, every land-owner of prudence and taste, who resides on his property, will furnish himself with shelter and shade near his habitation, in such situations as local circumstances shall direct; but large plantations should not be made on good land, of much more value in this county for corn or pasture; but always (where such abounds), upon land impracticable to the plough, upon precipices, sides of hills, or in dingles, where the land unplanted is of little value. Plantations upon this system have been made by the late John Holliday, Esq. to a great extent. The Dilhorn woods form a chain of three or four miles in length, and those of Kingsley and Oakamoor, are upon steep uneven hills: the underwood will bear cutting every seven years for cratewood, and pay near £1. per acre annually, whilst large tracts of similar land unplanted are not worth one-fourth of that value.

Respecting the species of wood, oak is undoubtedly the first forest tree, and its acorn is valuable hog-food; but being slow of growth, the profit lies back. It should always be plentifully intermixed with other sorts, whose place it may supply when they are cut for use. Ash is valuable for many uses, but not very quick of growth, and its fruit is of no value; it should be planted in corners, or dingles, or coppices, and not in hedge-rows, where it is very injurious. Elm (the Worcestershire or narrow-leaved), is proper for groves and avenues, and does the least damage in hedge-rows of any timber tree, but requires a deep soil and warm situation. Eighty trees of this sort are now growing upon Tettenhall-green, planted less than a century ago, some of which contain upwards of 100 feet of timber each. The Witch-elm has in some instances been of very rapid growth. One of these has been known to grow to 60 feet of timber in 30 years from planting; but the poplar, in rapidity of growth, is superior to every other timber tree. The before-named gentleman has been known to plant and cut down poplars of 60 or 80 feet growth. Our own native poplars are thought to be superior to the Lombardy or Po poplar. Some of the willow species are remarkable for quick growth, and their loppings are useful for many purposes. Beech is a valuable wood, and its fruit is good hog-food: it will also destroy furze, and not suffer it to grow near it. The sycamore will answer well on high and bleak hills, and bear exposure to the cold winds. All our native forest-trees should be renewed at times from the seed; even the elm, and the poplar, are doubtless liable to degenerate, if raised from shoots or cuttings for successive generations.

Respecting the fir-tribe, planters have certainly been rather too partial here. The larch is believed to be one of the best sorts. Every gentleman making a fir plantation, should intermix at least an equal number of English forest-trees, and not forget oak; and in hedge-rows the sorts bearing useful fruit deserve more attention, as the chesnut and the walnut; and particularly on all deep mellow loams, the apple and the pear, as the raising of orchards, and increase of fruit, in this county, is a desideratum, and may be ranked amongst the greatest agricultural improvements of which it is capable.

Waste Lands.—These have been rapidly lessened within the last twenty years. In 1796, the practicable waste land of the county was estimated at 100,000 acres; since that time to 1815, Needwood Forest, Sutton Coldfield, and many other commons, have been inclosed, to the extent perhaps of one-third of the whole. The remaining land now lying waste may be calculated at from 60 to 70,000 acres.

Cannock-heath is the most extensive waste in the county. Some of the skirtings adjoining it are now under inclosure, particularly Teddesley-hay, within the manor of E. J. Littleton, Esq. M. P. the successor of the late Sir Edward Littleton; but the extent lying waste may be estimated at 25,000 acres. Large tracts of the north and west parts are a good light soil, fit for turnips and barley; the east and south parts are a colder gravelly soil, in many places covered with heath (*erica*), to a great extent, but the whole capable of improvement and cultivation.

Whittington-heath is a considerable tract of good sound land, and many similar ones lie in different parts of the county; but the most extensive wastes, besides those above-named, are in the north of the county, and in the Moorlands. Some of them are but cold barren spots, but they are all capable of being improved, and applied to useful purposes, either for tillage, grass land, or plantation. Waste land in general is thus best reclaimed: 1. Pare and burn, spread the ashes directly, and plant potatoes; 2. oats; 3. lime well, and make a clean turnip fallow; 4. barley, or spring wheat, with plenty of seeds; but draining must be the first operation, if necessary.

Improvement of Heath Land.—Waste lands are admirably adapted to the growth of potatoes. "The east side of Dilhorn-heath was cultivated with potatoes, after the heath and gorse had rotted, and been mixed with lime and compost; the crop of potatoes was so abundant as to admit of many waggon-loads being sent in the winter into the vicinity of the Pottery, about six miles from Dilhorn, which afforded a seasonable supply to many thousand manufacturers. The quantity was not only immense, but the quality of the potatoes was in so high repute, that the Dilhorn potatoes produced two-pence per bushel above the common market price. In this part of the Moorlands the potatoe harvest is of great consideration, and the 30,000 artificers and yeomanry there eat less wheaten bread than in most other places with the same numbers. Give a cottager in the Moorlands, with a wife and large family, a cow, and a few roods of potatoe-ground, and you make him a happy man; he goes to his daily labour, earns money to purchase clothing, &c. for his large family, the younger children collect the dung and soil from the public roads for the improvement of the potatoe-ground, and the industrious dame, with her stouter children, keep the ground clean, and assist to get in the potatoe harvest, the chief support of their family about nine months in the year." A pig might be added, to eat up the offal, which, when fat, would add to the general plenty.

Many instances have occurred of great success in raising potatoes on waste land, but the shortest way is first to pare and burn. Two day-labourers gave a guinea for an acre of waste land to plant with potatoes; they pared and burnt it by moonlight after their daily labour, spread the ashes, and paid for ploughing them in; the crop proved so good, and the price of potatoes so high, that they shared £40. between them, besides reserving plenty of potatoes for their families.

A peat-bog on waste land was drained, then pared and burnt; the ashes immediately regularly spread, and the land ploughed in twelve furrow ridges (it could not be ploughed in narrow ones from toughness), the furrows were hacked and levelled with heavy hoes, then planted across the ridges with potatoes in rows, and owing to the large quantity of ashes, produced an abundant crop. The land afterwards produced the two next years two very strong crops of oats in succession: it was then well limed, and clean fallowed, and is now a good meadow.

IMPROVEMENTS: 1. *Draining.*—Of this most essential improvement of all land that requires it, much remains to be done, though great exertions have been made in it, and are now making. The practice generally adopted is, after having found a sufficient fall, to open ditches and main drains to a proper depth, and to cut from and into them, through all wet and unsound spots, a number of drains of about fourteen inches wide at top, four or five at bottom, and of a proper depth, three feet or more; these last-mentioned drains to be filled up with proper materials, so as to let a current of water pass along the bottom, thus at all times keeping the moisture in circulation, and freeing the land from the chill of stagnant water.

The materials now used for filling up drains are most generally draining tiles, made on purpose at the brick kilns, of a semi-elliptical form, about four inches deep, two inches and a half wide within, and a foot long, placed end to end along the bottom of the drains. On the top of these is laid some straw, or heath, then turf, with the turfy side downwards, and then filled up with loose earth: this is the practice in arable land. In grass-land the natural turf may be reserved, and replaced on the surface.

In countries where stone is plentiful, two side stones, and a covering stone may be used instead of tiles: close drains thus executed will last a great length of time if properly done, and are a very capital improvement to all lands injured by springs or stagnant water. The open drains and ditches must at times be cleaned out, to let the water pass off freely.

It is also a common practice to bore the bottom of hollow drains, with a view of tapping the spring, and sometimes with great success. After boring through a clayey stratum into one of gravel beneath, the water has been known to rise in a constant stream up the hole, and thus to drain considerable breadths of land above which the water had injured it for want of a vent, which is now given along the bottom of the hollow drain.

2. *Irrigation*, or the watering of land, is a very important part of agriculture, the advantages of which are generally admitted, and well known by all intelligent farmers; but it is by no means carried to the extent of which it is capable. The omission is in part owing to the jealousy of millers and others interested in the streams. The industry of many individuals has, however, been successfully exerted in this valuable species of improvement.

Respecting a system for irrigation, no general one can apply to particular cases; but in all cases where a stream naturally falls down a valley, and the sides of such valley consist of easy and regular declivities, the best way undoubtedly is to draw the water nearly upon a level along a main carrier, as far as it is intended to use it, this water to be let out of the said main carrier at pleasure by sluices constructed in different places in its sides, into floating gutters, such floating gutters being cut nearly on a level along the sides of the declivities, one below another. These floating gutters marked out by a level, will collect the water from the spaces of land above them, and deliver it upon the spaces below with a little attendance and attention. The land should be watered in rotation, different parts successively, and not too long in a place: the gutters will require an annual cleansing or scouring out, otherwise they will choke up with grass.

Watering will not have the same good effect upon all soils: it makes the greatest improvement upon warm soils, but on strong cold clay it should be used sparingly; and every one knows the necessity of draining the land properly previous to watering.

Reservoirs.—The greatest benefit would be derived from watering in dry seasons, when all kinds of land would be most improved by it. This might be effected by reservoirs constructed in valleys, to retain the water of rainy seasons, or violent showers; and however visionary the idea may appear, it is economically practicable where the fall is not already occupied by mills or public works. A reservoir of a few acres might be constructed at the head of a vale on entering an estate, at (in general) less than twenty pounds per acre, to receive violent rains, or the melting of snow; the usual current, if such there be, might pass along side it in common, if claimed by others; and the flood water thus collected might be distributed at pleasure upon any part of the estate below its level. A very large proportion of the rain water now runs away in waste, which might thus be collected for use, and the mischief of floods prevented. Such reservoirs, as a fishery, might be equal or superior in value to the same breadth of land, and might be drawn down to a certain depth for watering at pleasure without injuring the fishery; water might thus be applied to such land as it would most benefit, or in dry seasons to any land below its level. Reservoirs for watering the land have been constructed in Spain, and other hot countries, and it is surprising they have not been more attended to in England. Many of the ancient water corn-mills are now of very little use. The natural situation of a corn-mill, under the present circumstances of the country, is not in a remote situation upon a stream, where a team must be employed to carry in and out, but upon a canal, and near a large town, where coals can be commodiously delivered to keep the machinery in motion, and where the consumption of the flour ground is upon the spot. Upon this principle, when the machinery of mills upon streams is worn out, they will be removed to more convenient situations, and the streams be left for the improvement of the land; in which case the construction of reservoirs in proper places, for public use, in proportion to every person's quantity of land, might effect a great public improvement.

Paring and Burning has not been much practised in this county, and opinions vary much as to its propriety. It is most applicable to waste lands, as the shortest way to get rid of rubbish, and to correct the acids in the soil. Excellent potatoes, as well as grain, have been grown on such land, after paring, burning, spreading the ashes, and then ploughing them in: some few persons have tried it upon inclosed land, and it is found to stimulate the soil to a crop, but it must waste the vegetable substance of the soil, and thus lower its staple. It is most applicable to deep peaty soils, which being composed of inert vegetable matter, may, by the ashes, be stimulated to action: the ashes should be spread immediately in a live state. Burning of couch-grass is practised by some active farmers, but most people think even *that* is better destroyed by tillage and sunshine upon the land, if the season will admit; if not, by carting off and mixing with quick lime.

Manuring.—The manures in general use in Staffordshire are, 1. The muck of the farm-yard, composed of the dung and urine of animals, littering straw, and rain water: most farmers turn this over once before laying it on the land, to promote fermentation, or draw it to a heap. 2. Town manure, which can be only used by those who are conveniently situated for conveying it: when of a good quality, it is the best of all manures, either for tillage or grass-land. 3. Lime; this is a manure much used upon all sorts of land: it is supposed to dissolve strong soils, and to bind light ones, and to correct the acidity of old uncultivated land. When laid on hot from the kiln, it is destructive to grubs, snails, and caterpillars: it is laid on tillage land, either for barley, turnips, or wheat, from six to ten quarters per acre. It is best laid on in a quick state, soon after slacking, and should be harrowed-in when the land is dry. It is laid on grass land with good effect, sweetening the herbage, by forming sugar in the joints, so that cattle bite it close with avidity: it is sometimes fetched fifteen miles by land-carriage for manure. 4. Marl has been much used on arable land for ages back, as is evident from the very extensive marl-pits to be found in different parts of the county. In many places complaints are made that the land has been over-marled, and rendered harsh; this may be remedied by well-fallowing and liming. It is in general only in the loamy soils that a true marl is found; in the mixed gravelly soils is often found a clay marl, which has been laid on such soils with good effect; the quantity laid upon an acre about 128 cubic yards, which adds nearly an inch to the staple of the soil. The improvement by marl is believed in this county to be much more adapted to tillage than pasture, and its good effects on corn land are long visible; indeed, in the case of thin, light, unimproved lands, the benefit of marling must endure for ever, as it not only consolidates the soil, but adds to its thickness and staple. It has formerly been much practised in this county, but the high price of labour has of late years been a check to its progress.

The above are the staple manures of the county, to which may be added composts, consisting of shovelings of soil, scrapings of roads, scouring of ditches, and mud from pits and ponds, either used alone or mixed with lime or muck, and generally used as a top-dressing for meadow and pasture land. The ashes of our culinary fires are used as top-dressings for coarse pastures, but are improved by mixing with night soil. Soot is used as a top-dressing on wheat, or for turnips; the land should be free from weeds, as soot will force the weeds as well as the crop. It is also a good top-dressing for mown grass land. The soot merchant always undertakes the sowing of the soot into the bargain of sale.

Pigeon's or fowl's dung is a good top-dressing for grass, turnips, or any other crop, to be sown on by hand. To increase the quantity, strew malt-dust over the floor every time the dove-house or hen-roost is cleared, and keep this compost in a dry place till used. Before sowing it, it should be turned over and well mixed: enough of this has been collected upon a farm in one year to top-dress ten acres equal to a dressing of other manure.

Bone or horn-shavings from Birmingham, is esteemed a good manure for light tillage land, as is ground bones and chopped rags.

Manuring Waste Land.—Some persons have contended that waste land can only be improved by robbing the land already in tillage of part of its manure; but waste land ought to be improved from its own resources, or foreign aids, as from paring and burning, lime and marl, with clean fallowing, turnips, clover, sheep, and other stock, with the manure they make from the product of the land in question, without robbing the old inclosed land.

Embankments.—Connected with the improvement of our valleys and low land, is the embanking them from external waters, to secure them from the damage, depredation, and ruin, that may be caused by inundation. The water of many rivers and brooks is liable to overflow its banks after excessive rains, and if this happens in summer, when mowing grass is growing on the land, such grass becomes greatly injured, by being fouled with mud, sand, and extraneous matter; it is difficult to mow, and of little value for hay when mown. If such flood occur in hay harvest, as is sometimes the case, the damage is still greater; the hay is sometimes totally ruined, or floated away. In some valleys of rich meadow land, this misfortune is expected once in three years upon an average; in others, the meadows are always grazed as the safest way.

It is very certain that such meadows may in most, or all cases, be secured from floods by embankments, and at the same time greatly improved by drainage, and by watering or not at pleasure. If the business were taken up in a general way, embankments should be raised parallel to the river or brook at a proper distance from it on either side, and carried up the collateral streams to the same level; these embankments to be formed of materials dug out of drains on their outside, such drains to be continued by culvers under the collateral streams, which would at all times drain the land. The distance between the embankments and their height must be proportioned to the swell of water, or flood, that it is proposed to guard against. The land between the embankments would be subject to floods as before, and should therefore be grazed; that on the outside the embankments and drains may be mown or watered at option; also in floods wholly, or partially, by trunks and paddles laid in the embankments, and over the drains. Every work of this kind, where different interests are concerned, must of course be left to a commissioner, to award to each his due share of expence, as well as benefit; and the distribution of waters for floating should be duly attended to, particularly from the collateral streams, and each one ought to be at liberty at all times, injuring none other, to take in water through the embankment in floods for the purpose of soaking or floating his land when required.

LIVE STOCK.—The principal profitable stock of the Staffordshire farmer consists of, 1. horned cattle; 2. sheep; 3. horses; 4. hogs; to which may be added, 5. rabbits; 6. poultry; 7. pigeons; 8. bees.

Horned Cattle.—Those of this county are generally of the long-horned breed, but with a few of other varieties; differing in value and quality in proportion to the attention of the breeder, and his resources for keeping. It was a maxim of the old farmers that "all breed is put in at the mouth," but it is now established upon the principles of Bakewell, that the soft handling beast, with a sound constitution, well proportioned and light in offal, is inclined to fatten, and will thrive better in any reasonable keeping than a coarse beast with thick hide and large bone. That blood and breed will shew itself, has been proved by well-bred beasts paying double for their keep to what others have done. The cattle stock of the county hath been for some years improving; but as it is not in the power of the lower order of farmers to pay the prices for prime stock, it would be a good plan if gentlemen would procure for the use of their tenants the best bulls, rams, stallions, and boars, which might be thrown upon a plan to indemnify expences, and would facilitate the improvement of the stock of the smaller farmers.

The great object in the produce from horned cattle being first, milk, and second, beef, the uniting of these two products in the greatest quantity from the least food or quantity of land, is the ultimatum of breed. It has been observed that cows with the best disposition to fatten give the least milk and soonest go off their milking, but this has been contradicted by others, who say that cows of a good sort that are good milkers, will, on being dried, make themselves fat in less time than a loose, open, ill-made beast; that bad milkers often occur in the unimproved breed; but that a well-bred beast being a bad milker, is sooner marked for it. A good or average cow should give, in the height of the season, from eight to twelve quarts of milk at a meal, and two meals per day; but they soon decline to three-fourths, and one-half of that quantity and less. A good cow, it is said, will make five hundred weight of cheese in a season; but three hundred weight is a good average. We have a well-authenticated account of a small farmer adjoining Needwood Forest, who kept seven cows, began and ended with the same, and sold on an average two tons of cheese annually. Sixteen pounds of butter have been made weekly from a prime cow, in the height of the season. The value of a good milch cow and calf at prime age, is nearly the same as the cow will be worth when fat. The long-horned cows of this county vary in size, when fat, from seven to twelve score pounds the quarter, and sometimes more. Many instances have occurred of superior cattle stock. The late Mr. Princep, of Croxall, on the borders of the county, brought the long-horned cattle to their highest perfection. An ox of his breed was fattened at the Marquis of Donegal's, in 1794: weight of the four quarters 1988 lbs. tallow 200 lbs. hide 177 lbs. This ox, when beef was at 9d. per pound, as was the case a year or two ago, would have been worth £80. or guineas. Very superior cow stock are kept at Lord Bagot's, and in that neighbourhood; by E. J. Littleton, Esq.; by Lord Anson, Lord Talbot, and many others. At Oxley, cows were fed by the late Mr. Huskisson to 374 lbs. the quarter, which, when beef fetched 9d. per pound, would have been worth £60. or guineas each.

The late Mr. Miller, of Dunstall, near Wolverhampton, had, perhaps, the largest and best dairy in the county. The number of his milkers were once upwards of seventy; but they have since been lessened, and the sheep stock increased. The main object was cheese, of which at least 4 cwt. was made per cow. Early in the spring, calves were fattened to 8 or 10 weeks old. The cows, when taken from the dairy, or when accidentally barren, were fattened for the butcher upon the farm, to 10, 12, or 13 score the quarter, and seldom any sold in store order, or in any other way. His successors are pursuing the same plan. Mr. Miller, in his cow management, had two main objects in view, milk and beef; in pursuit of which, the carcass and the milk bag were equally attended to, and the cows were bred of good size, good milkers, and well made, with a disposition to fatten. The dairy too has always been managed in a first-rate style. A large breadth of land, capable of irrigation, has been attended to, which, by furnishing an abundant supply of winter keep, maintains not only the dairy cows, but also a number of young cattle for succession stock.

Mr. Miller, by length of time and perseverance, succeeded in raising a new and very valuable variety of the breed of cattle. Their peculiar characteristics are, hornless, milk white, ears sometimes tipt with red, size and weight large, producing well to the pail, and weighing, when full grown and fat, the cows 12 score per quarter, and the male 15 score and upwards. One of these cows, six years old, was weighed in Wolverhampton market, June 26, 1811, whose weight was 11 score 10 lbs. or 230 lbs. per quarter; she had been milked three summers, and to Michaelmas 1810, but having missed being in-calf, was then let dry, and fattened with turnips and hay. This cow had probably produced 480 lbs. weight of cheese per annum from her milk, besides a calf 120 lbs. and whey butter from the dairy, &c. equal to 120 lbs. more; this would be 2160 lbs. weight of human food, besides its own carcass (920 lbs. more), and the eatable parts not weighed may be reckoned 80 lbs. making in all 3160 lbs. weight of human food, and principally of the richest and best quality. This cow had not very probably consumed more than the produce of 12 acres of land, which gives the produce of the dairy at 263 1-3d lbs. per acre per annum of the best and richest of human food, and is much superior to what can be produced by breeding oxen, or even sheep.

This breed of cattle is well formed, and disposed to fatten, handles sleek and well, with great weight in the more valuable joints, and are of singular and beautiful appearance. They are not derived from any particular country, breed, or descent: the originals were selected at fairs and markets, and Mr. Miller bred the rest himself, still selecting for rearing what he most approved, and particularly the bull calves. There is now a full dairy, and succession stock, and any gentleman or farmer may be supplied with calves of either sex for rearing.

It appears, by some well-authenticated experiments, that dairy cows would be the most profitable, if fed constantly in the cow-shed or yard: in summer with cut-grass, vetches, lucern, or other green food; and in winter, with turnips, cabbages, and hay, or with grains at any time. They produce more milk from being less harassed in hot weather in procuring their food, and their dung, thus collected, turns to better account as manure.

When dairy cows are fed with turnips, if butter be made, a taint or strong flavour is given to it from the turnips. To prevent this, the following is recommended: Pick off the decayed leaves, and then dissolve saltpetre in water, half a pound to the gallon; bottle the liquor for use, and put a glass of it (containing 1-12th of a pint, or from 2½ to 3 cubic inches), into each pan of milk, warm from the cow, containing a gallon: one ounce of nitre is thus applied to 12 gallons of milk, and will, it is said, effectually prevent the unpleasant flavour. Some persons recommend an ounce of nitre to 8 pans of milk, stirring the milk when set up, and skimming it at every meal. A lump of nitre put in the vessel that holds the cream has the same effect.

Feeding.—Staffordshire is not a great feeding county, at least not comparable with the neighbouring one of Leicester; yet there are some gentlemen, and many considerable farmers, that think much of the trouble of a dairy, who fatten a good many cattle and sheep. Calves and hogs are fattened upon most farms. The horned cattle fattened here are principally heifers and cows, with a few bullocks or oxen. This county breeds more cattle than it feeds, as a good many partly fattened are sold to dealers, who drive them into other counties.

Cattle are sometimes fattened at grass, at others only brought forward, and finished by stall-feeding. A heifer brought into good plight at grass, will generally, on the approach of winter, come on well in a stall, with plenty of turnips and straw only. Potatoes have been given with success when not wanted at market, and ground barley, when plentiful and cheap; but otherwise only tail-end or damaged barley. Linseed oil-cake is also used; also pollard and bran, cabbages, Swedish turnips, and hay. Cattle stall-fed are generally sold to the butcher, without being put again to the spring grass.

Dried or barren cattle are often kept through the winter in a turnip-field, with a little hay or straw given in severe weather, or kept in a turnip-field by day, and in a yard by night; or on a turf by day, with an allowance of turnips drawn to them, and hay or straw given in the field or in the yard at night. With a tolerable allowance of turnips in this way, they will thrive and improve through the winter, and may be either fattened on the spring grass, or sold to dealers at the early spring fairs, who buy up with avidity all such cattle, being dry and in an improved state. The system of fattening cattle is not supposed to pay so well as the dairy, nor indeed ought it, as the labour is considerably less.

Sheep are fattened with the summer's grass or clover, and what are not sold during that season are kept in winter on turnips: a little hay is given in severe weather. From the native or common breeds of sheep, the lambs are often sold fat, but seldom from the improved breed. Calves are oftentimes well fattened, especially in the winter seasons, until six or eight weeks old or more, and to 30 or 40 lbs. per quarter, and upwards.

A good many hogs are fattened in this county, both for pork and bacon: the food generally used in the country for fattening them, is the refuse of the dairy, boiled potatoes mixed with barley or bean-meal, and grey pease or pease-meal. In towns various kinds of offal, and the refuse of the butcher's slaughter-house, with pollard, bran, barley-meal, pease, and bean-meal, are the general food.

A small pamphlet has been published by a Scotch farmer, called the Grazier's Ready Reckoner, directing how the weight of fat animals may be found by measure. We shall here give a short dissection of the principle.

The girt of a beast, calf, or sheep, is to be taken just behind the shoulder and fore legs, and the length from the fore part of the shoulder to a plumb line from the hind part of the buttock, allowing a little for the protuberance; the solid content in feet is then to be found, and reckoning 42 lbs. to the foot gives the weight of the animal. The operation may be abridged as follows:

Square the circumference; multiply that product by the length all in inches; multiply the last product by 46, and point off 6 figures, and you have the true cylindric content of the animal, weighing 42 lbs. each foot.

Example.—Suppose a beast 80 inches circumference, and 60 inches long; required its weight of butcher's meat—80×80×60×46= 17,664000, or 17,664 cubic feet, the true cylindric content of the animal.

17,664×42=741,888 lbs. or 185,472 lbs. per quarter, the weight of butcher's meat, or 185½ lbs. per quarter.

Rule 2: find the content timber measure thus: square one-fourth of the circumference; multiply by the length in feet, and divide by 144, gives the feet timber measure, two-thirds of which is the weight per quarter.

Again: 20×20×5 feet=2000, divide by 144, gives 13,888 feet content timber measure, two-thirds of which is 9 score 5 lbs. and a fraction per quarter, as before. If the animal be not fully fat, deduct one-twentieth part, and the same for a cow that has had calves.

SHEEP.—Of the different varieties to be found in this county, three only are deserving of particular attention. These are, 1. the grey-faced, hornless, with fine wool, natives of Cannock-heath and Sutton Coldfield; 2. the white-faced, hornless, with long or combing wool, introduced from Leicestershire; 3. the South Down breed, introduced by Lord Bradford, Lord Anson, and some other public-spirited gentlemen.

The grey-faced, without horns, are the native common sheep: their characteristics are grey faces, lighter or darker, varying from white to black in different shades; the legs of the same colour with the face; wool fine, closely and compactly covering the carcass, of a moderate size. Those from a sound walk, and a good managing master, are well disposed to fatten, and produce mutton equal to that from any other breed. The better breed of these sheep are similar to the South Down, and not inferior: their general fault has been the want of thickness in proportion to their length. The late Sir Edward Littleton, with a patriotism highly commendable, for many years set the example of improving this breed, by crossing with Ross rams, by which he produced what he called the Teddesley breed, improved in form, increased in weight, and the wool refined in a very high degree. Sir Edward kept several hundreds of these sheep, the ewes fattening from 12 to 16 lbs. the quarter, and the wethers 16 to 20 lbs. The ewes of this breed are good milkers, and, if taken into pasture at lambing time, will feed every lamb before Midsummer fit for the butcher, and themselves the same by Michaelmas. These sheep, taken into pasture and crossed with strong rams, make a heavy breed of sheep, and have been brought to weigh, the ewes to 20 lbs. and the wethers to 30 lbs. the quarter. The fleeces of those of the original size will average about 3 lbs. each.

2. The white-faced polled sheep, with combing wool. The best flocks of this breed now in the county have been introduced or crossed from Leicestershire, and refined and improved in a high degree upon principles introduced by the late Mr. Bakewell, of Dishley. They are fine and light in the bone, thick and plump in the carcass, broad across the loin, with the back bone well covered, and not rising into a ridge, but sinking in a nick; fine and clean in the neck and shoulder; not too short in the leg, and of a sufficient bulk to weigh, when fat, the ewes 20 to 25 lbs. per quarter, and the wethers at from 18 months to two years old, 24 to 30 lbs. per quarter; but if kept another year, or till full grown, select sheep will rise to 40 and even 50 lbs. the quarter, and upwards. The ewe fleeces weigh about 7 lbs. each, and the rams and wethers 8 to

10 lbs. or sometimes more. The advocates of this breed say that pasture may be stocked harder and closer with them than with any other of equal weight, as they are always in good condition, even when suckling lambs, or hard kept. They are not, however, famous for fattening lambs for the butcher, which is no defect, as lambs of this breed are much better kept on for mutton and wool, and it would be a public loss to slaughter them prematurely. Many public-spirited gentlemen and farmers have paid great attention to this breed, and brought it to high perfection. Mr. Princep, of Croxall, had a very numerous, choice, and select flock; Richard Dyott, Esq. of Freeford, near Lichfield, kept of this breed between 200 and 300 breeding ewes, rearing annually 300 lambs. Excellent flocks of this breed were also many years ago upon Lord Bagot's estate, belonging to Mr. Harvey, and others; and they are now common in various parts of the county, in the hands of the most respectable and intelligent farmers. Mr. Miller, of Dunstall, near Wolverhampton, with great spirit, crossed his ewes of this breed several years successively with first-rate Leicester rams, procured from Dishley, and from Messrs. Green, Farrow, Stone, and Buckley, at from 50 to 200 guineas each ram: he also purchased ewes up to ten and twelve guineas each, and raised his flock to the first-rate new Leicester blood: he has had in his possession 40 rams at a time, which were let out for the season at from five to fifty guineas each. His successors continue the same plan, and have prime rams to let of this breed to any gentleman or farmer desirous of improving their flock.

Many farmers, however, in this populous manufacturing county, finding a demand for lamb as well as mutton, have been induced to keep none other than an annual stock of sheep consisting of ewes and wethers, bought in at Michaelmas, from Cannock-heath, Shropshire, Leicestershire, and from other counties and places. The ewes are immediately put to a ram, the lambs suckled in spring and sold to the butcher, and the ewes fattened in pasture and sold after them: the wethers are wintered in turnips and sold to the butcher, and the whole flock generally cleared off within the year, thus making room for the same rotation to succeed.

South Down Sheep.—The chief flocks of these sheep in this county are those upon the farms of the Earl of Bradford, Lord Anson, and Sir George Pigot. They came originally from the south downs of Sussex, where they have been greatly improved in their size, form, and feeding qualities, by Mr. Ellman, of Glynd, who has been to this stock a second Bakewell. They are a grey-faced, active sheep, hardy, and of sound constitution. The wool is short and fine, and nearly of the same quality as the Ross Ryeland. Of late years they have been so much improved by attention in breeding them, as to become, in some instances, rivals of the Leicesters in shape, weight, and early maturity. But between them, in fact, there should be no rivalry. The long-woolled Leicesters are adapted to the rich pastures of the county, from whence they have originated, and to other counties equally rich and fertile. The short-woolled South Downs are best suited to the lighter arable districts, which is the character of the country from whence they have come. They are both excellent in their respective qualities; and in the choice of either of these, the circumstances of soil and situation should be the chief guide of the agriculturist.

Disorders of Sheep.—Those most common in this county are the rot, the scab, the foot-rot, and the striking or dying in the blood. The rot is supposed to be occasioned by feeding on wet lands, where the animal licks up the spawn of the fluk-worm (*fasciola hepatica*), which insect creeps up the gall-ducts from the intestines, and preys upon the livers of sheep. A spoonful of spirits, or oil of turpentine, is a specific; and to prevent the disease, drain the lands where wanting, and let no sheep graze on water meadows in summer. No danger is to be apprehended from the first frost to the return of warm weather, the water in winter being too cold for these insects.

The scab is a very troublesome disorder, and when got to any considerable height, requires a good deal of pains and attention to cure. The following is given as an effectual remedy:—Take green broom two good handfuls, tobacco, stone brimstone, saltpetre, and bay salt, of each one pound, common salt two pounds, terebinthinated balsam of sulphur half a pound; boil all in five gallons of soft water for half an hour, and rub a little on the parts affected.

The foot-rot is to be cured by paring the hoof, cleaning out the dirt and matter, and anointing it with butter of antimony.

The striking or dying in the blood, is a very serious complaint, and generally happens to sheep in a growing or fattening state, often in turnips in winter, in clover in spring, and sometimes to young fat lambs. To prevent this complaint, bleed the sheep in the nostril about a week after they are put to good keep, and again about a fortnight or three weeks after, and give to each two or three spoonfuls of salt and water, made rather strong; to lambs one spoonful: the ewes suckling lambs to be omitted.

For maggots from the fly, apply tar water, or spirits of turpentine, to dislodge the maggots, and white lead to dry the parts. The best shepherds dissolve one ounce of sublimate in one gallon of spring water, and then add three spoonfuls of spirits of turpentine: this applied to the fly-blown part will destroy the maggots.

Horses.—The draught horses are generally of the Flemish breed, in colour black or brown, and some valuable stallions are kept of either colour, but mostly black. There are some considerable shows or fairs, where a great many excellent colts and horses are offered to sale, at Stafford, Burton, and Rugeley: many of them are bred in the county. The poorer sort of land in the Moorlands answers well for breeding; and a few acres of rich feeding land, on the banks of the Dove, pushes up the young stock surprizingly. They produce a good price at these fairs at two years old, so as to give a fair profit to the farmer who breeds them, and an inducement to attend to this discrimination of poor and rich land for the breed and improvement of live stock.

Lord Talbot has introduced upon his farm the Suffolk breed of draught horse. They have good bone, are clean legged, active, and very well adapted either for the waggon or the plough. For economical ploughing, an active, quick-stepping horse, is quite essential. In the slow pace in which that operation is too generally performed in this county, the resistance is much increased by the want of impetus, afforded by the more active step of the Suffolk horse. His Lordship has two stallions of that breed, and several mares. With the same object in view, Mr. Tollet has introduced upon his farm the breed of Cleveland horses, so called from a district in the north-east part of Yorkshire. They are a native breed, colour bay, large in size, and equal muscular power, clean in the leg as a blood horse, but with sufficient bone; and are good for the coach or the plough. By crossing with blood stallions, in order to breed the lighter sort of carriage horses and hunters, the pure race of Cleveland bays had nearly become extinct. Mr. Mason, of Chilton (the celebrated breeder of Durham cattle), with some other spirited agriculturists in the north, have taken great pains in preserving the small remnant of them. Mr. Tollet has procured his by the exertions of Mr. Mason, and has in his possession two stallions, six mares, and two foals.

The consumption of landed produce by horses has been much complained of, and they are doubtless in many cases great consumers, but it is very questionable whether their numbers can be much lessened, consistent with the convenience, or even the welfare of the nation. The greatest consumers are not those used in agriculture, (they are indeed the least), but those kept for pleasure and luxury, for harness, the saddle, and the road, as well as for the army; and of which the farmers are generally the breeders, and their stock in hand the source from whence they proceed, and without which they could not be supplied.

It has been calculated, that the number of horses kept in the kingdom is about 24 upon a square mile; of these one-fourth are colts and young horses in the breeder's hands, one-half are working horses employed in agriculture, and one-fourth are used for the saddle, harness, draught on the road, for the army, and for all other uses; and that these horses consume one-fourth of the whole produce of the land, leaving three-fourths for mankind, and for animals used as human food. If we allow this county its average proportion, the number kept of all sorts will be 29,280, which may be divided in the above proportion.

Many economical practices in keeping horses have been proposed, but as they are generally attended with more trouble and labour than giving hay and corn, they have not been generally adopted. It is very certain that horses working easy need not be much pampered; and those not working at all, will get their living and improve, in any sour rough pasture even in the winter; but if they are closely and hard worked, they must be well kept, or they will sink both in flesh and strength.

Vetches or green clover, or mown grass, are sometimes given in the stable, which is an excellent practice, as it prevents waste by treading. Carrots, or Swedish turnips, are good food for horses, but the latter require to be washed and sliced by a machine. Boiled potatoes are said to be excellent for horses. These kinds of food would maintain horses from less land than giving them hay and corn. Horses turned to pasture waste a considerable proportion of their food by treading, and by their dung, and when full-bellied are apt to range, and do much mischief: they are such great consumers, that every means of keeping them well with economy deserves the utmost attention.

A person who kept nine cart horses at jobbing, in constant heavy work, gave them his oat crops cut in a straw-cutting engine all together, corn and straw, but mixed with beans, about 6lbs. weight of beans to a heaped bushel of the former to each horse per day. The oat straw is thus wholly eaten up instead of hay, and the horses were in good working condition, and equal to their work.

A hard-working horse, if kept on hay and corn, requires daily 28 lbs. of hay, and 8 lbs. of corn, suppose 6 lbs. oats, 2 lbs. beans.

	A.	R.	P.
Hay, at a ton and a half per acre, will require of land …………	3	0	0
Oats 2190 lbs. at 40 lbs. the bushel, 54¾ bushels, …………	1	2	0
Beans 730 lbs. at 12 lbs. the bushel, at 4 quarters per acre, ………	0	1	20
Land to maintain a working horse the whole year, ……	4	3	20

An acre of oats, corn, and straw together, ought to weigh 3 tons; if we allow one-half of this to a horse, and add to it the produce of half an acre of oats or beans, it will be a good allowance for a horse in corn,

	A.	R.	P.
Being about 12 lbs. daily, (cut oat-straw included,) per annum, …	1	0	0
Vetches, June, July, and August, or mown clover, 1 cwt. per day: no hay would be necessary, but the above allowance of corn, ….	0	3	0
September, October, November, at grass or aftermath clover; each horse no hay still, but corn as above, …………	0	2	0
For the other six months, with corn as above, give half the allowance of hay, or 14 lbs. per day each horse, 22¾ cwt. from land, ….	0	3	0
Swedish turnips washed and sliced, or carrots to each horse, ….	0	1	0
On this plan, land to maintain a working horse the whole year, ….	3	0	0

A quarter of an acre of Swedish turnips, or in part carrots, would afford a good allowance for six months; for at half a cwt. per day it would only require 4 tons 11 cwt. 1 qr. or 18 tons, 5 cwt. per acre, which is a common crop.

That the above saving might be made, or horses kept upon less than two-thirds of the land they require at present, must be evident to every one who will consider the subject; and whoever keeps his team with the oat crop wholly cut, with green food in the stable, or upon carrots and Swedish turnips in part only, will approximate towards this system of economy, which no doubt would bear to be carried to the extent above proposed: whoever shall put it in practice will find their account in it.

Horses, and Oxen, for Draught.—The former are generally used by most farmers, being superior in quickness and dispatch, and for fetching coals, lime, and other articles, from a distance; and where a horse team can do the whole business, oxen are not kept, as they are very inferior in point of profit to heifer or cow stock, the latter producing annually a calf, besides milk, butter, and cheese, and their own carcass at last; whereas the ox, if not kept to labour, produces only his carcass.

But a good many large farmers, and gentlemen who farm, and

have more work than a team of horses can do, keep a team of oxen; and either rear or buy in succession stock, fattening off a proportion every year: they are very docile and useful in ploughing, carting manure, or other articles about the farm, but not for long journeys. The number of oxen thus kept for use has increased of late years.

Hogs.—The breed most esteemed, is not the large slouch-eared breed, but a cross between them and a smaller dwarf breed. They should be fine in the bone, thick and plump in the carcass, with a fine thin hide, and of a moderate size; large enough to fat, at from one to two years old, to the weight of from 300 lbs. to 400 lbs. each. These, if well bred, will keep themselves in good plight, with moderate feeding, and will soon grow fat with a plentiful allowance of proper food. Hogs of the large breed have been fattened here to from 600 lbs. to 800 lbs. weight exclusive of the entrails; but requiring much time and food, have given way to a smaller sized, finer boned, thick, plump animal. Hogs are generally fattened by farmers with the refuse of the dairy, and with pease either whole or ground: some throw a few dry old beans daily into the troughs for the pigs; bean-meal is also given. The most economical food for fattening hogs is boiled refuse potatoes and a little barley-meal mixed, or when potatoes are plentiful and cheap, the saleable size may be thus used. It was found by experiment that 28 lbs. of potatoes was equal to 7 lbs. of bean or barley-meal as hog food, and that either one or the other would produce 1 lb. of pork or bacon; that 2 cwt. of boiled potatoes, with 14 lbs. of barley-flour mashed-in, was a good weekly allowance for a moderate-sized hog, who would grow 10 lbs. weight with it, consequently that potatoes will pay a farthing a pound in feeding hogs, when pork is 7d. per pound, and so in proportion, with the hog manure for labour.

In an experiment made upon a labourer's hog, it was found, that one hundred and a half of potatoes boiled, with 28 lbs. of bean-meal mashed-in, was a good weekly allowance. The owner said, the pig, which cost 56s. would at first have eaten half a hundred per week of bean-meal, but was then satisfied with a quarter, with the above potatoes; said the bean-meal made him lay down to sleep, and thought a quarter of a hundred per week, with a hundred and a half of potatoes, better than more potatoes and less meal. The potatoes were boiled as wanted, and given with the meal mashed-in warm, or with a little warm wash. He thought the pig grew 2 lbs. per day, and was then large and fat enough to weigh 16 to 18 score lbs. having had him four months, but not all the while at fattening keep.

By this proportion, 168 lbs. of potatoes weekly should add of pork, .. 6 lbs.
28 lbs. of bean-meal ditto, ... 4

Growth per week, ... 10 lbs.

But in the above case the owner believed his pig grew 14 lbs. per week. These particulars seem sufficient to prove that hogs will pay well for feeding on potatoes and bean-meal, when those articles are at a moderate price, and that a farmer may profitably grow them with that view, and thereby greatly increase his manure.

The coarser and more productive kind of potatoes, will, when boiled and mashed with meal, be equal in hog feeding, to the more delicate flavoured sorts, which are less productive, and should therefore be reserved for human food. Of the former 12¼ tons. per acre have been grown (Mr. Curwen): these boiled and mashed with 2 tons of bean-meal, which may be grown upon two acres, would produce at the lowest estimate as above, from the potatoes 1000 lbs. of pork, and from the beans 640 lbs. total 1640 lbs. from three acres, or 546 2-3ds lbs. per acre. An acre of potatoes is thus proved superior in nutriment to either man or beast, to an acre of the best grain or pulse in the proportion of at least 3 to 1, but no preparation of potatoes alone is good enough for the staple food of mankind; nor will potatoes alone, in any form, answer for fattening hogs, without additions from grains, pulse, or the dairy.

Hogs are fattened by millers with pollard, with inferior grain and pulse ground down, and by butchers with the refuse and offal of slaughtered animals. A medical gentleman has suggested, that in his opinion, tallow or rough fat boiled away in water, and made into a stir-pudding with ground oats, barley, or other meal, or with pollard, merits trial. Bacon and pork are articles of considerable consumption in this county, and it is supplied in part from Shropshire, Cheshire, and North Wales.

Rabbits are but little attended to, except in warrens on waste land. They may be kept to some advantage in sandy or rocky precipices, impracticable to the plough, which may at the same time be planted; and, when properly fenced in, and thus stocked, such land seems to be in the highest state of improvement of which it is capable. Rabbits are well known to be extremely prolific, breeding almost every month. Some tame ones are kept, and some persons occupying land have bought the young ones to turn down, and hunt with dogs for amusement, and for sale when grown.

Poultry.—The profits of these are of but little consideration in or near the populous parts of Staffordshire, from the risk of their being carried off by thieves, which is done in some places two or three times a-year; and these depredations being committed in the night, are seldom detected. To prevent this, the poultry-house should be under the command of a lodging-room window, where fire-arms are kept in readiness, and a dog chained near. If they could be kept in security they would deserve more attention, and those who are at the pains to rear them early in the spring, find it worth their while, the chickens selling at a good price, and at an early growth. Eggs also produce considerable profit, being sometimes sold at 2d. each. A poor woman, who reared some chickens and ducks, took them to market and sold them well, and bought a pig with the money: this is commendable management, and worthy of imitation. The gallinæ or Guinea-fowls, are very prolific; their eggs are much admired; their appearance is singular, and adds to the variety; their flesh is also much esteemed by some people, who fancy its taste and flavour resembles the pheasant.

Turkies are in higher estimation than any other sorts of tame fowls. They require more attention in rearing, and do best where there is plenty of growing wood, being fond of acorns: their eggs are esteemed delicious. Since they have become common, the capon (though a fine fowl), is not so much sought after: these latter, however, might be very profitable, on account of their rarity.

Geese are become more scarce in the neighbourhood of commons lately inclosed, where the poor people used to rear them, and sell them lean to farmers to fat in their stubbles for the table. In such cases, they must in future be bred on the farmer's premises. Geese are reckoned salutary and wholesome to horned cattle; and attentive people assert, from their own observation, that where geese are kept to graze with horned cattle, such cattle are less subject to disorders than where no geese are kept.

Ducks.—The breeding of ducks is not so much attended to as it ought to be, particularly when their utility is considered in clearing grass lands and cultivated vegetables from worms, snails, and insects. It is the custom of the Chinese to keep large flocks of ducks to devour certain reptiles common in that country, and which would otherwise destroy the crops of rice. They are employed in some parts of England for similar purposes. A field of flax of about six acres was in danger of being destroyed by the slug-worm. To put a stop to their depredations, the owner procured an hundred ducks of the common sort, and engaged a man to drive them as soon as day-light appeared into the flax-field, where they devoured great quantities of these slug-worms: they were then pent up till the following morning, when they were again turned out; and this was repeated for about nine days, by which time the insects were destroyed, and the flax was preserved, and produced a good crop. The same practice is said to have been applied with success to the caterpillar, and to slugs, on turnips. Ducks thrive very well in thick water, but if there is plenty of clearer water so much the better, as both ducks and geese are fond of a change. The Muscovy duck is by much the largest and finest flavoured, as well as most beautiful and various in colour; and, what is singular in this variety, it will sometimes perch.

Pigeons are kept by gentlemen resident in the country from choice, and for the supply of their own tables; by farmers who happen to rent old mansions where pigeon-houses have been formerly erected; and by whom the markets are supplied; and by pigeon-fanciers, who keep tame ones. They are very prolific, breeding almost every month: the young are readily sold at market, and their dung is valuable as a top-dressing for crops. They certainly deserve some attention. They do best in dove-houses secured from vermin, especially rats, as boxes fixed on the outside of buildings subject them to many inconveniences. They sometimes require to be fed, particularly in the benting time: ten dozen of young pigeons have been taken in a short time from a moderate-sized dove-house.

Pigeons will sometimes forsake their habitations for want of being fed, or kept clean: in which case boil assafœtida in water, and wash the holes with it; their feathers will receive the scent, which pleases their companions so much that you will soon have the flock restored. Cummin-seed is also reckoned a great enticer of pigeons, by washing the holes with its decoction, or feeding them with grain steeped in such water. They are also very fond of salt: a lump laid on a plate on the dove-house floor, is very salutary to them: or in a large dove-house this may be done in separate places.

Bees are kept occasionally by farmers, and cottagers: it is matter of surprise that they are not more attended to, and that every farmer has not an apiary. A number of bee-hives is attended with little expence, and considerable profit may be made by selling the honey, which may be used instead of sugar. It also makes excellent liquor, under the names of mead and metheglin. The blossoms of fruit are very kindly to bees, and much honey is produced in the fruit counties around the Malvern-hills. Were the number of fruit-trees therefore increased in this county, bees might be increased in in proportion. It has been said; they are so much attended to in Flanders, that a farmer near Louvain sold a thousand stocks annually at five shillings each.

PRODUCE *and* CONSUMPTION *of* PROVISIONS: *Poor and Population.*—The number of acres of the different kinds of grain grown in this county was attempted to be ascertained about 20 years ago, by returns made by the high constables to the Quarter sessions, in answer to certain queries proposed. These answers were not general, but promiscuous from all quarters, and the parts omitted were supplied by analogy, and may be brought down to the present time, by adding a certain proportion for land since inclosed and brought into cultivation. The result is, supposing the enclosed and cultivated land of the county 640,000 acres, as follows:

Sorts of Grain, &c. grown.	Acres.	Produce per Acre Winches. measure. Bushels.	Amount. Bushels.	Deduct Seed. Bushels.	Remains for Consumption. Bushels.
Wheat, ...	60,000	24	1,440,000	140,000	1,300,000
Barley, ...	40,000	30	1,200,000	140,000	1,060,000
Oats, ...	50,000	30	1,500,000	240,000	1,260,000
Beans, Peas, &c.	20,000	30	600,000	80,000	520,000
Wheat Fallow, ..	30,000				
Turnips, ...	20,000				
Grass, ...	420,000				
	640,000				

The rest of the county, besides homeage and garden-ground, may be reckoned, waste land, woodland, roads, rivers, plantations, or unimproved land.

Poor.—The increase of paupers is no more than a natural consequence of the increase of population, and the alteration in the money-price of provisions. If we calculate by analogy from the population acts, the poor, as well as all others, have increased in that time as two to three; and the money-price of provisions of all sorts, corn, butcher's-meat, cheese, as well as malt liquor for beverage, has nearly doubled. These circumstances will account for the nominal advance of the expence of maintaining the poor. Respecting aids to the poor, this county is by no means remiss. The property of the poor, under the management of ministers and churchwardens, left them by charitable donations, is very considerable, and properly applied to the relief of those who keep themselves from being chargeable to the parish. Voluntary subscriptions have often been set on foot, and very liberally supported, in scarce and inclement seasons, when people of all ranks have come forward in proportion to their means and ability.

The population of the country, by the returns in 1811, was 295,153 persons; it may now be taken in round numbers at 300,000; and reckoning every acre of wheat grown to supply four persons, the growth at present will supply 240,000, or four-fifths of the whole, leaving the remaining one-fifth to be supplied from elsewhere, which is the case at present, chiefly from Shropshire and Worcestershire, both of which counties grow more than their own consumption.

If the tillage were improved so as to add one-fourth to the produce, the county would supply itself. This would easily be done by a cleaner culture, rooting-out grass and weeds, and putting wheat-plants in their places. The present net produce seems very low at 21 2-3ds Winchester bushels per acre. If it were raised to 27, the county would grow its own wheat, and further improvements, with the gradual cultivation of unimproved lands, might make it the case for many years to come.

Barley.—The consumption of malt, at four bushels per head, upon 300,000 persons, is 1,200,000 bushels, which is considerably more than the growth of the county in malting barley: the deficiency is principally made up from the Rylands of Shropshire. This, and any future increased demand, might easily be supplied by the raising of orchards, and cultivation of fruit-trees for cyder and perry.

Oats, Beans, and Pease.—The county, it is conjectured, supplies itself with horse-keep and hog-food, a good deal of the land being inclined to cold and moist, adapted to the growth of oats: a good porportion is suited to beans. It, however, does not breed and feed hogs enough for its own consumption, having a considerable supply of fat pigs, as well as bacon, from other counties, and some from Ireland. Oat-meal is much used as human food (from oats grown in the county), in oat-cake, and thickening soups, &c.

Turnips and Grass Land.—The county more than supplies itself with fresh butter, but receives salt butter from North Wales and Ireland. In cheese it has a considerable surplus, which is sent down the Trent, for the navy, the London, and other markets. It breeds cattle enough for its own consumption, but does not feed its own supply, numbers being sold when lean, or in part fattened, and many bought-in fat from Leicestershire. It is much the same with respect to sheep, supplying itself with lamb, but not wholly with mutton. Of wool, part is manufactured in the county, but the greater part sold out of it. It breeds horses enough for its own supply, and some to spare, a good deal of the cold grass land being more adapted to breeding than feeding, and perhaps paying better at grass than it would under the plough.

It is very possible for an improved cultivation to add very shortly one-fourth to our average products per acre, to effect which nothing more is wanting than for the improved management to become general; this quantity, and more, being already raised by our best farmers, independent of any advantage from goodness of soil. This circumstance, united with the additional cultivation of Cannock-heath, and the other waste lands, would increase the product of grain nearly in the proportion of two to three; and the production from pasture-land might, by a similar improvement, be increased in as high proportion.

Increase of Animals and Vegetables.—We shall add a short memoir on this subject, as a matter of curiosity and some utility; tending to shew how soon a new breed of animals, or variety of vegetables, may be introduced and extended, by application and perseverance.

The cattle species, though not very rapid in increase, will be four-fold in 8 years, beginning with 1 bull and 1 cow, - - 2

In 8 years they may be - -	8
16 years, - - - -	32
24 years, - - - -	128
32 years, - - - -	512
40 years, - - - -	2048

and continue in that proportion, though slaughtered at eight years old.

Sheep.—Suppose the ewes each rear annually one lamb only, over and above losses, the young ewes to bring forth at two years old, half males, and the old stock cut off at five years old, all the others preserved,—

One ram and one ewe at first,	- - - -	2
In one year may be	1 ram and 2 ewes, - - -	3
Two years,	2 rams and 3 ewes, - - -	5
Three years,	3 rams and 5 ewes, - - -	8
Four years,	6 rams and 7 ewes, - - -	13
Five years,	10 rams and 10 ewes, - - -	20
Suppose the original pair now cut off, deduct	-	2
	Remains, -	18

The increase is nine-fold in five years.

If 2 sheep in 5 years increase to	18
In 10 years, they may be	162
15 years,	1,458
20 years,	13,122
25 years,	118,098
30 years,	1,062,882
35 years,	9,565,938
40 years,	86,093,442

And so on in proportion.

Swine.—This species may increase five-fold annually, at a very reasonable calculation: suppose ten at one or two litters brought forth by each dam annually, half males, half females, and the old stock annually cut off,—

Thus two, a boar and sow, may in one year be	10
In two years,	50
Three years,	250
Four years,	1,250
Five years,	6,250
Six years,	31,250
Seven years,	156,250
Eight years,	781,250
Nine years,	3,906,250
And in Ten years,	19,531,250

Rabbits are still more prolific, and would, if permitted, soon increase in numbers sufficient to overrun the earth. It seems to be in the system of Nature, that a large proportion of animals should be prematurely cut off as food for others; otherwise, their tendency to increase would so exceed their means of subsistence, as to occasion a general scarcity of food.

Vegetable Increase is still more rapid than that of animals. It is providentially ordered, that animals as the food of man, should have a tendency to increase faster than man himself; and vegetables as food for both, have a tendency to increase faster than either, insomuch as in a very short time to cover the whole surface of the earth by their natural power of accumulation.

We shall conclude these theoretic calculations by the often-repeated example of the increase of a grain of wheat in a ten-fold proportion:

One grain of wheat may be, in 1 year,	10 grains.
In 2 years,	100
3 years,	1,000
4 years,	10,000
Which, suppose 1 pint in 5 years,	10 pints.
6 years,	100
7 years,	1,000

Or 125 gallons, or 15⅝ bushels, which is enough to sow		8 acres.
In 8 years would sow	-	80
9 years,	- -	800
10 years,	- -	8,000
11 years,	- -	80,000
12 years,	- -	800,000
13 years,	-	8,000,000
14 years,	-	80,000,000
15 years,	-	800,000,000
16 years,	-	8,000,000,000
17 years,	-	80,000,000,000
18 years,	-	800,000,000,000

which is more than six times the superficies of the whole earth and sea.

COMPARATIVE VIEW

OF THE

Present State of the Agriculture of Staffordshire.

UPON a comparative view of the present state of the Agriculture of this County with what it was thirty or forty years ago, it may be confidently asserted that few districts of the kingdom have improved more than this has done. The live stock of the county, particularly in the southern part of it, has been very much attended to. The long-horned cattle and the Leicester sheep have been the sorts most generally selected, and the county can vie with most others in stock of these valuable breeds.

The cultivation of the Turnip, the key-stone as it were of our arable agriculture, has been greatly extended, so much so, that several individuals now grow upon their respective farms a larger breadth of turnips than could be found fifty years ago in the whole county. Draining, the foundation of all other improvements, when from the nature of the soil it is necessary, has been practised upon a very extensive scale, and executed in a scientific and masterly manner. Since the introduction of that incomparable article the Draining Tile, the art of laying the land dry, whether it be incommoded by springs or by surface water, has been brought to great perfection in this and in the neighbouring counties of Warwick and Leicester; an example that should be followed upon all retentive sub-soils throughout the kingdom. The Tile is greatly preferred to any other material. They are light of carriage, the drains completed with them are less likely to become out of repair, and if by any accident they get out of order, the work admits of easy reparation. Since they were exempted from duty (an advantage for which the agriculturist is chiefly indebted to the exertions of Sir John Wrottesley, when he represented Lichfield in Parliament:) it will be generally found that tile-draining is the cheapest of any permanent mode. Irrigation has been extensively adopted in various parts of the county, according to the most approved methods, and has been found of infinite advantage. The use of lime at the period before alluded to, was little known. Of late years it has been extensively applied, both upon the arable land, and, as a top-dressing, upon the grass lands; and under judicious management has been found to be highly beneficial.

The implements of husbandry have been much attended to. The Staffordshire Plough is an efficient implement, and the construction has been in many parts improved by imitating the Swing Ploughs, that have been introduced from Leicestershire, from the northern counties, and from Scotland. The Twin Harrow, perhaps the best of all implements for eradicating weeds and bringing the soil into good tilth, is coming into very general use.

The waggons and carts are constructed upon a less heavy and a less cumbrous plan. At Tamworth, one and two-horse carts are made as good and as cheap as in any part of the United Kingdom.

At Wolverhampton the various sorts of drilling machines are manufactured, a proof that this excellent system of husbandry is making gradual advances amongst us.

A more economical attention is paid to the making and collecting of various kinds of composts and manures than was formerly done; and they are certainly applied to the land in a much more scientific manner than they used to be. The gates and fences are kept in better repair; the mode of laying and pleaching hedges has very much improved of late. But considerable as the improvements in agriculture may have been in this county, yet it must be confessed that very much remains to be done.

The arable lands are not yet brought in general, to that clean and perfect tilth, which is essential for the securing of first-rate crops. The drill husbandry has not made that progress which it ought to have done, from the examples which have been set in different parts of the county. The practice of getting white strawed crops in succession is by far too general; and the clover seed is almost always sown upon land that is exhausted and foul from these successive white-straw crops, nor is it sowed in sufficient quantity. The turnip husbandry does not yet prevail over all the lighter soils; and much of the stronger soils might be made to bear them to great advantage, by being previously well drained, and by growing the turnips upon the Northumberland one-bout ridge. Upon the turnip lands the Norfolk system should be pursued. Upon the strong soils, instead of the present course of 1. fallow, 2. wheat, 3. oats, 4. clover, it would be better to substitute the following rotation: 1. fallow, 2. wheat, 3. clover, 4. oats upon the layer. By this mode of culture, the land would be kept more clean; the wheat and the oats would be at least as good, if not better; and the clover layer would in general be of double value to what it now is.

The next course might consist of the following shifts: 1. fallow, 2. wheat, 3. beans or pease, 4. oats, by which change the clover would only be put in the land once in eight years, as it is well known that the land will not bear to advantage a more frequent repetition of that crop. These we presume are principal objects to be attended to in our future progress in improvement.

Farms of Land-owners.—Among the various causes which have led to the improvement of the Agriculture of the County, the partiality of many of the landed proprietors for this most interesting and useful of arts, may justly be considered as a principal one. Scattered about in different parts of the county, the demesnes of these gentlemen may be looked upon as a sort of school in which the practices of other districts may be learnt, the best implements may be inspected, and the merits of the various stock of different counties may be duly estimated by the practical farmer. Among persons devoted to the same pursuits, the most lively and lasting friendships are usually formed; and this effect has been produced among the gentlemen of this county. To inspect the improvements carrying on upon each other's farms, has led to a social intercourse, which could scarcely have been formed under any other circumstances. In these friendly meetings, the paltry jarrings of politics and party are forgotten in the contemplation of those far more interesting pursuits, which have for their object to better the condition of our fellow-creatures, and to place the wealth and prosperity of our country upon the surest foundation. Among the first of these may be placed the Lord Lieutenant of the county.—From the period that he came of age, Lord Talbot has devoted his time and his talents to the practice of agriculture: and the demesne of Ingestrie will be a conspicuous proof of the success of his pursuits.

Ingestrie, Farm, &c.—The principal entrance or avenue to the demesne of Ingestrie is by the south lawn, through a very magnificent gateway of ancient architecture. The whole parish, manor, lordship, and township, belongs to Earl Talbot. It consists of a capital mansion and demesne, gardens, plantations, park, lawns, meadow and arable land, pleasantly situated at a moderate elevation, on the south-west side the river Trent: the Grand Trunk Canal is parallel to the Trent here on the opposite side. The demesne in hand contains upwards of 1600 acres, of which about 300 acres are contained in a walled park, 200 in plantations, 500 of arable land in tillage, and the rest, about 600 acres, permanent grass in lawns and meadows, on the Trent and elsewhere.

The park and plantations are well stored with timber trees of every sort, amongst which are some very fine oaks, and beeches in great numbers, remarkable for size and vigour of growth. The park is well stocked with deer, and shews marks of ancient cultivation, but it has long been at grass.

The tillage land is managed nearly upon the Norfolk system, in four divisions; green crops, spring corn, clover, and wheat, from 120 to 130 acres of each, according to the size of the inclosures. The green crops comprehend the common and Swedish turnips, cabbages, and cole. The turnips are cultivated in the Northumberland system, in rows, and horse-hoed, and afterwards hand-hoed till not a weed remains; the cabbages in about three-feet ridges in the usual way. Mr. Ginders, who has the care of the whole, is satisfied by repeatedly weighing large breadths to half an acre, that 35 tons per acre is a full crop of Swedish turnips: they are used for sheep and stall-feeding cattle. A considerable lot of Swedish were (Nov. 1815,) topped and rooted, and laid along a dry ditch upon a brushwood bottom, and meant to be covered with straw, and thus stored for use, the row being from 100 to 200 yards in length, and the ground might thus be cultivated for any early crop.

The spring corn is of different varieties, barley on the dry land, oats on the colder lands, a few beans sometimes on the stronger soils, and buck wheat on a moderate scale for the pheasants and poultry. It has been found very advantageous to lay land to grass with buck wheat, as the straw, standing firm and upright, never smothers the grass seeds.

On laying land to permanent grass, which has been done upon some of the lawns, which have been cultivated for improving the herbage or levelling the surface, the land is effectually under-drained; if necessary, lime is freely used, and the grass seeds always sown with the first crop after turnips, cow grass and white clover to about 20lbs. per acre, and Buxton hay seeds half a bushel, the latter collected by the ostlers of innkeepers at or about Buxton, and bought from them by seedsmen, and believed to be free from weeds and couch grass. Some parts of the south lawn thus sown last season, is now extremely promising for pasture.

The wheat and barley is always sown with Cooke's drill, in rows at nine inches, pease of course wider, the crop always clean hoed, and no weeds suffered to grow therein.

A complete farmery has been erected, consisting of a farm-house and offices, barns, stables, very extensive cow sheds, hoggery, roomy farm yards, store-houses, smith's and carpenter's shops, coach and cart houses, rick yards, and every convenience, within a compact square; and the whole business is managed by a regular system. Only one man is permitted to cut the hay-ricks, who deals out the fodder to the grooms, carters, and cow-keepers, without waste or loss.

The grass land, besides clover, consists of about 600 acres, of which about 70 acres is water meadow, principally on the Trent: the morassy parts have been effectually drained, and the herbage improved; and the irrigation is attended to in various ways upon both the catch and pane work, some of the land having been so laid out. Mr. Ginders is convinced, from close observation, that no improvement can answer better than watering meadows with due attention, but they will not admit of neglect; the gutters must be annually cleansed, and the water spread in rotation and periodically. By such attention, some of the meadows on the Trent have been twice annually mown, the first time the end of May, or early in June, producing two tons of hay per acre, the second time in August, a ton and an half per acre, and an aftermath to graze in autumn: no manure required, but due attention to the water, by which the ground is kept fully in heart, and further improved. An immense stock of hay is thus raised, some of the single hay ricks being supposed to contain 100 tons of hay. The live stock kept on this farm consists of from 300 to 500 breeding ewes, of the best new Leicester breed, from the stock of Mr. Buckley: these rear annually an equal number, the twins making up for casualties. The lambs are not shorn, and the wethers are sold as shear hogs, before a second shearing; they produce wool on an average 7lbs. to the fleece, or four to the tod of 28lbs., the best shearlings having produced 13lbs. to the fleece: the aged ewes when fat average 24lbs. per quarter, and shear hog wethers about the same, but have sometimes risen to 35lbs. the quarter.

The cow stock is of the Hereford breed, from the stock of Mr. Price, of Worcestershire, and Mr. Tomkins, of Herefordshire. The preference is given to that breed on account of their aptitude to fatten, the land here being of a good staple for feeding: they are generally red in colour, with white faces.

This stock consists of about 200 head of cattle: the bull calves are reared upon cows, and the oxen fattened when grown without being worked, and some Hereford oxen are bought-in of four or five years old, for fattening at grass or in the stalls upon hay, turnips, and sometimes oil-cake, very little corn being thus given.

The cultivation is by horses, principally of the Suffolk punch breed, colour light chesnut, of which are kept two stallions, and about forty mares and geldings.

The hogs are chiefly pure Suffolk, of a good size and kindly; the stock consists of about ninety, from six to eight being breeding sows: they are fed principally with potatoes, Swedish turnips, and refuse from the oxen.

There is a walled garden of seven acres, furnished with all the varieties of vegetables and fruit; the demesne is thus rendered highly productive of human necessaries, timber, grain, fat cattle, sheep and wool, and of the comforts and luxuries of life.

Three threshing-mills are employed; the first fixed and permanent in the farm yard, a six-horse power, by Forrest, of Shiffnal; a second, of less power, at an out barn; and a third, portable, moving from place to place as wanted: the large one will fetch wheat out clean at the rate of 120 bushels for the day's work.

Lord Anson's *Farm and Demesne.*—The family-seat of Shugborough is an elegant and magnificent building, covered with white stucco in the best style. It is delightfully situated in the Vale of Trent, at the confluence of the Penk and Sow with that river, and lies between Stafford and Lichfield. The demesne is a rich plain of several hundred acres, well wooded and watered by the above rivers, and skirted by distant hills, variously disposed by nature and art, part in arable culture, part plantation, and some in a state of native wildness. On the eminences are an Obelisk and a Triumphal

Arch, erected in honour of the great naval commander George Lord Anson. The shrubberies and pleasure-grounds are extensive, and laid out in the best modern taste, well decorated with fancy buildings and sculpture, and kept in the highest style of neatness. A kitchen garden of several acres is walled and subdivided; the walls well stored with the choicest fruit-trees, with very extensive ranges of hot-houses, in which the pine-apple, the grape, the peach, the fig, and other varieties of hot-house fruits, flowers and plants, are cultivated in the highest perfection.

One of the hot-houses is heated with steam, in which melons and cucumbers are produced in perfection at all seasons. These gardens are a kind of Academy for the study of Horticulture, in which young men enter themselves to assist without pay, for the purpose of improving themselves, and gaining knowledge in the art.

His Lordship's farming is to the extent of 2000 acres. A farm-yard has been constructed at a convenient distance from the family-seat, under the direction of Mr. Wyatt the architect, consisting of the farming steward's house on one side: a range of buildings on another contains a brewhouse, upon a large scale, a water corn-mill for the use of the family and farm, and in which corn is ground for the neighbouring poor gratis, and also a malt-house. The opposite side and end are occupied by stalls for feeding cattle, store-rooms, stables, and other appendages. In the middle of the yard is a very complete hoggery, built of large stones set edge-wise, and covered with slate, with a boiler for heating hog-food, and a cold bath, supplied by the mill stream, for giving an occasional washing to the pigs, to promote their health and cleanliness. Here a number of hogs are fattened on dairy refuse, boiled roots or vegetables, pulse, ground barley or bran, supplied by the mill near at hand.

At some distance above the farm-yard is the stack-yard and barns, where a powerful threshing machine is worked by the same stream that afterwards supplies the garden and works the corn-mill. This machine will thresh out 200 bushels of grain per day. The erecting it cost less than £100.; but in this expence the barns are not reckoned, having been before built; nor the mill-wheels, they having been erected by a millwright, separate from the threshing-machine.

Arable Cultivation.—This has, we believe, been upon so large a scale, that £1100. has been paid in one year for the purchase of lime as manure; and the growth of barley in 1804 had been near

300 acres, and a full crop, a large quantity of which was then in hand in the stack, but going to a good market, being contracted for at 9s. per bushel, 2000 bushels to be delivered immediately, which the threshing-mill can get out in ten days. With the barley is always sown plenty of the usual seeds, red and white clover, trefoil, and ray grass, and it was intended in future to confine the cultivation generally within less compass; about 100 acres of turnips, and the same breadth of barley and grass seeds each year; the turnips were then extremely promising for a full crop.

The usual course of crops: 1. wheat; 2. turnips; 3. barley and seeds; sometimes, 1. pease; 2. wheat; 3. turnips; 4. barley and seeds: or on weaker land, 1. pease; 2. turnips; 3. barley and seeds, and wheat omitted. Manure laid on for the turnips: lime, farm-yard dung, or rape cake; the barley and wheat were then full crops, a considerable breadth of pease and vetches, good. The soil is generally a sandy or gravelly loam, with plenty of pebbles: the original spontaneous produce, heath and furze. The site of the upland generally varied and uneven; the knobs or brows of banks, generally poor and thin of soil; many of them are very properly reserved for clamps and plantations. Lord Anson's ploughing is uniformly done with a swing plough, with two horses a-breast, without a driver. Cooke's drill machine has been much used both for wheat and barley, putting in two bushels of barley per acre, three being sown broad-cast. When barley is drilled, the ground is harrowed for the grass seeds and rolled, without any hoeing. Turnips have been sown by Cooke's drill, at both nine and twelve inches distant with success, but they are now chiefly sown on the Northumberland ridge, and astonishing crops have been grown upon the gravelly hills.

For cleaning the land, both horse-hoeing and hand-hoeing have been used, as well as hand-weeding; the most prevalent weeds in the turnip crops being corn lake-weed (*polygonum persicaria*), corn spurry (*spergula arvensis*), and chickweed.

The carting, ploughing, and team-work, was then done by horses, of which about thirty were kept for draught, including the drawing of fuel, building materials, &c. which are sometimes considerable. Some time back, seventeen oxen were kept for draught; and the reason assigned for the change is, the land being strong and uneven, and on those accounts inconvenient to cultivate with oxen.

Of *Grass Land* the quantity is very considerable, consisting of water meadows on the Trent and Sow rivers, subject to inundation; grass land, that can be watered at pleasure by the mill stream, after passing through the farm-yard; and upland pasture, out of the reach of water; this last being occasionally refreshed by top-dressing with dung or compost.

The meadows on the rivers are subject to occasional damage from summer floods, whereby the hay may be fouled, spoiled, or carried away; but this does not often occur. An inundation in May, 1805, continued several days, and left the ground slimy, and apparently chilled and starved, but it soon recovered, and the grass ran to a great crop. The hay harvest of 1805 had been upon a very large scale, and from the showery season, got together with great difficulty. Near 250 acres having been mown for hay, eleven very large stacks had been made from it, besides a great quantity of old hay in hand. One of the old stacks was estimated to contain 100 tons of hay; and Lord Anson, we believe, had then the largest stock of hay of any individual in the county.

The land that can be watered at pleasure (of which there is a considerable quantity), is forced to an early spring of grass, and reserved for the ewes and lambs. The unsound parts having been drained, no inconvenience has been found, or is apprehended, as likely to arise from this practice.

This great farming concern was, in 1805, managed under the direction of Mr. Wheelock, the farming steward, by 22 labourers and two men servants, and one man managed the mill and malt-house: the labourers had then 10s. per week each the year round, and a good allowance of beer, with ale in harvest, and now and then a gratuitous dinner, at the option of their employer; and all the regular work let by the piece. Additional hands are employed in harvest, and in improvements, as a great deal of under-draining had been done, and the drains filled up with rock stone, heath, and earth, after boring in different places in their bottom. The farm is now under the management of Mr. Stamford, and affords a brilliant example of the good effects of the Norfolk system, and of the Drill Husbandry.

Live Stock.—The cow stock, in 1805, were of the very best long-horned sort: Lord Anson's cattle are now of the Devon breed. His Lordship's Devon bull is a very good one, and his stock of cows by constant attention and selection, will, ere long, be of the first-rate character: about 20 dairy cows are kept. An elegant dairy has been fitted-up, and finished with marble, which is appropriated to the immediate use of the family. A number of young cattle are annually reared for succession stock, which, with feeding cattle, made the number wintered generally exceed 100 head. The last season upwards of 60 head had been stall-fed, entirely on the produce of the farm (having stall-room for so many at once). They were all finished-up together, and sent to Smithfield.

Sheep.—The flock was very considerable, amounting to near 1700. Lord Anson had fixed upon the South Down breed as the staple flock, of which he possesses a capital flock, of good symmetry and sound constitutions. At the show of live stock by the Staffordshire Agricultural Society, at Lichfield, Lord Anson has obtained gold medals for the best short-wool rams, as well as ewes. These ewes, we are informed, will feed when full grown from 18 lbs. to 20 lbs. the quarter, and the wethers at two years old to 20 lbs. or more: produce of wool, yearling hoggets (not having been shorn as lambs) 4 lbs. average fleece; ewes ditto, 3¼ lbs. average, and some of the ram fleeces 6 lbs. each. The wool has been sold at 2s. 6d. per lb. The fat wethers of this flock have found their way to Smithfield, where they have been sold for first-rate prices.

The EARL of BRADFORD has made very great improvements upon his demesne, at WESTON, both in an ornamental and in an agricultural point of view. He has drained the land, and introduced a spirited and successful system of agriculture upon his farm.

The turnip husbandry is managed upon the one-bout Northumberland ridge, and his Lordship raises abundant crops of this most useful plant.

The ploughing is wholly done with two horses a-breast, and is conducted by an able Norfolk ploughman. The arable land is managed according to the Norfolk system, for which sort of husbandry the soil of the farm is for the most part very well adapted. Lord Bradford's Devon cattle are in the highest repute, having been judiciously selected, at a great expence, and they have answered all the expectations that were formed of them. The sheep stock are South-downs, bred from the flocks of the Duke of Bedford, and Mr. Coke, and are remarkably neat and good. The Pigs are of the Suffolk kind, having a great aptitude to fatten, with very little offal.—The whole farming concern is conducted in a masterly style, and the noble Earl, both by his example and influence, has largely contributed to the improvement of the Agriculture of the County.

The demesne at WROTTESLEY affords an excellent specimen of

practical and substantial agricultural improvement. We have before stated that the public are much indebted to Sir JOHN WROTTESLEY, when he represented Lichfield in Parliament, for obtaining an exemption of duty for the Draining Tiles. He has amply availed himself of the benefits arising from his own exertions in that important measure, having for a great number of years annually made use of from forty to seventy thousand tiles. The draining at Wrottesley is executed very scientifically. The worthy Baronet has himself laid out the plan of all the drains, and is so particular in their arrangement, as to use the spirit level to ascertain the exact slope of the land, that the drains may be formed with the greatest exactness, and have the fullest effect. His usual method is to carry up the tailing drain, so as to form a right angle with the slope, and to carry the side drains or herring bones (as we call them), as nearly as the situation will allow at right angles, into the tailing drain. The surface-drains range from four to two feet in depth. The spring draining of course depends upon the depth of the water. These drains Sir John Wrottesley has carried to very great depth; in some instances he has been obliged to keep up the sides with a well-formed scaffolding, and under such circumstances, that the getting-in of a single tile was considered a great achievement. The effects of these deep drains have been wonderful, and have amply repaid the cost and trouble incurred in making them. Any other attempt to drain such land would have been fruitless, and the money expended would have been thrown away. By the scale upon which this most substantial of all agricultural improvements has been carried on upon the demesne of Wrottesley, that which was for the most part a cold wet backward soil, is now made perfectly dry; and the turnip husbandry has been extended to lands that formerly produced nothing but rushes and other noxious aquatic plants. The arable farm is chiefly managed upon the Norfolk rotation. The turnips are sown upon the Northumberland ridge. The corn is for the most part drilled. The land is ploughed with two horses a-breast, without a driver. The implements are of the most improved sorts; and the whole farm, which consists of about 6 or 700 acres, is cultivated in an exemplary manner. The stock consists of short-horned cows, Leicester sheep, and pigs of a small but very good sort.

At PATTESHULL, Sir GEORGE PIGOT cultivates his demesne upon a magnificent scale. His farm consists of about 1,200 acres. The land is for the most part light and well adapted to the Norfolk husbandry, which system is adopted with great success by the worthy Baronet.

The implements upon the farm are particularly well selected. The farm buildings are as commodious as possible. A most copious spring of water has enabled Sir George to thresh his corn, &c. &c. by means of a water-wheel, the power best adapted to that operation. The cattle are mostly of the galloway breed. The flock of sheep is South-Down. Patteshull, as a gentleman's residence, may be justly considered a great ornament to the county.

Sir ROBERT LAWLEY has long been fond of agriculture, and has carried on considerable farming concerns, both at CANWELL and upon his Shropshire estate. His Leicester sheep have frequently carried away the Sweepstakes, both in the Staffordshire and Shropshire Societies.

Mr. INGE, of THORPE, has in his hands a large farm, which is very well managed. He has made great improvements in draining, having thereby turned cold unsound land into good turnip soil. His stock of cattle are of the long-horned breed. His sheep are Leicesters; and both of them of first-rate character.

Sir JOHN F. BOUGHEY, of AQUALATE, has directed his attention to the improvement of that valuable breed of sheep the Ross Ryeland.

His object is to assimilate the carcass as much as possible to that of the Leicester, at the same time preserving the fineness and character of the wool. The progress he has made is quite surprising, and there is little doubt but he will attain the object he has in view. He will thereby establish the reputation of one of our most useful varieties of sheep. He has a good dairy of Devon cows, and always works a team of Devon oxen, of his own breeding. After working them for a certain time, he feeds them, and has sold some of them when fat for upwards of fifty guineas a-piece.

Mr. BLOUNT, of BELLAMORE, the worthy Vice-President of the Agricultural Society of the County, has in hand a very considerable quantity of land, which is cultivated with great ability in the best modern style. He has made a surprising improvement upon part of his farm by irrigation. His stock consists of short-horned cows, and Leicester sheep.

Mr. TOLLET, at BETLEY, has much contracted his farming concerns, having let those farms which he had in hand, merely for the sake of improving them.

He has now only a small flock of merinos, as the overwhelming importation of fine wool from Spain and Germany, has rendered the British merino wool very difficult of sale. His farm, from the extent of water-meadow he has made upon it, is better calculated for the dairy than for sheep stock. These meadows have been made with considerable cost and contrivance. From an ornamental piece of water near the Hall, the stream is conducted to thresh, grind, and perform many useful purposes, at the farm-yard; afterwards it irrigates a large extent of meadow, and produces an abundant crop of hay. In short, wherever irrigation is practicable, the water upon this farm is not suffered to run to waste. Much under-draining has been done, the effects answering the most sanguine expectations: tiles only are used, which are made upon the farm. The dairy cows are of the Devon breed, and have a considerable mixture of the blood of Mr. Childe's well-known stock at Kinlet. Mr. Tollet has for his team, and for the plough, introduced the Cleveland breed of horses, from the north of Yorkshire, and Durham. They are a remnant of the native English breed, before the introduction of the Arabian blood for the turf and the field, and of the Flemish horse for the waggon and the plough. The colour of them is bay; they are active, with great bone and much muscular strength. They are well calculated for the plough and the coach; and for the use of the cavalry would perhaps be unrivalled. From the breed being constantly crossed with blood-horses, for the purpose of breeding hunters and coach-horses, the race was in danger of becoming extinct. It will probably now be continued: if that is the case, it will chiefly arise from the exertions of Mr. Mason, of Chilton, (the celebrated breeder of Durham cattle), and some other spirited agricultural gentlemen in Durham. It is chiefly owing to the friendly attentions of Mr. Mason, that Mr. Tollet has procured this breed. His stock at present consists of six mares, a filley, and three colts.

Mr. Grove, of Shenstone Park, has for a long time paid great attention to agricultural pursuits. His farm is well managed, and his flock of Leicester sheep are in high repute.

Mr. Tennant, of Little Aston, is also a great improver. The draining upon the lower part of his estate at Aston, has been executed at vast labour and expence. It has fully answered its purpose, having reclaimed a considerable tract of land, now highly fertile, that was before of little or no value. His improvements upon the new inclosure on Aston common, are equally conspicuous. Part of it he has judiciously selected for planting; the other he will soon turn into productive turnip and barley land.

There are many other Gentlemen, not here enumerated, who are fond of farming, and set good patterns in their respective neighbourhoods. With such examples scattered over the county in every direction, it cannot be at all surprising that great progress has been made in the improvement of its agriculture.

AGRICULTURAL SOCIETIES.

The prevailing taste for Agricultural improvement, which has been so long manifested in this county, has caused the establishment of two Societies, viz. one at Lichfield, and the other at Newcastle.

The former is denominated the Staffordshire General Agricultural Society.

It is liberally supported by the Nobility, Gentry, and the principal farmers of the county, and possesses ample means for attaining the useful and patriotic objects of the institution.

The premiums are confined to the practical farmer, as there cannot be expected to be a fair state of competition between the owner of the soil and one who rents it only. Therefore, the Landed Proprietors in all matters of competition, either make Matches among themselves, or enter into different sets of Sweepstakes, which of course are open to the tenants also. It is a gratifying circumstance to have to record, that notwithstanding the great pains, skill, and expence which have been bestowed by many of the principal Staffordshire proprietors upon the selection and improvement of their stock, several of the Sweepstakes have been carried off by the tenants.

The principal premiums of this Society are given to encourage a general good cultivation, rather than that of particular crops, or specific modes of cultivation and improvement. In the latter cases it was often found that the chief attention was drawn to the single object held out by the premium, to the detriment of the rest of the farm. Thus there is offered to the tenant who shall cultivate, and improve in the best manner, the farm he occupies, consisting of not less than 200 acres, a piece of plate of the value of £50.

To the second best, a piece of plate of the value of £25.

To the tenant who shall cultivate and improve in the best manner the farm he occupies, consisting of not less than 100 acres, or more than 200, a piece of plate of the value of £30.

To the second best, a piece of plate of the value of £15.

The judges, who are always persons of the first-rate character for their agricultural attainments, and from some other districts, are directed in looking over the farms of the claimants, to pay particular attention to the system of cultivation as adapted to the nature and quality of the soil to which it is applied; to the quantity and description of the several manures made use of (or in reserve), with the application of each; to the quantity, quality, and management of the stock; to the degree of cleanliness of the arable, meadow, and pasture land; and to the best execution and mode of conducting the whole concern. A detailed report of the state of the several farms is given in by the judges, and read by the president to the meeting.

To gain one of these silver cups is a high agricultural honour, and will no doubt in future times be considered as such by the family of him by whose merits it was gained.

Besides these, there are the usual premiums given for the various sorts of horned cattle, sheep, and pigs; and to the person who shall invent and exhibit the best tool or implement of husbandry, found to be better adapted to the intended purpose than any now in use, a premium of £10.

There are also several premiums offered to day-labourers who have brought up the greatest number of children without parish relief; and also to those who, producing a good character, shall have continued the greatest number of years upon the same farm.

The candidates for these premiums are required (in addition to the usual certificates of the persons they have served), to produce one from the Clergyman, or some other respectable householder of the parish in which they reside, not only of good character, but of the cleanly state and condition of the cottage and garden they occupy. The latter stipulation has an excellent effect in encouraging the cottager to attend to those circumstances which administer so much to his own comforts, and which do so much credit to the district in which they are practised.

There are also premiums for domestic servants in husbandry, of both sexes, who have continued the greatest number of years in the same service. The certificates produced at the annual meeting of the characters of these persons, and the accounts given of their faithful services, reflect the highest credit to these humble individuals, and are an honour to the county in which they live. Of this excellent institution, Earl Talbot, the Lord Lieutenant of the County, is the President, and Mr. Blount, of Bellamore, is the Vice-president. The annual meetings are well attended, and the shew of stock of every description of the first-rate quality; affording a convincing proof of the improved state of the county in this important branch of agriculture.

The *Newcastle and Pottery Agricultural Society* was established about 20 years ago by some gentlemen of that neighbourhood, who devote their time to the study and practice of agriculture. The limits are confined to a district of 15 miles round that town. The premiums are directed to promote the best modes of husbandry; and the objects of the patriotic founders of the institution have been answered, by the various improvements that have been introduced into the district. This Society also offers premiums to that useful class of men, the labourers and servants in husbandry. The meetings are annual, and are held in the month of November.

BOTANY OF STAFFORDSHIRE:

ACCORDING TO THE CLASSES OF LINNÆUS.

CLASS I.—MONANDRIA.

MONOGYNIA.

Chara tomentosa, (brittle stonewort) on *Needwood-Forest*
vulgaris, (common stonewort) ditches and pools
hispida, (prickly stonewort) ditches in *Forton-moors*
Aphanes arvensis, (parsley piert) cultivated land, on dry soil
Hippuris vulgaris, (mare's tail) near *Stafford* in ditches, adjoining the foot-road to *Aston*.—Dr. WITHERING

DIGYNIA.

Callitriche aquatica, (star grass) ditches and shallow waters

CLASS II.—DIANDRIA.

MONOGYNIA.

Ligustrum vulgare, (privet) hedges, common on moist land
Circæa Lutetiana, (enchanter's nightshade) woods, marsh-lands, *Pendeford*
Veronica arvensis, (corn speedwell) fallow fields and walls
officinalis, (male speedwell) heaths and barren ground
serpyllifolia, (smooth speedwell, Paul's betony) pastures
Becabunga, (brook-lime) watery places
Anagallis, (long-leaved brook-lime) ditches and watery places
scutellata, (narrow-leaved speedwell) on swampy ground
montana, (mountain speedwell) hedges and moist woods
Chamædrys, (germander speedwell) hedges and road sides
agrestis, (procumbent speedwell) cultivated lands
hederifolia, (ivy leaved speedwell) often abundant amongst corn
Pinguicula vulgaris, (butterwort) bogs and moors
Utricularia vulgaris, (hooded water milfoil) bogs near *Blymhill*
Lycopus europæus, (water horehound) moist places
Salvia verbenaca, (wild clary) about *Kinver*, and ruins of *Tutbury Castle*
Fraxinus excelsior, (ash tree) woods and hedge-rows

DIGYNIA.

Anthoxanthum odoratum, (sweet-scented vernal grass) meadows and pastures

CLASS III.—TRIANDRIA.

Monogynia.

Valeriana dioica, (small valerian) moist meadows

officinalis, (great wild valerian) moist hedges and marshes

Valeriana Locusta, (lamb's lettuce, corn salad) corn-fields and ditch-banks

Crocus nudiflorus, (naked flowering crocus) field near *Wolstanton*, in great abundance

Iris Pseudacorus, (yellow flag) marshes and banks of rivers

Nardus stricta, (small mat-grass) heaths and marshes, common

Eriophorum angustifolium, (many-headed cotton grass) bogs, common

vaginatum, (single-headed cotton grass) *Moreton-moors*

Scirpus sylvaticus, (millet cyperus grass) marshes and bogs, *Weston-under-Lizard*

palustris, (club-rush) lakes and ditches

cæspitosus, (dwarf-rush) heaths and bogs, *Weston-under-Lizard*

fluitans, (floating club-rush) shallow waters

lacustris, (bull-rush) rivers and lakes

setaceus, (least club-rush) *Blymhill*, in wet sandy land

Cyperus longus, (sweet cyperus, English galingale) wet ditches, *Hamstall-Ridware*

Schænus Mariscus, (long rooted bog-rush) moors near *Moreton*

nigricans, (round black-headed bog-rush) *Moreton-moors*

Digynia.

Phalaris arundinacea, (reed Canary grass) banks of rivers and moist ditch banks, common. Stem grows to the height of six feet

Phleum pratense, (timothy grass) common

Alopecurus pratensis, (meadow fox-tail grass) meadows and pastures

agrestis, (field fox-tail grass) in wet arable land

geniculatus, (spiked float-grass) in shallow waters frequent

Milium effusum, (millet grass) in the Park, *Weston-under-Lizard*

Agrostis palustris, (bent grass) cold arable land

canina, (dog's bent) wet pastures and peat bogs

vinealis, (short-bearded bent) moist pastures, *Blymhill*

alba, (white couch) the root will perforate a potatoe; in ditches and moist lands

nigra, (black couch) in cold, clayey, arable land

stolonifera, (black couch) moist meadows

vulgaris, (common bent) dry land and road sides

Holcus lanatus, (soft grass) common in meadows

mollis, (creeping soft grass) corn-fields and hedges

Aira aquatica, (water hair grass) shallow waters

cæspitosa, (turfy hair grass) moist land

flexuosa, (heath hair grass) on heaths

Aira præcox, (early hair grass) formerly common on hedge banks at *Weston-under-Lizard*

caryophyllea, (silver hair grass) in light sandy lands, *Blymhill*

Melica cærulea, (purple melic) in *Restlar's meadow*, *ditto*

uniflora, (wood melic grass) in the Park, *Weston-under-Lizard*

Poa aquatica, (reed meadow grass) sides of rivers, six feet high; it frequently occurs on the banks of the Canals near *Newcastle*

pratensis, (smooth stalked meadow grass) meadows

annua, (annual meadow grass) foot-paths and gravel walks

trivialis, (rough stalked meadow grass) meadows

nemoralis, (wood meadow grass) in woods, in the *Hide Rough*, near *Brewood*

rigida, (hard meadow grass) walls and lime-stone, *Tutbury Castle*

Briza media, (quaking grass) meadows

Dactylis glomerata, (rough cock's foot grass) pastures

Cynosurus cristatus, (crested dog's tail grass) pastures

Festuca bromoides, (barren fescue grass) walls and dry pastures

myurus, (capon's tail grass) dry sandy soil, near *Blymhill*

ovina, (sheep's fescue) heaths and high pastures

duriuscula, (hard fescue) in meadows and pastures

dumetorum, (pubescent fescue) in hedges, in *Restlar's meadow*, *Blymhill*

tenuifolia, (fibthorpe) in *Moreton-moors*

elatior, (tall fescue) common

fluitans, (flote fescue) wet ditches, commons

It is well known that old horses have been bogged in searching for this grass.

decumbens. (decumbent fescue) moist barren pastures

sylvatica, (wood fescue) hedges and thickets

Bromus mollis, (brome or oat grass) common in meadows

secalinus, (field brome grass) culm three feet high, in corn-fields

The seed is called *long-tail* by farmers, when in their corn samples

racemosus, (cluster brome grass) hedges, *Hamstall-Ridware*

asper, (hairy stalked brome grass) hedges and thickets

sterilis, (barren brome grass) hedges

arvensis, (corn brome grass) corn-fields, *Hamstall-Ridware*

giganteus, (tall brome grass) hedges in moist places

pinnatus, (spiked brome grass) dry situations, *Hamstall-Ridware*

Avena elatior, (tall oat grass) moist places

nuda, (naked oat, pilcorn) said by Dr. Plot to have been cultivated

fatua, (bearded oat, wild oat) a weed in corn-fields

flavescens, (yellow oat grass) meadow and pastures

Arundo Phragmites, (common reed) rivers and lakes

epigejos, (small reed) south side of *Aqualate-mere*, *Streetway* side

Arundo Calamagrostis, (wood reed) in moist woods and ditches, *Aqualate-mere*
Lolium perenne, (ray grass) pastures
temulentum, (darnel) a very pernicious weed amongst wheat
Hordeum murinum, (wall-barley) against old walls and sides of bridges
Triticum repens, (dog's couch grass) very common

Trigynia.

Montia fontana, (water chick weed or blinks) shallow waters.

CLASS IV.—TETRANDRIA.

Monogynia.

Dipsacus sylvestris, (wild teasel) ditch banks, and in wet situations
pilosus, (small teasel, or shepherd's staff) moist hedges, *Blymhill*, frequent
Scabiosa succisa, (devil's bit scabious) marshy meadows
arvensis, (field scabious) a common corn weed
columbaria, (small scabious) dry pastures
Sherardia arvensis, (little field madder) light arable soils, common
Asperula odorata, (sweet woodroof) woods, *Brockhurst coppice*, *Blymhill*
Galium cruciatum, (cross-wort) borders of fields and hedges
palustre, (white ladies bed-straw) moist hedges and pastures
montanum, (mountain goose grass) high parts of *Handsworth heath*
procumbens, (creeping goose grass) heaths and commons
uliginosum, (marsh goose-grass) wet pastures and heaths
Mollugo, (madder goose-grass) heath and hedges
verum, (yellow ladies bed-straw) hedges
Aparine, (catchweed, cleavers or eriff) hedges and gardens
Plantago major, (broad leaved plantain) road sides, frequent
media, (hoary plantain) pastures and lanes
variety 2. (lamb's tongue, with straw-coloured stripes,) *Dudley Castle*
lanceolata, (rib-grass) pastures
Coronopus, (buck's horn plantain) lanes in a gravelly soil
Sanguisorba officinalis, (meadow burnet) meadows in upland. *Bentley*
Cornus sanguinea, (dogberry tree) in hedges, very common
Parietaria officinalis, (pellitory of the wall) old walls, common
Alchemilla vulgaris, (ladies mantle) meadows and pastures.

Digynia.

Cuscuta europæa, (greater dodder) a parasitical plant on flax, and heath
Epithymum, (lesser dodder) on heath and thyme.

Tetragynia.

Ilex Aquifolium, (holly) woods and hedges

Potamogeton natans, (broad-leaved pond-weed) pits and rivers
perfoliatum, (perfoliated pond-weed) ponds, frequent
lucens, (shining-leaved pond-weed) in ponds, frequent
compressum, (flat-stalked pond-weed) *Blymhill*
crispum, (curled pond-weed) stem more than six feet long, in *Aqualate Mere*, *White-Pool*, *Weston*
pectinatum, (fennel-leaved pond-weed) *Forton Moors*
Sagina procumbens, (trailing pearl-wort) dry walls, and also bogs
apetala, (annual pearl-wort) walls and gravel walks.

CLASS V.—PENTANDRIA.

Monogynia.

Myosotis palustris, (water mouse-ear scorpion-grass) marshy ground, and springs
arvensis, (field mouse-ear scorpion-grass) sandy grounds and walls
Lithospermum officinale, (gromwell) *ruins of Tutbury Castle*
arvense, (corn gromwell) corn-fields
Cynoglossum officinale, (hound's tongue) road-sides, common
Symphytum officinale, (comfrey) hedge-sides and banks of rivers
Borago officinalis, (borage) corn-fields near *Cheadle*—Mr. Bourne
Lycopsis arvensis, (bugloss) corn-fields and road-sides
Echium vulgare, (viper's bugloss) corn-fields, lime-stone hills, *Dudley-Castle*
Primula vulgaris, (primrose) hedge-sides and thickets, common
elatior, (oxlip) hedge-sides
officinalis, (cowslip) meadows and pastures
Menyanthes trifoliata, (buckbean) pits, common
Hottonia palustris, (water violet) ditches, *Aqualate-Mere*
Lysimachia vulgaris, (yellow loosestrife) *Motty Meadow-hedge*, *Blymhill*
nemorum, (yellow pimpernel) moist shady places
Nummularia, (money-wort) marshy meadows
Anagallis arvensis, (pimpernel) corn-fields
tenella, (bog pimpernel) moist meadows and bogs
Convolvulus arvensis, (corn bindweed) corn-fields
sepium, (great bindweed) in hedges
Campanula rotundifolia (bell-flower) hedges and ditch banks
patula, (field bell-flower) road-sides and hedges
Rapunculus, (rampions) hedge banks, *Enville*—Dr. Stokes
latifolia, (giant throat-wort) *Rowley*, *Crowden Abbey*, *Moorlands*—Mr. Bourne; at *Darlaston*, near *Stone*, with white flowers—Mr. Foster
Trachelium, (Canterbury bells) *Rowley*; also at *Wood Eaton*
Lonicera Periclymenum, (honey-suckle) in hedges, common

Jasione montana, (hairy sheep's scabious) hedge banks and pastures
Verbascum Thapsus, (mullein) sandy hedge banks, common
Lychnitis, (hoary mullein) about *Kinver* and *Wombourne*
nigrum, (black mullein) at *Hampstead*, near *Walsall*
Blattaria, (yellow moth mullein) *Hill Ridware*
Hyoscyamus niger, (henbane) amongst rubbish, *Four-crosses*, *Streetway*
Atropa Bella-donna (dwale, deadly nightshade) *Dudley Castle*
Solanum Dulcamara, (woody nightshade) moist hedges, common
nigrum, (garden nightshade) gardens and rubbish
Chironia Centaurium (lesser centory) pastures
Rhamnus catharticus, (purging buck-thorn) hedges, *Blymhill*
Frangula, (alder buck-thorn) woods, *Marsh-lane*, *Pendeford*
Evonymus europæus, (spindle tree) in a hedge at *Stretton*—E. Grove, Esq.
Ribes alpinum, (sweet mountain currant) in a hedge near *Ilam*
nigrum, (black currant) in moors near *Moreton*
Grossularia, (rough gooseberry) hedges, *Croxden Abbey*, *Moorlands*—Mr. Bourne
Hedera Helix, (common ivy) woods and hedges
Viola odorata, (sweet violet) hedge banks
canina, (dog's violet) hedge banks
palustris, (marsh violet) moist meadows
tricolor, (pansies, heart's ease) corn-fields, in light sandy soil
lutea, (yellow violet) corn-fields, *Moorlands*
grandiflora, (great yellow violet) *Wever-hills*, *Moorlands*
Glaux maritima, (sea milk-wort) in a salt-marsh near *Tixall*.—R. Wolseley, Esq.
Vinca minor, (periwinkle) hedges, *Holloway*, *Gosbrook* near *Wolverhampton*, *Smethwick*.

Digynia.

Chenopodium Bonus Henricus, (English mercury) road-sides, *Church-yard*, *Blymhill*
album, (white goose-foot) gardens and cultivated land
viride, (green goose foot, john a neal) frequent on cultivated ground
Ulmus campestris, (elm-tree) hedges and groves
montana, (wych-hasel) woods and hedges
Gentiana campestris, (field gentian) on bur-walls, near *Wolseley-bridge*
Amarella, (autumnal gentian) dry pastures, *Moorlands*—Mr. Bourne
Hydrocotyle vulgaris, (marsh penny-wort) marshes and bogs, common
Sanicula europæa, (sanicle) woods and hedges
Caucalis Anthriscus, (hedge parsley) hedges, common
Daucus Carota, (wild carrot) on neglected land
Bunium flexuosum, (pig-nut) pastures, common

Conium maculatum, (hemlock) waste grounds
Heracleum sphondylium, (hog-weed, cow parsnip) pastures and hedges
Angelica sylvestris, (wild angelica) marshes and by sides of rivulets
Sium angustifolium, (water parsnip) rivulets and ditches
nodiflorum, (creeping water parsnip) rivulets, very common
Sison inundatum, (least water parsnip) *White-sitch-pool*, *Weston-under-Lizard*
Oenanthe fistulosa, (water drop-wort) watery places, frequent
crocata, (hemlock drop-wort) at *Willowbridge*
Phellandrium aquaticum, (water hemlock) ditches, *Church-Eaton*
Cicuta virosa, (long-leaved water hemlock) sides of pools and rivers
Æthusa Cynapium, (fool's parsley) a common garden weed
Scandix odorata, (sweet chervil) *Tixall*, near *Stafford*; *Dimsdale* and *Bradwell*, near *Newcastle*
Pecten, (shepherd's needle) common on corn-land
Anthriscus, (rough shepherd's needle) rubbish, hedge-sides
Chærophyllum sylvestre, (cow weed) orchards and pastures
temulum, (hedge chervil) hedges, common
Pimpinella saxifraga, (lesser saxifrage) dry pastures, *Dudley Castle*
magna, (saxifrage) *Moorland pastures*, *Wednesbury-field*
Ægopodium Podagraria (ground-ash) common in gardens.

Trigynia.

Viburnum Opulus, (water-elder) moist hedges
Sambucus Ebulus, (dwarf elder) *Tutbury-Castle*, *Tamworth-Castle Hill*
nigra, (common elder) hedges, common
Var. 2. White berries, *Blymhill*, *Essington*
Alsine media, (chick-weed) in land in a high state of cultivation.

Tetragynia.

Parnassia palustris, (grass of Parnassus) *Motty Meadows*, *Wever-hills*, *Blymhill*.

Pentagynia.

Linum usitatissimum, (common flax) common
catharticum, (purging flax) dry meadows and pastures
Radiola, (all-seed flax) in light sandy soil, frequent
Drosera rotundifolia, (round-leaved sun-dew) wet heaths, *Blymhill-heath*
longifolia, (long-leaved sun-dew) on a bog at *Willowbridge*, *Balterley*.

Polygynia.

Myosurus minimus, (mouse-tail) corn-fields, *Elford* & *Hamstall Ridware*.

CLASS VI.—HEXANDRIA.

Monogynia.

Galanthus nivalis, (snow-drop) hedge-sides
Narcissus Pseudo-narcissus, (daffodil) meadows

Allium ursinum, (ramsons) near brook-sides
Fritillaria Meleagris, (fritillary) meadows near *Blymhill*; in a meadow near *Wolseley-bridge*, on the road to *Stafford*
Narthecium ossifragum, (Lancashire asphodel) bogs near *Moreton*; bogs in the *Moorlands*
Convallaria majalis, (lily of the valley) *Rough-Park-Woods* at *Belmont*
multiflora, (solomon's seal) *Needwood-Forest*, *Moorlands*
Hyacinthus non scriptus, (hare-bell hyacinth) woods and hedges
Acorus Calamus, (sweet-smelling flag) *Tamworth*
Juncus conglomeratus, (round-headed rush) wet places, common
effusus, (common rush) marshes and miry places
inflexus, (hard rush) moist pastures
squarrosus, (moss rush) wet heaths
articulatus, (jointed rush) meadows, woods, and pastures
bufonius, (toad rush) moist places
bulbosus, (bulbous rush) wet heaths, *Moorlands*
uliginosus, (rush-grass, proliferous-headed) *Petmoor-pool*, *Weston*
pilosus, (hairy rush) woods, common
maximus, (wood rush) in a dingle at *Cotton-hall*, *Burnt-wood*, near *Mucclestone*
campestris, (hairy field rush) pastures
Berberis vulgaris, (barberry bush) hedges
Colchicum autumnale, (meadow saffron) wet meadows
Peplis Portula, (water purslane) near *Russell's Hall*, *Sedgley Old Park*, *Dudley*—Mr. Wainwright.

Trigynia.

Rumex sanguineus, (blood wort) near *Cheadle*—Mr. Bourne
crispus, (curled dock) pastures
acutus, (sharp-pointed dock) moist grounds, by sides of rivulets
Hydrolapathum, (water dock) pools and river sides
pulcher, (fiddle dock) dry meadows near *Cheadle*, Mr. Bourne
obtusifolius, (broad-leaved dock) rubbish and pastures
Acetosa (meadow sorrel) meadows and pastures
Acetosella, (sheep's sorrel) barren sandy soil
Triglochin palustre, (arrow headed grass) marshy meadows, *Blymhill*
maritimum, (sea arrow-grass) in a salt marsh near *Tixall*.

Polygynia.

Alisma Plantago, (water plantain) pits and wet places
ranunculoides, (lesser thrumwort) pits and pools, rare
lanceolata, (narrow-leaved thrumwort) shallow waters.

CLASS VIII.—OCTANDRIA.

Monogynia.

Epilobium angustifolium, (rose bay willow herb) *Cotton-wood*, *Moorlands*

Epilobium hirsutum, (codlins and cream) moist ditches and brooks
parviflorum, (woolly willow herb) in rivulets, common
montanum, (smooth leaved willow herb) woods and ditches
palustre, (marsh willow herb) marshes and sides of lakes
Acer Pseudo-platanus, (sycamore tree) hedges and plantations
campestre, (common maple) hedges
Chlora perfoliata, (yellow-wort) limestone hills near *Dudley Castle*, abundant
Vaccinium Myrtillus, (bilberry) woods and heaths
Vitis-idæa, (red whortle-berry) dry heaths, summits of barren hills
Oxycoccos, (cranberry) bogs near *Aqualate-mere*, *Cannock-heath*
Erica tetralix, (cross-leaved heath) common
vulgaris, (common heath) moors and commons
cinerea, (fine-leaved heath) common on heaths
Daphne Mezereum, (spurge olive, mezereon) *Needwood forest*, woods, rare
Laureola, (spurge laurel) woods and hedges, *Walsall-wood*.

TRIGYNIA.

Polygonum amphibium, (amphibious lakeweed) pits and moist places
Hydropiper, (lakeweed, water pepper) watery places
Persicaria, (spotted lakeweed) moist arable land
Bistorta, (bistort) meadows
aviculare, (knotgrass) road sides, common
Fagopyrum, (buckwheat) corn-fields
Convolvulus, (black bindweed) corn-fields and gardens.

TETRAGYNIA.

Paris quadrifolia, (herb paris) hedges and woods, *Pendeford*, *Bradwell-wood*
Adoxa Moschatellina, (moschatel) hedge banks and woods.

CLASS IX.—ENNEANDRIA.

HEXAGYNIA.

Butomus umbellatus, (flowering rush, water gladiole) in the *Penk*, at *Richerscote* near *Stafford*—Mr. J. H. DICKENSON.

CLASS X.—DECANDRIA.

MONOGYNIA.

Monotropa Hypopithys, (primrose scented bird's-nest) *Earl of Stamford's woods*, *Enville*,—WITHERING, STILLINGFLEET
Andromeda polifolia, (marsh cistus, wild rosemary) *Chartley moss*
Pyrola rotundifolia, (winter-green) *in a wood near Cotton-hall*, *woods at Belmont*, *Moorlands*

Digynia.

Chrysosplenium alternifolium, (alternate-leaved golden saxifrage) shady woods, *Moorlands*

oppositifolium, (golden saxifrage) shady moist ditch banks

Saxifraga granulata, (white saxifrage) meadows near *Tamworth*

tridactylites, (rue-leaved saxifrage) *Burton Abbey walls*

Scleranthus annuus, (german knotgrass) sandy corn-fields

Saponaria officinalis, (soapwort) *Blymhill*, *Moorlands*, in hedges near *Hedgford*, on the road to *Rugeley*

Dianthus Armeria, (Deptford pink) *Morry hills* near *Yoxall*

deltoides, (maiden pink) *Trysull*, and *Swindon meadows*—Mr. Wainwright.

Trigynia.

Cucubalus Behen, (bladder champion) in corn-fields, frequent

Stellaria holostea, (stitchwort) hedge sides, common

uliginosa, (bog stichwort) springs and rivulets

graminea, (grass-leaved stichwort) moist grounds

Arenaria trinervia, (plantain-leaved sandwort) woods and moist hedges

serpyllifolia, (least chickweed) walls and sandy hedge-banks

marina, (sea spurry) in a salt marsh near *Shirleywich*

rubra, (purple spurry) corn-fields, in a light sandy soil.

Pentagynia.

Cotyledon Umbilicus, (navelwort) rocks under *Heyley Castle* bank, rare

Sedum Telephium, (orpine, livelong) hedges and pastures, on roofs of houses, in the *Moorlands*

acre, (pepper stone-crop) walls and dry hedge banks

reflexum, (yellow stone-crop) walls and roofs

Oxalis Acetosella, (woodsorrel, cuckoo-meat) woods and hedge-banks

Agrostemma Githago, (cockle) in arable land

Lichnis dioica diurna, (red-flowered campion) in hedges, common

vespertina, (white campion) amongst corn and clover

Lychnis flos cuculi (cuckoo flower, ragged robin) in meadows, common

Var. *flore pleno* (double-ragged robin) near *Norton in the Moors*

Cerastium viscosum (clammy mouse-ear) walls, dry banks, and ant hills

vulgatum (mouse-ear) pastures and hedge banks

Spergula arvensis (corn spurry) in light arable lands

nodosa (knotted spurry) marshy places, *Moorlands*.

CLASS XI.—DODECANDRIA.

Monogynia.

Lythrum Salicaria (purple-spiked willowherb) banks of rivers, common.

DIGYNIA.

Agrimonia Eupatoria, (agrimony) hedges and borders of corn-fields.

TRIGYNIA.

Reseda luteola, (dyer's weed) road sides—*Dudley* and *Tutbury Castles*, near *Red-Street*
lutea, (base rocket) rough ground, near *Russell's Hall*, *Dudley*
Euphorbia Peplus, (petty spurge) in cultivated land, common
exigua, (dwarf spurge, devil's milk) corn-fields
helioscopia, (sun spurge, wart spurge) gardens
amygdaloides, (wood spurge) *Needwood forest*
Characias, (red spurge) *Needwood forest*.

DODECAGYNIA.

Sempervivum tectorum, (house-leek) roofs and walls.

CLASS XII.—ICOSANDRIA.

MONOGYNIA.

Prunus Padus, (bird cherry) hedges at *Pendeford*, *Byshbury*, and in the *Moorlands*
insititia, (bullace) hedges, frequent
spinosa, (sloe, black thorn) hedges, common.

DIGYNIA.

Cratægus Aria, (white-beam) rocks and mountains, in the *Moorlands*
torminalis, (wild service) in a hedge at *Pendeford*
Oxyacantha, (white thorn, hawthorn) hedges.

TRIGYNIA.

Sorbus aucuparia, (quicken tree, mountain ash) hedges and barren hills
Sorbus domestica, (service tree) in the *Moorlands*.

PENTAGYNIA.

Pyrus communis, (wild pear) woods and hedges
Malus, (crab-tree) in hedges, common
Spiræa Filipendula, (drop wort) *Wever Hills*, very rare
Ulmaria, (meadow sweet) meadows and banks of rivers.

POLYGYNIA.

Rosa arvensis, (white dog-rose) in hedges
spinosissima, (burnet rose) *Betley*
tomentosa, (downy-leaved dog-rose) in hedges
canina, (dog-rose) in hedges, common
rubiginosa, (sweet-briar rose) hedges, *Belmont*, *Moorlands*
Rubus ideus, (raspberry) in hedges and woods
fruticosus, (blackberry) hedges, common
Fragaria vesca, (strawberry) woods and hedge-banks
sterilis, (barren strawberry) hedge-banks, *Moorlands*

Potentilla anserina, (silver weed) road-sides, in wet places
reptans, (common cinquefoil) hedge-banks, pastures
Tormentilla officinalis, (tormentil) wet pastures
Geum urbanum, (herb bennet) hedges, frequent
rivale, (water avens) *Moreton-Moors*, meadows at *Stretton*, meadow at *Shenstone*
Comarum palustre, (marsh cinquefoil) in old pits, common

CLASS XIII.—POLYANDRIA.

Monogynia.

Chelidonum majus, (celandine) old hedges, *Sedgley* and *Rowley*
Papaver Rhœas, (red poppy) corn-fields, in a sandy soil
dubium, (smooth-headed poppy) in sandy fields—Rev. Samuel Dickenson
Argemone, (rough-podded poppy) corn-fields
Nymphæa lutea, (yellow water lily) rivers and pools
alba, (white water lily) *Aqualate-Mere*, *River Sow*, near *Stafford*
Tilia europæa, (lime or linden-tree) groves and plantations
Cistus Helianthemum, (dwarf cistus) *Moorlands*, near *Thor's-house cavern*

Pentagynia.

Aquilegia vulgaris, (columbine) in thickets, *Needwood-Forest*, *Moorlands*.

Polygynia.

Anemone nemorosa, (anemone) hedges and meadows, common
Thalictrum flavum, (meadow rue) marshy meadows
Ranunculus Ficaria, (pilewort) pastures and hedges
Flammula, (spearwort) moist ditches, common
Lingua, (great spearwort) pits and ditches
auricomus, (sweetwood crowfoot) woods and hedges
reptans, (narrow-leaved crowfoot) moist places, *Moorlands*
sceleratus, (round-leaved water crowfoot) wet places
acris, (upright crowfoot, butter-cups) meadows and pastures
parviflorus, (small-flowered crowfoot) hedge-banks on dry soil
hederaceus, (ivy-leaved crowfoot) watery places
aquatilis, (water crowfoot) on ponds in May
hirsutus, (pale hairy crowfoot) corn-fields, near *Newcastle*
bulbosus, (bulbous crowfoot) meadows and pastures, common
repens, (creeping crowfoot) pastures, common
arvensis, (corn crowfoot) in corn-fields
Trollius europæus, (globe flower) woods at *Belmont*
Helleborus fœtidus, (bear's foot) *Moorlands*, woods at *Belmont*
Caltha palustris, (marsh marigold, meadow-boot) meadows.

CLASS XIV.—DIDYNAMIA.

GYMNOSPERMIA.

Ajuga reptans, (common bugle) moist meadows and woods
Teucrium Scorodonia, (wood sage) dry hedge banks and woods
Nepeta cataria, (cat mint) *ruins of Croxden abbey, Moorlands*
Verbena officinalis, (vervain) waste places about villages
Mentha viridis, (spear mint) watery places, near *Cheadle*—Mr. Bourne
    Var. 2 villosa, near *Cheadle*—Mr. Bourne
    aquatica, (water mint) sides of rivulets
    piperita, (pepper mint) watery places near *Wolstanton*
    gentilis, (red mint) watery places
    arvensis, (corn mint) frequent in stubbles on moist land
    Pulegium, (penny royal) moist heaths, *Blymhill*
    hirsuta, (round headed mint) watery places
Lamium album, (white dead-nettle) hedges and amongst rubbish
    purpureum, (red dead-nettle) gardens and rubbish
    amplexicaule, (great henbit) in cultivated land
Galeopsis Tetrahit (nettle-hemp) corn-fields and hedges
    cannabina, (bee-nettle) corn-fields, near *Newcastle, Blymhill*, not uncommon
Galeobdolon luteum, (yellow dead-nettle) shady places
Betonica officinalis, (wood betony) sides of roads amongst bushes
Stachys sylvatica, (hedge nettle) hedges, common
    palustris, (clown's woundwort) moist places
    arvensis, (corn woundwort) corn-fields, frequent
Ballota nigra (stinking horehound) hedges and waste places
Marrubium vulgare (white horehound) road-sides, frequent
Clinopodium vulgare, (wild basil) pastures in calcareous soil
Origanum vulgare, (wild marjoram) road-sides, *ruins of Tutbury-Castle, Moorlands*
Thymus Serpyllum, (wild thyme) sandy heaths and dry road sides
Melissa Calamintha, (calamint) hedge-banks, near *Wolseley-bridge, Dudley-castle*
Scutellaria galericulata, (hooded skull-cap) marshy ground, *Pendeford*
Prunella vulgaris, (common self-heal) pastures, common
Leonurus Cardiaca, (motherwort) hedges, *Gornall-wood*.

ANGIOSPERMIA.

Rhinanthus Cristagalli, (yellow rattle) meadows, common
Euphrasia officinalis, (common eye-bright) pastures, frequent
    Odontites, (purple eye-bright) moist pastures and corn-fields
Melampyrum pratense, (cow wheat) woods and pastures.
Lathræa Squamaria, (great toothwort) *Longley meadows, King's Bromley*
Pedicularis palustris, (marsh lousewort) wet land, near *Betley*

Pedicularis sylvatica, (common lousewort) marshy meadows
Antirrhinum Cymbalaria, (ivy-leaved snapdragon) walls at *Lapley*
minus, (least snapdragon) corn-fields, *Blymhill*, rare
linaria; (toad flax) dry hedge banks
majus, (great snapdragon) walls at *Rushall-hall*
Scrophularia nodosa, (great figwort) sides of moist hedges
aquatica, (water betony) sides of rivulets
Digitalis purpurea, (fox glove) road-sides and hedge banks
Orobanche major, (broom rape) in the *Pyehill-lane*, *Blymhill*.

CLASS XV.—TETRADYNAMIA.

Siliculosa.

Draba verna, (whitlow-grass) walls and pastures
muralis, (speedwell-leaved whitlow-grass) *Wever-hills*, very rare
Lepidium petræum, (mountain dittander) *Wever-hills*
Thlaspi arvense, (French mustard) in fields, *Tittensor*
campestre, (mithridate mustard) in corn-fields
Bursapastoris, (shepherd's purse) amongst rubbish and in corn
Iberis amara, (bitter candy-tuft) *Moorlands*, between *Cheadle* and *Oak-a-moor*
nudicaulis, (rock cress) hedge banks, *Blymhill*, and near *Betley*.

Siliquosa.

Cardamine impatiens, (impatient ladies smock) *Barrow-hill, at the quarries of Rowley-Regis*, and at *Sedgley*
pratensis, (ladies smock) meadows
hirsuta, (hairy-leaved ladies smock) near springs and ditches
flexuosa, (zig-zag ladies smock) moist places, *Blymhill-marsh ditch*
amara, (bitter cress) moist places, *Blymhill*
Sisymbrium Nasturtium, (watercress) ditches and rivulets, common
sylvestre, (water rocket) watery places
amphibium, (water radish) banks of rivers and ditches
terrestre, (land rocket) near *Stafford*, towards *Castle-hill*
Sophia, (flix-weed) walls and rubbish, *Tutbury-castle*
Erisymum officinale, (hedge mustard) hedges and rubbish, common
Barbarea, (winter cresses) wet places
Alliaria, (jack by the hedge) hedges, near houses, common
cheiranthoides, (treacle wormseed) in fields and by hedges, *road-sides near Mathfield*,—Mr. Bisby
Arabis thaliana, (codded mouse-ear) dry cultivated land
Turritis glabra, (tower mustard) hedges, near *Lichfield*, *Weston-under-Lizard*
hirsuta, (hairy towerwort) stony places, *ruins of Tutbury-castle*

Brassica Napus, (wild rape) corn-fields and ditch-banks
muralis, (wild rocket) walls, *Lichfield Close*
Sinapis arvensis, (rough charlock, wild mustard) in corn laud
alba, (white mustard) corn-fields and roads
nigra, (common mustard) corn-fields, near *Tutbury*
Raphanus Raphanistrum, (wild radish,) in arable land.

CLASS XVI.—MONADELPHIA.

Decandria.

Geranium dissectum, (jagged-leaved geranium) hedge banks and meadows
columbinum, (long-stalked geranium) pastures, near *Barr-beacon, on the road from Birmingham*
molle, (dove's-foot geranium) hedges and dry banks, common
pratense, (crow-foot geranium) moist pastures
maritimum, (sea geranium) *Wombourne* and *Orton*,—Mr. Wainwright
lucidum, (shining geranium) walls, and hedge banks, *Alton Castle* and *Croxden Abbey*, stone walls, *Rowley*
robertianum, (herb robert, stork bill) hedges and amongst rubbish
moschatum, (musk geranium) pastures, *Rowley*—Mr. Wainwright
cicutarium, (hemlock-leaved geranium) light arable land.

Polyandria.

Malva rotundifolia, (round-leaved mallow) road sides, common
sylvestris, (common mallow) rubbish near villages
moschata, (jagged-leaved mallow) hedges and pastures, road-sides.

CLASS XVII.—DIADELPHIA.

Hexandria.

Fumaria bulbosa, (bulbous fumitory) shady places, *Perry-hall Park*, near the river *Tame*
officinalis, (common fumitory) gardens and corn-fields
capreolata, (ramping fumitory) cultivated land, *Wolstanton*
claviculata, (climbing fumitory) hedges, common.

Octandria.

Polygala vulgaris, (milkwort) meadows and pastures, frequent.

Decandria.

Spartium scoparium, (broom) dry preserves
Genista tinctoria, (dyer's broom) pastures and hedge-sides
pilosa, (hoary green-weed) marshy ground, *Pendeford*
anglica, (petty whin, hengorse) moist spungy ground

Ulex europæus, (gorse) heaths and barren neglected land
Ononis spinosa, (thorny rest-harrow) pastures and hedge-banks
arvensis, (hairy rest-harrow) pastures, more common in this county than the preceding species
Orobus tuberosus, (wood peas) common in pastures
Lathyrus Nissolia, (crimson grass-vetch) *Cotton-fields, woods at Belmont*
pratensis, (yellow vetchling) hedges and meadows
sylvestris, (narrow-leaved pea-everlasting) *Redhill*, near *Stone-Park*
Vicia sylvatica, (wood vetch) woods and hedges near *Stone*,—Mr. Foster
Cracca, (perennial tufted vetch) meadows and hedges
sativa, (common vetch) hedges and borders of land
sepium, (bush vetch) hedges, pastures and woods
Ervum tetraspermum, (smooth-podded tare) corn-fields
hirsutum, (wild tare) corn-fields
Ornithopus perpusillus, (common bird's foot) on dry sandy land
Astragalus glycyphyllos, (liquorice vetch) on ditch banks
Trifolium Melilotus officinalis, (melilot) in hedges *about Tutbury*
repens, (white clover) pastures, common
striatum, (soft-knotted trefoil) pastures, near *Stafford*
medium, (cow clover, perennial clover) pastures
pratense, (common clover) pastures, common
procumbens, (hop trefoil) pastures
filiforme, (least hop-trefoil) sandy pastures
Lotus corniculatus, (bird's foot trefoil) meadows and pastures
Medicago lupulina, (trefoil) light soil.

CLASS XVIII.—POLYADELPHIA.

Polyandria.

Hypericum Androsæmum, (tutsan, St. John's wort) woods and moist hedges
quadrangulum, (St. Peter's wort) moist and shady places
perforatum, (common St. John's wort) thickets and hedges
dubium, (imperforate St. John's wort) thickets, *Stretton*,—E. Grove, Esq.
humifusum, (trailing St. John's wort) pastures, *Blymhill*
elodes, (marsh St. John's wort) watery places on *Calf-heath*
hirsutum, (hairy St. John's wort) hedges near *Tutbury*
pulchrum, (heart-leaved St. John's wort) heaths.

CLASS XIX.—SYNGENESIA.

Polygamia Æqualis.

Tragopogon pratense, (yellow goat's beard) pastures and hedges, on a clayey soil

Sonchus arvensis, (corn sow-thistle) corn fields and hedges
palustris, (marsh sow-thistle) banks of rivers and watery places
oleraceus, (common sow-thistle) gardens and cultivated land
Lactuca virosa, (strong-scented lettuce) near *Chartley Castle*, *Hamstal Ridware*
Prenanthes muralis, (wall ivy leaf) walls and hedge banks
Leontodon officinale, (dandelion) pastures
autumnale, (autumnal dandelion) grass land
hispidum, (rough dandelion) meadows and pastures
Hieracium Pilosella, (mouse-ear hawkweed) hedge banks on a dry soil
subaudum, (shrubby hawkweed) woods and hedges
paludosum, (marsh hawkweed) *on the banks of a rivulet near Biddulph*
murorum, (French lungwort) hedge banks, *Blymhill*
sylvaticum, (wood hawkweed) woods, *Hatch-lane*, *Weston-under-Lizard*
umbellatum, (bushy hawkweed) hedges and woods
Crepis tectorum, (smooth succory) road sides and among oats
Hypochæris radicata, (long-rooted hawkweed, cat's ear) meadows and pastures
Lapsana communis, (nipplewort) gardens and cultivated land, frequent
Cichorium Intybus, (wild succory) road sides and borders of fields
Arctium Lappa, (burdock) waste places and on rubbish
Serratula tinctoria, (sawwort) woods and marshy meadows
arvensis, (common thistle) road sides, corn fields, and pastures
Carduus palustris, (marsh thistle) hedge sides and marshes
acanthoides, (welted thistle) near *Chartley Castle*, *in the road*
lanceolatus, (spear thistle) road sides and pastures
nutans, (musk thistle) pastures in calcareous soil
eriophorus, (woolly-headed thistle) road sides and pastures
pratensis, (gentle thistle) wet meadows
Carlina vulgaris, (wild carline thistle) heaths, *dry hills on the Moorlands*
Bidens cernua, (water hemp, agrimony) watery places
tripartita, (trifid leaved water hemp) agrimony) pits
Eupatorium cannabinum, (hemp agrimony) about rivulets, common.

POLYGAMIA SUPERFLUA.

Tanacetum vulgare, (tansy) borders of fields and banks of rivers
Artemisia Absinthium, (wormwood) on rocks and light soil
vulgaris, (mugwort) hedges and waste places
Gnaphalium margaritaceum, (pearl everlasting) meadows, *Longdon*, *Flor. Brit.* 3, 1405
dioicum, (cat's foot cudweed) on hills, *Moorlands*
rectum, (upright cudweed) pastures, *Blymhill*
uliginosum, (black-headed cudweed) moist grounds

Gnaphalium montanum, (least cudweed) light sandy pastures
germanicum, (common cudweed) pastures and road-sides
Erigeron canadense, (canada flea-bane) road-sides, *Ashwood*,—Mr. Wainwright
acre, (blue flea-bane) *Dudley-castle*, *Stretton-bridge*, on *Street-way*
Tussilago Farfara, (coltsfoot) in arable land
Petasites, (butter-bur) sides of rivulets
Senecio vulgaris, (common groundsel) cultivated ground
sylvaticus, (bushy groundsel) dry heaths and sandy ditch banks
tenuifolius, (hoary groundsel) pastures, *Blymhill*, near *Etruria*
Jacobæa, (ragwort) road sides and pastures
aquaticus, (water ragwort) marshes and watery places
saracenicus, (broad-leaved groundsel) hedges near *Compton-mill*, *Bradwell*
Aster Tripolium, (sea starwort) salt marshes, *Tixall*, and *Shirleywich*
Solidago virga-aurea, (golden-rod) woods
Inula dysenterica, (elecampane) moist ground, road-sides
Bellis perennis, (common daisy) meadows and pastures
Chrysanthemum Leucanthemum, (greater daisy, oxeye) meadows and pastures
segetum, (corn marigold) on light soil
Matricaria Parthenium, (feverfew) hedges and waste places
Chamomilla, (chamomile feverfew) road sides and farm yards
Anthemis arvensis, (corn chamomile) corn fields, common
Cotula, (stinking Mayweed) a corn weed
nobilis, (sweet chamomile) heaths and pastures, *Blymhill-road* near *Hedgford*
Achillea Ptarmica, (sneezewort) in moist places
Millefolium, (yarrow) pastures and road sides.

FRUSTRANEA.

Centaurea Cyanus, (blue bottle) corn fields
nigra, (black knap weed) road sides and pastures
Scabiosa, (greater knap weed) a corn weed, frequent.

CLASS XX.—GYNANDRIA.

DIANDRIA.

Orchis bifolia, (butterfly orchis) *Restler's meadow*, *Blymhill*, *Wever hills*, near *Bradwell*
Morio, (meadow orchis) moist meadows, common
mascula, (early orchis) meadows and woods
latifolia, (broad-leaved orchis) marshy meadows
maculata, (spotted orchis) meadows, common
conopsea, (red-handed orchis) *Blymhill*, *Moorlands*

Satyrium viride, (frog satyrion) *Restlar's moor, Blymhill*
Ophrys ovata, (tway-blade) woods and marshy meadows, *Blymhill*
Serapias latifolia, (helleborine) *Pipe marsh, Lichfield, Pendeford.*

CLASS XXI.—MONŒCIA.

Diandria.

Lemna trisulca, (ivy-leaved duck meat) in pits, *Blymhill*
minor, (common duck meat) ditches and ponds
gibba, (gibbous duck meat) ditches and ponds, *Blymhill*
polyrhiza, (greater duck meat) ditches and ponds, near *Brewood, in a pit by the road.*

Triandria.

Sparganium ramosum, (greater bur weed) pits and banks of rivers
simplex, (lesser bur weed) sides of pools
natans, (floating bur weed) ditches, near *Aqualate-mere*
Carex dioica, (small sedge) in boggy ground
pulicaris, (flea sedge) marshes and wet meadows
stellulata, (star sedge) marshy meadows, *Blymhill*
curta, (white sedge) marshes, *Hide Rough*, near *Brewood*
ovalis, (oval spiked sedge) meadows and pastures
remota, (remote sedge) moist ditch banks
vulpina, (great sedge) banks of rivers and pits
paniculata, (panicled sedge) bogs, frequent
pendula, (pendulous sedge) moist woods and hedges, *Big Hide Rough*, near *Brewood*
præcox, (vernal sedge) marshy meadows, common
flava, (yellow sedge, marsh hedge-hog grass) wet meadows
distans, (loose sedge) marshes, *summit of Wever-hills*
panicea, (pinky sedge) moors, *Blymhill*, frequent
sylvatica, (wood sedge) woods, *Blymhill, in a hag or grove at the Restlar's*
recurva, (heath sedge) moist places, *Blymhill pool meadow*
pallescens, (pale sedge) moist places, *Blymhill*
limosa, (brown sedge) boggy meadows
pilulifera, (pill bearing sedge) moist places *Far Restlar's meadow, Blymhill*
cæspitosa, (turfy sedge) marshes and old pits
paludosa (acute sedge) moorish ground and wet ditches
vesicaria, (bladder sedge) bogs and pits, *Aqualate-mere*
hirta, (hairy sedge) meadows, common.

Tetrandria.

Betula alba, (birch tree) woods and hedges on thin wet gravelly soil

Betula Alnus, (alder) on wet land, common in moist fences
Urtica dioica, (common nettle) hedges and amongst rubbish
urens, (lesser nettle) in similar situations.

PENTANDRIA.

Bryonia dioica, (wild vine) hedges, near *Lichfield* and *Burton-upon-Trent*.

POLYANDRIA.

Myriophyllum spicatum (spiked water milfoil) ditches and ponds
verticillatum, (verticillate water milfoil) *Forton-moors*
Sagittaria sagittifolia, (arrow head) ditches and banks of rivers
Arum maculatum, (wake robin cuckoo pint) hedges and shady places
Poterium Sanguisorba, (upland burnet) pastures, *Wever-hills*, *Moorlands*
Quercus Robur, (oak tree) woods and hedges
Fagus Castanea, (chesnut tree) hedges and plantations
sylvatica, (beech tree) in similar situations
Carpinus Betulus, (horn beam) woods and hedges, rare
Corylus Avellana, (hazel nut tree) hedges and woods.

CLASS XXII.—DIŒCIA.

DIANDRIA.

Salix pentandra, (sweet willow) woods and hedges, *without the East-gate, Stafford*, *Moorlands*
fragilis, (crack willow) woods and banks of rivers
repens, (creeping willow) marshy meadows, frequent
caprea, (sallow withy) hedge rows, pits, and woods, very common
acuminata, long-leaved sallow) moist places, *Blymhill*
viminalis, (osier) willow beds near rivers
alba, (white willow) meadows near rivers.

TRIANDRIA.

Empetrum nigrum, (black-berried heath, crake berries) heaths.

TETRANDRIA.

Viscum album, (misletoe) on apple trees
Myrica Gale, (sweet gale) moist heaths, *Forton* and *Mereton moors*.

PENTANDRIA.

Humulus Lupulus, (hops) hedges near villages, not unfrequent.

HEXANDRIA.

Tamus communis, (black bryony) hedges, frequent.

OCTANDRIA.

Populus alba, (white poplar, abele) common
tremula, (aspen) in moist ground, frequent
nigra, (poplar) in moist ground, common.

ENNEANDRIA.

Mercurialis perennis, (dog's mercury) woods and moist ditch banks, common.

MONADELPHIA.

Taxus baccata, (yew tree) hedge rows, and village church-yards, *woods at Himley*, very large, and numerous.

CLASS XXIII.—POLYGAMIA.

MONŒCIA.

Atriplex hastata, (wild orach, john-a-neal) gardens
patula, (narrow-leaved orach) on rubbish.

CLASS XXIV.—CRYPTOGAMIA.

FILICES.

Equisetum sylvaticum, (wood horsetail) in ditches and woods, *dingle at Coton-hall, and near Burslem*
arvense, (corn horsetail) on arable land
palustre, (marsh horsetail) marshes
fluviatile, (river horsetail) moist woods, rivers, and pools
limosum, (smooth naked horsetail) shallow water
hyemale, (shave-grass) marshes, *Prestwood-farm, Wednesfield*
Lycopodium clavatum, (club-moss) *Hogley Hey, Cannock-heath*
Ophioglossum vulgatum, (adder's tongue) *meadows near Blymhill*
Osmunda lunaria, (moonwort) *woods at Belmont, Moorlands*
regalis, (osmund royal) *woods near Aqualate-mere*
spicant, (rough spleenwort) moist heaths, frequent, *on the dam of White-sitch-pool*
Pteris Aquilina, (common fern) on heaths, [burnt to make ash balls]
Asplenium Scolopendrium, (spleenwort) fissures of rocks, *Stapenhill, near Burton-upon-Trent*
Trichomanes, (maidenhair) old walls, clefts of rocks, *on Wever-hills, Moorlands*
Ruta muraria, (wallrue, white maiden hair) moist walls, rocks
Adianthum nigrum, (black maiden hair) rocks and old walls
Polypodium vulgare, (common polypody) walls, roots of trees, frequent
Filix mas, (male polypody) woods and hedges
Filix femina, (female polypody) moist places, *Blymhill, Pitmoor-pool dam, Weston-under-Lizard*
Cristatum, (crested polypody) moist hedges and roots of oaks
fragile, (brittle polypody) walls, *Moorlands*
Dryopteris, (branched polypody) shady woods, *Trentham-park, Moorlands.*

MUSCI.

Sphagnum palustre, (grey bog moss) in a bog near the *White-sitch-pool*

Phascum subulatum, (subulated earth moss) ant-hills, on a wet clayey soil, *Blymhill*

Fontinalis antipyretica, (great water moss) deep waters, roots and rocks

Polytrichum commune, (besom moss, goldilocks) woods and boggy ground

subrotundum, (dwarf hair moss) *upon a sandy-hill, Weston-under-Lizard*

aloides, (aloe-leaved hair moss) rocky banks, near *Dawford-brook*

Mnium pellucidum, (transparent marsh moss) moist shady places

scoparium, (broom marsh moss) hedge banks by the side of heaths

fontanum, (fountain marsh moss) mountainous commons

heteromallum, (hair pencil mnium) *stone quarry, Weston-Park*

glaucum, (bristle marsh moss) *moors north of Aqualate-mere*

palustre, (common marsh moss) bogs and wet heaths

purpureum, (purple marsh moss) walls, rocks, and hedge banks

hygrometricum, (yellow bulbed marsh moss) on stones, on the ground

hornum, (swan's neck marsh moss) moist places

annotinum, (long-leaved marsh moss) *turf bogs, near Bishop's woods*

punctatum, (dotted-leaved marsh moss) *Patteshull-park, by a fountain*

undulatum, (waved-leaved marsh moss) woods and moist hedge banks

cæspititium, (pendulous-matted marsh moss) walls, and hedge banks

Bryum murale, (wall thread moss) walls and rocks, very common

pyriforme, (pear-headed thread moss) pastures and ditch banks

viridulum, (green thread moss) wet marshy grounds, and hedge banks, common

cirrhatum, (star-topped thread moss) *Weston-park-wall*

striatum, (curled-leaved thread moss) trunks of trees, rare

subulatum, (subulated thread moss) hedge banks

flexuosum, (soft thread moss) fruit stalks, twisted like a cork-screw, very rare

rurale, (hoary rustic thread moss) roofs, and trunks of trees, common

undulatum, (curled thread moss) shady banks and woods

argenteum, (silver thread moss) rocks, and dry hedge banks

pulvinatum, (pincushion thread moss) roofs, walls, very common

Hypnum complanatum, (flat feather moss) trunks of trees

trichomanoides, (feather moss) roots of trees

taxifolium, (yew-leaved feather moss) ditch banks in a moist clayey soil, and woods, *Blymhill*

ruscifolium, (knee-holly-leaved feather moss) stones in rivulets

Hypnum triquetrum, (triangular feather moss) in woods, frequent
rutabulum, (toothed feather moss) woods, roots of trees, and on the ground
proliferum, (fern feather moss) woods, frequent
parietinum, (wall feather moss) woods, heaths, and shady places
filicinum, (bog feather moss) wet moors, and boggy ditches
crista castrensis, (crested feather moss) amongst the turf on marly soil
prælongum, (trailing feather moss) rotten wood, and grounds
squarrosum, (scurfy feather moss) frequent in soils of the most opposite quality
aduncum, (hooked feather moss) bogs, and wet pastures
cupressiforme, (cypress feather moss) woods, at the roots of trees
dendroides, (tree feather moss) woods, at the roots of trees
alopecurum, (foxtail feather moss) wet moorish meadows
cuspidatum, (spear-pointed bog feather moss) moorish and boggy grounds
purum, (neat feather moss) wet meadows, and dry shady banks
sericeum, (silky feather moss) trunks of trees, and on the ground
myosuron, (mouse-tail feather moss) woods and hedge banks
serpens, (creeping feather moss) ground, sticks and stones
velutinum, (velvet feather moss) trees, stones, and ground, very common.

ALGÆ.

Jungermannia epiphylla, (broad-leaved star-tip) wet rocks
bidentata, (clover star-tip) moist rocks and ditch banks
sphærocephala, (powdered star-tip) *wet rocks near Windford mill*
complanata, (flat star-tip) trunks of trees, frequent
resupinata, (curled star-tip) wet rocks
dilatata, (scaly star-tip) trunks of trees
Marchantia polymorpha, (stone liverwort) sides of wells, and rivulets
conica, (conic livergreen) *side of Dawford-brook, Blymhill*
Lichen incanus, (powdery lichen) hedge banks, and trunks of trees
sulphureus, (sulphur-coloured lichen) *walls of Blymhill church*
scriptus, (lettered lichen) trunks of young trees, *Heath-birches-wood, Weston-under-Lizard*, not unfrequent
canescens, (hoary lichen) trunks of aged trees and walls
ater, (black-cupped lichen) walls and trees, frequent
concentricus, (concentric lichen) *Weston-park-wall*
niger, (black lichen) lime-stone, *Dudley-castle*
fusco-ater, (black-knobbed lichen) on the bark of young pear trees, *Blymhill*
fagineus, (white lichen) bark of beech and other trees

Lichen immersus, (sunk lichen) *walls of Dudley-castle*
pertusus, (porous lichen) bark of trees, and walls
coccineus, (scarlet lichen) *walls of Blymhill church*
parellus, (crab's eye lichen) rocks, walls, and trunks of trees
scruposus, (hollowed lichen) *barr-walls near Wolseley*
atrocinereus, (grey-bordered lichen) *walls of Blymhill church*
subfuscus, (brownish lichen) walls, and bark of trees
pallescens, (pale lichen) walls, and trunks of trees, frequent
Psora, on stones, *rag-stone at Rowley*
candelarius, (yellow lichen) rocks, pales, and trunks of trees
stellarius, (starry lichen) bark of trees
physodes, (inflated lichen) trunks of birch trees
parietinus, (yellow wall lichen) walls and trees
olivaceus, (olive-coloured lichen) trunks of apple and sycamore trees
pyxidatus, (cup lichen) hedge banks and heaths
cocciferus, (scarlet cupped lichen) *Cannock*, and other heaths
cornutus, (horned lichen) heaths, frequent
digitatus, (branched scarlet-headed lichen) moist heaths and roots
rangeferinus, (reindeer lichen) *bogs near Aqualate-mere*
uncialis, (short lichen) *Cannock*, and other heaths
hirtus, (rough lichen) trees and posts
chaliebformis, (wiry lichen) posts, *Weston-park*
plicatus, (stringy lichen) oak trees, *wood near Mucclestone*, rare
floridus, (flowering lichen) *Bishop's wood and Burnt wood*
farinaceus, (mealy lichen) apple and sloe trees
Prunastri, (ragged hoary lichen) old plum trees and old pales
ciliaris, (hairy lichen) apple and elm trees
tenellus, apple trees, *Blymhill*
fraxineus, (wrinkled lichen) trunks of timber trees
caperatus, (rose lichen) trees, pales, and stones
caninus, (grey ground liverwort) hedge banks and woods
Tremela arborea, (tree tremella) dead sticks in woods
Nostoc, (star slough) meadows and gravel-walks, after rain
Conferva ægragropila, (moor balls) *White-sitch-pool, Weston-und.-Lizard*
rivularis, (crow silk) brooks and rivers.

FUNGI.

Agaricus campestris, (common mushroom) dry meadows and pastures
Cantharellus, (yellow champignon) *dam of the White-sitch-pool, Weston-under-Lizard*, very rare
deliciosus, (saffron-juiced agaric) *fir plantations, at Barr, in a plantation near the White-sitch-pool, Weston-under-Lizard*, very rare
orcades, (fairy-ring agaric) upland pastures

Agaricus giganteus, (great agaric) [four to fourteen inches diameter,] *Blymhill*
terreus, (dirt-coloured agaric) meadows and woods
dentatus, (orange agaric) meadows, pastures, parks, and forests
integer, (crimson agaric) pastures, common
caseus, (cream-cheese agaric) *grove at Little Wyrley*
procerus, (tall agaric) hedge banks and dry pastures
muscarius, (fly agaric) *borders of White-sitch-pool*
Listeri, (bitter agaric) *Brockhurst coppice, Blymhill*
necator, (deadly agaric) *borders of White-sitch-pool*
piperatus, (pepper agaric) *fir plantation, Blymhill-heath*
latus, (broad agaric) *borders of White-sitch-pool*
floccosus, (shag agaric) *Bishop's-wood and Burnt-wood.*
fascicularis, (clustered agaric) near decayed timber, in clusters, very frequent
cumulatus, (crowded agaric) hedge banks, frequent
cinereus, dung-hills and meadows
ovatus, (egg agaric) hedge banks and woods
semi-ovatus, (half egg-shaped agaric) dung-hills and pastures
semi-globatus, (hemispherical agaric) pastures, frequent
plicatilis, (plaited agaric) decayed roots
campanulatus, (bell agaric) woods and pastures, common
varius, (variable agaric) roots of trees
stercorarius, cow-pastures
æruginosus, (verdigris agaric) *in the grove at Wyrley-hall*
araneosus, *Hatch-lane plantation, Weston-under-Lizard*
clypeatus, (long-stalked agaric) *birch plantation, White-sitch-pool*
cæspitosus, *side of White-sitch-pool*
castaneus, (chesnut agaric) *birch plantation, White-sitch-pool*
calyciformis, *grove at Wyrley-hall*
ericeus, near *White-sitch-pool*
eburneus, (ivory agaric) pastures, common
fulvus, (tawny agaric) pastures
farinaceus, (mealy agaric) *Weston-park*
fusco-flavus, *Weston-park*
lacer, (lacerated agaric) near *White-sitch-pool*
livido-purpureus, moist woods, *Brockhurst coppice, Blymhill*
nemoralis, *wet ground, near White-sitch-pool*
subcæruleus, *fir plantation, Blymhill-heath*
violaceus, (violet agaric) *grove at Wyrley-hall*
ardesiaceus, (black-stalked agaric) *birch plantation, by the side of the Whitesitch-pool*
velutipes, (velvet-stalked agaric) foot of decayed posts, in clusters, *Weston-under-Lizard*
lateralis, (short-stalked agaric) decayed trees and posts

Agaricus quercinus, (stemless oak agaric) decayed trees and posts, common
Boletus flavus, (yellow boletus) *side of Weston-park-wall, on the road to Tonge*
aurantiacus, (orange boletus) near *White-sitch-pool*, frequent
squamosus, (honey-comb boletus) *trunk of an ash-tree, Weston*
versicolor, (striped boletus) trunks of decayed trees and posts
igniarius, (touchwood boletus) trunks of cherry trees
Helvella mitra, (mitre mushroom) *Weston-under-Lizard-park*
Peziza cochleata, (wreathed peziza) *Weston-under-Lizard-park, birch plantations near the side of White-sitch-pool*
Phallus impudicus, (stinking morel) shady hedge banks
Clavaria Hypoxylon, (horned club top) rotten sticks, common
ophioglossoides, (black club top) *road-side, Blymhill-heath*
pistillaris, (hollow club top) *Chartley and Ingestre parks, Weston-under-Lizard*
muscoides, (yellow pointed club top) *Heyley Castle-bank*
Lycoperdon Bovista, (great puff ball) drained peaty ground
equinum, (horse puff ball) *Weston-under-Lizard*
stellatum, (star puff ball) *hedge-bank, Blymhill*
Sphaeria scripta, (lettered sphaeria) *on the trunks of trees in Heath-birches-wood, Weston-under-Lizard*
tremelloides, (red-knobbed sphaeria) rotten sticks
tuberculosa, (knobbed sphaeria) dried hazel-sticks.

A SYNOPSIS

OF THE

MINERALOGY OF STAFFORDSHIRE.*

EARTHS, including stones, form aggregates: the surface-soil is composed of clay, colorific earth, marl, and mould. Clay is a decomposition of the primitive stones with other matter, silica alumina, carbonate of lime, oxyd of iron, carbonate of magnesia, water, and air. The colorific earths are composed of clay and oxyd of iron. Marl is a mixture of carbonate of lime and clay. Mould, the soil on which vegetables grow, contains silica alumina, lime, magnesia, sometimes iron, but always carbon, derived from decayed vegetable and animal substances, and carbonic acid and water.

Giobert the chemist, found a fertile soil near Turin to contain 78 parts silex (sand), 12 alumina (clay), and 8 to 9 lime, the whole being supposed 100 parts: the annual rains there 30 inches. Near Paris, where the rains are 20 inches, M. Tillet found 46 parts silica, 16½ alumina, 37½ lime: total, 100.

The four Earths, silica, alumina, lime, and magnesia, are the principal basis of soils. If silica most abounds, the soil is dry; if alumina preponderates, the soil is wet: lime and magnesia are correctors; they increase the moisture of land, and diminish the wet-

* In compiling this article, it will be seen that the excellent letter of Mr. Keir, communicated by that gentleman to Mr. Shaw, and published in his "Antiquities of Staffordshire," has been had recourse to; and it may be proper to add, with the entire concurrence of the venerable writer, who would cheerfully have gone into the subject a second time, and furnished us with whatever suggested itself, if age and bodily infirmity did not prevent him from accomplishing so laborious a task. Some apparent inaccuracies will be accounted for in the altered circumstances of the times, but these do not in the least affect the substance of this very able and elaborate view of the mineral treasures of this county.

ness of clay. Different proportions of these four earths will form soils of every degree of dryness or moisture, and their fertility is derived from carbon: dung-hill water contains much carbon. A fertile soil should contain one-sixteenth its own weight of carbon, but it must exist in the soil in a particular state of combination to promote fertility; it must be in a state to be soluble in water. Manures, on putrefaction, part with their carbon, which is absorbed by the soil, and by the plants growing thereon.

Mr. Keir conceives that the coarse bluish clay-soil of Rowley is principally decomposed rag-stone, which stone, from its analysis, is of the kind called basaltes or trapp: it first decomposes into a reddish powder. Marl is clay mixed with carbonate of lime; sand consists of small grains of siliceous stones, not cohering nor softened by water; clay-soils are formed of common clay mixed with decayed vegetable and animal substances. Loam, not cohering so strongly as clay, is of several varieties: the stiff, strong, heavy loam, is composed of clay and coarse sand; calcareous loam is a mixture of lime, clay, and coarse sand, the lime predominating; sandy loam, the same ingredients, the sand predominating to eight or nine-tenths of the whole. Thus are all the upland soils formed and composed, and by degrees impregnated with carbon by vegetation.

Peat-earth, or the meadow and lowland soils, have been formed by the decomposition of aquatic vegetables, root and branch, mixed and impregnated with earthy particles, brought from the upland by wind and water, during a long course of succeeding ages.

The mineral clays of this country are a mixture of alumina and silica, the alumina in powder, the silica in small stones: besides these, clay often contains lime, magnesia, and oxyd of iron. The varieties are common brick-clay, potter's-clay, indurated clay, and shistose clay, the latter found in strata in coal mines, and when impregnated with bitumen, called shale, roach, and clunch, by the miners.

The aggregates of stone are small grains of sand, consisting of quartz, flint, hornstone, siliceous shistus, felspar, and mica, all or in part cemented by lime, alumina, silica, and iron.

The southern part of Staffordshire, to a great extent, contains valuable mines of coal, iron-stone, lime-stone, and clay, to which are owing the foundation and prosperity of the populous villages of this part of the county, and of the neighbouring towns, Birmingham, Dudley, Wolverhampton, Stourbridge, Walsall, Wednesbury, Bilston, &c.; and the surrounding counties are, by means of numerous canals, supplied with the comfort and benefit of fuel, as well as lime, iron, and the other productions of these mines.

By an examination of the country, from actual survey, under the direction of experienced miners, the coal country was found to extend over about 60 square miles, of which one-half, including Bilston, extends southward to Brierley Hill, and Amblecot, near Stourbridge, which may be about seven or eight miles in length, and four in breadth. This tract of land contains beds of coal ten yards in thickness, through which the Birmingham canal, with its several branches, passes; also the Netherton and Stourbridge canals, the Dudley tunnel, and Lord Dudley's canal, including the townships of Bilston, Darlaston, Wednesbury, Dudley, Rowley, Oldbury, Tipton, Sedgeley, Netherton, part of Westbromwich, and the collieries about Dudley-wood, Brettel-lane, Amblecot, Brierley Hill, and the Lye, near Stourbridge. The abundance of coal, iron-stone, lime-stone, and clay, together with the intercourse opened by canals to distant parts, and sea-ports, has induced the establishment of iron furnaces, forges, founderies, and other extensive manufactories, and given employment to a vast number of smiths, and other workmen, in making guns, locks, screws, and above all, nails, of which the quantity manufactured here exceeds all others in the known world.

Besides the above, there are other thinner beds of coal, of four, six, and eight feet in thickness, extending northwards from Bilston and Darlaston, through Essington-wood, Pelsall, and Brown-hills, to Cannock-heath, and are at least of equal extent to the thicker beds before-described. It was calculated, about the year 1800, that the whole coal land of this district was 40,000 acres, and that the consumption from the beginning had not been more than a tenth of the whole.

The thin mine of coal last-mentioned commences near the termination of the ten-yard coal, which there ceases and crops-out, that is, rises abruptly to the surface of the ground, and appears no more. This district of thin coal is intersected by the line of the Wyrley and Essington Canal, which, with its different branches, passes through its whole length, and into the lime-stone strata of Rushall, and Hay-head, north-east of Walsall.

To return to the district of thick coal: A range of lime-stone mountains rise on the west side of this district, extending, from their commencement, two miles south of Wolverhampton, in a southern direction to Dudley Castle, being about four miles in length, and one in breadth. A second remarkable feature, is another range of mountains, which begin to rise from the side of Dudley opposite to the lime-stone, and inclining to an easterly direction: these are a basaltic rock, proceeding from Dudley through Rowley, called Rowley-hills, and, dividing into two branches, terminate between Oldbury and Hales Owen. There are two other detached hills, on one of which stands Wednesbury Church, and the other near the village of Netherton. These hills are supposed to be lime-stone elevations, although lime-stone does not appear on the surface.

The range of lime-stone mountains between Dudley and Wolverhampton, is formed of beds of lime-stone elevated to a high pitch on each side, and inclining to each other so as to form an oblong ridge near their tops, not unlike the roof of a house. In a similar manner the coal, and its accompanying strata, lying on the sides of these mountains, rise or crop-out in directions corresponding with the neighbouring lime-stone elevation; so that, generally, the nearer a pit is sunk to a lime-stone hill, the coal is found at less depth. Most of the coal is got to the east of the lime-stone.

In the same manner that the coal rises or crops-up to the sides of the great ranges of lime-stone mountains, so it also follows the direction of the detached hills. In some intermediate space between the lime-stone range and Wednesbury on the east side, and between that range and Netherton, the coal lies nearly level, and from thence rises on both sides towards the respective hills, forming what the colliers call a *trough*; but though the coal follows the direction of the lime-stone hills, it does not always follow the direction of the surface of the ground, the ordinary rising or falling of which it seems to have no correspondence with. How the coal is affected with relation to the rising of the basaltic hills, has not been fully ascertained, excepting that it certainly does not crop-out along the skirts of these hills as it does along the lime-stone range, but is to be found at the foot of them at moderate depths.

Various opinions have been formed respecting the formation and origin of coal, as well as of other minerals, and which is certainly a fair subject of speculation, though probably above the reach of human genius. Some consider coal as an argillaceous schistus, impregnated with bitumen, which indeed may be the case with some species; but in the best kinds there is too little argillaceous matter to consider it as the basis of the composition. The most generally received opinion is, that coal is formed from vegetable matter. Dr. Darwin says there is reason to believe that bitumen or petroleum, with jet, amber, and all the fossil coal in the world, owes its inflammable part to the recrements of destroyed forests of terebinthinate vegetables—so important to the present race of mankind has been this vegetable secretion. The incumbent strata of coal, sand, iron, clay, and marl, are chiefly the products of vegetable organization, changed in the long progress of their decomposition; and all these solid parts of the earth have been thus fabricated from their simpler elements by vegetation, and by animal life. Mr. Keir says, "but we know no similar fact or experiment, from the analogy of which we can infer the possibility of such conversion." He believes, that in all or most coals, the vestiges of vegetable fibres are to be seen; "not, however, forming the whole substance of the coal, but interspersed as thin laminæ between the thicker and more shining bituminous layers of the coal." We know also, that in all coal-mines, the superincumbent strata of clunch and rock, contain abundance of vegetable impressions, chiefly of reeds and broad leaves, like the stems and foliage of aquatic plants. The substance or body of the reed is sometimes rock and sometimes iron ore, and its surface only is covered with a thin coat of coal, as if the rocky or ferruginous matter which filled up the space which had been occupied by the decayed vegetables, had, by the contraction of drying, left an interval which was afterwards filled up by a bitumen, whether that bitumen came in a liquid form or in the state of a distilled vapour, or whether it was the remaining oil and resin of the plant hardened by age, or by absorption of air. Some kinds of coal are entirely free from vegetable vestiges, and are an uniform, compact bitumen; it therefore seems most probable, that the vegetables have done little more towards the formation of coal, than to furnish a convenient space for the insinuation of the ejected, or distilled bitumen. But in whatever manner coal was formed, it is very certain that it was formed all over the country together, (where now existing) and that its irregularity has been caused by internal convulsions of the earth, which convulsions broke through and raised the strata of lime-stone into the form of mountains, and which also raised the superincumbent strata of which coal is one, and left it in the form it now appears, except the softer parts near the surface, which have been carried off by the action of air and water. The formation of coal must therefore be referred to a very remote period in the history of the earth, as it must have preceded the existence of mountains. The shells of marine animals seem to have furnished the substance of these immense beds of lime-stone which envelope the earth, raised into their present form and appearance. It appears also, from the rearing of the beds of lime-stone, and their broken and irregular position, that their present state could not have been the effect of any gradual operation of nature, but of some sudden and violent effort or earthquake.

When the elastic vapour and air, that were the cause of the earthquakes and elevation of strata, began to lose their force, either by escape at the fractured tops, or by condensation from cold, the weight of the incumbent strata would incline them to subside; and this subsidence being irregular, would occasion still greater irregularities in the fractures, and be the cause of the *faults*, *falls*, and *slips*, which now appear. Many irregular vacuities must also have been left, which since have been filled up by argillaceous and rocky substances, that have fallen or been washed into them, more or less consolidated.

The most singular and extensive vacuity in the coal of this country, now filled up with clay and argillaceous rock, is that occasioned by the separation of the two upper beds of the main coal, called the *flying reed*, and which begin to part from the lower beds of the coal at Bloomfield colliery, from whence they continue to diverge for several miles, till they crop-out at the surface, and are lost before they arrive at Bilston. The separation of strata originally contiguous, and the subsequent interposition of adventitious matter, is a very curious fact in the history of the earth.

The coal does not crop-out on its approach to the Rowley-hills, as it does to the Dudley lime-stone hills, but continues its course some way under them. The formation of these basaltic hills seems more difficult to explain than that of the lime-stone hills, for we know that the latter are part of strata which extend themselves horizontally over a great tract of country, but which in some places are elevated from their inferior situation by some violent convulsion, of which we see manifest indications, both in their own dislocated state, and that of the superincumbent strata. But, with regard to the stone composing Rowley-hills, we do not know of any such inferior stratum, by the elevation of which these hills could have been formed. If it be pretended that basaltic hills are of the kind called primitive, that is, anterior to all other formation or change on the earth's surface, this pretension will be defeated by the consideration that coal lies under them; and, consequently, that their formation is posterior to that of coal. Under all circumstances, (says Mr. Keir,) "I do not know any other mode of formation that can be supported by analogy, excepting that by means of matter ejected in a fluid state from the bowels of the earth, through a chasm in its surface, whether that fluidity was the consequence of fusion by fire, or of water mixed with the smaller particles of earth; after which ejection, the fluid matter may have been consolidated, either by cooling, or by gradual evaporation and drying."

The formation of basaltic rocks has lately been a subject of controversy with mineralogists, some of whom consider them as a lava thrown out of volcanoes, and others as the produce of a watery deposition; and the advocates for these two opinions have been called Vulcanists and Neptunists. Bergman ascribes the basaltic matter to ejection from volcanoes, but not always as a lava melted by fire, but a mass of earthy particles softened and diluted with water, which afterwards has become dry and consolidated.

The Rowley-stone is in large masses in the mountains, of a quadrilateral form, with perpendicular and horizontal joints or cracks, standing on their edges; but on the surface, and declivities, it is confused in all directions. It has been accurately analysed by Dr. Withering, who found that 1000 parts of it contained 475 parts of siliceous earth, 325 argillaceous earth, and 200 calx of iron; but this iron seems to be in a very small degree of calcination from the dark-blue colour of the stone, and from the rusty colour it assumes on being exposed to air and water.

The range of lime-stone hills extends from Dudley in the direction nearly of N.N.W. It consists of oblong hills of which the west sides are the steepest: the most conspicuous are Dudley Castle hill, Wrens-nest hill, and Sedgeley hill. The construction of the two former is, that of several large beds of lime-stone standing at a very steep inclination, corresponding with the surface in opposite directions, till they meet and rest against each other along the summit or ridge of the hills: the elevation of these lime-stone-hills is less than that of the hills of Rowley. The Birmingham canal summit, Wyrley and Essington canal, and Dudley tunnel, form a level (varying only a few inches) of more than 40 miles in length; this level is 600 feet perpendicular above the mean tide of the ocean. The lime-stone hills may be estimated at 200 to 300 feet above this level, and Corney-hill Rowley at 400 feet: a considerable part of the lime-stone is upon the level of the canal, and some beneath it, that to the N. E. of Walsall being nearly on the same level, but none there so much elevated as the Dudley and Sedgeley hills. The coal-pit shafts are generally sunk at near the canal level, but none have, we believe, yet penetrated so deep as to the level of the sea.

The lime-stone beds consist of thin layers of the thickness from three to eight inches. In the centre of the hills are large indefinite masses, called *crog*, of good lime-stone. The beds are separated from each other by substances called *ratch* and *bavin*, which are mixtures of calcareous and argillaceous earth. Immediately under the surface of the ground, detached masses of good lime-stone are found in loose earth, which masses are called turf-stone. Some of these beds contain abundance of petrified shells, amongst which is a representation of an animal called by the workmen a locust, by others the Dudley fossil.

Strata above the Coal.—These are very various in different places, without uniformity; and the depth of the pits are very different. In some places the coal has been got at the surface of the ground in open quarries, in others the pits are 140 yards deep, though the beds of coal are similar in a considerable degree in thickness, quality, and relative position. The strata containing iron ore (called iron-stone,) and those of the finer kind of clay, called fire-clay, pipe-clay, and pot-clay, from its power of resisting heat and its fitness for making tobacco-pipes and glass-house pots, are the next in order with regard to uniformity. As to the other intervening masses of rock, bind, clunch, and especially the upper earths, they vary frequently in the same field, or even at a few yards distance.

The following is a list of measures found in digging a coal-pit in Tividale colliery:

	Feet	In.
1. Soil,	1	0
2. Brick clay,	5	6
3. Brown-coloured roach, (1)	6	0
4. Blue clay,	1	0
5. Red-coloured roach,	5	0
6. Rock, with coal interspersed, (2)	5	0
7. Clunch and iron-stone, (3)	8	3
8. Smutt, (4)	0	10
9. White clunch,	3	3
10. Grey clunch,	7	3
11. Red wild stuff,	41	6
12. Greenish rock,	4	0
13. Red wild stuff,	7	6
14. Binds, with balls of grey rock,	11	0
Carried forward,	107	1

	Feet	In.
Depth brought forward, ..	107	1
15. Wild stuff	16	8
16. Rocky black stuff, ...	1	0
17. Smutt, ...	0	3
18. Black rocky stuff, ...	8	0
19. Kind clunch, with iron-stone, ...	11	2
20. Clunch binds, ...	12	0
21. Soft clunch, ...	8	9
22. Coal called the two-foot coal, (5)	1	6
23. Fire clay, (6)	4	0
24. Fine clunch, ...	12	0
25. Rock, ...	2	0
26. Rock-binds, ...	2	0
27. Parting, (7) emitting inflammable damps, ...	0	3
28. Rock-binds, ...	7	0
29. Kind clunch, ...	1	0
30. Broach coal, (8)	3	9
31 Black clunch, ...	7	0
32. Coal, ...	1	3
33. Penny-earth, (9) with iron-stone, ...	7	0
34. Clunch-binds, ...	26	0
35. Clunch, with iron-stone, ..	2	9
36. Rock-binds, ...	17	0
37. Clunch, with iron-stone, ..	14	0
38. Rock-binds, ...	16	0
39. Strong rock, ...	4	0
40. Rock, with laminæ of coal	4	0
41. Strong rock, ...	4	0
42. Clunch parting, ...	0	10
43. Chance coal, (10)	0	9
44. Clunch-binds, ...	4	0
45. Rock, or rock-binds, ...	8	10
46. Clunch and iron-stone, ..	2	9
47. Black batt, (11)	6	0
48. Chance coal, ...	0	10
49. Catch earth, a clunch, ..	2	9
50. Black batt, ...	0	7
51. Main coal, ...	31	6
Total, ...	360	3

(1) *Roach* is ferruginous clay, of a coarse quality, differently coloured and veined.

(2) The *rock* is white, composed of siliceous earth, with a smaller proportion of argillaceous. Next to this are the rock-binds, which, having more argillaceous earth, are softer than the rock; then the clunch-binds, which have still less siliceous earth; and lastly, clunch, which has the least, and is the softest. The clunch and clunch-binds shiver into flakes when exposed to the weather. The rock-binds, and still better, the rock, retain their texture. The rock is subject to cracks or fissures, through which the water flows, and it is chiefly from these fissures in rock that the water in mines issues: thin laminæ of coal often lay horizontally in the rock, and frequently there is thin coal in the form of broad leaves of aquatic plants running in all directions through the rock.

(3) *Clunch* is a smooth soft earthy matter, which soon shivers into flakes. When of a reddish or yellowish colour, the colliers call it *wild*; but when of its proper bluish or greyish colour, it is said to be *kindly*, by which the colliers mean the latter indicates coal. Those called wild, being irregular and accidental clunch, generally contain balls of iron stone.

(4) *Smutt* is a mixture of coal and clunch.

(5) This coal is too thin to get for use.

(6) *Fire-clay*, called also pipe-clay, from its having been made into tobacco-pipes, for which it is not now used, not being white enough.

(7) A *Parting* is a small quantity of clunch, or soft earth, separating thicker beds from each other: from these interstices between the strata generally proceed the inflammable gas, or air, that incommodes the miners.

(8) *Broach-coal* is of very good quality, and is sometimes got, but generally neglected, not being thought thick enough to pay the expence.

(9) *Penny-earth* is a clunch which contains a good many nodules of iron-stone, for the sake of which pits are sunk at Wednesbury.

(10) *Chance-coal* is a name given to accidental masses of coal, which are not regular strata.

(11) A smooth schistus, rendered black from its vicinity to the coal.

The main coal consists of divisions, and indeed is a number of beds, differing regularly in quality and thickness, and separated by very thin partitions, which are sometimes wanting; so that it is generally named as a single bed of coal. These divisions, with their thickness, are as follows:

	Feet	In.
1. Roof floor, top floor, (*Dr. Plot,*)	4	0
2. After a parting of 4 inches of soft and dark earth—top-slipper; over-slipper, (*Plot,*)	2	2
3. Jays,	2	0
White stone, called Patchel, 1 *inch.*		
4. Lambs,	1	0
5 Tough kitts, or heath,	1	6
6. Benches,	1	6
7. Brassils or corns,	1	6
Foot coal parting (sometimes only)		
Thickness,	13	8

	Feet	In.
8. Foot coal, or bottom slipper	1	8
John coal parting 1 *inch.*		
9. John coal, or slips or veins,	3	0
Hard stone 10 *inches, or less.*		
10. Stone coal, or long coal,	4	0
11. Sawyer, or springs,	1	6
12. Slipper,	2	6
Humphrey parting.		
13. Humphreys, or bottom-bench—Omfray floor, (*Plot,*)	2	3
	14	11
	13	8
Total main coal,	28	7

The coal, including the partings, which vary in thickness, generally exceeds ten yards. A very extraordinary variation occurs in one instance; the two upper beds of the main coal, namely, roof, floor, and top slipper, separate from the rest of the coal at the Bloomfield colliery, and the separation grows wider and wider, in a northern direction, till at length those two beds which, when thus separated, acquire the name of the flying reed, crop-out to the surface, and are lost, whilst the lower part of the coal proceeds on to Bilston, where consequently it has only the thickness of eight yards.

The interval between the flying reed and the main coal, is filled up with soft clunch at the place of separation, which by degrees assumes a harder texture, and changes to a rock that will strike fire with steel.

The measures which are generally known, under the main coal, which in some places are dug for the sake of the iron-stone and glass-house pot-clay, and sometimes for the sake of the coal called heathing coal, are as follows; but here it may be observed, that the main coal is not in such places at so great a depth from the surface as in the instance heretofore given:

	Feet	In.
1. Dark clunch, generally about	1	6
2. Light-coloured clay, with small round iron-stone, called white grains, ...	2	6
3. Main iron-stone, in balls or nodules, involved in clunch: of this mine there are three distinct measures, ...	3	0
4. Table batt, a smooth level-faced schistus, sometimes thicker,	3	0
5. White clay, containing white iron-stone,....................	1	6
6. Heathing coal, in three distinct layers: good coal, ...	6	0
7. Measures of clunch and white-rock to the pot-clay, which is about 16 or 20 yards below the main coal at the lye, where it is principally got, the clay being there nearer the surface of the ground, and of a better quality than in most other places which have been tried. This is what is called Stourbridge clay, and, from its quality of resisting the most violent heat, has caused the establishment of the glass-manufactories in this neighbourhood.		

Iron-stone is found in several of the measures or strata, as before-named. Of these two only are worked for the ore, namely, that which lies immediately under the broach coal, and that which lies under the main coal. In the neighbourhood of Wednesbury, the former bed is worked; and in the other parts of the country, the latter is most considerable. The iron-stone is generally got in coal-works after the coal has been extracted, particularly where it lies at a moderate depth from the surface of the ground, that the expence of sinking pits, and raising it, may be less.

Iron-stone, when dug, is put up in masses called *blooms*, the dimensions of which are three feet by four feet, with a height of 22 inches, and the weight is estimated at 35 hundred, each hundred being 120 pounds. Sometimes 1000 or 1200 such blooms are good from one acre of good mine. The quantity of iron-stone now (1800) got, is sufficient to keep at work about 14 smelting furnaces in the coal country, which produce annually about 18,000 tons of pig-iron; all of which, and more, from other countries, is worked up in the founderies and forges of this neighbourhood.

Spars, calcareous and gypseous, chiefly the latter, in very thin plates, are sometimes found on the coal: when this abounds, its vitriolic acid forms with the coal in the fire a sulphur which smiths observe to be injurious to their iron, they therefore avoid using such coal.

The coal of this country in small pieces, does not cake or conglutinate in the fire, as that of Newcastle-upon-Tyne, and some other coals, do; but being stronger in texture, and not so apt to break, this quality is not requisite. It kindles more readily, and makes a pleasanter fire, requires less trouble in the management, and makes less dust then the caking coals, and is preferred for chamber-fires. It is also a good coal for all kinds of metallic processes.

Of the different beds of the main coal, the upper bed, called roof-floor, is generally left to support the earth or clunch above it from falling. The second bed, called top-slipper, and the third and fourth beds, which together are called the white coal, are reckoned the best for chamber-fires. Next to them in goodness are reckoned the eleventh and twelfth beds, called sawyer and slipper; after them come the eighth, ninth, and tenth, called foot-coal, john-coal, and stone-coal. The tows and benches are preferred for making coaks, and are generally reserved for the furnaces; they do not kindle and flame so vividly as the foregoing measures, but they give a more durable and stronger heat. The part of the brassil measure, which contains pyrites, is generally laid aside, or used for burning bricks or lime; the humphreys being the lowest measure, is cut away to let those above it fall down, and therefore most of it is reduced to small coal or slack.

The same beds of coal, however, vary in quality in different fields, or in parts of the same field. In general the collieries on the east side of the Dudley and Rowley-hills, yield better coal than those on the opposite side. The coal is liable to variations and accidental defects, and some is black without lustre, others appear broken or crushed, and some very bright and shining, burns with little or no flame.

In getting of this coal, in order to support excavations ten yards high, it is necessary to leave very large pillars of coal, eight or ten yards diameter. When the pits are sunk, and communication made with the engine-pit for the water to go off, and roads made for the conveyance of coal to the shaft of the pit, the colliers begin to get coal by working a *stall*. They first cut out the lower bed called the humphreys the length of their stall, ten, twelve, or fourteen yards; and when they have thus removed the foundation to a certain breadth, they loosen its adhesion to the sides by cutting as high as the beds called slipper and sawyer, which makes the first fall of coals, the stone-coal makes the second fall, the john-coal or slips makes the third, the foot-coal makes the fourth, the brassil makes the fifth, the benches and tow-coal make the sixth, the lambs make the seventh, the jays make the eighth, the top slipper the ninth, and the roof is seldom cut, but only as much of it got as drops of itself and can be safely collected. It is the facility of separation, by means of the partings, that principally determines these different falls. A large proportion of the coal is left ungot in these collieries. It is generally reckoned that one-third part is left in pillars, and about another third is small coal, part of which is made into coaks, and part consumed by the fire-engines; the rest is left in the pit. So there remains only about one-third of the mine to be sold as marketable coal, which, at one ton per cubic yard upon a ten-yard mine, is upwards of 16,000 tons per acre.

In the Marquis of Anglesea's park, (Beaudesert), there is a mine of *cannel coal*, which is reserved for the exclusive use of the Marquis's family. This coal has a brownish-black colour, and much less lustre than common coal. The fracture is flat conchoidal, and quite smooth; but the cross fracture is more rough, and on that account has a blacker appearance. This coal is hard, and does not soil the fingers. Interspersed through it are numerous specks of a brown matter, very similar in appearance to Bovey-coal.

Many *faults* occur in this coal-field. They are rents in the beds, which are usually filled with clay. Very frequently the height of the beds varies on the two sides of a fault. By a great fault which occurs near Bilston, the dip of the coal is reversed; that is to say, the coal beds on the south side of the fault dip south, and those on the north side dip north: but this is an unusual occurrence.*

The quantity of coals raised weekly on the banks of the Birmingham canal, and its several branches, has been computed at 15,000 tons, of which about 8,000 tons are sent to Birmingham and beyond, 2,000 tons to Wolverhampton and the Severn, and 5,000 tons consumed in manufactories, towns and villages, near the canal, and the neighbouring country. There are also about 1,200 tons carried weekly upon the Stourbridge and Dudley canals, which makes 16,200 tons of coal, or the produce of an acre raised weekly from these mines. Dr. Plot reckoned in his time (about 1680) twelve to fourteen collieries, each yielding from 2,000 to 5,000 tons annually, which is about 45,500 tons annually, not equal to three weeks' present consumption.

The consumption of coals in this part of England is prodigious.

* Thomson's Annals, Vol. VIII. p. 169.

All the neighbouring counties, to a considerable distance, are supplied by means of the numerous canals of which Birmingham constitutes the centre. Besides this, an immense quantity of coal is required for the iron-works, which are established in the neighbourhood of Dudley to the amount of 68. These smelted an immense quantity of iron; probably more than the quantity manufactured in all the rest of Great Britain. But the low price to which iron has of late sunk, (about £9. or £10. sterling per ton) has in a great measure destroyed this formerly lucrative manufacture. No less than 32 of the 68 furnaces have stopped, or *been blown out*, as the phrase is in Staffordshire. The Welsh iron manufacturers, it seems, produce a greater proportion of iron from their ore, and work with less coals than they can do in this county. They are able, in consequence, to undersell them. This opposition has been carried so far as to sink the price of iron much lower than it seems possible to manufacture it at. Before the late peace it sold at £18. per ton, which was almost double its present price.

There are also extensive tracts of coal country in the northern parts of this county, and collieries are now working at Kidcrew, Apedale, Silverdale, Cobridge, &c. to all of which the preceding remarks are generally applicable. A correct account of one of the principal collieries in this district, is given in the account of the POTTERIES.

A SKETCH

OF THE

ZOOLOGY OF STAFFORDSHIRE.

QUADRUPEDS.

ORDER I.——PRIMATES.

VESPERTILIO MURINUS—*The Common Bat:* Teeth sharp pointed, length 2½ inches: fly like birds.

ORDER III.——FERÆ.

CANIS FAMILIARIS—*The Dog:* Bends his tail to the left. Varieties many, as mastiff, hound, greyhound, pointer, spaniel, terrier, shock, lap-dog, cur, &c.

2. VULPES—*The Fox:* Tail a brush, tipped with white, legs white, fore-feet black.

FELIS CATUS—*The Cat:* Domesticated, tail long, annulated, body marked with spiral and three longitudinal stripes. Our tame cats are less in size, and vary in colour: the tortoise-shell most esteemed.

MUSTELA LUTRA—*The Otter:* Colour dark brown, with two white spots on each side the nose, length three feet three inches, tail compressed and taper, eyes small, ears short, long whiskers, feet webbed. The otter is found in the Trent, and smaller rivers communicating with it, and affords much diversion in hunting; no animal defends itself with greater obstinacy, and will make a desperate resistance to the hounds, who are frequently maimed in the conflict. In rivers where weirs are set, otters will force their way in, and devour the fish, and then effect their passage out by biting asunder the twigs. Dr. Plot says, "at Ingestre I ate potted otter, so well ordered by the cook, that it required a very nice palate to distinguish it from venison."

2. PUTORIUS—*The Fitchet or Polecat:* Length about two feet,

colour blackish tawny, muzzle and ears white, nose sharp. Its fierceness is remarkable. We have known one, when confined and unable to escape, attack a large greyhound: it has the faculty of stinking most abominably when irritated, to annoy and drive away its enemies.

3. Vulgaris—*The Weazel*: Upper part of the body and tail tawny, throat and belly white, length nine inches.

4. Erminea—*The Stoat or Ermine*: A beautiful white variety, with the tail tipped with black, is sometimes seen. These two last are so fierce, that they will devour rats, young hares, or poultry. A rat larger than themselves will fly instantly at their approach.

Ursus Meles—*The Badger*: Length two feet and a half, upper part of the body grey, under black, tail short, with long stiff hair, face black and white, fore claws long, a transverse orifice between the tail and the anus, three black bristles over each eye. In Manstey and other woods. This harmless animal is sometimes subjected to much cruelty in being baited by dogs for amusement.

Talpa Europæa—*The Mole*: Five toes on each foot, snout long, eyes small, fur remarkably soft, no external ears, fore legs short, with broad feet turned outwards: a cream-coloured variety has been taken near Blymhill. These creatures are great destroyers of earth worms, though injurious to the land and crops by their hillocks.

Sorex Araneus—*The Shrew Mouse; Nurserow*: Snout and tail long, above brown, beneath dirty white, toes five on each foot, eyes small, length four inches.

Erinaceus Europæus—*The Hedge Hog; Urchin*: Five toes on each foot, long snout, body covered with long sharp bristles, length 11 inches. This animal is capable of folding itself up in its rough prickly husk, as a retreat from danger, and appearing like a rough senseless ball, and thoughtless boys sometimes torture it with dogs, and otherwise, to make it appear. The peasantry charge it with sucking the teats of cattle.

ORDER IV.——GLIRES.

Lepus Timidus—*The Hare*: Ears tipped with black, most abounding on light dry soils.

2. Cuniculus—*The Rabbit*: Ears almost naked, pupil of the eyes red. They abound on Cannock-chase, and Bishop's-wood, near Breewood.

Mus Norwegicus—*The Brown Rat*: Length to the tail 9 inches, tail nine inches more, back tawny, belly dirty white, feet and legs almost bare, tail scaly. This species has extirpated the black rat, but they are a dreadful substitute; they swarm about farm-houses, destroying pigeons, poultry, leather, and all kinds of grain in great quantities. Though so rapacious and fierce, they are great cowards, being put to flight in a moment, or destroyed, if caught by the ferret, or by our native animal the weasel, either much less than themselves.

2. Amphibius—*The Water Rat*: Length to the tail seven inches, tail five, body covered with long hairs, brownish black, belly grey, tail black, white at the extremity, ears and eyes small.

3. Musculus—*The Common Mouse:* Differs very little from the common rat except in size; in other respects similar.

4. Sylvaticus—*The Field Mouse*: Larger than the common mouse, back brown, belly grey.

5. Agrestis—*The Short-tailed Field Mouse*: From nose to tail two inches and a half, tail two inches, ears naked, weight one-sixth of an ounce.

6. Avellanarius—*The Dormouse*: Found in oat-ricks.

Sciurus Vulgaris—*The Common Squirrel*: Upper fore-teeth like wedges, lower compressed, colour red brown, belly white, ears tufted, tail a brush. A lively little creature, and nut-cracker: said to do much injury to fir-plantations in severe winters.

ORDER V.——PECORA.

Cervus Elaphus—*The Stag*, *Red Deer*, or *Hart:* Horns round, branched, and turned backwards. A few are kept in parks; but their ferocity at certain seasons has caused their numbers to be reduced. Weston Park, Chartley, &c.

2. Dama—*The Fallow Deer*: Horns compressed, branched, turned backwards, and broad at the extremities. Many of those on Cannock-chase are of the deep brown kind, introduced from Norway.

Capra Hircus—*The Goat*: Horns hollow, erect, bending backwards, body covered with long hair, long beard: domesticated.

Ovis Aries—*The Sheep*: Horns compressed, rough, hollow, simple, spiral, turning outwards; face grey or black, legs of the same colour, wool short, fine: Cannock-wood sheep: improved in pasture.

2. Domestica: Face and legs white, wool longer, but fine in staple, hornless, fine in bone, thick, compact, inclined to fatten, quiet in pasture, not inclined to roam or leap: New Leicester sheep.

3. MERINO *or Andalusia*: Looser and lighter made, longer, and less compact, twirling horn, wool silky, extremely fine: Spanish sheep.

BOS TAURUS—*The Wild Ox*: Chartley-park is among the few places where the indigenous wild cattle of Britain are now found: the breed has been here preserved pure and unmixed, having been constantly confined in a separate enclosure. The following are their characteristics: colour white, muzzles and ears black, horns white, fine, tipped with black. Their native wildness appears at the first approach of man; the whole herd instantly sets off at full speed to a considerable distance, and then wheels round, at the same time staring with eager curiosity. The calves, even when very young, have a natural shyness; if disturbed, they bleat to alarm their dams, which will immediately attack the intruder: their flesh is finely marbled, and of excellent flavour. The domesticated varieties of this species are most generally the long-horned breed in different degrees and stages of improvement; and where due attention is paid to selection, this variety is very respectable both as milkers and feeders. Other varieties are

2. TAURUS ALBA: Hornless, colour milk-white, ears tipped with red, size and weight large, producing well to the pail, and weighing, when full grown and fat, the cows 10 to 12 score the quarter, and the male 12 to 15 score: they are well-formed, and disposed to fatten, and handle sleek and well, with great weight in the more valuable joints, and the cows of singular and beautiful appearance. To be found at Dunstall-hall, near Wolverhampton.

ORDER VI.——BELLUÆ.

EQUUS CABALLUS—*The Horse*: Tail with long hair, ears short, erect, mane long. The sort generally preferred for draught is the heavy black breed, having the greatest weight and strength; but some of the brown, and other colours, have doubtless more action and energy of motion, and are to be preferred in all cases where dispatch is more the object than strength. The Welsh mountain breed is often very hardy, and useful, for the saddle and road.

2. ASINUS—*The Ass*: Ears long, flaccid, a black cross on the top of the shoulders. This animal finds its value in the revolutions of time, being now kept by many respectable females as a donkey or pad, for short rides.

SUS SCROFA—*The Hog*: Hoof divided, two tusks in each jaw, ridge beset with strong bristles. The dividing of the hoof was formerly considered as an essential character, but a variety has been lately produced with the hoof undivided, having otherwise the same specific character.

BIRDS.

ORDER I.——ACCIPITRES.

FALCO CHRYSAETOS—*The Golden Eagle:* Ceré and legs yellow, feathered to the toes, wings reach to the extremity of the tail, beak lead colour, irides hazel, colour dark brown, weight 12lb. tail clouded with dark ash colour. Two birds of this species were discovered on Cannock-chase, preying upon the carcass of a sheep, one of which was shot by the late Sir Edward Littleton's game-keeper: they have also been seen at Beaudesert, and in Needwood-forest.

2. CYANEUS—*The Hen-harrier:* Male cere and feet yellow, above grey, beneath white, weight 12 ounces, length 17 inches. Female, a ruff round the head, under each eye a white spot, back dusky, rump white, belly reddish brown, weight 16 ounces, length 20 inches.

3. MILVUS—*The Kite-Glead:* Cere and legs yellow, head grey, back brown, tail forked. Flies remarkably steady; preys much on mice.

4. BUTEO—*The Buzzard:* Cere and feet pale yellow, beak lead colour, above ferruginous brown, beneath yellowish white, spotted with brown, tail barred with black and ash colour, and tipped with brownish white; weight 32oz. length 22 inches.

5. TINNUNCULUS—*The Kestrel or Standing Hawk:* Cere and feet yellow: male, head light grey, tail the same, with a black bar near the end tipped with white; back purple spotted with black, weight 6½ ounces, length 14 inches. Female, back less bright, head and tail pale brown spotted with black, weight 11 ounces.

6. NISUS—*The Sparrow Hawk:* Cere green yellow, feet yellow, bill blue, claws black, above brown or grey, beneath tawny white waved with brown, tail ash colour, barred and tipped with white, weight of the male five ounces; female nine. Most destructive to game.

7. ÆSALON—*The Merlin:* Cere and legs yellow, bill lead colour, above purplish ash colour, with ferruginous spots, beneath yellowish white with brown spots, tail barred with brown, length 6½ inches.

STRIX BRACHYOTUS—*The Short-eared Owl:* Horns or ears a single feather, above brown, beneath pale yellow, quill feathers barred with red, tip of the tail white, wings when closed reach beyond the tail; length 13¾ inches. Migrates with the woodcock. Does not perch on trees, and prefers wild solitary parts abounding with heath, amidst which it breeds on the ground. It flies by day, and destroys many mice. The ears being small and decumbent, are scarcely conspicuous in dead birds. Found in Chartley Park.

2. FLAMMEA—*The White Owl:* Above pale yellow with white spots; beneath white; interior sides of the quill feathers, four black spots on each side; edge of the middle claw serrated; weight 11 ounces, length 14 inches. The plumage is exquisitely pencilled with unrivalled delicacy.

3. ULULA—*The Brown Owl:* Above deep brown spotted with black and white, beneath pale ash colour mixed with tawny, with black strokes; disk round the eyes ash colour, with brown spots. Flies by night.

LANIUS EXCUBITOR—*The Cinereous Shrike or Ash-coloured Butcher Bird:* Bill and legs black, tail wedged, black in the middle, and white on the sides, crown and back ash colour, wings black with a white bar; cheeks white, with a black stroke; weight three ounces, length 10 inches: migrates. Dr. Plot says, there is of them in Needwood-forest, and in Bramshall-park, and that one was killed at Scierscot, near Tamworth, colour and size of a thrush. Feed on insects and small birds, and have sometimes been reclaimed by falconers to fly at small game.

ORDER II.——PICÆ.

CORVUS CORAX—*The Raven:* Above shining bluish black, beneath dusky; weight 3lb. length 26 inches.

2. CORONE—*The Common Crow:* The whole body bluish black, tail rounded; weight 20 ounces, length 18 inches. White crows are named by Dr. Plot.

3. FRUGILEGUS—*The Rook:* Colour bluish black, nostrils, chin, and sides of the mouth, white and bare, somewhat larger than the last. Multitudes of them resort to the sea-coast in severe winters, in search of periwincles; having found a shell, they raise it to the height of 40 feet, and drop it upon the rocks, instantly descending; if the shell should not break, the operation is repeated; the labour is immense for so small a recompence. A cream-coloured variety was taken in a nest at Wyrley-grove. Rooks abound in this

country to the great injury of corn-land. (*Mr. Dickenson.*) They may be serviceable in a moderate degree, in picking worms and caterpillars out of the earth, and will often examine each furrow after the plough.

4. Cornix—*The Royston Crow; Hooded Crow:* Back, breast, belly, and upper part of the neck ash colour, the rest bluish black; weight 22 ounces, length 22 inches. Migrates: rare.

5. Monedula—*The Jackdaw:* Back of the head, breast, and belly, ash colour, the rest black, irides white; weight nine ounces, length 13 inches.

6. Glandarius—*The Jay:* Head covered with long feathers, forehead white with black strokes, neck black, coverts of the wings fine blue, barred with black and white, back, breast, and belly purple, dashed with grey, rump white, tail black; weight six ounces, length 13 inches. Dr. Plot names a white popinjay.

7. Pica—*The Magpie:* Black and white tail, long, wedge-fashioned and richly tinged; weight nine ounces, length 18 inches. Build their nest with thorns.

Picus Viridis—*The Green Io or Woodpecker:* Bill angular, straight, finely squared, admirably adapted in all the species for the perforation of trees, crown crimson, back green, rump pale yellow, beneath pale green, tail tipped with black, bill lead colour, legs and feet greenish ash colour, weight six ounces and a half, length thirteen inches. Has a tone of voice resembling the human laugh.

2. Major—*The Greater Spotted French Magpie:* Irides red, crown of the head black, back part crimson, a black collar round the neck, back and rump black, cheeks and scapulars white, breast yellowish white. This bird is more common than the former, flying in the same manner from tree to tree. It is erroneously asserted to keep entirely in the woods.

3. Minor—*The Little Spotted Woodpecker:* Crown crimson, above black barred with white, beneath dirty white; weight one ounce, length six inches. In the female the crown is white, rare, and a miniature of the preceding, insomuch that some have esteemed it a young bird of that species.

Yunx Torquilla—*The Wryneck; Cuckoo's Mate:* Above ash colour, with black and brown strokes, beneath light brown, with black spots, tail ash colour, with four black bars, irides hazel, weight one ounce and a quarter, length seven inches: migrates with the cuckoo, with whom it has great affinity in other respects.

Alcedo Ispida—*The King Fisher:* Bill, upper mandible black, under yellow, wings dark green, spotted with blue; scapulars and coverts of the tail bright azure, beneath orange, tail deep blue, feet orange; weight 1¾ oz. length 7 inches. It skims along our rivers and brooks, and devours small fish.

CUCULUS CANORUS—*The Cuckoo:* Bill roundish, and curved a little; nostrils circular, with a raised margin, tongue entire, two fore, and two hind claws; above ash colour, beneath white, waved with transverse black lines, tail rounded and spotted with white; weight five ounces, length 14 inches: migrates. Zoologists have in all languages assigned to this harbinger of nature's fairest season, a name expressive of its singular note. Cuckoos were known in Greece in the time of the poet Hesiod. These birds are observed to rest twice annually on Malta, at times which prove them to be on their migration to and from Europe. In Staffordshire, a vulgar error prevails, that cuckoos feed entirely on the eggs of other birds; but this is disproved by anatomists, who have found in their stomachs worms and insects.

SITTA EUROPÆA—*The Nut Hatch or Wood Cracker:* Upper mandible black, lower white, crown black, black stroke across the eyes, cheeks and chin white, coverts of the wings bluish ash colour, beneath dull orange, legs pale yellow, twelve feathers in the tail; weight one ounce, length six inches: rare, but found in Earl Bradford's woods, at Weston.

CERTHIA FAMILIARIS—*The Creeper:* Bill arched, slender, and somewhat angular, tongue sharp, above brown, streaked with black, quills tipped with white, beneath white, tail of twelve long feathers; weight five drams, length 5¾ inches.

ORDER III.—ANSERES.

ANAS ANSER—*The Grey Lag Goose:* Weighs 10lb. length two feet nine inches, extent five feet, bill pale yellow, general colour grey, legs flesh colour. Origin of the domestic goose: originally from the fens. Other varieties are sometimes met with, as the *Anas Canadensis*, shot at Shenstone. The French Goose, kept about Amerton near Chartley, called, from the blackness of its bill, *Anser Guineensis*, and *Swan*, are met with near Rugeley, with legs flesh-coloured, like those of a goose, and may be termed *Anser Cygnoides*.

2. BOSCHAS—*The Mallard or Wild Duck:* Bill green yellow, tail of 24 feathers, four middle ones black. Origin of the tame duck. Found in vast numbers on Aqualate-mere, Chillington-pool, &c.

and sometimes stray off, and build their nests in rivers and pits; and in winter frequent the smaller brooks. They dive to a great depth in search of subaqueous plants.

3. Crecca—*The Teal:* Bill and vent feathers black, head bay, with a green and white line on each side, beneath dirty white, a green and a black spot on the wings, coverts of the wings brown, tail pointed; weight 12 ounces, of the female nine, length 17 inches. The male, when it has attained its true colours, is the most beautiful of our aquatic birds.

4. Cygnus—*The Wild Swan:* Cere yellow, bill black, plumage white, legs and feet brown. On the Trent.

5. Mansuetus—*The Tame Swan:* Cere knotty and black, bill red, tipped with black, feet lead colour. Larger than the former; long-lived.

6. Anas Penelope—*The Widgeon:* Bill lead-colour, tipped with black, head bright bay, back waved with black and white, breast purplish, belly white, vent black, middle quills green, tipped with black, tail two middle feathers black, longer pointed, the rest ash colour; weight 23 ounces, length 20 inches.

Mergus Merganser—*The Goosander; Sowgander:* Male; bill, irides, and legs red, head green black, neck, breast, and belly pale yellow, wings and tail black and white; weight four pounds, length two feet four inches.—*Dundiver:* female; head ferruginous, back and tail ash colour, less than the male. Aqualate-mere.

Pelecanus Carbo—*The Cormorant:* Bill ash colour, with a yellow naked skin round the base, head and neck black, chin white, a short pendant crest, above bluish green, beneath dusky, a white tuft on the thighs, tail long and rounded, legs black, the middle claw serrated; weight four pounds, length 32 inches. Frequent in winter about Aqualate-mere.

2. Bassanus—*The Solan Goose or Gannet:* Bill bluish grey, jagged at the sides, a long furrow in the upper mandible, eyes yellow, surrounded by a blue skin, a black pouch under the chin, neck long, crown buff-colour, wings and legs black, the rest of the plumage dirty white, tail rounded; weight 4½ pounds, length 37 inches. Aqualate-mere, in winter.

Larus Canus—*The Common Gull, Mew, or Sea Mall:* Bill yellow, back grey, the rest white; weight 12 ounces, length 17 inches. At Batchacre.

2. Tridactylus—*The Tarrock:* Bill, tips of the wings, and of the tail black, head, neck, breast, belly, and tail white, a black

crescent on the back of the neck; back bluish grey, legs dull ash colour, a small protuberance in lieu of a back toe; weight seven ounces, length 17 inches. At Weston-under-Lizard.

3. Ridibundus—*The Pewit-Gull, Black Cap or Sea Crow:* Bill and legs red, head, throat, and tips of the wings black; back and wings ash colour, the rest white, the wings long, and adapted to long flights its note a hoarse laugh. These birds formerly visited the Skrymsher estate, about half a mile south-west of Norbury church, to build and breed: some curious particulars of them from Plot will be found under "High Offley."

Colymbus Cristatus—*The Great-crested Grebe:* Killed at Comberford, near Tamworth: a pendant ruff under the chin, no tail, above black brown, beneath silver white, weight 2½lb. length 21 inches, legs near the rump, made for diving, not walking or flying.

2. Auritus—*The Little Grebe, Dobchick, or Doucker:* Above black brown, beneath white, head and neck black, ridge and tips of the wings white, a tuft of long feathers on each side the head, legs green, irides red, less than a teal. These birds are so shy, they are with much difficulty shot, having a most exquisite quickness of sight, with a power of remaining long under water. They emerge always towards the water's edge, for the purpose of sheltering in holes, or under the roots of trees; if taken alive they strike furiously with the bill, like the heron or bittern. Builds its nest in the water, but generally attached to a subaqueous twig: on approaching the nest, the bird darts from it, covers the eggs, and instantly disappears.

3. Stellatus—*The Grey Speckled Diver:* Bill turns a little upwards; above grey speckled with white, beneath white, tail of 20 black feathers, some tipped with white, feet fintaed. Vast flocks on Aqualate-mere, in winter.

ORDER IV.—GRALLÆ.

Ardea Major—*The Common Heron:* Bill greenish above, yellowish beneath, black at the end, a short pendant crest, above grey, beneath white, with longitudinal spots; a greenish naked skin between the bill and the eyes, irides yellow, quills black, legs greenish. The vulgar imagine their legs attract fish. The heron feeds on fish, but nests in the highest trees. "This bird in numbers is on the decline, having been marked out by man as a fish-stealer. Builds in Norbury Park."—*Plot.*

2. Stellaris—*The Bittern:* Bill a little arched, irides yellow,

crown black; a short pendant crest, a black spot at each corner of the mouth; plumage pale dun yellow, spotted with black, feathers of the breast long and loose, legs greenish, hind claw very long; makes a bellowing noise, hollow and loud, which may be heard afar, and formerly deemed ominous.

RECURVIROSTRA AVOSETTA—*The Avosette*. Bill 3½ inches long; above black and white, beneath white, legs blue and very long; thighs naked. Rare: Aqualate-mere.

SCOLOPAX ARQUATA—*The Curlew*: Bill arched, six inches long, plumage white, spotted with black, quills black, tail white, barred with black, legs bluish grey; weight from 22 to 37 ounces. Shot near Breewood.

2. RUSTICOLA—*The Woodcock*: Bill three inches long; above variegated with red brown, black, and grey, beneath dirty white, with transparent dark lines, chin yellow; weight 12 ounces, length 14 inches.

3. GALLENAGO—*The Snipe*: Bill three inches, four black and red longitudinal lines on the head; chin, breast, and belly white; neck red and brown; scapulars black and yellow; tail tipped with white; legs pale green: weight four ounces, length 12 inches.

4. GALLENULA—*The Jack Snipe; Beckett*: Bill 1½ inch, crown black, rump shining purple, over each eye a yellow stroke; weight scarce two ounces, length nine inches.

5. CALIDRIS—*The Redshank; Whistling Plover*: Bill two inches, red at the base, black at the end; head, back of the neck, and scapulars dusky ash colour, spotted with black; back and under side of the body white, spotted with black; tail barred with black and white; legs orange: weight 5½ ounces, length 12 inches. The note of this species is shrill, resembling the human whistle, and may be heard half a mile. Aqualate-mere.

TRINGA VANELLUS—*The Lapwing; Pewit*: Bill, crown, crest, and throat black; a black line under each eye, back purplish green, wings and tail black and white, legs red; weight eight ounces, length 13 inches. Has great powers of flight, and darts with great celerity; during nidification shew great anxiety, and to divert the attention of a man or dog approaching their nest, will fly so near as almost to come in contact.

2. HYPOLEUCOS—*The Common Sandpiper*: Bill, head, and quills brown; neck dark ash colour, back brown mixed with green, with transverse dusky lines; longitudinal black lines on the head; a

white stroke above each eye, beneath white; legs greenish; weight two ounces. In lakes and rivers.

3. ALPINA—*The Dunlin; Summer Snipe:* Bill, legs, and feet black, above red brown, beneath white, with black spots, wings ash colour, back toe very short; length eight inches.

FULICA ATRA—*The Coot:* A thick callus red membrane above the bill; bill and legs olive; garter yellow; belly ash colour; margin of the wings white; the rest black, or very dark ash colour; weight 26 ounces; toes scolloped.

2. CHLOROPUS—*The Water-Hen; Moor-Hen:* Callus, irides, upper part of the bill, and garters red; head, neck, and breast black; belly deep ash colour; margin of the wings white; the rest glossy olive brown; fins of the toes strait, narrow; weight 15 ounces, length 14 inches.

RALLUS AQUATICUS—*The Water Rail:* Bill slightly arched, slender, above black, beneath orange; above black, the feathers edged with olive brown; beneath ash colour; base of the wings white; tail short, legs reddish, and near the tail; weight 4¼ ounces, length 12 inches : rare.

2. CREX—*The Land Rail; Corn Crake:* Bill and legs grey; above black edged with light brown; beneath white; legs very long; weight seven ounces.

CHARADRIUS PLUVIALIS—*The Green Plover:* Bill, head, back and legs black, breast brown, belly white; upper part of the body and breast spotted with green; tail barred with black, green and brown; weight nine ounces, length 11 inches. Migrates; rare.

2. MORINELLUS—*The Dottrel:* Bill, head, shafts of the quills, belly, tip of the tail, and legs black; a broad white line above the eyes; cheeks, throat, and thighs white; breast and sides orange, the rest olive; weight four ounces. Migrates; rare.

ORDER V.——GALLINÆ.

PHASIANUS COLCHICUS—*The Pheasant:* Bill horn colour, cheek membrane bright red, a tuft of black feathers near the ears, head and neck tinged with rich green purple, beneath brown, tail of 18 feathers, long and of various colours, a short sharp spur; weight 45 ounces, length 36 inches.

TETRAO—*The Grouse:* Bare scarlet spot above the eyes, legs feathered to the feet.

2. TETRIX—*The Black Game or Heath-Cock:* Plumage black, a

white spot on the shoulders, tail of 16 feathers, forked; thighs and legs dark brown; weight near 4lb. length 22 inches: female, plumage mixed with red, weight 2lb. Cannock-chase, Burntwood, Bishop's-woods, and Chartley-park. Feed on whortle-berries, but in summer and autumn wander into corn-fields, being polygamous. Sportsmen aim at destroying the males only.

3. Scoticus—*The Red Grouse or Moor Game:* Bill black, irides hazel; a white spot on each side the lower mandible, throat red, head and neck tawny red, mixed with black; back red with black spots; beneath purplish brown; streaked with dark lines; tail of 16 feathers; thighs pale red; legs feathered to the claws; weight 19 ounces, length 15 inches: female, belly spotted with white, weight 15 ounces. These birds prefer the summits of hills, braving the utmost rigour of winter, and never seek shelter in woods; when other food fails, they will devour the tops of heath, and abound in the Moorlands of Staffordshire, and on Cannock-heath. Legs naked.

4. Perdix Cinerea—*The Partridge:* Bill white, plumage orange, black, and brown; a semilunar mark on the breast, tail of 18 feathers; weight 15 ounces, length 13 inches: female, less. Plentiful in cultivated lands: in Needwood-forest they subsist on ants' eggs. A brace of white partridges were killed at Wrottesley in 1796.

5. Coturnix—*The Quail:* Bill dusky, head black, a yellowish line from the crown to the back, chin and throat whitish, breast pale orange spotted with black, back black and yellow; tail of 12 feathers black and red; length seven inches. Migrates: rare. These birds are seldom seen, but their note has been heard most seasons in corn-fields at Pendeford.

ORDER VI.——PASSERES.

Loxia Coccothraustes—*The Cross Beak or Haw Finch:* Bill, under side the body, and legs reddish, head chesnut, chin and quills black, the ends of the middle quills curved outwards, a black line from the bill to the eyes, back brown; weight two ounces, length seven inches. Migrates. Feed on berries of yew and hawthorn.

2. Pyrrhula—*The Bullfinch:* Bill, crown, coverts of the wings, and tail black, back of the neck and back grey, coverts of the tail and vent white, cheeks, breast, and belly crimson, a white line across the wings: the breast and belly of the female dirty buff colour.

3. Chloris—*The Green Finch:* Plumage yellow green, exterior

webs of the outmost feathers of the tail yellow, lower belly white; length 5½ inches.

FRINGILLA CARDUELIS—*The Goldfinch; Redcap; Seven-coloured Linnet:* Circle round the bill red, breast, back, and rump brown, middle of the wings yellow, tip of the bill, line to the eyes, crown, wings, and tail black, the rest white. The most elegant of our small birds in form, plumage, and nidification.

2. CŒLEBS—*The Chaff Finch; Pied Finch:* Bill, crown, back, and sides of the neck bluish, forehead, wings, and tail black, three white lines across the wings, above the eyes, cheeks, and throat red; belly reddish white: female, without red feathers.

3. MONTIFRINGILLA—*The Brambling:* Back and crown black, edged with brown, chin, throat, breast, and less wing coverts orange, inner coverts yellow, quills edged with yellow, belly whitish; length six inches. The male is a bird of great beauty: they breed in Scandinavia, and a few migrate here in winter.

4. DOMESTICA—*The House Sparrow:* numerous, and injurious to corn in summer.

5. MONTANA—*The Tree Sparrow or White Cap:* Back of the neck, under the eyes, and belly white, rest of the body brown and black, legs pale red. Observed here only in autumn and winter, and supposed to breed elsewhere.

6. LINOTA—*The Common Linnet:* Back ash colour and brown, belly white.

7. CANNABINA—*The Red-headed Linnet:* Back, scapulars, and coverts red brown, on the forehead a dark red spot, breast pale red, rest black and white.

EMBERIZA MILIARIS—*The Bunting:* Above olive brown, with black spots, beneath yellow white, edges of the wings and legs yellow, sides of the tail white; length 7½ inches.

2. CITRINELLA—*The Yellowhammer; Goldfinch:* Crown, chin, throat, and belly yellow, back of the neck, less wing coverts, part of the quills, and part of the tail green, breast orange, rump brown red; length six inches.

3. SCHŒNICULUS—*The Reed Sparrow:* Above black, edged with red, beneath white, a white circle round the head of the male; length near six inches.

CAPRIMULGUS EUROPÆUS—*The Goatsucker:* Plumage black, white, brown, and ash colour intermixed, mouth purple, irides hazel, weight 2½ ounces, length 10 inches. Woods near Walsall, and in Sutton Park: migrates, and found here from April to September.

It sings at the close of day, and at times in the night; the noise resembles the whizzing of a spinning wheel, and when near gives a sensible vibration to the human frame, and it is difficult to discover whence the sound is emitted. This bird is a natural ventriloquist, has wonderful powers of wing, and may be termed the night swallow, never flying in the day unless disturbed; it is nearly as large as a cuckoo; its unfledged young squat amongst heath, and resemble a toad.

HIRUNDO RUSTICA—*The Chimney Swallow*: Above purplish black, beneath reddish white, a red spot on the forehead, and under the chin, mouth black; length seven inches. Two birds of this species, entirely white, were bred at Horsebrook, 1796.

2. URBICA—*The Martin*: Head and back purple black, breast, belly, rump, white, feet covered with white down, mouth yellow; length six inches.

3. RIPARIA—*The Sand Martin*: Above mouse colour, beneath white, feet black, a mouse-coloured ring round the neck, length five inches. Will penetrate several feet into the banks of sand-pits to deposit their eggs, the shell of which is white, and beautifully pellucid.

4. APUS—*The Martlet*; *Swift*; *Jack Squealer*: Black, except a white spot on the chin, all the toes stand forward; weight less than one ounce, length near eight inches. This species cannot rise from off the ground, from its shortness of legs and length of wing.

TURDUS VISCIVORUS—*The Misletoe Thrush*; *Thricecock*: Above brown grey, beneath yellow white, with dark spots, irides hazel, tail of 12 feathers, exteriors tipped with white; inner coverts of the wings white; weight five ounces, length 11 inches. Our largest singing bird: builds often in orchards, and propagates the misletoe.

2. PILARIS—*The Fieldfare*: Bill yellowish, tipped with black, head and rump ash colour, back deep brown, tail and legs black; beneath white, breast reddish, with dark spots; weight four ounces, length 10 inches. Migrates.

3. MUSICUS—*The Throstle*: Above brown grey, beneath white, with black spots, throat and breast reddish, inner coverts of the wings yellow; weight three ounces, length nine inches. Sings loudly, and early in spring mornings.

4. ILIACUS—*The Redwing*: Above brown grey, beneath whitish brown spots, sides and under coverts of the wings red yellow, a yellow white line athwart the eyes; weight 2½ ounces.

5. MERULA—*The Blackbird*: Plumage black, bill and edges of the eyelids yellow.

MUSCICAPA GRISOLA—*The Spotted Fly-catcher*: Bill, legs and feet black; above brownish grey, beneath white, quills edged with yellow, throat and sides tinged with red, mouth yellow, length 6½ inches, builds in the corner of walls, or on the end of beams, and lays five eggs. This bird frequents gardens, and devours cherries as well as flies.

MOTACILLA ALBA—*The White Water Wagtail*: Bill, mouth, head, neck, back, tail, and legs black, breast, belly, and sides of the tail white, tail and back claw very long; weight six drams, length eight inches.

2. FLAVA—*The Yellow Water Wagtail*: Above olive green, beneath yellow, a yellow line above the eyes, tail edged with white; length eight inches.

3. REGULUS—*The Golden-crested Wren*: Weight 70 grains, length four inches. One of our smallest birds: in a wood near Blymhill.

4. LUSCIANA—*The Nightingale*: Head and back tawny olive, tail tawny red, beneath pale ash colour, lower belly white, irides hazel, eyes large, legs deep ash colour; weight one ounce, length seven inches. They have been heard, though very rarely, in the woods at Weston. "I have (says Mr. Dickenson) no conception why this species is not more common in Staffordshire, for it inhabits much farther north, being found in plenty near Doncaster, Yorkshire." It is found in the south of Staffordshire, near Himley, Swindon, and Enville.

5. MODULARIS—*The Hedge Sparrow*: Head deep brown and ash colour, back and tail blackish, throat and breast bluish ash colour, belly dirty white, legs reddish; weight less than an ounce. A number of white ones were observed some years near Norton-under-Cannock.

6. PHŒNICURUS—*The Redstart*: Bill, cheeks, throat, and legs black, forehead white, crown, neck, and back blue grey, breast, rump, and tail red; weight six drams.

7. RUBICULA—*The Robin Redbreast*: Bill and and legs blackish, forehead, chin, throat, and breast yellow red, above green ash colour, belly white; weight half an ounce. It was formerly held impious to destroy this species, or the common wren, or the martin, or the swallow.

8. OENANTHE—*The Wheat Ear*: A black and a white line across

the face, head and back ash colour, tinged with red, beneath yellowish white, bill, mouth, quills, and half the tail black, other half and rump white; size of a sparrow. Migrates.

9. TROCHILUS—*The Willow Wren:* Weight two drams; very diminutive.

10. TROGLODYTES—*The Wren:* Weight three drams, length 4¼ inches; of a sprightly nature, perpetually in action, and flirting up its tail, and often observed to sing in its flight.

11. RUBETRA—*The Whinchat:* Above red brown, with black spots, beneath reddish yellow, a white stroke above the eyes, and a broad black one under them, tail black and white, bill, mouth, and legs black; weight one ounce.

12. RUBICOLA—*The Stone Chatter; Gorse Hatcher:* Bill, mouth, head, neck, back, tail, legs black, sides of the throat and rump white, beneath reddish yellow, white spots on the wings; size of a linnet.

13. SYLVIA—*The White Throat:* Above brown ash colour, beneath white, breast and belly tinged with red, tail edged with white; length six inches.

PARUS MAJOR—*The Tomtit; Greater Titmouse:* Bill, head, throat black, back green, beneath green yellow, rump bluish, quills tipped with blue and white, tail edged with white, legs lead colour: lays many eggs; nests in holes of trees.

2. CŒRULEUS—*The Blue Titmouse:* Crown, wings, tail blue, forehead, cheeks white, back yellow green, beneath yellow; weight 3¼ drams, length five inches.

3. ATER—*The Colemouse:* Bill, head black, with a white spot, above green grey, beneath white, wing coverts tipped with white, legs bluish; weight two drams.

4. PALUSTRIS—*The Marsh Titmouse:* Head black, cheeks white, above greenish, beneath white, legs lead colour; weight above three drams.

5. CAUDATIS—*The Long-tailed Titmouse; Canbottle:* Bill black, thick and convex, crown white, coverts of the wings black, tail black and white, remarkably long. The nest is very beautiful, and arched over, with the entrance on one side.

ALAUDA ARVENSIS—*The Sky Lark; Ground Lark:* Lower mandible spot above the eyes, soles of the feet, and under side of the body yellow, head and breast spotted with black, exterior web of the quills edged with white; weight 1¼ ounce, length seven inches. Sings flying, or soaring in the air.

2. Pratensis—*The Tit Lark:* Bill, spots on the head, breast, and back black; above greenish brown; breast yellow, throat and belly white, sides of the tail white.

3. Arborea—*The Wood Lark:* A white ring round the head; crown, throat, and back, spotted with black; breast tinged with red; belly white; tail black edged with white. Sings in the night soaring in the air, or in early morn.

Sturnus Vulgaris—*The Stare or Starling:* Bill yellow, depressed, the base of the lower mandible furrowed on each side, a prominent margin round the nostrils, tongue bifid, horny; irides hazel; plumage black, tinged with green purple; tips of the feathers yellowish, legs and feet reddish black.

Columba Œnas—*The Common Pigeon:* Plumage deep, bluish ash colour, breast tinged with purple green, two black bars on the wings, and one near the end of the tail, back white. All our tame pigeons are varieties of this species.

2. Palumbus—*The Ring Dove; Queest:* Above bluish ash colour, breast tinged with purple, belly dirty white; on the back of the neck a semicircular white line; weight 20 ounces, length 18 inches. In winter they congregate in vast flights, and feed on the shoots of turnips; in spring they seperate in pairs, and seek remote places: their nest is very slightly constructed of a few short sticks only.

AMPHIBIOUS ANIMALS.

ORDER I.—REPTILES.

Bufo—*The Toad:* Back broad, black, beset with tubercles, belly swollen, yellow with black spots; crawls: a loathsome creature. Many well-authenticated instances are recorded of toads found in solid stones, and timber trees: they are supposed to have crept in through an aperture, and growing, could not get out again, and so are grown round, in trees by vegetation, and in stones by petrifaction; and in that dormant state require little or no sustenance, having little or no stimulus to exhaust them: they generally die upon the access of fresh air.

2. Temperaria—*The Frog:* Less than the toad, back smooth, margin of the upper jaw white; leaps: torpid in winter: in ponds and

ditches. In March, enlivened by the warmth of spring, they rise to the surface in vast numbers, and soon make a great croaking: the young tadpole is a curious interesting little animal.

LACERTA AGILIS—*The Common Lizard or Swift*: Tail round, scaly, five toes and nails on each foot, belly black, spotted.

2. VULGARIS—*The Common Eft; Brown Lizard; Newt; Asker*: Fore-feet four toes, hind-feet five, without nails, a brown line on each side the head.

3. PALUSTRIS—*The Water Eft or Lizard*: Three or four inches long, above black brown, beneath deep yellow, spotted; feet like the former.

ORDER II.——SERPENTES.

COLUBER BERUS—*The Viper or Adder*: Numerous on Cannock-heath and Sutton Coldfield, and often seen on Morton-moors; the bite said to be very dangerous, yet swine and peacocks devour them greedily.

2. NATRIX—*The Snake*: Swim very swiftly, prey on lizards under water, very prolific if undisturbed; 70 have been destroyed in one dunghill.

ANGUIS FRAGILIS—*The Blind Worm*: Common heaths and the Moorland hills, perfectly harmless, though reckoned poisonous by the peasantry.

ORDER III.——NANTES.

PETROMYZON MARINUS—*The Lamprey*: In the Severn and Trent rivers.

2. FLUVIATILIS—*The Lampern*: In the Severn, about one foot long.

3. BRANCHIALIS—*The Nine Eyes*: In rivulets, length five or six inches.

FISHES.

ORDER I.——APODES.

Muræna Anguilla (common eel). Plot says they are viviparous, and make journies by land, stocking waters themselves, being night-walkers, and catch snails; and adds, they will live six days out of water.

ORDER II.——JUGULARES.

Gadus Lota, (burbot, eel pout) in the Trent; a great delicacy, but of hideous appearance; head like a toad, body like an eel.

ORDER III.——THORACICI.

Cottus Gobio, (bullhead) in rivers and rivulets.

Perca Fluviatilis, (perch) in ponds, pools, lakes and rivers.

2. *Cernua*, (ruff) length five inches; in the Staffordshire canal, and in rivulets.

Gastrosteus Aculeata, (stickleback, sharpling) in rivers, brooks, and rills.

ORDER IV.——ABDOMINALES.

Cobitis Barbatula, (groundling or loach) length three inches; brooks and rivers.

2. *Tænia*, (bearded loach) in the Trent.

Salmo Talar, (salmon) in the Severn, and sometimes in the Trent and Dove.

2. *Fario*, (trout) highly esteemed; in rivulets, brooks, and rivers.

3. *Thymallus*, (grayling) in the Churnet, Trent, &c.; have been taken four pounds weight.

Esox Lucius, (pike) grow to great size, will swallow young ducks; voracious: taken in Aqualate-mere 36 pounds weight.

Cyprinus Carpio, (carp) ponds and pools in Aqualate-mere; 33 inches long, weight 15 pounds.

2. *Barbus*, (barble) in the Trent; have been taken 11½ pounds weight.

3. *Tinca*, (tench) ponds, pools, and rivers.

4. *Gobio*, (gudgeon) in Black-brook; seven inches long and four about.

5. *Brama*, (bream) in large rivers; frequent: length 12 inches.

6. *Rutilus*, (roach) in brooks and rivers; length six or seven inches.

7. *Leuciscus*, (dace) length six or seven inches; in the Trent.

8. *Cephalus*, (chub) in the Trent; have been known five pounds weight.

9. *Alburnus*, (bleak) in the Trent; length five inches.

10. *Phoxinus*, (pink or minnow) in the Peak; length two inches.

INSECTS.

THE FOLLOWING ARE A FEW OF THE MOST REMARKABLE:

Scarabæus Melolontha, (free beetle humbuz) sometimes extremely numerous.

2. *Solstitialis,* (fern chaffer) supposed to eat off the roots of wheat and clover.

Coccinella, (lady-cow) seeks its food on the leaves of trees.

Crysomela Nemorum, (turnip beetle) skippers; prey on the young turnip plants.

Curculio Granarius, (weevil) long-snouted; devours corn in granaries.

Lampyris Noctiluca, (glow-worm) the light arises from two small specks under the tail.

Gryllus Domestica, (house cricket) sometimes very numerous, within and without doors.

2. *Grossus,* (common grasshopper) the larger sort greenish.

Cancer Astacus, (crayfish) in brooks and rivers.

Aphis Brassicæ, (cabbage louse) very minute, and very numerous on plants.

Papileo Brassica, (cabbage butterfly) caterpillar; breed from May to July.

Libellula Grandis, (great dragon fly): *Libellula Vulgatissima,* (small dragon fly) sport and fly over waters, and about in fine summer weather.

Ephemera, (short-lived insect) produced and exhausted in a day or two.

Tenthredo Rustica, (turnip fly) the caterpillar of this fly sometimes preys on turnips.

Vespa Crabro, (hornet) length one inch; sometimes gets into upper rooms.

2. *Vulgaris,* (common wasp) female larger than the male; sting well known.

Apis Mellifica, (common hive bee) well known for its excellent honey.

Formica Fusca, (common ant) proverbial for industry.

Oestrus Bovis, (breeze or gad-fly) breeds on the backs of cattle.

Tipula Oleracea, (long legs) one of our domestic insects.

Musca Carnaria, (common flesh fly): *Musca Domestica*, (domestic house fly) two very common well-known species.

Tabanus Bovinus, (horse fly) a plague to horses and men.

Culex Pipiens, (common gnat) often dancing in the air.

Apis Terrestris, (humble bee) builds deep in the ground.

Bumbylus Medius, (humble bee fly) hovers in the air, and darts with celerity.

Phalangium Opilio (long-legged spider) eyes on the top of the head.

Limax Ater, (black snail) in moist shady places.

MISCELLANEOUS PARTICULARS

CONNECTED WITH THE

NATURAL HISTORY OF STAFFORDSHIRE.

It has been observed in Leek church-yard of the setting of the sun at the summer solstice, near a hill called the Cloud, about six miles distant, (which hill being a precipice, and appearing almost perpendicular on the northern side,) that the sun seems so nicely to cut the edge of it at setting, it can plainly be perceived by the help of this hill that no two days are equal, but that there is a sensible variation of the solar motion. If the parts were well defined by solid work and painting, and one or more fixed stations established in Leek church-yard, this might be converted into a quadrant of six miles radius, and applied to the purpose of very accurately determining the obliquity of the ecliptic, and its variation.

Dr. Plot mentions a *Black Moist Rotten Earth*, that lies just under the turf in heathy soils, where he was told seriously by different gentlemen, near Beresford and Rudyerd in the Moorlands, and also at Bescote, near Walsall, "that if one ride in a dark night, in a moist season, when a horse breaks through the turf, and throws up this black, moist, spongy sort of earth, he seems to fling up so much fire, which lies shining upon the ground like so many embers; by the light whereof one horse may trace another, though at some distance, and never so dark, it continuing light upon the ground, and being gradually dying away for some minutes, or near a quarter of an hour." We have heard of similar phenomena occurring at Chartley-moss, or some moist ground thereabouts; and without personal experience can conceive it possible, upon the principle of some other bodies or substances shining in the dark, as rotten wood, glow-worms, &c.

Tobacco-pipe Clay was formerly got in different parts of the county, "as at Stile-cop, on Cannock-heath, whereof they made pipes at Armitage and Lichfield; and again at Darlaston, Wednesbury, and in Monway-field, towards Willingsworth, of a whitish colour, which makes excellent pipes; and again in Pennet-chace, south of Dudley. And Charles Riggs, of Newcastle, made very good pipes of three sorts of clay, which he had from Shelton, Hanley-green, and Grubber's Ash." We believe these have all given way to the superior whiteness of the Broseley clay, which now enjoys a monopoly of the pipe-trade of this county.

The *Amblecot Clay*, on the banks of the Stour, near Stourbridge, famous in the time of Plot, still maintains its superiority as a fire-clay, and will bear the most violent heat unmoved: it is used in the neighbouring glass-houses, and exported to all parts of the world.

Potter's Clay, for the more common wares, is found in Monway-field, and near Tipton, Wednesbury, and Bilston, where some small manufactures of this kind are now carried on. But the largest Potteries in this county, or perhaps in the world, are established at Burslem, Hanley, Stoke, Lane End, &c. which places occupy an extent of about ten miles, and contain upwards of 30,000 inhabitants, chiefly employed in this extensive manufacture.

Plot says, "of *Earths used in Colouring and Painting*, we may reckon the yellow and red ochres met with near Stansope, in the parish of Alston-field. Ruddle, or red ochre, they dig very good in the parish of Ipstones; and I was told there is also of it at Dilhorn." The ochres are, doubtless, an imperfect or decomposed ore of iron.

The *Peacock Coal*, dug at Hanley-green, near Newcastle-under-Lyme, is much softer than the cannel; it most vividly represents all the colours of the most glorious feathers of a peacock's train, whence it borrows its name: *Plot.*

"At Waterfall, in the Moorlands, the *Lime-stone* rises in tables or flats of about two inches thick, and of so very fine a grain and close texture, that it polishes like marble, whereof in the arbours up and down the country I find tables made of it."

Dr. Plot also relates that a *Sweet Liquor* is sometimes contained in the iron-stones of Staffordshire, particularly in those found in the "lands of Henry Legh, Esq. of Rushall; where, in the Mill-meadow, in the Moss-close, near the old Vicarage-house, and in the Furnace-leasow, it was frequently met with amongst the best sort of iron-stone called Mush, in round or oval, blackish or reddish stones, sometimes as big as the crown of one's hat, hollow and like a honey-comb within, and holding a pint of this matter, which, according to the colour of the comb within, is either red or white, of a sweet sharp taste, very cold and cutting, but greedily drank by the workmen." This may be matter of fact, but nothing is known of it at present, and no iron-stone is now got there, though there are evident traces of it formerly.

Building-stone is found plentifully in various parts of the county. One of the most remarkable quarries was formerly at Pendeford, in the lands of Mr. Fowler, where it lies in depth 14 to 20 yards; indeed the workmen found no bottom, and the deeper they went the finer the grit, the colour whiteish at the top, then reddish to 10 or 12 yards, and then whiter than at top for five or six yards more; it being all good weather stone, but not enduring the fire, as the stone in the quarry in Purton-park will, which is both a good weather and fire stone too, and so free from clefts that a stone may be had of any magnitude. Specimens of this stone may be seen in the new Church, Wolverhampton, which was built from this quarry.

The quarry at Bilston is also remarkable, it laying in beds in horizontal planes one under another, twelve beds deep, every bed being thicker than that above it an inch or more; so that the lower bed is about a yard thick, of which they make troughs, cisterns, &c. The grind-stones dug here are of a peculiarly fine grit, and frequent exportations of them to America have been made. The stone is also excellent for building-stone, and fit for cornices and mouldings. Grind-stones of a coarser quality are got at Gornall and Upper Areley, in this county.

A very large facing of stone is now open in a quarry at Coton, on the north bank of the Trent, facing Rugeley; and very good stone is also raised upon the Brockton estate of Sir George Chetwynd. The Tixall stone is also famous for canal buildings, well resisting the action of water; very large quantities of it are used in canal locks, raised in a quarry upon the Tixall estate, about half a mile west of the Staffordshire and Worcestershire canal.

Mill-stones have been made out of great round pebbles, found in different parts of the county, and used for grinding wheat. At Mole-Cop they dig them as in a quarry, which they cleave from the rock with a great number of small wedges, driven with slight strokes lest the stone should crack or flaw. This stone lies well in building, the parish church of Biddulph being built with it.

Between Swithamley and Wharnford, in the north-west Moorlands, is a stupendous cleft in the rock, called Lud-Church, 208 yards long and from 30 to 60 feet deep, the sides steep and so hanging over, that it sometimes preserves snow all the summer. Snow has been brought from thence to Leek July fair.

Alabaster is contained in the whole bank of red marl between Needwood-forest and the river Dove, from Marchington to Tutbury; but that at Castle-Hayes is the best, of which they make grave-stones, tables, chimney-pieces, &c. and many smaller utensils.

In the lordship of Grindon are whole mountains of *Marble*. Yelpersley Torr, and most of the hills thereabouts, are (says Dr. Plot) a rance marble, of a white hard shining grit, striped red, and capable of a good polish; and at Stansope, in the parish of Alston-field, is a grey marble not inferior in its kind, and in quantities sufficient for any purpose.

Timber-trees of great bulk have been produced in this county: the following are upon record. On Needwood-forest, *Swilcar Oak*, the father of the forest, girths at five feet high 21 feet, the lower stem is 10 feet clear; the whole height about 65 feet, the extent of the arms about 45 feet. The great oak at Norbury was six yards in the girth; and one standing on Canneck-wood nine yards ⅔ girth. An oak was felled in the Grefield-coppice, Chillington, about 20 years ago, of nine yards in circumference. The trunk of an oak in Ellenhall-park was of so vast a bulk, that though sunk in the ground, two men on horseback could not see each other over it. A prodigious *Witch-elm* grew at Field, upon the Bagot estate, of which it is attested that it fell 120 feet in length, was 17 yards in circumference at the butt-end, and 25 foot ½ girth in the middle, containing 96 tons of timber; that 14 loads of fire-wood, as much as six oxen could draw, broke off in the fall, and 47 such loads more were cut from the top; that they were forced to piece two saws together, and put three men to each end, to cross-cut the body; that there were cut out of it 80 pair of naiths for wheels, and 8000 feet of sawn timber in boards and planks.—*Plot.*

A *Fir-tree* grew at Walton, in the parish of Norbury, six yards about, and 47 yards high, according to admeasurements of it by three distinct persons at different times.

Trees are often found beneath the earth in a fossil state, and Dr. Plot has made long and elaborate dissertations to account for their being there; but nothing can be more simple and natural than to suppose that they may have fallen down by natural decay, or in consequence of high winds, or have been cut down to clear the ground, and not taken away, and that the natural accumulation of earth over them, from decayed vegetable growth, or from inundations, or other causes, has left them beneath the surface in the state they are found.

From the instances given of *Fruit-trees* growing formerly in this county, it should appear that more attention has heretofore been paid to their culture than has unfortunately been the case in later times. An apple-tree grew in the moat of the parsonage-house at Leigh that spread its boughs to 17 or 18 yards diameter, and 54 circumference, and having borne some years 50 strike of apples. All sorts of fruit-trees were then cultivated (about 1680) much more than they anciently were, and they have been too much neglected since. Many good orchards have been suffered to decay for want of a succession stock of young trees, and the crab-mills of many villages where they occasionally made perry, have been disused for want of fruit. The parish of Areley was planted in the Worcestershire manner, having all sorts of pippins of the best kinds. These were afterwards neglected, but were again partially reinstated. Dr. Plot says, for fruits of the parterre or house-gardens, they were as choice at Ingestre as in most gardens of the south; and in the gardens at Okeover were then growing 60 different sorts of apples, 30 sorts of pears, 16 of cherries, 35 of apricots and plums, and seven sorts of nectarines and peaches.

Some instances of remarkable *Echos* are given in the particulars of parishes. Dr. Plot mentions one at Norbury, north-easterly from the manor, near a little bank under the wood side, at about 440 yards distant, where in a still day the echo will repeat ten to twelve syllables distinctly, or more if spoken quick, the object whereof must needs be the manor. A similar remarkable echo is from the front of Pendeford Hall, at about 200 yards distant: this was weakened some years ago by cutting down the side groves, which gave room for the air to expand. The theory of echos seems to depend upon the vibration of the air being stopped or confined by surrounding objects.

Among remarkable *Pools*, Dr. Plot mentions one at Pendeford, which, though a standing lake, is seldom dry, and tolerably clear in settled fair weather, only against rain it becomes troubled, rising full of bubbles, and thickening at the top into a yellow scum, which presently as it rains vanishes away, and the water recovers its former colour and clearness. Mr. Fowler, the owner, said that he had often been admonished by it in time of harvest to fetch-in his corn. Similar circumstances, we believe, occur upon most standing waters in hot seasons previous to a change of weather, the variation of heat putting the insects, and other creatures preying upon them, in motion: thus, upon the approach of rain in hot weather, ducks and other water-fowl clap their wings, and as it were rejoice, and are busy with their heads under water, which is a symptom of a change of weather.

"As some pooles and ponds (says Plot) thus foretell rain, so there are other ponds and springs that prognosticate dearth, by rising or overflowing; such is the Moss-poole near Mearton, in the parish of Forton, not far from Newport; and Druid-meer, in the parish of Aldridge, the rising or exudation of which being taken for a certain sign of a dearth of corn. But what excels all others in this feat is Hungry-pit, situate in a field below the old fortification not far from Billington, but in the parish of Seighford, which predicts scarcity by its rise upon sticks set upright in the mud, which people place there for observation, fore-judging the rise or fall of corn in the mercats by the rise or fall of the water on these sticks, and keeping or vending it accordingly; yet it little regards the quality of the weather, overflowing sometimes in droughts, and having nothing in it after falls of rain; yet these superstitious notions have been profitable to some, who, whether by casualty or not, had thereby advanced their fortunes."

These circumstances are less noticed now than in former times, though probably founded on just principles, for dearths of corn in England are most frequently or always the consequence of wet seasons; and though this spring might not be much affected by a sudden shower, or partial drought, yet it probably follows very accurately the general tenor of the season. The cause of springs is doubtless the atmospherical moisture condensing on the summits of hills, which water, thus condensed, descends between the strata of the incumbent soil, till it settles in the earth, or bursts forth on the surface: the whole process is a work of time, and it is only the exuberance of the season that appears in these springs. The Nileometer of Egypt is well known, as foretelling plenty or scarcity by the rise of the Nile; but their greatest plenty is when water is most abundant, by spreading the farther to fertilize the naturally barren sand.

Dr. Plot mentions another kind of springs he met with in this county, "that have this surprising quality, that though they are cold, they never freeze in the hardest weather; such is the spring that feeds the mill-pool at Overholm, near Leek, and the spring which supplies the mill at Tittensor; whence it is that the mills of both these places never fail going in the severest frost. There are springs also about Hints that do not freeze; also Clayer's-pool, in the corner of the park at Enville, which is fed by a spring (coming out of a spout) called Shadwell, under which if you set a vessel of water frozen, it will certainly thaw it." The Brown-hills engine coal-pit water, in a very severe frost, kept the Wyrley canal fluid for a considerable distance from its place of supply, though the canal was hard frozen over in general. This is to be accounted for, because the supply of spring water comes from a degree of heat within the earth higher than the freezing point, which consequently keeps fluid till cooled by the external air to a degree below 32. We remember some years ago observing the effects of water, from a spring of sensible warmth near Mothershall, early in the season, in fertilizing the grass, and meadow-bouts in blossom, a fortnight or three weeks before other places: water within the earth is often warmer than the external air, generally so in winter, though not in summer.

Of the *Salt Springs* in divers parts of the county, (says Plot) the most considerable are in the parish of Weston-upon-Trent, belonging to Earl Ferrers, of Chartley, where they make as good white salt for all uses as any in England, though not to so great advantage as in Worcestershire and Cheshire, where the brine is so strong as to yield from a fourth to a sixth part salt; whereas here in Staffordshire it affords but a ninth: 16 hogsheads of brine made nine strike of salt after 16 hours of evaporation.

Besides these salt springs, there are other weak brines that gently rise out of the earth about Enson, St. Thomas, and in the parish of Ingestre, in a ground called the Marsh, where the brine of itself breaks out of the ground, and frets away the grass, and the very earth also, so that it lies in a plash half a foot lower than the turf about it: the cattle standing in it in the summer time, and throwing it on their backs with their tails, the sun so candies it upon them, that they appear as if covered with a hoar frost. Brine lately appeared in the meadows of Rickerscote, near Stafford, upon digging into the earth, but was soon diluted by the river water; and a water has been since found there strongly impregnated with salt and sulphur.

In Pensnet-chase, south from Dudley, there was a brine found upon the estate of Lord Dudley, where they once attempted to make salt, but the brine proving too weak, they thought proper to desist.

The following information respecting the MEDICINAL WATERS in this county, is extracted from Plot (ch. ii. secs. 115—129) without strictly regarding, in some places, either his precise words or arrangement, though such are for the most part observed.

Whatever may be the opinion of some, most certain it is, that divers strange unaccountable cures have been performed at various Wells in Staffordshire: and more there might be were the waters attended (as some others are) with a skilful physician to prepare the body before-hand, direct the use of the waters, and how to order the body after drinking and bathing. Lord Bacon well observes, that some medicines, if unadvisedly administered, will do no cures, but being orderly applied, will do great ones; even so the success of using *medicinal waters* depends much upon method. Various waters may be taken inwardly, or applied outwardly, for the *prevention* or *cure* of divers distempers: both drinking and bathing in such waters have been attended with beneficial effects. And no one need to wonder that there should be any such thing as a cold sulphur bath, or one in waters which hold an oily substance and salt.

The *Well of St. Erasmus*, situate in the parish of Ingestre, and grounds of Walter Chetwynd, Esq. was formerly of so great esteem, that there was a *chapel* built near it, and endowed; where the offerings were so considerable, that the rector of Ingestre pays at this day (1686) for the dismes (tenths) of them more than for the profits of the whole parish beside; notwithstanding its revenue, with all other such lands given to pious uses, were taken into the king's hands at the Reformation, and the chapel be now demolished, and all applications thither long since ceased. Although the well is wholly neglected and overgrown with weeds, the water is clear, remains as good, and might be as beneficial as ever it was. It has no very eminent smell or taste, and is exactly of the colour of sack.

Codsall Well, situate near Codsall-wood, is such another. Though its water is much clearer than at Ingestre, yet it most certainly is sated with sulphureous particles; for it always emits a sulphureous smell: and in winter, and sometimes against rain, the odour is so strong, that, with the advantage of the wind, one may smell it now and then at least 23 yards off. Moreover, so volatile is it, and so little restrained, that when set over the fire, it flies away so fast, that the water quickly loses its smell.

In ancient times, when leprosies were frequent, this water was accounted a sovereign remedy for such as were troubled with that foul distemper; and for whose better accommodation there was a house built near it, which retains to this day the name of the *Leper House.* This water is in use at present against scabs and itch, both in man and beast, and purges both by siege and urine. It not only rakes the body within, but most effectually drives forth all ill humours, and sometimes it vomits, according to the constitutions of the patients, who commonly drink about three quarts at a time. Less, scarce works except by vomit, where it meets with weak stomachs.

The inhabitants hereabout brew their drink with this water, especially at that which they call the *Brimstone Alehouse;* and boil their meat with it. Upon which it is observed, that none of them are ever troubled either with scabs or itch, or such like cuticular diseases.

Willowbridge Well, situate in Willowbridge Park, near to Ashley, is a water that carries with it the most rectified sulphur, it being hardly visible in the water itself, which appears, not like others, of a yellow, but clear chrystalline colour; only on the sides of the glasses, after they have been used awhile, one may perceive with good attention a bright oilyness. So abundant was this water in Plot's time, that he computed no less than sixty springs of it, of a most uninterrupted profluence, all rising within the space of ten yards square.

It cures many *diseases* by its balsamic virtue and great subtilty and volatility, easily permeating the closest texture, and most inaccessible parts of the body, when once heated by the stomach if taken inwardly; or by the external heat of the skin, if applied outwardly by way of bath.

The Right Honourable Jane Lady Gerard, Baroness Gerard of Gerard's Bromley, was the first who discovered the restorative virtues of these waters; and at her charitable expence several of the springs have been inclosed with squared stone, to preserve them pure and fit for bathing and drinking; and divers apartments built for lodging the poorer sort of diseased impotent people. By which deeds she was justly entitled to the praise of a pious and charitable lady.

This lady's son, Digby Lord Gerard, had a stately seat at Bromley, within two miles and a half of these wells, which was considered the most magnificent structure in the county.

The above Wells, whose properties are thus enumerated by Plot, he terms sulphureous oleagenous waters. He then proceeds to give an account of other *Sulphur Waters*, which, he says, are at many places in this county, though not so clear and well-concocted as these: being generally thick, of a bluish colour, and emitting unpleasant fetid odours. Such as that at

TATENHIL, on the highway-side near the pound, which in the summer-time, if undisturbed by fowl, has been observed to lay down a sediment almost of all sorts of colours, and might possibly have the same use, that other stinking spaws have, were it but kept clean. And so might that stinking water, which crosses

WATLING-STREET way, not far from Horsebrook, and another of this kind betwixt

WILLENHALL and BENTLY, could it be kept from a mixture with other water. There is another of these in a watery lane not far from

ECCLESHALL, and another was said to be near

HARTLEY GREEN, besides that, which also petrifyes, between

SANDON and GAYTON. And there is another at

BUTTERTON, in the parish of Mathfield by How-brook side, which, like the baths of Banca in Hungary, will tinge silver of a blackish colour in an hour's time.

There are other *Sulphur waters combined with Vitriol* in this county, whereof there is one in a ditch in the Park meddow under

BROUGHTON Park pale, and another at

MONMOORE green near WOLVERHAMPTON. Another at

GRINDLESTONE edge, about a quarter of a mile eastward of

HORTON church.

One *purely Vitriolic water* is in Needwood Forest, about a mile and a half south easterly from HANBURY. Another is said to be at

BURSLEM, near Newcastle-under-Lyme: but respecting the quality of this last Plot entertains a doubt.

Among what he terms *milky waters*, that hold somewhat of vitriol in them, he notices one at HAMPSTED, the seat of Sir John Wyrley, knt.

Aluminous waters are found at

DRAYCOT in the Clay, in the parish of Hanbury, and near

HORECROSS, westerly from the hall by the brook side in a ground called Broadfield, belonging to Robert Howard Esq.

The water of the well at Hore-cross Hall seems of this kind, and the well water of the house of Mr. John Cumberlege, mayor of Walsall, is also aluminous; and so is the fountain head of the rivulet called Stichbrook, between Lichfield and Elmhurst, which is so apparently such, that the very *alum slat* may be seen sticking in the bank side, whence the water issues.

Besides these, there are many other *waters not mineral*, which no doubt carry with them some more subtile steams whereby they perform unaccountable cures. Such is the water of the well near Gawton Stone, in Knypersley Park, which has some reputation for the cure of the *king's evil;* and so has the spring called *Saltar's Well*, near Newcastle-under-Lyme; and such is the water of the three wells near *Shuston House*, and of *St. Edith's* well, both in the parish of Church Eyton; of the two *St. Modwen's* wells, at Burton and Cannal, and all the *holy wells* in the country, which the people still adorn, at some certain times of the year, with green boughs and flowers, in grateful memory of the good they have formerly done. And among these must be reckoned all sorts of *eye-waters*, such as that of the Elder well betwixt Blymhill and Brineton, and many others of the kind all over the country. And so must a spring in a narrow lane about mid way betwixt Wolverhampton and a house called Sea well, which was anciently of such repute, that it still retains the name of the *Spaw*.

The above were all the wells of note in this county in the time of Plot; and the names of the owners above-mentioned, were proprietors in his day. The following observations appear in another part of his work, (ch. viii. sec. 89) and may aptly conclude this subject.

" They have also *a custom* in this county, which I observed on holy Thursday at Breewood and Bilbrook, *of adorning their wells* with boughs and flowers: this it seems they do at all *gospell-places*, whether wells, trees, or hills; which being now observed only for decency and custom sake, is innocent enough. Heretofore too it was usual to pay their respect to such wells as were eminent for cureing distempers, on the saint's day whose name the well bore, diverting themselves with cakes and ale, and a little musick and danceing; which, whilst within these bounds, was also an innocent recreation.

" But whenever they began to place sanctity in them, to bring alms and offerings, or make vows at them, as the ancient Germans and Britons did, and the Saxons and English were too much *inclined* to, for which *St. Edmund's Well*, without St. Clements, near *Oxford*, and *St. Laurence's* at *Peterborough*, were famous heretofore; I doe not find but they were forbid in those times, as well as now, this superstitious devotion being call'd (in English) *well-worship*, and was strictly prohibited by our Anglican councills, as long ago as king Edgar; and in the reign of Canutus; not long after again in a councill at London, under St. Anselm, archbishop of Cant. ann. 1102; as it was also particularly at those two wells, near Oxford, and at Peterborough, by Oliver Sutton, bishop of Lincoln."

Dr. Darwin relates that a complete cylinder of bark, an inch in length, was cut from the branch of a pear tree against a wall in Mr. Howard's garden at Lichfield, about five years after the circumcised part was not above half the diameter of the branch above and below it, yet the branch had been full of fruit every year since, when the other branches bore only sparingly; the leaves of the wounded branch were smaller and paler, and the fruit less, but ripe sooner than on the rest of the tree: another branch had the bark taken off not quite all round with similar effect.

The following instance of longevity was transmitted by a gentleman who was a neighbour to the individual referred to, who died a few years ago at the age of 116 years. The industry of his early years acquired a competence with which he purchased a small farm called Lapley Hayes, near Wednesfield. His name was William Hyven: he lived on this farm a great many years, and at the age of 105 married a third wife, and in defence of his marriage used the words of St. Paul: "It is better to marry than to burn." He was rather under the middle size, but robust and strong, of a lively disposition, and retentive memory, which he retained to the last; he took for breakfast a gruel of oatmeal and leeks, the latter in plenty, as he considered them as conducing to health; he was an early riser, but observed no rigid discipline of abstinence, not rejecting that stimulus by which mirth and vivacity are augmented, nor indeed at times refusing larger potations to dispel the cares attendant on mortality. He died at Little Bloxwich at the age above mentioned, and was buried in Wednesfield chapel-yard.

BIOGRAPHICAL SKETCHES

OF

EMINENT NATIVES OF STAFFORDSHIRE.

LORD ANSON.

PERHAPS no other country in the world has produced so many eminent and memorable men as England; and among the great names which adorn the history of this celebrated spot, several natives of STAFFORDSHIRE will be found, whose genius or heroism were conducive to national greatness. This midland county is remarkable for having been the birth-place of several naval heroes, particularly an ANSON, a GARDNER, and a ST. VINCENT; names which will be frequently repeated by every admirer of British enterprise, skill, and intrepidity.

GEORGE ANSON, the fourth son of William Anson, Esq. and Elizabeth, eldest daughter of Robert Carrier, Esq. was born at Shugborough manor-house, in the parish of Colwich, in Staffordshire, on the 23d of April 1697. The bent of Mr. Anson's genius was directed rather to an active than a studious profession, and having been instructed in the rudiments of a classical education, he was entered as a midshipman on board a man of war, which sailed with a squadron to the Baltic in the spring of 1716. On the 16th of May in that year, Mr. Anson was promoted to the rank of Second Lieutenant of the Hampshire; and thus at the early age of nineteen, the hero who was afterwards destined to exalt the naval character of his country, had the honour to receive a commission as an officer in one of his Majesty's ships of the line. That his ardour for distinction was stimulated by this early preferment will not be doubted by the reader of his eventful history, yet no particular instance of his superiority of juvenile talent is on record.

In 1717, a British fleet was sent to cruize in the Baltic, under the command of the unfortunate Admiral Sir George Byng, and in this fleet our hero served in the Hampshire.

On the 15th of March 1718, Lieutenant Anson was appointed Second Lieutenant of the Montague; that ship was one of the squadron which sailed under Admiral Byng to protect the island of Sicily from the intended invasion of the Spaniards; and in the course of the summer, the British brought the Spanish fleet to action, which terminated in the total destruction of the enemy, and the consequent relinquishment of the King of Spain's ambitious projects against the independence of the Sicilians. Lieutenant Anson was highly distinguished for his skill and bravery on this occasion. On the 19th of June 1722, he was appointed commander of the Weazle sloop of war; and on the 1st of February 1724, he was raised to the rank of Post Captain in the Royal Navy, and to the command of the Scarborough man of war.

Captain Anson was immediately ordered in the Scarborough to the South Carolina station, where he continued about three years and a half. During his residence on the shores of that colony, Captain Anson was from time to time actively employed in the improvement of a tract of land which he had purchased, in which he built a small town which he called Anson-Brough, and gave name to a division of the province, still called Anson county.

In consequence of orders from the Admiralty, Captain Anson returned to England in October 1727, and the Scarborough was paid off in May 1728. On the 11th of October 1728, our hero was appointed Captain of the Garland man of war, and sent again to the coast of South Carolina, to watch the motions of the French cruisers. He was ordered home in December 1729, and the ship was paid off at Sheerness. In May 1731, he was appointed commander of the Diamond, but that ship was paid off in August, and in January 1732, he was again sent to South Carolina, as commander of the Squirrel man of war. Captain Anson continued on this station three years, and had an opportunity of pursuing his favourite plan for the improvement of Anson-Brough, doubtless with the honourable ambition to perpetuate a name which will ever adorn the annals of his country. It does not appear, however, that his wishes and expectations respecting this town were ever realized; perhaps his pecuniary resources were insufficient to procure the requisite supplies of furniture, utensils, &c. from the mother country for the accommodation of settlers; and from the uncertainty of his own profession as a naval officer, he was unable to bestow that uninterrupted attention to this infant establishment, which would have insured its permanence. In his public employment he behaved with that firmness and discretion which were among his most conspicuous characteristics: when ordered home in the spring of 1735, he left South Carolina for the last time, and on his return to England his ship was put out of commission.

Our hero now enjoyed the calm pleasures of retirement at his birth-place for some time, but like all active and aspiring men, he was still emulous of distinction. His former services were not forgotten, and on the 9th of December 1737, he was appointed Captain of the Centurion, a ship in which he afterwards added new lustre to the naval glory of his country, and signalized himself by his fortitude and achievements during his celebrated voyage round the Globe.

In February 1738, Captain Anson was ordered in his ship to the coast of Guinea, to prevent the French cruisers from annoying our merchantmen in their trade to the Gold Coast; and during a cruise of eighteen months, he effectually protected the British traders without the necessity of coming to actual hostilities with our ambitious rival. His vigilance and prudence were highly approved by Government.

But the era now approached which was to eternize Captain Anson's fame as an enterprising and skilful navigator. At the commencement of the war with Spain in 1739, it was determined by the British Government to attack the Spanish American Settlements in the Pacific Ocean, and our hero was appointed Commodore of the fleet designed for that purpose.

The expedition, however, was delayed till September 1740. Commodore Anson in the Centurion doubled Cape Horn in March 1741, and arrived at the island of Juan Fernandez in June, with only two ships of his squadron, and 335 men. During this disastrous voyage, the seamen on board the Centurion were afflicted with the scurvy, which proved fatal to many; on their arrival at Juan Fernandez, Commodore Anson by exemplary humanity and attention to the sick excited the emulation of his officers, who actually lent their aid in carrying the infirm sailors ashore in their hammocks to the temporary infirmary erected for their comfort and accommodation. During the three months that he remained in Juan Fernandez, he sowed the seeds of several culinary herbs and roots, and a variety of plum, apricot, and peach stones, for the purpose of producing salutary herbs, roots, and fruits, for the use of such seamen as might touch there. He also employed his officers in examining the creeks and havens along the coast of the island, and in comparing their observations with their charts, to facilitate the progress of future voyagers.

Commodore Anson sailed from Juan Fernandez in September 1741, and in his voyage towards the Spanish settlements in Quito he took some prizes. He afterwards took the town of Paita, which was but feebly defended by the Spanish garrison, and burnt it, because the governor refused to pay a sufficient ransom. He continued cruising off the American coast till May 1742, and then proceeded on his voyage to China. All his squadron were now dispersed or lost except the Centurion. During his voyage across the Southern Ocean, he found it expedient to touch at the island of Tinian for wood, water, and fresh provisions. Here he found numbers of cattle and fowls, but no inhabitants. There was no harbour along the shore of this island sufficient to afford safe anchorage for the Centurion, and that ship was driven out to sea, leaving Commodore Anson, several of his officers, and part of the crew, on shore. The fortitude of our hero was now put to the test; hopeless of the return of his ship, which was now considered as lost, he animated the exertions of his men by his example, and assisted in the construction of a vessel. The return of the Centurion revived his hopes of ultimate success, and he continued his voyage to China, where he arrived in Nov. 1742. The Chinese at Macao are sufficiently vigilant on the approach of armed Europeans, and it required the utmost prudence and address in Commodore Anson to procure the necessary aid from them. When this was accomplished, and a fresh stock of provisions got on board, he left China in the spring of 1743, and cruised off the Philippine islands, in expectation of meeting with the Acapulco galleon. This Spanish ship, laden with treasure, came in sight on the 20th of June 1743, and was captured by the Centurion, after a smart action. Some ladies on board the prize, who were passengers for Spain, were treated with the greatest propriety by the victor, whose moderation on this occasion was only exceeded by his equanimity when in the hottest part of the engagement; the Centurion was on fire near the powder-room, and the instantaneous destruction of the ship and her crew was expected.

On his arrival at Canton with his prize, Commodore Anson firmly maintained the honour of the British flag; his remonstrance was heard with respect, and the requisite supplies of provisions, &c. granted. He removed the treasure taken in the Spanish ship on board the Centurion, sold his prize to the Chinese, and sailed for England in December 1743. After a voyage of about six months he entered the English Channel, where he narrowly escaped being captured by a French fleet, from which he was fortunately concealed by a fog. He arrived at Spithead on the 15th of June 1744, and was received on shore with acclamations.

His Sovereign, as a reward for his services, promoted him to the rank of Rear-admiral of the Blue; and he was soon afterwards chosen Member of Parliament for the borough of Heydon, in Yorkshire.

Thus Admiral Anson, after having been exposed to great perils and hardships, was publicly rewarded for his patience and successful perseverance in the performance of his duty as a naval commander. A series of auspicious circumstances now succeeded his temporary adversity, and in his life the assertion of the poet was completely illustrated:

> There is a tide in the affairs of men,
> Which taken at the flood leads onto fortune;
> Omitted, all the voyage of their lives,
> Is bound in shallows and in miseries!

He was now popular, and his preferment kept pace with his popularity. On the 27th of December 1744, he was appointed one of the Commissioners of the Admiralty; and on the 23d of April 1745, promoted to the rank of Rear-admiral of the White.

In the winter of 1746-7, Admiral Anson commanded the British fleet which cruised in the Channel, to intercept the French fleet, commanded by the Duke D'Arville; but the object of the cruise was frustrated by intelligence communicated by the captain of a Dutch ship to the French admiral, who was thus apprized of his danger. Our hero suffered the hardships of a long and tempestuous winter navigation with his characteristic fortitude, and his vigilance was crowned with success on the 3d of May 1747, when with his fleet, consisting of fourteen sail of the line, he intercepted, off Cape Finisterre, a French fleet bound to the East and West Indies, and laden with treasures, merchandise, and warlike stores. On this occasion, all the French men of war, consisting of six sail of the line, were taken, besides four East Indiamen, with valuable cargoes. By this fortunate achievement,

Admiral Anson defeated the purpose of two hostile expeditions, fitted-out at a great expence to annoy our settlements in the East and our West India islands, and for this important public service he was promoted to the rank of Vice-admiral of the Blue, and raised to the dignity of the peerage, by the title of Lord Anson, Baron of Soberton, in the county of Hants. The motto chosen by his Lordship, on this elevation, was *Nil Desperandum*, which was a very appesite allusion to his former perils and eventual success.

On the 25th of April 1748, Lord Anson married Elizabeth, eldest daughter of Lord Hardwicke, then Lord Chancellor. This lady died without issue on the 1st of June 1760. On the 12th of July 1749, Lord Anson was appointed Vice-admiral of Great Britain; and on the 12th of June, he was made First Lord of the Admiralty. During the absence of King George II. in the years 1752 and 1755, his Lordship was appointed one of the Lords Justices of the Kingdom, a proof of the high opinion entertained by his Sovereign of his knowledge and prudence.

On a change of the Ministry in 1756, Lord Anson resigned his office in the Admiralty; on the 24th of February 1757, he was made an Admiral; and on the 2d of July the same year, he resumed his place at the head of the Board, and retained that honourable post during the remainder of his life.

In 1758, he hoisted his flag as Admiral of the White, on board the Royal George of 110 guns, and sailed from Spithead on the 1st of June, with a formidable fleet. The celebrated Sir Edward Hawke served under him; and these two naval heroes, by their skilful manœuvres while they cruised before Brest, effectually protected the descents made by the British troops and seamen that summer, at St. Maloes, Cherburgh, and other places along the French coast. While their shores were thus insulted by invaders, the French fleet did not venture to come out of Brest, and Lord Anson kept his fleet and seamen in constant exercise, for the purpose of thus acquiring a superiority in naval tactics.

The highest dignity which could be conferred upon him as a naval officer was the reward of these important services; for on the 30th July 1761, his Lordship was appointed Admiral and Commander-in-Chief of the British Navy. A few days afterwards he sailed from Harwich, in the Charlotte yatch, to convey Her present Majesty to England.

In 1762, Lord Anson accompanied the Queen's brother, Prince

Charles of Mecklenburgh, to Portsmouth, to shew him the arsenal, and the fleet which was ready to sail under the command of Sir George Pocock, for the Havannah. On this occasion his Lordship caught a cold, which was accompanied with a fit of the gout, under which he languished two or three months, and died at his seat at Moor-park, in Hertfordshire, on the 6th of June, 1762, in the 65th year of his age, and was buried in the family-vault at Colwich. He bequeathed his fortune to his brother, Thomas Anson, Esq. who was then Member of Parliament for Lichfield. On the decease of this gentleman, the united fortunes of the family devolved to his nephew, George Adams, Esq. who, in consequence, assumed the name of Anson. This gentleman was father to the present proprietor of Shugborough, who was raised to the dignity of the peerage on the 17th of February 1806, by the titles of Baron of Soberton, in the county of Hants, and Viscount Anson, of Shugborough and Orgrave, in the county of Stafford.

The public character of the first Lord Anson is illustrated in the naval annals of his country, where his enterprize, skill, perseverance, and intrepidity, are recorded, to excite the emulation of other heroes. By the discoveries made during his perilous and celebrated voyage, he enlarged the boundaries of nautical science, and was consequently a benefactor to all the maritime states of the world; and by his firmness in sustaining toils, his skill and bravery in action, and the strictness of his discipline, he greatly contributed to that superiority which the British navy has since so gloriously obtained over all others. As a man of business, in his official character, he was attentive and indefatigable, and as a member of the Senate, an honest and incorruptible patriot, equally devoted to his King and Country. Of his private character, little has been recorded: a habit of gaming has been mentioned among his foibles, but it appears from the best authority, that he merely had recourse to cards as an amusement. Though a prudent man, he was not avaricious, and the gamester is generally under the influence of that ignoble passion. But whatever may have been his errors or defects, they are lost in the contemplation of his merit, and his fame as a circumnavigator will be transmitted through the successive generations of posterity to the latest period of time.

THOMAS ALLEN.

Thomas Allen was born at Uttoxeter, on the 21st of Dec. 1542, and, according to Camden, was a descendant through six generations of Henry Allen, of Bucknall. In 1561, he was admitted a Scholar of Trinity College, Oxford, became a Fellow in 1565, and, in 1567, took his Master's degree. He continued to pursue his studies for three years afterwards in this college, but in consequence of his disinclination to enter into holy orders, as required by the statutes, he resigned his Fellowship, and went to Gloucester Hall (now Worcester College) in the year 1570. He now pursued the study of the mathematics with great attention and success, and in consequence of his attainments, acquired a high reputation for his superior knowledge of his favourite branch of learning.

Henry Earl of Northumberland, a nobleman much devoted to mathematical science, now patronized Mr. Allen, invited him to his house, and introduced him to those celebrated mathematicians, Thomas Harrison, John Dee, Walter Warner, and Nathaniel Torporley. In their society Mr. Allen enjoyed the highest gratification, by the discussion of topics most congenial to his habits of thinking, and his friends were no less pleased and instructed in their intercourse with a young philosopher, whose demonstrations of science were so complete and conclusive.

Among other distinguished characters who respected the talents of Mr. Allen, Robert Earl of Leicester was emulous to patronise him, and offered to confer a bishopric upon him, but our philosopher declined this clerical preferment, and continued in that retirement which was so agreeable to his unostentatious character, and his simple and temperate habits of life. Devoted to the studious pursuits of science, Mr. Allen continued in the University, and availing himself of the advantages of his situation, he collected many valuable manuscripts relating to Antiquities, History, Philosophy, Mathematics, and Astronomy. A catalogue of this collection is preserved in the Ashmolean Museum.

Mr. Allen published in Latin the second and third books of Ptolemy, "Concerning the Judgment of the Stars," with an Exposition. He also wrote notes on many of Lilly's books, and some on Bale's work, entitled "De Scriptoribus Maj. Britanniæ;" and was doubtless misled by the belief in judicial astrology, so prevalent in that age. His skill as a mathematician induced the vulgar to suspect him of practising the art of magic; and the author of a book entitled "Leicester's Commonwealth," accused him of exercising his necromantic art to promote the Earl of Leicester's ambitious schemes, and effect a match between that nobleman and Queen Elizabeth. This absurd assertion doubtless originated in the well-known confidence which existed between Mr. Allen and his patron, between whom a constant correspondence was kept up; insomuch, that nothing important respecting the State was transacted without the cognizance of the philosopher, who, in return, informed the Earl of what passed in the University.

From the uniformity of a collegian's life, few interesting incidents are to be expected; and Mr. Allen was content with the esteem of a few select friends in preference to emolument or fame. He was highly respected by several celebrated contemporaries, particularly Mr. Camden, Sir Thomas Bodley, Sir Henry Savile, Sir Robert Cotton, Sir Henry Spelman, and Mr. Selden.

Mr. Allen died at Gloucester-hall, Sept. 30, 1632, in the ninetieth year of his age, and was interred with great solemnity. Mr. Burton, who delivered a funeral oration on that occasion, called him "not only the Coryphæus, but the very soul and sun of all the mathematicians of his time;" and Mr. Selden, who was his intimate friend, mentions him as "a person of most extensive learning and consummate judgment, the brightest ornament of the University of Oxford." Camden says, "he was skilled in most of the best arts and sciences." These high panegyrics from such distinguished men are certainly honourable memorials of the learning of Mr. Allen, yet he does not seem to have been ambitious of transmitting his name to posterity by any literary production, which might have promoted the progress of science. In fact, it has long been the practice of the learned to compliment each other most hyperbolically; and in the ardour of their admiration, and wish to shine as encomiasts, they overpraise the abilities and attainments of men of real merit.

The following sketch of Mr. Allen's character is from a manuscript in the library of Trinity College, Oxford: "He studied polite literature with great application, he was strictly tenacious of academic discipline, always highly esteemed both by foreigners, and those of the University, and by all of the highest stations in the Church of England, and the University of Oxford." Yet with all this boasted knowledge how much is it to be regretted by the world, that Mr. Allen was so secret a lover of the Muses, for we have not a single scrap to illustrate his taste for polite literature, and very few articles indeed from his pen respecting even his favourite mathematics. The majority of those students in the academic shades are either very selfish or very indolent, for few of them communicate knowledge to others. In this respect they resemble the Monks, who seemed to consider religion as their peculiar study, and amused the people with a number of absurd or unmeaning ceremonies. Our Universities have indeed been censured with some appearance of truth for the laxity of their discipline and the indifferent progress made by the majority of students. This defect has been accounted for in the following epigram, which however, like all witticisms, is rather to be admired for its point than its truth:

> No wonder that Oxford and Cambridge profound,
> In wisdom and science should ever abound;
> Since each one takes thither some knowledge each day,
> And we meet with so few that bring any away.

Mr. Allen founded and endowed a free-school in Uttoxeter, his natal town.

ELIAS ASHMOLE,

Was the only son of Simon Ashmole, saddler, and Anne, the daughter of Mr. Boyer, of Coventry, woollen draper. He was born in Lichfield, May 23, 1617, received the rudiments of his education at a grammar-school in his native city, and afterwards, at his own request, instructed in the science of music, to which his genius was strongly inclined. In consequence of his proficiency in music he was admitted a chorister in the cathedral; but a more favourable opportunity for his advancement soon afterwards presented itself, in the patronage of James Paget, esq. Puisne Baron of the Exchequer, who married his mother's sister. Accordingly in 1633 he went to London. His father died in 1634, leaving him in the seventeenth year of his age, with little property, and consequently dependant on his uncle, with whom he lived about four years, during which period he made a considerable progress in the study of the law.

In March 1638, he married Eleanor, daughter of Mr. Mainwaring, of Smallwood, in Cheshire. In Michaelmas term the same year, he became a Solicitor in Chancery, and subsequently an Attorney in the Court of Common Pleas. On the 5th of December 1641, his wife died, of whom he left an affectionate memorial; and London at this period being in a very disturbed state, he retired into Cheshire, where he remained till 1645, and on the 9th of May that year, he joined the Royal standard in opposition to the Parliament, and was appointed one of the engineers or gentlemen of the ordnance in the garrison of Oxford.

Mr. Ashmole had now an opportunity for the full exertion of his active corporeal and mental powers, and as he was a zealous loyalist, he was advanced to places of considerable trust. From Oxford he was removed to Worcester, where he was appointed commissioner, receiver, and register of the excise, a very difficult and troublesome office, at a time when a civil war raged in the land, and the royal revenue was so uncertain. Ardent in the cause of his Sovereign, he was desirous to devote his life to his service, and was appointed captain in Lord Ashley's regiment of infantry, and comptroller of the ordnance. Yet amid the perils of civil warfare, he found moments of leisure, in which he applied himself vigorously to the study of natural philosophy, mathematics, and astronomy. His intimacy with Mr. Wharton, who was an astrologer, diverted his attention from more useful pursuits to the whimsical theories of that science as it was then called, and highly prized by many learned men in Britain, and on the Continent.

In July 1646, his mother died. She had always been a kind parent, and the object of his filial affection. On the 16th of October, the same year, he was elected a free mason, and wrote an account of the lodge established at Warrington.

After the surrender of the garrison of Worcester, and the fatal termination of the war, Mr. Ashmole again retired into Cheshire, where he continued a few months, and then went to London, where he became acquainted with Sir Jonas Moore, Mr. Lilly, and Mr. Booker, at that time considered as the most eminent astrologers in the world. From these his philosophical friends Mr. Ashmole imbibed all the ridiculous absurdities of judicial astrology. An anniversary feast was then held in honour of the science, and persons of high distinction thought it an honour to associate with these English magi.

In 1647, Mr. Ashmole retired to a pleasant seat in Berkshire, where he studied botany, with the aid of able masters. But a more agreeable pursuit soon engaged his attention in the rural shades, for he was introduced to Lady Mainwaring, widow of Sir Thomas Mainwaring, an attachment took place, and they were married on the 16th of November 1649. He now returned to London, and by the acquisition of a considerable fortune with his wife, he was enabled to live in a splendid style, and his house became the resort of several of the most ingenious and learned men of that age. Mr. Ashmole's passion for occult philosophy seems now to have returned with renovated force, the golden dreams of alchymy took possession of his imagination, and under the influence of the illusion, he published Dr. Dee's Treatise on the Philosopher's Stone. He also occasionally amused himself by engraving seals, and learning the trade of goldsmith and jeweller, but his principal employment during two years, was in preparing the unpublished works of English alchymists for the press, with illustrative engravings produced under his immediate direction. Whilst engaged in this work, he found that a competent knowledge of the Hebrew was requisite for the complete elucidation of the arcana of the hermetic philosophy; and, indefatigable in the pursuit of his favourite science, he studied the rudiments of this ancient language, and towards the close of the year 1652, published his "*Theatrum Chemicum Britannicum.*" This work advanced his reputation in the learned world, and obtained him the friendship of the celebrated Selden, Mr. Oughtred the mathematician, and Dr. Wharton.

But whatever eclat he might have obtained by his marriage with Lady Mainwaring, the litigation in which he was involved respecting her property must have been a great diminution of his domestic happiness; and to render these circumstances still more disagreeable, the lady entered a suit against her husband in the Court of Chancery, which came to a hearing in October 1657, nearly eight years after their union. On this occasion, the Counsellor employed by Mr. Ashmole observed, that in eight hundred sheets of depositions taken on the part of the lady, not a word of censure nor a serious charge was brought against his client. According to the decision of the Chancellor, the lady's bill was dismissed, and she was required to return to her husband, but it does not appear that they were cordially reconciled.

Mr. Ashmole, however, found solace in his studious pursuits, and he now turned his attention to antiquities and records, and became intimately acquainted with Sir William Dugdale, whom he accompanied in his Survey of the Fens. In a letter to Sir William, he gives an account of the Roman road from Weedon to Lichfield, mentioned in the Itinerary of Antoninus.

He now discontinued his "*Theatrum Chemicum*," which according to his original plan was to have consisted of several volumes; yet he still retained such a regard for alchymy that he published a short treatise on the Philosopher's Stone, with a well-written preface, in which he bids farewell to the philosophical gold-manufacturers.

In the spring of the year 1658, Mr. Ashmole began to collect materials for his great work, "*The Institution, Laws, and Ceremonies of the Most Noble Order of the Garter*," which occupied his attention nearly fourteen years, and was published in folio in the year 1672.

In September 1658, Mr. Ashmole went to Oxford, where he was well received, and engaged to write a description of the Coins and Medals presented to the public Library by Bishop Laud. On his return to London, he lodged and boarded at Mr. Tradescant's, in Lambeth; this gentleman was a botanist, and had collected many rare and curious subjects in natural history, which he presented to his lodger on the 16th of December 1659.

Soon after the Restoration of Charles II. in 1660, Mr. Ashmole was introduced to his Majesty, who, as a reward for his zeal and attachment, bestowed on him the office of Windsor Herald. The King afterwards appointed him to give a description of his Majesty's Medals, which were delivered to him for that purpose. Henry the Eighth's closet was appropriated to the use of our antiquary, and he was allowed to reside and diet in the palace.

Mr. Ashmole was now a courtier, and high in the estimation of his Sovereign, to whom he presented the three books which he had published, and who received them very graciously. Preferments now poured rapidly upon him. He was appointed commissioner of excise, secretary of Surinam, one of the commissioners for recovering the King's goods, a commissioner of the white office, and accomptant-general, and country accomptant of the excise.

In the beginning of the year 1662, he presented a set of Services and Anthems to the Cathedral church of Lichfield, and twenty pounds towards repairing that edifice. In 1666, he presented a silver bowl to the Corporation of Lichfield.

On the 1st of April 1668, his second wife died, and although he does not appear to have enjoyed much conjugal felicity from that connexion, he made another matrimonial experiment, and on the 3d of November following, was married at Lincoln's Inn Chapel, to Elizabeth, the daughter of his friend Sir William Dugdale, Knt. Garter King at Arms.

The academical distinction of M. D. was conferred on Mr. Ashmole by the University of Oxford in the year 1669; the diploma was dated the 16th of July, and presented to him on the 3d of November, by Dr. Yates, Principal of Brazen Nose College. As he had never regularly studied the medical art, this compliment was doubtless paid to him for his works on chemistry.

Dr. Ashmole was now visited at his Chambers in the Temple by some of the most eminent men in the kingdom, both with respect to title and merit. In a summer excursion he visited his native city, and was splendidly entertained by the Corporation.

On the 8th of May 1672, he presented his work on the Most Noble Order of the Garter to Charles II. who, as a proof of his approbation, granted the author £400. out of the duty on paper. This curious production, the work of years of research, is in its kind one of the most valuable books in our language.

The infirmities of old age were now creeping on, and in January 1675, Dr. Ashmole resigned his office of Windsor Herald, which, by his influence, was bestowed on his brother-in-law, Mr. Dugdale. As a proof that his learning and abilities were not forgotten at Court, on the death of Sir Edward Walker, Garter King at Arms, February 20, 1677, on a contest arising between the King and the Duke of Norfolk about the right of disposing of the place, the matter was left to the decision of Dr. Ashmole, who, like a true courtier and loyal subject, decided in favour of his Majesty. This high office was then offered to Dr. Ashmole, who modestly declined it, but employed his interest in favour of his father-in-law, Sir William Dugdale, on whom it was conferred.

In October 1677, he made an offer to the University of Oxford of all his coins, medals, and manuscripts, on condition that a building should be erected for their reception: to this proposal the University assented. About the close of the same year, a proposal was made to Dr. Ashmole to become a candidate for the city of Lichfield; to this flattering proof of the esteem of his fellow-citizens he could have no objection, but on application to several of the burgesses, he found himself so poorly supported that he prudently withdrew his pretensions.

In January 1679, his Chambers in the Temple were consumed by an accidental fire, which destroyed the noble library which he had been collecting for thirty-three years, together with 9000 coins, and a vast collection of seals, charters, and antiquities. His gold medals, and his MSS. escaped, being at his house in Lambeth.

On the 15th of May 1679, the first stone of the Ashmolean Museum was laid on the west side of the Theatre in Oxford: the building was finished in March 1682, and in 1683 Dr. Ashmole sent thither his collection of rarities.

In the beginning of 1685, he was invited by the Magistrates and the Dean of Lichfield to become a candidate for the representation of that city in Parliament, and he probably would have been elected without opposition, but King James sent an intimation to him that he wished another individual to be returned, on which Dr. Ashmole not only declined the honour intended by his fellow-citizens, but successfully exerted his influence in favour of Mr. Lewson, the person nominated and approved by his Majesty.

On the demise of Sir William Dugdale, the office of Garter King at Arms was a second time offered to Dr. Ashmole, and again declined by him. From this time he spent the remainder of his life in honourable retirement, and died on the 18th of May 1692, in the 76th year of his age; after having witnessed a remarkable variety of political vicissitudes in his native land, particularly the dethronement and death of Charles I.; the usurpation of Oliver Cromwell; the restoration of Charles II.; the abdication of James II.; and the final re-establishment of the Protestant religion, and the British constitution, by the glorious Revolution of 1688, which placed William III. on the throne.

His remains were interred in Lambeth church on the 26th of May 1692, and a tombstone of black marble, with an encomiastic Latin inscription, placed over his grave.

On a general view of the life of Elias Ashmole, his character will be found irreproachable, his learning considerable, and notwithstanding his early predilection for the absurdities of astrology and alchymy, his subsequent attention to the antiquities of his country, entitle him to respect. His liberality to indigent men of letters may also be mentioned to his praise, particularly the care he took in the education of Dr. George Smalridge.

Dr. Ashmole's published and unpublished works, including translations, are seventeen in number:

1. "Fasciculus Chemicus; or Chemical Collection, expressing

the ingress progress, and egress, of the secret hermetick science, out of the choicest and most famous authors. Whereunto is added, the Arcanum, or grand secret of hermetick philosophy."

2. "Theatrum Chemicum Britannicum, containing several poetical pieces of our famous English Philosophers who have treated the Hermetique mysteries, in their own ancient language. Faithfully collated in one volume, with annotations thereon, by Elias Ashmole, Esq. qui est Mercuriophilus Anglicus." London, 1652, quarto.

3. "The Way to Bliss," in three books, made public by Elias Ashmole, Esq. qui est Mercuriophilus Anglicus." London, 1658, quarto. This was the treatise in which he took leave of the astrologers and alchymists.

4. "The Institution, Laws, and Ceremonies of the Most Noble Order of the Garter. Collected and digested into one body by Elias Ashmole, of the Middle Temple, Esq. Windesore Herald at Arms. A work furnished with variety of matter relating to honour and noblesse." London, 1672, folio.

5. "The Arms, Epitaphs, Inscriptions, with the Draughts of the Tombs, &c. in all the Churches in Berkshire." It was written in 1666, and published under the title of The Antiquities of Berkshire, in three volumes octavo, in 1717.

6. "Familiarum Illustrium Imperatorumque Romanorum Numismata Oxoniæ in Bodleinæ Bibliothecæ Archivis descripta et explanata." This work was published by the author in 1659, and given by him to the Public Library in Oxford, in 1666, in three volumes folio, prepared for the press.

7. "A Description and Explanation of the Coins and Medals belonging to King Charles II." A MS. in folio.

8. "A Brief Ceremonial of the Feast of St. George, held at Whitehall, 1661, with other Papers relating to the Order."

9. "Remarkable Passages in the year 1660."

10. "An Account of the Coronation of our Kings, transcribed from a MS. in the King's private Closet."

11. "The Proceedings on the Day of the Coronation of King Charles the Second."

12. "The Arms, Epitaphs, &c. in some Churches in Staffordshire."

13. "The Arms, Epitaphs, Inscriptions, &c. in Cheshire, Shropshire, Derbyshire, Nottinghamshire," &c.

14. "Answers to the Objections urged against Mr. Ashmole's being made Historiographer to the Order of the Garter."

15. "A Translation of John Francis Spina's Book of the Catastrophe of the World; to which is subjoined, Ambrose Merlin's Prophecy."

16. "Collections, Remarks, and Notes, on Books and MSS."

17. "Memoirs of the Life of that learned Antiquary Elias Ashmole, Esq. Written by Himself, by way of Diary, with an Appendix of Original Letters." Published by Charles Burnum, Esq. London, 1717, 12mo.

DR. SAMUEL JOHNSON.

Individuals pre-eminent for their genius, learning, and moral excellence, undoubtedly confer a degree of celebrity upon their birth-place, which it could never otherwise have attained. Like beautiful and odoriferous plants, they diffuse their influence around, cheering and invigorating all within the sphere of their attraction; for their productions, and their lives, are equally conducive to the public weal, by affording superior examples of social utility, refined manners, and exalted rectitude. Lichfield, with its numerous pretensions to the approbation of the visitor and the resident, owes much of its consequence to the great men which it has produced, and among them Samuel Johnson claims undoubted pre-eminence.

Samuel Johnson, the eldest son of Michael Johnson, bookseller and stationer, in Lichfield, was born in that city, on the 7th of September, o. s. 1709. His father was a native of Cubley, in Derbyshire, and his grandfather was an honest and industrious husbandman in that village. This circumstance has been mentioned by some biographers as a derogation of the dignity of our great Moralist, as if the ancestry of all mankind could not be traced to one common stock; and certainly it is more honourable to have among our more immediate forefathers men of sober and industrious habits of life, than profligate, though more refined beings, who can only display the mere glitter of heraldic distinction without personal merit.

Michael Johnson was a meritorious and successful adventurer, who, by application to business, rose to respectability as a citizen of Lichfield. He kept his shop as a bookseller in a very conspicuous situation, in the corner house of the Market-square, opposite to the Market-house, and was considered a man of intellect, as appears from the following extract of a letter, dated Trentham, St. Peter's Day, 1716, and written by the Rev. George Plaxton, chaplain to Lord Gower: "Johnson, the Lichfield librarian, is now here; he propagates learning all over this diocese, and advanceth knowledge to its just height; all the clergy here are his pupils, and seek all they have from him; Allen cannot make out a warrant without his precedent, nor our quondam John Evans draw a recognizance *sine directione Michaelis*." This quotation indeed seems to be written in a strain of irony, but Michael Johnson was undoubtedly held in some estimation by his fellow-citizens, for in 1718 he was chosen Under-bailiff, and in 1725 was appointed Senior-bailiff of Lichfield. He died in December 1731, aged 76. His wife, the mother of our great lexicographer, was sister of Dr. Ford, a physician, and died in 1759, of a gradual decay, at the advanced age of 89. Nathaniel, their youngest son, died at the age of 27.

Samuel, the eldest son, was afflicted with the king's evil from his earliest infancy. When two year's old, he was touched by Queen Anne,* but with little effect, for the disorder continued with him, and is supposed to have deprived him of his left eye, and impaired his hearing. His mother, who was a woman of understanding, instilled moral and religious sentiments into his infant mind; and the son, whose filial gratitude was strong, often acknowledged how much he owed to maternal instruction. His father, a man of integrity and very industrious, was a steady high-churchman and a warm adherent to the house of Stuart, and was so solicitous to inspire his son with congenial opinions, that when Dr. Sachaverel, in his memorable tour through England, came to Lichfield, he carried the boy, not quite three years old, to the Cathedral, and placed him on his shoulders that he might see the preacher.

When eight years old, Samuel Johnson was placed under the tuition of Mr. Hawkins, at the free grammar-school in Lichfield, where he was more remarkable for his capacity, especially a retentive memory, than for diligence. When Mr. Hunter was appointed master of this school, however, young Johnson underwent severe discipline from him for his inattention, but he acknowledged that he required correction, and was so well convinced that severity is necessary in public education, that he was ever of the opinion of

* The piece of gold given by the Queen on this occasion, was preserved by the Doctor, and after his death it fell into the hands of his biographer, Sir John Hawkins.

Cowper, that "man's coltish disposition needs the thong." Even when a boy his memory was uncommonly tenacious, and on the advantages of this powerful faculty he prided himself through life. The weakness of his sight prevented him from joining in the sports of his school-fellows, for which he was naturally well qualified by personal strength and courage; and he spent his hours of recreation and leisure in rambles in the numerous public walks adjacent to the city, or in reading old romances.

In his fifteenth year, Johnson was indulged with a visit to his mother's brother, Dr. Ford; he continued in Warwickshire beyond the usual time allowed for the vacation, and on his return to school was refused admission by Mr. Hunter, who was offended at his pupil for having addressed some complimentary verses to his niece, Lucy Porter.

Johnson was afterwards sent to the grammar-school in Stourbridge, where he remained a year, during which he wrote several poems, some of which were printed in the Gentleman's Magazine, when he became a contributor to that periodical miscellany. On his return to Lichfield, he continued in his father's house without any determination respecting his future life. It has been asserted indeed, but without sufficient authority, that he was for some time employed in learning the trade of a bookbinder. The strong bias, however, of the youth's mind for classical attainments, doubtless induced his father to send him to the University of Oxford, and in his nineteenth year he was entered of Pembroke College.* He continued about three years at College, where his vigorous intellect, under the tuition of Dr. Adams, imbibed that knowledge at the fountain-head, with the diversity and excellence of which he afterwards delighted and instructed his countrymen. While at College he was mostly in a state of seclusion, occasioned in a great degree by his indigence, and to this circumstance may possibly be attributed that successful application to study, and the consequent acquisition of classic lore, which enabled him to outshine his contemporaries. It has been ascertained indeed, that Johnson was compelled by a want of pecuniary aid to quit College without a de-

* His apartment in Pembroke College was upon the second floor over the gateway. One day, when he was sitting in it quite alone, Dr. Ponting, then Head of the College, overheard him uttering this soliloquy in his strong emphatic voice. "Well, I have a mind to see what is done in other places of learning. I'll go and visit the Universities abroad. I'll go to France and Italy; I'll go to Padua ——, and I'll mind my business. For an Athenian blockhead is the worst of all blockheads."

gree, and this circumstance has been lamented by some of his biographers, but perhaps in the life of no other individual has the truth of the old adage, "Necessity is the mother of invention," been better demonstrated than in that of Samuel Johnson; his poverty therefore was an advantage to that community of which he became an ornament.

Soon after his return to Lichfield in 1731, his father died, leaving a small property scarcely sufficient for the maintenance of the widow. Thus left afloat on the great ocean of life without any resource but his classical acquirements, he in the 23d year of his age obtained the place of Usher at the grammar-school in Bosworth, Leicestershire. In this situation he continued but a few months; for indignant at the insolence of Sir Wolstan Dixie, the patron of the institution, he left Bosworth, and went to Birmingham on a visit to Mr. Hector, his former school-fellow, and then an eminent surgeon in that town. He continued an inmate in the house of Mr. Hector about six months, and under that hospitable roof commenced his literary career by a translation of Father Lobo's Voyage to Abyssinia, for which he received five guineas from the publisher.

In February 1734, Mr. Johnson returned to Lichfield, where in August following he issued proposals for publishing the Poems of Politian, &c. with his Life, by the editor. The book was to be printed in thirty sheets octavo, price five shillings, but this project was relinquished for want of encouragement. This failure in his first attempt as a professed author, for the translation of Lobo's Voyage was anonymous, was in consequence of the unestablished claim of a literary adventurer. Nor was his birth-place a well-chosen place for the early public efforts of his genius, for many learned and accomplished residents doubtless thought themselves equal if not superior in literary attainments to so young a man, and probably considered his proposals a mere effusion of vanity.

After this disappointment, Mr. Johnson offered his assistance to Mr. Cave, the proprietor of the Gentleman's Magazine. In an anonymous letter to Mr. Cave, dated Nov. 25, 1734, he offers to supply that publisher with original poems and inscriptions, &c. and concludes by saying, "Your letter, by being directed to S. Smith, to be left at the Castle, in Birmingham, Warwickshire, will reach your humble servant." This letter was immediately answered by Mr. Cave, who engaged our author as a contributor to his miscellany. The remuneration paid monthly is unknown, but doubtless it was insufficient for the maintenance of an individual.

Mr. Johnson in 1736, offered to assist Mr. Budworth, then Head master of the grammar-school at Brewood, Staffordshire, but his offer was declined. He then engaged as tutor in Mr. Whitby's family at Haywood, where the superiority of his memory was frequently displayed, on his return from Colwich church on Sunday, by his repeating the greatest part of the sermon, with critical remarks and illustrations.

Conscious of the superiority of his natural endowments and acquirements, Johnson, who was ever emulous of distinction, did not continue long in this obscure and secluded state. While in Birmingham he had been introduced to Mrs. Porter, the widow of a mercer in that town, and became captivated by her personal charms and accomplishments. The lady was flattered by the addresses of a man of merit, who was nearly twenty years younger than herself, and consented to a matrimonial union. Johnson, always independent in his sentiments, resolved to assert the dignity of a husband, even on his wedding-day, and came off triumphant. "She had read," says he, "the old romances, and had got into her head the fantastical notion that a woman of spirit should use her lover like a dog. So, at first she told me that I rode too fast, and she could not keep up with me; and when I rode a little slower, she passed me, and complained that I lagged behind. I was not to be made the slave of caprice; and I resolved to begin as I meant to end. I therefore pushed on briskly till I was fairly out of her sight. The road lay between two hedges, so I was sure she could not miss it; and I contrived that she should soon come up with me. When she did, I observed her to be in tears."

His wife's fortune was £800. with part of which he rented Edial-hall, and fitted it up as an academy, of which notice was given by a public advertisement, that "At Edial, near Lichfield, young gentlemen are boarded, and taught the Latin or Greek languages, by Samuel Johnson." His warm friend, Mr. Gilbert Walmsley, exerted his influence to promote this plan, but only three pupils could be obtained, one of whom was the celebrated David Garrick, then about eighteen years of age.

Johnson continued at Edial about a year and a half, during which, besides the instruction of his three pupils, he wrote Irene, a tragedy, which met the approbation of his judicious friend, Mr. Walmsley, who advised him to go to London and offer it to one of the managers of the principal theatres. In March 1737, he left Mrs. Johnson at Lichfield, and went to London, accompanied by his friend Garrick. Johnson presented a recommendatory letter from Mr. Walmsley to the Rev. Mr. Colson, whose exertions in his favour were unsuccessful. In July 1737, Johnson went to Greenwich, where he lodged for some time next door to the Golden Hart, Church-street, as appears by a letter from him to Mr. Cave, to whom he sent a specimen of the History of the Council of Trent. Mr. Cave consented to publish the work, and printed twelve sheets of it, for which Johnson received £49.; but another translation being advertised by a Mr. Samuel Johnson, librarian, of St. Martin's-in-the-Fields, Cave never completed his volume. Johnson now returned to Lichfield, where he continued about three months with his wife, and corrected and prepared Irene for the stage. About this time, he offered himself a candidate for the Mastership of the free-school at Appleby, in Leicestershire, the salary of which was £60.; but as it was a requisite to that appointment that the person chosen should be M. A. he was rejected. It required no common degree of fortitude to bear up against these frequent disappointments, but Johnson was yet in possession of part of his wife's fortune, and had great expectations of the success of his tragedy.

On his return to London with Mrs. Johnson, he offered Irene to Fleetwood, the patentee of Drury Lane Theatre, but was unsuccessful, and obliged to resume his literary pursuits. He was now in the thirtieth year of his age, unpatronized and unknown as an author, except by Mr. Cave, for whose Magazine he produced a number of essays, biographical sketches, poems, and translations. The Gentleman's Magazine now became a popular publication; and gradually rose in the estimation of men of taste, from the time that Johnson became a regular writer for it. He was now considered by Cave as the editor, and received from him £100. per annum for his contributions and superintendance.

From November 19, 1740, till February 1743, he revised the Parliamentary debates for the Gentleman's Magazine, under the name of the Senate of Lilliput. The original manuscript of these debates was for some time supplied by Guthrie the geographer and historian, and underwent the revision of Johnson. But when Guthrie obtained more advantageous employment, our author was obliged to fill up that department of the Magazine from the fertile sources of his own invention, with the occasional aid of scraps supplied by persons who attended the debates. Thus not only the style, but even the sentiments, were Johnsonian. In his old age our great Moralist expressed his disapprobation of this practice, as inconsistent with truth.

In the spring of 1738, he offered his "London," a poem, to Mr. Cave, as the production of an author who was under very disadvantageous circumstances of fortune. "I will," says he in his letter to Cave, "if you please to transmit the sheets from the press, correct it for you, and will take the trouble of altering any stroke of satire which you may dislike." Cave sent a present to Johnson's poor friend, and recommended Dodsley as a purchaser, who paid our author ten guineas for the copyright of the poem. It was published in May 1738, on the same morning with Pope's Satire of "Seventeen Hundred and Thirty-eight;" and such was the rapidity of the sale, that a second edition was called for by the public in less than a week. The approbation of Pope contributed not a little to the success of Johnson's London; it was certainly a happy prelude to the author's greater productions, and as such is memorable in his history.

Besides his "London," published in 1738, Johnson in 1739 published two satirical tracts, one entitled "A complete Vindication of the Licensers of the Stage from the malicious and scandalous aspersions of Mr. Brooke, author of Gustavus Vasa;" and the other "Marmor Norfolciense," &c. The latter was a severe attack on the administration of Sir Robert Walpole.

Among other literary adventurers with whom Johnson became acquainted at this time, the unfortunate Richard Savage was distinguished for the peculiar circumstances of his life, particularly the unnatural cruelty of his mother, the infamous Countess of Macclesfield. Johnson, naturally compassionate, sympathized with this ingenious but imprudent young gentleman, and even joined him in some of his nocturnal frolics, which were scarcely reconcileable to the moral principles by which our author was generally guided. On the demise of Savage, his friend Johnson wrote his Life in 1744, and a more masterly as well as instructive piece of biography, is perhaps not to be found in the English language.

The confidence of Johnson seemed now to increase with his increasing reputation. In 1745 he published "Miscellaneous Observations on the Tragedy of Macbeth, with Remarks on Sir Thomas Hanmer's edition of Shakspeare;" and at the same time offered proposals for a new edition of that dramatic poet. To this plan, which for some time met with but little encouragement, he seems to have devoted his attention.

In 1747, at the suggestion of Dodsley, he issued his plan for "A Dictionary of the English Language," and entered into an agreement with the booksellers to produce the work for fifteen hundred guineas, to be paid in sums proportionate to the quantity of manuscript prepared for the press. The plan of this great national work, which conferred immortality on the author, was addressed to the celebrated Earl of Chesterfield. Our author now obtained "a local habitation and a name," and removed from his obscure lodgings to a house in Gough-square. He now collected materials for his Dictionary, engaged six amanuenses, and commenced a work which engaged his attention for a considerable part of eight years. He varied his literary labours however during that time, by the production of some of his best original works.

In 1749 he published "The Vanity of Human Wishes," a very beautiful ethic poem, for the copyright of which he received fifteen guineas. His friend Garrick, who was now exalted to the very summit of theatrical fame, offered to bring forward Irene, and suggested some judicious alterations in the tragedy, to render it more fit for representation, to which the author reluctantly agreed. The tragedy had a run of nine nights, which entitled the author to the profits of three nights, and he received £100. from Dodsley for the copyright. For his only dramatic production, it may therefore be safely stated, that Johnson received £500. yet Irene, however admirable for purity of moral sentiment, and elegance of language, is no longer acted: a farce or comedy would perhaps have been more successful.

The permanent and exalted reputation obtained by Addison from the Spectator, probably stimulated Johnson to aspire to equal distinction as an essayist, and, on Tuesday the 20th of March, 1750, he published the first number of "The Rambler," and the work was continued in regular numbers every Tuesday and Friday till Saturday the 17th of March 1752, on which day it closed. Mr. John Payne, a bookseller in Paternoster-row, engaged to pay the author two guineas for each number, besides a share of the future profits of the work. Many of the numbers of this masterly production, which at once raised the author to the highest place among our ethical writers, were written in haste, as the moment pressed, and without being read over by the author before they were printed. Such is the account given by Johnson himself, and his veracity is unquestionable. When Sir Joshua Reynolds asked him by what means he had attained his extraordinary accuracy and flow of language, he told him that "he had early laid it down as a fixed rule to do his best on every occasion, and in every company to impart whatever he knew in the most forcible language he could put it in; and that by constant practice, and never suffering any careless expressions to escape him, or attempting to deliver his thoughts without arranging them in the clearest manner, it became habitual to him." It does not appear that the Rambler gained that popularity to which, from its excellent moral tendency, animated illustrations of manners, and superior elegance of diction, it was entitled; for the author, in the last number, says with a degree of candour, "I have never been much a favourite of the public." He corrected the two first editions, however, with that critical exactitude, in which, when he exerted himself, he was known to excel, and lived to see his great ethical work pass through successive editions with increasing celebrity.

The cessation of this periodical work was probably occasioned by the illness of Mrs. Johnson, who expired three days after the last number was published. This was the most severe affliction which Johnson ever experienced, as his conjugal affection was extreme, and he never, at the latest period of his life, recollected the death of Mrs. Johnson without emotion. She was buried in Bromley church-yard, and her husband placed an epitaph in Latin on her tomb, expressive of her beauty and amiable qualities. His affection for this estimable woman was doubtless increased by the fortitude with which she shared all his vicissitudes. During their intercourse their attachment was put to the test by adversity; a trial which, while it more firmly unites those who love with genuine feeling, has a tendency to produce strife and estrangement among the indifferent. Mrs. Johnson was not only the sharer of her husband's joys and sorrows, but a friend to whose judgment and taste he frequently submitted his productions. Of this there is one instance on record. After a few numbers of the Rambler were published, she expressed her unqualified approbation, and said, "I thought very well of you before, but I did not imagine you could have written any thing equal to this." The memory of her excellence was cherished by Johnson to the latest period of his life; his affection extended beyond the grave, and he often mentioned her in his prayers and meditations, which gave occasion to his enemies to ridicule his superstition, as they called it, and to accuse him of a bias to popery. How unfounded these malignant aspersions were every one who has read his excellent moral productions must know. A more sincere Churchman, and friend to Protestantism, perhaps, never existed; and so unimpeachable was his integrity, that if he had been convinced of the necessity of abjuring his Protestant faith, and adopting that dictated by the Church of Rome, he would have publicly avowed his sentiments.

In one instance Johnson was the dupe of an impostor. William Lander, a Scotchman, had with unparalleled impudence represented Milton as a plagiarist, and produced some passages from Grotius, and other modern Latin authors, which had a faint resemblance to some passages in "Paradise Lost." In 1750 Lander published some scraps of Hog's Latin Translation of the immortal production of our great Epic Poet, and averred that the mass thus fabricated was the archetype from which Milton copied, and to this pamphlet Johnson wrote a preface, in full persuasion of Lander's integrity. In the postscript we have a specimen of the powerful and persuasive eloquence of our great ethical writer, who recommended a subscription for Milton's grand-daughter: "It is yet in the power of a great people," says he, "to reward the poet whose name they boast, and from alliance to whose genius they claim some kind of superiority to every other nation of the earth; that poet whose works may possibly be read when every other monument of British greatness shall be obliterated; to reward him not with pictures or with medals, which if he sees, he sees with contempt, but with tokens of gratitude, which he, perhaps, may even now consider not unworthy the regard of an immortal spirit." From this passage we may justly defend Johnson from having been a party to the forgery of Lander, which was afterwards detected and exposed to popular execration by Dr. Douglas.

Soon after the conclusion of The Rambler, Dr. Hawkesworth and Dr. Warton began a periodical work entitled "The Adventurer," to which Johnson contributed several papers, the profits of which were, by his desire, given to Dr. Bathurst, a physician, whom he highly respected.

Johnson's reputation as a man of genius and literature was now high, and the circle of his friends extensive. Among others, Sir Joshua Reynolds, Dr. Goldsmith, and Mr. Burke, were the most distinguished as well as the highest in his estimation. Reynolds had from the first reading of Johnson's Life of Savage conceived a very high admiration of the literary powers of the author, and when they accidentally met at the house of a friend, an intimacy commenced which terminated only with life. Johnson at this period also became acquainted with Beauclerk and Langton, two gentlemen of rank, wit, and genius, in whose society he sometimes indulged in freaks of gaity, scarcely reconcileable with his general character for morality. Yet it does not appear that in any instance the gay Beauclerk seduced him into the purlieus of vice; but certain it is, that he sometimes indulged himself in nocturnal festivity with that witty companion, which was incompatible with the character of the author of the Rambler.

In 1754 Lord Chesterfield, who had formerly treated Johnson with neglect when he applied to him for patronage, was now induced by a desire of reconciliation, to publish two papers in "The World," in recommendation of the Dictionary, which his Lordship foresaw would confer a new accession of popularity on the author. But Johnson now felt conscious of his own superiority, and could no longer brook the idea of receiving favour from a nobleman who had so far deviated from his characteristic politeness as to treat with neglect the first application of a man of genius. He therefore thought it necessary, in his own defence, to undeceive the public, by publishing the following letter to his Lordship:

"*To the Right Hon. the* EARL OF CHESTERFIELD.

"MY LORD, "*February*, 1755.

"I have been lately informed by the proprietor of The World, that two papers, in which my Dictionary is recommended to the public, were written by your Lordship. To be so distinguished, is an honour which, being very little accustomed to favours from the great, I know not well how to receive, or in what terms to acknowledge.

"When, upon some slight encouragement, I first visited your Lordship, I was overpowered, like the rest of mankind, by the enchantment of your address; and could not forbear to wish that I might boast myself *Le vainqueur du vainqueur de la terre*;—that I might obtain that regard for which I saw the world contending; but I found my attendance so little encouraged, that neither pride nor modesty would suffer me to continue it. When I had once addressed your Lordship in public, I had exhausted all the art of pleasing which a retired and uncourtly scholar can possess. I had done all that I could; and no man is well pleased to have his all neglected, be it ever so little.

"Seven years, my Lord, are now past since I waited in your outer rooms, or was repulsed from your door; during which time I have been pushing on my work through difficulties of which it is useless to complain, and have brought it at last to the verge of publication, without one act of assistance, one word of encouragement, or one smile of favour. Such treatment I did not expect, for I never had a patron before.

"The shepherd in Virgil grew at last acquainted with Love, and found him a native of the rocks.

"Is not a patron, my Lord, one who looks with unconcern on a man struggling for life in the water, and when he has reached the ground, encumbers him with help? The notice which you have been pleased to take of my labours, had it been early, had been kind; but it has been delayed till I am indifferent, and cannot enjoy it; till *I am solitary*, and cannot impart it; till I am known, and do not want it. I hope it is no very cynical asperity not to confess obligations where no benefit has been received, or to be willing that the public should consider me as owing that to a patron, which Providence has enabled me to do for myself.

"Having carried on my work thus far with so little obligations to any favourer of learning, I shall not be disappointed though I shall conclude it, if less be possible, with less; for I have long been wakened from that dream of hope in which I once boasted myself with so much exultation.

"My Lord,

"Your Lordship's most humble, most obedient servant,

"Samuel Johnson."

This spirited and contemptuous rejection of the untimely aid of Lord Chesterfield, was probably very mortifying to the vanity of a nobleman who was ambitious to be looked up to as the Mecænas of the age; but Johnson having thrown down the gauntlet, expressed his opinion of his Lordship with his strong sarcastic humour. "This man," said he, "I thought had been a Lord among wits, but I find he is only a wit among Lords." When Lord Chesterfield's celebrated Letters to his natural son were published, Johnson censured them for the laxity of principle with which they are imbued. "They teach," said he, "the morals of a whore, and the manners of a dancing-master;" and it is remarkable that later writers on morality, particularly Cowper, have condemned that work in still severer terms.

Mr. Cave died in 1754, and Johnson recorded the merit of his old friend in a very elegant biographical sketch of him, which was published in the Gentleman's Magazine. From that time Johnson discontinued his regular contributions to that miscellany, though this was by no means a time of prosperity with him, for he was actually arrested for a debt of £5. 18s. which was paid by Richardson, the author of Sir Charles Grandison, &c. This circumstance is magnified by some of Johnson's biographers into a calamitous proof of the miseries of authors, but it probably originated in the improvidence of our author, whose mind was so deeply engaged with his literary pursuits that he forgot the debt till reminded by the severity of his creditor. It was one of those trifling casualties to which men of genius are continually liable among the pitfalls of civilized life, which no where abound more than in London, and a moment's reflection must convince us that if a friend had not been at hand, the booksellers who engaged him to write his Dictionary would have discharged the debt.

In the spring of 1755, the degree of M. A. was conferred upon Johnson by the University of Oxford, and immediately afterwards his Dictionary was published in two volumes folio. A few days prior to that event Johnson transmitted a part of the work to Dr. Birch, with the following cautionary note:

"Sir, "*March* 29, 1755.

"I have sent some parts of my Dictionary, such as were at hand, for your inspection. The favour which I beg is, that if you do not like them, you will say nothing.

"I am, Sir, your most affectionate humble servant,

"Samuel Johnson."

Dr. Birch's answer is too remarkable to be omitted, as it anticipated that universal approbation with which this production of our great philologist has been received by a grateful nation.

Sir, "*Norfolk-street, April* 3, 1755.

"The part of your Dictionary which you have favoured me with the sight of, has given me such an idea of the whole, that I most sincerely congratulate the public upon the acquisition of a work long wanted, and now executed with an industry, accuracy, and judgment, equal to the importance of the subject. You might perhaps have chosen one in which your genius would have appeared to more advantage, but you could not have fixed upon any other in which your labours could have done such substantial service to the present age and to posterity. I am glad that your health has supported the application necessary to the performance of so vast a task; and can undertake to promise you as one (though perhaps the only) reward of it, the approbation and thanks of every well-wisher to the honour of the English language.

"I am, with the greatest regard, Sir,
"Your most faithful and most affectionate humble servant,
"Thomas Birch."

To render his Dictionary of more general utility, he abridged it in 1756 into one volume octavo; and the same year he engaged as Editor of a monthly publication entitled "The Literary Magazine." Among the numerous articles which he wrote for this Magazine, his reviews of new books were most remarkable for elegance of style, justness of criticism, and propriety of sentiment. His review of Soame Jenyns's "Free Inquiry into the Nature and Origin of Evil," was particularly distinguished; it was reprinted separately to gratify the public, and two large editions of it were rapidly sold. The Magazine was discontinued at the expiration of two years from the want of encouragement. In 1756 he also wrote several essays for the "Universal Visitor," a Magazine which lasted only one year. His proposals for an edition of Shakspeare were revived in June 1756, and subscription tickets issued, but the work was not published till the year 1765.

In 1758 Johnson was engaged to write a short paper weekly for "The Universal Chronicle, or Weekly Gazette." This he entitled The Idler, and when the newspaper was discontinued, these papers were collected and published in two volumes, 12mo. which he corrected for the press. The first number of The Idler appeared April 14, 1758, and the last April 5, 1760. In No. 41, he alludes to the death of his mother, who died at Lichfield in the 89th year of her age, and for whom he had ever evinced the most grateful affection, having for many years contributed liberally to the support and comfort of her old age. He attended her funeral at Lichfield on the 23d of January 1759, and afterwards wrote a Latin epitaph expressive of her virtues and his filial love. On his return to London he wrote "Rasselas," in one week, to defray the expences of his mother's funeral, and pay some debts she had left. For the copyright of this philosophical fiction, which the Monthly Reviewers termed "a literary gem," Johnson received £125. which fully answered the purpose for which it was written. He also produced some smaller pieces in 1759, particularly an Introduction to the "World Displayed," and a translation of a "Dissertation on the Greek Comedy."

In 1762, while Johnson was proceeding with his edition of Shakspeare, he was agreeably surprised with the information that his Majesty had been pleased to grant him a pension of £300. a-year, as a reward for his literary merit; and certainly royal munificence could not have been more honourably employed. No proviso accompanied this annuity, for the independence of Johnson's mind was so well known, that any intimation of his services being required by the Government, would have occasioned his indignant rejection of a pension offered on such terms. That he afterwards defended some of the measures of Government is well known, but that was his own spontaneous and voluntary act; and it was not till 1770 that his first political pamphlet, "The False Alarm," was published. Soon afterwards he received a diploma from Trinity College, Dublin, which conferred upon him the highest academical honour, that of Doctor of Laws. This unsolicited mark of distinction from that learned body, was very gratifying to Johnson, who was thus unexpectedly raised to a degree of literary eminence and competence, the due reward of his successful exertion of great and useful talents. He now rented a house in a court in Fleet-street, which has since been called "Johnson's Court," and allotted an apartment to Miss Williams, a blind woman, of considerable talents and agreeable colloquial powers. She had been left in a state of indigence by her father, who was a physician, and on the recommendation of Miss Lucy Porter, Johnson took her under his protection, procured her a benefit play from Garrick, and assisted her in publishing a volume of poems; she thus obtained about £300. and with this fund became the inmate of our author's house, where she passed the remainder of her days, protected and cheered by his friendship. In 1762 Doctor Johnson received another inmate under his hospitable roof, a Mr. Levett, a medical man, who acted in the capacity of surgeon and apothecary to his friend till the time of his death, which happened on the 17th of January 1782, nearly three years before the demise of Dr. Johnson.

Thus, in the 53d year of his age, and the zenith of his reputation, Dr. Johnson had no longer occasion to "provide for the day that was passing over him;" and like all great men who, like the lion described by Milton, had vigorously disencumbered themselves of surrounding obstructions, he now arose to that eminence which shed a lustre on the literature of his country. While in a state of comparative indigence, no patron came forward to distinguish, cherish, and establish his merit; but when fortune smiled, the ingenious, the gay, and the great, thought it a privilege to associate with the formerly neglected Sam. Johnson. Such is human nature, such has it ever been, and ever will continue, while self-love is the predominant passion of man.

Notwithstanding the avowed partiality of Dr. Johnson to Oxford, he now visited Cambridge; and the following amusing account of that visit, narrated by Dr. Sharp, contains several illustrations of the most remarkable traits of our author:

"*Cambridge, March* 1, 1765.

"As to Johnson, you will be surprised to hear that I have had him in the chair in which I am now writing. He has ascended my aërial citadel. He came down on a Saturday evening, with a Mr. Beauclerk, who has a friend at Trinity. Caliban, you may be sure, was not roused from his lair before next day noon, and his breakfast probably kept him till night. I saw nothing of him, nor was he heard of by any one, till Monday afternoon, when I was sent for home to two gentlemen unknown. In conversation I made a strange *faux pas* about Barnaby Greene's poem, in which Johnson is drawn at full length. He drank his large potations of tea with me, interrupted by many an indignant contradiction, and many a noble sentiment. He had on a better wig than usual, but one, whose curls were not, like Sir Cloudesley's, formed for 'eternal buckle.' Our conversation was chiefly on books you may be sure. He was much pleased with a small Milton of mine, published in the author's life-time, and with the Greek epigram on his own effigy, of its being the picture, not of him, but of a bad painter. There are many manuscript stanzas, for aught I know, in Milton's own hand-writing, and several interlined hints and fragments. We were puzzled about one of the sonnets, which we thought was not to be found in Newton's edition, and differed from all the printed ones. But Johnson cried, No! no! repeated the whole sonnet instantly, *memoriter*, and shewed it us in Newton's book; after which he learnedly harangued on sonnet-writing, and its different numbers. He tells me he will come hither again quickly, and is promised 'an habitation' in Emmanuel college. He went back to town next morning; but, as it began to be known that he was in the University, several persons got into his company the last evening at Trinity, where, about twelve, he began to be very great, stripped poor Mrs. Macauley to the very skin, then gave her for his toast, and drank her in two bumpers.

"J. S."

In 1765, Dr. Johnson became an inmate in the family of Mr.

Thrale, then one of the most eminent brewers in London, and Member of Parliament for the Borough of Southwark. Arthur Murphy was then high in his reputation as a dramatic poet and essayist, and was intimate with Mr. Thrale and his lady, who was distinguished for her literary taste; by his praise of Johnson he so strongly excited the curiosity of Mrs. Thrale, that the literary colossus received an invitation to dinner, when his conversation proved so agreeable to Mr. and Mrs. Thrale, that reiterated invitations terminated in our author's acceptance of apartments in the town-house of his new friends, and their country-house at Streatham, where he was completely at home. Dr. Johnson became an inmate at the hospitable board of Mr. Thrale in 1765, and spent a considerable portion of several subsequent years in the society of the elegant, the witty, and the wise, who visited his friends in their pleasant country residence during the summer months. Indeed so great was his influence over the mind of Mr. Thrale, that he had a suit of apartments for himself both at the town and country-house, formed a library principally of his own selection, and directed the education of the young ladies. so much to the satisfaction of their father, that he bequeathed £200. to his learned friend, and appointed him one of his executors. His edition of Shakspeare was published in October 1765; and previous to a convivial meeting of his friends, on the night before the publication, Tonson the publisher, desired a gentleman to ask the Doctor for a list of the subscribers. "Why, Sir," said Johnson, "I have two material reasons against it: in the first place, I have lost all their names; and in the second, I have spent all the money." In his criticisms on Shakspeare, he displayed that original and vigorous conception, and beautiful diction, for which he was so remarkable, and which constitutes the excellence of his great critical work, "The Lives of Eminent English Poets," written several years afterwards. In his masterly preface to the edition of Shakspeare, he has displayed the excellencies and defects of the Avonian bard with a power of discrimination never surpassed, and a beauty and harmony of style never equalled. Yet such was the manliness and candour of his mind, that when a lady told him she thought it excelled Pope's Preface to the Iliad, the Doctor replied, "I'm afraid not, Madam; the little fellow did wonders!"

In 1766, he wrote the preface to Miss Williams's poems. This lady still continued an inmate of his house, of which she might be said to be absolute mistress, for such was Johnson's compassion for her infirmities, that he overlooked her peevishness. He was now, in fact, seldom at home, and when he visited Johnson's Court, the perverse tempers of some of his dependants drove him from it to the more attractive social circle at Mr. Thrale's.

When the Royal Academy of Arts was established in 1767, Dr. Johnson was appointed Professor in Ancient Literature; and the same year he had the honour of an interview, by appointment, with the King, at the Queen's library in Buckingham House. In the course of conversation his Majesty asked him "Why he *did* not continue to write?" 'Why, Sire,' said Johnson, 'I thought I had done enough!' "So should I too, Doctor," replied the King, "*if* you had not written so well."

In 1770, he wrote his first political pamphlet, "The False *Alarm*," to justify the Ministry and the House of Commons for expelling the celebrated Wilkes; and in 1771, he produced "Thoughts *on* the late Transactions respecting Falkland Islands." In the latter pamphlet he eloquently pointed out the absurdity of going to war with another nation about a barren island.

Dr. Johnson was now calumniated by his enemies as a ministerial hireling; but when some writers of the Whig party answered his pamphlets, he merely said, "These fellows are only advertising my book; it is surely better a man should be abused than forgotten."

In 1771, an attempt was made by Mr. Thrale and Mr. Strahan to get the Doctor returned as a representative for some borough, and Lord North, then Premier, had two meetings with them for that purpose. But the Minister finally declined the idea, which the Doctor could never forgive. "That fellow," said he, speaking of Lord North, "has a mind as narrow as the neck of a vinegar cruet."

From the year 1765, when he first entered into a familiar intercourse with Mr. and Mrs. Thrale till 1781, when that intercourse was interrupted by the death of the former, Dr. Johnson enjoyed all the gratifications communicable by elegant society and accommodation. His fame as the first writer of the age was established on the firm basis of critical approbation, and his conversation admired by persons of high rank and undoubted talents.

Mrs. Thrale, after the death of her first husband, married Piozzi, an Italian singer, and in 1785 published anecdotes of Dr. Johnson, in which his peculiarities are minutely described:

"The first time I ever saw this extraordinary man," says she, "was in the year 1754, when Mr. Murphy, who had been long the friend and confidential intimate of Mr. Thrale, persuaded him to wish for Johnson's conversation, extolling it in terms which that of no other person could have deserved, till we were only in doubt how to obtain his company, and find an excuse for the invitation. Mr. Johnson liked his new acquaintance so much, that from that time he dined with us every Thursday through the winter. In 1766, his health grew so exceedingly bad, that he could not stir out of his room in the Court he inhabited for many weeks together. Mr. Thrale soon after prevailed on him to quit his close habitation in the Court, and come with us to Streatham, where I undertook the care of his health, and had the honour and happiness of contributing to its restoration."

This lady related several amusing anecdotes of Johnson, and other celebrated men. Soon after Beattie's Essay on Truth was published, Dr. Goldsmith exclaimed, "Here is such a stir about a fellow that has written one book, and I have written many!" 'Ah, Doctor,' replied Johnson, 'there go two and forty sixpences, you know, to one guinea.' When talking of Richardson, the author of Sir Charles Grandison, "You think I love flattery," says Dr. Johnson, "and so I do; but a little too much always disgusts me: that fellow Richardson, on the contrary, could not be contented to sail quietly down the stream of reputation, without longing to taste the froth from every stroke of the oar." When Sir Joshua Reynolds left the room one day, "There goes a man," said the Doctor, "not to be spoiled by prosperity." He had a high opinion of the talents of Burke, and said that "you could not stand five minutes with that man beneath a shed while it rained, but you must be convinced you had been standing with the greatest man you had ever yet seen."

The Doctor had a custom of reading nightly in bed, and the top of all his wigs were burned by the candle to the net-work. Mr. Thrale's valet, for that reason, kept one always ready with which he met him at the parlour-door when the bell called him down to dinner, and as he went up stairs to sleep in the afternoon, the same attendant followed him with another. He told Mr. Thrale that he never sought to please till past thirty years old, as he considered the matter hopeless, but he had always been studious not to make enemies by apparent preference of himself. Vanity was always the object of his aversion, and his reproof to the vain was sometimes successful. A lady who had dressed herself fantastically for church at Streatham one Sunday morning, was censured by the Doctor, who expressed his disapprobation of her hat, her gown, &c. in such strong terms, that she hastened to change them, and when she returned quite another figure, received his applause, and thanked him for his reproof, to the amazement of her husband, who could scarcely believe his own ears. This anecdote is related by Mrs. Piozzi, who probably was herself the fantastical lady thus reformed by the animadversions of our great Moralist.

The death of Mr. Thrale was a subject of the deepest regret to Dr. Johnson, who in a letter to a friend, written on that occasion, pathetically says, "I saw for the last time that face which for eighteen years had never been turned towards me but with looks of complacency and affection." That the moral precepts of Johnson were highly conducive to the instruction of Mr. Thrale's daughters cannot be doubted by whoever will read his excellent letters to those young ladies, but their mother, after the demise of her husband, hastened with a degree of levity to Bath, for the purpose, as she has acknowledged, of a separation from the instructor of her children, and the friend of her husband. "I was forced," said she, "to take advantage of my lost law-suit, and plead inability of purse to remain longer in London or its vicinity. I had been crossed in my intentions of going abroad, and found it convenient for every reason of health, peace, and pecuniary circumstances, to retire to Bath, where I knew Mr. Johnson would not follow me, and where I could for that reason command some little portion of time for my own use; a thing impossible while I remained at Streatham, or at London, as my horses, servants, and carriage, had long been at his command, who would not rise in the morning till twelve o'clock perhaps, and oblige me to make breakfast for him till the bell rung for dinner, though much displeased if the toilette was neglected, and though much of the time we passed together was spent in blaming or deriding, very justly, my neglect of economy, and waste of that money which might make many families happy." By such ingenious excuses does this lady endeavour to palliate her abandonment of her venerable friend; but the Monthly Reviewers bestowed on her a sufficient portion of just castigation, and her subsequent matrimonial connection proved that it was not so much the Doctor's overbearing manners, as her own volatility, that induced her to violate that friendship, without which she would have lived and died unmentioned by the biographer.

The year 1773 is memorable in the life of Johnson for his Journey to the Western Isles of Scotland. He set out from London in company with his obsequious friend Mr. Boswell, and arrived at Edinburgh on the 18th August. In the course of his tour he visited the Universities of Edinburgh, St. Andrews, Aberdeen, and Glasgow, and after having enjoyed all the romantic beauties of the Hebrides, he returned to the Scottish capital in November. Like most patriotic Englishmen at that period, he had an aversion for the Scotch on account of their time-serving and assiduous endeavours to grasp at places under Government. This aversion he took no pains to conceal, and it may account for the coldness of his reception in some parts of Scotland. But his "Tour" is an instructive performance, and contains numerous and orignal remarks on men and manners, philosophical reflections, and passages of superlative beauty and sublimity. The defects in the domestic arrangements, accommodations, and want of cleanliness among the people, were not described by him for the purpose of censure but improvement; and it must be a wretched national pride indeed, which will cling to its errors, and persevere in ignorance, because a man of genius from another country has pointed them out. Every honest Scotchman should be grateful to Dr. Johnson for having mentioned inconveniencies which a little ingenuity and industry might remove.

Mr. Boswell wrote an entertaining Journal during this tour with his friend, in which he recorded with manly impartiality the Doctor's real sentiments on various subjects. On the first night of their arrival in Edinburgh, as they walked up High-street arm in arm, Boswell said, "Well now, Doctor, we are at last in Scotland." 'Yes, Sir,' cried the Doctor, 'I smell it in the dark.' The Doctor lost his oak stick in the isle of Mull; he suspected his guide had stolen it, but Boswell endeavoured to persuade him that it would be restored to him again. "No, Sir," says Johnson, "it is not to be expected that any man in Mull who has got it, will part with it, *consider the value of such a piece of timber here!*"

On his return to London, a Scotch resident in the metropolis asked him with a firm tone of voice, what he thought of his country? "That it is a very vile country, to be sure, Sir," said the Doctor. 'Well Sir,' said the other, somewhat mortified, 'God made it.' "Certainly he did," replied Johnson, "but we must always remember that he made it for Scotchmen." At this time "Ossian's Poems," published by Macpherson, excited much controversy in the literary world, and Dr. Johnson, always the friend to veracity, made frequent inquiries during his tour in Scotland respecting the authenticity of the work; but he could get no information of the original poems, and from the evasive answers which he received from some individuals, and the unsatisfactory report of others, the Doctor declared his opinion that the supposed production of an ancient Caledonian bard was the fabrication of Macpherson, who thus wished to impose on the public. Macpherson was exasperated at this detection of his imposture, and wrote a menacing letter to Johnson, which the English critic answered in a composition expressive of dignified contempt.

In 1775, when the dispute between the English Government and the American Colonists ran high, but prior to the actual commencement of hostilities, Dr. Johnson wrote his "Taxation no Tyranny," in which he endeavoured to prove that the mother country, through the medium of her legislature, had a right to tax her colonies, though they had independent legislative assemblies of their own. An appeal to arms, however, soon set aside all the sophistical ingenuity of literary men; the American revolt ended in a revolution and the establishment of the independence of the United States, and a mutual intercourse between them and Great Britain, is, at this moment, conducive to the prosperity of both countries. In autumn, the Doctor went on a tour of two months, accompanied by Mr. and Mrs. Thrale, to Paris, and other parts of France; and while at Versailles, where our tourists were shewn the machinery of the theatre, Johnson's patriotism was excited by a remark of one of his companions, who said, "Now we are here, what shall we act? The Englishman at Paris?" 'No, no,' replied Johnson, 'we will try to act Henry the Fifth.'

In the summer of 1776 he visited Lichfield, in company with his friend Boswell, who gives a most interesting account of the excursion. "I wished to have staid at Birmingham to-night," says he, "but my friend was impatient to reach his native city, so we drove on that stage in the dark. When we came within the focus of the Lichfield lamps, "Now," said he, "we are getting out of a state of death." We put up at the Three Crowns, not one of the greatest inns, but a good old-fashioned one; which was the very next house to that in which Johnson was born and brought up, and which was still his property. We had a comfortable supper, and got into high spirits. I felt all my Toryism glowing in this old capital of Staffordshire.

"Next morning he introduced me to Mrs. Lucy Porter, his step-daughter. She was now an old maid, with much simplicity of manner. Her brother, a captain in the navy, had left her a fortune of

£10,000. about one-third of which she had laid out in building a stately house, and making a handsome garden, in an elevated situation in Lichfield. Johnson, when here by himself, used to live at her house. She reverenced him, and he had a paternal tenderness for her. He expatiated in praise of Lichfield and its inhabitants, who, he said, were "the most sober, decent, people in England, the genteelest in proportion to their wealth, and spoke the purest English."

"Very little business appeared to be going forward in Lichfield. I found, however, two strange manufactures for so inland a place, sailcloth, and streamers for ships; and I observed them making some saddle-cloths and dressing sheep-skins; but, upon the whole, the busy hand of industry seemed to be quite slackened. "Surely, Sir," said I, "your are an idle set of people." 'Sir,' said Johnson, 'we are a city of philosophers: we work with our heads, and make the boobies of Birmingham work for us with their hands.' We went and viewed the museum of Mr. R. Greene, apothecary here, who told me he was proud of being a relation of Dr. Johnson's."

Dr. Johnson had now for some years relaxed in his literary labours, probably from a consciousness that he had performed his duty as a moralist and public writer; but a circumstance occurred in 1777, which again called into full exertion all his intellectual powers. At a meeting of about forty of the most opulent booksellers of London, who were proprietors of the copyrights in the works of our most eminent poets, it was agreed that an elegant and uniform edition of the whole should be printed, with a concise account of the Life of each author by Dr. Johnson, and that a deputation of three publishers should wait upon him with proposals. The Doctor readily engaged in a task so congenial to his taste, added the names of Watts, Blackmore, Pomfret, and Yalden, to the list proposed to him, demanded two hundred guineas for his biographical sketches; and in April the new edition of the poets was sent to the press.

In May 1777, he was applied to by the friends of Dr. William Dodd, a clergyman, who was convicted of forgery; and willingly contributed whatever could be suggested by humanity to prevent the disgraceful public execution of a minister of the church of England. Dr. Johnson wrote nine different papers on this subject, in a style of pathetic, but ineffectual, eloquence; for the delinquent underwent the sentence of the law, and afforded by his fall a memorable example of the danger of deviating from the path of rectitude.

Dr. Johnson commenced his great biographical work in the 68th year of his age, and concluded it in the 74th: it is now universally allowed to be the most entertaining, as well as the most elegant, of his literary productions, and demonstrated the truth of the assertion that his genius suffered no diminution by the advance of old age. The first four volumes were published in 1779, and the remainder in 1781, when his employers presented him with £100. more than the stipulated sum. The avidity with which his Lives of the Poets were read, and the general approbation of the public, rendered it necessary to print them detached from the poetry: a second edition was published in 1781, and a third edition, enlarged and corrected by the author, was published in 1783, and thus terminated the literary labours of the most elegant writer in English prose that adorned the eighteenth century. Envy of his established fame, and partiality for particular poets, induced several writers to publish animadversions on his Lives of the Poets. He was accused of prejudice for his unqualified condemnation of Milton's democratic principles, and the occasional praise and censure with which he criticised the poems of Gray, Hammond, and Prior; but the gratification afforded to the true admirers of poetry by his masterly and unparalleled delineations of character, and eulogiums on poetic genius, in his Lives of Cowley, Milton, Dryden, and Pope, at once stamped his reputation as the most faithful and elegant biographer that had ever written in the English language. Since the appearance of his Lives of the Poets, indeed, biography, always popular, has become the favourite study of the intelligent part of the community, and numerous writers have produced the Lives of eminent men, but none of them have caught the inspiring mantle of Johnson.

After this conclusion, and it may be said consummation of his various and useful literary labours, Dr. Johnson appeared but little in public. The infirmities of old age increased; several of his intimate friends were dead, particularly Mr. Thrale, Miss Williams, and Dr. Goldsmith. He had removed from his former residence to a house in Bolt-court, Fleet-street, where, in June 1783, he had a paralytic stroke, which deprived him of the powers of speech. He awoke with the attack, and immediately rung the bell; but on the approach of his servant could not articulate a syllable. He made signs for pen, ink, and paper, and wrote the following note to Mr. Allen, a printer, who lived next door to him; an honest, virtuous man, who had been his intimate and confidential friend for many years:

"DEAR SIR,

"It hath pleased Almighty God this morning to deprive me of the powers of speech; and as I do not know but that it might be his further good pleasure to deprive me soon of my senses, I request you will on the receipt of this note, come to me, and act for me, as the exigencies of my case may require.

"I am sincerely yours,

"*To Mr. Edmund Allen.*" "S. JOHNSON."

Mr. Allen immediately attended him, and sent for his physicians, Drs. Heberden and Brocklesby, by whose aid he recovered so far as to be able to take the air and visit his friends as usual. In August 1783, he took another journey to Lichfield and Oxford, and on his return to London he received a visit from Mrs. Siddons. He gives an account of this interview in a letter to Mrs. Thrale, with whom he kept up a correspondence. "Mrs. Siddons, in her visit to me, behaved with great modesty and propriety, and left nothing behind her to be censured or despised. Neither praise nor money, the two powerful corrupters of mankind, seem to have depraved her. Her brother Kemble calls on me, and pleases me very well." When Mrs. Siddons came into the room there happened to be no chair ready for her, which the Doctor observing, said with a smile, "Madam, you who so often occasion a want of seats to other people, will the more easily excuse the want of one yourself."

Towards the close of the year 1783, as he found his spirits much cheered by society, it was proposed by some of his friends to establish a club in the neighbourhood, which would answer that purpose. The Doctor was highly pleased at the proposal, and a club consisting of twenty-four friends met at the Essex Head, Essex-street, on the 10th of December, 1783, and continued to meet three evenings every week, except the week before Easter-day.

In the spring of 1784 he was confined for some weeks to his house by a dropsy, brought on by the frequent bleeding which he was obliged to undergo for his alleviation from an asthma. In the summer of the same year he grew so much better, that supposing the air of Italy might renovate his health, a tour through that country was recommended by his friends. But this scheme was never much encouraged by his physicians, though Sir Joshua Reynolds applied to the Lord Chancellor to procure some addition to his pension to enable him to defray the expence of the intended tour; and his Lordship, when his application proved unsuccessful, offered to advance £500. for that purpose, and Dr. Brocklesby offered an equal sum. These amicable offers, however, were not accepted; the termination of his useful life approached, and his dropsy and asthma were pronounced incurable.

That terror of death which is said to have preyed on the spirits of Doctor Johnson, has been attributed to his constitutional melancholy, but there is also reason to suppose that the estimation in which he was held by numerous and respectable friends, the comfortable accommodations supplied by a competent income, and the gratification arising from a well-earned fame, all contributed to heighten the natural dread of dissolution. Yet he endured the pains of disease with becoming fortitude; and felt a revival of his filial affection, as the moment approached when he might hope to see his parents in a higher state of existence.

This appears by the following letter written by him eleven days before his death:

"*To* Mr. Greene, *Apothecary, at Lichfield.*

"Dear Sir,

"I have enclosed the epitaph for my father, mother, and brother, to be all engraved on the large size, and laid in the middle aisle in St. Michael's Church, which I request the clergyman and churchwardens to permit.

"The first care must be to find the exact place of interment, that the stone may protect the bodies. Then let the stone be deep, massy, and hard; and do not let the difference of ten pounds or more defeat our purpose.

"I have enclosed ten pounds, and Mrs. Porter will pay you ten more, which I gave her for the same purpose; and I beg that all possible haste may be made, *for I wish to have it done while I am yet alive.* Let me know, dear Sir, that you receive this.

"I am, &c.

"*Dec.* 2, 1784." "Sam. Johnson."

On the Thursday before his death, when Dr. Brocklesby gave his final opinion that he could never recover without a miracle: "Then," said he, "I will take no more physic, not even my opiates; for I have prayed that I may render up my soul to God unclouded."

The following authentic and edifying account of the death of Dr. Johnson, was published by his intimate friend Dr. Brocklesby:

"For some time before his death, all his fears were calmed and absorbed by the prevalence of his faith, and his trust in the merits and propitiation of Jesus Christ.

"He talked often to me about the necessity of faith in the sacrifice of Jesus, as necessary, beyond all good works whatever, for the salvation of mankind.

"Johnson having thus in his mind the true Christian scheme, at once rational and consolatory, uniting justice and mercy in the Divinity, with the improvement of human nature, while the Holy Sacrament was celebrated in his apartment, fervently uttered this prayer:

"'Almighty and most merciful Father, I am now as to human eyes it seems, about to commemorate, for the last time, the death of thy Son Jesus Christ, our Saviour and Redeemer. Grant, O Lord, that my whole hope and confidence may be in his merits and thy mercy; enforce and accept my imperfect repentance; make this commemoration available to the confirmation of my faith; the establishment of my hope, and the enlargement of my charity; and make the death of thy Son Jesus Christ effectual to my redemption. Have mercy upon me, and pardon the multitude of my offences. Bless my friends; have mercy upon all men. Support me, by thy Holy Spirit, in the days of weakness, and at the hour of death; and receive me, at my death, to everlasting happiness, for the sake of Jesus Christ. Amen.'

"The Doctor, from the time he was certain his death was near, appeared to be perfectly resigned, was seldom or never out of temper or fretful, and often said to his faithful servant, 'Attend, Francis, to the salvation of your soul, which is an object of the greatest importance.'

"On Monday the 13th of December, the day on which he died, Miss Morris, the daughter of a particular friend of his, called, and said to Francis, that she begged to be permitted to see the Doctor, that she might earnestly request him to give her his blessing. Francis went into the room, followed by the young lady, and delivered the message. The Doctor turned himself in the bed and said, 'God bless you, my dear!' These were the last words he spoke. His difficulty of breathing increased till about seven o'clock in the evening, when Francis Barber, and Mrs. Des Moulins, who were sitting in the room, observed that the noise he made in breathing had ceased, went to the bed, and found he was dead."

On Monday the 20th December 1784, his body was interred in Westminster Abbey, at the foot of Shakspeare's monument, in the Poet's Corner, near the grave of his friend Garrick. His pall was supported by the Right Hon. Edmund Burke, Right Hon. William Wyndham, Sir Joseph Banks, Sir Charles Bunbury, George Colman, and Bennet Langton, Esqs. His executors and a considerable number of his friends attended, and paid this last tribute of affection to his memory. His schoolfellow, Dr. Taylor, performed the mournful office of reading the funeral service, and over the grave was placed a large blue flag stone, with this inscription:

SAMUEL JOHNSON, LL. D.
Obiit XIII. Die Decembris.
Anno Domini,
M.DCC.LXXXV.
Ætatis suæ, LXXV.

By the first clause of his will he left an annuity of seventy pounds to his old faithful black servant Francis Barber, who lived with him near forty years. He also left two hundred pounds to a descendant of Mr. Innys, a bookseller, to whom his father was indebted thirty pounds at the time of his death, about thirty years before; and several smaller legacies to distant relations and friends. Dr. Johnson not only advised others to be charitable, but was charitable himself. He gave away all he had, except the two thousand pounds he left behind; he had numerous dependants out of doors, as well as under his own roof, and for these people he used to raise contributions on his richer friends.

The following fact is strongly illustrative of his characteristic humanity. "Coming up Fleet Street alone about two o'clock in the morning, he was alarmed with the cries of a person seemingly in great distress. He followed the sound of the voice for some time, when, by the glimmer of an expiring lamp, he perceived an unhappy female, almost naked, and perishing on a truss of straw, who had just strength enough to tell him, she was turned out by an inhuman landlord in that condition, and to beg his charitable assistance not to let her die in the street. The Doctor melted at her story, desired her to place her confidence in God, for that under him he would be her protector. He accordingly looked about for a coach to put her into, but there was none to be had. He then kneeled down, raised her in his arms, wrapped his great coat about her, and carried her on his back to his own house. Next day, from the nature of her disorder, he was advised to abandon her, but he nobly replied, 'I am determined to give her the chance of a reformation.' Accordingly she was attended by a physician during thirteen weeks, and restored to health; and Dr.

Johnson having learned from her story, that she was one of those unhappy women who are impelled to prostitution more from necessity than inclination, he set on foot a subscription for her, and established her in a small milliner's shop in an inland town, where she lived in considerable repute." Such philanthropy is above encomium; it is doubtless that 'charity which shall cover a multitude of sins,' and it is not presumptuous to believe that the subject of these memoirs is now enjoying the reward promised to the merciful and benevolent. As for any encomium on his talents, his own works are a sufficient proof of their superlative excellence and popular utility. Among the numerous honours paid to the memory of Dr. Johnson, by his admiring countrymen, the following epitaph by the celebrated Cowper, is equally remarkable for its truth and energy:

Here JOHNSON lies---a sage by all allow'd,
Whom to have bred may well make England proud;
Whose prose was eloquence, by wisdom taught,
The graceful vehicle of virtuous thought;
Whose verse may claim---grave, masculine, and strong,
Superior praise to the mere poet's song;
Who many a noble gift from Heav'n possess'd,
And Faith at last, alone worth all the rest.
O man, immortal by a double prize!
By fame on earth---by glory in the skies!

WILLIAM MOUNTFORD.

This eminent Actor was born in Staffordshire in 1659. There are no records illustrative of his parentage or education, but he undoubtedly went early on the stage, and was distinguished as a good comedian and tragedian. Jacob, one of his contemporaries, says, that after he had attained a degree of excellence in his profession, he left the stage, and became an inmate in the family of Lord Chancellor Jefferies. This cruel and iniquitous Judge, was, it appears, fond of buffoonery; for we are informed by Sir John Reresby, "that at an entertainment of the Lord Mayor and Court of Aldermen in the year 1685, the Lord Chancellor called for Mr. Mountford to divert the company. He being an excellent mimic, my Lord made him plead before him in a feigned cause, in which he aped all the great lawyers of the age, in their tone of voice, and in their action and gestures of body, to the very great ridicule not only of the lawyers but of the law itself; which to me did not seem altogether prudent in a man in his lofty station in the law: diverting it certainly was, but prudent in the Lord High Chancellor I shall never think it."

After the fall of his patron, Mountfort returned to the stage, and continued in that profession till his tragical death in 1692. Whatever may have been the levity of this dramatic writer and performer, the manner in which he perished by the hands of unmanly assassins is disgraceful to the age in which the crime was perpetrated. The following narrative of the circumstance is interesting:

Lord Mohun and Captain Hill, two young men of dissolute lives, had entered into close intimacy. Hill had long felt a passion for Mrs. Bracegirdle, then the heroine of the English stage, and a woman of great beauty and accomplishment, but she rejected his suit with contemptuous disdain. This repulse was attributed by the unsuccessful lover to the lady's predilection for some favoured rival, and as Mountfort was a man of a very agreeable person, and frequently performed the part of a lover on the stage it seems with Mrs. Bracegirdle, his apparent assiduity and respect induced Capt. Hill to fix on him as the obstacle to his happiness, and he secretly vowed revenge.

He also laid a plan to carry Mrs. Bracegirdle off, in concert with his friend Lord Mohun, and for that purpose they went to the play-house, but not finding her there, they got intelligence where she was, and having hired a number of soldiers to assist them, they waited near the door till she came out, seized her, and endeavoured to force her into a coach, but were prevented by the resistance of her mother and a gentleman who struggled with the ruffians, and assisted by several passengers, rescued the actress, and conducted her in safety to her own house.

Enraged at this disappointment, Lord Mohun and Captain Hill now directed their vengeance against the unsuspecting and unfortunate Mountfort; and with loud imprecations avowed their intention to destroy him. Mrs. Bracegirdle's mother and the gentleman who was with her heard their threats, and immediately went to inform Mrs. Mountfort of her husband's danger. She sent a messenger to warn him not to come home that night, but unfortunately he could not find him. In the mean time Lord Mohun and Captain Hill paraded the streets with their drawn swords, and about midnight met Mr. Mountfort in Norfolk-street, in the Strand. He was accosted in an amicable manner by Lord Mohun, who treacherously held him in conversation till the cowardly assassin Hill came behind him, first struck him on the head with his left hand, and before he could draw his sword and stand on his defence, the villain ran him through the body. This circumstance was declared by Mr. Mountfort while expiring. Hill made his escape, but Lord Mohun was seized, tried for the murder as an accomplice, and acquitted. He was afterwards killed in a duel with the Duke of Hamilton about the Sandon estate, in which similar treachery was practised against him to that which he so villainously resorted to for the destruction of poor Mountfort.

The body of Mr. Mountfort was interred in the church-yard of St. Clement Danes. His dramatic works consist of six pieces, none of them remarkable for excellence. As a player, however, he had great merit. He is described by Colley Cibber as "tall, well-made, fair, and of an agreeable aspect. His voice was clear, full, and melodious; a most affecting lover in tragedy; and in comedy gave the truest life to the character of a fine gentleman. In scenes of gaiety he never broke into that respect that was due to the presence of equal or superior characters, though inferior actors played them, nor sought to acquire any advantage over other performers, by finesse or stage tricks, but only by surpassing them in true and masterly touches of nature. He might perhaps have obtained a higher degree of excellence and fame, had he not been untimely cut off by the hands of an assassin in the 33d year of his age."

ELIJAH FENTON.

Elijah, the youngest son of John Fenton, was born at Shelton, in this county, on the 20th day of May, 1683. His father was an Attorney, and Coroner for the County of Stafford: he was descended from an ancient family, but the chief means for the support of eleven children were derived from his professional labours. He died in the year 1694, consequently the care of the education of Elijah devolved on his guardians. From the early indications of intellectual superiority evinced by young Fenton, he was intended by his friends for the clerical profession, and removed from school to Jesus College, Cambridge, where he took a Bachelor's degree in the year 1704, but removed in 1706 to Trinity Hall.

Having refused to take the oaths to qualify him for public employment, he left the University without proceeding, A.M. but it is not certain whether he became a separatist from the established Church. He had the decency and discretion not to censure or ridicule the religion of his fathers, if he even disapproved of some of its tenets; and hence, he did not, like those malignant apostates whose pride or other violent passions prompt them to an opposition to the faith of their friends, vilify with base ingratitude the venerable institutions of ages. Elijah Fenton seems to have acted from conscientious scruples, and however censurable his imprudence may appear in thus counteracting the amicable endeavours of his friends to obtain an honourable employment for him, his integrity was not only respectable but unimpeachable.

Elijah Fenton left the University with a well-instructed mind, but without any prospect of obtaining a livelihood. The small subdivision of his father's property which fell to his share had been expended on his education, and he was consequently cast afloat upon the ocean of life to pursue his progress according to fortuitous circumstances. There is considerable obscurity in this part of his life, and none of his biographers have been able to clear away the mist.

It is however ascertained that he was employed for some time as Usher, by Mr. Bonwicke, of Headley, in Surrey, but was soon afterwards patronized by the representative of the noble family of Boyle, and appointed Secretary by Charles Earl of Cork and Orrery. Under the auspices of a family which had produced so many men of genius and piety, Mr. Fenton's abilities were happily matured, and so great was the confidence of Lord Orrery in the integrity and prudence of his secretary, that he appointed him tutor to his only son Lord Boyle. Mr. Fenton remained with this noble family for some time in Flanders.

On his return to England he opened a grammar-school at Seven Oaks, in Kent, which, though not very productive, increased his reputation as an able instructor of young gentlemen, a task for which he seems to have been well qualified by his classical knowledge, and the amenity of his manners, which had doubtless received a higher polish by travel, and an intercourse with the visitors of his noble patron while on the Continent. But notwithstanding the approbation of Lord Orrery, and the high estimation in which Mr. Fenton was held by his pupils, it does not appear that his pecuniary resources were improved during his residence with that family, nor was there an annuity bestowed on him as a reward for his services. Hence he was obliged to avail himself of his knowledge of the classic languages for a subsistence.

In the year 1707 he published a collection of poems entitled "Oxford and Cambridge Verses," but as the copies of this publication have long since disappeared, the poems written by himself, which appeared in that volume, cannot be ascertained. This publication seems to have obtained some approbation among the wits and critics of the age, for in the year 1710 Mr. St. John, afterwards Lord Bolingbroke, took him under his patronage, in consequence of which he broke up his establishment at Seven Oaks.

The principal advantage derived by Fenton from the patronage of Mr. St. John, was an introduction to Pope, then in the zenith of his reputation. An intercourse with the bard of Twickenham undoubtedly improved the poetical taste of Fenton, and roused him from his habitual or constitutional indolence to the exertion of his mental powers. In 1717 he published a volume of original poems, containing an Ode to the Sun, a Paraphrase on Isaiah, and several short pieces, of which it is no small commendation to say, that they were favourably received by the public at a time when the harmonious productions of Pope, and the humourous effusions of Swift and Gay, had rendered criticism rather fastidious.

In 1720, Fenton was appointed private secretary to Mr. Craggs, when that statesman was exalted to one of the most honourable situations under government. Thus a fair opportunity for advancement in public life presented itself to our poet, but the death of his patron cut short these hopes, and he was again thrown into the obscurity of retirement. He seems to have borne his disappointment with great equanimity.

On the death of Mr. Craggs, Pope did not, like too many men of the world, neglect the man whom he had once patronized, but recommended Fenton to the patronage of Lady Trumbull, who appointed him tutor to her eldest son. Thus happy in an employment which did not preclude his favourite pursuit, our poet continued for several years to enjoy that learned ease so congenial to his love of quiescence and seclusion. His chief poetical effort while under the roof of Lady Trumbull, was the tragedy of Mariamne, which was performed at one of the London Theatres in 1723 with considerable approbation. When this tragedy was shewn to Cibber, who was then not only Poet Laureate, but Manager of a Theatre, he rejected it as unfit for representation, and with the petulance commonly arising from "the insolence of office," he advised the author to engage himself in some handicraft employment by which he might obtain that subsistence which he could never receive from his poetry. The event was more propitious than was foretold by the critic, for the profits arising from the tragedy amounted to nearly one thousand pounds, which enabled the poet to pay some debts which he contracted during his attendance at Court.

It would be superfluous to review a tragedy which is no longer represented; Mariamne has several beauties as a poem, but there is a degree of uniformity in the construction of the verse which tires by its stiffness, and is unlike that familiarity of dialogue and redundancy of measure, so happily adopted by Shakspeare and Otway.

About this time Fenton translated the eleventh book of Homer's Odyssey into English blank verse, and as Pope had now completed his justly-celebrated translation of the Iliad, and engaged to translate the Odyssey, he employed our poet, and Mr. Broome, as his coadjutors in the work. The books allotted to Fenton were the first, fourth, nineteenth, and twentieth, and he performed his task to the entire satisfaction of his friend and employer, and all the admirers of English poetry. Some critics have observed in approbation of the abilities of Fenton and Broome respecting this masterly translation, that the books translated by them are not inferior to nor distinguishable from those of Pope; but there can be no doubt that a poet who was so elegant and exact as the sweet bard of Twickenham, carefully corrected the productions of his associates, as well as his own, before they were committed to the hands of the printer. From the specimen given by Dr. Johnson in his admirable life of Pope, of the care and exactitude of that great poet in his corrections of the translation of the Iliad, he was undoubtedly as careful to maintain his merited reputation undiminished in his subsequent version of Homer's inferior work. Hence that regularity, harmony, and vigour, which pervades both the translations, were derived from the masterly touches of the great harmonist of the poetic lyre.

It was indeed highly honourable to the genius of Fenton to be considered a worthy consociate of Pope, and that he was a man of elegant taste and considerable abilities as a critic, and an original writer in prose as well as verse, was proved by a brief biography of our greatest poet, Milton, which he prefixed to a new edition of the Paradise Lost and other poems, the punctuation of which was carefully corrected by our author.

When Fenton had prepared his pupil, Mr. Trumbull, for an introduction as a student in the University, he accompanied him from his mother's country-seat to Cambridge. He was afterwards employed by Lady Trumbull as auditor of her accounts, and he enhanced the innocent pleasures of elegant retirement by occasional visits to his friends in London. He also paid a yearly visit to his relations in Staffordshire, and from his high reputation and amiable manners, he was received with respectful affection by his kindred. In one of those visits, an elder brother made an entertainment for the family, and our poet observing that one of his sisters was absent, enquired the reason. He was informed, that in consequence of poverty, she was thought unworthy to make one of the family party, on which he refused to sit at the table till she was invited, and, as she lived in the neighbourhood, she soon came. During the entertainment he treated her with particular attention, and although the circumstance is not mentioned, there can be little doubt that he also gave her some pecuniary aid.

In the year 1729 Fenton published a splendid edition of Waller's Poems, with illustrative and entertaining notes. But the life of this blameless bard was now drawing towards a close. Habitually indolent, and when employed, chiefly engaged in the sedentary pursuits of study, he was attacked by the gout, of which he died at Lady Trumbull's seat at East-Hamstead, in Berkshire, in the year 1730.

As Fenton had lived beloved by cotemporary authors, he died lamented by them. But the most durable record of his virtues is the following Epitaph, written by Pope, and inscribed upon his tomb in East-Hamstead church-yard:

> This modest stone, what few vain marbles can,
> May truly say, Here lies an honest man,
> A Poet, blest beyond the poet's fate,
> Whom Heaven kept sacred from the proud and great:
> Foe to loud praise, and friend to learned ease,
> Content with science in the vale of peace.
> Calmly he look'd on either life, and here
> Saw nothing to regret, or there to fear;
> From Nature's temp'rate feast rose satisfied,
> Thank'd Heaven that he had lived, and that he died.

The person of Fenton is described as tall and bulky, inclining to corpulence. This "tendency to fat," as Walcot calls it, was undoubtedly increased by his habitual enjoyment of ease. He had no family or household to provide for, no children to feed, clothe, or educate. His life passed in an even tenor for many years before his death, insomuch that it might be said he could scarcely perceive the lapse of time except by the succession of the seasons:

Far from the mad'ning crowd's ignoble strife,
His sober wishes never learn'd to stray;
Along the cool sequester'd vale of life
He kept the noiseless tenor of his way.

This quiescence of character and life may be thought enviable; but without greater activity than that of Fenton, we should never have made those discoveries in arts and sciences, nor have arrived at that high degree of refinement as a civilized community, which we now possess. Yet in the solitary instance of Elijah Fenton, little blame is imputable to the individual, for as he was in no ostensible public capacity, his indolence could only be injurious to himself.

His character as a friend and companion was highly praised by Pope, his pupil the Earl of Orrery, and several other individuals who were fully competent to form a just estimation of his merit.

In his public character of poet, he must rank among the minor bards. He has produced no original piece which entitles him to stand in the same rank with such men as Milton, Shakspeare, Dryden, Pope, Young, Otway, Rowe, Akenside, Gray, or Cowper; but may be fairly classed with Gay, Prior, Collins, and many other bards who have contributed to the amusement if not to the edification of the reader.

After the masterly critique of Dr. Johnson, that greatest of English critics and biographers, it would be ridiculous to enter into an analysis of Fenton's original poems. "To examine his performances one by one," says that energetic writer, "would be tedious. His translation from Homer into blank verse will find few readers while another can be had in rhyme. The piece addressed to Lambarde is no disagreeable specimen of epistolary poetry; and his Ode to Lord Gower was pronounced by Pope the next ode in the English language to Dryden's Cecilia. Fenton may be justly styled an excellent versifier and a good poet."

JOHN ROBYNS.

This learned mathematician seems to have been one of those writers, whose works are preserved rather as curiosities of literature than for their intrinsic value. He was born in Staffordshire about the close of the fifteenth century, and entered a student at Oxford in

1516. In 1520 he was elected a Fellow of All Souls College, where he took his degree in arts, and was ordained. His studies, however, were more devoted to the sciences than theology, and as he lived in a superstitious age, his favourite pursuit was judicial astrology.

We are informed by Wood, that Robyns made such a progress in "the pleasant studies of mathematics and astrology, that he became the ablest person of his time for those studies, not excepting his friend Record."

In the year 1531, he took the degree of B. D. and was appointed one of the Chaplains to Henry VIII. In 1532 he was made by the King one of the Canons of his College in Oxford, and in 1543 preferred to the dignity of Canon of Windsor.

On the accession of Queen Mary, he was appointed Chaplain to her Majesty, and highly venerated by her for his great learning. He died at Windsor, on the 25th of August 1558, and was buried in St. George's Chapel.

Wood enumerates five productions of Robyns, all in Latin, and in manuscript. They are chiefly on Astrology and Astronomy. "All which books," says Wood, "were in MS. in the choice library of Dr. Thomas Allen, of Gloucester-hall. After his death, coming into the hands of Sir Kenelm Digby, they were by him given to the Bodleian Library, where they yet remain. It is also said, that the said Robyns hath written a book entitled De Portentosis Cometus, but such a thing I have not yet seen."

Whatever gratification the obsolete tracts of visionary philosophers respecting astrology may give the curious admirers of ancient books and manuscripts, mankind will be no loser if they should remain in the Bodleian Library, Ashmolean Museum, or even the British Museum, till consumed as the proper food of the book-worm.

ISAAC HAWKINS BROWNE.

Few individuals of any age or nation have obtained literary fame upon easier terms than the subject of the following biographical sketch. But the reason is obvious. His principal production was written in Latin, a language which has been sanctioned by the custom of ages as the great essential in the acquisition of knowledge, and hence, mere smatterers, ambitious to be thought men of classical taste, have lavished their encomiums on what they did not understand.

Isaac Hawkins Browne, the son of the Rev. William Browne, minister of the parish of Burton-upon-Trent, was born in that town, January 21, 1706. His father was possessed of a small inheritance, but his fortune was increased by his marriage with Anne, daughter and heiress of Isaac Hawkins, Esq. and thus, happily for our author, he was from his earliest infancy blest with all the aid which opulence could give to facilitate his studies. He received the rudiments of a classical education at the grammar-school in Lichfield, and was afterwards sent to Westminster-school, where he made a commendable proficiency in scholastic attainments.

In the seventeenth year of his age, Mr. Browne was sent to Trinity College, Cambridge, where he applied with success to the various branches of mathematical science and the principles of the Newtonian philosophy. He also cherished his bias for classical literature, and obtained the degree of Master of Arts.

In May 1724 George I. established a foundation at both the Universities for the study of modern languages and history, to qualify the students for employments at Court and on Foreign Embassies. Mr. Browne was one of the young gentlemen selected to be scholars upon this foundation; and on the death of the King, he testified his gratitude by a copy of elegiac verses, for which he was complimented by the Professor of Poetry at the University.

In 1727 Mr. Browne went to London, and commenced the study of the law at Lincoln's Inn. In the course of several years he acquired considerable professional knowledge; but he never arrived at any eminence as a practitioner, and in consequence of the possession of a competent income, he relinquished the profession many years before his death.

During his residence at Lincoln's Inn, he wrote his Poem on Design and Beauty, which he inscribed to Mr. Highmore the painter, who was his particular friend. His ideas of Beauty and Order are borrowed from the principles of the Platonic Philosophy; but though a poem has been called "*a speaking picture*," the practical artist will derive little instruction from the futile illustrations of Mr. Browne on a subject which admits of such diversity. In the *jeu des esprit*, which he entitled "The Pipe of Tobacco," he imitated the manner of no less than five authors then living, with considerable humour and effect. That he might successfully imitate the peculiarities in the style of Cibber, Thomson, and Swift, whose poetical poems were little superior to his own, will not be denied, but Pope and Young were far above his imitation. This innocent and comic effusion was however amusing, and consequently popular.

In the twenty-ninth year of his age, and on the 10th of February, 1734, Mr. Browne married Jane, daughter of the Rev. Dr. Trimnell, Archdeacon of Leicester. His union with this lady, who is described as of a most amiable disposition, was highly conducive to his domestic felicity.

Mr. Browne continued in the uninterrupted enjoyment of his favourite study of the classic authors for several years after his marriage; and as he lived with unostentatious economy, his income was fully adequate to supply all the genteel accommodations of life. The birth of his son Hawkins also contributed to his happiness, while an intercourse with several learned and ingenious friends gave an additional zest to his elegant pursuits. Thus he lived till the year 1744, when he was introduced to public life, and elected a representative in parliament for the borough of Wenlock, in Shropshire, by the interest of Mr. Forester, a gentleman of fortune and influence in that county.

On the dissolution of parliament in 1748, Mr. Browne was re-elected to represent Wenlock by a great majority of the electors, who expressed their entire approbation of his senatorial integrity.

As Mr. Browne had been admired for his colloquial powers in the social circle, his friends expected that he would have been distinguished for his eloquence in the House of Commons. But in that celebrated assemblage of the greatest orators and statesmen of the age, Mr. Browne felt his comparative inferiority. It is one thing to flourish the pen in the privacy of retirement, when thoughts flow freely without interruption or embarrassment; and another to expatiate on an important subject in the presence of the representatives of a great nation. Mr. Browne consequently, like many other men of great intellectual powers, contented himself with expressing the monosyllables *aye* or *no*, according to the dictates of judgment and patriotism.

In the year 1754, our author published his great work entitled "De Animi Immortalitate." This Latin poem certainly excited a strong emotion among the classical scholars of the day, for in a few months after it was published, two English translations of it in rhyme and one in blank verse, gratified the public. Mr. Browne's poem on "The Immortality of the Soul," is divided into two books, to which he intended to have added a third, and of which he left a fragment. In these three books, it is said, that he meant to give a full explanation of Natural Religion, and thus to lay the foundation of Christianity. But a candid and even critical examination of the two books now before the public, must convince the reader, that a passion for classic elegance rather than a zeal for Revelation, animated the author. There is nothing indeed in the poem contrary to Christianity, but little in illustration of its divine truth. The author keeps up that philosophical tone which he borrowed from Lucretius, Seneca, and Lucan; and his reasoning, even in the most interesting passages, does not seem superior to that of Plato, Socrates, or any wise heathen, on the important subject of Immortality. In this spirit he says in the first book:

Well worth our search discoveries may be made
By *Nature*, void of the celestial aid:
Let's try what her conjectures then can reach,
Nor scorn plain *Reason*, when she deigns to teach.
* * * *
That mind and body often sympathize
Is plain; such is this union Nature ties:
But then as often too they disagree,
Which proves the soul's superior progeny.
Sometimes the body in full strength we find,
While various ails debilitate the mind;
At others, while the mind its force retains,
The body sinks with sickness and with pains:
Now did one common fate their beings end,
Alike they'd sicken, and alike they'd mend.
But sure experience, on the slightest view,
Shows us that the reverse of this is true;
For when the body oft expiring lies,
Its limbs quite senseless, and half-closed its eyes,
The mind new force and eloquence acquires,
And with prophetic voice, the dying lips inspires."

These are pleasing illustrations of the immortality of the human soul, yet they are merely such as any sagacious and ingenious heathen might have derived from reflection and observation.

At the conclusion of the second book, Christianity has thrown some irradiations on his view of a future state:

If ev'ry rule of equity demands
That vice and virtue from th' Almighty's hands
Should due rewards and punishments receive,
And this by no means happens whilst we live,
It follows, that a time must surely come
When each shall meet their well-adjusted doom:

Then shall this scene, which now to human sight
Seems so unworthy, wisdom infinite;
A system of consummate skill appear,
And ev'ry cloud dispers'd, be beautiful and clear.

From these two most unexceptionable passages of this over-praised poem, selected from the translation of it by Soame Jenyns, which is allowed to be the most accurate, the English reader will be enabled to conceive an idea of its general merit. If a mere author by profession had produced such a work, it would scarcely have been mentioned, but as it was the production of a scholar, a senator, and a gentleman of independent fortune, it was sure to find admirers. Mr. Browne undoubtedly deserves the approbation due to good intentions, and his own comparatively blameless life afforded the best proof of his sincerity as a professing christian. In all the social duties too, of son, husband, father, friend, and master, he was exemplary and amiable. His principal attention during the latter part of his life was directed to the education of his only son, and some portion of his time was also devoted to a correspondence with his friends, many of whom were men of high character and influence as statesmen and literati. Towards the close of his life Mr. Browne, whose constitution never was very vigorous, felt the infirmities of premature old age stealing upon him, and after languishing some time in a lingering illness, he expired on the 14th of February 1760, in the 55th year of his age, at his house in Great Russell Street, Bloomsbury-square.

The public as well as the private character of Mr. Browne was unimpeachable. He was an unostentatious patriot, who ever voted in the Senate according to the dictates of his conscience. In private life he was both an estimable and an amusing companion, possessing great cheerfulness and equanimity; and his conversation was occasionally enlivened by spontaneous sallies of genuine but inoffensive humour. As a poet, his productions entitle him to no higher rank than that of a minor author; his short and light pieces are playfully trivial, but his didactic poems, especially his poem on the Immortality of the Soul, are calculated to inspire the reader with elevated ideas of futurity, and of the justice and philanthropy of the Deity.

Since his death, "De Animi Immortalitate" has been translated by the Rev. Mr. Grawley, of Huntingdonshire; and another, and what is considered by most critics the best translation of that poem, was published by Soame Jenyns, Esq. a poet who, like himself, devoted his talents to truth and virtue. An edition of Mr. Browne's poems was published in the year 1768 by his son, Mr. Hawkins Browne. This edition is adorned with a very fine portrait of the author, engraved from an original painting by his friend Highmore.

BISHOP HURD.

If Staffordshire can boast of having given birth to characters who have upheld the honour of their country by their prowess and gallantry, she can likewise exult in her sages and philosophers, whose talents and acquirements have illumined the page of modern history. Independent of her Colossus of Literature, and other profound scholars, she may proudly recount the virtues and unostentatious wisdom of a HURD, who, whilst he added lustre to his high calling, thought it no condescension to regard with complacency the virtuous, though less exalted, among his fellow-men, to deal abroad that brotherly love so forcibly inculcated by the doctrines of Christianity, and who, through a long and honourable course of years, was indefatigable in illustrating, both by his preaching and living, the important truth, that earthly honours are only valuable in proportion as they are seconded by the virtues of their possessors, and that not even the dignified assumption of the mitre can shed one ray of glory on the brow of him who is not a good man and a practical Christian.

This accomplished and erudite Prelate was born at the village of Congreve, in this county, on the 13th of January, 1719-20, at which place, at the time of his birth, his father rented a considerable farm, but soon after removed to a larger at Pendeford, about midway between Wolverhampton and Breewood. He received the rudiments of his education at Market-Bosworth, in the county of Leicester, under the tuition of Anthony Blackwall, the well-known editor of the Sacred Classics; and was afterwards removed to the free grammar-school at Breewood, under the Rev. Mr. Hillman, and, upon his death, under his successor, the Rev. Mr. Budworth, whose memory Mr. Hurd affectionately embalmed so early as 1757, in a Dedication to Sir Edward Littleton, who had also been educated at Breewood school. He continued under the care of this gentleman till he went to the University; and was admitted at

Emmanuel College, Cambridge, October 3, 1733, but did not take up his residence till nearly two years afterwards. In this college he received the countenance, and was permitted to attend the Lectures, of that excellent tutor, Mr. Henry Hubbard, although he had been admitted under another person. He took the degree of B. A. in 1738-9; proceeded A. M. and elected Fellow in 1742; was ordained Deacon in St. Paul's Cathedral, on the 13th of June, in the same year; and took Priest's orders in the Chapel of Gonville and Caius College, Cambridge, on the 20th of May, 1744. About this time he formed an intimate friendship with the celebrated Dr. Warburton, Mr. Mason, and several other celebrated literary characters. In 1749, he took the degree of B. D. and the same year published his Commentary on Horace's Ars Poetica, with a complimentary preface to Mr. Warburton, who soon after returned the eulogium in his edition of Pope's Works, wherein he speaks of Mr. Hurd's Commentary in terms of unqualified approbation. This work was "distinguished by a new consideration of the subject, and a fancied discovery of a systematic plan, which he supposed to have been adopted by the Roman poet. Mr. Hurd was dissatisfied with the opinion entertained by Scaliger and Heinsius, that the Ars Poetica was a collection, though not a system, of criticism on poetry in general. His dissatisfaction resulted from an idea, that the purpose of the author was, neither to abridge the Greek critics, nor to amuse himself with composing a short critical system for the general use of poets, but simply to criticize the Roman drama." Extensive erudition and refined taste are conspicuously displayed in this work, but the views of the author upon the subject have not been generally assented to. Though this was avowedly the first performance of Mr. Hurd, yet he is known to have been the author of a "fine pamphlet," as Warburton calls it, entitled Remarks on an Enquiry into the Rejection of the Christian Miracles by the Heathens, which appeared in 1747; and a copy of verses, by Mr. Hurd, on the Peace of Aix-la-Chapelle, is printed in the University Collection of 1749. In 1750, he was, by the recommendation of Mr. Warburton, appointed one of the Whitehall Preachers; and next year appeared his excellent Commentary on the Epistle to Augustus, a work every way worthy the talents of the author; and two pamphlets on the subject of some internal divisions in the government of the University, which took place at that period. In 1752, he published the Mischiefs of Enthusiasm and Bigotry, an assize-sermon preached at Norwich;

and the following year he was presented to the donative curacy of St. Andrew's the Little, in Cambridge, commonly called Barnwell, and published a Sermon delivered at Trinity Church, Cambridge, March 28, being the time of the annual meeting of the children educated in the charity-schools of that town. On the 27th of November, 1755, he had to lament the death of his father, at the age of 70, which circumstance is noticed in an affecting letter printed in Mr. Hurd's correspondence with Bishop Warburton.

The friendship which had taken place between Dr. Warburton and Mr. Hurd, had, from its commencement, continued to increase by the aid of mutual good offices; and an opportunity now offered for the latter to shew the warmth of his attachment. Dr. Jortin having spoken of Mr. Warburton with less deference and submission than the claims of an overbearing and confident superiority seemed to demand, was called to account for it in a pamphlet entitled The Delicacy of Friendship, "which, while it gave the most heartfelt satisfaction to his friend, severely hurt the feelings of Dr. Jortin. This circumstance Mr. Hurd afterwards so much regretted, that he expressed the most earnest wish that the essay should be suppressed."

Though Mr. Hurd's reputation for genius and learning had been long established, we do not find that his merit had attracted the notice of the great, or that any disposition had appeared at this time to advance his fortune. He still continued to reside at Cambridge in learned and unostentatious retirement, till, in December 1756, he became, on the death of Dr. Arnald, entitled to the rectory of Thurcaston, as senior Fellow of Emmanuel College, on Mr. Hubbard's declining it, and was instituted in it Feb. 16, 1757: he accordingly took up his residence at this place, perfectly satisfied with his situation. In the same year appeared A Letter to Mr. Mason on the Marks of Imitation; and Remarks on Hume's Essay on the Natural History of Religion. This little tract was occasioned by some passages in Hume's life of himself, and is a most excellent ridicule and display of Hume's bad logic and reasoning; but Dr. Warburton is supposed to have had a considerable share in it. Hume appears to have felt its severity; for he declares that it is written "with all the illiberal petulance, arrogance, and scurrility which distinguishes the Warburtonian school." In 1762, he was presented to the sinecure rectory of Folkton; and two years afterwards appeared his celebrated Dialogues, which are said to have laid the foundation of his preferment in the Church.

He was soon afterwards made Preacher of Lincoln's Inn; and, on the death of Dr. Gekie, received the Archdeaconry of Gloucester, at the recommendation of his friend Warburton. On July 5, 1768, he was admitted D. D. at Cambridge, and on the same day appointed to open the Warburtonian Lecture. In the same year he published an edition of Cowley's Works, with a preface and notes, in which he is accused of having omitted some of the most distinguishing beauties of that poet. In 1770, he lost his steady friend and patron Mr. Yorke; and in 1773, his mother died at the advanced age of 88. About this time an intimacy took place between Dr. Hurd and Lord Mansfield, who was a great admirer of the Doctor's talents; and on February 12, 1775, he was consecrated Bishop of Lichfield and Coventry, which dignified preferment is said to have been reserved for him by the King, as a reward for the pleasure which his Majesty derived from the perusal of one of his Dialogues. In the same year the new Bishop gave to the world a republication of Dr. Jeremy Taylor's Demonstration of the Truth of the Christian Religion, and the following year he was appointed Preceptor to the Prince of Wales and Prince Frederick.

The Bishop, having employed the interval between 1776 and 1781 in the discharge of his episcopal functions, and the publication of a number of excellent sermons, he was, in the latter year, on the death of Dr. Thomas, Bishop of Winchester, translated to the Bishopric of Worcester, on the collation of the Hon. Dr. Brownlow North to the vacant See of Winchester. About two years afterwards he had the honour, in conjunction with Dr. Louth, Bishop of London, of refusing the Archbishopric of Canterbury, it having been offered him on the death of Archbishop Cornwallis, but was subsequently given to Dr. Moore, Bishop of Bangor. On his arrival at Hartlebury Palace, he thoroughly repaired the castle, built a new library, and furnished it with the greater part of Dr. Warburton's collection, which he had purchased on the death of that prelate, and to which he made considerable additions, at various times. Bishop Hurd had on different occasions received tokens of the Royal regard; and in 1788, he received a visit at Hartlebury from their Majesties and the Princesses, who were then taking the waters at Cheltenham. In 1807 the Prince of Wales and Prince Frederick visited their venerable preceptor. The Bishop passed the remainder of his days at Hartlebury, in the constant discharge of his pastoral duties, and the exercise of those virtues which exalt and dignify human nature; and, after having filled the See of Worcester twenty-seven years, he expired in his sleep on the 28th of May 1808, having completed four months beyond his eighty-eighth year. He was, according to his own directions, buried in Hartlebury church-yard.

Of Bishop Hurd's character (says his biographer) little need be said. Where calumny had not ventured even to insinuate a fault, and where respect and reverence were his constant attendants, it would be unnecessary to expatiate on good qualities. As a writer, his taste, learning, and genius, were universally confessed. His sermons are read with no less advantage than they were delivered. With his friends and connections he had obtained their best eulogium, their constant and warm attachment; and with the world, a kind of veneration, which in times like the present could neither be acquired nor preserved but by the exercise of great virtues.

A complete edition of Bishop Hurd's Works, in eight volumes 8vo. was published in 1811, as were also editions of Addison, and his friend Warburton, all prepared for the press by the Bishop.*

RICHARD CALDWELL.

Richard Caldwell was born at Overholm, in Staffordshire, in the year 1530. He was educated at Brazen-nose College, Oxford, of which he was elected fellow. He took his degree of M. A. in 1539, became one of the senior students of Christ Church in 1547, and having studied physic, and passed through the gradations of that faculty, he became an eminent physician. His reputation for medical skill was so great, that he was elected Censor of the College of Physicians, soon afterwards chosen one of the Elects of that Institution, and in 1570 raised to the honourable station of President. We are informed by Wood that he wrote several pieces on the Art of Healing, but none of them have been preserved.

In 1585, Dr. Caldwell translated and published a medical work, written by Moro, a Florentine physician. He also founded a Chirurgical Lecture in the College of Physicians, and endowed it with a salary; and it appears from the few documents respecting him

* Nichols's Literary Anecdotes, Vol. VI. p. 468, et seq.

which have been preserved, that he was not only a skilful but a benevolent physician. It does not appear that Dr. Caldwell was married. He died in London in 1585, aged seventy-three years, and was buried in the Church of St. Bennett, near Paul's Wharf.

CHARLES COTTON.

This eccentric genius was the son of Charles Cotton, Esq. of Beresford or Barrisford, a manor on the banks of the Dove, in Staffordshire. He was born at Beresford on the 28th of April 1630, and was educated at the University of Cambridge. His acquirements while he continued at College, are stated by different biographers to have been very great, particularly his knowledge of the Greek and Roman classics. He also became master of the French and Italian languages. He did not however take a degree, or study with a view to any of the learned professions, but contented himself with those acquisitions as a scholar which should enable him to shine as a gentleman. The improvidence or imprudence of his father, who had wasted a considerable part of his patrimony in unsuccessful litigation, prevented him from the full indulgence of youthful curiosity by making the grand tour; but when he completed his education, he travelled into France and some other parts of the Continent. He did not continue long abroad; and soon after his return to England in 1656, he married Isabella, daughter of Sir Thomas Hutchinson, knt. of Owthorpe, in Nottinghamshire. With this lady he resided in the family-mansion during the remainder of his father's life, as his income did not enable him to keep up a separate establishment. His father died in 1658, and he came into possession of the estate of Beresford, in the twenty-eighth year of his age; but a considerable part of this estate had been mortgaged, and he was not blest with sufficient prudence and self-denial to recover it.

Mr. Cotton's first original production was "A Panegyric to the King's Most Excellent Majesty," written in the year 1660, at the time of the Restoration. It is preserved in the British Museum. In 1664, he published a translation from the French of M. de Vaix's "Moral Philosophy of the Stoics;" and in 1665, he translated the works of Horace from the French of Corneille, for the amusement of his sister.

During several years Mr. Cotton confined his literary efforts to translations from the French, or imitations of celebrated authors who wrote in that language, and in this respect he was allowed to excel his contemporaries. His translation of "The History of the Life of the Duke d'Espernon," was printed in folio and published in the year 1670, but notwithstanding the partiality of Charles II. and his polite and witty courtiers to every French production of genius, Mr. Cotton obtained but little profit from this work.

His agreeable manners and literary attainments had however increased the number of his friends, one of whom presented him with a Captain's commission in a regiment of infantry. He went over to Ireland with his regiment in 1670, and some adventures which he met with during his march, and while on his voyage, gave rise to his first humourous poem, which he entitled "A Voyage to Ireland." This ludicrous production consists of three cantos written in the anapestic measure, or what the Monthly Reviewers have not inaptly termed "*wheelbarrow verse*." In this poem he characterizes the Mayor of Chester as a superficial man, easily caught by shew. This head of an ancient Corporation was particularly struck, on coming out of Church, with the richness of Captain Cotton's regimentals, and especially with a gold belt which encircled the loins of the hero. In consequence of the lustre of the Captain's habiliments, the admiring Mayor invited him to supper, and regaled him sumptuously. The military poet gives a brief account of part of their conversation in the following lines:

> I answer'd, My country was fam'd Staffordshire,
> That in deeds, bills, and bonds, I was ever writ squire;
> That of lands I had both sorts, some good and some evil,
> But that a great part on't was pawn'd to the Devil.

As his appointment to a captainship was conferred in a time of profound peace, when the nations of Europe, after being harassed by ruinous wars, wished for a continuance of repose, there was no reasonable prospect of preferment, emolument, or fame. To a gentleman of Mr. Cotton's liberality, his pay was inadequate to his expenditure, he therefore resigned his commission, and returned to the calm retirement of Beresford-hall, and his pleasant recreations of alternate composition and angling. His residence on the banks of a beautiful river which abounded with fish, was peculiarly adapted to the solitary amusement of angling, and as an author can always render his most trivial recreations interesting to others by the graces of fine writing, Mr. Cotton gratified the admirers of the finny tribes by his "Instructions how to Angle for Trout or Grayling in a clear Stream."

In the year 1674, soon after his return to Beresford, he published his translation of "The Fair One of Tunis," a romantic French novel; and of the "Commentaries of Blaise de Montlieu, Marshal of France."

Mr. Cotton appears to have devoted himself to rural pursuits about this period. Indeed, the situation of his paternal mansion was calculated to inspire a taste for the beauties of nature. It stood on a fertile eminence, on the picturesque bank of the winding Dove, and commanded an extensive prospect of part of the counties of Stafford and Derby. His attachment to his natal spot is marked by the circumstance of his giving the name of the place to his only son Beresford.

Ever desirous to combine literary improvement with rural recreations, Mr. Cotton published his Planter's Manual, in 1675. This tract contains instructions for the cultivation of fruit-trees.

We now come to that period in his literary life in which he acquired more popularity, but less esteem, than by his former attempts with the pen. In imitation of some of the prurient wits of the French school, he wrote his "Scarronides, or Virgil Travestie; a mock poem, on the first and fourth books of Virgil's Æneis, in English burlesque." This precious piece of ribaldry was published in 1678, when licentiousness and profaneness were at the height, not only in the Court of Charles II. but in London, and throughout England; a time when decency of manners, and a regard for truth or the sacred Scriptures, were stigmatized with the opprobrious names of puritanism and fanaticism: in short, when Rochester, Sedley, Dryden, and Mrs. Behn, contributed to the utmost of their abilities to pervert the public mind. This appeared to Cotton the auspicious moment, when a witty and profligate country-squire might hope to redeem his mortgaged lands by contaminating the public morals.

The success of Mr. Cotton's burlesque poem was at least equal to his expectation; and from the rapid sale of several editions, there can be no doubt that it perverted the taste and the morals of tens of thousands. His success in turning to ridicule the greatest epic poem of which ancient Rome could boast, encouraged him to attack another classic author, and publish his last comic performance, entitled "Burlesque upon Burlesque, or the Scoffer Scoffed; being some of Lucian's Dialogues newly put into English fustian."

This little work also passed through several editions; but if we may judge from the event, both these humourous productions did not realize the philosopher's stone to the author.

There is no record of the demise of Mr. Cotton's first wife; no monument consecrated by conjugal affection marks the spot where her relics rest, no elegiac strain from her poetical spouse celebrates her domestic virtues and the anguish and regret of the survivor. This neglect of the virtuous mother of his children on his part was succeeded by a retributive oblivion of his own remains a few years afterwards.

Mr. Cotton afterwards paid his addresses to a noble widow. He was a successful suitor, and was married to Mary Countess Dowager of Ardglass, widow of Wingfield Lord Cromwell, second Earl of Ardglass. This lady was possessed of a jointure of fifteen hundred pounds a-year, which was secured by law from the imprudence of our poet.

In the year 1681, Mr. Cotton published "The Wonders of the Peak," an original poem. This descriptive piece did not add to his poetical reputation. What poet indeed can hope, after Dante's Inferno, to gratify the imagination of the reader with ideal views of subterranean horrors. What imagery can the bleak and barren hills of Derbyshire, and their hideous caverns, afford for the gratification of the lover of nature? Mineralogists indeed may profit by the exploration of valuable strata, but the muses turn away in disgust from subjects so little suited to the regions of fancy.

The last great effort of Mr. Cotton's mind, was a translation of Montaigne's Essays, which he inscribed to George Saville Marquis of Halifax, a courtly son of Apollo. This polite nobleman was highly gratified with such a mark of respect from the celebrated author of "Virgil Travestie," and expressed his satisfaction in strong terms of approbation. The translation is allowed by the best critics to possess the merit of expressing the sense of the original with great felicity.

After his second marriage he spent much of his time in the Metropolis; and the fortune possessed by his lady doubtless contributed to his comfortable accommodation in the decline of life. But his personal estate still continued in a state of temporary alienation, and he died insolvent in the parish of St. James, Westminster, in 1687, in the fifty-eighth year of his age. It is recorded that Elizabeth Bludworth, his principal creditor, administered to his effects, his widow and children having previously renounced the administration. In 1694 Mr. Beresford Cotton published the "Memoirs of the Sieur de Pontis," translated by his father from the French.

From the foregoing narrative of the principal events of Mr. Cotton's life, and of his literary productions, it is sufficiently evident that his character was inconsistent. At one time devoted to the innocent pleasures of retirement, and at another emulous of distinction among the wits and humourists of the most immoral age upon record in English history. With his natural gaiety, and that suavity of manners which he acquired during his tour on the Continent, and his occasional visits to London, Mr. Cotton became what was then termed a man of wit and pleasure; he was consequently profuse in his expenditure, prodigal of his time, and careless of his morals. Morality was in that age considered by the fashionable part of the community as a relic of puritanism unworthy of an accomplished gentleman and are fined scholar; consequently, dissipation became the theme of panegyric with poets and buffoons: and all the men of spirit of the day laughed at religion. That poor Mr. Cotton sailed with the stream is but too clearly illustrated by his life, his writings, and his death. In an epistle to one of his poetical friends, he thus characterises himself:

> He always wants money, which makes him want ease;
> He's always besieg'd, though himself of the Peace,
> By an army of duns, who batter with scandals,
> And are foemen more fierce than the Goths or the Vandals.

Such is the inconsistency of the human character, that Mr. Cotton's best productions were not published till after his decease. In the volume of his poems published in 1689, there are indeed some pieces containing a heterogeneous mixture of broad humour with delicacy of sentiment, and some of them are moral and devotional. These smaller poems, by far the best productions of his mind, had never been re-published till they were admitted into the late edition of the English Poets, probably to fill up the volumes for the benefit of the booksellers. But his ludicrous compositions, particularly the "Virgil Travestie," passed through *fifteen editions* from the year 1678, in which it was first published, to the year 1771; a strong proof of the depravity of those ages, the moral taint of which is from time to time revived by the disgusting ribaldry published by a description of men well described by one of our reviewers by the characteristic epithet of "*dirty publishers.*"

Mr. Cotton's burlesque poems are founded upon the model of

Butler, but have neither the humour nor the point of Hudibras. Butler only wished to turn the laugh against fanaticism, and he succeeded; but Cotton seems to have written for no other purpose than to gratify the mere sensualist. When we see a man of parts like Cotton employing his powers for the mere gratification of his own lowest passions, or those of his fellow-creatures, we feel regret at this unhallowed prostitution of talents, which properly applied, might have enlightened and improved the possessor, and millions yet unborn! Like a baneful meteor, his genius blasts all that come within the sphere of its pernicious influence, instead of shining like a benignant star, to guide the wanderer through the mazes of life to the path of virtue and happiness.

DR. GEORGE SMALLRIDGE

Was born in Lichfield, in the year 1663. He was educated at Westminster School, where he distinguished himself by his classical acquirements, and in the sixteenth year of his age, was admitted as a student in Christ Church College, Oxford, where he took the degree of A. M.

At a very early age he ventured into the labyrinth of controversy, and in the year 1687 published "Animadversions on a Piece upon Church Government." He soon afterwards went into orders; and on the 12th of June 1693, he was admitted a prebendary in Lichfield Cathedral. By a regular gradation of preferment, Mr. Smallridge was appointed minister of the new chapel in Tothill-fields; Dean of Carlisle, November 29, 1711; thence removed to the Deanery of Christ Church, Oxford, in July 1713; and consecrated Bishop of Bristol, April 4, 1714.

On the accession of George I. to the throne, he was appointed Lord Almoner to his Majesty, but lost that situation for refusing to sign the declaration of the Bishops against the rebellion in 1715. From his intimacy with Whiston he was suspected of Arianism; but he vindicated himself from this charge of heresy, in a letter to the Bishop of Winchester, which he published. He also published some volumes of sermons, equally remarkable for the purity of style, and soundness of morality. The Bishop died on the 17th of September 1719, in the 53d year of his age, and was buried in the north aisle of Christ Church, near the choir, without any inscription on his tomb.

BISHOP NEWTON.

THOMAS NEWTON, the son of John Newton, wine-merchant, was born in Lichfield, on the 1st of January 1707. He received the rudiments of his education in the free grammar-school of his native city, and in the tenth year of his age was removed to Westminster School, where he continued six years, five of which he studied as a King's scholar in the College. In 1723 he went to Cambridge, and entered at Trinity College, where he resided eight months in the year, till he had taken his degree of Bachelor of Arts. He was afterwards chosen Fellow of his College, and ordained Deacon in December 1729, and Priest in February 1730. He then went to London, the great scene for the display of genius, and was appointed Curate of St. George's, Hanover Square. In this church he continued for several years to officiate as assistant preacher to Dr. Trebeck, and at length obtained the preferment of reader and afternoon preacher at Grosvenor Chapel, in South Audley Street, where he soon became distinguished for the eloquence and perspicuity of his sermons, and was patronized by the Earl of Bath, who nominated him his first chaplain, and presented him to the rectory of St. Mary-le-Bone in 1744. In the memorable year 1745, when the whole kingdom was agitated by the Scotch rebellion, and party-spirit assumed a fierce and malignant character, Mr. Newton distinguished himself as a strenuous champion for the House of Brunswick, and an enemy to popery, and obtained much celebrity by the explicit and determined manner in which he asserted the cause of his Church and King. His enemies had recourse to the dastardly expedient of anonymous threatening letters, which must have given him some uneasiness, as he shewed them to his patron Lord Bath, who advised him to transmit them to the Secretary of State. Here the matter rested, for the ill-concerted rebellion was soon quelled, and public tranquillity restored; but the zeal of Doctor Newton, for he had taken the degree of D. D. was not forgotten. In 1747 he was chosen Lecturer at St. George's, Hanover Square, and the same year he married a daughter of his former coadjutor Dr. Trebeck. He preached and published a pathetic sermon on the death of Frederick Prince of Wales, which obtained him the appointment of Chaplain to the Princess Dowager.

Thus Dr. Newton gradually and steadily advanced in the estimation of all who knew him; equally respected as a zealous and eloquent pastor, a man of learning, and an ornament to the Protestant Church. In 1750, he was appointed Chaplain to the King, Prebendary of Westminster, and Precentor of York. His wife died in 1754; and in September 1761, he married the widow of the Rev. Mr. Hand. He had been elevated to the Deanery of Salisbury; and in 1761, he received a still higher proof of his Sovereign's esteem by being appointed Bishop of Bristol. In 1764, he was offered the Primacy of Ireland, which he modestly refused, being content with the high clerical honours already conferred on him.

Bishop Newton continued in the conscientious discharge of his important duties till the year 1769, when his health declined, and during the last thirteen years of his life he was a valetudinarian, and bore his indisposition with pious fortitude. On the 14th of February 1782, after a long and well-spent life, this virtuous prelate expired without a sigh or the least emotion.

Bishop Newton was the author of several works, the principal of which, entitled "Dissertations on the Prophecies," is held in high estimation for its orthodoxy.

THOMAS ASTLE.

Thomas, the son of Daniel Astle, keeper of Needwood-forest, was born at Yoxall in the year 1734, and having received a good education, was bound apprentice to an attorney in his native town. At the expiration of his apprenticeship, his father wished him to settle at Yoxall, and practise as an attorney, but young Astle, who had, in consequence of his being employed in decyphering ancient records, acquired a taste for antiquities, went to London in order to improve in his favourite study.

In 1763, he was so fortunate as to obtain the patronage of Mr. Grenville, First Lord of the Treasury and Chancellor of the Exchequer; and under the auspices of that celebrated statesman he was employed as an assistant to Sir Joseph Ayliffe and Dr. Ducarel, in the regulation of the Public Records at Westminster. He was afterwards patronized by Lord Townshend, and became acquainted with the Rev. Philip Morant, author of the History of Essex, a gentleman of considerable property in that county. In consequence of his intimacy with Mr. Morant, he was introduced to the family, soon afterwards married his daughter and heiress, and on the demise of his father-in-law, came into possession of an ample estate.

In the year 1765, Mr. Astle was appointed Receiver-General of six-pence in the pound on the Civil List; and in 1766, he was employed in the preparation of the Harleian Catalogue of Manuscripts in the British Museum, a work which engaged his principal attention during five years.

On the death of Henry Rooke, Esq. his Majesty's Chief Clerk in the Record Office, in the Tower of London, Mr. Astle was appointed his successor in that office, and was soon afterwards appointed Keeper of the Records. His character as an antiquary was now established, and he became a member of the Royal and Antiquarian Societies, and an honorary member of several learned societies on the Continent. He was for several years a distinguished member of the Antiquarian Society, during which he contributed several articles to the Archæologia. He was also a Trustee of the British Museum.

In 1775, Mr. Astle published "The Will of King Henry the Seventh," in quarto; "A Catalogue of the Manuscripts in the Cottonian Library, to which are added many emendations and additions, with an appendix containing an Account of the Damage sustained by the Fire in 1731; and also a Catalogue of the Charters preserved in the same Library, prepared by him for the press, and published by S. Hooper, in one volume octavo, in 1777. In 1784, Mr. Astle published, in quarto, "The Origin and Progress of Writing, as well hieroglyphic as elementary; illustrated by Engravings taken from marbles, manuscripts, and charters, ancient and modern: also, some Account of the Origin and Progress of Printing." A new edition of this valuable work, which was indeed the author's masterpiece, was published in 1783, with an additional copper-plate engraving from a manuscript in the British Museum, marked Nero, D 4. This edition was also embellished with a portrait of the author, engraved by Shelton, from a painting by Howard, in which the accidental loss of an eye while at school is concealed.

The Will of King Alfred, found in a Register at Winchester, and which was given by Dr. Lort to Mr. Astle in 1769, was printed in 1788, at Oxford, with illustrations, under the superintendence of Sir Herbert Croft. An Account of the King's Royal Burghs and Magnates of Scotland, was prepared for the press by Mr. Astle, and published with fine plates in one volume folio, in 1798. The Calendar to the Patent Rolls in the Tower of London, from the third year of the reign of John to the twenty-third of Edward IV. was printed from four MS. volumes, procured in 1775, by Mr. Astle, for public use, from the executors of Henry Rooke, Esq. his predecessor in the office of Keeper of the Records in the Tower, and collated with two Manuscripts in the Cottonian Library, marked Titus c 2 and 3.

Mr. Astle's principal residence was at Battersea-rise, a beautiful eminence contiguous to Clapham-common. His library was richly furnished with a large collection of books and manuscripts, which afforded the highest gratification to the most celebrated British antiquaries of the age, who were his frequent visitors. Among his most curious manuscripts was a series of original Saxon Charters, unequalled in number, beauty, and preservation. He died of a dropsy, on the 1st day of December 1803, in the 69th year of his age, leaving eight sons and daughters, by whom he was deservedly venerated and beloved.

Mr. Astle's library was purchased by the Royal Institution for £1000. His manuscripts were, according to his will, to be offered first to the Marquis of Buckingham, and on his declining the purchase, to the British Museum.

Mr. Astle was not only a learned antiquary but a liberal contributer to the history and antiquities of his country; and very willing to give assistance to gentlemen employed in historical investigation. His principal work is "The Origin and Progress of Writing." He was fortunate in his matrimonial connexion, and his antiquarian researches were facilitated by the possession of opulence. In private life his manners were irreproachable, and he was much esteemed by his learned contemporaries.

DR. JOHN LIGHTFOOT.

John, the second son of the Rev. Thomas Lightfoot, was born at Stoke-upon-Trent, on the 29th of March 1602: his mother's maiden name was Bagnall, a gentlewoman of a good family. The Rev. Thomas Lightfoot was in holy orders fifty-six years, thirty-six of which he was vicar of Uttoxeter, and he died July the 21st 1658, in the eighty-first year of his age. He had five sons: the eldest, named Thomas, was brought up to trade; the second, John, the subject of this memoir; the third, Peter, a physician, who practised at Uttoxeter; the fourth Josiah, who succeeded his brother John in the living of Ashley, in Staffordshire; and the fifth and youngest, Samuel, was brought up to the church.

Our author was instructed in the elements of classical learning by Mr. Whitehead, at his grammar-school on Moreton-green, near Congleton, in Cheshire. He was removed in the sixteenth year of his age to Cambridge, and placed under the tuition of Mr. William Chappel, Fellow of Christ's College. His preceptor, who was also the instructor of Henry More, Milton, and other eminent men, was afterwards Bishop of Cork, in Ireland. At college, our author applied himself to eloquence with such success, that he was considered the best orator of the under-graduates in the University. He was also distinguished for the extraordinary proficiency which he made in the study of Latin and Greek, but he neglected the Hebrew language.

When he had taken his degree of B. A. he left the University, and became an assistant to his former master, Mr. Whitehead, who then kept a school with great reputation at Repton, in Derbyshire. He continued at Repton about two years, and then entered into orders, and began his auspicious career as a divine, as curate of Norton-under-Hales, in Shropshire. This curacy lay in the vicinity of Bellaport, the seat of Sir Rowland Cotton, who was his constant hearer, nominated him chaplain, and received him as an inmate in his hospitable mansion. Sir Rowland was a proficient in the Hebrew language, and persuaded his chaplain to engage in that study, who applied himself to it with vigour; and when his patron removed with his family to London, at the request of Sir Allan Cotton his uncle, who was then Lord Mayor of that city, Mr. Lightfoot accompanied him thither. He had not been long in the metropolis before he resolved to go to the Continent for improvement, and he consequently went down into Staffordshire to take leave of his father and mother. In the course of his journey he came to Stone, which was then destitute of a minister, and by the persuasions of the parishioners he consented to preach there. His idea of going abroad was now relinquished; and having in the year 1628 obtained possession of the living, he married the daughter of William Crompton, Esq. of Stone Park. Soon after this event, he quitted his charge at Stone, and removed to London with his family, that he might the more successfully pursue his rabbinical studies, with the aid of the numerous books in the Hebrew language in Sion College. He now settled at Hornsey, in the vicinity of Islington, and in 1729 published his first work, entitled "Embhim; or Miscellanies, Christian and Judaical." As a proof of his gratitude to his patron and friend Sir Rowland Cotton, he dedicated his work to him. This compliment was not misapplied; for in 1631, Sir Rowland presented him to the rectory of Ashley, in Staffordshire.

In this comfortable situation, and blessed with competency, he now considered himself settled for life; and built a study in his garden, at a convenient distance from his mansion. In this happy seclusion he continued for twelve years to increase his biblical knowledge, and apply himself to the study of the Holy Scriptures. Thus employed the time passed agreeably away, but his studious serenity was interrupted by the political change which took place in 1642, and he was nominated a Member of the Assembly of Divines appointed to settle a new form of ecclesiastical government. He accepted this honourable trust from the most patriotic motives, and although he contended on some points with many of the most able innovators in the assembly, he had a favourable opinion of the Presbyterian form of church government.

In consequence of his nomination, he was obliged to reside in London, and resigned his rectory, but obtained the presentation for his brother Josiah. Soon after his arrival in London he was chosen minister of St. Bartholomew's, near the Royal Exchange.

The Assembly of Divines met in 1643, and our author made a distinguished figure in the debates, and gave many proofs of his learning and intrepidity in his opposition to many tenets which some of the divines endeavoured to establish. His abilities recommended him to the notice of the Parliament; and when Dr. William Spurstow was ejected from the mastership of Catherine Hall, in Cambridge, Lightfoot was appointed in his place, and presented to the living of Much Munden, in Hertfordshire. His residence at this time was at the upper end of Moor-lane. As he had, like other favourite ministers, preached before the Parliament, most of his sermons were printed, and in those eloquent pieces we see him warmly pressing the speedy settlement of the Church in the Presbyterian form, which he cordially believed to be superior to any other. When not engaged in his public duties, he employed his time in preparing and publishing the several branches of his "Harmony."

In 1652 he took the degree of Doctor in Divinity, and in 1655 he entered upon the office of Chancellor of Cambridge. He performed all the regular exercises for his degree with distinguished approbation, supplied the place of Professor of Divinity, at an act which was kept for a doctor's degree in that professorship, and was exemplary for the diligence and fidelity with which he discharged his duty as Vice-Chancellor. He was now engaged with other learned professors in completing the celebrated Polyglot Bible, and was patronized by Oliver Cromwell, then Lord Protector. Dr. Lightfoot expressed his satisfaction at this high patronage, in his speech delivered at the commencement of the work, and at the same time manfully expressed his commiseration for the oppressed clergy of the Church of England, and extolled their learning, zeal, and confidence in God.

On the restoration of Charles the Second, Dr. Lightfoot offered to resign the mastership of Catherine Hall to Dr. Spurstow, who declined it. Another person would have been preferred by the Crown, in which the right of presentation lay, but as Dr. Lightfoot's compliance with the measures of the Republican government had arisen rather from necessity than choice, Gilbert Shelden, Archbishop of Canterbury, procured him a confirmation both of his place and of his living. This attention on the part of the Archbishop arose entirely from his respect for the high character of Dr. Lightfoot for learning, abilities, and virtue, for he was not personally acquainted with him; and this circumstance was equally honourable to both parties, as it proved the liberality of the one and the merit of the other.

After the Restoration, Dr. Lightfoot was appointed one of the assistants at the Conference upon the Liturgy which was held in the beginning of 1661, but he attended only once or twice, being more intent upon the completion of his "Harmony," and his attention to his parishioners at Munden, among whom he lived in a hospitable and charitable manner, much beloved and respected. So great was his attachment to his flock, and his love of retirement, that he seldom left his rectory, except to perform the requisite residence at Cambridge and Ely. During his absence he frequently said "he longed to be with his russet coats."

Dr. Lightfoot was of a strong healthy constitution, and remarkably temperate; which enabled him to pursue his studies with unabated vigour and attention to the end of his life. He continued to publish, though in a letter to Buxtorf he declares "that he could scarce find any bookseller in England who would venture to print his works, and that he was obliged to print some of them at his own expence." But he was not the only author who found difficulties in publishing, for Frederic Miege, in a letter, informed him "that there was not a bookseller in Germany who would freely undertake the impression of his Commentary upon the First Epistle to the Corinthians." A short time before he died, Dr. Lightfoot promised some booksellers that he would collect his works, in order that they might be printed uniformly, but his death, which happened at Ely on the 6th of December 1675, in the 74th year of his age, prevented the fulfilment of this promise. His remains were interred at Great Munden, amid the tears and regret of his parishioners.

Dr. Lightfoot was well proportioned and comely; his complexion was ruddy, and from the temperature of his habits he enjoyed excellent health. His manner was grave, yet he was easy of access, affable, and communicative. His countenance was expressive of his disposition, which was uncommonly mild and tender. In the performance of his clerical duties he was strict, conscientious, and pathetic; the care of his flock was the principal motive of his actions, and he was a faithful representative of Goldsmith's Country Clergyman:

At Church, with meek and unaffected grace,
His looks adorn'd the venerable place;
Truth from his lips prevail'd with double sway,
And fools who came to scoff, remain'd to pray.
The service past, around the pious man
With ready zeal each honest rustic ran;
E'en children follow'd with endearing wile,
And pluck'd his gown to share the good man's smile.
His ready smile a parent's warmth express'd,
Their welfare pleas'd him, and their cares distress'd;
To them his heart, his love, his griefs were given,
But all his serious thoughts had rest in heav'n.
As some tall cliff that lifts its awful form,
Swells from the vale, and midway leaves the storm,
Tho' round its breast the rolling clouds are spread,
Eternal sunshine settles on its head.

Dr. Lightfoot was twice married. He had four sons and two daughters by his first wife. His eldest son, John, was bred to the clerical profession; he was Chaplain to Bryan Walton, Bishop of Chester, and died soon after that prelate. His second son, Anastatius Cottonus Jacksonus, was minister of Thundridge, in Hertfordshire; his third son, Anastatius, was brought up to trade in London; and his fourth son, Thomas, died young. His eldest daughter, Joice, was married to the Rev. John Duckfield, rector of Aspeden, in Hertfordshire, and his youngest daughter, Sarah, married Mr. Colclough, a native of Staffordshire. The Doctor's second wife was a relict of Mr. Austin Bregrave. He had no issue by her; she also died before him, and was buried in Munden church.

In 1684, Dr. Lightfoot's works were collected and printed, first in two volumes folio, containing his Latin writings, with a translation into Latin of those which he wrote in English. At the end of both those editions, there is a list of such pieces as he left unfinished. A third edition, in three volumes folio, was published by John Leusden, at Utrecht, in 1699. This edition contained several of his unfinished pieces in Latin, which were communicated by Mr. Strype to the publisher. In 1700, Mr. Strype published another collection of these papers, entitled "Some genuine Remains of the late pious and learned Dr. John Lightfoot." This publication contains some curious and interesting particulars of his life.

In rabbinical learning Dr. Lightfoot was equalled by few of his contemporaries, and excelled by none; and foreigners who came to England to prosecute those studies, visited him for the purpose of instruction. The most distinguished of these scholars was Frederic Miege, who, with Theodore Haak, was recommended to Dr. Pocock, with whom Dr. Lightfoot kept up a correspondence. Among the number of his correspondents were also Dr. Marshall, of Lincoln College, Oxford; Samuel Clarke, Keeper of the Bodleian Library; Dr. Barnard, of St. John's; and the learned Buxtorf.

THOMAS BLAKE

Was born in Staffordshire in the year 1597, and in the 19th year of his age he was entered of Christ Church, Oxford, where he passed through different gradations of academic distinction, and afterwards went into the Church. In 1648 he promoted the cause of the Republicans, and became an eminent puritanical divine. After he subscribed to the Covenant, he was appointed pastor of St. Alemonds, in Shrewsbury, and was afterwards minister of Tamworth, in his native county.

Mr. Blake was one of the Committee appointed by Parliament to examine and eject "ignorant and scandalous ministers and schoolmasters," and exerted himself with much zeal on that occasion. He died at Tamworth in June 1657, in the 60th year of his age, and was buried in the church of that town. Mr. Anthony Burgess, of Sutton Coldfield, an eloquent minister, preached his funeral sermon.

There are but few documents respecting the private life of Mr. Blake. He published several tracts and sermons besides his "Vindiciæ Fœderis, a Treatise of the Covenant of God with Mankind, &c." This work was published in London, 1653, 4to. "The Covenant Sealed," quarto, 1656; Meditations entitled "Living Truths in Dying Times;" and Controversial Tracts on Infant Baptism.

RICHARD MEADOWCOURT,

An English critic, was born in Staffordshire in 1697, and was educated at Merton College, in Oxford, of which he became a Fellow. In 1732, he published Notes on Milton's Paradise Regained, and in the following year was promoted to a canonry in the Church of Worcester. He was author of several small tracts, containing critical remarks on the English poets: and his Notes were not neglected by the late Bishop Newton in publishing his edition of Milton. He was greatly esteemed by the learned in general, and died at Worcester in 1769, aged 72. Dr. Newton thus speaks of him in his preface to the Paradise Regained. After enumerating the assistance given by friends, he adds, "I had the honour of all these for my associates and assistants before, but I have been farther strengthened by some new recruits, which were the more unexpected, as they were sent me by gentlemen with whom I never had the pleasure of a personal acquaintance. The Rev. Mr. Meadowcourt, Canon of Worcester, in 1732, published a Critical Dissertation, with Notes, upon the Paradise Regained, a second edition of which was published in 1748; and he likewise transmitted to me a sheet of his manuscript remarks, wherein he hath happily explained a most difficult passage in Lycidas better than any man had done before him." The passage alluded to is the 106th line of that poem, in which Mr. Meadowcourt explained the words "Bellerus" and "Bayona's hold." He was author also of eleven printed sermons, which are enumerated in Cook's Preacher's Assistant.*

* Chalmers's Biographical Dictionary.

REV. STEBBING SHAW.

This gentleman is well worthy of a brief memorial in the History of Staffordshire, both as the Topographer of a considerable part of his native county, and for his private virtues. He was the son of the Rev. Stebbing Shaw, Rector of Hartshorn, near Ashby-de-la-Zouch, and was born in the year 1762, at Stone, where his mother inherited a small estate, which descended to him. He received the rudiments of a classical education at the grammar-school of Repton, near Hartshorn, and under the tuition of the Rev. William Bagshaw Stevens, who was an ingenious poet and scholar, he early imbibed a love of literature.

In the eighteenth year of his age, he was admitted a student in Queen's College, Cambridge, where, as is usual with youthful and ardent minds, his attention was first directed to English poetry; but before he made a proficiency in that elegant study, the science of music prevailed in his estimation over the sister art of verse, and he devoted much of his time to it. In the meantime he did not neglect his studies, but took the degree of Bachelor of Arts, was elected to a Fellowship, and went into holy orders.

A most cordial intimacy had long subsisted between Mr. Shaw's father and his neighbour, Sir Robert Burdett, of Foremark, in Derbyshire, in whose hospitable mansion the subject of these memoirs passed many happy days; and on his return from College, he was engaged as tutor to the present Sir Francis Burdett. Mr. Shaw accompanied his pupil on a tour to the Highlands of Scotland in 1787, of which he kept a diary, which he afterwards published anonymously. "It fell," to use the words of Pope, "still-born from the press;" and a much superior production on the same subject by an anonymous author would doubtless have been neglected at a time when Dr. Johnson's Journey to the Hebrides had obtained such general and deserved popularity.

In 1788, Mr. Shaw made a tour to the West of England, of which he published a more elaborate account, with his name. This work met a better reception from the public than his first essay. This Tour was published in 1789; when he went to London, and obtained access to the topographical and genealogical stores deposited in the British Museum, of which he availed himself by the successful study of those branches of History. About this period Mr. Shaw, in conjunction with a learned friend, published a periodical work, entitled "The Topographer," which was continued for more than two years, during which many useful facts, illustrative of English topography, were communicated in its pages to the public. While engaged in those researches, Mr. Shaw spent part of the summer of 1790 in Sussex, where he visited several parishes, and collected many documents, of which only a small part was published when the Topographer was discontinued.

In 1791, Mr. Shaw retired to his father's house in Hartshorn, where he amused himself with topographical researches. He soon afterwards conceived the idea of writing the History of Staffordshire. The project seemed bold and difficult of execution even to his most partial friends; but his enthusiasm for his favourite branch of literature stimulated his perseverance in the laborious task. He had already written on the subject of topography with success, and he was ambitious to transmit the History of his native county to posterity. Animated by this motive he set about his work with ardour, and his progress was much facilitated by the increase of the number of his friends during his excursions through different parts of Staffordshire in quest of materials. Many enlightened and liberal individuals supplied him with valuable documents of the places he visited, while his mild and inoffensive manners engaged their esteem and confidence. Thus he acquired a vast fund of valuable materials illustrative of the Natural History, Agriculture, Manufactures, and local Scenery of Staffordshire, while his own taste for drawing, and the aid of his companion, Mr. James, whom he employed as draughtsman, soon enriched his portfolio with numerous sketches, which were afterwards engraved for the work.

While engaged in this pleasing and praise-worthy pursuit, Mr. Shaw discovered and obtained the manuscript collection of the Antiquities of Staffordshire which had long been supposed to be *lost*, and from the moment these valuable papers came into his possession, he received countenance and assistance from the principal nobility, gentry, and literati of the county. He had made a great variety of drawings of antiquities, churches, and mansions, and many of the latter were engraved at the expence of the proprietors.

Mr. Shaw devoted four years to the augmentation and arrangement of his collections, and in 1796 he began to print the first volume of his work, which was published in August 1798. This volume not only answered but even exceeded public expectation, by the curious retrospection of past ages intermixed with modern facts, the correct and elaborate genealogical tables which it contains, the descriptions of local scenes, and the variety of engravings with which it is enriched.

Encouraged by his success, and the reputation which he so justly obtained, Mr. Shaw published the first part of his second volume in 1801. It was in all respects equal to the former, and the lovers of British Topography looked eagerly forward to the completion of a work which would add to the stock of national information; but this expectation was disappointed by an unforeseen event. Health, which "is above all gold and treasure," forsook Mr. Shaw, and the capacity to perform his task was destroyed.

Mr. Shaw's father died at the close of 1799, and he succeeded him in the rectory of Hartshorn. At this village he spent the summer, and found some relaxation from his clerical duties and topographical studies, in improving his house and the pleasure-grounds around it. He was, however, subject to a bilious complaint, an attack of which was much aggravated by the heat of the season. On his recovery, he returned to London in the winter of 1801, and resumed his work. But his constitution had received an alarming shock, and early in the ensuing spring he was unfit for his usual studies. While sitting at dinner with his confidential friend Mr. James, and his housekeeper, he was seized with a paroxysm of insanity, which was succeeded by a fever, from which, by medical aid, he was at length restored, but all application to books was prohibited. In June it was deemed advisable for him to pay a visit to the Kentish coast, for the benefit of a change of salubrious air, and thither he was accompanied by his half-sister, the daughter of his father by a second wife. They went first to Ramsgate, and thence for more perfect seclusion to Sandgate, where the amiable and unfortunate historian of Staffordshire passed the autumn, and was so far recovered, that he was permitted to join a party of friends in a trip to Boulogne. But in October 1802, his disorder returned with redoubled violence, and after a struggle of ten days he was removed to London, where he died on the 28th of the same month, aged 41, "deeply lamented by all who knew him, and leaving a chasm in the department of literature which he had embraced not easy to be supplied." Such was Stebbing Shaw. As an author he is more to be esteemed for his indefatigable perseverance in the accumulation of facts, than his skill or taste in presenting them to the reader, for his style is harsh, his sentences perplexed and often obscure, and his attention was more powerfully attracted by the vague assertions and conjectures of antiquaries, than by the infinitely more interesting and valuable progress of society, and the improvement of those arts and sciences which humanize and enlighten civilized man. Perhaps the most valuable parts of Shaw's History of Staffordshire consist of the communications respecting agriculture, which were transmitted to him by intelligent farmers from different parts of the county. Still his work is truly valuable for reference, and it is much to be regretted that he did not live to finish it.

JOHN DUKE OF NORTHUMBERLAND.

This accomplished statesman was the son of the unfortunate Edmund Dudley, and was born near Okeover, in 1502. By the influence of his friends, his father's attainder was reversed by an act of parliament passed in 1511, and himself restored in blood. He received a classical education, and was introduced at the Court of Henry the Eighth in the 22d year of his age, where he soon was distinguished for a handsome person and various accomplishments. Naturally emulous of distinction, young Dudley accompanied Charles Brandon Duke of Suffolk, in his expedition to France, and received the honour of knighthood for his bravery in the field. On his return to England, he became a partizan of the celebrated Cardinal Wolsey, whom he attended in his embassy to France, and on the disgrace of that minister, Sir John Dudley was so fortunate as to obtain the confidence of Lord Cromwell. When that eminent statesman fell, Sir John Dudley, like an adroit courtier, by a ready submission to the will of the king, continued in favour, and in 1542 he was exalted to the dignity of the peerage by the title of Viscount L'Isle, and at the next festival of St. George was elected Knight of the Garter.

This nobleman owed his fortune, and many of his preferments to King Henry the Eighth, who appointed him Lord High Admiral of England during life, a post for which he was well qualified by his abilities and courage. The King towards the close of his reign also conferred upon Viscount L'Isle large grants of church lands, which excited the envy and hatred of the clergy, and many other enemies. Lord L'Isle was also nominated by the King as one of his sixteen executors, and received a legacy of £500. bequeathed to him by his Majesty.

On the demise of King Henry the Eighth, which happened January 28, 1547, his son, then in his minority, was proclaimed by the title of Edward the Sixth, and a regency appointed to manage the affairs of the state. But the Earl of Hereford, uncle to the King, incited by ambition, exerted his influence so successfully in the cabinet, that he was declared Protector of the kingdom, and exercised little less than regal authority. Jealous of the abilities of Lord Viscount L'Isle he deprived him of the office of Lord High Admiral, and appointed his own brother Sir Thomas Seymour to that place of high trust and responsibility. As a compensation to Lord L'Isle, he was created Earl of Warwick, and Great Chamberlain of England, but this did not abate his aversion to the Protector.

Insurrections soon afterwards broke out in several parts of the kingdom, the most remarkable of which was the rebellion in Norfolk, where Robert Ket, a tanner, appeared at the head of ten thousand rebels. The Earl of Warwick marched with the Royal army against the insurgents, and in a general engagement defeated them, and killed about one thousand men. But the rebels resolutely rallied their scattered forces, and offered him battle a second time. The Earl willing to spare the lives of his countrymen, sent a flag of truce, with a message "that he was sorry to see so much courage exerted in so bad a cause, but notwithstanding what had passed, they might depend upon the King's pardon, if they delivered up their leaders, and dispersed quietly to their homes." To this they answered, that "he was a nobleman of so much worth and generosity, that if they might have the assurance from his own mouth, they were willing to submit." On receiving this answer, the Earl gallantly rode forward, and exhorted the insurgents to lay down their arms, which they immediately did, and having delivered up their leader, Robert Ket, and his brother William, they were immediately dispersed, and their chiefs tried by a court-martial and executed.

The disgrace and public execution of Sir Thomas Seymour for treasonable practices in 1549, opened the way for the preferment of the Earl of Warwick, who was reinstated in the office of Lord High Admiral, with very extensive powers. He was now the favourite of his Prince, and in such friendship and estimation with the other Lords of the Privy Council, that nothing was done without his advice and concurrence. The Duke of Somerset, who in those times of precarious power had been committed to the Tower on some unfounded charges, was released by the influence of the Earl of Warwick, and restored to a share of power. The King was pleased with this instance of magnanimity in the Earl towards his former rival, and in order to establish a lasting friendship between these two great men, he proposed the marriage of the Earl of Warwick's eldest son with the Duke of Somerset's daughter, which was solemnized on the 3d of June 1550, in his Majesty's presence. In April 1551, the Earl of Warwick was nominated Earl Marshal of England, and in October in the same year he was advanced to the dignity of Duke of Northumberland. These honours, however, were not unattended with danger; a conspiracy to assassinate him was said to have been concerted between the Duke and Duchess of Somerset, and they were committed to the Tower. The King being persuaded of the criminality of the Duke of Somerset, he was tried by his peers, condemned, and beheaded on the 22d of February, 1552.

The Earl of Warwick, as Regent, had obtained a complete ascendancy over the mind of the young King, and with much dexterity and address brought most of the nobility over to his interests, and humbled and depressed his enemies. On the decline of the King's health, the Earl directed his principal attention to the aggrandizement of his own family, and for that purpose the marriage of Lady Jane Grey, and his fourth son Lord Guilford Dudley, was solemnized in May 1552, about two months before the death of the King. By this matrimonial connection, Lord Guilford Dudley had a distant claim to the Crown of England, for Lady Jane Grey's mother, the Duchess of Suffolk, was sister to Henry VIII.

In order to facilitate his ambitious project, the Duke of Northumberland persuaded King Edward to set aside his sister Mary from the succession, in consequence of her attachment to Popery; and so great was the ascendancy of the wily Regent, that he even prevailed on the young Sovereign to pass by his sister Elizabeth, though a Protestant, and to nominate his cousin Lady Jane Grey as his successor. Edward the Sixth died on the 6th of July 1553, in the 16th year of his age, and seventh of his reign, and the Duke of Northumberland having conveyed his daughter-in-law from Durham-house to the Tower for greater security, he on the 10th of the same month proclaimed her Queen of England. But though Lady

Jane Grey had thus been invested with regal dignity and power with the concurrence of the Council, the submission of the Princess Mary was considered indispensable to prevent the evils of civil war. The Council therefore wrote to her requiring her to resign her claims, but they were informed that she had retired to Norfolk, where many of the nobility and their adherents had taken up arms in support of her right to the Crown. It was then resolved to subdue her party by force, and for that purpose the Duke of Northumberland marched from London on the 14th of July, with 2000 cavalry and 6000 infantry. The Duke seems to have felt a presentiment of his fall, for as he marched through Bishopsgate-street, he said to Lord Grey, "The people press to see us, but no one says God speed us."

The Duke advanced to St. Edmundsbury, in Suffolk, but was deserted by part of his army on the march, and not receiving the expected reinforcements and supplies from London, he retired to Cambridge. In the meanwhile, the Council, unwilling to be any longer the tools of his inordinate ambition, and convinced that Queen Jane's assumed authority was unconstitutional, had Mary proclaimed Queen. When the news of this event reached the ears of the Duke of Northumberland, he joined in the popular sentiment, and throwing up his cap, cried "God save Queen Mary!" and caused her to be proclaimed at Cambridge. But his submission and affected loyalty were unavailing, for he was soon afterwards arrested, tried for high treason, and condemned. August the 21st, 1553, was the day fixed for his execution, and a vast concourse of people assembled upon Tower-hill to behold the tragical spectacle of a great man's fall; but after waiting some hours, the people were ordered to depart, for the criminal had received a respite. But this semblance of clemency was only for the purpose of persuading the Duke to make a formal recantation of the Protestant religion, that he might be received into the bosom of the Catholic Church, a common trick in that age, by which malignant Papists at once insulted and deceived their victims. Accordingly, the Duke of Northumberland, a man once famed for his heroism and political abilities, was induced by the love of life to gratify his treacherous enemies by the abjuration of his religion for the superstitions of Popery. This he did in the presence of a temporizing Lord Mayor of London, the Aldermen, and some of the Council; and next day he was beheaded! At the place of execution he made a long speech in favour of Popery, and behaved with great firmness and composure. Such was the end of the Duke of Northumberland, one of the most powerful noblemen that ever lived in England, a man highly endowed, yet, like most great courtiers, his virtues were counteracted by vices or errors. He had a numerous issue, eight sons and five daughters, some of whom died before him, and others survived and lived to see a great change in their fortunes. His eldest son, John Earl of Warwick, was condemned to die with his father, but reprieved, and released out of the Tower, on which he went to his brother's house at Penshurst, in Kent, where he died two days afterwards. Lord Guilford Dudley, and his unfortunate consort, the virtuous, learned, and ingenious Lady Jane Grey, were both executed upon the scaffold on the 12th of February 1554, thus affording a memorable example of the danger of ambition and usurpation.

ARCHBISHOP SHELDON.

Gilbert, the youngest son of Roger Sheldon, was born at Stanton, in Staffordshire, on the 19th of July 1598. At the time of his birth his father was a menial servant to the Earl of Shrewsbury, who stood sponsor for him at his baptism, and gave him the name of Gilbert. Under the patronage of this nobleman, young Sheldon received a classical education, and took the degree of Bachelor of Arts in Trinity College, Oxford, November 27, 1617, in the twentieth year of his age, and that of Master of Arts, May 20, 1620. He was elected Fellow of All Souls College, in the year 1622, entered into holy orders, and having attracted the notice of Lord Keeper Coventry, he was appointed domestic Chaplain by that statesman, and in consequence of his abilities, employed in various affairs relating both to Church and State. Lord Clarendon observes, that Sheldon soon became distinguished for his uncommon abilities and attainments, and was considered by competent judges to be fully qualified to shine in any ecclesiastical preferment.

Lord Coventry recommended Sheldon to Charles the First, as a person versed in political as well as theological knowledge. He was soon afterwards appointed rector of Ickford, in Buckinghamshire, and also presented to the rectory of Newington by Archbishop Copland. On the 11th of November 1821, he was honoured with the degree of Bachelor of Divinity, and May 2, 1632, he was presented by the King to the vicarage of Hackney, in Middlesex. On the 25th of June, 1634, he compounded for his degree of D. D. and in March, 1635, was elected Warden of All Souls College.

In 1635, Dr. Sheldon was appointed Chaplain in Ordinary to the King, and was afterwards Clerk of the Closet. During the Civil War between the King and Parliament, Dr. Sheldon adhered conscientiously to the cause of his Royal Master; and in 1644, he was one of the King's Chaplains, sent by his Majesty to attend his Commissioners at the Treaty of Uxbridge, where the learned Doctor argued earnestly and eloquently in favour of the Church.

In April 1646, he attended the King at Oxford, and was witness to a vow made by his Majesty, that if it should please God to re-establish his throne, he would restore to the Church all lands, impropriations, &c. which were taken from any Episcopal See, Cathedral, &c. A copy of this vow was preserved by Dr. Sheldon, and afterwards published in the Appendix to Echard's History.

Dr. Sheldon attended the King at Newmarket, and afterwards in the Isle of Wight, in the year 1647; and on the 30th of March, 1648, he was ejected from his Wardenship of All Souls College by the Parliamentary Visitors, and imprisoned by order of Parliament, with Dr. Hammond, at Oxford. He was confined about six months, and then liberated, on the hard conditions that he should not come within five miles of Oxford, nor go to the King in the Isle of Wight, and that he should appear before the Reforming Committee at fourteen days' warning.

Dr. Sheldon now retired to Snelston, in Derbyshire, where he collected money by contribution among his friends, and sent it from time to time to the exiled King. He pursued his studies without interruption till the approach of the Restoration. Dr. John Palmer, who had been placed in the Wardenship of All Souls College, died on the 4th of March, 1660, and there being a prospect of his Majesty's immediate return to England, there was no election of a successor, bnt the place was left vacant for Dr. Sheldon, who, however, never re-possessed it, for he now had higher prospects of preferment.

On the return of Charles the Second, Dr. Sheldon met him at Canterbury, where he was most graciously received by that Sovereign, to whom he had been so faithful. He was soon after made Dean of the Royal Chapel, and when Bishop Juxon was translated to the See of Canterbury, Dr. Sheldon was elected Bishop of London, October 9, 1660, and consecrated on the 28th of the same month. Dr. Sheldon also held the Mastership of the Savoy, where the famous Conference between the episcopal and presbyterian clergy was held at his lodgings in 1661. At this conference Dr. Sheldon exerted himself with his usual zeal and ability in favour of the established Church, and upon the death of Archbishop Juxon in 1663, he was elected to the See of Canterbury. Thus, by a series of preferments most honourably obtained, this eminent prelate arrived at the very summit of episcopal power and authority.

In the year 1665, Archbishop Sheldon gave unequivocal proofs of his magnanimity and charity by continuing at his palace at Lambeth during the plague, and exerting himself to the utmost of his power in aid of many afflicted and necessitous individuals.

On the 21st of December 1667, Archbishop Sheldon was elected Chancellor of the University of Oxford, but he did not hold that highest academical distinction long, for he resigned the office on the 31st of July, 1669. After a long, active, and well-spent life, this venerable prelate died at Lambeth, November 9, 1677, in the eightieth year of his age. His remains were interred in Croydon Church, in Surrey, where a monument was erected to his memory, by his nephew and heir, Sir Joseph Sheldon, the son of his eldest brother, Ralph Sheldon, of Stanton, in Staffordshire.

From an impartial review of contemporary writers respecting the public and private character of this eminent man, it appears that he was more distinguished as a politician than a divine. His zeal for the Church, and his resentment of personal injuries, made him take a decided and severe part in the enactment of penal laws against the non-conformists.

Parker says, "Archbishop Sheldon was a man of undoubted piety; but though he was very assiduous at prayers, yet he did not set so great a value on them as others did, nor regarded so much worship as the use of worship, placing the chief point of religion in the practice of a good life. His advice to young noblemen and gentleman, who by their parents' command resorted daily to him, was always: 'Let it be your principal care to be honest men, and afterwards be as devout and religious as you will. No piety will be of any advantage to yourselves or any body else, unless you are honest and moral men.' His worthy notions of religion meeting with an excellent temper in him, gave him that even tranquillity of mind by which he was still himself, and always the same, in adversity as well as in prosperity; and neither over-rated nor despised life, nor feared nor wished for death, but lived agreeably to himself and others." This is the character of a great and good man, and the more praise-worthy, as he lived in an age remarkable for hypocrisy at one period, and for profaneness at another.

On the Restoration of Charles II. when the former members of the University of Oxford returned to their places, and re-established the ancient institutions, Archbishop Sheldon munificently founded a Theatre for the acts, exercises, &c. of the students, laid the first stone, and defrayed the whole expence of the edifice, which amounted to above £12,000. He also bestowed large sums in benefactions to other public institutions, especially those connected with the established Church.

DR. ROBERT JAMES.

This ingenious gentleman was born at Kinverston, in this county, in the year 1703. His father was a major in the army, his mother a sister of Sir Robert Clarke. He received his education at St. John's College, Oxford, where he took the degree of A. B. and afterwards became a licentiate in the Royal College of Physicians. In 1743, he published the "Medicinal Dictionary," in three volumes folio; and shortly afterwards a translation, with a supplement by himself, of "*Ramazzini de Morbis Artificum*," to which also he prefixed a piece by Frederic Hoffman, upon "Endemical Distempers." In 1746 appeared "The Practice of Physic," in two volumes octavo; in 1760, the Treatise "On Canine Madness," octavo; and in 1764, the "Dispensatory," also in octavo. On the 25th of June, 1755, while the King was at Cambridge, Mr. James was admitted by mandamus to the Doctorship of Physic. Three years after this, were published "A Dissertation upon Fevers;"—"A Vindication of the Fever Powder;" and "A Short Treatise on the Disorders of Children," with a good print of the author. This was the eighth edition of the "Dissertation;" the first having been published in 1751. The "Vindication" was in fact a posthumous publication, the author dying before it was completed, in 1776, while he was employed upon it.

Dr. Johnson, in his Life of Smith, makes an affectionate mention of Dr. James, for whom he had a great regard. It appears also that Johnson acquired from Dr. James some knowledge of physic.

Speaking of Dr. James, Johnson on one occasion said, "No man brings more *mind* to his profession than James." This character, from such a judge of mind, must ever stamp the memory of Dr. James with respect. Dr. James was somewhat rough and unpolished in his manners. Some whimsical stories are told of him; particularly of his evening prescriptions. However this may be, and whatever might be Dr. James's failings, there can be no doubt of his having been a skilful and experienced physician.

For a long time his Fever Powder was violently opposed by the Faculty; an opposition which all secret nostrums ought to experience, but, in this instance, it subsequently appeared, unmerited. Dr. Pearson took great pains in analyzing it, and concluded that "by calcining bone ashes, that is, phosphorated lime, with antimony in a certain proportion, and afterwards exposing the mixture to a white heat, a compound may be formed containing the same ingredients, in the same proportion, and containing the same chemical properties."* The London Pharmacopœia now contains a prescription under the title of *Pulvis Antimonialis*, which is intended to answer the same purposes. "It is well known," says Dr. Pearson, "that this powder cannot be prepared by following the directions in the specification in the Court of Chancery."

It has long been doubtful whether Dr. James was really the inventor of this powder. Dr. Pearson remarks, that "the calcination of antimony and bone ashes produces a powder called Lile's and Schawenberg's fever powder; a preparation described by Schroeder and other chemists 150 years ago."—"According to the receipt in the possession of Mr. Bromfield, by which this powder was prepared forty-five years ago, and before any medicine was known by the name of James's Powder, two pounds of hartshorn shavings must be boiled, to dissolve all the mucilage, and then, being dried, be calcined with one pound of crude antimony till the smell of sulphur ceases, and a light grey powder is produced. The same preparation was given to Mr. Willis, above forty years ago, by Dr. John Eaton, of the College of Physicians, with the material addition, however, of ordering the calcined mixture to be exposed to a given heat in a close vessel, to render it white."—"Schroeder prescribes equal weights of antimony and calcined hartshorn; and Poterius and Michaelis, as quoted by Frederic Hoffman, merely order the calcination of these two substances together (assigning no proportion) in a reverberatory fire for several days." It has been

* Philosoph. Trans. for 1791, p. 367.

alleged, that Dr. James obtained the receipt of this powder of a German baron named Schawenberg, or one Baker, to whom Schawenberg had sold it.

However all this might be, it is certain that Dr. James gave a credit and currency to this valuable powder, which it otherwise would probably never have obtained. It is to this day prescribed by the Faculty; and is almost the only secret medicine, or medicine bearing the name of the inventor, which our cautious physicians have introduced into their own practice. It has proved a noble fortune, (says the editors of the General Biographical Dictionary,) to Dr. James's family.*

EDMUND DUDLEY.

The life and death of this celebrated lawyer and statesman may afford a moral lesson to posterity, by offering an illustration of the danger attendant upon a deviation from rectitude, and the propriety and wisdom of a strict adherence to the principles of common honesty. Edmund, the son of Sir John Dudley, second son of John Dudley, Baron of Dudley, and Knight of the Garter, was born in Staffordshire in the year 1442. After having received the first rudiments of education, he was sent in the sixteenth year of his age to the University of Oxford, and was afterwards removed to Gray's Inn, in London, to pursue the study of the law, for which he had a great aptitude, and from his natural acuteness it was foreseen by his friends that he would rise into eminence.

Young Dudley made so rapid a progress in jurisprudence, that he obtained the reputation of an able lawyer when scarcely out of his minority, and was taken into the service of Henry the Seventh. This was only the prelude to farther advancement, for the King was so well pleased with his prudence, knowledge, and fidelity, that he was sworn of the Privy Council in the 23d year of his age.

In 1492 he went with the King to France, and was one of the commissioners who were chiefly instrumental in making peace with the French monarch. In the year 1494, Mr. Dudley was married to Elizabeth, daughter of Edward Grey, Viscount L'Isle; and in 1499, he was one of the British commissioners who signed the ratification of a peace with France. He was now in high favour with

* Beauties, Vol. XIII. p. 1196.

his Prince, whose coffers he replenished, with the aid of his coadjutor Empson, at the expence of the opulent, contrary to equity and justice. This fact is fully elucidated by Lord Bacon in the following passage : "As kings do more easily find instruments for their will and humour, than for their service and honour, he had gotten for his purpose two instruments, Empson and Dudley, bold men, and careless of fame, and that took toll for their master's grist. Dudley was of a good family, eloquent, and one that could put hateful business into good language, but Empson, that was a son of a sieve-maker, triumphed always in the deed done, putting off all other respect whatever. These two persons being lawyers in science, and privy councillors in authority, turned law and justice into wormwood and rapine. For first their manner was to cause divers subjects to be indicted for sundry crimes, and so far forth to proceed in form of law, but when the bills were found, then presently to commit them: and, nevertheless, not to produce them in any reasonable time to them; but to suffer them to languish long in prison, and by sundry artificial devices and terms to extort from them great fines and ransoms which they termed compositions and mitigations."

The activity of Dudley in these nefarious transactions soon made him very odious to the people, but he retained his consequence at Court, and in the Senate too; for in 1504 he was Speaker of the House of Commons, and two years afterwards he obtained from the King a grant of the Stewardship of the Rape of Hastings, in Sussex. This was the last favour which he received from his royal master, who is said to have felt great compunction for his arbitrary proceedings against the liberties and unjust exaction of the property of his people.

Soon after the demise of Henry the Seventh in 1509, the popular clamour against Dudley was so loud that it reached the ears of Henry the Eighth, who imprisoned him and Empson in the Tower, and commanded that their subordinate instruments should be brought to trial and punished.

In July 1509, after about three months imprisonment, Dudley was arraigned before commissioners assembled at Guildhall, and found guilty of high treason, but no sentence was passed upon him. When the Parliament met in January 1510, Dudley and Empson were both attainted of high treason, but Queen Catherine having interposed in favour of Dudley, the execution of both the state criminals was deferred. This circumstance excited much indignation among the people, and the King was at last obliged to resign these unfortunate statesmen to the executioner. They were both beheaded upon Tower-hill, August 18, 1510.

During his confinement, Dudley amused himself with writing a curious piece, entitled "The Tree of the Commonwealth, by Edmund Dudley, Esq. late Counsellor to Henry the Seventh, the same Edmund being at the compiling thereof prisoner in the Tower, in 1. Henry VIII." "This treatise," says the author, "consists of three especial points. First, remembrance of God and the faithful of his holy Church, in which every Christian prince had need to begin. Secondly, of some conditions and demeanour necessary in every prince, both for his honour, and a surety of his continuance. Thirdly, the Tree of the Commonwealth, which toucheth people of every degree, of the conditions and demeanour they should be of." This work was never published, but several copies of it are still in existence in manuscript. It displays much knowledge of the world, and some courtly panegyrics on royalty; but as it never reached the King's hand, it failed of the intended influence on the royal clemency. Like many other unfortunate statesmen who have been the passive instruments of despotism, Dudley debased his talents by injustice, and with all his eloquence and learning he never knew the truth and value of the honest but important adage, "*Honesty is the best policy.*"

ISAAC WALTON.

This amiable man and useful writer was born at Stafford, August 9, 1593, but no records exist that throw any light upon the early transactions of his life, which indeed seems to have afforded few materials for the biographer. His father died when Isaac was little more than two months old. The first notice of Walton fixes him in business as a linen-draper under the Royal Bourse or Exchange, in London, where he continued to reside till the year 1624. He afterwards removed to a house in Fleet-street, "two doors west of the end of Chancery-lane, and abutting on a messuage known by the sign of the Harrow," which answers to the situation of an old half-timbered house till within these few years known by that name. About 1632 he married Anne, daughter of Mr. Thomas Ken, of Furnival's Inn, and niece of Dr. Ken, Bishop of Bath and Wells, one of the seven Prelates sent to the Tower: the Doctor was afterwards ejected, and died in retirement. Mrs. Walton appears to have been a very accomplished and pious woman, and greatly contributed to the domestic happiness of her husband. Death deprived Walton of his amiable helpmate in the year 1662, and he has recorded her virtues in an affectionate and well-written epitaph.

After the death of his wife, he retired from business, and left London for scenes more congenial to his contemplative turn of mind; "living (according to Wood) sometimes at Stafford and elsewhere, but mostly in the families of the eminent clergymen of England, by whom he was much beloved."

Whilst Mr. Walton resided in London, his favourite recreation was angling, in which art he is said to have excelled almost all men living. Langbaine calls him "the common father of anglers." His most frequented river for this primitive amusement seems to have been the Lea, which has its source above Ware, in Hertfordshire.

While Mr. Walton lived in the neighbourhood of St. Dunstan's in the West, he had frequent opportunities of attending the ministry of Dr. John Donne, who was vicar of that Church. To this prelate's sentiments, he became, as he himself expresses it, a convert; and upon the doctor's decease, in 1631, Sir H. Wotton requested Walton to collect materials for his life, which Sir Henry had undertaken to write. Sir Henry died before he had completed his work, and Walton undertook it himself, and in 1640, finished and published it, with a collection of the Doctor's sermons, in folio.

On the death of Sir Henry, which took place in 1639, Walton was importuned by the King to write his life also; and it was accordingly finished about 1644. In 1655, he published that most pleasing and highly interesting book, for which he will be ever remembered by all lovers of the art it is designed to recommend, "The Complete Angler, or Contemplative Man's Recreation," in duodecimo, adorned with some very excellent cuts of most of the fish mentioned in it. It is not known exactly who was the artist on this occasion; but it is generally believed to have been Lombart, who is mentioned by Mr. Evelyn in the "*Sculptura*," and also that the plates were of steel. Walton had not the advantage of a classical education; yet he frequently cites Latin and other learned authors, as Gessner, Camden, Aldrovandus, Rondeltius, and even Albertus Magnus. He of course had made use of such translations as he could meet with; yet some of the authors whom he quotes, we believe, have never been translated.

There has not been a writer who has had occasion to make mention of the art of angling, since Walton's time, that has not referred to The Complete Angler, as of undoubted authority on the subject; and it has been read, even as a parlour companion, by men of taste, both at home and abroad.

About two years after the Restoration, Walton wrote the Life of Mr. Richard Hooker, long known by the honourable, and in many respects deserved, appellation of "*Judicious*." This life of Hooker appears to have been written with great care and faithfulness. It is the life of one of the most learned and excellent men of his age—the author of a book which has placed a large portion of what are usually called religious people in such a situation as to render it almost criminal in them to dissent from the church establishment of this country till they have carefully read it through, and seriously weighed its reasonings.

In 1670 he published The Life of Mr. George Herbert, brother to the celebrated Lord Herbert, of Cherbury, in Shropshire. That part of "The Complete Angler" which treats of *fly-fishing*, was chiefly communicated by Mr. Thomas Barker, an ingenious and highly facetious person, and a very expert angler.

In his eighty-third year, a period when, to use his own words, he might have claimed "a *writ of ease*," he undertook the Life of Bishop Sanderson. The concluding paragraph of this book has been particularly noticed by Dr. Johnson as a specimen of nervous sentiment and pious simplicity. This paragraph informs us, that Walton was then in the eighty-fifth year of his age. Besides these books, for which the memory of Walton will ever be venerated and esteemed, he was somewhat of a poet; but in this department of literature he did not excel. He collected materials for some other lives, but did not live to finish them; yet in his *ninetieth* year he published "Thealma and Clearchus, a pastoral history, written long since by John Chalkhil, Esq, an acquaintance and friend of Edmund Spenser." To this he wrote a preface, containing a character of the author. He lived but a short time after this. He died on the 15th of December 1683, during the great frost, at Winchester, in the cathedral church of which a large black flat marble stone, with a miserable poetic epitaph, marks the place of his interment. Such was the life of this excellent man; and such the useful nature of his studies and labours. As a biographer he will always be respected, and as an angler, will never cease to be consulted and referred to. To one of the most pleasing and useful branches of literature, and one of the most primitive of recreations, he devoted a long and valuable life.* The Complete Angler is indeed unique in its kind; and while it simply professes to initiate in the mysteries of angling, by a happy combination of ideas, apt allusions, and beautiful imagery, seduces the reader from the low and unsatisfactory amusements of the throng to the more stable and rational enjoyments of primitive time, that involuntarily lead to a train of thought embracing the contemplation of the Creator, and filling the mind with admiration of the stupendous works of creation. Its praise-worthy aim seems to have been to impress man with a just estimate of himself; and in the pursuit of an innocent recreation, to "look through Nature up to Nature's God."

We have now concluded our Biographical Sketches; and it will be seen that Staffordshire is rich in Worthies. In the nobler walks of life, and in almost every department of science, Staffordshire has her representative. She has been illustrious in the Field, eloquent at the Bar, persuasive and dignified in the Pulpit: the graces and essentials of our native language, the flowers of poetry, the intricacies of science, the soundest maxims of morality, the philosophy of faith, and the benign and healing influence of Christianity—have been expounded, adorned, and illustrated by her combined and luminous intellect,—the works of her academic masters will descend like "precious ointment" upon posterity,—to the edification of the mind, the improvement of the heart—to the well-being of Man and the exaltation of Deity. It will be the duty of some future biographer to add other exalted names to this phalanx of wisdom; and, whilst pausing to admire, we feel at once unequal to the task of fully estimating their worth, and an adequate sense of the emptiness of human praise.

* See Sir John Hawkins's Life of Walton, and the Gen. Biog. Dict. in Beauties.

PEERAGE OF STAFFORDSHIRE:

HISTORICAL & GENEALOGICAL ACCOUNT OF NOBLE FAMILIES

Who derive Titles from the Names of Places in this County.

[FROM THE BEST AUTHORITIES.]

PELHAM, DUKE OF NEWCASTLE-UNDER-LYME.

HENRY PELHAM CLINTON, *Duke of Newcastle-under-Lyme*, and *Earl of Lincoln*, Lord Lieutenant of the County of Nottingham, K. G. *born* Jan. 31, 1785, succeeded his father Thomas, the late Duke, May 17, 1795; *married* July 18, 1807, Caroline Georgiana, daughter and sole heiress of Edward Miller Mundy, Esq. M. P. for the county of Derby, (by his second wife the Dowager Lady Middleton, of Wollaton, co. Nottingham,) and has issue, Anna Maria, *b.* Aug. 6, 1808; the Earl of Lincoln, *b.* May 22, 1811; a daughter, *b.* Sept. 1812.

This family derives its name from a place called Clinton in the Valley (now Glimpton), near Woodstock, and is descended from William de Tankerville, Chamberlain of Normandy. He had issue three sons, who accompanied William the Conqueror in his expedition: 1. Osbert; 2. Renebald; 3. William; of which, the 2d had by gift from William the Conqueror the Lordship of Clinton, in Oxfordshire, and his descendants assumed that name; from whom descended John de Clinton, who in 1298 was created Baron Clinton; he married Ida, daughter and co-heir of William de Odingsells, Baron of Maxtoke.

Henry Clinton, the 9th Earl of Lincoln, and 21st in paternal descent from Renebald de Clinton, became Duke of Newcastle by intermarriage into the family of Pelham. He was born April 20, 1720; on February 6th, 1742, appointed Gentleman of his Majesty's Bedchamber, and July 24th following, Lord Lieutenant and Custos Rotulorum of the county of Cambridge. On December 25th, 1746, his Grace was appointed Cofferer of his Majesty's Household; and on April 1st, 1751, Auditor of the Exchequer. On March 13th,

1752, he was elected Knight Companion of the Most Noble Order of the Garter, and chosen High Steward of Westminster in 1759. His Grace was also vested with the offices of Comptroller of the Customs in the Port of London, Master of Geddington Chase, co. Nottingham, and President of the Westminster Infirmary. He was a Lord of the Bedchamber to the late King, and for some time in the same station to the present, at whose coronation he carried the sword called *Curtana;* soon after which he resigned all his places dependant on the Crown, except those of Auditor and Comptroller, which are for life. His Grace Thomas Pelham, Duke of Newcastle-upon-Tyne, being by patent, dated November 13, 1756, created Duke of Newcastle-under-Lyme, with remainder on default of issue male to Henry Earl of Lincoln, and his heirs male by Catherine his then wife, was at his death, on November 17th, 1768, succeeded in the said title of Duke of Newcastle-under-Lyme by the said Henry Earl of Lincoln, who likewise obtained his Majesty's licence to take and use the name of Pelham; and on December 16th following, his Grace was sworn of his Majesty's Privy Council. On the same day he was also appointed Lord Lieutenant and Custos Rotulorum of the county of Nottingham, and of the town of Nottingham and county of the same town; also Steward, Keeper, and Guardian of his Majesty's Forest of Sherwood, and Park of Folewood, co. Nottingham. His Grace was also Fellow of the Royal Society, Doctor of Laws, and President of the Westminster Hospital. His Grace *m.* Oct. 16, 1744, Catherine, eldest surviving daughter and co-heir of the Right Hon. Henry Pelham, brother to the before-mentioned Thomas Duke of Newcastle, and by her (who died July 27, 1760, and was buried at Bamber, co. Lincoln) he had issue, 1st. George Lord Clinton, *b.* Nov. 26, 1745, and *d.* Aug. 19, 1752; 2d. Henry Earl of Lincoln, *b.* Nov. 5, 1750, and *m.* May 22, 1775, Lady Frances Seymour Conway, sister to the Marquis of Hertford, and *d.* Oct. 22, 1778: by his Lady he had a son, Henry Pelham, who *d.* in 1779; and a daughter, Catherine, *b.* April 6, 1776, *m.* Oct. 2, 1800, Viscount Folkstone, and *d.* May 17, 1804: his Lordship *d.* in France Oct. 18, 1778. 3d. Thomas, *b.* July 1, 1752, who afterwards succeeded his father; 4th. John, *b.* Sept. 13, 1755, and *d.* Nov. 10, 1781. His Grace *d.* Feb. 22, 1794, and was succeeded by his third son,

Thomas, third Duke, who was a Major-General in the Army and Colonel of the 17th regiment of Light Dragoons. He *m.* Lady Anna Maria Stanhope, sister to the present Earl of Harrington, by whom he had, 1st, Lady Anna Maria, *b.* Aug. 1, 1783, *m.* Jan. 1, 1801, Lieutenant-General Sir Stapleton Cotton, Bart. (now Lord Combermere), and *d.* May 31, 1807; 2d, Henry Pelham, the present Duke; 3d, Thomas, *b.* Feb. 27, 1786, a Lieutenant in the 1st regiment of Life Guards, *d.* of a malignant fever at Gibraltar, Oct. 8, 1804; 4th, Lady Charlotte, *d.* May 23, 1811. His Grace dying May 17, 1795, was succeeded by his eldest son Henry Pelham, the present and fourth Duke.

Creations.---Earl of Lincoln, May 4, 1572; and Duke of Newcastle-under-Lyme by patent, Nov. 13, 1756.

Heir Apparent.---[illegible] Earl of Lincoln.

Arms.---Argent, six Crosslets fitchy, Sable, on a chief Azure, two mullets, round pierced, Or.

Crest.---In a ducal coronet Gules, five ostrich feathers, proper, banded Azure.

Supporters.---Two greyhounds Argent, plain collared and Gules.

Motto.---*Loyalte na honte* (Loyalty is never ashamed).

Chief Seat.---Clumber Park Lodge, co. Nottingham, formerly part of the Holles estates.

GOWER, MARQUIS OF STAFFORD.

George Granville Leveson Gower, *Marquis of Stafford, Earl Gower, Viscount Trentham, Baron Gower of Sittenham*, and a *Baronet*, Knight of the Garter, Recorder of Stafford, and (*jure uxoris*), High Sheriff of the county of Sutherland, *b.* Jan. 9, 1758, succeeded his father the late Marquis Oct. 26, 1803. In 1790 he was sent Ambassador Extraordinary to Paris, and recalled in 1792. In 1798 he was called up to the House of Peers by the title of his father's Barony, as Lord Gower. In Jan. 1803, on the death of his uncle Francis, last Duke of Bridgewater, he became his heir-general, and succeeded to that part of his immense property which consisted in shares in those numerous and splendid inland Canals of which his Grace was the illustrious founder, and by which he has immortalized his name. His Lordship *m.* Sept. 4, 1785, Elizabeth, Countess of Sutherland and Baroness of Strathnaver, in Scotland, in her own right, and has issue, 1st, George Granville, Earl Gower, *b.* Aug. 8, 1786, one of the representatives in Parliament for this county; 2d, Lady Charlotte, *b.* June 8, 1788, *m.* Dec. 27, 1814, Henry Howard, Esq. son of Bernard Edward Howard, Esq. (now Duke of Norfolk); and has issue, Henry Granville, Lord Fitz Alan, *b.* Nov. 7, 1815; 3d, Lord William, *b.* June 4, 1792, and *d.* in

1793; 4th, Lord Henry Leveson, *d.* June 17, 1804; 5th, Lady Elizabeth, *b.* Nov. 8, 1797; and 6th, Lord Francis, *b.* Jan. 1, 1800.

This family is of considerable antiquity in the county of York. Sir Alan Gower, of Sittenham, at the time of the Norman Conquest, was Sheriff of Yorkshire (1066), and from him in a direct line the family have continued their seat at Sittenham to the present time.

In 1257 was living Sir Walter, the father of Sir Robert de *Guer*, living in 1289 and 1293, about which time probably the name began to be differently written; for in 28th Edward I. Sir John Gower was one of those persons of note summoned to be at Carlisle to march against the Scots; and the year after he had summons to be at Berwick with horse and arms on the same account. In 1313 Lawrence Gower had the King's pardon for being concerned with Thomas Earl of Lancaster in the murder of Piers Gaveston Earl of Cornwall, the great favourite of Edward II. at Blacklow-hill, co. Warwick, June 19, 1312.

Sir John Leveson Gower, 5th Baronet, was created Baron Gower March 16, 1703, *m.* Catherine, eldest daughter of John 1st Duke of Rutland, and had by her four sons and two daughters. John, the eldest son, was by George II. July 8, 1746, advanced to the dignity of Viscount Trentham and Earl Gower; he *m.* 1st, March 18, 1712, Evelyn, youngest daughter of Evelyn Duke of Kingston, and by her had four sons, viz. John, who *d.* July 15, 1723; William, who *d.* April 4, 1739; Granville (the 1st Marquis); and Richard, who *d.* Oct. 19, 1753; also seven daughters. His Lady dying June 27, 1727, his Lordship *m.* on Oct. 31, 1733, secondly, Penelope, daughter of Sir John Stonehouse, Bart. by whom he had a daughter Penelope, who *d.* in 1741, and her mother dying in the same year, his Lordship *m.* thirdly, the Lady Mary, daughter and co-heir of Thomas Tufton Earl of Thanet, by whom he had a daughter and three sons, 1st, Thomas, who was *b.* Aug. 23, 1738, and *d.* when about two years old; 2d, John Leveson, *b.* in 1743, appointed a Captain in the Navy June 30, 1763, and an Admiral in 1787. He distinguished himself while he commanded the Quebec in the former war, and during the last while he commanded the Valiant; also while he was first Captain of Lord Howe's ship at the relief of Gibraltar in 1782. In 1785 he hoisted his broad pennant on board the Hebe frigate, and sailed round the coast of Great Britain and Ireland, having on board his Royal Highness Prince William Henry (Duke of Clarence), who served as Lieutenant on board the ship. In 1788 he commanded a squadron of observation in the Channel.

He was a Lord of the Admiralty from 1783 to 1789, and *d.* Aug. 15, 1790. 3d, Thomas, *b.* July 4, 1743, and *d.* an infant. His Lordship dying Dec. 25, 1754, was succeeded by his eldest son,

Granville, second Earl and first Marquis, who *m.* in 1744 Elizabeth, daughter of Nicholas Fazakerly, Esq. of Prescot, and by her, who *d.* May 19, 1745, had a son John, who *d.* an infant. His Lordship in 1744 was elected M. P. for Bishop's Castle, Salop, and in the next parliament in 1747 was unanimously chosen for Westminster; but in Nov. 1749 being constituted a Lord of the Admiralty, he was opposed by Sir George Vandeput, but succeeded. At the General Election in 1754 he was returned for the city of Lichfield, and sat for the same till the death of his father. On succeeding to the Earldom, he was appointed Lord Lieutenant and Custos Rotulorum of Stafford. On Dec. 19, 1755, he had a grant of the office of Lord Privy Seal, and in Jan. following was sworn a Privy Counsellor. Resigning the Privy Seal, he was, July 2, 1757, made Master of the Horse, and on Nov. 25, 1760, was nominated Keeper of the Great Wardrobe. On April 23, 1763, he was declared Lord Chamberlain of his Majesty's Household, which his Lordship resigned July 10, 1765. On Dec. 23, 1767, he was appointed President of the Privy Council, but resigned in Nov. 1779, appointed again Dec. 19, 1783, and again resigned in 1784, and was appointed in Nov. of that year Lord Privy Seal, which he held till 1794. On Feb. 28, 1786, he was advanced to the title of Marquis of Stafford. On Feb. 11, 1771, he was elected a Knight Companion of the Order of the Garter. His Lordship *m.* secondly, Lady Louisa Egerton, daughter of Scroop, first Duke of Bridgewater, by whom he had issue, 1st, George Granville, Viscount Trentham, *b.* Feb. 9, 1758; 2d, Lady Louisa, *b.* Oct 22, 1749, *m.* to Sir Archibald Macdonald; 3d, Lady Carolina, *b.* Nov. 2, 1753, *m.* to Frederick Earl of Carlisle; 4th, Lady Anne, *b.* Feb. 22, 1761, *m.* the Hon. Dr. Edward Vernon, now Archbishop of York. His Lordship *m.* thirdly, May 25, 1768, Lady Susan Stewart, daughter of John Earl of Galloway, by whom he had issue three daughters; 1st, Lady Georgiana Augusta, *b.* April 13, 1760, *m.* the Hon. William Eliot; 2d, Lady Charlotte, *b.* Jan. 11, 1771, *m.* Henry Charles now Duke of Beaufort; 3d, Lady Susanna, *b.* in 1772, *m.* in 1795, Dudley now Earl of Harrowby; and one son, George Granville Leveson, *b.* Oct. 12, 1773, now Lord Viscount Granville, and Ambassador in 1807 to St. Petersburgh, who *m.* Dec. 24, 1809, Lady Henrietta Cavendish, second daughter of William, 5th Duke of Devonshire, by whom he has issue. The Marquis dying Oct. 26, 1803, was succeeded by his eldest son, George Granville, present and second Marquis.

Creations.---Baronet, June 2, 1620; Baron Gower of Sittenham, co. York, by letters patent March 16, 1702-3; Viscount Trentham and Earl Gower, July 8, 1746; and Marquis of Stafford, Feb. 26, 1786.

Arms.---Quarterly, 1st and 4th, barry of eight, Argent and Gules, over all, a cross fleury, Sable, for Gower; 2d and 3d Azure, three laurel leaves erect, Or, for Leveson.

Crest.---On a wreath, a wolf passant, Argent, collared and chained, Or.

Supporters.---Two wolves, Argent, each having a collar and chain, Or.

Motto.---*Frangas non Flectes*---(You may break, but shall not bend me.)

Heir-Apparent.---George Granville, Earl Gower.

Chief Seats.----Trentham Hall, in this county; and Dunrobin Castle, Scotland.

PAGET, MARQUIS OF ANGLESEY.

Henry William Paget, *Marquis of Anglesey, Earl of Uxbridge, Lord Paget of Beaudesert*, co. Stafford, a Lieutenant-General in the Army, Colonel of the 7th regiment of dragoons, succeeded his father, Henry Bayley, the late Earl, March 13, 1812; *b.* March 17, 1768, *m.* July 25, 1795, Caroline Elizabeth, daughter of George, fourth Earl of Jersey, (which marriage was dissolved by the law of Scotland), and had issue, Lady Caroline, *b.* June 6, 1796; Lord Henry (Earl of Uxbridge) *b.* July 6, 1797; Lady Jane, *b.* Oct. 13, 1798; Lady Georgiana, *b.* Aug. 20, 1800; Lady Augusta, *b.* Jan. 26, 1802; Lord William, *b.* March 1, 1803; Lady Agnes, *b.* Feb. 11, 1804; Lord Arthur, *b.* Jan 31, 1805. His Lordship *m.* secondly, Charlotte, daughter of Charles Earl of Cadogan, and has issue Lady Emily Caroline, *b.* March 4, 1810; Lord Edward Clarence, *b.* June 17, 1811; Lady Mary, *b.* June 16, 1812; a still-born daughter, June 25, 1813; Lord Alfred, *b.* May 4, 1815, *d.* May 7, 1815; Lord Alfred Henry, *b.* June 29, 1816.

The first who attained to the dignity of the Peerage was William Paget, a person of humble origin, but of very great and eminent abilities. On Dec. 3, 1549, he was called up to the House of Peers by the title of Lord Paget of Beaudesert, co. Stafford. His Lordship *m.* Anne, daughter and sole heir of Henry Preston, Esq. by whom he had issue four sons and six daughters. His Lordship was succeeded by his eldest son,

Henry, second Lord Paget, who *m.* Catherine, daughter of Sir Henry Knevet, of Buckenham, co. Norfolk, by whom he had issue Elizabeth, who *d.* June 29, 1571. His Lordship *d.* Dec. 28, 1568, and was succeeded by his brother,

Thomas, third Lord Paget, who *m.* Nazaret, daughter of Sir John Newton, of Barr's-court, co. Somerset, and had issue William, his son and heir. In the Parliament holden at Westminster in 29th Eliz. he was attainted with his brother Charles, and their lands and possessions confiscated: he died at Brussels in 1589, and was succeeded by his only son,

William, fourth Lord Paget: he *m.* Lettice, daughter and coheir of Henry Knollys, Esq. of Kingsbury, co. Warwick, and had issue three sons and four daughters. His Lordship dying Aug. 29, 1629, was succeeded by his eldest son,

William, fifth Lord Paget, who *m.* Lady Frances Rich, eldest daughter of Henry Earl of Holland, and by her had issue, William; Henry, who *m.* a daughter of —— Sandford, Esq. of Sandford, co. Salop, and had issue a daughter Dorothy; Thomas, one of the Grooms of the Bed-chamber to his late Majesty, and a Brigadier-general, who had issue Caroline, *m.* in April 1737, to Sir Nicholas Bayley, of Plasnewydd, Bart. grandfather to the present Marquis; Thomas; and seven daughters. His Lordship dying Oct. 19, 1678, was succeeded by his eldest son,

William, sixth Lord Paget, who *m.* Frances, daughter of the Hon. Francis Pierpoint, and had issue, William, who *d.* young; Henry, created Earl of Uxbridge. His Lordship *d.* Feb. 26, 1712-13, and was succeeded by

Henry, first Earl of Uxbridge, who *m.* Mary, eldest daughter of Thomas Catesby, Esq. of Whiston, co. Northampton, by whom he had issue Thomas Catesby, his only son, who died in January 1741. His Lordship dying in August 1743, was succeeded by his grandson,

Henry, second Earl of Uxbridge, who died unmarried on Nov. 16, 1769, when the Earldom became extinct, but the title of Baron Paget, of Beaudesert, devolved on Henry Bayley Paget, eldest son of Sir Nicholas Bayley, *b.* Jan. 18, 1744, and created Earl of Uxbridge May 19, 1784. His Lordship *m.* Jane, eldest daughter of Arthur Champaigne, Dean of Clonmacnoise, in Ireland, and had issue, 1st, Henry William, now Marquis of Anglesey; 2d, William, *b.* Dec. 22, 1769, *d.* in 1794; 3d, Sir Arthur, *b.* Jan. 15, 1771; 4th, Caroline, *b.* Feb. 6, 1773, *m.* John Thomas, second son of the Earl of Essex; 5th, Jane, *b.* Sept. 1, 1774, *m.* April 18, Admiral George present Earl of Galloway; 6th, Sir Edward, K.B. and K.T.S. Lieutenant-General in the Army, *b.* Nov. 3, 1775, *m.* May 21, 1804, the Hon. Louisa Bagot, daughter of William first Lord Bagot, who *d.* May 30, 1806; 7th, Louisa, *b.* March 26, 1777, *m.* Lieut.-General Sir James Erskine; 8th, Charles, *b.* Oct. 7, 1778, a Captain in the Royal Navy, *m.* Elizabeth Araminta, second daughter of Henry Monck, Esq.; 9th, Berkeley, *b.* Jan. 2, 1780, *m.* Nov. 22, 1804, Sophia, daughter of the Hon. William Bucknall Guinston; 10th, Charlotte, *b.* Oct. 27, 1781, *m.* Oct. 15, 1805, John Willoughby Earl of Enniskillen; 11th, Mary, *b.* April 9, 1783, *m.* June 27, 1803, Thomas Lord Graves. The Earl dying March 13, 1812, was succeeded by his eldest son,

William Henry, third Earl of Uxbridge, and *first* MARQUIS OF ANGLESEY. His Lordship was early destined to the army, and after a series of honourable services, embarked for Lisbon at the beginning of the Peninsular war, having received the command of the cavalry attached to the memorable expedition under the lamented Sir John Moore. Much mystery still attaches to this unfortunate affair, but there can be no doubt that Sir John was deceived by the artifices of the enemy, and the information of pretended friends, as to the strength and character of the troops with which he had to contend, composed, as it afterwards appeared, of the flower of the French army, commanded by Buonaparte in person, and numbering upwards of 70,000 men, whilst the British consisted of less than 30,000. Sir John marched from Lisbon in four divisions, and assembled his troops at Salamanca on the 23d of November 1808: here he soon learned that all hope of assistance from the Patriots was at an end. Belvidere, Castanos, and Blake, had been successively defeated, and Burgos and Valladolid were in the hands of the French, the latter town only 60 miles north of him. Thus deprived of the aid which he had been led to expect, prudence dictated retreat; but a transitory burst of patriotism in the populace of Madrid, and a remonstrance from the Junta against such a measure, determined Sir John to attempt something, though without the least probability of success. He accordingly resolved to attack Soult, who was at Saldanna, and thus make a diversion in favour of the Spaniards in the south. The whole army united at Majorga on the 20th December, where the head-quarters were fixed. At this place Lord Paget had a gallant affair with a detachment of the enemy; and having gained information that a strong party of cavalry were posted at Sahagun, his Lordship resolved to make an attempt to cut them off. The weather was intensely cold, and the ground covered with snow. Lord Paget began his march at two o'clock in the morning, and before day-light approached the town, and surprised a picquet, but two or three of the men escaped, and gave the alarm. He pushed forward, and discovered the enemy already formed. The ground was particularly unfavourable for Lord Paget, and both parties attempted to flank each other: his Lordship however completely out-manœuvred the enemy, charged, overthrew, and dispersed them in every direction. The British loss was trifling: the French had many killed, and two lieutenant-colonels, and a number of prisoners, fell into the hands of the English. On the following day the army moved to Sahagun, 12 miles from Saldanna, the object of attack: Soult was strongly posted behind the Carrion with 18,000 men. Sir John wrote to Romana, who was on the enemy's further flank, to prepare for the united attack on the morrow's dawn. In this state of things, messenger after messenger arrived, with intelligence that the French were rapidly advancing on all sides, that Soult had been considerably reinforced, that a corps was advancing from the south; and, to crown all, an express from Romana announced the advance of Buonaparte in person, with an overwhelming force from Madrid. This information immediately determined Sir John; the former orders were countermanded, and immediate preparations made for retreat. This accordingly commenced on the following day, December 24, and on the 25th, the whole army were in full retreat upon Astorga. This retreat, however, was not so precipitate as has been represented; it was both skilfully and deliberately arranged. On the 29th, at day-break, Sir John followed the divisions of Gens. Fraser and Hope with the reserve; Lord Paget bringing up the rear with the cavalry. But at nine o'clock in the morning, before his Lordship had moved, a party of the enemy's cavalry were observed trying a ford near the bridge that had been blown up, and immediately between 5 and 600 of the Imperial Guards plunged into the river, and crossed over. They were opposed by the British picquets under Colonel Otway, with a small party of the 3d dragoons; and the colonel charged them with so much fury that the front squadron broke through, and was for a short time surrounded by the enemy's rear squadron wheeling up. They extricated themselves by charging back again through the enemy. Lord Paget now galloped up with some of the rear-guard, and found brigadier-general Stuart very sharply engaged, and the squadrons on both sides sometimes intermixed. The 10th hussars having arrived, his Lordship wheeled it into line in the rear of the picquets. The latter then charged the enemy, supported by the 10th hussars; but before they could close, the French wheeled round, fled to the ford, and plunged into the river. They were closely pursued, and left on the field 55 killed and wounded and 70 prisoners, including General Lefebvre, the commander of the Imperial Guards. They formed upon the opposite bank, but a few shots from the horse artillery, who now came up, soon dispersed them, and they fled with the greatest precipitation. It was learned from the prisoners that Buonaparte was only twenty miles behind. Having so brilliantly concluded this affair, Lord Paget, with the cavalry, followed the reserve to La Banissa. It is not to our purpose to follow the retreat in detail. In the course of it, Sir John repeatedly offered the enemy battle, without effect; and the several affairs in which they were engaged arose more from necessity than choice on their part, and in these the British were always victorious. The steady gallantry of the infantry, and the impetuosity of the cavalry, under their intrepid commander, were not to be resisted. The retreat was continued till the 10th of January; and on the following morning the army marched from Betanzos for Corunna, their last day's march. Here the gallant Moore received his death-wound, after a life of honourable vicissitude, and resigned his last breath with the consoling reflection that he had done his duty. Thus, with a "most admirable degree of military skill and patience on the part of the lamented Sir John Moore, and with a spirit, a heroism, and an alertness on the part of Lord Paget, the British army was at length brought to Corunna, having thus made a retreat through nearly three hundred miles of the most difficult country in Europe, and in a succession of mountains, ditches, and rivers, in daily and almost hourly contact with the enemy, and this enemy more than double its number! In all this long interval of war, and against this vast superiority of force, the English had invariably the advantage in every encounter; the rear guard and covering cavalry were never broken, and the main army, secure behind this invincible screen, proceeded almost unassailed towards their ultimate destination. As long as military skill is understood and valued, so long will the military name of Lord Paget be united with that of Moore, and both be recorded together as having conducted and successfully concluded the most masterly retreat on modern record."

Lord Paget shortly afterwards returned to England, and passed some years in the bosom of his family. His father dying on the 13th of March 1812, he succeeded him in the Earldom of Uxbridge. It is probable that his Lordship felt somewhat chagrined at the public discussion which the retreat of Sir John Moore had created: but no blame could possibly attach to either of the commanders upon that occasion, who, as we have already seen, conducted themselves with a degree of heroic firmness under the most trying difficulties, that has perhaps never been surpassed. Much ill-humour was apparent at the time, and the obloquy intended to be fastened upon Ministers was unjustly extended to those gallant men who had so nobly fulfilled their trust. It is well known that the Peninsular war was at this time very unpopular in England: it had commenced under unfavourable auspices, and much of its equivocal character might undoubtedly be traced to an ill-calculated policy at home. It is the more unnecessary for us to enlarge upon this subject, as the leading events of the campaign are fresh in the memory of all. To whatever cause attributable, we do not find Lord Paget again employed on foreign service during the war. But a yet more noble distinction awaited him: the Earl of Uxbridge was destined to become an instrument in the final overthrow of that frightful despotism which he had been one of the first to curb and chastise in the desolated fields of Spain. The important and almost unparalleled train of events that preceded the day of Waterloo need not be re-capitulated. The nation as to a man viewed with silent awe the gathering storm that darkened the horizon, and threatened to engulph the fairer part of Europe in one common ruin. The Moloch of devastation was abroad, and England, in conjunction with her allies, was driven to wage with him a war of existence. The whole moral and physical strength of the country was called forth to meet the exigency of the moment. The Sovereign summoned his legislators and his warriors to their posts: they threw the gauntlet to the tyrant, and the people applauded their wisdom. After Wellington, the Prince Regent evinced his sense of the former services of the Earl of Uxbridge by selecting him to second that great commander in the tremendous struggle about to be commenced. He cheerfully obeyed the mandate; and the legions of Britain were again marshalled to punish those traitorous and guilty Frenchmen who had betrayed their King, and delivered over their country to the merciless sway of an Usurper. Little did Buonaparte calculate upon being compelled a second time to feel the vengeance of an offended and stigmatized people, to fly before her victors, and to sue for safety even at the gates of Paris. The Earl of Uxbridge joined the army at Brussels, where the British head-quarters were fixed. Whilst the most extensive preparations were making in Flanders for the commencement of hostilities, all was anxious suspense in England. The chances of success and defeat were nicely calculated: if the Allies came up in time, all would be well; if not, it was feared that the result might prove unfavourable. In the midst of these calculations at home, and preparations abroad, the storm suddenly burst, and the demon of war stalked forth in all his gigantic fury. Buonaparte had left Paris at the head of an immense army, and was already on the frontiers. The Prussians were immediately attacked, driven back, and ultimately compelled to re-cross the Sambre. Buonaparte quickly forced Charleroi, and dealt destruction in their discomfited ranks: they fled with precipitation. The four Pussian corps on the line of the Sambre fell back upon Fleurus, Namur, Cincy, and Hannut: one of these, under Ziethen, had engaged the enemy with partial success. Blucher being apprized of these movements, ordered the other three corps to make a forced march to Sombref, where he intended to give the enemy battle next day. Early on the morning of the 16th, the French army moved forward, and soon discovered the Prussian army drawn up in condensed columns, and occupying a strong position, with its right on the village of St. Amand, its centre at Ligny, and its left extending towards Sombref: a most sanguinary contest ensued, and each of these places were taken and re-taken several times. This was a day of blood: it is called the battle of Ligny. Buonaparte hoped to accomplish the annihilation of the Prussian army at Ligny, and afterwards bring his united strength against the English. These movements led to the obstinate struggle at Quatre Bras. It was at one o'clock on the morning of this day, that the bugle sounded through the streets of Brussels to summon every soldier to his rendezvous. On the repulse of the Prussians on the 15th, Blucher had sent off intelligence of the state of things to the Duke of Wellington, who had previously received information to the same effect, but he did not consider it authentic. The troops were immediately put in motion, and before ten o'clock in the morning the whole of them had left Brussels. The Duke of Brunswick and the Prince of Orange had already marched, with their corps. On their arrival at the scene of action, they became immediately opposed to Marshal Ney. He attacked the Belgians and Brunswickers in their position at Quatre Bras with all his force. It was at this time that the Duke of Brunswick was mortally wounded: the Prince of Orange also received a musket-ball in his shoulder. It was near mid-day when the British began to arrive: the Duke of Wellington rode into the field near the post of Quatre Bras, which now fell to the British to maintain. The English became engaged as they successively arrived. The French attacked the British repeatedly with a large body of infantry and cavalry, supported by a powerful artillery, but they were repulsed in the "steadiest manner." This dreadful day was sustained with great loss, owing, perhaps, to the want of cavalry, which had not yet arrived. Night put an end to the contest. Blucher had suffered dreadfully; and, as his fourth corps had not yet arrived, he fell back in the night, and concentrated his army upon Wavre. This rendered a correspondent movement on the part of the British necessary: the Duke of Wellington retired from Quatre Bras to Genappe, and thence, on the morning of the 17th, upon Waterloo. The enemy made no effort to pursue Blucher, nor did he attempt to molest the march of the British to the rear, although made in the middle of the day, excepting by following with a large body of cavalry the cavalry under the Earl of Uxbridge. This gave his Lordship an opportunity of charging them with the 1st Life Guards in an effective and gallant manner. Indeed the cavalry were constantly fighting with the enemy whilst covering the retreat of the infantry, and the Duke of Wellington declared that he never saw finer charges, or more ably conducted. Both armies now appear to have been exhausted: a sort of respite ensued, but this was only to prepare for the dreadful 18th, when all was to be consummated. Whilst engaged in the arduous task of covering the retreat on this day, the Earl of Uxbridge, being in the rear of the last troop of the cavalry, he observed a French regiment form across the road to charge. He instantly turned round, *and alone*, galloped back towards the enemy, waving his hat to his soldiers, who had advanced some way in their retreat. Major Kelly joined his Lordship at full gallop. The regiment soon came up, and dashed amongst the enemy, who were entirely overthrown. A violent thunder-storm raged at this time, accompanied with torrents of rain. *Sunday* the 18th arrived: this was the day, "the great the important day." We shall content ourselves with giving the Duke of Wellington's account of this awful battle—the "grave of France, the deadly Waterloo." It would be an endless task to point out individual traits of heroism; but we must advert to the cool and steady manner in which the *Earl* of Uxbridge conducted himself on this never-to-be-forgotten occasion. Where-ever danger appeared, he was to be found, enspiriting his troops, and alluring them to victory by his courageous and intrepid example in the hottest of the fight. His Lordship headed every squadron that was ordered forward. His division was literally annihilated! Some accounts say that he had formed the determination of taking Buonaparte in person, and that he was nearly accomplishing his purpose, when he received the unfortunate wound which compelled him reluctantly to quit the field—but the work of Victory was complete!——"The position which I took up, in front of Waterloo, (we use the words of the dispatches from Wellington) crossed the high roads from Charleroi and Nivelles, and had its right thrown back to a ravine near Merke Braine, which was occupied, and its left extended to a height above the hamlet Ter-la-Haye, which was likewise occupied. In front of the right centre, and near the Nivelles road, we occupied the house and garden of Hougoumont, which covered the return of that flank; and, in front of the left centre, we occupied the farm of La Haye Sainte. By our left we communicated with Marshal Prince Blucher, at Wavre, through Ohain; and the Marshal had promised me, that in case we should be attacked, he would support me with one or more corps, as might be necessary. The enemy collected his army, with the exception of the third corps, which had been sent to observe Marshal Blucher, on a range of heights in our front, in the course of the night of the 17th and yesterday morning; and at about ten o'clock he commenced a furious attack upon our post at Hougoumont. I had occupied that post with a detachment from General Byng's brigade of Guards, which was in position in its rear; and it was for some time under the command of Lieutenant-Colonel Macdonald, and afterwards of Colonel Home; and I am happy to add, that it was maintained throughout the day, with the utmost gallantry by these brave troops, notwithstanding the repeated efforts of large bodies of the enemy to obtain possession of it. These attacks were repeated till about seven in the evening, when the enemy made a desperate effort with the cavalry and infantry, supported by the fire of artillery, to force our left centre near the farm of La Haye Sainte, which after a severe contest was defeated; and having observed that the troops retired from this attack in great confusion, and that the march of General Bulow's corps by Frichermont upon Planchenoite and La Belle Alliance, had begun to take effect, and as I could perceive the fire of his cannon, and as Marshal Prince Blucher had joined in person, with a corps of his army to the left of our line by Ohaim, I determined to attack the enemy, and immediately advanced the whole line of infantry, supported by the cavalry and artillery. *The attack succeeded in every point; the enemy was forced from his position on the heights, and fled in the utmost confusion, leaving behind him as far as I could judge* ONE HUNDRED AND FIFTY PIECES OF CANNON, with their ammunition, which fell into our hands. I continued the pursuit till long after dark, and then discontinued it only on account of the fatigue of our troops, who had been engaged during twelve hours, and because I found myself on the same road with Marshal Blucher, who assured me of his intention to follow the enemy throughout the night; he has sent me word this morning that he had taken 60 pieces of cannon belonging to the Imperial Guard, and several carriages, baggage, &c. belonging to Buonaparte, in Genappe. The Earl of UXBRIDGE, after having succesfully got through this arduous day, received a wound, *by almost the last shot fired,* which will, I am afraid, deprive his Majesty for some time of his services." The illustrious Wellington's fears were but too fatally realized. The Earl of Uxbridge's wound was so severe as to require the almost immediate amputation of his right leg*—and thus his country was deprived of his active services for ever! His Lordship underwent a tedious and painful operation with great composure; and expressed his gratitude to God for having brought him through that dreadful day in comparative safety. What a noble struggle! What an incalculable victory! that at once averted the dreadful calamities which threatened Europe,

* Lord Uxbridge's leg is buried in a garden opposite to the inn, or rather public-house, at Waterloo. The owner of the house in which the amputation was performed considers it as a relic which has fallen to his share. He had deposited it at first behind the house, but as he intended to plant a tree upon the spot, he considered, that as the ground there was not his own property, the boys might injure or destroy the tree, and therefore he removed the leg into his own garden, where it lies in a proper sort of coffin, under a mound of earth about three or four feet in diameter. A tuft of Michaelmas daisies was in blossom upon this mound when we were at Waterloo; but this was a temporary or-

and altered the destinies of nations. It was a day both of joy and of grief: England rejoiced in her triumph, but mourned for her heroes. The flower of her fine army was no more. The unconquerable descendants of Cressy and Agincourt, the bold assertors of their country's rights and the gallant avengers of her wrongs—who had traversed the burning sands of Egypt in pursuit of their deadly foe—who had encountered and beaten him on the mountains of Portugal—who had chased and discomfited him in the devoted fields of Spain—who had scaled the Pyrennees, conquered him in his native rocks, and planted the British standard on the towers of Bayonne—and who, four centuries after the soil of France had been immortalized by the victories of their ancestors, attacked and defeated him almost in the very heart of his kingdom. Peace to thy manes, departed warriors! Thou art avenged: the day of Waterloo dated the existence of England's avowed enemy. There he fought, and there he fell: thence he flew before the victorious arms of Britain, and soon lay crushed even under the walls of his usurped capital. The names of Alexandria, of Maida, of Vimiera, of Corunna, of Talavera, of Barossa, of Albuera, of Salamanca, of Vittoria, of Orthes, of the Pyrennees, of Thoulouse, and lastly, of WATERLOO, no time shall obliterate: they will form the rallying cries of distant posterity, and animate many a future race of warriors whilst hurling the thunders of Britain upon her enemies in either hemisphere, responding to the patriotic war-cry of some unborn chief whose genius shall have grasped the inspiring mantle of WELLINGTON!—As the Prince Regent cherished and brought into action the military talents of Lord Uxbridge, so was he ready to reward their developement: on the 23d of June, 1815, whilst the Earl was suffering under the effects of his severe wound, his Royal Highness was pleased to raise him to the dignity of a Marquis of the United Kingdom by the name, style, and title of MARQUIS OF ANGLESEY; and on the same day he received the thanks of Parliament for his services at Waterloo. The Emperor of Austria

nament: in November the owner meant to plant a weeping willow there. He was obliging enough to give me a copy of an epitaph which he had prepared, and which, he said, was then in the stone-cutter's hands. It is as follows:

Ci est enterres la Jambe de l'illustre, brave, et vaillant Comte Uxbridge, Lieutenant General, Commandant en Chef la Cavalerie Angloise, Belge, et Hollandoise; blesse le 18 Juin, 1815, a la memorable battaille de Waterloo; qui par son heroisme a concouru au triomphe de la cause du Genre humain, glorieusement decidee par l'eclatante victoire du dit jour.—SOUTHEY's WATERLOO, p. 213.

and the Emperor of Russia sent him the Cross of a Commander of the Order of Maria Theresa, and the decoration of St. George, in testimony "of their Imperial Majesties' approbation of his services in the late battles fought in the Netherlands." The Marquis of Anglesey is also a Knight Grand Cross of the Most Honourable Order of the Bath. As soon as he could be removed with safety, the Marquis returned to England; and Beandesert received the mutilated warrior, who had so lately left its noble domain in all the plumy pride of chivalry, and in the full vigour of manhood. In November following he received a visit from the Prince Regent and other branches of the Royal family. The Noble Marquis may henceforth recline under "his vine and his fig-tree," in the proud recollection of having fulfilled the expectations of his Prince and the hopes of his country: long, long may he live to enjoy the confidence of the one, and the admiration of the other. Long may he wear his hard-earned laurels, and continue to be a living monument of the pre-eminence of his native county in Worth and Valour. In the peaceful haunts of Beandesert he may recount his many hair-breadth escapes—"shoulder his crutch and shew how fields were won:"—and were we called upon to select the proudest feature in his heraldic honours, (we speak with something of latitudinarianism) we should point to the *crutch-head*, and exclaim, "Here is Waterloo, the noblest emblem in the field of the Pagets!"

Creations.—Baron Paget, of Beaudesert, in co. Stafford, January 19th, 1550, 4 Edward VI.; Earl of Uxbridge, May 19th, 1784; and MARQUIS of ANGLESEY, June 23, 1815.

Arms.—Quarterly, first and fourth, Sable, on a cross engrailed, between four eagles displayed, Argent, 5 lions passant of the first, for Paget; second and third, Azure 9 Estoiles, 3, 3, 2, and 1, Argent for Bayley.

Crest.—On a wreath, a demi tiger, Sable, tufted and maned, Argent, and ducally gorged, Or.

Supporters.—Two tigers, Sable, tufted and maned, Argent, and ducally gorged, Or.

Motto.—*Per il suo contrario.* (By the reverse of it.)

Heir-Apparent.—The Earl of Uxbridge.

Chief Seats.—At Sinai Park, and Beaudesert, both in Staffordshire; and Plasnewydd, in the county of Anglesey.

TOWNSHEND, MARQUIS TOWNSHEND.

GEORGE FERRARS TOWNSHEND, *Marquis of Townshend, Earl of Leicester, Viscount Townshend, Baron Townshend of Lynn, Baron de Ferrars of Chartley, Baron Bouchier, Lovaine, Basset, and Comp-*

ton, and a *Baronet*, *born* Dec. 13, 1778, succeeded his father George Ferrars, the late Marquis, July 27, 1811; *married* May 12, 1807, —— daughter of W. D. Gardner, Esq.

This family have long flourished in the county of Norfolk. John Townshend, son of Sir Roger, was of Raynham, in Norfolk, (the present seat of this family), as appears by his will, dated 1465. His son Roger was returned for Calne, and in the 1st. Edward V. was constituted a Serjeant at Law; the following year he was appointed Justice of the Common Pleas, which office he retained till his death, which happened in 1492; he *m.* Anne, daughter and co-heir of Sir William de Brewse, Knt. and had a numerous issue; from whom descended Roger, who in 1617 was created a Baronet; *m.* Mary, second daughter of the celebrated Horatio Lord Vere, of Tilbury, and had issue two sons and five daughters. Sir Roger *d.* Jan. 1, 1636. From this marriage descended Sir Horatio, 1st Viscount; in 1660 he was elected for the county of Norfolk, and was deputed with several other members to wait on the King at the Hague, and invite him "to return to his dominions, and take the government thereof into his own hands;" in consequence of his loyalty he was advanced to the Peerage by the title of Baron Townshend of Lynn Regis, in Norfolk, and in 1682 was created Viscount Townshend of Raynham; by his second wife, Mary, daughter of Sir Joseph Ashe, Bart. he had three sons; 1st, Charles, second Viscount; 2d, Roger, *d.* May 22, 1709; 3d, Horatio, *d.* Oct. 24, 1751. He was succeeded by his eldest son,

Charles, second Viscount, *m.* first, Elizabeth, daughter of Thomas Lord Pelham, father of Thomas Duke of Newcastle, who *d.* May 11, 1711, leaving issue a daughter, Elizabeth, *m.* Charles, first Earl of Cornwallis, and four sons; 1st, Charles, the third Viscount; 2d, Thomas, *b.* Jan. 2, 1701, *d.* in May 1780; 3d, William, *b.* Jan. 9, 1702, *d.* Jan. 29, 1738; 4th, Roger, *d. unm.* Aug. 7, 1760. His Lordship *m.* secondly, in July 1713, Dorothy, daughter of Robert Walpole, of Houghton, co. Norfolk, and by her, who *d.* March 29, 1726, had issue, 1st, George, a Rear Admiral of the White, *d.* Aug. 9, 1762; 2d, Augustus, Captain in the East India Service, *d.* in 1746; 3d, Horatio, *d.* in Feb. 1764; 4th, Richard, *d.* young; 5th, Edward, Prebend of Westminster, *b.* May 11, 1726, *d.* Jan. 27, 1765; 6th, Dorothy, *m.* in 1743 to Dr. Spencer Cowper, *d.* in 1779; 7th, Mary, *m.* 1753, to Lieutenant-Gen. Edward Cornwallis, *d.* Dec. 29, 1776. His Lordship dying in June 1738, was succeeded by

Charles, third Viscount, who *m.* in May 1723 Audrey, daughter of Edward Harrison, Esq. and had issue, 1st, George, first Marquis; 2d, Charles, *d.* Sept. 4, 1767; 3d, Edward, *d.* June 29, 1731; 4th, a son, who *d.* young; 5th, Roger, killed at Ticonderago July 25 1759; 6th, Audrey, *d.* Feb. 1781. His Lordship dying March 12, 1764, was succeeded by his eldest son,

George, first Marquis, *b.* Feb. 28, 1724. His Lordship served under George I. at the battle of Dettingen, and was second in command at the taking of Quebec, which city surrendered to him after the death of General Wolfe, and *m.* Dec. 1, 1751, Charlotte, Baroness de Ferrars in her own right, and by her had issue, 1st, George, the second Marquis; 2d, Lord John, *b.* Jan. 19, 1757; 3d, Lord Frederick Patrick, *b.* Dec. 30, 1767; 4th, Lord Charles Patrick Thomas, *b.* Jan. 6, 1768, *d.* May 27, 1796, also four daughters, Ladies Charlotte, Caroline, Frances, who *d.* young, and Lady Elizabeth, *b.* in Aug. 1766, *d.* March 21, 1811. His Lordship *m.* secondly, Anne, daughter of Sir William Montgomery, Bart. and had issue, 1st, Lady Anne, *b.* Feb. 1, 1775; 2d, Lady Charlotte, *b.* March 19, 1766; 3d, Lady Honora Maria, *b.* July 6, 1777; 4th, Lord William, *b.* Sept. 5, 1778, *d.* in 1794; 5th, Lady Henrietta, *b.* April 20, 1782; 6th, Lord James Nugent Boyle Bernardo, *b.* Sept. 11, 1785. His Lordship dying Sept. 14, 1807, was succeeded by his eldest son,

George, second Marquis, *b.* April 18, 1755, who, on the death of his mother, succeeded to the titles of Baron de Ferrars of Chartley, Bourchier, Lovaine, Bassett, and Compton, created Earl of Leicester, May 18, 1784, *m.* Dec. 24, 1777, Charlotte, second sister and co-heir of Roger Mainwaring Ellerker, Esq. and by her had issue, 1st, George, the present Marquis; 2d, Lord Thomas Compton, *b.* March 29, 1780, *d.* Jan. 7, 1787; 3d, Lord Vere Charles, *b.* Sept. 16, 1785; 4th, Lady Charlotte Barbara, *b.* June 26, 1781, *d.* Oct. 3, 1807; 5th, Lady Harriet, *b.* May 23, 1782; 6th, Lady Elizabeth Margaret, *b.* Aug. 16, 1784; 7th, Lady Arabella, *b.* April 2, 1787. His Lordship dying July 27, 1811, was succeeded by his son George Ferrars, the present Marquis.

Creations.—Baron de Ferrars, of Chartley, in the county of Stafford, originally by writ of summons to Parliament, 27 Edw. I. Baron Bourchier, by writ also 16 Edw. III. Baron Lovaine, by writ 22 Edw. I. Baron Bassett, of Drayton Bassett, in the county of Stafford, by writ 49 Hen. III. and lastly, Baron Compton, of Compton, in the county of Warwick, by writ 14th Elizabeth; Baron Townshend, of Lynn Regis, by letters patent April 20th, 1661, and Viscount Townshend of Rainham, December 2d, 1682, Baron of Lynn, by writ May 24th, 1723; Earl of Leicester, May 18th, 1784, and Marquis Townshend, October 27, 1787.

Arms.—Quarterly, of six, 1st Azure, a chevron ermine, between three escallop shells, Argent, (Townshend) 2d, France and England, quarterly, within a border, Argent; (Plantagenet of Woodstock, Duke of Gloucester) 3d, Sable, a lion of England, between three helmets, proper, garnished, Or, (Compton) 4th, Paly of six, Or, and Azure, a canton ermine, (Shirley) 5th, quarterly, 1 and 4, Argent, a fess, Gules, three torteaux in chief, (Devereux) 2 and 3, Varre, Or and Gules, (Ferrars of Chartley) 6th, Gules, seven Mascles conjoined, 3, 3, and 1, (Ferrars of Groby and Tamworth.)

Crest.—On a wreath, a buck tripping sable, a hind proper. Also out of a ducal coronet, Or, a swan rousant Argent, ducally collared and chained of the 1st, holding in its beak Sable, an ostrich feather of the 2d; (which crest was granted to his Lordship, by sign manual of his present Majesty, in regard to his being the direct lineal heir to Thomas Plantagenet, surnamed of Woodstock, Duke of Gloucester, sixth son to Edward III. whose cognizance was a swan, with an ostrich feather.) Also on a wreath, an unicorn passant ermine, armed, hoofed, maned, and tufted, Or, the crest of the family of Ferrers, of Tamworth. Also on a wreath, a peacock's tail erect, proper; the crest of the ancient Lord de Ferrars, of Chartley.

Supporters.—Dexter, a buck Sable, attired Argent. Sinister, a greyhound Argent.

Motto.—*Hæc generi incrementa fides* (Ennobled for our Fidelity.)

Heir-Presumptive.—Lord Charles Vere Townshend, brother to the Marquis.

Chief Seats.—At Tamworth Castle; and Rainham, in Norfolk.

SHIRLEY, EARL FERRERS.

Robert Shirley, Earl Ferrers, *Viscount Tamworth*, and a *Baronet*, F. S. A. *born* Sept. 21, 1756, succeeded his father the late Earl, April 18, 1787; *married* 1st, March 13, 1778, Elizabeth Prentise, by whom, who *d.* Sept. 14, 1799, he had issue Robert Sewallis, Viscount Tamworth, *b.* Nov. 9, 1778, *m.* Sept. 5, 1800, Sophia Caroline, daughter of Nathaniel Curzon, Lord Scarsdale. His Lordship *m.* 2dly, Elizabeth, youngest daughter of the late Wrightson Mundy, Esq. of Markeaton, co. Derby.

The family of Shirley is descended from Sewallis, whose residence at the time of the Conquest was at Ettington, co. Warwick. His descendant, James, of Ettington, first assumed the name of Ettington, *temp.* Henry III. His grandson, Sir Thomas Shirley, *m.* a daughter and co-heir of the Bassets, of Drayton, *temp.* Richard II. His descendant, Sir Ralph Shirley, was created a Knight-banneret for his distinguished gallantry at the battle of Stoke, in 1417. Sir George Shirley, his great grandson, was created a Baronet in 1611, from whom descended,

Sir Robert Shirley, Knt. first Earl Ferrers, *b.* at East Sheen, co. Surry, during his father's confinement in the Tower for his loyalty to Charles I. where he *d.* during his imprisonment, on Dec. 14, 1677. His Majesty was pleased to confirm unto him and his heirs the ancient Baronies of Ferrars, of Chartley, Bourchier, and Lovaine, which honours had been in abeyance between the ladies Frances and Dorothy Devereaux and their descendants from the decease of their brother, the Earl of Essex, without issue. His Lordship was in the 1st of Anne advanced to the dignity of Viscount Tamworth. He *m.* first Elizabeth, daughter of Lawrence Washington, Esq. of Caresden, co. Wilts, by whom, who *d.* Oct. 2, 1693, he had issue ten sons and seven daughters; and by his second wife, Selina, daughter of George Finch, Esq. he had five sons, 1st, Robert, *d.* young; 2d, George, *d.* an infant; 3d, George, *b.* in 1705; 4th, Sewallis, *b.* in 1709, *d.* Oct. 31, 1765; 5th, John, *b.* in 1752, *d.* Feb. 15, 1768; and five daughters. His Lordship dying Dec. 25, 1717, was succeeded by his second son,

Washington, second Earl, *b.* June 22, 1677. He *m.* Mary, daughter of Sir Richard Levings, Bart. by whom, who *d.* in Jan. 1740, he had issue three daughters, his co-heirs: 1st, Elizabeth, *m.* June 24, 1725, Joseph Gascoigne Nightingale, of Enfield, co. Middlesex, *d.* Aug. 1731; 2d, Selina, *m.* June 23, 1728, Theophilus, Earl of Huntingdon, and *d.* in 1791; 3d, Mary, *m.* June 29, 1730, Thomas Needham, Lord Viscount Kilmorey, and *d.* Aug. 4, 1767. His Lordship dying April 14, 1729, and leaving no heir male, the title devolved on

Henry, third Earl, his next brother, who was *b.* April 14, 1691, and dying in Aug. 1745 unmarried, the title devolved on his nephew, Lawrence, son and heir of Lawrence Shirley, tenth son of Robert first Earl Ferrars.

Lawrence, fourth Earl, *b.* Sept. 16, 1752, *m.* Mary, youngest daughter of Amos Meredith, Esq. son and heir of Sir William Meredith, Bart. of Hanbury, in Cheshire, from whom he was separated by act of Parliament. His Lordship, though he was at times a very intelligent person, and a nobleman conversant in the Constitution of his country, yet, on several occasions, exhibited symptoms of a constitutional insanity of mind. In one of these fits of disorder he shot his steward, Mr. Johnson, in Jan. 1760; in April following he was tried by his Peers, in Westminster-Hall, and suffered death on the 5th of May following. His title and estate descended to his next brother,

Washington, fifth Earl, *b.* May 26, 1722, *m.* April 19, 1746,

Miss Anne Elliot, of Plymouth, was made Captain in the Royal Navy, in which he gave eminent proofs of courage, and was promoted to the rank of Vice-Admiral of the Blue. His Lordship dying Oct. 1, 1778, was succeeded by his brother,

Robert, 6th Earl, *b.* July 10, 1773, *m.* Catherine, daughter of Rowland Cotton, of Etwall, co. Derby, by whom, who *d.* March 16, 1786, he had issue, 1st, Robert, the present Earl; 2d, Lawrence Rowland, *d.* young; 3d, Washington, *b.* Nov. 13, 1760, *m.* Frances, only daughter of the Rev. William Ward. His Lordship dying April 18, 1787, was succeeded by his son Robert, the present and seventh Earl.

Creations.—Baronet, May 22, 1611, Viscount Tamworth, co. Stafford, and Earl Ferrars, Sept. 3, 1711.

Arms.—Quarterly, 1st and 4th, Paly of six, Or, and Azure, a Canton Ermine; 2d and 3d, France and England, quarterly, within a border, Argent.

Crest.—On a Wreath, the Bust of a Saracen, side-faced, and couped, proper, wreathed about the Temples, Or, and Azure.

Supporters.—Dexter a Talbot Ermine, eared Gules, and gorged with a ducal Collar, Or, Sinister, a Rein-deer of the second, attired and gorged with a ducal Collar, Or, and charged on the Shoulder with a Horse-shoe, Argent.

Motto.—*Honor Virtutis Præmium,* (Honour is the reward of Virtue).

Heir-Apparent.—Viscount Tamworth, his Lordship's eldest son.

Chief Seats.—At Stanton Harold, in Leicestershire; at Chartley-castle, in Staffordshire; and at Shirley, in Derbyshire.

CHETWYND, EARL TALBOT.

Charles Chetwynd Talbot Chetwynd, *Earl Talbot, Viscount Ingestre, in Staffordshire, and Lord Talbot, Baron of Hensal, co. Glamorgan,* Lord Lieutenant of the county of Stafford, *b.* April 25 1777, succeeded his father the late Earl, May 19, 1793; *m.* Aug. 28, 1800, Frances Thomasine Lambart, eldest daughter of Charles Lambart, of Beanpark, Ireland, Esq. by whom he has issue, 1st, Lady Frances *b.* May 17, 1801; 2d, Charles Thomas, Viscount Ingestre, *b.* July 11, 1802; 3d, the Hon. Henry John, *b.* Nov. 8, 1803; 4th Arthur, *b.* Jan. 12, 1805; 5th, John, *b.* May 31, 1806; 6th, Lady Cecil, *b.* April 18, 1808; 7th, George Gustavus *b.* March 19, 1810; 8th, Walter Chetwynd, *b.* Sept. 29, 1812; 9th, William Whitworth, *b.* Jan. 17, 1814; and 10th, Gilbert, *b.* April 28, 1816.

This noble branch of the ancient and illustrious house of Talbot, is lineally descended from Sir Gilbert Talbot, of Grafton, Worcestershire, third son of John 2d Earl of Shrewsbury, by his wife Elizabeth, daughter of James Butler, Earl of Ormond.

Charles first Lord Talbot, was the eldest son of Dr. William Talbot, Bishop of Durham. On May 31 1717, he was nominated Solicitor-General to George Prince of Wales, the late King; and on April 22, 1726, was constituted Solicitor-General to George I. in which post he continued till Nov. 29, 1733, when he was appointed Lord High Chancellor of Great Britain, and at the same time sworn of his Majesty's Privy Council. He was by letters patent, dated Dec. 5, 1733, created a Peer of Great Britain, by the title of Lord Talbot Baron of Hensol. His Lordship *m.* Cecil, daughter of Charles Matthews, Esq. of Castle-y-Menich, co. Glamorgan, by whom he had issue, 1st, Charles Richard, who *d.* a bachelor in 1733; 2d, William Earl Talbot; 3d, John, *m.* Henrietta Maria, daughter of Sir Matthew Decker, Bart. who dying in Sept. 1746, he *m.* secondly, Catherine, eldest daughter of John Lord Viscount Chetwynd, in Ireland, by whom he left issue four sons; 4th, Edward, who *d.* an infant; 5th, George D. D. *m.* Anne, eldest daughter of Jacob, late Lord Viscount Folkstone, by whom he had issue two sons and four daughters. His Lordship dying Sept. 23, 1756, was succeeded by his second son,

William, advanced to the dignity of Earl Talbot March 21, 1761; *m.* Mary, sole daughter of Adam de Cardonnel, of Bedhampton-park, co. Southampton, Esq. by whom he had a son, William, who *d.* an infant; Cecil, *b.* in July, 1735, *m.* Aug. 16, 1756, George Rice, Esq. of Newton, co. Carmarthen, by whom she had issue. His Lordship *d.* April 27, 1782, on which the Earldom became extinct; but having been created Baron Dinevor Sept. 29, 1780, with remainder to his daughter Lady Cecil, she succeeded to that Barony, and the Barony of Talbot descended to his nephew and heir male, John Chetwynd Talbot.

John Chetwynd, eldest son of the Hon. John Talbot, succeeded his uncle as third Baron Talbot, and was created Viscount of Ingestre, co. Stafford, and Earl Talbot of Hensol, co. Glamorgan, by patent. His Lordship *m.* May 7, 1776, Lady Charlotte, daughter of Wills Hill, first Marquis of Downshire, and by her, who *d.* Jan. 1804, left issue Charles Chetwynd, the present Earl; John, *b.* April 4, 1779. His Lordship dying May 19, 1793, was succeeded by his eldest son Charles Chetwynd, the present Earl.

Creations.—Baron Talbot, of Hensol, in the county of Glamorgan, December 5th, 1733; Viscount Ingestre and Earl Talbot, July 3, 1784.

Arms.—Gu. a lion ramp. within a border engrailed, Or, a crescent difference.

Crest.—On a chapeau, Gules, turned up ermine, a lion, Or, his tail extended.

Supporters.—On each side, a Talbot, Argent, collared with a double treasure fleury, counter fleury, Gules.

Motto.—*Humani nihil Alienum.*—(Nothing is foreign to me which relates to man.)

Heir-Apparent.—Viscount Igetre.

Chief Seat.—Ingestre, Staffordshire.

JERVIS, EARL OF ST. VINCENT.

JOHN JERVIS, *Earl of St. Vincent, Viscount St. Vincent of Meaford*, and *Baron Jervis, of Meaford, in Staffordshire*; Admiral of the Red, G. C. B. General of the Royal Marines, an Elder Brother of the Trinity House, and one of the Council of State for the Prince of Wales in Cornwall, *b.* at Meaford, Jan. 9, 1734, and baptized at Stone the 26th of the same month; *m.* his first cousin Martha, daughter of the Right Hon. Sir Thomas Parker, who *d.* February 8, 1816, and was buried in Caverswall church, in the county of Stafford, but has no issue.

He is descended from James Jervis, of Chatkyll, in the parish of Eccleshall and county of Stafford, who left two sons, 1st, Robert, who died without issue; 2d, William, who had issue, John, who *m.* Ellen, daughter of —— Whittington, by whom he had issue, Thomas; John, who *m.* Elizabeth, daughter and sole heir of John Jervys, by her he had issue (beside other children) John Jervys, *b.* Sept. 26, 1681, *m.* Elizabeth, daughter of Nicholas Wakelin, of Gentleshaw, in the parish of Longdon, co. Stafford, and had issue, John Jervis, *b.* at Chatkyll, and buried at Darlaston, in the parish of Stone, *m.* Mary, only daughter and heir of John Swynfen, heir-apparent of John Swynfen, of Swynfen, co. Stafford, Esq. By her he had issue, John; William, *b.* in 1695; Benjamin, *b.* 1696; Thomas, *b.* in 1699; Swynfen, in Nov. 1700; Matthew, *b.* in 1704.

Swynfen, fifth son of John Jervis, before-mentioned, was *b.* at Darlaston in Nov. 1700; he was Auditor of Greenwich Hospital. He *m.* in 1727, Elizabeth, daughter of George Parker, Esq. of Park Hall, and sister of the Right Hon. Sir Thomas Parker, and *d.* at Meaford in March 1784, leaving issue, William; *b.* in May 1728, *m.* Jane, only surviving sister of Thomas Hatrell, late of Newcastle, Esq. and *d.* in 1813; John, now Earl of St. Vincent; Elizabeth, *b.* in 1729, *m.* the Rev. Mr. Batewell; Margaret and Mary *d.* in-

-fants; Mary, m. April 19, 1769, William Henry Ricketts, and by him had issue, Capt. William Ricketts, R. N. b. Nov. 4, 1764, who took the name of Jervis by royal sign manual, m. Nov. 9, 1793, Elizabeth Jane Lambert, daughter of Richard Earl of Cavan, (which marriage was dissolved by Act of Parliament in 1799), by whom he had issue, Edward Jervis Ricketts, Barrister-at-Law, who m. Jan. 29, 1790, the Hon. Cassandra Twisleton, daughter of Thomas, late Lord Say and Sele, from whom he was divorced in 1799, and by whom be had issue, one son and two daughters, William Jervis, b. April 11, 1794; Eliza, d. Aug. 29, 1805; and Maria, b. in 1797. The Captain was drowned by the upsetting of his barge, Jan. 26, 1805.

John Jervis, second son of Swynfen Jervis aforesaid, now EARL of ST. VINCENT, was born at Meaford, in this county, Jan. 9th, 1734, and baptized at Stone 29th of the same month. Being early brought up to the navy, he was promoted to the rank of Post Captain, October 13th, 1760, and early distinguished himself in his profession, particularly in the action with the Pegase, when he commanded the Foudroyant, April 20, 1782, for which he was made a Knight of the Bath. In 1787 he obtained the rank of Admiral.

In the latter end of 1793, his Majesty having determined to send a formidable armament to the West Indies, to reduce the French Islands in that quarter, and to secure his own from any attack of the enemy, under the command of Lieutenant-General Sir Charles Grey, K. B., Vice-Admiral Sir John Jervis, K. B. was nominated Commander in Chief of the naval force on the same expedition. On Monday Jan. 6, 1794, they arrived off Barbadoes, and immediately prepared for the expedition against Martinique; for which the total military force embarked was 6085 effectives, besides 224 sick. The fleet sailed from Barbadoes on Febuary 3d, and approached Martinique on Wednesday February 5th. On the following day the General effected a landing. After some intermediate operations, the town of St. Pierre was attacked by sea and land, and carried. The enemy's strength was then concentrated in one point at Forts Louis and Bourbon. These the Admiral now attacked; and on the 20th of February, these strong holds were completely invested. On March 25, Fort Bourbon was surrendered; Fort Louis was also taken by storm. The island of Martinique being thus conquered, the fleet and army proceeded on March 30th for St. Lucia, which also surrendered on the 4th of April.

On Tuesday April 6th, the commanders proceeded against Guadaloupe; Fort Fleur D'Epee was taken on the 12th, and with this fell Hog Island, and Fort Louis, an old fortification, commanding the entrance into the harbour of Point-a-Petre, which town was also taken possession of at the same time by Sir Charles Grey. On the 13th, the army was re-embarked, and landed on Basse-Terre at Petit-Bourg, in another part of the island. The Admiral anchored in Ance-de-Bailiff. The Palmiste was taken by assault, and the island of Guadaloupe was surrendered to the Commander in Chief on the 23d of April. In June an armament arrived from France, made good their landing on Grand Terre, and retook Fort Fleur D'Epee by storm. The Admiral was at this time with the General at St. Christopher's, on his return to England. The commanders instantly determined to return to Guadaloupe, and pushed for Basse-Terre, where they arrived on June 7th. Sir Charles Grey took the heights of Mascot, and several actions took place; but the General found it now in vain to attempt any thing against Fleur D'Epee, at this season, with an army so greatly reduced by the yellow fever, as well as service. Having therefore made the best arrangements to enable him to renew his attacks after the hurricane months, he embarked on board the Boyne, which proceeded to Fort Royal Bay, where she was laid up during those months in a strong harbour called Trois Islet Bay, and the sick and wounded were landed for the benefit of fresh air. The General then sailed with the Admiral to Martinique, and established his head-quarters at St. Pierre. On the 30th of September, the enemy having been too successful against the little remnants of the army in Guadaloupe, the Admiral embarked, and sailed from Martinique, and anchored off Grozier, in the bay of Point-a-Petre. But our troops there, under General Graham, reduced by fatigue and sickness, and no longer able to undergo duty, were forced to capitulate. Sir John Jervis, who had made every attempt to succour General Graham's camp at Berville, and had been an unwilling spectator from the fleet of the surrender of that camp to the enemy, now made sail for Basse-Terre, to render every assistance in his power to General Prescot; and on October 9th, anchored within half a cable's length of the town. However, after various attempts to assist General Prescot, the Admiral, worn down by long and severe exertions, the fatigues of which were augmented by the anxiety for the welfare of the service, that not all his exertions could promote without the arrival of a strong reinforcement, together with the unhealthiness of the climate, found himself no longer able to continue on this station, and therefore, to the great grief of General Prescot, was obliged to give up his command to Admiral Caldwell, and embarking his seamen, under Lieutenant James, from Fort Matilda, sailed for St. Pierre, when every thing being arranged between the several commanders, Sir Charles Grey and his suite embarked once more with Sir John Jervis on board the Boyne. On November 27th, they sailed for England, and after a tedious voyage (being for near a month tossed about in the Channel by contrary winds) arrived at Spithead, January 21st, 1795.

In February 1797, he fought the famous battle off Cape St. Vincent, for which he was rewarded with a Peerage. The *New Annual Register* gives the following account of this glorious victory: —" The British fleet, or to speak more correctly, the British squadron, under the command of Admiral Sir John Jervis, amounted to no more than fifteen sail of the line, four frigates, a sloop of war, and a cutter. Of these six were three-deckers, eight were of 74 guns, and one of 64. The Spanish fleet consisted of twenty-seven sail of the line, one of which was a four-decker, and carried 136 guns, six were three-deckers of 112 guns each, two of 84 guns, and eighteen of 74. The Spanish Admiral, Don Josef de Cordova, had sailed from Carthagena on the 4th of February, and passed Gibraltar on the following day; having left in that bay three line-of-battle ships, supposed to be laden with military stores for the Spanish troops before that garrison. On the night of the 11th, this fleet had been discovered by the Minerva frigate, which carried the broad pennant of Commodore Nelson, then on his way from the Mediterranean to join Admiral Jervis. Captain Foote, of the Nigre, also kept company with them for some days previous to the 13th, and that night they approached so near the British fleet, that their signal guns were distinctly heard. The signals were, therefore, made that night for the British fleet to prepare for battle, and at day-break on the 14th they were in complete order. The morning was dark and hazy, but about half-past six the Culloden made the signal for five sail in the south-west quarter; at eight o'clock the squadron was ordered to form in close order, and in a few minutes after, the signal was repeated to prepare for battle. At a little after ten, the Minerva made the signal for twenty sail in the south-west quarter; and in about half-an-hour after, the enemy's fleet were visible to all the British squadron. The ships first discovered by the Culloden were at this period separated from their main body, which was bearing down in some confusion to join the separated ships. It appeared to have been the British Admiral's intention at the first, to cut off these vessels from the enemy's fleet before the main body could arrive to their assistance, and with this view the fast-sailing ships were ordered to chase, but observing the near position of their main body, he afterwards formed his fleet into a line of battle a-head and a-stern as most convenient. At about twenty-six minutes past eleven, the Admiral communicated his intention to pass through the enemy's line, and immediately after the signal was made to engage. At about half-past eleven, the action commenced by the van ship, the Culloden, commanded by Captain Trowbridge, firing against the enemy's headmost ships to the windward. As the squadron advanced, however, the action became more general; and it was soon apparent that the British Admiral had accomplished his design of passing through the enemy's line. In the mean time, the regular and animated fire of the British fleet was but feebly returned by the enemy's ships to windward, which were also completely prevented from joining their companions to leeward, and obliged to haul their wind on the larboard tack. Thus a part of the Spanish fleet was effectually cut off from the main body, and they were reduced to the necessity of also forming on their larboard tack, apparently with the intention of passing through or to the leeward of the British line; but such was the reception they experienced from the centre of the British, that they were obliged to put about, and did not appear again in the action till the close of the day. The British Admiral having thus fortunately obtained his first object, now directed his whole attention to the enemy's main body to windward, which was reduced at this time, by the separation of the ships to leeward, to eighteen sail of the line. At a little after twelve o'clock, the signal was made for the British fleet to tack in succession, and soon after, the signal for again passing the enemy's line; while the Spanish Admiral's design appeared to be to join his ships to leeward by wearing round the rear of the British line. The intention of the enemy was, however, soon perceived by Commodore Nelson, whose station in the rear afforded him an opportunity of observing the manœuvre. In order to frustrate the design, therefore, his ship, the Captain, had no sooner passed the Spanish rear, than he ordered her to wear, and stand on the other tack towards the enemy. In executing this bold manœuvre, the Commodore found himself alongside of the Spanish Admiral, the Santissidma Trinidada, of 136 guns, which is said to be the largest ship at present in existence. Notwithstanding this immense disparity (the Captain being only a seventy-four), this brave officer did not shrink from the contest; though the Spaniard was also warmly supported by her two seconds a-head and a-stern, which were each of them three-deckers. While he sustained, however, this unequal conflict, his friends were eagerly pressing to his assistance; the enemy's attention, therefore, was soon directed to the Culloden, Captain Trowbridge; and the Blenheim, Captain Frederick; and the able support afforded by these vessels to Commodore Nelson, and the approach of Rear-Admiral Parker, with four others of the British line, determined the Spanish Commander to relinquish his design of rejoining his ships to leeward, and to make the signal for his main body to haul their wind, and make sail on the larboard tack. The advantage was now evidently on the side of the British, and while the advanced division warmly pressed the centre and rear of the enemy, the Admiral meditated with his division a co-operation which might effectually compel some of them to surrender. In the confusion of their retreat, several of the Spanish ships had doubled on each other. It was therefore Admiral Jervis's plan to reach the weathermost of those ships, then to bear up and take them all in succession, with the seven ships composing his division. The casual position of the rear ships in his own division, however, prevented the executing this design. He therefore ordered the leading ship, the Excellent, Captain Collingwood, to bear up, while with his own ship, the Victory, he passed to leeward of the rearmost ships of the enemy. Captain Collingwood, in obedience to the Admiral's orders, passed between the two rearmost ships of the enemy, and gave one of them, the San Isidro, so effectual a broadside, that having been much injured before, she was obliged to submit. The Excellent then passed on to the relief of the Captain, which was engaged with a three-decker, carrying a flag; but before she could arrive, the vessel became entangled with the second, a two-decker. In this state they were both boarded by the Captain, and the smaller of them, the San Nicholas, was in a short time in the possession of her opponents. The three-decker, the San Joseph, followed the fate of her second, and became immediately a prize to Commodore Nelson, who headed the party which boarded her from the San Nicholas. In the mean time Admiral Jervis ordered the Victory to be placed on the lee quarter of the rearmost ship of the enemy, the Salvador del Mundo; and threw-in so effectual a discharge, that her Commander seeing the Barfleur carrying Vice-Admiral Waldegrave's flag, bearing down to second the Victory, thought proper to strike. Thus four of the enemy's ships were in possession of the British, while the van ships continued to press hard on the Santissima Trinidada, the Spanish Admiral's ship, and the others which composed the rear of the flying fleet. The career of victory was however stopped by circumstances not in the power of the British Commander to control. The ships which in the morning had been separated from the main body of the Spanish fleet, were now able to make their approach; two fresh ships which had not appeared in the action, bore down from windward, and two of the flying ships tacked about to support their chiefs. These circumstances, therefore, with the lateness of the hour, and the necessity of securing the prizes, determined the conquering Admiral to bring-to. A little after four in the afternoon the signal was made to this effect; and a strong line was formed for the protection of the prizes and disabled vessels. The enemy's fresh ships, on approaching, opened a fire on the covering ships; but though superior in number, and fresh for action, they contented themselves with a few irregular broadsides, and left the British Admiral to sail off triumphantly with his prizes, which the reader will remember amounted to four; viz. two, the Salvador del Mundo, the San Josef, of 112 guns; the San Nicholas, of 84, and the San Isidro, of 74 guns. The Spanish Admiral, which was greatly the object of attention to the British ships, was rendered a perfect wreck; her firing had ceased before the close of the action, and some even affirm that she had struck her colours. The loss of the British in this engagement, in killed and wounded, was exactly 300 men. The loss of the Spaniards which were captured amounted to 693; and the ships which escaped must also have suffered considerably. So important a victory with so decisive a disparity of force, is, perhaps, unparalleled in our naval annals. The ability displayed by the Commander was only to be equalled by the valour and adroitness of the seamen; indeed, we have been informed by an eye-witness that the fire of the British was superior to that of their opponents, in the proportion of five or six to one, during the whole of the action; and the expenditure of ammunition was consequently beyond example. The Culloden, it is said, expended 170 barrels of powder; the Captain, 146; and the Blenheim, 180. The Spaniards fought bravely, but with little skill; and it is but fair to remark, that their fleet was ill-equipped, and very indifferently manned, and in no respect fit for action; their flag-ship had not more than sixty or eighty seamen on board: the rest consisted of impressed landsmen, or soldiers, of their new levies. As the port of Cadiz had been their original destination, and as many of their ships were disabled, the Spaniards manifested no inclination to renew the action, but took shelter in Cadiz."

For this victory Sir John Jervis was rewarded with an Earldom, being created Baron Jervis of Meaford, in the county of Stafford, and Earl of St. Vincent, to him and the heirs male of his body, by patent dated June 23d, 1797. He afterwards obtained a patent for a Viscountcy, with a collateral limitation, being created by patent dated April 27, 1801, Viscount St. Vincent of Meaford, in the county of Stafford, to him and the heirs male of his body lawfully begotten, and in default to William Henry Ricketts, Esq. Captain in his Majesty's Navy, son of Mary Ricketts, by William Henry Ricketts, Esq. late of the Island of Jamaica, (which said Mary Ricketts is the sister of the said John Earl of St. Vincent,) and the heirs male of his body lawfully begotten, and in default of such issue, to Edward Jervis Ricketts, Esq. Barrister-at-Law, another son of the said Mary Ricketts, by the said late William Henry Ricketts, and brother of the said William Henry Ricketts, and the heirs male of his body lawfully begotten; and in default of such issue, to Mary Countess of Northesk, daughter of the said Mary Ricketts; and after the decease of the said Mary Countess of Northesk, and in default of such issue, as well of the said John Earl of St. Vincent, as of the said William Henry Ricketts, and Edward Jervis Ricketts, as aforesaid, to the heirs male of the body of the said Mary Countess of Northesk, lawfully begotten. In March 1801, when the reins of Administration were committed to Mr. Addington, the Earl of St. Vincent was appointed First Lord of the Admiralty. His Lordship had afterwards for some time the command of the Channel fleet.

Creations.—Earl of St. Vincent and Baron Jervis of Meaford, by patent, June 23, 1797; and Viscount St. Vincent of Meaford, April 27, 1801.

Arms.—Sable, a chevron Ermine, between three martlets, Or.

Crest.—A demi-Pegasus issuing from a wreath.

Supporters.—On the dexter side, an eagle grasping in its left talon a thunderbolt; on the sinister side, a Pegasus charged on the wing, with a fleur-de-leuce.

Motto.—*Thus.*

Seat.—Rochetts, Essex.

Heir-Presumptive.—Edward Jervis Ricketts, Esq. to the Viscountcy only.

RYDER, EARL OF HARROWBY.

Dudley Ryder, *Earl of Harrowby, of Harrowby, in the County of Lincoln, Viscount Sandon, of Sandon, in the County of Stafford, Lord Harrowby, of Harrowby, in Lincolnshire;* High Steward of Tiverton, President of the Council, L.L.D., F.S.A.; *born* Dec. 22, 1762, succeeded his father, Nathaniel, the late Lord, June 20, 1803; *married* in 1795, Lady Susan Leveson Gower, daughter of Granville, first Marquis of Stafford, and has issue, 1st, Susan, *b.* June 20, 1796, 2d, Dudley, Viscount Sandon, *b.* May 20, 1797; 3d, a son, *b.* Oct. 27, 1809; 4th, a daughter, *b.* Jan. 28, 1811.

The name of Rythre, Ryther, or Ryder, is local, being derived from Ryther, in the hundred of Barkston, in the county of York, in which county, as well as Kent, respectable families of that name have at various times settled.

Robert Ryder, whose son was the Rev. Dudley Ryder, of Bedworth, in the county of Warwick, *m.* Anne, fourth daughter of Richard Bickley, of Halloughton, co. Warwick, by whom he had issue four sons, and two daughters, Richard; Benedi; Francis; Dudley Ryder, who *m.* Catherine, daughter of ——— Shiers, and had several children, of whom John Ryder, D.D. was Archbishop of Armagh, and left issue Richard, who left by his first wife two daughters. He *m.* secondly, Elizabeth Marshall, by whom he had three sons, Richard; Sir Dudley; William, who by Mary his wife, daughter of ——— Burton, left an only child, Elizabeth, married to Dudley Baxter, Solicitor of Excise, who died without issue, at Farnham, co. Surrey.

Sir Dudley Ryder, Knt. second son, father of the late Lord Harrowby, was born in the year 1691. After receiving a good school education, he went to the University of Edinburgh, and from thence to that of Leyden. On his return to England he settled in the Temple, where he studied the laws of his country with that assiduity and attention which laid the foundation of his future rise in his profession. He was made Solicitor-General in 1733; was in 1736 advanced to the office of Attorney-General, and discharged with integrity and ability the duties of that laborious situation during a period of about eighteen years. In 1754, he was appointed Chief Justice of the Court of King's Bench, and in 1756, his Majesty, in reward for his long and faithful services, determined to raise him to the dignity of the peerage; for which purpose the King signed a warrant on May 24th of that year, but Sir Dudley dying the next day, before the patent was completed, it did not take effect. He married Anne, daughter of Nathaniel Newnham, of Streatham in Surrey, Esq. and by her, who *d.* May 9th, 1774, left one son,

Nathaniel, first Lord Harrowby, who represented the borough of Tiverton in 1768 and 1774, till his Majesty was pleased to advance him to the dignity of a Peer of Great Britain, by the style and title of Baron Harrowby, of Harrowby, in Lincolnshire (with the like dignity to the heirs male of his body lawfully begotten) by letters patent, bearing date May 20th, 1776. His Lordship *m.* in January 1762, Elizabeth, daughter of the Right Reverend Richard Terrick, D. D. Lord Bishop of London, by whom (who *d.* August 24th, 1804), he had issue, Dudley, the present Peer; Henry, *d.* young; the Right Hon. Richard Ryder, *b.* July 5th, 1766, late Judge Martial and Advocate General to the Army, and First Justice of Cardigan, Pembroke, and Carmarthenshires; *m.* August 1, 1799, Frederica Skymer, daughter of Sir John Skymer, Knight, and has issue, of which Frederica-Anna, the eldest, *d.* December 12, 1808. Elizabeth, *b.* November 22, 1777; Nathaniel, *d.* young; Nathaniel, *b.* June 7, 1775, *d.* young; Henry, M. A. Vicar of Lutterworth, in Leicestershire, *m.* in December 1802, Sophia, daughter of Thomas March Phillips, Esq.; Anne, *b.* Dec. 1, 1779, *d.* June 26, 1801. His Lordship dying June 20, 1803, was succeeded by his eldest son,

Dudley, second Baron, and first Earl of Harrowby, who was *b.* Dec. 22, 1762, and was educated at St. John's College, Cambridge. While a Commoner he sat in Parliament for Tiverton. In 1790, he was Under Secretary of State to the Duke of Leeds. He then became Comptroller of the Household; and, in 1791, was Joint Paymaster of the Forces, in which office his brother Richard was his deputy. In 1801 he was appointed Treasurer of the Navy, and a Member of the Board of Controul; and in 1804, held for a short time the place of Secretary of State for the Foreign Department. In 1805 his Lordship was appointed Ambassador Extraordinary and Plenipotentiary to the Court of Berlin, when the plans of our government were suddenly defeated by the battle of Austerlitz. His Lordship was elevated to an Earldom on the 18th of July 1809, by the titles of Viscount Sandon of Sandon, in Staffordshire, and Earl of Harrowby.

Creations.—Baron, by patent, May 20, 1776; Earl and Viscount, by patent, July 18, 1809.

Arms.—Azure, three crescents, Or, each charged with an ermine spot, Sable.

Crest.—In a mural crown, Or, a Wyvern's head, Argent, thereon an Ermine spot, Sable.

Supporters.—Two Griffins, Argent, charged on their breast with an Ermine spot, Sable, collared Azure, thereon three crescents, Or, and chained Azure.

Motto.—*Servata fides cineri*—(The promise made to my ancestors is now kept.)

Heir-Apparent.—Dudley Viscount Sandon.

Chief Seat.—Sandon, Staffordshire.

WHITWORTH, EARL WHITWORTH.

CHARLES WHITWORTH, *Earl Whitworth, Viscount Adbaston, of Adbaston, in Staffordshire, Baron Whitworth, of Newport-Pratt, in Ireland;* Lord-Lieutenant of Ireland, and Grand Master of the Order of St. Patrick, a Privy Counsellor, a Lord of Trade and Plantations, High Steward of Stratford-upon-Avon, a Lieutenant-General, G. C. B. and D. C. L. *born* in 1760, appointed Minister Plenipotentiary at Warsaw in 1786; Envoy Extraordinary and Minister Plenipotentiary to the Court of Russia, in September 1788; invested with the Military Order of the Bath in 1793; Minister Plenipotentiary to the Court of Denmark in 1800; created a Peer of Ireland, March 21, 1800, by the title of Baron Whitworth, of Newport-Pratt, in the county of Galway; Ambassador Extraordinary to the Court of France in 1802; appointed a Lord of the Bedchamber in 1813; created a Peer of Great Britain by the title of Viscount Whitworth of Adbaston, sworn of the Privy Council, and appointed Viceroy of Ireland in 1813; *married* April 7, 1801, Arabella Diana Duchess of Dorset, widow of John Frederick third duke of Dorset, and eldest daughter of Sir Charles Cope, Bart. of Orton Longueville, by Catherine Countess of Liverpool.

The Viscount's grandfather, Francis Whitworth, Esq. of Leyborne Castle, in Kent, was younger brother of Charles Lord Whitworth, Baron of Galway, who was created a Peer of Ireland in 1721, but *d.* without issue in 1725, when the Barony became extinct. His Lordship had filled many important diplomatic situations, being successively appointed Envoy Extraordinary to the Court of St. Petersburgh in 1707, in the reign of Queen Anne;

Minister Plenipotentiary to the Diet of Augsburgh and Ratisbon in 1714; Envoy Extraordinary to the Court of Prussia in 1716; Envoy to the Hague in 1717; Ambassador Extraordinary to the Congress of Cambray in 1724. His Lordship's estates devolved to his brother, Francis Whitworth, of Leyborne, who *d.* in 1743, leaving issue, Sir Charles Whitworth, Knt. M. P. *m.* Miss Shelley, and *d.* in 1772, having had issue, 1st, Charles, the first Viscount; 2d, Sir Francis, Lieutenant-Colonel in the Army, *d.* Jan. 16, 1805; 3d, Richard, Captain R. N., lost at sea; 4th, Diana, *m.* 1st, Sir Bellingham Graham, Bart. of Norton Conyers, co. York, and 2dly, in Jan. 1800, Francis Gerard, second and present Viscount Lake. 5th, Catherine, *m.* Feb. 23, 1774, Henry, fourth Lord Aylmer; 6th Mary, *m.* Thomas Lloyd, Esq.; 7th, Anne, *m.* Sir Henry Russell, Chief Justice of Bengal.

The extraordinary circumstances attending Lord Whitworth's celebrated mission to Paris in 1802, deserve recapitulation. At this time, Buonaparte finding himself firmly established in power, took every opportunity of provoking a rupture of the ill-advised peace of Amiens, which had given him leisure to recruit his armies, and to renew with increased vigour his long-cherished designs upon Great Britain. Buonaparte had been re-elected First Consul for ten years, and soon afterwards attained the Chief Consulate for life. Finding himself thus elevated, he took occasion to question the right of this country to discuss with freedom the actions of his public life, and intimated a desire that a kind of censorship should be established for inspecting all writings of this description prior to their appearance in print. The noble and manly reply of the British Government upon this occasion, ought to be written in letters of gold. "His Majesty cannot, and never will, in consequence of any representation, or any menace from a foreign power, make any concession which can be in the smallest degree dangerous to the liberty of the press, as secured by the Constitution of this country, which admits of no previous restraints upon publications of any description." In the mean time Lord Whitworth repaired to Paris, and General Andreossi was replaced by M. Otto at the Court of St James's. One demand succeeded another in a tone of arrogant superiority and insulting irony. France complained that one of the principal articles of the Treaty, which stipulated for the evacuation of Egypt, had not been complied with. This delay arose from a misunderstanding on the part of General Stuart, who held the command there, but instructions were immediately sent out for withdrawing the King's troops from that country. Buonaparte could no longer dissemble his ambitious projects. The shameful violation of the treaty of Luneville, in overrunning and subjecting the Swiss Cantons, the annexation to France of Piedmont, Parma, Placentia, and the Isle of Elba, confirmed the suspicions entertained by the British Government that Buonaparte was preparing for the worst. Indeed, the First Consul little less than avowed his intentions in a conference with Lord Whitworth on the 17th February. A tedious negociation for the giving up of Malta was interrupted by an insulting demand, accompanied by a threat on the part of the French Government, for its immediate evacuation; and an official note was delivered to their Legislature, rudely asserting "that Great Britain was not able to contend *single-handed* against France." With the avowed intention of reducing the refractory West India colonies, immense naval and military preparations were making in the ports of France and Holland. In this state of things it behoved his Majesty's Ministers to resort to something like precautionary measures, especially as they were said to be in possession of certain intelligence, which left no doubt of Buonaparte's ulterior object. On the 8th of March, therefore, the following Message from his Majesty was delivered to the House of Commons, and another of a similar import to the Lords: "His Majesty thinks it necessary to acquaint the House of Commons, that as very considerable military preparations are carrying on in the ports of France and Holland, he has judged it expedient to adopt additional measures of precaution for the security of his dominions. Though the preparations to which his Majesty refers are avowedly directed to colonial service, yet as discussions of great importance are now subsisting between his Majesty and the French Government, the result of which must be uncertain at present, his Majesty is induced to make this communication to his faithful Commons, in the full persuasion, that whilst they partake of his Majesty's earnest and unvarying solicitude for the continuance of peace, he may rely with perfect confidence on their public spirit and liberality, to enable his Majesty to adopt such measures as circumstances may seem to require for supporting the honour of the Crown, and the essential interests of his people." This Message was immediately transmitted to Lord Whitworth, who was instructed to assure the French Government of his Majesty's earnest desire for the preservation of peace. M. Talleyrand asserted that there were no grounds for his Majesty's apprehensions, and assured Lord Whitworth of the pacific intentions of the First Consul; adding, that he should always consider a refusal to evacuate Malta as a commencement of hostilities, and as we had hitherto hesitated upon that point, his Government was justified in the precautionary measures which it had taken. At a subsequent conference, M. Talleyrand informed his Lordship of the extreme displeasure of the First Consul at his Majesty's suspicions, yet he (Buonaparte) would not suffer his passions to master him so far as to entail upon humanity the sufferings which this discussion might lead to. M. Talleyrand then delivered to Lord Whitworth a sort of Proces Verbale, which stated in substance, that if his Britannic Majesty in his Message meant to speak of the expedition of Helvoetsluys, all the world knew that it was destined for America, but that in consequence of his Majesty's Message it must be countermanded; that if the armaments in England actually took place, it would be natural for the First Consul to march 20,000 troops into Holland; to form an encampment on the frontiers of Hanover; to embark additional troops for America; to form several camps at Calais; to continue a French army in Switzerland; to send a French force into Italy; and as England was arming, and arming with so much publicity, to put the armies of France on the war establishment—a step so important, as could not fail to agitate all Europe. Two days afterwards, the British Minister had a conference with the First Consul himself, which is thus described by Lord Whitworth, in his dispatch to Lord Hawkesbury, dated March 14, 1803:—" The messenger Mason, went on Saturday with my dispatches of that date, and until yesterday (Sunday) I saw no one likely to give me further information such as I could depend on, as to the effect which his Majesty's Message had produced on the First Consul. At the Court which was held at the Thuilleries on that day, he accosted me, evidently under very considerable agitation. He began by asking me if I had any news from England. I told him that I had received a letter from England two days ago. He immediately said, " And so you are determined to go to war." 'No,' I replied, 'we are too sensible of the advantages of peace.' " We have," said he, " already waged war these 15 years." As he seemed to wait for an answer, I observed only, 'That is already too long.' " But," said he, " you wish to carry it on for 15 years more, and you force me to do it." I told him that was very far from his Majesty's intention. He then proceeded to Count Markoff and the Chevalier Azara, who were standing together at a little distance from me, and said to them, "The English wish for war; but if they be the first to draw the sword, I shall be the last to sheath it. They have no regard for treaties. We must henceforth cover them with shame." He then went his round. In a few minutes he came back and resumed the conversation, if such it can be called, by something personally civil to me. He began again, "For what reason are these armaments? Against whom are these measures of precaution? I have not a single ship of the line in the ports of France; but if you will arm I shall arm likewise; if you will go to war I shall go to war also. You may perhaps be able to destroy France, but never to intimidate her." 'We do not desire,' said I, 'either the one or the other: we wish to live in good understanding with her.' It is requisite to pay regard to treaties," replied he, "woe to those who pay no regard to treaties: they will be responsible for it to all Europe." He was too much agitated to make it advisable for me to prolong the conversation, I therefore made no answer; and he returned to his apartment repeating the last phrase. It is to be remarked, that all this passed loud enough to be overheard by 200 people that were present; and I am persuaded that there was not a single person who did not feel the extreme impropriety of his conduct, and the total want of dignity as well as of decency on this occasion." All attempts at conciliation between the Governments of England and France proved abortive. Andreossi received his passports, and quitted England on the 16th of May. On the 13th of the same month, Lord Whitworth demanded his passports, and arrived in London on the 19th, his Majesty's declaration of war having been issued on the preceding day. The First Consul, with a degree of mean perfidy unknown in former wars between civilized states, detained all those persons who were so unfortunate as to trust themselves in the French dominions after the departure of the British Ambassador. The number of English thus detained as prisoners in France, Italy, and Holland, was estimated at 7500, among whom were several Generals, Colonels, and officers of all ranks. Thus commenced the "long ten years' war," that terminated in the downfal of Buonaparte, and the re-establishment of the Bourbon dynasty, after an exile of nearly twenty-five years.

Creations.—Baron Whitworth, of Newport-Pratt, March 21, 1800; Viscount Whitworth, of Adbaston, 1812; and Earl Whitworth, Sept. 30, 1815.

Arms.—Argent, a bend, Sable, in the sinister chief point, a garb, Gules.

Crest.—Out of a ducal coronet, Or, a garb, Gules.

Motto.—*Dum spiro spero*—(While I breathe I hope.)

Supporters.—Two eaglets, wings erect, Sable, ducally gorged, Or, and a shield pendant, Argent, charged with a garb, Gules.

Heir-Apparent.—None.

Chief Seats.—Knowle, Kent; Burghurst, Sussex; and Batchacre, Staffordshire.

ANSON, VISCOUNT ANSON.

Thomas Anson, *Viscount Anson, of Shugborough and Orgreave, in the county of Stafford, Baron Soberton, of Soberton, co. Hants,* by patent; born Feb. 17, 1767, LL. D.; married September 15, 1794, Anne Margaret, second daughter of Thomas William Coke, Esq. of Holkham Hall, in the county of Norfolk, and has issue, 1st, Thomas William, *b.* Oct. 20, 1795; 2d, George,* *b.* Oct. 13, 1797; 3d, Charles-Littleton,† *b.* Jan. 28, 1799; 4th, William,‡ *b.* Feb. 26, 1801; 5th, Henry, *b.* May 15, 1804; 6th, Edward, *b.* June 21, 1805, *d.* an infant; 7th, Anne Margaret, *b.* Oct. 3, 1796; 8th, Georgiana, *b.* June 6, 1802, *d.* an infant; 9th, Georgiana, *b.* Jan. 3, 1807; 10th, Edward Harcourt, *b.* Aug. 20, 1808; 11th, Frances Elizabeth, *b.* Jan. 9, 1810; 12th, Frederica Sophia, *b.* Aug. 24, 1814; 13th, Eliza Jane, *b.* Feb. 26, 1816.

The family of Anson have been seated in this county for many generations; first, at Dunstan, in the parish of Penkridge, till William Anson, Esq. having purchased, in the reign of James I. the manor of Shugborough, made it his chief residence, and it has continued to be the seat of the family to the present day.

William Anson, Esq. his son and heir, *m.* Elizabeth, daughter of Thomas Stafford, of Botham Hall, Derbyshire, Esq. and by her had issue, three daughters, Hannah, Elizabeth, and Mary; also a son, William, his heir, who *m.* Isabella, daughter of Charles Carrier, Esq. of Wirkworth, co. Derby, (sister to Janette Countess of Macclesfield) from whom is descended the present Earl. He *d.* in Aug. 1720. Thomas, the eldest son, *b.* in 1695, and dying unmar-

* A Lieutenant in the 3d Regiment of Foot Guards, and served with his regiment at the memorable battle of Waterloo.

† Served as a Midshipman on board H. M. S. Bacchante, commanded by Capt. (now Sir William) Hoste, and was unfortunately killed by the bursting of a cannon off Lissa.

‡ A Midshipman, and served on board H. M. S. Glasgow, commanded by the Hon. Capt. A. Maitland, under Lord Exmouth, in the late severe engagement with the Algerines.

ried in 1773, left his estate to his nephew, Mr. Adams; George, the celebrated Admiral, who for his signal naval services was rewarded by George II. with a Peerage, by the title of Lord Anson, Baron of Soberton, co. Hants, June 13, 1747; he *m.* the eldest daughter of the first Earl of Hardwicke, and *d.* June 6, 1762, without issue, (for an account of the eminent services of this great man, see article "Anson," in the Biography of this work); Janette, *m.* Sambrook Adams, Esq. by whom she left issue George, (who took the name of Anson,) father of the present Peer; Isabella, Anne, and Johanna.

George Adams, Esq. his Lordship's nephew, (already mentioned) succeeded to his Lordship's property, as well as to that of his uncle Thomas Anson, Esq. He *m.* Jan. 5, 1763, Mary, daughter of George Venables Vernon, first Lord Vernon, and had issue, Thomas, the present Viscount; George, *b.* Aug. 12, 1769, who received the unanimous thanks of the House of Commons three different times for his gallant services in the Peninsular war, a Major-General in the Army, M.P. for Lichfield, Aid-de-Camp to the King, and K.C.B. *m.* May 27, 1809, Frances, sister of Sir Francis Hamilton, and has issue; Charles, in holy orders, M.A. *b.* Aug. 20, 1770; Sir William, *b.* Aug. 13, 1772, a Major-General in the Army, Knight Commander of the Most Hon. Order of the Bath, who likewise received the unanimous thanks of both Houses of Parliament, (well-earned rewards of his military services;) Henry, in holy orders, M.A. *b.* Dec. 19, 1793; Sambrook, *b.* Feb, 18, 1779, a Captain in the King's Own Militia, *m.* in Jan. 1801, Henrietta, daughter of James Ramsbottom, Esq.; Frederick, in holy orders, M.A. *b.* March 23, 1779, *m.* in May 1807, the only daughter of the late Rev. Richard Levett, of Milford, co. Stafford; Mary, *b.* Dec. 8, 1763, *m.* Jan. 22, 1785, Sir Francis Ford, Bart.; Anne, *b.* Feb. 22, 1768, *m.* Bell Lloyd, Esq. of Bodfach, co. Montgomery; and Catherine Juliana, *b.* Dec. 27, 1780.

Creations.—Viscount Anson, of Shugborough and Orgreave, co. Stafford, and Baron of Soberton, February 17th, 1806.

Arms.—Argent, three bends, engrailed, Gules, a crescent for difference.

Crest.—Out of a ducal coronet, Or, the top of a spear, Argent.

Supporters.—On the dexter side, a sea-horse, Argent; on the sinister, a lion guardant, collared.

Motto.—*Nil Desperandum*—(Fear not).

Heir-Apparent.—The Hon. Thomas William Anson.

Chief Seat.—Shugborough, Staffordshire.

GOWER, VISCOUNT GRANVILLE.

Granville Leveson Gower, *Viscount Granville, of Stone-Park, Staffordshire*, and a Privy Counsellor; *born* October 12, 1773, *married* December 24, 1809, Harriet, second daughter of the late Duke of Devonshire, by whom he has issue, 1st, Susan, *b.* October 25, 1810; 2d, Georgiana, *b.* September 23, 1812; 3d, Granville-George, *b.* May 11, 1815; and 4th, William, *b.* October 2, 1816. His Lordship is youngest son of Granville-Leveson, late Marquis of Stafford, by Susan, daughter of Alexander Earl of Galloway. Viscount Granville was appointed Minister Plenipotentiary to the Court of St. Petersburgh, in 1807. [For an account of this noble family, see "Gower, Marquis of Stafford."]

Creation.---Viscount, June 15, 1815.

Arms.---Quarterly, 1st and 4th, barry of eight, Argent and Gules, over all a cross flory, Sable, for Gower; 2d, Azure, three laurel leaves erect, Or, for Leveson; 3d, Gules, three clarions, Or.* In the fess point a crescent for difference.

Crest.---On a wreath, a wolf passant, Argent, collared and chained, Or.

Supporters.---Two wolves, Argent, collared and chained, Or, bearing on the shoulder a shield charged with a clarion.

Motto.---*Frangas non Flectes*---(You may break but shall not bend me.)

Heir-Apparent.---Granville-George, his Lordship's eldest son.

Chief Seat.---Tigall, Staffordshire.

GARDNER, VISCOUNT GARDNER.

Alan-Legge Gardner, *Viscount and Baron Gardner, of Uttoxeter, in Staffordshire, and Baron Gardner, of Uttoxeter, in Ireland; born* January 29, 1810, and succeeded his father, Alan-Hyde, the late Viscount, Dec. 27, 1815.

William Gardner, grandfather of the late Lord Gardner, was born at Coleraine, in Ireland, March 24, 1691, and *d.* at Uttoxeter, in Staffordshire, August 14, 1764. He was a Lieutenant-Colonel in the 11th regiment of dragoons, to which he, with distinguished honour, rose from the rank of a Cornet; and *m.* at Preston, in Lancashire, Dec. 27, 1729, Elizabeth, eldest daughter and co-heir of Valentine Farrington, of Preston aforesaid, M. D. (youngest son of William Farrington, of Werden, in Lancashire), by his wife Agnes,

* This is a correct description of the Stafford Arms: Sir E. Bridges', in Collins, and the other Peerages, describe the 3d quartering inaccurately. We copied Collins, in our previous account of it.

sole daughter and heir of ——— Prickett, of Nutland Abbey, co. Westmoreland, Esq. and by her (who *d.* Aug. 16, 1783,) had issue, William, of Liverpool, merchant, *b.* 1734, *d. unm.* Oct. 27, 1788; Anne, *b.* May 1733, *m.* Thomas Dixon, Esq. *d.* Nov. 1803; Agnes, *b.* Nov. 1735; Elizabeth, *d. unm.* August 30, 1738; Valentine, late a Major in the Army, *b.* May 16, 1739, *m.* in 1762, Alida, third daughter of Colonel Robert Livingston, of Livingston Manor, New York, and *d.* September 1791, leaving issue, William Linnæus, a Captain in the Army, and *m.* 2dly, Dec. 1, 1792, Frances, 2d daughter of Samuel Holworthy, of Elsworthy, co. Cambridge, widow of Framingham Thruston, of Weston, co. Suffolk, Esq. and had issue Valentine, *b.* in 1794; Mildred, *d. unm.* Dec. 29, 1758; Henry, *d. unm.* in 1740; Alan, the first Lord; Sarah, *b.* May 26, 1745, *m.* Nov. 9, 1771, Henry Humphreys, Esq. of London, merchant, and by her (who *d.* May 27, 1778) had issue, John-Montague, *b.* Oct. 7, 1772; Henry, *b.* April 29, 1774; Sophia, who *d.* in 1776; Harry-Farrington, *b.* in 1746, Colonel Commandant of the 20th regiment of dragoons, *d. unm.* at Jamaica, July 20, 1792; Margaretta, *d. unm.* Nov. 8, 1777; Dorothy, *b.* September 29, 1751, *m.* at St. Augustine, in East Florida, May 10, 1772, Robert Barrie, Esq. in the Army, who *d.* in 1775, having had issue William-Oughton, *d.* June, 1773, and Robert, May 5, 1774, she *m.* 2dly, Sept. 22, 1784, George Clayton, of Lostock Hall, in Lancashire, Esq. and has issue Frances, *b.* July 29, 1785; George, *b.* Nov. 9, 1787; William, *b.* June 17, 1790; and Elizabeth, *b.* Oct. 15, 1792.

Alan Gardner, fourth son, commenced his naval life, May 1, 1755, by entering on board his Majesty's ship, the Medway, Captain Peter Dennis, of Harwich, which ship, under Lord Anson's orders, conveyed George II. to Helvoetsluys; he was Admiral of the White, Major-General of Marines, and a Baronet by patent, September 1794. His Majesty, in consideration of a series of gallant services to his country,* was pleased to advance him to the dignity of the Peerage, by the title of Baron Gardner of Uttoxeter, in Ireland, Dec. 23, 1800; and further advanced him to the Peerage of England, Nov. 27, 1806; *born* April 12, 1742, *married*, in Jamaica, May 20, 1769, Susannah Hyde, sole daughter and heiress of Francis Gale, of Liguania, in Jamaica, Esq. and had issue, 1st, Alan-Hyde, the late Lord; 2d, Francis-Farrington, *b.* June 21, 1772, R. N., *m.* in 1809, C. V. Straubenzee, third daughter of C. S. V.

* See Memoir of his Lordship, p. 313.

Straubenzee, of Yorkshire, Esq.; 3d, William Henry, *b.* Oct. 6, 1774, a Captain in the Royal Horse Artillery, *m.* Feb. 20, 1803, Eliza Lydia, third daughter of Colonel Fyers, Commanding Royal Engineer at Gibraltar, by whom he had issue, Anne Europa, *b.* Dec. 10, 1804; Eliza, *b.* June 17, 1806; and Susan, *d.* Aug. 7, 1809, and a daughter, *b.* Oct. 30, 1811; 4th, Henry Crosby, *d.* Nov. 6, 1792; 5th, Herbert, *b.* Aug. 20, 1781, a Lieutenant in the Royal Artillery, *m.* Nov. 7, 1804, Miss Cornwall, youngest daughter of the late John Cornwall, Esq. and has a son, *b.* in Dec. 1808; 6th, Edward, *b.* March 9, 1784; 7th, Valentine, *d.* Feb. 1, 1786; 8th, Valentine, *b.* March 20, 1787; 9th, Samuel Martin, *b.* on board his Majesty's ship Europa, Aug. 10, 1789; 10th, Susannah Hall, *b.* July 1, 1773; *m.* Jan. 7, 1794, John Cornwall, of Hendon, in Middlesex, Esq.; and by her, who *d.* June 20, 1802, left issue, John, *b.* Jan. 22, 1795; 11th, Susan Caroline, *d.* in Jan. 1797; 12th, Susannah, *b.* May 20, 1797; 13th, Alan, *b.* July 16, 1798; 14th, William Henry, *b.* Sept. 4, 1799; 15th, Augusta, *b.* Jan. 13, 1801; and 16th, Sophia, *b.* April 2, 1802. His Lordship *d.* Dec. 30, 1808, and was succeeded by his eldest son,

Alan-Hyde, the second Lord and first Viscount Gardner, Vice-Admiral of the Blue, *b.* Feb. 8, 1772, and succeeded his father Alan, the late Lord, Dec. 30, 1808; *married*, 1st, in March 1796, Maria-Elizabeth, only daughter of Thomas Adderley, Innishannon, in Ireland, by Margaret Baroness Hobart, (which marriage was dissolved by Act of Parliament in 1805); his Lordship *m.* 2dly, April 10, 1809, Charlotte, daughter of Robert Smith, Lord Carrington, and sister to Catharine-Lucy Viscountess Mahon, and by her (who *d.* March 27, 1811,) had issue, Alan-Legge, *b.* January 29, 1810; and Charlotte-Susan. His Lordship dying December 27, 1815, was succeeded by Alan-Legge, the present and second Viscount.

Creations.—Baron Gardner, in Ireland, Dec. 13, 1800; Baron Gardner, Nov. 15, 1806, Viscount Gardner, both in England, Sept. 30, 1815.

Arms.—Argent, on a chevron, Gules, between three griffins' heads erased, Azure, an anchor, between two lions combatant, Or.

Crest.—A demi-griffin, Azure, collared and lined, holding an anchor, Or.

Supporters.—Two griffins, Azure, murally gorged, and resting their hind off-leg upon an anchor, Or.

Heir-Presumptive.—Francis-Farrington, his Lordship's uncle.

Motto.—*Valet anchora virtus*—(Virtue is a sheet-anchor).

TUCHET, BARON AUDLEY.

GEORGE THICKNESSE TUCHET, *Baron Audley, of Heleigh Castle, in the county of Stafford; born* Feb. 4, 1758, and succeeded his uncle John, the last Earl of Castlehaven, as Baron Audley, April 28, 1777. His Lordship, April 3, 1784, assumed the name of Tuchet, by permission of his Majesty; *married*, 1st, May 19, 1781, Elizabeth, co-heiress of John late Lord Delaval, by whom, (who *d.* July 11, 1785, he had issue, Elizabeth Susannah, *b.* April 3, 1782; *m.* in Nov. 1805,) John Coffin, Esq.; George John, *b.* in March 1783. His Lordship *m.* 2dly, May 2, 1792, Mrs. Moorhouse, relict of Colonel Moorhouse.

This family is of great antiquity, William Tuchet having summons to Parliament among the Barons from 1296 to 1307. Thomas, his son, was beheaded, being concerned in the insurrection of Thomas, Earl of Lancaster, in 1321. Sir John Tuchet *m.* Joan, daughter of James, and sister-co-heir of Nicholas Audley, Baron Audley, of Heleigh, and had issue, Sir John, who was, in 1391, found to be heir to Joan, his grandmother, and assumed the title, and was possessed of many lordships and inheritances. He *d.* Dec. 19, 1409. He left issue, James, the second Baron Audley, who distinguished himself in the wars of France in 1418; but, in 1428, being sent by Queen Margaret to oppose Richard Neville Earl of Salisbury, then in arms in favour of the House of York, he was defeated and killed at Blore-heath, co. Stafford, with 2400 of his companions. He *m.* Margaret, daughter of William Lord Roos, of Hamlake, (ancestor of Charlotte, the present Baroness de Roos,) leaving issue, John third Baron; and Anne, *m.* Sir Thomas Dutton, of Dutton, co. Chester, whose 5th daughter, Eleanor, *m.* Richard Cholmondeley, Esq. ancestor of the Earls of Cholmondeley.

James, third Earl of Castlehaven in Ireland, and twelfth Baron Audley, *m.* Elizabeth, daughter and co-heir of —— Bard, Esq. by Anne, daughter of Charles Villiers, Earl of Anglesea, and widow of Thomas Ratcliffe, Earl of Sussex, and by her had James, fourth Earl of Castlehaven, and Baron Orrier in Ireland, and Baron Audley in England. James, fourth Earl of Castlehaven, *m.* Elizabeth, only daughter of Henry, fifth Lord Arundel of Wardour, and by her, who *d.* 1743, had issue, 1st, John, his successor; 2d, Mary, *m.* Nov. 20,

1749, Philip Thicknesse, Esq. late Lieutenant-Governor of Languard Fort, and *d.* 1762, leaving issue, George, the present Baron; Philip, R. N. *b.* June 23, 1760; James, *d.* an infant; Elizabeth, *b.* Aug. 4, 1750; Joyce, *b.* March 26, 1753; Charlotte, *m.* July 15, 1786, Mr. Wilder. His Lordship *d.* at Paris, November 1740, and was succeeded by his only son,

John, fifth Earl of Castlehaven, *m.* 1776, Mrs. Cracraft, and *d.* without issue, April 22, 1777, when the Earldom became extinct; but the Barony of Audley descended to his nephew, George, present and sixteenth Baron Audley.

Creations.—Baron Audley, of Heleigh Castle, in the county of Stafford, by writ of summons, Oct. 20, 1403; but originally Jan. 26, 1296.

Arms.—Ermine, a chevron, Gules, quartering, Gules, a fret, Argent, for Audley.

Crest.—In a ducal coronet, Or, a swan rising, Argent, ducally gorged of the first.

Supporters.—Two wyverns, with wings expanded, Sable.

Motto.—*Je le tiens*—(I hold it.)

Heir-Apparent.—George John, his Lordship's son.

Chief Seat.—Grovely, near Sarum, Wiltshire.

BAGOT, BARON BAGOT.

WILLIAM BAGOT, *Baron Bagot, of Blithfield, in Staffordshire*, and a *Baronet*; F.L.S.; *born* Sept. 11, 1773; succeeded his father, William, the late Lord, Oct. 22, 1798; *married*, 1st, May 30, 1799, Emily Fitzroy, daughter of Charles first Lord Southampton, (who *d.* June 8, 1800), leaving a daughter, who *d.* Jan. 1, 1801; and, 2dly, Feb. 17, 1807, Louisa Legge, eldest daughter of George, third Earl of Dartmouth, K. G. and has issue by her, (who *d.* Aug. 13, 1816,) Louisa, *b.* March 1, 1808; Agnes, *b.* August 10, 1809; William, son and heir, *b.* March 27, 1811; Harvey, *b.* Dec. 17, 1812; Eleanor; and Alfred.

This noble family is of great eminence and antiquity; and, as appears by Domesday-Book, were possessors of Bagot's Bromley, in Staffordshire, at the time of the Conqueror's Survey, which they held of the Baron Stafford, whose male line became extinct, 1195, and they became possessed of it by marriage with the heiress of Robert de Stafford with Hervey Bagot, a younger branch of this family, who thereupon assumed the surname and title of Stafford and became progenitor to the succeeding Barons and Earls of Stafford, and Dukes of Buckingham. From Bagod Domians Medietatis de Bramale, descended Sir John Bagot, Knt. Lieutenant of Calais, and Privy Counsellor to Henry IV. ancestor of Sir Lewis Bagot, Knt. who was of the body to Henry VII. and from whom descended Walter Bagot, who m. Elizabeth, sister to William Lord Burleigh, by whom he had four daughters and five sons, the eldest of which died in the life-time of the father, and the second, Hervey, ancestor of the present Lord, was created a Baronet May 30, 1627. His Lordship's grandfather, Sir Edward Bagot, the fourth Baronet of the family, and lineally descended from Henry the first Baronet, m. Frances, daughter and sole heir of Sir Thomas Wagstaffe, Knt. of Tachbrooke, in Warwickshire (who surviving him, m. the late Sir Adolphus Houghton, Bart.). By this Lady he had issue two sons and two daughters; Hervey, *d.* an infant, 1699; Walter Wagstaffe, his successor; Frances, m. Sir John Chester, Bart. of Chichley, in Bucks; Jane, m. Craven Kynnersley, Esq. of Loxley, in Staffordshire. Sir Edward *d.* May 1712, and was succeeded by his only surviving son,

Sir Walter Wagstaffe Bagot, Bart. m. in 1724 Barbara, eldest daughter of William Earl of Dartmouth, by whom (who *d.* Oct. 29, 1765) he had eight sons and eight daughters: the sons were Edward and John, *d.* infants; William, his successor, the first Lord; Walter, in holy orders, *b.* Nov. 2, 1731, m. September 7, 1773, Anne, daughter of W. Swinnerton, Esq. of Staffordshire, and has five children, one of whom, Honora, m. Dec. 15, 1795, Augustus-George, brother to the present Earl of Dartmouth; Charles, *b.* Sept. 1, 1730, who took the name of Chester, on the death of Sir Charles Bagot-Chester, Bart. m. Catharine, daughter of Heneage Legge, Esq. son of William Earl of Dartmouth, and has twelve children; he *d.* April 2, 1792; Richard, *b.* Nov. 13, 1733, m. February 26, 1783, Frances, sister to the twelfth Earl of Suffolk, and, by the King's permission, changed his name to Howard; Lewis, D.D. Dean of Christ-Church, Oxford, and late Bishop of St. Asaph, *b.* January 1, 1740, and *d.* June 4, 1802, having m. Mary, daughter of Edward Hay, brother to Thomas eighth Earl Kinnoul, and his Lady *d.* August 17, 1799; Thomas, *b.* Feb. 14, 1746, and *d.* unm. The eight daughters were, Barbara, m. Ralph Sneyd, Esq. of Keel, in Staffordshire; Anne; Frances, *d.* an infant; Jane; Elizabeth; Maria, m. Richard Wingfield, Esq. Catharine; and Harriet. Sir Walter was one of the trustees for executing the will of the learned Dr. Ratcliffe, and was a member of Parliament for six Parliaments of Great Britain, and dying January 28, 1768, was succeeded by his son, William, the first Lord, created a Baron Oct. 17, 1780, *b.* February 20, 1728, *m.* August 20, 1760, Louisa, only surviving daughter of John St. John, second Viscount Bolingbroke, by whom (who *d.* June 7, 1800) he had issue, Edward, *b.* January 23, 1763, and *d.* 1773; Walter, *b.* March 28, 1776, and *d.* 1763; William, the present Lord; Hervey, *b.* March 14, 1777, *m.* July 22, 1806, Miss Pole, daughter of the Hon. William Pole; Louisa, *b.* September 30, 1764, *m.* Walter Sneyd, Esq. of Keel Hall, by whom she has issue; Barbara, *b.* October 25, 1768, and *d.* 1773; Henrietta, *b.* February 14, 1780, and *d.* January 13, 1797; Charles, *b.* September 22, 1781; Richard, *b.* November 22, 1782; Frances, *b.* November 9, 1785, *m.* May 21, 1805, Major-General Edward Paget, third son of Henry Earl of Uxbridge, and she *d.* May 30, 1806. His Lordship died October 22, 1798, and was succeeded by his son William, the present Lord.

Creations.—Baronet, May 30, 1627; Baron Bagot, of Bagot's Bromley, in the county of Stafford, Oct. 17, 1780.
Arms.—Ermine, two chevronels, Azure.
Crest.—In a ducal coronet, Or, a goat's head, Argent, attired, Or.
Supporters.—Two goats, Argent, attired and unguled, Or.
Motto.—*Antiquam obtinens*—(Possessing antiquity.)
Chief Seat.—Blithfield, near Lichfield, in the county of Stafford.

CHICHESTER, BARON FISHERWICK.

GEORGE AUGUSTUS CHICHESTER, *Baron Fisherwick, in the county of Stafford, and Marquis of Donegal, Earl of Belfast, Viscount Chichester, and Baron of Belfast, in the kingdom of Ireland;* *born* Aug. 13, 1769, *married*, Aug. 8, 1795, Anne, daughter of Sir Edward May, of Mayfield, in the county of Waterford, Bart. and has issue, 1st, George Hamilton, Earl of Belfast, *b.* Feb. 8, 1797; 2d, Edward, *b.* June 11, 1799; 3d, Spencer Augustus, *b.* Nov. 27, 1805; 4th, Arthur, *b.* Sept. 30, 1808; 5th, Hamilton Frances, *b.* March 9, 1810; 6th, John, *b.* Nov. 12, 1811.

This family seems anciently to have borne the name of Cirencester, and was one of the most eminent in the county of Devon, for its antiquity, estate, employments, and alliances, having flourished for several generations at South Poole, not far from Kingsbridge.

Arthur, Viscount Chichester, was created, in 1647, Earl of Donegal, and was succeeded, pursuant to the limitation of the patent, by his nephew, Arthur, second Earl. He *m.* Jane, daughter and sole heir to John Itchingham, of Dunbrody, co. Wexford, Esq. by whom he had four sons, Arthur; John; Edward, in holy orders, who, by Elizabeth, daughter of Captain John Chichester, had issue the Rev. Arthur Chichester, *b.* Jan. 6, 1716; Charles; and three daughters: Anne, *m.* Dacre-Lennard Barret, of Bell-house, in Essex, Esq. by whom she had three daughters and one son, viz. Jane *m.* John Ranby, Esq.; Dorothy, *m.* Hugh Smith, of Weald Hall, in Essex, Esq. and had issue, Lucy, mother of Edward, the present and twelfth Earl Derby; Henrietta, *d. unm.*; and Richard, the son, *m.* Anne Lennard, daughter and heir of Thomas Earl of Sussex, Lord Dacre, and dying in 1716, four months after his marriage, left her (who became Baroness Dacre) with child of a son, Thomas, Lord Dacre, who *m.* Anne Pratt, sister of the first and aunt of the present Marquis Camden, and a daughter, who *d.* in 1749, and her father *d.* Dec. 13, 1785: her Ladyship *m.* 2dly, the Hon. Robert Moore, sixth son of the third Earl of Drogheda, and next brother of William, whose son *m.* one of the daughters and co-heirs of M. Cassan, of the Queen's County, Esq.; Catharine, *m.* John Taylor, Esq.; Mary *m.* Sir Robert Newcomen, Bart. by whom she has issue, Sir Arthur, and five daughters.

Arthur, the eldest son, succeeded his father, and was the third Earl; in 1705 he was Major-general of the Spanish forces, and was killed on April 10, 1706, at the fort of Montjuick. He *m.* 1st, Barbara, youngest daughter of Roger, first Earl of Orrery, by whom he had a son, Charles, who *d.* young; and his lady dying, he *m.*, 2dly, Catharine, daughter of Arthur, first Earl of Granard; and by her (who *d.* in June 1743) left two sons and five daughters; Catharine, *m.* Clotworthy, third Viscount Masserene; Jane, Frances, and Henrietta, were unfortunately burnt at the house in Belfast; Mary *d. unm.*; Anne, *m.* James, fourth Earl of Barrymore: the sons were, Arthur; John, *b.* in 1700, who, in 1726, *m.* Elizabeth, eldest daughter of Sir Richard Newdigate, of Arbury, co. Warwick, Bart. He *d.* June 1, 1746, and his lady in 1747, leaving two sons, Arthur, 1st Marquis; and John *b.* Dec. 26, 1750, who *d.* Feb. 6, 1783; and a daughter. Arthur, eldest son of Arthur, third

Earl, *b.* March 28, 1695, was, in 1716, *m.* to Lucy, one of the two daughters and co-heirs of Robert Ridgway, Earl of Londonderry; but by her (who *d.* July 16, 1732) he had no issue; and dying in 1756, was succeeded by his nephew,

Arthur, fifth Earl, who was created Marquis of Donegal and Earl of Belfast, June 27, 1791, having been created Baron Fisherwick, an English Peer, July 9, 1790; *b.* June 13, 1739, *m.* 1st, Nov. 16, 1761, Anne, eldest daughter of James, fifth Duke of Hamilton, by whom (who *d.* in 1780) he had Charlotte-Anne, Henrietta, and Amelia, who all *d.* infants; Elizabeth Juliana, *d.* April 24, 1787; George-Augustus, the present Marquis; Arthur, *b.* May 3, 1771, *d.* Sept. 11, 1788; Spenser-Stanley, *b.* April 20, 1775, *m.* August 8, 1795, (on the same day as his brother,) Lady Harriet Stewart, daughter of John eighth Earl of Galloway, K.T. and has issue a son and a daughter. His Lordship *m.* 2dly, Oct. 24, 1788, Mrs. Moore, who *d.* without issue, Sept. 18, 1789; and, 3dly, Oct. 12, 1790, Barbara, daughter of the Rev. Dr. Godfrey. The Marquis died Jan. 5, 1799, and was succeeded by

George, second Baron Fisherwick.

Creations.—Lord Chichester, Baron of Belfast, in the County of Antrim, Viscount Chichester, April 1, 1625; Earl of Donegal, March 30, 1647; Marquis of Donegal in Ireland, 1791; and Baron Fisherwick, in England, July 9, 1790.

Arms.—Quarterly, the first and fourth cheque, Or, and Gules; a chief, vair; the second and third Azure, frettee, Argent.

Crest.—On a wreath, a stork, proper, holding in its beak a snake, Argent, the head, Or.

Supporters.—Two wolves, Gules, ducally gorged and chained, Or.

Motto.—*In vitum sequitur honos*—(Honour follows against his will.)

Heir-Apparent.—George Hamilton.

Chief Seats.—Belfast and Carrickfergus, in the county of Antrim. [The mansion, formerly at Fisherwick, in Staffordshire, is now taken down.]

ERRATA.

DEPARTMENT I.

Page.	Line.	
23	7 and 8,	for 'Thestenhall,' r. Thotenhall.
58	(note,)	for 'and may be reckoned three lives to a century,' r. that four may be reckoned to a century.
103	29	for 'Hamerwich,' r. *Hammerwich.*
130	9	for 'Charles Bearman, Esq.' r. *John Beaman Player, Esq.*
134	4	for 'Oakenhill,' r. *Ocker-hill.*—Also, same page, the statement that the 'road to Birmingham passes through West-Bromwich,' is incorrect.
168	6	from the bottom, for 'Walton', r. *Weston*, part of which is in the parish of Norbury, and part in Forton parish, and not wholly in the latter, as stated in page 272. l. 18.
181	8	for 'liberty of Bushbury,' r. *parish of Bushbury*. [There does not seem any good authority for including the Goughs as owners of this manor.]
184	1 and 2,	after 'inhabitants about 6000,' r. *which may be increased in some degree*; expunging 'from 5 to 13 per cent.'—Same page, line 30, for 'Joseph Pearson, Esq.' r. *John Pearson, Esq.*
186	7	for 'Cannock-wood is an extensive waste,' r. *a moderate-sized waste.*
255	last line,	after 'Rev. Richard Slaney,' omit *Vicar*; Penkridge is a royal peculiar.
256	28	for 'the principal manufacture of Penkridge is iron,' r. *Penkridge is not remarkable for any trade or manufacture.*
294	26	for 'wild geese and ducks', r. *wild ducks and geese, occasionally, in severe seasons. The ducks, &c.*
313	29	for 'ten,' r. *after.*
338	30	In the list of benefactions, Mr. Brock's bequest should have been thus stated:

"Mr. J. Brock left to the poor twenty shillings a-year, charged upon lands in *Barlaston* parish, which Mr. T. Brock afterwards increased to one-fourth of the rent of the land, and which, at the present rent, amounts to two guineas a-year."

DEPARTMENT II.

Page.	Line.	
17	26	for '81,' r. 183 parishes.
38	13	for 'seed 20lbs. is drilled to an acre,' r. 8lbs. or less.
150	4	for 'Hidlbundus,' r. *Ridibundus.*
164	20	for 'contain upwards of 30,000 inhabitants,' r. 60,000 inhabitants.
260	last line,	for '1821,' r. 1621.
275	26	for 'Feb. 9,' r. *Jan.* 9.
291	30	after 'Lawrence, fourth Earl, b.' insert in *August*, 1720.

[*The two last errata occur in the* PEERAGE, *in which, we fear, there are many similar blunders in the several accounts compiled from the standard "Peerages." These are very incorrect, and will sufficiently apologize for our having been drawn into mistakes, which we had no means of avoiding; nevertheless, we have endeavoured, as far as our information enabled us, to render this part of the Work correct. But it is a subject of all others the most liable to error from the constant changes making in the different branches of Noble Families by marriage, intermarriage, &c.* COLLINS, LONGMATE, *and* DEBRETT, *are the writers that we have chiefly consulted. The judicious reader will also make allowance for inaccuracies in other parts of this Work, to whom the nature of such an undertaking, and its concomitant difficulties, are well known;—to the injudicious, all apology were useless.*]

INDEX,

TO DEPARTMENT THE FIRST.

TOPOGRAPHY, &c.

A.

ABBEYS, ancient, 32, Burton, 42, Crozden, 225, Alveton, 228
Abbot's Bromley, ancient market town, 309, large parish, free school, market and fairs, ibid. hobby horse dance, 310
Acton, hamlet of Swinnerton, 386
—— Trussell, village and chapel, 267
Adbaston, parish and manor, 322
Agardsley, manor, chapel, Earl Talbot's, 58
Alabaster, at Hounhill, 58
Aldershaw, near Lichfield, 129
Aldridge, large village, soil, canal, and christmas customs, 161, church, monuments, distillery, and population, 162
Alfred, English monarch, 13
Alrewas, a large and ancient village on the Trent and Mersey Canal, 64, ancient church and manor, Turton family there, 65, now Lord Anson's with Orgreave estate, 66, church register extracts, 67
Alstonefield, parish on Dove, 242, church, hamlet, and population, 243
Alveton, fortress, 9, a large populous parish on Churnet, its church and castle, 226, wire and cotton works, corn mill and tan-yard, 227, abbey, Lord Shrewsbury's, 228, tower, free-school, earthquake and hail storm, ib.
Amblecote, coal pits, 193, fire clay, 194, glass making, 194—5
Ancient and modern names of places in the county, 20 to 23
Antiquities, 1 to 33
Areley, on Severn, its soil and fruit, 201, perish and church, mansion, Roman camp, hamlets & mines, 202
Armitage, south of the Trent, its church and hamlets, 78—79
Ashley, parish, village, and church, 324
Aston, Little, hamlet of Shenstone, an elegant mansion of Mr. Tennant's, 163
Audley, parish, church, &c. 345.

B.

Bage Robert, anecdotes of, 136
Bagnall, parish on the Caldon canal, manor-house and church, 246
Bagnall, township and chapel, 373
Ball, Rev. John, his piety, 377
Barlaston Church, hall and village, 388
Barr Beacon or Hill, 1, village and parish, 160, mansion and chapel, 161
Barons' wars, 14
Barrows or lows, 9—10
Barton-under-Needwood, its antiquity, manors and chapel, 61, curious origin of its chapel, 62, school and school-house, ibid.
Bastard, curious custom, (note) 11
Baswich, church and hamlets, 266, population and mansions, 267
Beamhurst, ancient village, 222
Beaudesert, Marquis of Anglesea's, 80, mansion and family, 81—4, visited by the Prince Regent, 86
Bednall, township and chapel, 267
Beggar's Petition, original, 195
Bent, Esq. and Dr. Bent, mansions of, 401
Bentley Hall, formerly Lane's, 150
Betley, town, mansions, church, monuments, &c. 341—344
Biddulph, parish and hamlets, 350, antiquities, church, &c. 351
Billington, ancient camp, 9, 279
Bilston, large and populous, on the Birmingham canal, 173, vast mines of coal and ironstone, quarry stone and clay, 174, chapel and meeting houses, ibid. stratum of the mines, old and new, 174 to 178, derangement and stagnation of trade, 178, efforts of the workmen, 179, excellent grindstones and sand, population, 179

Birmingham Canal, 173
Bishops of Lichfield, 118 to 127
Blithfield, village and mansion, Lord Bagot's, 310, church, paintings, and park, 311
Blore, small parish on Dove, formerly of the Bassetts, its church, &c. 237
Blore-heath, battle, 12, 328, pedestal & inscription, 330
Bloxwich, large village and chapel, 147, common field, and Wyrley canal, ib.
Blurton, chapelry of Trentham, 384
Blymhill, village, church, and hamlets, extent and population, rector, &c. 281
Blythbury, church and priory, 77
Bobbington, village, soil, 36, and church, 200
Bonehill, hamlet of Tamworth, 145
Bradley colliery, Sir Joseph Scott's, 180
Bradley-in-the-Moors, parish, church, and village, 226
Bradley Moor, hamlet of Bilston, wild-fire there, and extensive iron works, 180
Bradley, village and hamlets, 279
Bramshall, parish, church and village, 216
Brewood, large parish, and ancient market town, 253, church spire and grammar-school, 254, manufacture and hamlets, 255
Brierly-hill Chapel, 195, vast trade in glass, 196
Brindley, engineer, his life, bred a mill-wright, 429, paper and silk mills, 430, inland navigation, Duke of Bridgewater's patronage, 431, tunnels and aqueducts, canal to Manchester and the Mersey, 432, Trent and Mersey grand trunk or Staffordshire canal, 433, Harecastle-hill, tunnels, aqueducts over Trent and Dove, Staffordshire and Worcestershire canal, 434, also Birmingham, Droitwich, Coventry, Oxford, &c. canals, 434, his character, and death, 435—436
Brineton, hamlet of Blymhill, "John of Brineton," 281
Brockton Hall, Sir Geo. Chetwynd's, 266, village, and lodge (Mr. Chetwynd's), 267
Brome, village, church, &c. 204
Bromley Regis, on Trent, 68, its ancient manor and demesne, now Mr. Lane's, its soil, canal, mills, church, and charities, 69
Broughton, parish, mansion, church, and plantations, 323
Bucknall, church, population, &c. 373
Bunbury, ancient fortress, 9
Bunster, a limestone mountain, 38
Burnet, indigenous, 37
Burslem, origin of the Pottery, 395, extensive works carried on, church, meetings, population, 396, schools, bath, machinery, canal, clay and coal, Wedgwood family, 397, schraggers, horse gins, steam engines, coal mines, 398, varieties of coal, population how increased, 399
Burston, hamlet of Stone, 391
BURTON-ON-TRENT, antiquities, its abbey, and patron saint, 41—2, bridge, burgess rights, town and hamlets, 43—44, market, fairs, chronology, manufactures, navigation, and farther hamlets, 45—9
Bushbury, village and parish, 180, church and monuments, 181; epitaph, hill, mansions, and lows, hamlets, Fordhouses, Moseley, and Essington, 182
Butterton, village and church, 240.

C.

Caldon, a small moorland parish, barren and dreary, but in limestone inexhaustible, 235, church, oat bread, ibid.
Callingwood, near Needwood, 60, Roman coins found.
Calton, township and chapel, 235
Camps, 5, 6, 9, 10, 11
Cannock, an ancient market town, very extensive parish, 261, waterworks, 262, hamlets, church, and heath, 263—4, Mr. Gilpin's manufactory, ibid. 448
Canwell, Sir Robert Lawley's, 164
Castles, 8, 10, 11, 14, 190, 227, 268, 308
Castle Church, parish near Stafford, 267, hamlets and church, 268, ancient castle, demolished and new erected, ibid.
Caverswall, parish and village, 232, castle and hamlets, modern nunnery, ibid. church and monuments, 233
Charles II. his escape, 15, 16
Chartley Castle, 11, Earl Ferrers, 308, Chartley Park, and wild cattle, 309
CHEADLE, market town, population and manufactures, 229, brass and copper works, church, chapel, &c. 230
Cheapness of former times, 379

Chebsey, church and village, 318
Checkley, battle, 12, on Tean river, 220, church and monumental stones, 221
Cheddleton, a parish and village on Churnet and the Caldon canal, 247, neat church and good houses, ibid.
Cherry wake, Enville, 36
Chorlton, village and chapelry, 387
Church Eaton, village, parish, and hamlets, 278, church and manor, Earl Talbot's, Wood Eaton, High & Little Onn, ibid.
Civil wars, 14
Clayton, liberty of Stoke, 388
Clent hills, 35, parish church and manors, 203, St. Kenelm's church, and cheese fair, ibid.
Clifton Campville, large village and manor, 137, No Man's Heath, ibid. hamlets Clifton & Haunton, 137-138
Cobridge, large pottery village, and Catholic chapel, 399
Codsall, small village and parish, 185, church and monuments, ibid. hamlets, the Birches, Oaken, Woodhall, and Bilbrooke, 186
Collegiate churches, 32
Colton, parish, village, church, and mansions, 311
Colton Clanford, Wollaston's birthplace, 300
Colwich, village and church, 312, monuments and hamlets, Oakedge Hall, Great Haywood, Wolseley, &c. 113, Staffordshire and Trunk Canals.
Comberford, ancient seat & manor, 144
Commons or wastes, 36, 40, 264, 267, &c.
Congreve, hamlet of Penkridge, 257
Convulsions of nature, 40
Coppenhall, chapel and township, 261
Cotes family, & hamlet of Standon, 386
Cotes-under-Needwood, hamlet, mansion, Mr. Adderley's, 57
Cows, breed of, 37, (note)
Creswell, mansion, estate & land, 314
Croxden, parish and village, the abbey a picturesque ruin, 225, the church and hamlets, 226
Cuddlestone Hundred, 19, 253.

D.

Danes on Severn, their customs, 10, (note) 12
Darlaston, a populous manufacturing parish, with good mines, and canal, 149, church, meeting-houses, and school, ibid.
Darlaston Hall, & bridge on Trent, 388
Declivities, should be planted, 37
Dilhorne, parish and village, 231, improvements, coal-mines, schools, church, curious register, &c. ibid.
Divisions of the county, 13
Dove river, 38
Draycott-in-the-Moors, small parish & village, tape manufacture, church, longevity, &c. 230
Draycott-under-Needwood, manor, Ld. Vernon's, ruins of an old mansion at, 57
Drayton Bassett, mansion and church, 165
Druids, their customs, 1, 2
Dudley Castle, antiquity and romantic appearance, 190, cannon balls and civil wars, 191, Lord Dudley owner, limestone, fossils, and subterraneous canal, ibid.
Dunstall, 60, 144, 173
——— manor, Mr. Meek's, 60
Dunston, township and chapel, 261.

E.

Eaves, township of Bucknall, 374
ECCLESHALL, its antiquity & extensive parish, 315, church and market, 316, fairs and castle, 317, Bishop's seat and woods, 318
Edingale, village and church, 137
Elford, village, church, mansion, and low, 132, enclosure, 133, agriculture, 134, bridge, paper & corn mills, 135
Elkstone, moorland parish, 243
Ellastone, a large parish containing the Wever Hills, 223, church and hamlets, school, &c. 224
Ellenhall, church and parish, 315
Elmhurst, pleasant mansion, 128
Endon, church, village and hamlets, 246
Entrenchments, 11
Enville, its soil, cherries, and magnificent mansion, 36, parish, manor, and estate, Earl Stamford's, 199, church ornaments and hamlets, 200.
Etruria, Mr. Wedgwood's pottery village, seat and family mansion, named from Italy, 399, chapel, Josiah Wedgwood's memoirs, 400.

F.

Fairwell, village and church, 88
Fazeley, Watling Street leaves the county here, 145, Tame river, canal,

cotton mills, large factories, and population, ibid.
Fisherwick, dilapidated, 130—131
Fish, petrified, 38
—— in Trent river, 74
Flash, moorland church & village, 245
Forton, parish and hamlets, 272, Sutton and Mereton, 273, Aqualate Hall and Mere, Sir John Boughey's, 274, battle, 275, church, tombs and inscriptions to 278
Footherley, hamlet of Shenstone, 163
Fradswell, village and chapel, 308
Freeford, manor near Lichfield, 129
Fulford, village, church, Fulford Hall, white building stone found, 392
Furnaces and ancient forges, 32—33.

G.

Gayton, village, parish and church, 307
Gnosall, large parish, 269, church and hamlets, 270, extended prospect, 271
Golden-hill, mines and prospects, 393
Grand Trunk Canal, 429-438
Gratwich, parish, church & village, 217
Grindon, ancient manor, parish and church, school, &c. 239.

H.

Hagley, near Rugeley, Lord Curzon's, 265
Hampstead, mansion of Handsworth, 159
Hammerwich, village and chapel, 128
Hamstall Ridware, north of Trent, and on Blythe, 69, ancient manor house, watch tower and curiosities, 70, bridle for a scold, 71
Hanbury, ancient nunnery, church, vicarage house and mansion, 56, hamlets and manors, 57
Hanch Hall, near Lichfield, 87
Handsacre, moat and old manor, 78-9
Handsworth, extensive village and parish, 157, church, rectory, and population, ibid. Soho manufactory, coining mill, and elegant mansion, 158, hamlets, 159
Hanley, large modern pottery town, 401, canal and market, 402, population, church, meetings, &c. ibid.
Harbourne, west of Birmingham, church, charity school, longevity, &c. 159
Horecross, now a hunting-seat of Lord Scarsdale's, 59
Harlaston, village and chapel, 138
Haselor Hall, village and chapel, 137
Hatherton Hall, Mr. Walhouse's, 263
Haughton, village and church, 269
Haywood, on two canals & Trent, 313, remains of a priory, ibid. remarkable bridge formerly at, 443
Heyley Castle, remains of, 11, 346
Heptarchy, 13
Hides of land, 13
High Offley, parish, church, and hamlets, 319, Shebben-pool, sea crows or pewits, 320
Hilderstone, manor, Mr. Bourne's, 442
Hilton Hall and park, 259
Himley, pleasant village, church, and noble mansion, Lord Dudley's, 188, extensive park and great variety, 189, Baggeridge-wood and pheasants, 190
Hints, mansion, Mr. Floyer's, 164, church, pig of lead found, ibid.
Hixon, ancient village, 309
Hopton Heath, battle, 14, 300
Hopwas, west of the Tame, on the Birmingham and Fazeley canal, 144, bridge over Tame, ibid.
Horton, village and hamlets, 245, church, and canal reservoir, 246
Hounhill, manor, alabaster, Lord Vernon's, 58
Hundreds of the county, 18, Offlow, 41, Seisdon, 167, Totmanslow, 205, Pirehill, 283, Cuddlestone, 253.

I.

Ilam, romantic villa, 38, parish, hamlet, church, &c. 237—8
Improvements proposed, (note) 36—37
Infirmary, North Staffordshire, 401
Ingestre, parish and manor, 297, new church, 298, magnificent mansion, Earl Talbot's, 299, battle, 300
Insects, singular application of to cure the tooth-ache, 449
Ipstones, sharp cliffs, 39, parish and village, church, and singular custom, 234, Belmont, Mr. Sneyd's, ibid.

K.

Keel, parish, village, and mansion, Mr. Sneyd's, 337, church register, wake, &c. to 341
Kinfare Edge, camp, 9, its soil, 35, village, formerly a market, 196, church, manufactures, hamlets, 197
Kingsley, parish and village, 233,

church, trees on the roof, school, and tape manufactory, 234
Kingston, parish, church, and fertile land, 217
Knave's Castle, 6.

L.

Lapley, priory and church, 258, soil and situation, formerly a market, 279, fortifications, land proprietors, and hamlets, 280
Lane Delph, suburb of Lane End, 410
Lane End, modern populous town, 408, population, church, and market, church tower, longevity, charity school, &c. 409
Lanes, of Bentley, protect Charles II. 16
Leigh, large parish, 217, fertile, on Blythe, 218, church and hamlets, 219, schools, &c. 220
Loam, 85—86
Longcroft, mansion and park, &c. 59
Longevity, instances of, 88, 208, 224, *et passim*
Longdon, large village, 79, ancient manors and families, 80, Beaudesert, ibid. cannel coal, 81, Paget family, 82, 83, 84, visited by the Prince Regent, 85, hamlets, 86—7, Gothic church, 88, longevity, ibid.
Lows, 9, 10
LEEK, Roches, 39, parish and township, 247, population and inclosures, 248, market and fairs, church, meetings, schools, charities, 249, eminent men, 250, rebel army, 251
LICHFIELD, city and Bishop's see, 89, 91, sheriffalty, 92, perambulations, 93, Greenhill bower, 95, burgage tenures, 96, parishes, the close and wards, houses and population, 97, state of the poor, wages and rents, churches, 98, Borrowcop Hill, St. Chadd's well, 99, popish customs, mansions, gardens, 100, Johnson's willow, Queen Elizabeth, and a great storm, 101, the plague, ibid. almshouses, manufactures, markets, conduits, soil, streets, public buildings, canal, &c. 102—103, close and cathedral, civil wars, population, monuments, Bishops, hamlets, &c. to 131
Limestone and lime, 34 to 37, 146, 199
Longnor, moorland market town, 244, church and monuments, ibid.
Longport, manufactories and Methodist chapel, 395

M.

Madeley, parish, church and village, 336, park, Lord Crewe's, 337
——— Heath, Roman coins, &c. found, 447
Maer, camp, 8, 9, parish and village, lake, &c. 332, church, sunday school, heath, &c. 335
Mansions of Nobility, &c. 32, 33
Manyfold river, 89
Marchington, 57, woodlands, 58, Houndhill hamlet contains alabaster, ibid.
Market towns, 17, 33
Marston, 279, village and chapel, 313
Maveston Ridware, south of the Blythe, family feud, 72, now Mr. Chadwick's, ancient manor house, ibid. extent and pleasant country, 73, Vale of Trent, fish, and swans, 74, otters ib. quails, old and new church and cemetery, parish register, 75, 76, hamlets and priory, &c. 77
Mayfield, on Dove, pleasant and fertile, 224, stone bridge, cotton works, church, Dove-dale, prospects, &c. 225
Mediolanum, Roman station, 319
Meerbrook, moorland parish, 245
Members of Parliament for Staffordshire, 30, 31
Mercia, 8, 15, Kings of, 33
Mill Dale, 39
Milstone Green, hamlet, 386
Milwich, large parish and village, 308
Moated houses, 12
Modern farming, suggestions for improving, 450
Moorlands, 36, great elevations of, 37-38, peculiarities of, 89
Morredge, peat mosses, 40
Mucclestone, parish, village, &c. 332.

N.

Naseby, battle of, 16
Needwood Forest, 53, 54, its lodges and dingles, 55, Mr. Mundy's poem on, 56
Newbold, manor, Mr. Webb's, curious property ascribed to the soil of a place called the Clots, 60
Newborough, its chapel, 58, linen weavers, ibid.
NEWCASTLE-UNDER-LYME, its origin, 11, borough town, 351, markets, fairs, manors, castle, 352, waters, antiquities, charters, confiscations, burgage tenures, church, meetings, schools, manufactures, roads, canals, inclosures, population, ancient customs, eminent men, &c. to 369

New-chapel, village and chapelry, 349, Brindley's tomb, 350
New-field, coal mines, &c. 393
Norbury, village, parish, and manor, 271, its lakes or meres, and church, 272
Norman conquest, 13
Norton-on-the-Moors, parish, hamlets, church, 370, meetings, population, &c. 371
Norton-under-Cannock, manor, and church, 165
Nunneries, ancient, 32, modern, 232.

O.

Oaken, hamlet of Codsall, 186
Oakover, 38, parish, Roman barrows, and hall, 235, paintings and church, 236, monuments, &c. 237
Offley, or High Offley, elevated situation, 319, manor, church, and parish, 320
Offlow Hundred, 18, boundaries, 41
Onecote, township of Leek, 240, Narrowdale noon, chapel, &c. ibid.
Outlands, hamlet of Adbaston, 323.

P.

Paget family, account of, 82, 84
Parishes of Staffordshire, 18, 20
Patteshull-house, taken by Capt. Stone, 15, village, church, demesne, and mansion, 188, park, and Pigot diamond, ibid.
Pattingham, a pleasant village, parish church, and vicarage, 187, of great antiquity, Roman reliques, 188
Peat, for fuel, 40, peat ashes for manure, 320
Peculiars, list of, 441
Pelshall, on the Walsall water and Wyrley canal, 165, chapel and population, ibid.
Pendrills, save King Charles II. 15
Penkridge, large parish and ancient market town, on the Penk, collegiate church, and stone bridge, 255, charity school, market, and fairs, 256, Littleton family owners, hamlets, 257, Cuttlestone bridge, ibid. population, &c.
Pennocrucium, 7
Penn, Over, a pleasant village and parish, charity school and alms-houses, hamlets, and resident families, 186
Pewits, their natural history, 320
Pipe Ridware, small parish, but ancient manor, 70, ancient church and pleasant church yard, 71
Pirehill Hundred, 19, its boundaries, fertility, and population, 283
Plantations on impracticable land recommended, (note) 37, 40
Population, see different parishes; also Totmanslow Hundred, 205, Pirehill Hundred, 283, and the whole county.
POTTERIES OF STAFFORDSHIRE, 393, history of, 411, materials analysed, ingenuity, and science, 412—13, manufacturing process, increased population, 414, butter pottery, Burslem, 415, modern improvements, glazing, flowering, female employment, schragers, white stone ware, crouch ware, tough tom, candlesticks, plaster moulds, saggars, pottery ovens, volumes of smoke, to 419, cream-coloured ware, 421, the Wedgwoods, 422, chemical improvements, and moderate wages, Queen's and other wares, clays from Devonshire, varieties of porcelain, clay and ground flint, &c. to 428
Powerful machinery, at Bilston coal-pit field and Ocker-edge, 445, the "Neptune and Vulcan," and mud chain-pump, 446
Priories, ancient, 32
Public charities, account of one at Stafford for the relief of clergymen's widows and orphans, 438.

R.

Rebellion 1745, 17
Rickerscote, village and spa water, 269
Rocester, a parish on Dove, 222, monastery, church, and village, 223, cotton manufactory, Churnet, Cowbridge, Mince-pie Hall, ibid.
Rolleston, its soil, church, and mansion, 48
Roman Conquest, 3, roads, 4 to 7, stations, 4, camps, 5, 6, 193, 202, coins, 60, 148, 208, 447
Ronton, parish, village, & abbey, 314
Rowley Regis, hills, 202, church, 203
RUGELEY, market town & parish, 264, Trent river and canal, ibid. church, vicarage, grammar school, Brindley's bank and coal mines, 266
Rushall, manor and village, 147, castellated mansion, and cannon balls, 148, church, population, ruins, 149
Rushton Spencer, township and parish, church, tombs, &c. 251.

S.

Saline spring, analysis of one near Stafford, 448
Sandon Hills, remains of tumuli there, 449
Sandon, village and manor, 305, mansion, formerly Erdeswicke's, now Lord Harrowby's, church and monument to Erdeswicke, 306
Sandwell, park and mansion, 157
Saredon, Great and Little, 259
Saxons in Britain, 7, 8, camp, 9, 10
Sedgley, large and populous parish, high ground and lofty spire, 191, Sedgley park, popish seminary, hamlets, and thunder storm, 192
Seighford, parish, church, and hamlets, 314
Shareshill, village and land owners, 258, church, monuments, & hamlets, 259
Sheen, small moorland parish, 243
Shelton, its excellent porcelain, 403
Shenstone, pleasant village and large parish, 162, longevity, gentlemen's seats, hamlets, &c. 163
Sheriffs of Staffordshire, 24
Sheriff Hales, parish, population, church, and hamlets, 282
Shirleywich, salt works, 303—4
Shugborough, vale of, 443, scenery, seat of Lord Anson, statues, monuments, arch of Adrian, 444, Chinese house, ancient mansion, 445
Silkmore, Mr. Mottershaw's, 269
Smethwick, manor and population, 159, chapel, iron foundry, and canal, 160
Smith-field, coal mines, 394
Soho manufactory, Bolton & Co.'s, 157
Soils of Staffordshire, 34
Standon, parish, church, & village, 386
Stanton Moor, 37
STAFFORD, county town, antiquities, 283, burgess rights, markets, Mr. Sheridan, 288, fortifications, edifices, 290, manufactures, population, customs, eminent men, &c. to 296
Stafford castle demolished, 14
Statfold, the manor, Mr. Wolferston's, manor house, observatory, &c. 138
Stoke, hamlet of Stone, 391
Stoke-upon-Trent, parish, and Pottery town, 403, Trent river, canal, church and monuments, ibid. sexes separated, national school, meeting houses, machinery, to 407
Stonall, hamlet of Shenstone, 163
STONE, town and parish, 389, church, market, fairs, canal, & population, 390
Stone, ancient monuments, 2-3
Stone wall fences, 37
Stoniwell, Bishop, 86
Stow-heath manor, 173
Stow, large parish and hamlets, 308
Street-hay manor, 130
Swinfen, its church and mansion, 163
Swinford Regis, church and monuments, mansions and village, 192, Roman camp, brine spring, and hamlets, 193
Swinnerton, parish and village, 384, mansion, church and park, 385.

T.

Talk-o'th'-hill, township & chapel, 346, explosion, sulphur-well, 347
TAMWORTH, on the Anker and Tame rivers, 138, antiquities, ancient foss and castle, 139, town, borough and charter, 141, fairs, population, manufactures, church, and meeting houses, 142, school, public buildings, and moat-house, ibid. the plague, invasion of Ireland, 143, hamlets, Sciersecote, &c. ibid.
Tatenhill, east of Needwood, 59, its church, manors, and hamlets, 60, remarkable echo, screw mill, ibid.
Tean, Upper and Lower, populous villages, 221, tape manufactory, 222, chapel and Sunday schools, ibid.
Teddesley Hall, Mr. Littleton's, 257, late Sir Edward's, his hospitality, ibid.
Terley Castle, 11
Tettenhall, a great battle, 12, 183, the church a free chapel, ibid. in four prebends, hamlets, ditto, population and register, 184, resident gentry and good villas, 184, Tettenhall Wood picturesque, ditto, Elm Grove, hamlets, thunder storm, great degree of cold, 185
Thorp Constantine, village and church, 138
Thor's-house cavern, 39
Tipton, on the Birmingham canal, rich iron mines and works; also manufactures of iron, Keir's chemical works, great population, church and bells, 150
Tittensor Heath, lunatic asylum, 388
Tixall, manor and parish, 297, hall rebuilt, canal, &c. ibid.
Totmanslow Hundred, 205, village, 230
Trentham, parish and hamlets, 382, magnificent mansion, the Marquis of Stafford's, grounds, bridge, mausoleum, &c. 384

Trysull, small village and parish on Smestall, stone church and tower, 187, Seisdon, an hamlet giving name to the hundred, ibid. Apewood Castle, ditto
Tubber Patrick, a well in Ireland visited by the superstitious, 437, St. Canice's well, 438
Tunstall, excellent tiles, 394, Methodist chapel, ibid.
Tutbury Castle, 14, 48, ancient customs, 49, Mary Queen of Scots confined there, 50, besieged in the civil wars, 51, market and manufactures, Dove meadows and bridge, ibid. church and ancient priory, abstinence of Ann Moore, 52, Needwood Forest, 53, 55.

U.

Urns, Roman, 447
Uttoxeter, antiquity and population, 205, market, fairs, church, 206, meeting houses, free-school, and meadows, 208, chronology, 209, Lord Gardner, to 213.

W.

Walford, Mr. his farming, 320
Wall, hamlet of Lichfield, the Roman Etocetum, 128
Walsall, ancient borough and market town, mayor and aldermen, 145, population, manufactures, parish and church, 146, limestone, canals, meeting-houses, school, &c. 147
Walter Parsons, the strong man, 150
Walton, hamlet of Stone, 391
Waterfall, moorland parish, river Manifold disappears, 239
Warnslow, village and chapel, 243
Watling-street, 128
Wedges Mill, and Church Bridge manufactory, 448
Wedgwood family, their seat and mansion, 400, their important scientific improvements, 401
Wednesbury, manor house, 153, coal mines, manufactures, wild fire, 154, steam engines, furnaces, powerful machinery, 155, Methodist riots, church and monuments, ibid.
Wednesfield, battle, 12, 152, parish, church, population, 153
Weeford, village and parish, on Black Brook, 162, a low called Offlow, church, and hamlets, ibid.
West Bromwich, village and parish, manufactures, and population, 156, valley of the Tame, and villas, ibid. Sandwell park and mansion, 157
Weston-on-Trent, village and church, 303, canal, &c. 304
Weston-under-Lizard, mansion, Earl of Bradford's, 281, church and village, Watling-street way, 282
Wetton, moorland parish, 240, Thor's-house cavern, Ecton Hill, and Warnslow, 241, copper and lead mines, church and village, 242
Wever Hills, 37
Wheaton Aston, village and church, 210, fire there, a May-pole, ibid.
Whichnor manor, and bacon flitch, 62, ancient moated mansion, and the lodge, 63, church, visited by King James, 64, hail storm and vorago, mills burnt, ibid. stone bridges, large iron forge, and slitting mill, 64
Whitgreave, Esq. *Moseley*, loyalty to Charles II. 16
Whitmore, *parish*, church, and village, 375
Whittington, parish and race course, 131, mansion, 199
Wiggington, near Tamworth, village, and chapel, 143, Low Flat and Robin Hood's Butt, money lands, ancient coin, ibid. instance of great abstemiousness, 144
Wilkes, Dr. View of Staffordshire, 32
Willenhall, populous manufacturing place, church, &c. 150, Dr. Wilkes' account of, 151
Willows, account of some species found at Blymhill, 447, 448
Wolseley, parish and bridge, 266
Wolstanton, extensive parish, 342, marriages under Cromwell, 348, coal mines and minerals, 349
Wolverhampton, antiquity and population, 167, deanery, mansions, churches, 168, meeting houses, springs, market, situation, and manufactures, 171, grammar school, 172, thoroughfare and hamlets
Wombourn, ancient village, and spire church, 186, early soil and Smestall water, 187, hamlets, Woodhouses, Heath Forge, Orton manor, Sir John Wrottesley's.
Worcester, battle of, 16
Wyrley bank, Great Wyrley, 263
Wyrley Grove, or Little Wyrley, 165, Mr. Hussey's, rookery, 166
Yoxall, formerly a market, 58, church and rectory, 59, hamlets, ibid.
Yoxall lodge, Needwood Forest, 55.

INDEX,

TO DEPARTMENT THE SECOND.

AGRICULTURE, NATURAL HISTORY, &c.

A.

AGRICULTURAL Societies, 98, 100
Agriculture of Staffordshire, 17
Alabaster, 166
Anson Lord, his farm and demesne, 90
Aqualate, Sir John Boughey's, ditto, 96
Arable land, cropping and cultivation, 29, 86, 96.

B.

Barley, its culture, drilling, &c. 34, 35
——, consumption of, 81, 82
Basaltic rocks, formation of, 133
Beans, setting by hand recommended, 29
——, culture, 35, setting and drilling, 36
Bees, 80, 81
BIOGRAPHY OF STAFFORDSHIRE, 175 to 270
Birds of Staffordshire, 145 to 158
Blount, Mr. of Bellamore, his farming, 96
BOTANY OF STAFFORDSHIRE, 101 to 126
Bradford, Earl, his demesne, 94
Brewster, threshing machine maker, 27
Buck wheat, its culture, 37
Buildings, 22, 23
Burnet, 50, indigenous, 51.

C.

Cabbages, in field cultivation, 41
Canals, 21
Carrots recommended, 42
Chambers, drill machine maker, 29
Clays, fire clay, potter's clay, 164
Cleft in a rock remarkable, 165, 166
Climate, 17
Coal mines, thick and thin, 128 to 140
——, formation of, conjecture on, 130
——, strata above, intervening and beneath, 134
——, qualities, and means of working, 138
——, quantity raised, 139, peacock coal, 164
COINAGE OF STAFFORDSHIRE, 1 to 8
Comforth, winnowing machine maker, 27
Comparative View of Agriculture, 85
Cottages, 23
Courses of crops, 29, 30, 87, 92
Cow stock, 66, sundry owners of capital cattle, 67, new breed introduced by Mr. Miller, their size and qualities, ibid.

D.

Dairying, 67, great produce from a cow, 68, superior to oxen and sheep, and improvements proposed, ibid. turnip butter to cure, 69
Divisions of the county, 17
Dove river, 9-10
Draining, the first of improvements, 85
Ducks, their utility, 79.

E.

Earths, analysis of, 127, 128
Echos, 167
Elevation above the sea, 18, 19
Embankments proposed, 65
Estates, 21.

F.

Fallows, sometimes necessary, 30, 46
Farms, size of, 24, of land-owners, 87
Feeding cattle, 69, sheep, 70, hogs, ib.
—— weight to ascertain, 70
Fencing on good land, 28
Fertile soil, analysis of, 127
Fishes of Staffordshire, 159
Flax culture, 37
Fossil wood, 166, 167
Fruit trees, neglect of, 55
—— ——, their importance, 167
—— ——, their ripening promoted, 174.
Gardens and orchards, 53, great importance to country houses, 64, 65
Geese, 79

INDEX.

G.

Grass land, 46, 89, 93
Grasses, recommended for cultivation, 50, 89
Grove, Esq. Shenstone, his farming, 97.

H.

Harvest, ancient, 14, 15
Hay harvest, making and stacking, 51, 52, produce of, 53
Heathy land, 11, 12
Hogs, breed most esteemed, 77
——, experiments and economy on, 78
——, powers of increase, 84
Horses, the Flemish breed, black, or brown, Suffolk, at Lord Talbot's, and Cleveland, at Mr. Tollet's, 74, 97, consumption of landed produce, and numbers kept, 75, economy in keeping, 76
Horses and oxen for draught, 76
Houses and offices, 22, 23
Hundreds of Staffordshire, 17.

I.

Implements of husbandry, 26, 86
Increase of animals and vegetables, 83
Inge, Esq. of Thorp, his farming, 96
Ingestrie, farm and demesne, 88
Insects, 161
Iron gates, hurdles, and ploughs, 27
——stone strata, and quantity made, 137
Irrigation of land, 63, 89, 97.

L.

Labourers' gardens, importance of, 54
Lamprey, and nine eyes, 159
Lawley, Sir Robert's, demesne & farm, 96
Leases, 22, usual covenants, ibid.
Leek coinage, 7
Lichfield coinage, 3, 5
Lime-stone, 128, 130, with petrified shells, 134, rises in large tables or flats, 164
Live stock, 66, 90, 98
Longevity of William Hyven, 174
Lucern, 51, will bear transplanting, ib.

M.

Mangel wurzel, a variety of beet, its culture and utility, 41, 42
Manuring, 63 to 65, and 86
Marble, 166
Meadow land, 46, 47, rich on Trent and Dove, ibid. meadow herbage, 48, 49, neutral plants and weeds, 50
Miller, Mr. his wheat culture, 32, meadow grass, 53, cattle stock, 68, sheep, 72
Millstones, 165
MINERALOGY OF STAFFORDSHIRE, 127 to 140
Miscellaneous natural history, 163
Moorlands, 9.

N.

Newcastle coinage, 6, 7.

O.

Oat culture, 35, compared with wheat as human food, ibid.
Ochre or raddle, 164
Orchards neglected, 54, importance of, 55, landlords should *promote* and enforce *their culture*, 56, more cultivated formerly, 167.

P.

Paring and burning, 63
Parks, 9
Patteshull, demesne and farm, 95
Pease, culture, 36
Peat earth, its formation, 168
PEERAGE OF STAFFORDSHIRE, 271 to 319
Pigeons, 80
Pools, foretelling weather and dearth, 167
Poor, 91
Poor rates, 25
Population, 82
Potatoes, culture and use, 40, the curl, ibid. superior in nutriment to most other plants per acre, 41, cautions in giving to cattle, 42
Poultry, 78, 79
Produce and consumption of provisions, 81, 82.

Q.

Quadrant of six miles radius, 163
Quarries of stone, 165.

R.

Rabbits, 78, increase of, 84
Rape or cole seed, 41
Rents, 24
Reptiles, 158
Reservoirs for irrigation proposed, 62
Rivers, 9

Rowley stone analyzed, 133
Rye, its culture, 34.

S.

Saintfoin, an excellent plant, 51
Salt springs, 169, 170
Serpents, 159
Sheep, 10, varieties of, 70, 143, sheep masters, eminent ones, 72, disorders of, 73, increase of, 83
Shining earth, 163
Shugborough demesne, 90
Springs that do not freeze, 169
Stafford coinage, 1 to 3
STATISTICAL TABLES, [*see end of "Topography*."]
Stone for building, 165
Strata of earths in mines, 134, 137
Soils, 19, 20, analysis of, 127, 128
Superstition of former times, 15.

T.

Talbot, Earl, his agriculture, 87
Tamworth coinage, 5, 6
Tennant, Esq. his farming, 97
Tenures, 21
Tettenhall pears, 55
Tillage, ancient, 11 to 13
Timber and woodlands, 56 to 59
Tithes, 24
Tobacco-pipe clay, 163, 164
Tollett, Esq. of Betley, his farming, 96
Trees, of great bulk, 166
Turf, for fuel, 16
Turkies, 79
Turnips, common, and Swedish, 30, 31, weight grown per acre, ibid.
———, fly, mildew, and caterpillar, 38-40
———, Northumberland culture, 37, 38
———, Sir John Wrottesley's method, 38.

U.

Uttoxeter coinage, 8.

V.

Vale of Trent, its beauties, 19
Vegetable increase astonishing, 84
Vetches, their culture, 36.

W.

Waste land, improvements, 59 to 62
Water, 20, 21, medicinal ditto, 170 to 173
Weeds, a great evil, 42, different varieties, and means of destroying, 43, 46
Weights and measures, 25
Wheat, ancient culture, 13, seed and produce, 31, spring wheat, weight and number of grains, 32, varieties, 33, steeping of seed, mildew, &c. 34, astonishing increase of, 84, 85
Woodlands, 9
Wool, 10
Wrottesley, demesne and farm, 95.

Z.

ZOOLOGY OF STAFFORDSHIRE, 141.

NAMES OF PERSONS IN DEPARTMENT THE FIRST.

ABNETT 338
Aikin 428
Adams 66, 388
Adderley 57, 351
Addenbrooke 113
Adey 374
Adie 265
Aysheton 142
Aitkens 402
Allen 298, 220
Allports 207
Amphlett 204
Anglesea, M. 43, 48, 80, 84, 86, 88, 89, 131, 261, 264, 338
Anson, L. 270, 271, 272, 313
Anson, V. 66, 150,
Anson 444, 445
Anwyll 327
Arden 71
Argyle, D. 268
Ash 249
Ashburn 295
Ashmole 107
Astle 217
Astley 188
Aston 291
Austrian Archdukes, John and Lewis, 85

B.

Baddeley 159
Bagnall 360
Bagot, L. 55, 218, 219, 311
Bagot 220, 309, 311, 361
Bailey 237
Bailye 56
Baker 98
Bakewell 388, 390
Ball 376, 379, 381
Balderstone 245
Bamford 231
Barber 410
Barker 258
Barlow 335, 336
Barnes 230
Barrett 185
Bate 186, 344
Bateman 351
Bath, M. 141, 145
Beane 44
Bedford 185
Bedford, D. 288
Beetham 244
Belcher 225
Betteson 272, 276
Benson 288
Bent 401
Bentley 249

INDEX.

Berwick, L. 162
Best 167
Biddulph, 127, 262, 280
Bill 230, 388
Billinge 244
Birch 156, 157, 159
Blackey 227, 235
Bloomfield 85
Blount 265
Blunt 361, 384
Bonney, 305
Booth 389
Boughey 274, 278, 335, 359, 403
Bourne 132, 408, 410, 442
Bowyer 333, 351
Boynton 65
Bradburne 128, 284
Bradford, E. 279, 281, 282
Brasier 376
Breck 339
Brett 232, 339
Bridgwater, D. 431
Bridgman 282
Bridone 223
Bright 278
Brindley 429
Broadhurst 374
Bromfield 61
Bromley 291
Bromwich 161
Brook L. 108, 109, 137
Broughton 323
Buckeridge 312, 390
Buller 231
Buonaparte 84
Burke 287
Burne 186
Brewster 284
Butler 98
Butt 382

C.

Caldwell 359, 397
Carless 347
Carlisle 234
Carrington 296
Carteret, L. 362
Caswell 201
Chadwicke 93, 106, 109
Chetwode 329, 359
Chetwynd, V. 326
Chetwynd 267, 298, 299
Child 393
Clarke 269, 297, 309
Clare 153, 182
Clarence, D. 85
Clayton 389
Cleyton 333, 381
Clements 313
Clinton, B. 88, 90, 106
Clutterbuck 182
Cobb 87
Collins 266
Congreve 257
Cook 410, 419
Corns 297, 385
Cornwallis, B. 318
Cotes 296, 387
Cotterill 409
Cotton 243, 361
Courtenay, V. 201
Cowper 265
Cox 102, 333
Coyney 232, 233
Crewe, L. 332, 337, 447
Crockett 270, 278, 279
Cross 69
Croxall 161
Cumberland, D. 17, 103, 390
Curzon, V. 265

D.

Dale 333
Daniel 418
Darlington E. 165, 167
Dartmouth, E. 156
Darwall 277
Darwin 240
Davenport 224
Degge 167, 207, 208, 310
Devonshire, D. 234, 241
Dickenson 6, 260, 269, 281, 292, 447
Donegal, M. 130, 144, 319
Dolman 161
Dolphin 162
Downing 142
Drayton 141
Dudley and Ward, V. 188, 190, 191, 192, 193, 194, 202
Dunn 231
Dyott 129, 131

E.

Evans 218, 219, 220
Ensor 238
Enfield 195
Ellerton 266, 293, 314
Eliot 120
Egerton 336
Egginton, 158, 161, 186
Eddowes 296, 233

F.

Falconer 94, 100
Farmer 243
Feyld 48
Fenton 403
Fereday 191, 192
Fernyhough 408
Ferrars E. 308
Fieldon 294
Fitzgerald 189
Fitzherbert 2, 95, 385
Fleetwood 241
Fletcher 341
Floyer 164
Foley 152, 193, 196
Foley, L. 193, 194
Forster 182
Fowke 183
Fowler 46, 88, 185, 307
Fox, C. J. 216
Fox 287

G.

Galway, V. 254
Gardner 208
Gardner, L. 213
Garrick 117
Giffard 254, 278
Gilbert 234
Gildart 163
Giles 338
Gilpin 263
Girdler 137
Gisborne 55, 62
Goodwin 366-369
Gough 153, 159, 180, 182, 270
Gould 242
Gower 383, 384, 393
Gower, L. 360
Granville, Cts. 360
Granville 224, 361, 391
Granville, V. 288, 297, 337
Graves, L. 85
Green 146, 270
Gresley 351, 362, 396
Grove 117, 162
Grundy 143
Guy 142

H.

Haden 155, 292, 332
Hake 223, 224
Hall 280
Harding 314, 319, 322
Harper 243
Harris 293
Harrison 366
Harrowby, E. 305, 307
Harvis 219
Hussey 165, 166
Hastings 287
Hatrell 361
Heath 338
Heathcote 410
Hellier 187, 259
Henderson 53
Henshall 434
Hewett 277
Heyrick 173
Hicken 214, 315, 345
Hickman 194
Higgins 271, 272
Hill 346, 360
Hilditch 300
Hilridge 217
Hinckes 181, 184, 185
Holbrook 75
Holiday 231
Holland, L. 289
Holland 336
Hollins 410

Holyoake 184
Honeybourn 186
Hoo 153, 179, 180
Hordern 167, 173, 182, 259
Horton 61, 182, 337
Hough, B. 262, 318, 387
Howard 131, 132
Howe, E. 215
Howe 154
Hulme 248, 251
Hunt 56, 57
Huntingdon 335
Hurd, B. 254, 257
Huskisson 181, 182
Hutton 135

I. J.

Inge 265
Jackson 129
Jell 223
Jenkinson 318, 319, 388
Jerningham 268
Jennings 169, 172
Johnson, 100, 103, 117, 118, 128, 303, 308
Jones 100, 176

K.

Keen 269
Kempson 173, 254
Kemsey 254
Kenrick 327
Kinnersly 327
Kirk 62
Knightley 254

L.

Lane 69, 158, 176, 263
Langelly 98, 221
Lauderdale, E. 288
Lawley 110, 164
Lawton 346, 350
Lea 162, 389
Ledwick 438
Lees 150
Lefebvre, Gen. 84
Leigh 148, 149, 174, 360
Leveson 169, 170
Levett, 63, 127, 144, 266
Lewis 217
Lister 53, 78
Littler 422, 423
Littleton, 172, 234, 255, 256, 257, 258, 261, 263, 267, 279
London, B. of, 288
Lowe 270
Lyttelton, B. 197
Lyttelton 193, 201

M.

Macclesfield, E. 226, 247, 249, 398
Mainwaring, 375, 376, 377, 381, 386, 388, 407
Mander 173
Marsh 186
Mason 437, 441
Massarene, V. 130
Maude 292
May 88
Mayer 447
Meaykin 252
Meynell 326, 327
Milley 105
Mills 224, 248, 388, 410
Milton 369
Minor 207, 103
Mitchell 425
Molineux 158
Monckton, 85, 172, 254, 280
Montague 116
Moore 52, 288, 304, 305
Moore, Sir John, 84
Morgan 332, 334
Mosley 48, 60, 200
Moss 195
Mottershaw 269
Mulgrave, E. 288
Mundy 55

N.

Noble 250
Newton 69, 235
Neale 309

O.

O'Connor 438
Offley 63, 332, 336, 337
Okeover 235, 236
Oldfield 207
Orme 361
Orritt 385
Owens 163

P.

Paget, L. 83
Paget 85
Palmer 344
Parker 44, 410
Parkes 159, 278
Parkhurst 88
Parr 243
Parsons 156
Paulett 44
Pearson 184
Peel 141, 145
Pennant 443
Pennington 344
Peploe 267
Perkins 269
Perry 173
Pershouse 155
Peter 161
Petit 191, 258, 259
Pigot 188,
Philips 221
Phillips 182
Pitchford 313
Pitt 133, 287, 303
Plaxton 271
Pole, C. 198
Poole, 335, 447
Port 231, 239
Porter 230
Portland, D. 286, 287
Pountney 182
Powell 318
Powlett, L. 344
Powys 247
Pratt 259
Prince Regent, H. R. H. 85, 401
Printon 47
Pritchard 222
Pritchett 230
Proby 254
Pye 138, 149

R.

Rathbone 292
Rawlins 103, 104
Reid 169
Repington 142
Repton 407
Reynolds, 114, 159
Rice 308
Richardson 243
Roberts 337
Robinson 312, 385, 403
Rogers 396, 401
Rolleston 225
Ross 437
Roylance 393
Russell 238

S.

Salt 246, 396
Sands 159
Savary 154
Scarsdale, L. 59
Scott 153, 156, 160, 180, 193, 204
Sedgwick 186
Sell 308
Simeon 360, 391
Simpson 162, 337
Sewell 351
Shaw 196, 202
Shelburne, L. 287
Shenstone 161, 203, 204
Sheridan 286, 287, 288, 294
Shrewsbury, E. 227, 228
Skrymsher 271, 272, 273, 274, 276, 319, 329, 331
Slaney 255, 256
Smithe 393
Smith 231, 247, 258, 334, 388, 390, 394
Snape 329, 334, 340, 344
Sneyd 55, 217, 234, 247, 338, 340, 341, 347, 348, 357, 396
Southwell 162
Sparrow 246, 265, 259

Spencer, L. 288
Spode 408
Stafford 295
Stafford, M. 150, 186, 240, 282, 357, 361, 381, 388, 401
Stamford, E. 169, 199
Stretch 336
Stubbs 185
Sussex, D. 288
Sutherland 128
Swinfen 164, 280

T.

Talbot, E. 57, 217, 278, 299
Talbot 278
Tankerville, E. 185
Tarratt 186
Taylor 61, 273
Tenant 168
Tether 333
Thacker 186
Thomas 219
Tindall 173
Tollet 335, 341, 343, 344, 345, 385
Tomlinson 408
Tooke 216
Toplady 369
Townshend, M. 140, 141
Turner 245, 252, 372
Turton 65, 66

U. V.

Unwyn 345
Valentia, V. 201
Vaughan 319
Vaughton 143
Vernon, L. 51, 55, 57, 58
Vernon 183, 259, 260, 345
Vincent, E. 391
Vincent, C. 233

W.

Waldron 194
Walford 319
Walker 158, 386
Wallis 435
Walmsley 116
Warburton 421
Ward 242, 361
Watkins 137
Weld 391
Wesley 155, 369
Wedgwood 326, 332, 333, 334, 397, 399, 400, 403, 421, 422, 424, 426, 427
Wheeldon 235
Weymouth V. 142
Whieldon 396
Whitaker 240, 243
Whitby 153, 161, 180, 314
White 232, 319
Whitehurst 338
Whitehall 219, 221
Whitgreave 181, 183
Whitworth 275, 322
Wickliffe 295
Wigan 280
Wightman 46
Wilkes 147, 150, 200
Willatt 231
Williamson 227
Willoughby de Brooke, L. 216
Wilmot 130
Wilson 288, 334, 369
Wilton E. 342
Witrings 365
Wood 186, 234, 397, 418, 424, 425
Woodhouse 403
Wollaston 300, 302
Wolseley 266
Wright 59, 345, 409
Wrottesley 183, 184, 186, 187
Wyatt 66, 114
Wynn *186*

Y.

Yarmouth, L. 85
Yeoman 403
York, D. 288
Young 128.

NAMES OF PERSONS IN DEPARTMENT THE SECOND.

A.

ALLEN 183
Anglesea M. 31, 57, 139, 276
Anson L. 67, 71, 72, 90, 93, 94, 175, 309
Ashmole 184
Astle 244
Audley L. 312

B.

Bakewell 66, 71
Bagot L. 56, 67, 72, 315
Barker 4
Bergman 133
Blake 251
Blick 6
Blount 96, 100
Boughey 96
Bradford E. 71, 72, 94
Brewster 27
Browne 227
Broughton 58
Buckley 89

C.

Caldwell 236
Cartwright 8
Chambers 27
Childe 97
Coke 50, 94
Cornforth 27
Cotton 237
Curwan 53, 78

D.

Darwin 40, 131, 174
Dickenson 34
Donegal M. 57, 317
Drayton 10
Dudley L. 57, 129
Dudley 265
Dyott 72

E.

Ellman 72

F.

Fenton 221
Ferrers E. 169, 290
Fryer 58

G.

Gardner V. 311
Giffard 56
Ginders 88, 89
Granville V. 311
Greene 4, 5
Grove 97
Guy 6

H.

Hancock 5
Harding 6
Harrowby E. 302
Harvey 72
Holliday 58
Horton 3
Huskisson 67
Hurd, B. 232

I. J.

Inge 96
James 263
Johnson 5, 191

K.

Keir 127, 128, 131, 133
Kempson 3, 5

L.

Lawley 96
Lettsom 41
Lightfoot 246

INDEX.

Littleton 57, 59, 67, 71

M.

Mainwaring 58
Marshall 45
Mason 72, 97
Mead 6
Meadowcourt 252
Mellor 32, 53, 67, 68, 72
Mountford 219

N.

Newcastle-under-Lyme, D. 271
Newton, B. 243
NorthumberlandD. 256

P.

Pennant 97
Pigott, 37, 57, 72, 95
Plott 9, *et passim*
Pratt 33
Princep, 67, 72

R.

Rich 7
Robyns 226
Rudge 27

S.

Shaw 8, 253
Sheldon, A. 260
Smallridge, B. 242
Snelling 3, 7
Sinclair 52
Stamford E. 57
Stamford 93
Stafford, M. 58, 273

T.

Talbot, E. 67, 72, 87, 88, 100, 292
Taylor 447
Tollet 72, 96, 97
Townshend, M. 287

V. W.

Vagg 40
Vernon 58
Vincent, E. 294
Walton 267
Wakelin 8
Wheelock 93
Whitworth, E. 304
Willetts 3
Withering 37, 49
Wright 5
Wrottesley 37, 38, 57, 86, 95
Wyatt, 91.

FINIS.

Printed by JAMES SMITH, High Street, Newcastle-under-Lyme.

Milton Keynes UK
Ingram Content Group UK Ltd.
UKHW022009100823
426691UK00004B/40